Two Manuals Combined into One Book

Infantry Rifle Platoon and Squad

And

Special Forces Survival Evasion and Recovery

Department of Defense

Pentagon
Publishing

***FM 3-21.8 (FM 7-8)**

Field Manual
No. 3-21.8

Headquarters
Department of the Army
Washington, DC, 28 March 2007

The Infantry Rifle Platoon and Squad

Contents

		Page
PREFACE		**viii**
SUMMARY OF CHANGE		**ix**
Chapter 1	**FUNDAMENTALS OF TACTICS**	**1-1**
	SECTION I – FUNDAMENTALS OF INFANTRY PLATOON AND SQUAD OPERATIONS	**1-1**
	Infantry	1-1
	Offensive and Defensive Combat	1-1
	Warfighting functions	1-3
	Intelligence	1-4
	Movement and Maneuver	1-4
	Fire Support	1-4
	Protection	1-5
	Sustainment	1-5
	Command and Control	1-5
	Characteristics of Close Combat	1-6
	Section II – Doctrine and Training	**1-7**
	Infantry Doctrine	1-7
	Individual Infantry Skills	1-8
	Warrior Ethos and Army Values	1-9
	Every Soldier Is a Sensor	1-11
	Section III – Organization	**1-11**
	Infantry Platoon	1-11
	Infantry Fire Team	1-14
	Infantry Squad	1-17
	Infantry Weapons Squad	1-20
	Section IV – Combinations	**1-21**
	Complementary Effects	1-21
	Dilemma	1-22
	Reinforcing Effects	1-22
	Section V – Individual Leadership	**1-24**
	Leadership by Example	1-25
	Authority	1-25
	Mission Command	1-27
	Section VI – Tactical Decisionmaking	**1-27**

Distribution Restriction: Approved for public release. Distribution is unlimited.

*This publication supersedes FM 7-8, 22 April 1992.

		Page
	Supporting Concepts	1-28
	Troop-Leading Procedures	1-29
	Actions on Contact	1-32
	Risk Management	1-34
	After-Action Reviews	1-34
	Section VII – Combat Power	**1-35**
	Four Critical Functions	1-35
	Doctrinal Hierarchy of Operations	1-37
	Section VIII – Situation	**1-43**
	Human Dimension	1-43
	Law of War	1-44
	Operational Environment	1-46
Chapter 2	**EMPLOYING FIRES**	**2-1**
	Section I – Considerations for Employing and Controlling Fire	**2-1**
	Command and Control	2-1
	Mass the Effects of Fire	2-2
	Destroy the Greatest Threat First	2-2
	Avoid Target Overkill	2-3
	Employ the Best Weapon for the Target	2-3
	Minimize Friendly Exposure (Protection)	2-4
	Prevent Fratricide	2-4
	Plan for Limited Visibility Conditions	2-4
	Develop Contingencies for Diminished Capabilities	2-4
	Section II – Weapon and Munition Characteristics	**2-5**
	Common Weapons and Munition Characteristics	2-5
	Fire Team Weapons	2-7
	Complementary and Reinforcing Effects at the Fire Team Level	2-8
	Section III – Engaging the Enemy Without Endangering Friendly Troops	**2-9**
	Surface Danger Zone	2-9
	Minimum Safe Distance and Risk Estimate Distance	2-10
	Section IV – Employing Direct Fire	**2-11**
	Rules of Engagement	2-11
	Control Measures	2-11
	Engagement Techniques	2-20
	Fire Commands	2-21
	Range Cards	2-25
	Adjustments	2-26
	Closure Report	2-27
	Section V – Employing Indirect Fires	**2-27**
	Call for Fire	2-27
	Adjust Fire	2-34
	End of Mission	2-41
	Fire Support Coordination Measures	2-41
	Section VI – Close Air Support and Naval Gunfire	**2-43**
	Close Air Support	2-43
	Naval Gunfire Support	2-46
Chapter 3	**TACTICAL MOVEMENT**	**3-1**
	Section I – Overview	**3-1**
	Movement Formations	3-2
	Movement Techniques	3-2
	Route and Navigation	3-3
	Danger Areas	3-3

		Page
	Security	3-3
	Other Considerations	3-3
	Section II – Movement Formations	**3-5**
	Primary Formations	3-6
	Fire Team Formations	3-8
	Squad Formations	3-9
	Platoon Formations	3-12
	Section III – Movement Techniques	**3-19**
	Squad Movement Techniques	3-21
	Platoon Movement Techniques	3-24
	Section IV – Route Selection and Navigation	**3-28**
	Navigation Aids	3-28
	Route Planning	3-29
	Types of Navigation	3-29
	Route Types	3-30
	Develop a Leg	3-31
	Execute the Route	3-32
	Section V – Actions At Danger Areas	**3-33**
	Crossing of Danger Areas	3-34
	Crossing of Linear Danger Areas (Platoon)	3-35
	Enemy Contact at Danger Areas	3-37
	Section VI – Movement With Combat Vehicles	**3-37**
	Combat Vehicle and Infantry Squad Formations	3-37
	Combat Vehicle and Infantry Platoon Formations	3-40
	Mounted Tactical Movement	3-41
	Convoys	3-42
	Actions at Danger Areas (Mounted)	3-42
	Section VII – Security	**3-44**
	Enemy	3-44
	Terrain	3-45
	Camouflage, Noise, and Light Discipline	3-45
	Security at Halts	3-45
	Section VIII – Other Movement Situations	**3-47**
	Administrative Movement	3-47
	Tactical Road Marches	3-48
	Movement by Water	3-48
	Movement During Limited Visibility Conditions	3-48
Chapter 4	**PROTECTION**	**4-1**
	Section I – Risk Management and Fratricide Avoidance	**4-1**
	Risk Management Procedures	4-1
	Implementation Responsibilities	4-5
	Fratricide Avoidance	4-5
	Prevention	4-7
	Section II – Air and Missile Defense	**4-8**
	Early Warning Procedures	4-9
	Section III – Chemical, Biological, Radiological, and Nuclear Defense	**4-10**
	Tenets of CBRN Defense	4-11
Chapter 5	**COMMAND, CONTROL, AND TROOP-LEADING PROCEDURES**	**5-1**
	Section I – Command and Control	**5-1**
	Leadership	5-1
	Mission-Oriented Command and Control	5-1
	Section II – Plans and Orders	**5-3**

		Page
	Mission Statement	5-3
	Combat Orders	5-4
	Section III – Troop-Leading Procedures	**5-6**
	Receive the Mission	5-6
	Issue a Warning Order	5-7
	Make a Tentative Plan	5-8
	Initiate Movement	5-17
	Conduct Reconnaissance	5-18
	Complete the Plan	5-18
	Issue the Operations Order	5-18
	Supervise and Refine	5-19
	Squad Orders	5-20
Chapter 6	**SUSTAINMENT**	**6-1**
	Section I – Individual Responsibilities	**6-1**
	Platoon Sergeant	6-1
	Squad Leader	6-1
	Trauma Specialist/Platoon Medic	6-2
	Combat Lifesaver	6-2
	Section II – Planning Considerations	**6-2**
	Planning	6-3
	Development of the Sustainment Plan	6-3
	Classes of Supply Considerations	6-4
	Maintenance	6-6
	Soldier's Load	6-6
	Combat Load and Basic Load	6-6
	Transportation	6-6
	Section III – Resupply Operations	**6-6**
	Routine Resupply	6-6
	Emergency Resupply	6-8
	Prestock Resupply	6-9
	Aerial Sustainment	6-9
	Section IV – Casualty Procedures	**6-10**
	Initial Care	6-10
	Movement	6-10
	Unit SOPs	6-14
	Casualty Report	6-14
	Killed in Action	6-15
	Section V – Enemy Prisoners of War and Retained/Detained Persons	**6-15**
Chapter 7	**OFFENSIVE OPERATIONS**	**7-1**
	Section I – Introduction to Offensive Operations	**7-1**
	Characteristics of Offensive Operations	7-1
	Types of Offensive Operations	7-2
	Forms of Maneuver	7-3
	Section II – Sequence of Offensive Operations	**7-8**
	Preparation in the Offense	7-8
	Assembly Area	7-9
	Reconnaissance	7-9
	Movement to the Line of Departure	7-9
	Maneuver	7-9
	Deployment	7-9
	Assault	7-10
	Consolidation and Reorganization	7-10

		Page
	Section III – Planning Considerations	**7-11**
	Intelligence	7-11
	Movement and Maneuver	7-11
	Fire Support	7-11
	Protection	7-11
	Sustainment	7-12
	Command and Control	7-12
	Section IV – Platoon Attacks	**7-14**
	Deliberate Attack	7-14
	Hasty Attack	7-17
	Section V – Other Offensive Operations	**7-18**
	Movement to Contact	7-18
	Exploitation	7-24
	Pursuit	7-24
	Attacks During Limited Visibility	7-25
	Section VI – Special Purpose Attacks	**7-26**
	Ambush	7-26
	Raid	7-29
	Counterattack	7-30
	Spoiling Attack	7-30
	Feint	7-31
	Demonstration	7-31
	Section VII – Offensive Tactical Tasks	**7-31**
	Seize	7-31
	Clear	7-32
	Suppress	7-34
	Support by Fire	7-35
	Attack by Fire	7-36
	Section VIII – Urban Areas	**7-36**
	Critical Tasks	7-36
	Find	7-37
	Isolate the Building	7-37
	Assault a Building	7-38
	Follow Through	7-47
	Section IX – Attacking Fortified Positions	**7-48**
	Find	7-48
	Fix	7-49
	Finish — Fighting Enemies in Fortifications	7-49
	Follow Through	7-53
Chapter 8	**DEFENSIVE OPERATIONS**	
	Section I – Characteristics of the Defense	**8-1**
	Preparation	8-1
	Security	8-2
	Disruption	8-2
	Massing Effects	8-2
	Flexibility	8-2
	Section II – Sequence of the Defense	**8-2**
	Reconnaissance, Security Operations, and Enemy Preparatory Fires	8-2
	Occupation	8-3
	Approach of the Enemy Main Attack	8-4
	Enemy Assault	8-4
	Counterattack	8-5
	Consolidation and Reorganization	8-5

		Page
	Section III – Planning Considerations	**8-6**
	Movement and Maneuver	8-6
	Fire Support	8-13
	Protection	8-13
	Sustainment	8-15
	Intelligence	8-15
	Command and Control	8-15
	Section IV – Engagement Area Development	**8-16**
	Identify Likely Enemy Avenues of Approach	8-16
	Identify Enemy Scheme of Maneuver	8-16
	Determine Where to Kill the Enemy	8-16
	Plan and Integrate Obstacles	8-17
	Emplace Weapons Systems	8-17
	Plan and Integrate Indirect Fires	8-17
	Conduct an Engagement Area Rehearsal	8-18
	Section V – Occupation and Preparation of Defensive Positions	**8-19**
	Occupation of the Defense	8-19
	Priority of Work	8-20
	Security in the Defense	8-21
	Establishment of Defensive Positions	8-22
	Weapons Placement	8-22
	Coordination	8-23
	Section VI – Defensive Techniques	**8-23**
	Defend an Area	8-24
	Defend a Battle Position	8-25
	Defend a Strongpoint	8-28
	Defend a Perimeter	8-30
	Defend a Reverse Slope	8-31
	Section VII – Fighting and Survivability Positions	**8-33**
	Principles	8-34
	Fighting Position Materials	8-37
	Types of Fighting Positions	8-38
	Trenches	8-43
	Section VIII – Retrograde	**8-44**
	Withdrawal	8-44
	Delay	8-48
	Stay-Behind Operations	8-49
	Retirement	8-50
Chapter 9	**PATROLS AND PATROLLING**	**9-1**
	Section I – Overview	**9-1**
	Patrols and Patrolling	9-1
	Purpose of Patrolling	9-1
	Types of Patrols	9-2
	Organization of Patrols	9-2
	Initial Planning and Coordination for Patrols	9-3
	Completion of the Patrol Plan	9-4
	Departure from Friendly Lines or Fixed Base	9-5
	Rally Points	9-6
	Section II – Combat Patrols	**9-7**
	Combat Patrol Planning	9-8
	Actions on the Objective – Raid	9-10
	Actions on the Objective – Ambush	9-10
	Security Patrols	9-22

		Page
	Section III – Reconnaissance Patrols	**9-22**
	Control Measures	9-24
	Actions on the Reconnaissance Objective	9-27
	Section IV – Area Reconnaissance	**9-29**
	Actions from the Objective Rally Point	9-29
	Observation Plan for an Area Reconnaissance	9-29
	Short Range	9-30
	Multiple Reconnaissance and Surveillance Teams	9-30
	Security Element	9-30
	Surveillance Teams	9-30
	Section V – Route Reconnaissance	**9-31**
	Conduct	9-31
	Section VI – Zone Reconnaissance	**9-33**
	Moving Element Techniques	9-33
	Stationary Element Techniques	9-36
	Multiple Area Reconnaissance	9-37
	Section VII – Patrol Preparations	**9-37**
	Preparations	9-37
	Section VIII – Post Patrol Activities	**9-41**
	Accounting for Patrol Members	9-41
Appendix A	**MACHINE GUN EMPLOYMENT**	**A-1**
Appendix B	**SHOULDER-LAUNCHED MUNITIONS AND CLOSE COMBAT MISSILE SYSTEMS**	**B-1**
Appendix C	**FIRE PLANNING**	**C-1**
Appendix D	**VEHICLE EMPLOYMENT CONSIDERATIONS**	**D-1**
Appendix E	**HELICOPTER MOVEMENT**	**E-1**
Appendix F	**OBSTACLE REDUCTION AND EMPLOYMENT**	**F-1**
Appendix G	**OTHER SMALL UNIT ORGANIZATONS**	**G-1**
Appendix H	**SECURITY**	**H-1**
Appendix I	**IMPROVISED EXPLOSIVE DEVICES, SUICIDE BOMBERS, AND UNEXPLODED ORDANCE**	**I-1**
	GLOSSARY	**Glossary-1**
	REFERENCES	**References-1**
	INDEX	**Index-1**

Preface

This field manual provides a doctrinal framework on how Infantry rifle platoons and squads fight. It also addresses rifle platoon and squad non-combat operations across the spectrum of conflict. Content discussions include principles, tactics, techniques, procedures, terms, and symbols that apply to small unit operations in the current operational environment (COE). FM 3-21.8 supersedes FM 7-8, *Infantry Rifle Platoon and Squad*, dated 22 April 1992 (with change 1, dated 1 March 2001). It is not intended to be a stand-alone publication. To fully understand operations of the rifle platoon and squad, leaders must have an understanding of FM 3-21.10, *The Infantry Rifle Company*, and FM 3-21.20 (FM 7-20), *The Infantry Battalion.*

The primary audiences for this manual are Infantry rifle platoon leaders, platoon sergeants, and squad and fire team leaders. Secondary audiences include, instructors in U.S. Army Training and Doctrine Command (TRADOC) schools, writers of Infantry training literature, other Infantry leaders and staff officers, and Reserve Officer Training Candidate (ROTC) and military academy instructors.

Infantry leaders must understand this manual before they can train their companies using ARTEP 7-8 *MTP*, and ARTEP 7-8 *Drill*. They should use this manual as a set along with the publications listed in the references.

The Summary of Changes list major changes from the previous edition by chapter and appendix. Although these changes include lessons learned from training and U.S. Army operations all over the world, they are not specific to any particular theater of war. They are intended to apply across the entire spectrum of conflict.

This publication applies to the Active Army, the Army National Guard (ARNG)/Army National Guard of the United States (ARNGUS), and the United States Army Reserve (USAR) unless otherwise stated.

The proponent for this publication is TRADOC. The preparing agency is the U.S. Army Infantry School (USAIS). You may send comments and recommendations for improvement of this manual by U.S. mail, e-mail, fax, or telephone. It is best to use DA Form 2028, *Recommended Changes to Publications and Blank Forms,* but any format is acceptable as long as we can clearly identify and understand your comments. Point of contact information follows:

E-mail: doctrine@benning.army.mil
Phone: COM 706-545-7114 or DSN 835-7114
Fax: COM 706-545-7500 or DSN 835-7500
US Mail: Commandant, USAIS, ATTN; ATSH-ATD, 6751 Constitution Loop
Fort Benning, GA 31905-5593

Unless otherwise stated, whenever the masculine gender is used, both men and women are implied.

SUMMARY OF CHANGE

FM 7-8, *Infantry Rifle Platoon and Squad*, has been updated and renumbered as FM 3-21.8, *The Infantry Rifle Platoon and Squad*. Following is an overview of the significant changes and updates in this new manual:

- Introduces the concept of the Contemporary Operational Environment (COE).
- Introduces the concept of the Warrior Ethos.
- Introduces the concept of Every Soldier is a Sensor (ES2).
- Updates the discussion of platoon command and control and troop-leading procedures.
- Adds an updated chapter on direct fire control and distribution.
- Updates the fundamentals of tactical operations in the COE.
- Updates the discussion of the construction of fighting positions.
- Adds a section on urban operations.
- Adds a section on convoy and route security operations, check points, and road blocks.
- Adds a chapter on sustainment.
- Updates the Patrolling chapter (now Patrols and Patrolling) by adding definitions and discussions on point reconnaissance, security, tracking, presence patrols, and pre- and post-patrol activities.
- Adds a chapter addressing risk management and fratricide avoidance.
- Updates the examples of platoon and squad warning orders (WARNOs) and operation orders (OPORDs).
- Replaces the term "Combat Service Support (CSS)" with "Sustainment".
- Updates the room-clearing drill. The platoon attack drill has been eliminated.
- Updates the discussion of range cards and sector sketches.
- Adds an appendix on AT section employment.
- Updates the discussion of armored vehicle employment with Infantry, tanks, and BFVs.
- Updates the discussion on hazards of unexploded ordnance (UXO), improvised explosive devices (IEDs), mines, and suicide bombers.
- Updates the discussion of CBRN defense operations.
- Updates the content on obstacle reduction.
- Adds an appendix on security.
- Adds an appendix on helicopter employment.
- Updates information on fire planning.
- Removes all Infantry Battle Drills.

Chapter 1
Fundamentals of Tactics

The mission of the Infantry is to close with the enemy by means of fire and maneuver in order to destroy or capture him, or to repel his assault with fire, close combat, and counterattack. The Infantry will engage the enemy with combined arms in all operational environments to bring about his defeat. The close combat fight is not unique to the Infantry.

SECTION I — FUNDAMENTALS OF INFANTRY PLATOON AND SQUAD OPERATIONS

1-1. The Infantry's primary role is close combat, which may occur in any type of mission, in any theater, or environment. Characterized by extreme violence and physiological shock, close combat is callous and unforgiving. Its dimensions are measured in minutes and meters, and its consequences are final. Close combat stresses every aspect of the physical, mental, and spiritual features of the human dimension. To this end, Infantrymen are specially selected, trained, and led.

INFANTRY

1-2. Of all branches in the U.S. Army, the Infantry is unique because its core competency is founded on the individual Soldier—the Infantry rifleman. While other branches tend to focus on weapon systems and platforms to accomplish their mission, the Infantry alone relies almost exclusively on the human dimension of the individual rifleman to close with and destroy the enemy. This Soldier-centric approach fosters an environment that places the highest value on individual discipline, personal initiative, and performance-oriented leadership. The Infantry ethos is encapsulated by its motto: Follow Me!

1-3. Although the battlefield may be entered from a differing range of platforms, all types of Infantry must be able to fight on their feet. To perform this role, each type possesses two distinguishing qualities. First, Infantry are able to move almost anywhere under almost any condition. Second, Infantry can generate a high volume of lethal well-aimed small arms fire for a short time in any direction. Neither movement nor fire are exclusively decisive. However, combined fire and movement win engagements. These two strengths reveal three distinct vulnerabilities to Infantry. First, once committed it is difficult to adjust the Infantry's line of advance due to its limited tactical mobility. Second, determining the Infantryman's load required to accomplish the mission is always in conflict with preserving his physical ability to fight the enemy. Third, Infantry are particularly susceptible to the harsh conditions of combat, the effects of direct and indirect fire, the physical environment, and moral factors.

OFFENSIVE AND DEFENSIVE COMBAT

1-4. Infantry platoons and squads have a distinct position on the battlefield—the point of decision. Their actions take place at the point where all of the plans from higher headquarters meet the enemy in close combat. This role requires leaders at all levels to quickly understand the situation, make decisions, and fight the enemy to accomplish the mission. Offensive close combat has the objective of seizing terrain and destroying the adversary. Defensive close combat denies an area to the adversary and protects friendly forces for future operations. Both types constitute the most difficult and costly sorts of combat operations.

BASIC ACTIONS

1-5. Whether operating on its own or as part of a larger force, the goal of Infantry platoons and squads remains constant: defeat and destroy enemy forces, and seize ground. To achieve this end state, Infantry platoons and squads rely on two truths.

(1) In combat, Infantrymen who are moving are attacking.

(2) Infantrymen who are not attacking are preparing to attack.

1-6. These two truths highlight another truth—offensive action and defensive action are reciprocal opposites that are found in all actions.

1-7. At the platoon and squad level it is necessary to make a clear distinction between these two basic actions of attacking and defending, and larger scale offensive and defensive operations. The difference is one of degree, not type. Offensive and defensive operations are types of full spectrum operations that are undertaken by higher-level units.

TACTICAL PRINCIPLES

1-8. To achieve the basic truths of offense or defense, Infantrymen rely on fundamental principles. From these they derive their basic tactics, techniques, and procedures used to conduct operations. The information in Table 1-1 is introductory and forms the basis for the remainder of this chapter.

Table 1-1. Tactical principles.

PRINCIPLE
Tactical Maneuver: Fire without movement is indecisive. Exposed movement without fire is disastrous. There must be effective fire combined with skillful movement. A detailed explanation of the supporting concepts is in Chapter 2.
Advantage: Seek every opportunity to exploit your strengths while preventing the enemy from exploiting his own strengths.
Combinations: The power of combination creates dilemmas that fix the enemy, overwhelming his ability to react while protecting your own internal weaknesses.
Tactical Decisionmaking: Close combat demands flexible tactics, quick decisions, and swift maneuvers to create a tempo that overwhelms the enemy.
Individual Leadership: Resolute action by a few determined men is often decisive.
Combat Power: The ability of a unit to fight.
Situation: Every military situation is unique and must be solved on its own merits.

Tactical Maneuver

1-9. Tactical maneuver is the way in which Infantry platoons and squads apply combat power. Its most basic definition is fire plus movement, and is the Infantry's primary tactic when in close combat. Fire without movement is indecisive. Exposed movement without fire is potentially disastrous. Inherent in tactical maneuver is the concept of protection. The principle of tactical maneuver is more fully explained in Chapter 3, and is further integrated in other sections of this manual.

Advantage

1-10. Leaders and Soldiers must look for every opportunity to gain and maintain an advantage over the enemy. In close combat there is no such thing as a fair fight. As much as possible, leaders must set the conditions of an engagement, confronting the enemy on his terms, while forcing the enemy into unsolvable dilemmas to defeat or destroy him. Important supporting concepts are doctrine and training, individual Infantry skills, and the organization of the Infantry platoon and its squads.

1-11. Surprise means taking the enemy when the enemy is unprepared. Leaders continuously employ security measures to prevent the enemy from surprising them. Infantry platoons and squads should be especially concerned with their own security. They should expect the unexpected while avoiding patterns. Tactical surprise is rarely gained by resorting to the obvious.

1-12. The ability to generate and apply combat power is a significant advantage of the Infantry platoon and squad. This advantage results from the training of the units' Soldiers; the Soldiers' organization into teams, squads, and platoons; Soldiers' collective training in tasks and drills; and Soldiers' ability to integrate other assets and units into their formations. Through these elements, leaders exploit strengths while mitigating vulnerabilities.

Combination

1-13. Based on the power of force and firepower combinations, combined arms is how Army forces fight. Leaders creatively combine weapons, units, and tactics using the principles of complementary and reinforcing effects to create dilemmas for the enemy. Making effective and efficient combinations puts a premium on technical competence. Leaders must know the characteristics of the weapons and munitions when employing fires. They must understand the inherent capabilities and limitations of their own and other unit formations.

Tactical Decisionmaking

1-14. Tactical decisionmaking is the ability to make decisions during all phases of the operations process (plan, prepare, execute, and assess). Within this framework, Infantry platoon and squad leaders exercise command and control (C2) to be both effective and efficient in accomplishing their mission. Effectiveness entails making accurate assessments and good decisions about how to fight the enemy. Control complements command by using the most efficient means available. Key supporting concepts are troop-leading procedures, actions on contact, and risk management.

Individual Leadership

1-15. Leadership at the Infantry platoon and squad level is comprised of three fundamental concepts: leadership by example, authority, and mission command. Leadership by example is simply and most powerfully expressed by the Infantry's motto: Follow Me! Authority is the power to act. Mission command is the Army's command philosophy that focuses on leaders telling subordinates what must be accomplished and why. Leaving the how to do it up to the subordinate.

WARFIGHTING FUNCTIONS

1-16. A warfighting function is a group of tasks and systems (people, organization, information, and processes) united by a common purpose that commanders use to accomplish missions and training objectives. The warfighting functions are intelligence, movement and maneuver, fire support, protection, sustainment, and command and control. These warfighting functions replace the battlefield operating systems.

1-17. Commanders visualize, describe, direct, and lead operations and training in terms of the warfighting functions. Decisive, shaping, and sustaining operations combine all the warfighting functions. No function is exclusively decisive, shaping, or sustaining. Figure 1-1 illustrates the warfighting elements of combat power.

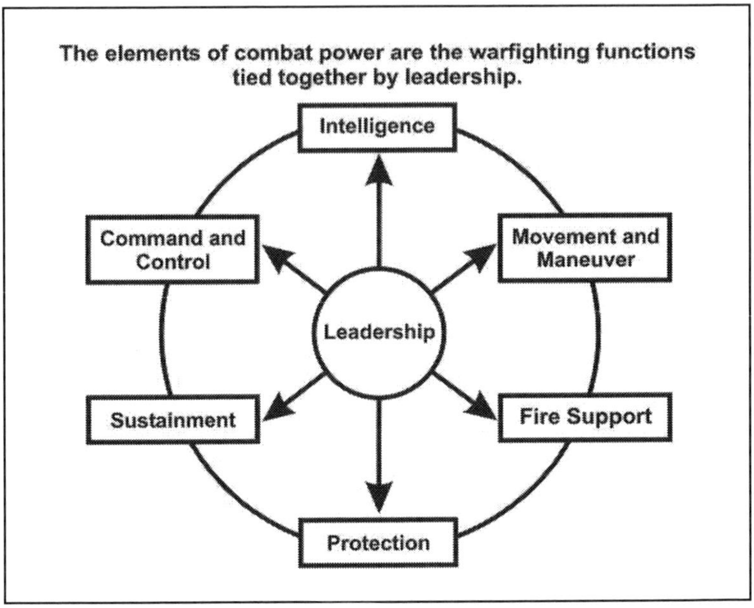

Figure 1-1. Warfighting elements of combat power.

INTELLIGENCE

1-18. The intelligence warfighting function involves the related tasks and systems that facilitate understanding of the enemy, terrain, weather, and civil considerations. It includes those tasks associated with intelligence, surveillance, and reconnaissance. The intelligence warfighting function combines a flexible and adjustable architecture of procedures, personnel, organizations, and equipment to provide commanders with relevant information and products relating to an area's threat, civil populace, and environment.

MOVEMENT AND MANEUVER

1-19. The movement and maneuver warfighting function involves the related tasks and systems that move forces to achieve a position of advantage in relation to the enemy. It includes those tasks associated with employing forces in combination with direct fire or fire potential (maneuver), force projection (movement), and mobility and countermobility. Movement and maneuver are the means through which commanders concentrate combat power to achieve surprise, shock, momentum, and dominance.

FIRE SUPPORT

1-20. The fire support warfighting function involves the related tasks and systems that provide collective and coordinated use of Army indirect fires, joint fires, and offensive information operations. It includes those tasks associated with integrating and synchronizing the effects of these types of fires with the other warfighting functions to accomplish operational and tactical objectives.

PROTECTION

1-21. The protection warfighting function involves the related tasks and systems that preserve the force so the commander can apply maximum combat power. Preserving the force includes protecting personnel (combatant and noncombatant), physical assets, and information of the United States and multinational partners. The following tasks are included in the protection warfighting function:

- Safety.
- Fratricide avoidance.
- Survivability.
- Air and missile defense.
- Antiterrorism.
- Counterproliferation and consequence management actions associated with chemical, biological, radiological, nuclear, and high-yield explosive weapons.
- Defensive information operations.
- Force health protection.

SUSTAINMENT

1-22. The sustainment warfighting function involves the related tasks and systems that provide support and services to ensure freedom of action, extend operational reach, and prolong endurance. Sustainment includes those tasks associated with—

- Maintenance.
- Transportation.
- Supply.
- Field services.
- Explosive ordnance disposal.
- Human resources support.
- Financial management.
- Health service support.
- Religious support.
- Band support.
- Related general engineering.

1-23. Sustainment allows uninterrupted operations through adequate and continuous logistical support such as supply systems, maintenance, and other services.

COMMAND AND CONTROL

1-24. The command and control warfighting function involves the related tasks and systems that support commanders in exercising authority and direction. It includes the tasks of acquiring friendly information, managing relevant information, and directing and leading subordinates.

1-25. Command and control has two parts: the commander; and the command and control system. Information systems—including communications systems, intelligence-support systems, and computer networks—back the command and control systems. They let the commander lead from anywhere in their area of operations (AO). Through command and control, the commander initiates and integrates all warfighting functions.

Combat Power

1-26. Combat power is a unit's ability to fight. The primary challenge of leadership at the tactical level is mastering the art of generating and applying combat power at a decisive point to accomplish a mission.

Leaders use the operations process (plan, prepare, execute, and assess) to generate combat power. They conduct operations following the Find, Fix, Finish, and Follow-through model to apply combat power.

1-27. At the core of a unit's ability to fight are three time-tested components of close combat:

(1) Firepower.

(2) Mobility.

(3) Protection/Security.

1-28. These components appear throughout military history under various names as the central elements required to fight and win against the enemy. Firepower consists of the weapons used to inflict casualties upon the enemy. Firepower alone is indecisive without movement. Mobility is the ability to move on the battlefield, dictating the speed, tempo, and tactical positioning of forces. Inherent in both firepower and mobility is the need for protection from the enemy's firepower and mobility. Leaders employ protection and security measures to preserve their unit's ability to fight. They deny the enemy protection through creative combinations of unit firepower and mobility.

Situation

1-29. Every combat situation is unique. Leaders do their best to accurately assess the situation and make good decisions about employing their units. The environment of combat, the application of military principles, and the desired end state of Army operations culminate with the close fight of Infantry platoons and squads. The leader should understand the larger military purpose and how his actions and decisions might affect the outcome of the larger operation.

CHARACTERISTICS OF CLOSE COMBAT

1-30. Close combat is characterized by danger, physical exertion and suffering, uncertainty, and chance. To combat these characteristics, Soldiers must have courage, physical and mental toughness, mental stamina, and flexibility.

COURAGE

1-31. Courage is the quality Soldiers must possess to face and overcome danger. Hazards, real or potential, are an ever-present aspect of the battlefield. Physical courage is necessary to deal with combat hazards. Physical courage results from two sources: mental conditioning that comes from demanding training; and motives such as personal pride, enthusiasm, and patriotism. Moral courage is necessary to face responsibilities and do what is necessary and right.

PHYSICAL AND MENTAL TOUGHNESS

> "The first quality of the Soldier is fortitude in enduring fatigue and privation; valor is only the second. Poverty, privation, and misery are the school of the good Soldier."
>
> *Napoleon, Maxim LVIII*

1-32. Physical and mental toughness are the qualities Soldiers must have to combat physical exertion and suffering. Physical toughness enables the Soldier to endure hardship and perform his rigorous duties. Mental toughness enables the Soldier to put the harshness of the environment and his duties into proper perspective. Mentally tough Soldiers can do what needs to be done to accomplish the mission.

MENTAL STAMINA

1-33. The individual's awareness during combat is never complete. There is no such thing as perfect awareness or understanding of the situation. Mental stamina is the quality Soldiers must have to combat this uncertainty. Mental stamina provides the ability to assess the situation based on whatever facts are at hand, to intuitively make reasonable assumptions about what is not known, and to make logical decisions based on that information.

FLEXIBILITY

1-34. Chance is luck, opportunity, and fortune, and happens to both sides in close combat. It is not predictable. However, it must be dealt with in that Soldiers must be flexible, resolute, and able to continuously look forward.

SECTION II — DOCTRINE AND TRAINING

INFANTRY DOCTRINE

1-35. Doctrine contains the fundamental principles by which the military forces or elements guide their actions in support of national objectives. It is authoritative, but requires judgment in application (FM 1-02, *Operational Terms and Graphics*). Infantry doctrine expresses the concise expression of how Infantry forces fight. It is comprised of principles; tactics, techniques, and procedures (TTP); and terms and symbols.

1-36. Infantry doctrine is based on hard-fought lessons from generations of combat Infantry Soldiers engaged in numerous conflicts. Doctrine is always evolving and adapting, yet its fundamental principles are as true today as they were generations ago.

1-37. Infantry doctrine facilitates communication among Infantry Soldiers regardless of where they serve, contributes to a shared professional culture, and serves as the basis for training and instruction. Infantry doctrine provides a common language and a common understanding of how Infantry forces conduct tasks and operations. To be useful, doctrine must be well known and commonly understood.

PRINCIPLES

1-38. Principles are fundamental concepts and facts underlying the conduct of tasks and operations. Principles are usually general, flexible, and apply across a broad spectrum. Because they are broad, they apply at the Infantry platoon and squad levels as well as at the higher levels with relatively the same meaning. Therefore, leaders at all levels need to remain aware of both the generic and specific aspects of doctrinal terms.

TACTICS, TECHNIQUES, AND PROCEDURES

1-39. One of the defining characteristics of war is chaos. TTP are the counterweight to this chaos. From the moment combat begins, plans often become obsolete, communications fail, Soldiers become casualties, and units fragment. Military tactics are the practical means armies use to achieve battlefield objectives. From this, "tactics" came to imply the deliberate control of military formation, movement and fire, and the attempt to impose order where there is disorder to defeat the enemy.

1-40. TTP are those generally-accepted practices used to conduct operations. "Generally accepted" means that the doctrine described is applicable to most operations, most of the time, and that there is widespread consensus about their value and usefulness. "Generally accepted" does not mean that doctrine should be applied uniformly on all missions. Leaders use their own standing operating procedures (SOPs) and judgment to determine what is appropriate based on the specific mission, enemy, terrain, troops-time, civil (METT-TC) conditions.

Tactics

1-41. Tactics are: (1) The employment of units in combat. (2) The ordered arrangement and maneuver of units in relation to each other or to the enemy to utilize their full potential (FM 1-02). Tactics are the ways that we engage in combat with an enemy force.

Techniques

1-42. Techniques are the general and detailed methods used by troops or commanders to perform assigned missions and functions, specifically the methods of using equipment and personnel. Techniques are the general methods used by the leader and his subordinates to perform the tactic. Techniques describe a way, not the only way (FM 1-02).

Procedures

1-43. Procedures are standard methods used by the leader and his subordinates to perform and accomplish a task or a portion of a task. For example, when the unit sustains a casualty, the leader or a radiotelephone operator (RTO) might use the 9-line medical evacuation (MEDEVAC) procedure to call for medical assistance.

Terms and Symbols

1-44. Doctrine provides a common language that professionals use to communicate with one another. Terms with commonly understood definitions are a major component of the language. Symbols are its graphical representation. Establishing and using words and symbols of common military meaning enhances communications among military professionals in all environments, and makes a common understanding of doctrine possible. (See FM 1-02.)

INDIVIDUAL INFANTRY SKILLS

1-45. Every Infantryman, from the private enlisted Soldier, to the general officer, is first a rifleman. As such, he must be a master of his basic skills: shoot, move, communicate, survive, and sustain. These basic skills provide the Soldier's ability to fight. When collectively applied by the fire team, squad, and platoon, these skills translate into combat power.

Shoot

1-46. Infantrymen must be able to accurately engage the enemy with all available weapons. Soldiers and their leaders must therefore be able to determine the best weapon-ammunition combination to achieve the desired effect. The best combination will expend a minimum of ammunition expenditure and unintended damage. To make this choice, they must know the characteristics, capabilities, and vulnerabilities of their organic and supporting assets. This means understanding the fundamental characteristics of the weapon's lay (direct or indirect), ammunition (high explosive [HE], penetrating, or special purpose), trajectory (high or low), and enemy targets (point or area). Properly applying these variables requires an understanding of the nature of targets, terrain, and effects.

Move

1-47. Tactical movement is inherent in all Infantry operations. Movement is multifaceted, ranging from dismounted, to mounted, to aerial modes, and is conducted in varying physical environments, including the urban environment. For the individual, movement is comprised of the individual movement techniques (IMT) of high crawl, low crawl, and 3-5 second rush; for the unit it is comprised of movement formations, movement techniques, and maneuver (fire and movement). Mastering the many aspects of tactical movement is fundamental. More importantly, Infantrymen must be thoroughly trained in the critical transition from tactical movement to maneuver.

1-48. Understanding the terrain is critical to applying the fundamental of the particulars of shoot and move. There are four basic terrain-related skills. First, the leader must know how to land navigate, mounted and dismounted, day and night, using the latest technology (global positioning systems [GPS], Falcon View). Second, leaders need to understand the basics of how to analyze the military aspects of terrain, Observation and fields of fire, Avenues of approach, Key and decisive terrain, Obstacles, Cover and concealment. (OAKOC). Third, once they understand how to look at the terrain in detail, leaders must understand how to integrate the aspects of fire (direct and indirect) and tactical movement to fit the terrain. Fourth, leaders must understand how to apply generic tactics and techniques to the unique terrain they are in, because understanding and appreciating terrain is an essential leader skill.

COMMUNICATE

1-49. Soldiers communicate to provide accurate and timely information to those who need it. Information is necessary to successfully execute combat operations. It enables leaders to achieve situational understanding, make decisions, and give orders. There are two aspects of communication: the technical means used to communicate; and the procedures used for reporting and disseminating information. The Soldier's and leader's ability to use information to assess the situation, make decisions, and direct necessary actions are also significant aspects in the communication process.

SURVIVE

1-50. To fully contribute to the mission, Soldiers must be able to survive. There are three aspects to surviving: the enemy; the environment; and the Soldier's body. Survival is both a personal responsibility and a unit responsibility. These aspects require Soldiers to discipline themselves in routine matters such as maintaining local security, maintaining field sanitation, caring for their bodies, and caring for their equipment. It also requires Soldiers to know how to respond to extraordinary circumstances such as dealing with casualties or functioning in a contaminated environment. Soldiers must know about the protective properties of their personal gear and combat vehicles, the effects of weapon systems and munitions, and how to build survivability positions. In short, Soldiers must do everything possible for the security and protection of themselves, their equipment, and their fellow Soldiers. In the same way, leaders must do everything possible to ensure the security and protection of their units.

SUSTAIN

1-51. Sustainment is an inherent feature in all operations. In order to shoot, ammunition is needed. Fuel and repair parts are needed for movement, and batteries are needed to communicate. To survive, the Soldier needs food and water. Soldiers and leaders need to forecast requirements before they need them, while at the same time managing the Soldier's load.

WARRIOR ETHOS AND ARMY VALUES

1-52. Warrior Ethos refers to the professional attitudes and beliefs that will characterize you. Developed through discipline, commitment to Army Values and knowledge of the Army's proud heritage, Warrior Ethos notes military service as much more than just a "job" — it is a profession with the enduring purpose to win wars and destroy our nation's enemies. Figure 1-2 displays the Warrior Ethos definition as embedded within the current Soldier's Creed:

Soldier's Creed

I am an American Soldier.

I am a Warrior and a member of a team. I serve the people of the United States and live the Army Values.

I will always place the mission first.

I will never accept defeat.

WARRIOR ETHOS

I will never quit.

I will never leave a fallen comrade.

I am disciplined, physically and mentally tough, trained and proficient in my warrior tasks and drills. I always maintain my arms, my equipment and myself.

I am an expert and I am a professional.

I stand ready to deploy, engage, and destroy the enemies of the United States of America in close combat.

I am a guardian of freedom and the American way of life.

I am an American Soldier.

Figure 1-2. The Soldier's Creed.

1-53. Warrior Ethos is the foundation for your commitment to victory in times of peace and war. While always exemplifying the tenets of Warrior Ethos — place the mission first, refuse to accept defeat, and never quit or leave a fallen comrade behind. You must have absolute faith in yourself. And you must have complete faith in your team, because they are trained and equipped to destroy the enemy in close combat.

1-54. The Army Values consist of the principles, standards, and qualities considered essential for successful Army leaders. They firmly bind all Army members into a fellowship dedicated to serve the Nation and the Army. Figure 1-3 lists the seven Army Values. It is not a coincidence that when reading the first letters of the Army Values in sequence they form the acronym "LDRSHIP".

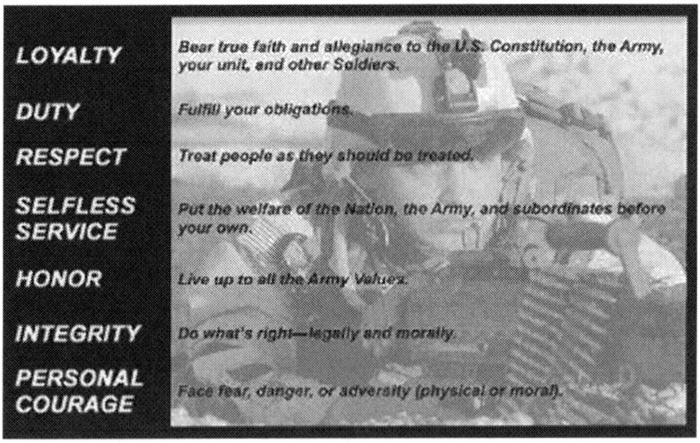

LOYALTY	Bear true faith and allegiance to the U.S. Constitution, the Army, your unit, and other Soldiers.
DUTY	Fulfill your obligations.
RESPECT	Treat people as they should be treated.
SELFLESS SERVICE	Put the welfare of the Nation, the Army, and subordinates before your own.
HONOR	Live up to all the Army Values.
INTEGRITY	Do what's right—legally and morally.
PERSONAL COURAGE	Face fear, danger, or adversity (physical or moral).

Figure 1-3. The Army Values.

EVERY SOLDIER IS A SENSOR (ES2)

1-55. Soldiers must be trained to actively observe details related to the commander's critical information requirements (CCIR) in an AO. They must also be competent in reporting their experience, perception, and judgment in a concise, accurate manner. Leaders who understand how to optimize the collection, processing, and dissemination of information in their organization enable the generation of timely intelligence. To accommodate this, leaders must create a climate that allows all Infantryman to feel free to report what they see and learn on a mission.

1-56. ES2 trains Soldiers and leaders to see intelligence development as everyone's responsibility. All must fight for knowledge to gain and maintain greater situational understanding. At the heart of the concept is the art of combat (tactical) collection. This process involves leaders directing and maximizing the collection of combat intelligence by patrols, and Soldiers who understand their vital role as collectors of combat information.

TACTICAL QUESTIONING

1-57. Tactical questioning involves the expedient initial questioning of an AO's local population to gather information of immediate value. Because tactical questioning applies to interaction with the local population, it is more "conversational" than "questioning" in nature. The Infantry Soldier conducts tactical questioning based on the unit's standing operating procedures, rules of engagement, and the order for that mission.

SITE EXPLOITATION

1-58. Site exploitation is defined as the search of a specific location or area to gain items of intelligence value. Locations may include apartments, buildings, multiple structures, compounds, or fields. Once a site has been cleared of enemy personnel, Infantry platoons will search for items of interest. Search items may include:

- Maps.
- Propaganda material.
- Phone or computer records.
- Photos.
- Weapons.

DEBRIEFING AND REPORTING

1-59. Once the platoon returns from the objective or site, a detailed debrief should begin. Everyone on the mission has a role to play in a debrief. A practical method for debriefing is to review all patrol actions chronologically. Leaders should not consider the mission complete or the personnel released until the debriefings and reporting are done.

1-60. All information collected by platoons in contact with the local population is reported through the chain of command. Upon return from the mission, photos should be downloaded. All material taken from the objective should be laid out.

1-61. Finally, as detailed a sketch as possible should be made for visual reference of debriefed patrol areas. For detailed information on debriefing, reporting, and tactical questions see FMI 2-91.4, *Intelligence Support to Operations in the Urban Environment.*

SECTION III — ORGANIZATION

INFANTRY PLATOON

1-62. The Infantry platoon is organized with three Infantry squads, a weapons squad, and a platoon headquarters. The headquarters section provides C2 of the squads and any attachments, and serves as the interface with the fire support and sustainment systems. Although all Infantry platoons use the same basic doctrinal principles in combat, application of those principles differs based on assigned organization or task organization (Figure 1-4).

1-63. One of the inherent strengths of the Infantry platoon is the ability to task organize. The Infantry platoon headquarters must expect to receive other Soldiers and units in command relationships, and direct other arms in support relationships.

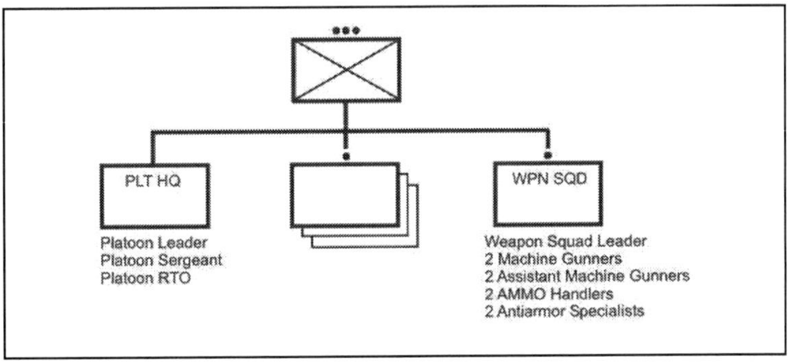

Figure 1-4. Infantry platoon.

PLATOON HEADQUARTERS

1-64. The platoon headquarters has three permanently assigned members: the platoon leader, the platoon sergeant, and the radiotelephone operator (RTO). Depending on task organization, the platoon headquarters may receive augmentation. Two traditionally-attached assets are the fire support team, and the platoon medic.

PLATOON LEADER

1-65. The platoon leader leads his subordinates by personal example. The platoon leader exercises authority over his subordinates and overall responsibility for those subordinates' actions. This centralized authority enables the platoon leader to act decisively while maintaining troop discipline and unity. Under the fluid conditions of close combat, even in the course of carefully-planned actions, the platoon leader must accomplish assigned missions using initiative without constant guidance from above.

Responsibilities

1-66. The platoon leader is responsible for all the platoon does or fails to do. In the conduct of his duties he consults the platoon sergeant in all matters related to the platoon. He must know his Soldiers and how to employ the platoon and its organic and supporting weapons. During operations, the platoon leader—

- Leads the platoon in supporting the higher headquarters missions. He bases his actions on his assigned mission and the intent and concept of his higher commanders.
- Maneuvers squads and fighting elements.
- Synchronizes the efforts of squads.
- Looks ahead to the next "move" for the platoon.
- Requests and controls supporting assets.
- Employs C2 systems available to the squads and platoon.
- Ensures 360-degree, three-dimensional security is maintained.

- Controls the emplacement of key weapon systems.
- Issues accurate and timely reports.
- Places himself where he is most needed to accomplish the mission.
- Assigns clear tasks and purposes to his squads.
- Understands the mission and commanders intent two levels up (the company and battalion).

Situational Understanding

1-67. The platoon leader works to develop and maintain situational understanding (SU). SU is a product of four elements. First, the platoon leader attempts to know what is happening in the present in terms of friendly, enemy, neutral, and terrain situations. Second, the platoon leader must know the end state that represents mission accomplishment. Third, the platoon leader determines the critical actions and events that must occur to move his unit from the present to the end state. Finally, the platoon leader must be able to assess the risk throughout.

PLATOON SERGEANT

1-68. The platoon sergeant (PSG) is the senior NCO in the platoon and second in command. He sets the example in everything. He is a tactical expert in Infantry platoon and squad operations, which include maneuver of the platoon-sized elements, and employment of all organic and supporting weapons. The platoon sergeant advises the platoon leader in all administrative, logistical, and tactical matters. The platoon sergeant is responsible for the care of the men, weapons, and equipment of the platoon. Because the platoon sergeant is the second in command, he has no formal assigned duties except those assigned by the platoon leader. However, the platoon sergeant traditionally—

- Ensures the platoon is prepared to accomplish its mission, to include supervising precombat checks and inspections.
- Prepares to assume the role and responsibilities of platoon leader.
- Acts where best needed to help C2 the engagement (either in the base of fire or with the assault element).
- Receives squad leaders' administrative, logistical, and maintenance reports, and requests for rations, water, fuel, and ammunition.
- Coordinates with the higher headquarters to request logistical support (usually the company's first sergeant or executive officer).
- Manages the unit's combat load prior to operations, and monitors logistical status during operations.
- Establishes and operates the unit's casualty collection point (CCP) to include directing the platoon medic and aid/litter teams in moving casualties; maintains platoon strength levels information; consolidates and forwards the platoon's casualty reports; and receives and orients replacements.
- Employs digital C2 systems available to the squads and platoon.
- Understands the mission and commanders intent two levels up (the company and battalion).

PLATOON RADIOTELEPHONE OPERATOR

1-69. The platoon radiotelephone operator (RTO) is primarily responsible for the platoon's communication with its controlling HQ (usually the company). During operations, the RTO will—

- Have communications at all times. If communication with the platoon's next higher element is lost, the RTO immediately informs the platoon leader or platoon sergeant.
- Conduct radio checks with higher (in accordance with unit SOPs) when in a static position. If the RTO cannot make successful radio contact as required, he will inform the platoon sergeant or platoon leader.
- Be an expert in radio procedures and report formats such as call for indirect fire or MEDEVAC, and all types of field expedient antennas.

- Have the frequencies and call signs on his person in a location known to all Soldiers in the platoon.
- Assist the platoon leader with information management.
- Assist the platoon leader and platoon sergeant employing digital C2 systems available to the squads and platoon.
- Determine his combat load prior to operations and manage his batteries during operations.

FORWARD OBSERVER

1-70. The forward observer (FO), along with a fire support RTO, is the unit's SME on indirect fire planning and execution. The FO is the primary observer for all fire support (FS) assets to include company mortars (if assigned), battalion mortars, field artillery, and any other allocated FS assets. He is responsible for locating targets and calling and adjusting indirect fires. He must know the mission and the concept of operation, specifically the platoon's scheme of maneuver and concept of fires. He works directly for the platoon leader and interacts with the next higher headquarters' fire support representative. The FO must also—

- Inform the FIST headquarters of the platoon situation, location, and fire support requirements.
- Prepare and use maps, overlays, and terrain sketches.
- Call for and adjust indirect fires.
- Operate as a team with the fire support RTO.
- Select targets to support the platoon's mission.
- Select observation post(s) (OP) and movement routes to and from selected targets.
- Operate digital message devices and maintain communication with the battalion and company fire support officer (FSO).
- Maintain grid coordinates of his location.
- Be prepared to back up the platoon leader's radio on the higher headquarters net if needed.
- Be prepared to employ close air support assets.

PLATOON MEDIC

1-71. The platoon medic is assigned to the battalion medical platoon and is attached upon order. His primary function is force health protection. As such, he is the unit's SME on treatment and evacuation of casualties. He works directly for the platoon sergeant. However, he also interacts heavily with the company's senior medic. During operations the medic—

- Treats casualties and assists the aid and litter teams with their evacuation.
- Advises the platoon leader and platoon sergeant on all force health protection matters, and personally checks the health and physical condition of platoon members.
- Reports all medical situations and his actions taken to the platoon sergeant.
- Requests Class VIII (medical) supplies for the platoon through the company medic.
- Provides training and guidance to combat lifesavers.

INFANTRY FIRE TEAM

1-72. The Infantry fire team is designed to fight as a team and is the fighting element within the Infantry platoon. Infantry platoons and squads succeed or fail based on the actions of their fire teams.

1-73. The Infantry fire team is designed as a self-contained team (Figure 1-5). The automatic rifleman (AR) provides an internal base of fire with the ability to deliver sustained suppressive small arms fire on area targets. The rifleman provides accurate lethal direct fire for point targets. The grenadier provides high explosive (HE) indirect fires for both point and area targets. A team leader (TL) who provides C2 through leadership by example ("Do as I do") leads this team.

Figure 1-5. Infantry fire team.

RIFLEMAN

1-74. The rifleman provides the baseline standard for all Infantrymen and is an integral part of the fire team. He must be an expert in handling and employing his weapon. Placing well-aimed, effective fire on the enemy is his primary capability. Additionally, the rifleman must—

- Be an expert on his weapon system—his rifle, its optics, and its laser aiming device. He must be effective with his weapon system day or night. He must be capable of engaging all targets with well-aimed shots.
- Be able to employ all weapons of the squad, as well as common munitions.
- Be able to construct and occupy a hasty firing position and know how to fire from it. He must know how to quickly occupy covered and concealed positions in all environments and what protection they will provide for him from direct fire weapons. He must be competent in the performance of these tasks while using night vision devices.
- Be able to fight as part of his unit, which includes being proficient in his individual tasks and drills, being able to fight alongside any member of the unit, and knowing the duties of his teammates and be prepared to fill in with their weapons if needed.
- Be able to contribute as a member of special teams to include wire/mine breach teams, EPW search, aid/litter, and demolitions.
- Be able to inform his team leader of everything he hears and sees when in a tactical situation.
- Be able to perform Soldier-level preventive medicine measures (PMM). (See Chapter 6.)
- Be able to administer buddy aid as required.
- Be able to manage his food, water, and ammunition during operations.
- Be prepared to assume the duties of the automatic rifleman and team leader.
- Understand the mission two levels up (squad and platoon).

GRENADIER

1-75. The grenadier is currently equipped with an M203 weapon system consisting of an M16/M4 rifle and an attached 40-mm grenade launcher. The grenadier provides the fire team with a high trajectory, high explosive capability out to 350 meters. His fire enables the fire team to achieve complementary effects with high trajectory, high explosive munitions, and the flat trajectory ball ammunition of the team's other weapons. The grenade launcher allows the grenadier to perform three functions: suppress and destroy enemy Infantry and lightly-armored vehicles with HE or high explosive dual purpose; provide smoke to screen and cover his squad's fire and movement; and employ illumination rounds to increase his squad's visibility and mark enemy positions. The grenadier must—

- Be able to accomplish all of the tasks of the rifleman.
- Be able to engage targets with appropriate type of rounds both day and night.

- Identify 40-mm rounds by shape and color. He must know how to employ each type of round and know its minimum safety constraints.
- Know the maximum ranges for each type of target for the grenade launcher.
- Know the leaf sight increments without seeing the markings.
- Know how to make an adjustment from the first round fired so he can attain a second-round hit.
- Load the grenade launcher quickly in all firing positions and while running.
- Be prepared to assume the duties of the automatic weapons gunner and the team leader.
- Understand the mission two levels up (squad and platoon).

AUTOMATIC RIFLEMAN

1-76. The AR's primary weapon is currently the 5.56-mm M249 machine gun. The M249 provides the unit with a high volume of sustained suppressive and lethal fires for area targets. The automatic rifleman employs the M249 machine gun to suppress enemy Infantry and bunkers, destroy enemy automatic rifle and antitank teams, and enable the movement of other teams and squads. He is normally the senior Soldier of the fire team. The AR must—

- Be able to accomplish all of the tasks of the rifleman and the grenadier.
- Be prepared to assume the duties of the team leader and squad leader.
- Be able to engage groups of enemy personnel, thin-skinned vehicles, bunker doors or apertures, and suspected enemy locations with automatic fire. He provides suppressive fire on these targets so his teammates can close with and destroy the enemy.
- Be familiar with field expedient firing aids to enhance the effectiveness of his weapon (for example, aiming stakes).
- Be able to engage targets from the prone, kneeling, and standing positions with and without night observation devices. Also understands the mission two levels up (the squad and platoon).

TEAM LEADER

1-77. The team leader leads his team members by personal example. He has authority over his subordinates and overall responsibility for their actions. Centralized authority enables the TL to maintain troop discipline and unity and to act decisively. Under the fluid conditions of close combat, the team leader must accomplish assigned missions using initiative without needing constant guidance from above.

1-78. The team leader's position on the battlefield requires immediacy and accuracy in all of his actions. He is a fighting leader who leads his team by example. The team leader is responsible for all his team does or fails to do. He is responsible for the care of his team's men, weapons, and equipment. During operations, the team leader—

- Is the SME on all of the team's weapons and duty positions and all squad battle drills.
- Leads his team in fire and movement.
- Controls the movement of his team and its rate and distribution of fire.
- Employs digital C2 systems available to the squad and platoon.
- Ensures security of his team's sector.
- Assists the squad leader as required.
- Is prepared to assume the duties of the squad leader and platoon sergeant.
- Enforces field discipline and PMM.
- Determines his team's combat load and manages its available classes of supply as required.
- Understands the mission two levels up (squad and platoon).

1-79. When maneuvering the team, the team fights using one of three techniques:

(1) Individual movement techniques (IMT, the lowest level of movement).
(2) Buddy team fire and movement.
(3) Fire team fire and movement (maneuver).

1-80. Determining a suitable technique is based on the effectiveness of the enemy's fire and available cover and concealment. The more effective the enemy's fire, the lower the level of movement. Because the team leader leads his team, he is able to make this assessment firsthand. Other leaders must be sensitive to the team leader's decision on movement.

INFANTRY SQUAD

1-81. There are several variations of Infantry, but there is currently only one type of Infantry squad (Figure 1-6). Its primary role is a maneuver or base-of-fire element. While the platoon's task organization may change, the organization of the Infantry squad generally remains standard.

1-82. The Infantry squad is a model for all tactical task organizations. It is comprised of two fire teams and a squad leader. It is capable of establishing a base of fire, providing security for another element, or conducting fire and movement with one team providing a base of fire, while the other team moves to the next position of advantage or onto an objective. The squad leader has two subordinate leaders to lead the two teams, freeing him to control the entire squad.

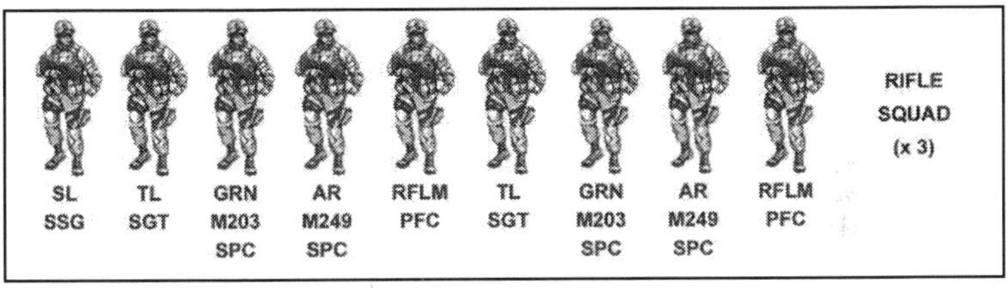

Figure 1-6. Infantry squad.

SQUAD LEADER

1-83. The squad leader (SL) directs his team leaders and leads by personal example. The SL has authority over his subordinates and overall responsibility for those subordinates' actions. Centralized authority enables the SL to act decisively while maintaining troop discipline and unity. Under the fluid conditions of close combat, even in the course of carefully-planned actions, the SL must accomplish assigned missions on his own initiative without constant guidance from above.

1-84. The squad leader is the senior Infantryman in the squad and is responsible for all the squad does or fails to do. The squad leader is responsible for the care of his squad's men, weapons, and equipment. He leads his squad through two team leaders. During operations, the squad leader—

- Is the SME on all battle drills and individual drills.
- Is the SME in the squad's organic weapons employment and the employment of supporting assets.
- Knows weapon effects, surface danger zone(s) (SDZ), and risk estimate distance(s) (RED) for all munitions.
- Effectively uses control measures for direct fire, indirect fire, and tactical movement.
- Controls the movement of his squad and its rate and distribution of fire (including call for and adjust fire).
- Fights the close fight by fire and movement with two fire teams and available supporting weapons.
- Selects the fire team's general location and sector in the defense.
- Communicates timely and accurate spot reports (SPOTREPs) and status reports, including—
  - Size, activity, location, unit, time, and equipment (SALUTE) SPOTREPs.
  - Status to the platoon leader (including squad location and progress, enemy situation, enemy killed in action [KIA], and security posture).
  - Status of ammunition, casualties, and equipment to the platoon sergeant.

- Employs digital C2 systems available to the squad and platoon.
- Operates in any environment to include the urban environment.
- Conducts troop-leading procedures (TLP).
- Assumes duties as the platoon sergeant or platoon leader as required.
- Understands the mission and commander's intent two levels up (the platoon and company).

Squad Designated Marksman

1-85. Squad designated marksmen are not squad snipers. They are fully integrated members of the rifle squad who provide an improved capability for the rifle squad. They do not operate as semi-autonomous elements on the battlefield as snipers, nor do they routinely engage targets at the extreme ranges common to snipers. The designated marksman employs an optically-enhanced general-purpose weapon. He also receives training available within the unit's resources to improve the squad's precision engagement capabilities at short and medium ranges

1-86. A rifleman may be assigned as the squad designated marksman (SDM). The SDM is chosen for his demonstrated shooting ability, maturity, reliability, good judgment, and experience. The SDM must be able to execute the entire range of individual and collective rifleman tasks within the squad (see FM 3-22.9, *Rifle Marksmanship M16A1, M16A2/3, M16A4, and M4 Carbine.*)

1-87. The designated marksman employs an optically-enhanced, general-purpose weapon and receives training available within the unit's resources to improve the squad's precision engagement capabilities at short and medium ranges. In contrast, snipers use specialized rifles and match ammunition, and are specially selected and trained to provide precision fire at medium and long ranges (normally from stationary positions).

1-88. The squad marksman engages visible point targets with target priorities of enemy leaders, personnel with radios, automatic weapons crews, enemy soldiers with rocket launchers or sniper rifles, or others as directed by his squad and platoon leaders. He is particularly effective against targets that are only partially exposed or exposed for only brief periods of time. A designated marksman delivers effective fire against very small targets such as loopholes or firing slits, bunker apertures, partially obscured and prone enemy snipers, crew-served weapons teams at close to medium ranges, and rapidly moving targets. He must be able to detect and engage targets rapidly from awkward or nonstandard firing positions while he, the target, or both are moving.

1-89. One designated marksman per fire team creates two highly flexible balanced teams with a squad automatic weapon, grenade launcher, and precision-fire rifleman in each. This combines increased situational awareness and target acquisition with precision point and area suppression. Integration of a designated marksman within each fire team allows the squad to suppress enemy individuals, support weapons, or small units while maneuvering to a position of advantage.

Equipment

1-90. The designated marksman uses an assigned weapon, normally an M16 or M4 equipped with optical sights. Optical sight magnification and wide field of view allow him to observe, detect, identify, range, and engage targets an iron sight or naked eye cannot. This provides the squad with improved situational awareness as well as increased lethality. The telescopic sight dramatically improves the probability of first-round hits on targets at unknown distances and greatly increases target identification capability for shadowed targets and during low light conditions.

Training

1-91. The designated marksman requires additional training on his new role and on the operation and maintenance of the optical sights. Additional training includes—

- Zeroing techniques.
- Target detection.
- Range, wind, and moving target estimation.

- Hold-off determination.
- Alternate and nonstandard shooting positions.
- Known distance field fire to 600 meters.
- Close combat firing techniques.
- Transition fire engagements.
- Rapid target identification and engagement.
- Night fire with and without additional night observation or aiming devices.
- Shooting while moving forward, sideways, and back.
- Shooting from vehicles.

Employment in Combat

1-92. The designated marksman moves and fights in combat as an integral part of the Infantry squad. He provides precision support fire in the offense during the assault and engages targets to the maximum effective range of his weapon in offensive, defensive, and retrograde operations. His ability to deliver lethal, precise, and discriminating fire during stability operations forms the basis of counterinsurgency combat. He enhances the squad's effectiveness and its ability to maneuver and accomplish its mission. When employed tactically, designated marksmen provide precision direct fire as directed by the squad leader. This fire limits fratricide, collateral damage, and noncombatant casualties.

1-93. The designated marksman is employed most effectively in combat situations where precision fire versus a volume of fires is required. Types of operations in which designated marksmen are most useful include:

- Situations in which the squad requires precision fires in an urban area containing an enemy mixed with multiple noncombatants or in those where the applicable ROE restricts the use of area-fire weapons.
- Close range engagements that have an immediate, critical need for precision rifle fire.
- Situations in which the unit is facing an enemy with trained marksmen or armed irregulars being used as snipers that must be countered.
- Civil disturbances involving armed rioters mixed with noncombatants.
- Vehicle and personnel checkpoint operations in which the squad needs an element in armed overwatch.
- Attacking specific targets identified by the platoon or squad leader.
- Covering the approach and entry of the assault element to the objective.
- Eliminating unexpected threats in and around the objective that appear and disappear suddenly and without warning.
- Covering specific avenues of approach into the unit's position and searching the area for signs of a counterattack.
- Isolating the objective area by fire.
- Providing diversionary fire for an assault element.
- Covering obstacles or other key installations with precision fire.
- Situations that require precision fire on apertures, exposed personnel, muzzle flashes, or other designated point targets.
- Situations with friendly troops on or near the objective when mortars, machine guns, and grenade launchers must cease or shift their fires to prevent fratricide. The designated marksman may be able to continue to fire in support of the assault.

INFANTRY WEAPONS SQUAD

1-94. The Infantry weapons squad provides the primary base of fire for the platoon's maneuver. It is comprised of two medium machine gun teams, two medium close combat missile (CCM) teams, and a weapons squad leader (Figure 1-7).

Figure 1-7. Infantry weapons squad.

MEDIUM MACHINE GUN TEAM

1-95. The two-man medium machine gun team is comprised of a gunner and an assistant gunner (AG). The weapons squad has two machine gun teams. These teams provide the platoon with medium-range area suppression at ranges up to 1,000 meters during day, night, and adverse weather conditions.

Gunner

1-96. The gunner is normally the senior member of the team. During operations, the gunner—

- Is responsible for his assistant gunner and all the gun equipment.
- Is responsible for putting the gun in and out of action.
- Is the SME for the information contained in FM 3-22.68, *Crew-Served Machine Guns, 5-56-mm and 7.62-mm.*
- When attached to a rifle squad, is the SME on employment of the medium machine gun. He advises the rifle squad leader of the best way to employ the machine gun.
- Enforces field discipline while the gun team is employed tactically.
- Knows the ballistic effects of the weapon on all types of targets.
- Assists the weapons squad leader and is prepared to assume his responsibilities.
- Understand the mission two levels up (the squad and platoon).

Assistant Gunner

1-97. The assistant gunner is the second member of the gun team. He is prepared to assume the gunner's role in any situation. During operations, the assistant gunner will—

- Constantly update the weapon squad leader on the round count and serviceability of the machine gun.
- Watch for Soldiers to the flanks of the target area or between the gun and the target.
- Report round counts of ammunition in accordance with the unit standard operating procedure.
- Obtain ammunition from other Soldiers who are carrying machine gun ammunition.
- Provide a supply of ammunition to the gun when employed.
- Spot rounds and report recommended corrections to the gunner.
- Immediately assume the role of gunner if the gunner is unable to continue his duties.
- Understand the mission two levels up (squad and platoon).

CLOSE COMBAT MISSILE TEAM

1-98. The two-man close combat missile team is comprised of a gunner and an ammunition handler. Currently, the team uses the Javelin missile system. The weapons squad has two close combat missile teams. This system provides the platoon with an extremely lethal fire-and-forget, man-portable, direct- and top-attack capability to defeat enemy armored vehicles and destroy fortified positions at ranges up to 2,000 meters. The Javelin has proven effective during day, night, and adverse weather conditions.

WEAPONS SQUAD LEADER

1-99. The weapons squad leader leads his teams by personal example. He has complete authority over his subordinates and overall responsibility for those subordinates' actions. This centralized authority enables the weapons squad leader to act decisively while maintaining troop discipline and unity and. Under the fluid conditions of modern warfare, even in the course of carefully-planned actions, the weapons squad leader must accomplish assigned missions using initiative without needing constant guidance from above.

1-100. The weapons squad leader is normally the senior squad leader, second only to the platoon sergeant. He performs all of the duties of the rifle squad leader. In addition, the weapons squad leader—

- Controls fires and establishes fire control measures.
- Recommends machine gun employment to the platoon leader.
- Coordinates directly with the platoon leader for machine gun base-of-fire effects and plans accordingly.
- Monitors ammunition expenditure.
- Coordinates directly with the platoon leader in placement of the Javelin close Combat Missile System (CCMS) to best cover armored avenues of approach in the defense and overwatch positions in the attack.
- Employs C2 systems available to the squad and platoon.
- Performs the role of the platoon sergeant as required.
- Understands the mission two levels up (platoon and company).

SECTION IV — COMBINATIONS

1-101. The Army's preferred method of fighting is combined arms. Combined arms warfare is based on the concept of strengths and weaknesses. All weapons, branches, and tactics have strengths and weaknesses, advantages, and disadvantages. Understanding this, leaders use the power of combinations to protect their weaknesses while using their strengths to create dilemmas for the enemy. There are two principles that guide leaders in fighting combined arms: complementary effects; and reinforcing effects. These two principles are separate and distinct, but are present in most situations.

COMPLEMENTARY EFFECTS

1-102. Leaders create complementary effects when they arrange elements with different characteristics together (Figure 1-8). Complementary effects enable leaders to protect friendly vulnerabilities or enhance effects on the enemy. For example, leaders can combine the effects of their direct fire weapons with those of mortars or artillery to produce an overall greater effect than if each were used separately. Combinations are created based on understanding the strengths and weaknesses of their weapons, the different branches and services, and tactical tasks.

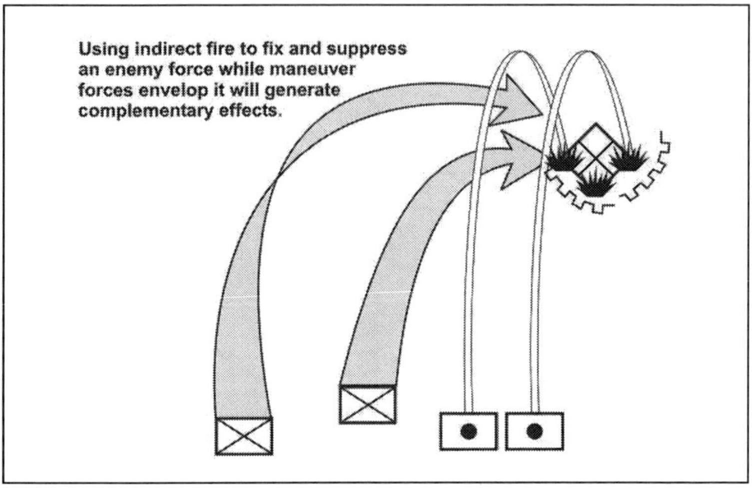

Figure 1-8. Complementary effects.

DILEMMA

1-103. A dilemma is a situation in which the enemy is presented with two or more equally bad alternatives. A problem is a situation in which the enemy is presented with only one bad alternative. Creative combinations allow the leader to create a dilemma for the enemy. When presented with a dilemma, an enemy has two reactions. The first reaction is not knowing what to do as he attempts to decide between equally bad options. This effect is commonly termed "fixed." When the enemy is fixed, the leader benefits from freedom of action. The second reaction is to simply choose one of the two equally bad options. Because the enemy's choice is an option in which the friendly force has the upper hand, the leader is able to exploit the enemy's decision.

1-104. Taking a single-tracked approach can lead to poor or unsuccessful results. Relying on one weapon type, on a single unit type, or a single tactical function does not present the enemy with a dilemma. Without a complementary effect, the enemy is exposed to a problem that can be resolved with a likely solution. Even if applied in rapid succession (sequentially), the enemy only needs to escape the problem at hand. Without a second or third stressor to impair his ability to make good decisions, the enemy is able to react and stay in the fight.

REINFORCING EFFECTS

1-105. Leaders create reinforcing effects when they combine the effect of similar capabilities (Figure 1-9). An example is a team leader reinforcing the effects of his squad automatic weapon with the fires of his rifleman. Leaders do this by either employing the elements simultaneously or sequentially to achieve focused, overwhelming effects at a single point. Simultaneous employment augments the effects of one element with that of another. Sequential employment sustains the effect longer than if just one element was used.

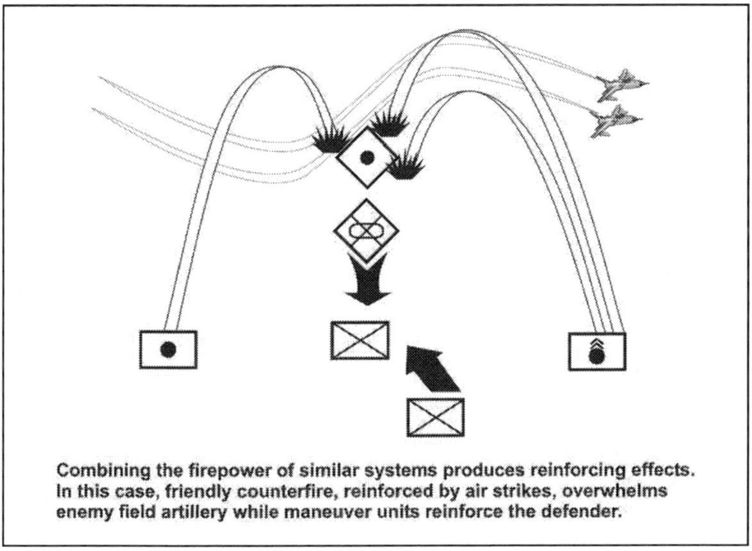

Figure 1-9. Reinforcing effects.

Effective Leaders Confront the Enemy with Dilemmas, Not Problems

1-106. Leaders always seek to present the enemy with a dilemma, not just with problems. There are many ways to do this including, using combinations of weapons, different types of units, tactics, and terrain.

1-107. In Figure 1-10 a moving enemy Infantry force makes contact with a stationary friendly Infantry force. There is an exchange of direct fire weapons. The direct fire contact poses a problem to which there is a solution. The universal reaction to direct fire contact is to get down and return fire. Once the situation develops, the direct fire effects, by themselves, tend to diminish as the enemy gets behind frontal cover and returns direct fire.

1-108. Instead of making contact with direct fire, the friendly force may call for indirect fire. This, too, poses a problem that can be solved with a solution. The universal reaction to indirect fire is for the receiving unit to move out of the indirect fire burst radius. Once again, as the situation develops, the indirect fire effects, by themselves, tend to diminish as the enemy moves out of the burst radius to an area with overhead cover.

1-109. Regardless of how lethal the effects of either direct fire or indirect fire are, by themselves they only pose problems that have solutions as their effects tend to diminish. Suppose the friendly force makes contact using both direct and indirect fire systems. What can the enemy do? He has a dilemma—if he gets up he gets shot, but if he stays down, he gets blown up. The enemy's dilemma results from the complementary effects of direct and indirect fire. This is the essence of combined arms warfare.

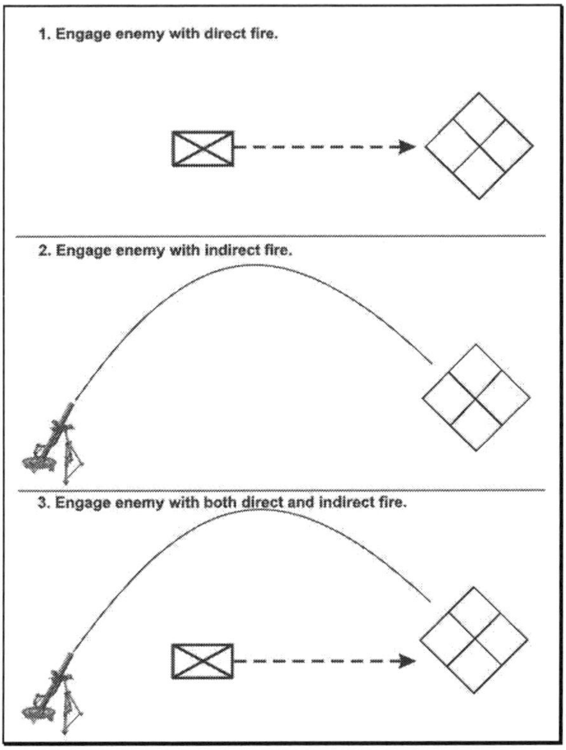

Figure 1-10. Example of problem versus dilemma.

1-110. To increase their effectiveness, leaders seek to combine both complementary and reinforcing effects. Continuing with the example from Figure 1-10, if the friendly Infantry has time, it can employ an obstacle to halt the enemy. The effects of the obstacle reinforce both the effects of direct and indirect fire. The synchronization of these three elements creates a "no-win" situation for the enemy. The engagement area development technique is designed using this as a foundation. Engagement area development combines the complementary effects of direct and indirect fire with the reinforcing effects of obstacles to produce an engagement area for killing enemy forces.

SECTION V — INDIVIDUAL LEADERSHIP

1-111. Tactical leadership is ultimately about one thing—leading Soldiers to accomplish the mission. Leadership is influencing people by providing purpose, direction, and motivation while operating to accomplish the mission and improve the organization (FM 1-02). Leaders need—

- Purpose: the *reason* to accomplish the mission.
- Direction: the *means* to accomplish the mission.
- Motivation: the *will* to accomplish the mission.

1-112. Leaders use command and control (C2) to influence their subordinates to accomplish the mission. Command is the authority leaders exercise over individuals in their unit by virtue of their assignment.

Control is the direction and guidance of subordinates to ensure accomplishment of the mission. Leadership is the art of exercising C2 to influence and direct men in such a way as to obtain their willing obedience, confidence, respect, and loyal cooperation to accomplish the mission. Leadership is the most vital component of C2.

1-113. Professional military leadership involves a combination of personal character and professional competence with a bias for the right action at the right time for the right effect. Leading Soldiers in combat is the Infantry leader's most important challenge.

1-114. There are three core principles that underlie the application of tactical leadership: leadership by example; authority; and mission command.

LEADERSHIP BY EXAMPLE

1-115. Follow me!—the Infantry motto—best summarizes the principle of leadership by example. This simple expression is further developed in the Army's leadership philosophy: **Be**, **Know**, **Do**. Character describes what a leader must be; competence refers to what leaders must know; action is what leaders must do (Figure 1-11). These concepts do not stand alone. They are closely connected and together make up who leaders seek to be (FM 6-22, *Army Leadership*).

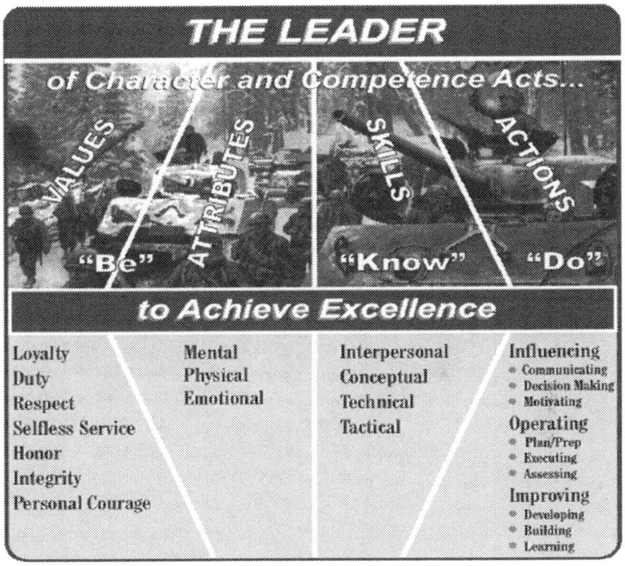

Figure 1-11. Leader by example "Be, Know, Do" principle.

AUTHORITY

1-116. Authority is the delegated power to judge, act, or command. It includes responsibility, accountability, and delegation (FM 6-0, *Mission, Command, and Control*). All Infantrymen in positions of authority are leaders. Leaders exercise authority as they make decisions to accomplish their mission and lead their Soldiers. Authority involves the right and freedom to use the power of position to carry out military duties. It carries with it the responsibility to act. Battle command is the exercise of authority against a hostile, thinking enemy.

1-117. Although commanders alone have the ability to enforce obedience under the Uniform Code of Military Justice (UCMJ), all leaders can expect subordinates to follow their orders. Commanders who delegate authority to subordinates are responsible to ensure their subordinates' lawful orders are followed. This authority to enforce orders by law if necessary is one of the key elements of military leadership, and clearly distinguishes military leaders from civilian leaders and managers.

1-118. Infantry leaders also have another source of authority: personal authority. It stems from values, attributes, personality, experience, reputation, character, personal example, and most of all, tactical and technical competence. Personal authority, freely granted to a leader by subordinates, ultimately arises from the actions of the leader, and the trust and confidence generated by these actions. It is often more powerful than legal authority and is the basis for leadership in the Infantry.

RESPONSIBILITY

1-119. Responsibility is the obligation to carry forward an assigned task to a successful conclusion. It includes the authority to direct and take the necessary action to ensure success (FM 1-02).

1-120. Leaders have three major responsibilities. First, leaders are responsible for accomplishing all assigned missions. Second, they are responsible for their Soldiers' health, welfare, morale, and discipline. Third, they are responsible for maintaining and employing the resources of their force. In most cases, these responsibilities do not conflict. However, the leader's responsibility for mission accomplishment can conflict with their responsibilities to the Soldier. In an irreconcilable conflict between the two, including the welfare of the leader himself, mission accomplishment must come first. However, leaders must understand that the excessive loss of Soldiers and resources can severely inhibit their ability to accomplish their mission.

1-121. Clear, legal, unambiguous orders are a responsibility of good leadership. Soldiers who receive illegal orders that clearly violate the Constitution, the Law of War, or the UCMJ must understand how to react to such orders. Soldiers who receive unclear or illegal orders must ask for clarification. Normally, the superior issuing the unclear or illegal order will make it clear, when queried. He should state that it was not his intent to issue an ambiguous or illegal order. If, however, the superior insists that his illegal order be obeyed, the Soldier should request the rescinding of that order. If the superior does not rescind, the Soldier has an affirmative legal obligation to disobey the order and report the incident to the next superior commander.

ACCOUNTABILITY

1-122. Leaders are accountable for their own decisions and for the actions, accomplishments, and failures of their subordinates. Accountability is non-negotiable and makes up the very backbone of the military chain of command. It is impossible to exercise authority without accountability. Accountability is included in the Army's core values and is what enables us to achieve and maintain legitimacy.

1-123. Accountability has two forms: the UCMJ, and personal accountability. Use of legal authority to enforce accountability at times may be necessary. However, it should not be used as a way of leading Soldiers. It is much more practical to foster a climate that uses the trust of personal authority as a basis for ensuring accountability. Leaders know that American Soldiers respond to trust as the stronger form of accountability, and that the power of the UCMJ is used only when personal accountability proves inadequate.

DELEGATION

1-124. Leaders delegate authority to allow subordinates to carry out their duties, and when necessary, decide and act on behalf of their commander. While leaders can delegate authority, they cannot delegate responsibility for the outcome of their subordinates' actions. Subordinates are accountable to their leaders for how they use their delegated authority.

1-125. When leaders delegate authority, they ensure subordinates understand the limits of their authority or their freedom of action. A leader's freedom of action includes his ability and responsibility to make decisions without the approval of the next higher headquarters. Disciplined initiative by subordinates can only occur when their freedom of action is clearly defined.

MISSION COMMAND

1-126. Mission command is the conduct of military operations through decentralized execution based upon mission orders for effective mission accomplishment. Successful mission command results from subordinate leaders at all echelons exercising disciplined initiative within the commander's intent to accomplish missions. It requires an environment of trust and mutual understanding (FM 1-02). A fundamental tenet of mission command is the importance of people over technology and equipment. There are too many variables, obstacles, and opportunities for leaders to attempt controlling everything. Therefore, mission command requires that leaders learn how to think rather than what to think. It recognizes that the subordinate is often the only person at the point of decision who can make an informed decision. Guided by the commander's intent, the mission, and the concept of the operation, the leader can make the right decision. A second fundamental tenet of mission command is that with the authority of freedom of action comes the subordinate's leader's responsibility to always accomplish his mission.

1-127. Mission orders that allow subordinates maximum freedom of planning and action to accomplish missions are an effective leadership technique in completing combat orders (FM 1-02). Mission orders leave the "how" of mission accomplishment to the subordinate. This way of thinking emphasizes the dominance of command rather than control, thereby providing for initiative, the acceptance of risk, and the rapid seizure of opportunities on the battlefield. Mission command is synonymous with freedom of action for the leader to execute his mission in the way he sees fit, rather than being told how to do it.

DISCIPLINED INITIATIVE

1-128. Execution of mission command requires initiative, resourcefulness, and imagination. Initiative must be disciplined because it should emanate from within the framework of the commander's mission, intent, and concept—not merely from a desire for independent action. Leaders must be resourceful enough to adapt to situations as they are, not as they were expected to be.

1-129. Disciplined initiative means that subordinates are required to make decisions, coordinate with their adjacent units, and determine the best way to accomplish their missions. This includes assuming responsibility for deciding and initiating independent actions when the concept of operations no longer applies, or when an unanticipated opportunity leading to achieving the commander's intent presents itself.

1-130. The amount of freedom of action afforded to his subordinates is a judgment call by the leader. New subordinates or an uncertain environment call for more detail and direction, while experienced subordinates familiar with the mission profile usually need less detail and direction.

NESTED PURPOSE

1-131. To integrate and synchronize all of their elements, leaders need to provide their subordinates with a nested purpose, or a common focus. Initiative, taken to the extreme, risks a dangerous loss of control. To correct this problem, leaders emphasize to subordinates the importance of their battlefield visualization as well as procedural controls for accomplishing tasks whenever possible.

SECTION VI — TACTICAL DECISIONMAKING

1-132. Tactical decisionmaking is one of the primary ways leaders influence subordinates to accomplish their mission. It is a process of the leader collecting information, employing a decisionmaking process, and giving an order to subordinates (Figure 1-12). The information leaders use to make decisions comes from the higher headquarters, the environment, and the common operating picture (COP). The processes used at the Infantry platoon and squad levels are troop-leading procedures (TLP) during planning and preparation, and actions on contact during execution. The combat order is the method of giving subordinates orders. Throughout this process of decisionmaking, leaders continuously assess the situation and their decisions using the risk management and after-action review (AAR) processes.

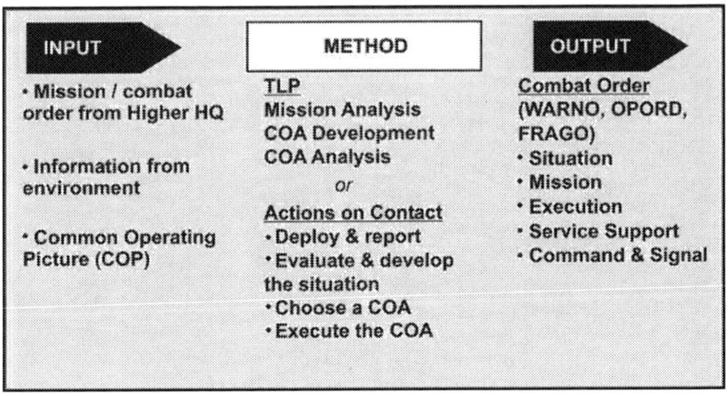

Figure 1-12. Tactical decisionmaking process.

1-133. Decisionmaking involves not only knowing how to make decisions, but knowing if to decide, when to decide, and what to decide. Understanding that once implemented, some commitments are irretrievable, leaders anticipate and understand the activities and consequences that follow their decisions.

SUPPORTING CONCEPTS

1-134. U.S. Army leaders use two decision making methods: visualize, describe, direct; and assess, decide, direct. Visualize, describe, and direct assists leaders in battlefield decisionmaking during planning and preparation. This method provides the underlying logic behind the TLP decisionmaking. The assess, decide, and direct method assists leaders in battlefield decisionmaking during operations. It provides the logic underlying the action-on-contact decisionmaking process.

VISUALIZE, DESCRIBE, DIRECT

1-135. The activities of visualize, describe, and direct are—

- Visualize the operation.
- Describe the visualization to subordinates.
- Direct subordinates with orders that make the visualization a reality.

Visualize

1-136. Effective battlefield leadership requires the leader to see through the fog and friction of military action and clearly articulate the mission. Visualizing the battlefield is a conceptual skill that requires the leader to imagine how to accomplish his mission based on the information he receives. Visualization requires critical reasoning and creative thought. Critical reasoning assists the leader in analyzing and understanding the situation. Creative thought enables the leader to merge his understanding of the unique situation with established tactics, techniques, procedures, and unit SOPs to produce a tailored solution to his tactical problem.

1-137. During operations one of the leader's primary responsibilities is to develop battlefield visualization. Four simple questions assist the leader in understanding the mission:

- Where do we want to be?
- Where are we now?
- How do we get from here to there?
- What will prevent us from getting there?

1-138. The leader's battlefield visualization is the basis for making sound decisions before, during, and after operations. However, it is important for the leader to know how much freedom of action he has in designing his visualization. If the platoon or squad is conducting independent operations, it is likely that he has the freedom to fully develop his visualization. If the leader's mission involves conducting platoon actions within the context of a larger unit's operations, the leader has less freedom to develop his visualization. Either way, the leader is always responsible for understanding the next higher leader's visualization.

Describe

1-139. Once leaders imagine the future and the means needed to achieve it, they influence their subordinates by describing their visualization. Their communication, in common doctrinal terms, concepts, and symbols, helps everyone understand what must be done and how each element contributes to the effort.

1-140. Leaders who communicate effectively—

- Display good oral, written, and listening skills.
- Persuade others.
- Express thoughts and ideas clearly to individuals and groups.

Direct

1-141. Leaders issue orders to direct subordinates. Examples include combat orders and fire commands. Orders can be oral or written.

Assess, Decide, Direct

1-142. Leaders assess by monitoring the situation through reports from subordinates and personal observation. The information they receive is then evaluated against how the operation or action was visualized. Leaders make many decisions during execution. Some are planned. Others are unforeseen; so leaders prepare for both. They use combat orders and procedural and positive controls to direct subordinates during execution.

1-143. Even when things are progressing satisfactorily, certain critical ongoing tasks must be accomplished. At the platoon and squad level these include—

- Focus on the decisive action.
- Ensure security.
- Monitor and adjust control measures.
- Perform battle tracking (control fires and control movement).
- Monitor sustaining actions.

TROOP-LEADING PROCEDURES

1-144. Troop-leading procedures (TLP) provide leaders a framework for decisionmaking during the plan and prepare phases of an operation. This eight-step procedure applies the logic of visualize, describe, and direct to the plan and prepare functions of the operations process. Steps in the TLP include:

- Receive the mission.
- Issue a warning order (WARNO).
- Make a tentative plan.
- Initiate movement.
- Conduct reconnaissance.
- Complete the plan.
- Issue the order.
- Supervise and assess.

1-145. For a complete discussion on making a tentative plan, see Chapter 6.

RECEIVE THE MISSION

1-146. Leaders receive their missions in several ways—ideally through a series of warning orders (WARNOs), operation orders (OPORD)s, and briefings from their leader/commander. However, the tempo of operations often precludes this ideal sequence, particularly at the lower levels. This means that leaders may often receive only a WARNO or a fragmentary order (FRAGO), but the process is the same.

1-147. After receiving an order, leaders are normally required to give a confirmation briefing to their higher commander. This is done to clarify their understanding of the commander's mission, intent, and concept of the operation, as well as their role within the operation. The leader obtains clarification on any portions of the higher headquarters' plan as required.

1-148. Upon receiving the mission, leaders perform an initial assessment of the situation (mission, enemy, terrain, troops-time, civil [METT-TC] analysis), focusing on the mission, the unit's role in the larger operation, and allocating time for planning and preparing. The two most important products from this initial assessment should be at least a partial restated mission, and a timeline. Leaders issue their initial WARNO on this first assessment and time allocation.

1-149. Based on their knowledge, leaders estimate the time available to plan and prepare for the mission. They issue a tentative timeline that is as detailed as possible. In the process they allocate roughly one-third of available planning and preparation time to themselves, allowing their subordinates the remaining two-thirds. During fast-paced operations, planning and preparation time might be extremely limited. Knowing this in advance enables leaders to emplace SOPs to assist them in these situations.

ISSUE A WARNING ORDER

1-150. Leaders issue the initial WARNO as quickly as possible to give subordinates maximum time to plan and prepare. They do not wait for additional information. The WARNO, following the five-paragraph field order format, contains as much detail as available. At a minimum, subordinates need to know critical times like the earliest time of movement, and when they must be ready to conduct operations. Leaders do not delay in issuing the initial WARNO. As more information becomes available, leaders can—and should—issue additional WARNOs. At a minimum the WARNO normally includes:

- Mission or nature of the operation.
- Time and place for issuing the OPORD.
- Units or elements participating in the operation.
- Specific tasks not addressed by unit SOP.
- Timeline for the operation.
- Rehearsal guidance.

MAKE A TENTATIVE PLAN

1-151. Once he has issued the initial WARNO, the leader continues to develop a tentative plan. Making a tentative plan follows the basic decisionmaking method of visualize, describe, direct, and the Army standard planning process. This step combines steps 2 through 6 of the military decisionmaking process: mission analysis, COA development, COA analysis, COA comparison, and COA selection. At the Infantry platoon level, these steps are often performed mentally. The platoon leader and squad leaders may include their principal subordinates—especially during COA development, analysis, and comparison.

1-152. To frame the tentative plan, Army leaders perform mission analysis. This mission analysis follows the METT-TC format, continuing the initial assessment performed in TLP step 1. This step is covered in detail in Chapter 6.

INITIATE MOVEMENT

1-153. Movement of the unit may occur simultaneously with the TLPs. Leaders initiate any movement necessary to continue mission preparation or position the unit for execution. They do this as soon as they have enough information to do so, or when the unit is required to move to position itself for the upcoming mission. Movements may be to an assembly area, a battle position, a new AO, or an attack position. They may include movement of reconnaissance elements, guides, or quartering parties. Infantry leaders can initiate movement based on their tentative plan and issue the order to subordinates in the new location.

CONDUCT RECONNAISSANCE

1-154. Whenever time and circumstances allow, leaders personally conduct reconnaissance of critical mission aspects. No amount of planning can substitute for firsthand assessment of the situation. Unfortunately, many factors can keep leaders from performing a personal reconnaissance. However, there are several means available to the leader to develop and confirm his visualization. They include: internal reconnaissance and surveillance elements, unmanned sensors, the higher unit's intelligence, surveillance, reconnaissance (ISR) elements, adjacent units, map reconnaissance, imagery, and intelligence products. One of the most difficult aspects of conducting reconnaissance is the process of identifying what the leader needs to know (the information requirements [IR]).

COMPLETE THE PLAN

1-155. During this step, leaders incorporate the result of reconnaissance into their selected course of action (COA) to complete the plan and order. This includes preparing overlays, refining the indirect fire target list, coordinating sustainment and C2 requirements, and updating the tentative plan as a result of the reconnaissance. At the platoon and squad levels, this step normally involves only confirming or updating information contained in the tentative plan. If time allows, leaders make final coordination with adjacent units and higher headquarters before issuing the order.

ISSUE THE ORDER

1-156. Infantry platoon and squad leaders normally issue verbal combat orders supplemented by graphics and other control measures. The order follows the standard five-paragraph field order format. Infantry leaders use many different techniques to convey their orders (see Chapter 6). Typically, platoon and squad leaders do not issue a commander's intent. They reiterate the intent of their company and battalion commanders.

1-157. The ideal location for issuing the order is a point in the AO with a view of the objective and other aspects of the terrain. The leader may perform reconnaissance, complete the order, and then summon subordinates to a specified location to receive it. At times, security or other constraints make it infeasible to issue the order on the terrain. In such cases, leaders use a sand table, detailed sketch, maps, aerial photos and images, and other products to depict the AO and situation.

SUPERVISE AND ASSESS

1-158. This final step of the TLP is crucial. Normally unit SOPs state individual responsibilities and the sequence of preparation activities. After issuing the OPORD, the platoon leader and his subordinate leaders must ensure the required activities and tasks are completed in a timely manner prior to mission execution. It is imperative that both officers and NCOs check everything that is important for successful mission accomplishment. The process should include:

- Ensuring the second in command of each element is prepared to execute in their leader's absence.
- Listening to subordinate operation orders.
- Checking load plans to ensure Soldiers are carrying only what is necessary for the mission and or what was specified in the OPORD.
- Checking the status and serviceability of weapons.
- Checking on maintenance activities of subordinate units.
- Ensuring local security is maintained.
- Conducting rehearsals.

1-159. Platoons and squads use five types of rehearsals:

(1) Confirmation brief.

(2) Backbrief.

(3) Combined arms rehearsal.

(4) Support rehearsal.

(5) Battle drill or SOP rehearsal.

ACTIONS ON CONTACT

1-160. Actions on contact involve a series of combat actions, often conducted simultaneously, taken upon contact with the enemy to develop the situation (FM 1-02). Leaders use the actions-on-contact process as a decisionmaking technique when in contact with the enemy. This process should not be confused with battle drills such as Battle Drill "React to Contact." Battle drills are the actions of individual Soldiers and small units when they come into contact with the enemy. Action on contact is a leader tool for making decisions while their units are in contact. The process assists the leader in decisionmaking concurrent with fighting his unit and assessing the situation.

1-161. The logic of assess, decide, and direct underlies the actions-on-contact decisionmaking process. As the leader evaluates and develops the situation, he assesses what is currently happening and its relation to what should be happening. The following four steps must be taken in the actions on contact process.

STEP 1 – DEPLOY AND REPORT

1-162. This step begins with enemy contact. Figure 1-13 details the forms of contact. This contact may be expected or unexpected. During this step, subordinates fight through the contact with the appropriate battle drill. While this is occurring, leadership has the following primary tasks:

- Fix the enemy.
- Isolate the enemy.
- Separate the enemy forces from each other by achieving fire superiority.
- Report to higher.
- Begin "fighting" for information—actively pursue and gather it.

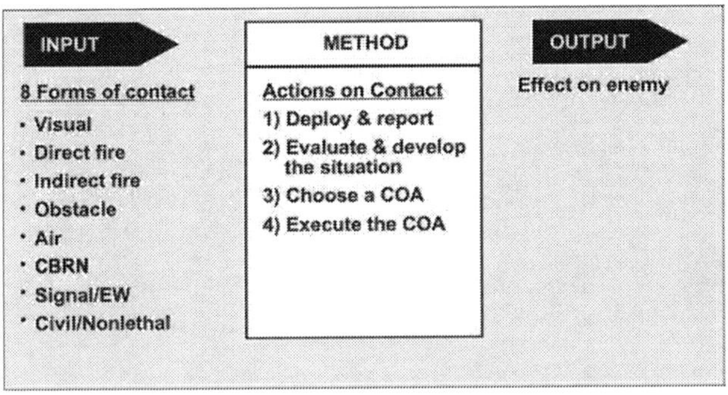

Figure 1-13. Enemy contact decisionmaking model.

1-163. During the TLP, leaders develop a vision of how their operation will unfold. Part of this process involves the leader anticipating where he expects the unit to make contact. This enables him to think through possible decisions in advance. If the leader expects contact, he will have already deployed his unit by transitioning from tactical movement to maneuver. Ideally, the overwatching element will make visual contact first. Because the unit is deployed, it will likely be able to establish contact on its own terms. If the contact occurs as expected, the leader goes through the procedure making decisions as anticipated and minor adjustments as required.

1-164. Regardless of how thorough the leader's visualization, there will always be cases in which the unit makes unexpected contact with the enemy. In this case, it is essential that the unit and its leader take actions to quickly and decisively take back the initiative.

STEP 2 – EVALUATE AND DEVELOP THE SITUATION

1-165. This step begins with the leader evaluating and developing the situation. The leader quickly gathers the information he needs to make a decision on his course of action. He does this through either personal reconnaissance or reports from subordinates. At a minimum, the leader needs to confirm the friendly situation and determine the enemy situation using the SALUTE format (size, activity, location, unit, time, and equipment), and enemy capabilities (defend, reinforce, attack, withdraw, and delay). During this analysis, the leader should look for an enemy vulnerability to exploit.

1-166. As part of developing the situation, the leader seeks a position of advantage to maneuver his force. During this process, the leader considers the following:

- Mutually supporting enemy positions.
- Obstacles.
- The size of the enemy force engaging the unit. (Enemy strength is indicated by the number of enemy automatic weapons, the presence of any vehicles, and the employment of indirect fires.)
- A vulnerable flank to the position.
- A covered and concealed route to the flank of the position.

1-167. If after his initial evaluation the leader still lacks information, he may attempt one or all of the following to get the information he needs:

- Reposition a subordinate(s) or a subordinate unit.
- Reconnaissance by fire.
- Request information from adjacent units or from the controlling headquarters.

STEP 3 – CHOOSE A COA

1-168. After developing the situation, the leader determines what action his unit must take to successfully conclude the engagement. The leader then determines if the chosen task is consistent with the original COA. If it still applies, he continues the mission. If it is not consistent, he issues a FRAGO modifying the original COA. If the leader is unsure, he continues to develop the situation and seeks guidance from higher. In general, the following options are open to the leader:

- Achieve fire superiority by assault/attack (including standard Infantry battle drills).
- Support by fire for another unit.
- Break contact.
- Defend.
- Bypass enemy position.

1-169. The order of COAs listed above is relative to the effectiveness of fire and strength of the enemy position. If the enemy is an inferior force, the unit in contact should be able to achieve fire superiority and still have enough elements to conduct movement to attack the enemy force. If the entire unit is needed to gain and maintain fire superiority, the next feasible COA is to establish a support by fire so another element can conduct movement to attack the enemy. If the unit cannot achieve fire superiority, or there is no other element to conduct an assault, the unit breaks contact. If the unit is decisively engaged and cannot break contact, it establishes a defense until assistance from another unit arrives. In some instances, based on METT-TC, the unit may bypass the enemy position.

STEP 4 – EXECUTE THE COA

1-170. Following his decision, the leader gives the order. When describing his visualization, he uses doctrinal terms and concepts and the five-paragraph field order format. The leader only needs to state those directions and orders that have changed from the original order and emphasizes other items he deems essential.

1-171. During this step, the leader must direct the engagement. There are three key things that the leader needs to control: movement; fires; and unit purpose. These controls may be standard procedures or hands-on positive controls.

RISK MANAGEMENT

1-172. Risk management is the process leaders use to assess and control risk. There are two types of risk associated with any combat action: tactical hazards that result from the presence of the enemy; and accidental hazards that result from the conduct of operations. All combat incurs both risks. The objective is to minimize them to acceptable levels. The following four considerations will help the leader identify risk to the unit and the mission (see Chapter 4):

- Define the enemy action.
- Identify friendly combat power shortfall.
- Identify available combat multipliers, if any, to mitigate risk.
- Consider the risks: acceptable or unacceptable?

AFTER-ACTION REVIEWS

1-173. An after-action review (AAR) is an assessment conducted after an event or major activity that allows participants to learn what and why something happened, and most importantly, how the unit can improve through change. This professional discussion enables units and their leaders to understand why things happened during the progression of an operation, and to learn from that experience. This learning is what enables units and their leaders to adapt to their operational environment. The AAR does not have to be performed at the end of the activity. Rather, it can be performed after each identifiable event (or whenever feasible) as a live learning process.

1-174. The AAR is a professional discussion that includes the participants and focuses directly on the tasks and goals. While it is not a critique, the AAR has several advantages over a critique:

- It does not judge success or failure.
- It attempts to discover why things happened.
- It focuses directly on the tasks and goals that were to be accomplished.
- It encourages participants to raise important lessons in the discussion.
- More Soldiers participate so more of the project or activity can be recalled and more lessons can be learned and shared.

1-175. Leaders are responsible for training their units and making their units adapt. The AAR is one of the primary tools used to accomplish this. It does this by providing feedback, which should be direct and on the spot. Each time an incorrect performance is observed, it should be immediately corrected so it does not interfere with future tasks. During major events or activities, it is not always easy to notice incorrect performances. An AAR should be planned at the end of each activity or event. In doing so, feedback can be provided, lessons can be learned, and ideas and suggestions can be generated to ensure the next project or activity will be an improved one.

1-176. An AAR may be formal or informal. Both follow the same format and involve the exchange of observations and ideas. Formal AARs are usually more structured and require planning. Informal AARs can be conducted anywhere and anytime to provide quick learning lessons. The AAR format follows:

- Gather all the participants.
- Go through introductions and rules.

- Review events leading to the activity (what was supposed to happen).
- Give a brief statement of the specific activity.
- Summarize key events. Encourage participation.
- Have junior leaders restate portions of their part of the activity.

1-177. The art of an AAR is in obtaining mutual trust so people will speak freely. Problem solving should be practical and Soldiers should not be preoccupied with status, territory, or second guessing "what the leader will think." There is a fine line between keeping the meeting from falling into chaos where little is accomplished, to people treating each other in a formal and polite manner that masks issues (especially with the leader).

1-178. The AAR facilitator should—

- Remain unbiased throughout the review.
- Ask open-ended questions to draw out comments from all.
- Do *not* allow personal attacks.
- Focus on learning and continuous improvement.
- Strive to allow others to offer solutions rather than offering them yourself.
- Find solutions and recommendations to make the unit better.

1-179. To avoid turning an AAR into a critique or lecture—

- Ask why certain actions were taken.
- Ask how Soldiers reacted to certain situations.
- Ask when actions were initiated.
- Ask leading and thought-provoking questions.
- Exchange "war stories" (lessons learned).
- Ask Soldiers to provide their own point of view on what happened.
- Relate events to subsequent results.
- Explore alternative courses of actions that might have been more effective.
- Handle complaints positively.
- When the discussion turns to errors made, emphasize the positive and point out the difficulties of making tough decisions.
- Summarize.
- Allow junior leaders to discuss the events with their Soldiers in private.
- Follow up on needed actions.

SECTION VII — COMBAT POWER

1-180. Combat power is the ability of a unit to fight. To generate combat power, Army forces at all levels conduct operations. An operation is a military action or carrying out of a mission (FM 1-02). Leaders at the operational level of war develop operations in response to receiving strategic guidance. These operations consist of numerous component operations, tasks, and actions. Within these operations, leaders at the operational level assign their subordinate's missions. These subordinates, in turn, develop operations to accomplish their mission. They then assign the mission to subordinates as part of their overall operations. This chain of events continues until the Infantry platoon and squad receives its mission. Leaders at all levels use many tools to develop and conduct operations. Two of the most important tools are—

- The four critical functions.
- Full spectrum operations doctrine.

FOUR CRITICAL FUNCTIONS

1-181. Most combat actions follow the sequence of find, fix, finish, and follow-through. First, the unit must find the enemy and make contact. Second, they fix the enemy with direct and indirect fires. Third, the unit must finish the enemy with fire and movement directed towards a vulnerable point in order to fight through to defeat, destroy, or capture the enemy. Fourth, the unit must follow-through with consolidation, reorganization, and preparing to continue the mission or receive a new mission.

1. Find the Enemy

1-182. At the individual, crew, and squad and platoon levels, finding the enemy directly relates to target acquisition. Target acquisition is the process of searching for the enemy and detecting his presence; determining his actual location and informing others; and confirming the identity of the enemy (not a friend or noncombatant). The most common method of target acquisition is assigning sectors to subordinates. Once assigned, Soldiers use search techniques within their sectors to detect potential targets.

1-183. There are many different sources for finding the enemy. They include:

- Other Soldiers, crews, squads and platoons.
- Forward observers.
- Reconnaissance elements (scouts, reconnaissance units, cavalry, and long-range surveillance units).
- Aviation assets such as the OH58D.
- Unmanned aircraft system(s) (UAS).
- Lightweight Counter-mortar Radar (LCMR).
- Special Forces.

1-184. Finding the enemy consists of physically locating him and determining his disposition. Enemy strength, composition, capabilities, probable COA, and exploitable vulnerabilities are important determinations made in the location process. The leader seeks to develop the situation as much as possible out of contact with the enemy. Once in contact, he fights for the information he needs to make decisions.

Plan and Prepare

1-185. Finding the enemy begins long before the unit moves across the line of departure (offense) or occupies its battle position (defense). During planning, the leader's METT-TC analysis is essential to developing the clearest picture of where the enemy is located, the probable COA, and the most dangerous COA. When there is little information about the enemy, a detailed analysis of terrain will assist the leader in predicting enemy actions. During preparation, the leader sends out his reconnaissance or submits his information requirements to higher headquarters to develop the enemy picture as thoroughly as possible.

Execute

1-186. During execution, the unit's first priority is to find the enemy before the enemy finds them. This involves employing good cover, concealment, camouflage, and deception while denying the enemy the same. During tactical movement, the unit must have an observation plan that covers their entire area of influence. Additionally, the leader takes measures to detect enemies in the unit's security zone.

1-187. Once found, the leader has a decision to make. In the offense, the leader must determine if he has enough forces to fix the enemy or if he should pass the enemy position off to a separate fixing force. In the defense, he must determine if he has enough forces to disrupt the enemy or if he should pass the enemy force off to a separate fixing force.

2. Fix

1-188. Immediately after finding the enemy, the leader has to fix the enemy in place. Fixing the enemy holds him in position. When the enemy is fixed, the leader can maneuver to the enemy's vulnerable point without the fear of being attacked in an exposed flank, or of more enemy forces reinforcing. Fixing the enemy normally consists of one of the following tactical mission tasks: support by fire, attack by fire, suppress, destroy, or block. An enemy that is fixed is affected physically and or psychologically. The means to achieve this effect—lethal, nonlethal, and combinations thereof—are endless.

1-189. Fixing the enemy is accomplished through isolation. "Isolate" means cutting the adversary off from the functions necessary to be effective. Isolation has both an external aspect of cutting off outside support and information, and an internal aspect of cutting off mutual support. Isolating the adversary also includes precluding any break in contact.

1-190. External isolation stops any of the fixed enemy force from leaving the engagement while preventing any other enemy force from reinforcing the fixed force. Actions outside of the objective area prevent enemy forces from entering the engagement. Internal isolation occurs by achieving fire superiority that prevents the enemy from repositioning and interfering with friendly maneuver elements.

1-191. Isolating the objective is a key factor in facilitating the assault and preventing casualties. Isolating the objective also involves seizing terrain that dominates the area so the enemy cannot supply, reinforce, or withdraw its defenders. Infantry platoons and squads may perform this function as a shaping element for a company operation, or it may assign subordinates this function within its own organization. In certain situations, the squads or platoon may isolate an objective or an area for special operations forces. Depending on the tactical situation, Infantry platoons may use infiltration to isolate the objective.

1-192. The enemy is fixed when his movement is stopped, his weapons suppressed, and his ability to effectively respond disrupted. Once fixed, the leader has a decision to make. In the offense, the leader must determine if he has enough forces to assault the enemy, or if he needs to request a separate assault element from the controlling headquarters. In the defense, he must determine if he has enough forces to counterattack, or if he needs to request a separate counterattack force from the controlling headquarters.

3. Finish

1-193. After finding and fixing the enemy, the leader finishes the fight. In the offense, this is known as the assault; in the defense, this is known as the counterattack. Finishing the enemy normally consists of one of the following tactical mission tasks: clear, seize, or destroy. It is extremely important that leaders understand the necessity to "have something left" when finishing the enemy and for the next step—follow-through. Failure to have enough combat power at the decisive point or during consolidation puts the unit at risk to counterattack. The fight is finished when the enemy—

- No longer has the physical ability to fight (meaning he is destroyed).
- Has determined physical destruction is imminent.
- No longer believes he can resist (meaning he is in shock).

4. Follow-Through

1-194. Follow-through involves those actions that enable the unit to transition from close combat to continuing the mission. It includes conducting consolidation and reorganization and exploiting success. Transitioning the unit from the violence of close combat back to a state of high readiness is difficult. Units are most vulnerable at the conclusion of close combat, and decisive leadership is absolutely essential to make the transition. Continuing the attack or counterattacking may be a deliberate phase of the operation (a "be-prepared-to" or "on-order" mission). It may also be a decision made by the controlling commander based on a window of opportunity.

DOCTRINAL HIERARCHY OF OPERATIONS

1-195. Figure 1-14 shows the doctrinal hierarchy and the relationship between the types and subordinate forms of operations. While an operation's predominant characteristic labels it as an offensive, defensive, stability, or civil support operation, different units involved in that operation may be conducting different types and subordinate forms of operations. These units often transition rapidly from one type or subordinate form to another. While positioning his forces for maximum effectiveness, the commander rapidly shifts from one type or form of operation to another to continually keep the enemy off balance. Flexibility in transitioning contributes to a successful operation.

1-196. Infantry platoons and squads conduct all the types of operations listed in the doctrinal hierarchy. However, the Infantry platoon and squad will almost always conduct these operations and their subordinate forms and types as part of a larger unit. In fact, many of these types of operations are only conducted at the battalion, brigade, or division level. Only the types of operations applicable to Infantry platoons and squads are further covered in this manual.

OFFENSIVE OPERATIONS	OFFENSIVE OPERATIONS
Movement to contact • Search and attack • Approach march Attack Special purpose attacks • Ambush • Demonstration • Feint • Raid • Spoiling attack Forms of maneuver • Envelopment • Frontal attack • Infiltration • Penetration • Turning Movement	Reconnaissance operations • Area • Route • Zone • Reconnaissance in force Security operations • Screen • Guard • Area security (including route and convoy) • Local security Combined arms breach operations Passage of lines Relief in place Troop movement (road march)
DEFENSIVE OPERATIONS	**DEFENSIVE OPERATIONS**
Area Defense Retrograde Operations • Delay • Withdrawal • Retirement	Patrols • Combat patrols • Reconnaissance patrols • Security patrols Tactical movement Battle drills Crew drills

Figure 1-14. Doctrinal hierarchy of operations.

Offensive Operations

1-197. Offensive operations aim to destroy or defeat an enemy. Their purpose is to impose U.S. will on the enemy and achieve decisive victory (FM 3-0, *Operations*). Dominance of the offense is a basic tenet of U.S. Army operations doctrine. While the defense is the stronger form of military action, the offense is the decisive form. Tactical considerations may call for Army forces to execute defensive operations for a period of time. However, leaders are constantly looking for ways to shift to the offense. Offensive operations do not exist in a vacuum—they exist side by side with defense, and tactical enabling operations. Leaders analyze the mission two levels up to determine how their unit's mission nests within the overall concept. For example, an Infantry platoon leader would analyze company and battalion missions.

1-198. Effective offensive operations require accurate intelligence on enemy forces, weather, and terrain. Leaders then maneuver their forces to advantageous positions before contact. Contact with enemy forces before the decisive action is deliberate and designed to shape the optimum situation for the decisive action. The decisive action is sudden and violent, capitalizing on subordinate initiative. Infantry platoon and squad leaders therefore execute offensive operations and attack with surprise, concentration, tempo, and audacity.

1-199. There is a subtle difference between attacking and conducting an attack. Attacking in everyday usage generally means the close combat action of fire and movement on an enemy or position. Attacking occurs frequently on the battlefield in all types of operations. Conducting an attack is one of the four types of offensive operations with specific doctrine meanings and requirements.

Offensive Purposes

1-200. How a unit conducts its offensive operations is determined by the mission's purpose and overall intent. There are four general purposes for the offense: throw the enemy off balance; overwhelm the enemy's capabilities; disrupt the enemy's defense; and ensure their defeat or destruction. In practice, each of these purposes has orientation on both the enemy force and the terrain. The labels merely describe the dominant characteristic of the operation.

Enemy-Oriented

1-201. Leaders employ enemy-oriented attacks to destroy enemy formations and their capabilities. Destruction results in an enemy unit (Soldiers and their equipment) that is no longer able to fight. Not everything has to be destroyed for the force-oriented attack offense to be successful. It is usually enough to focus on an enemy capability or unit cohesion. These attacks are best employed against an enemy vulnerability. Once destruction occurs, a window of opportunity opens. It is up to the leader to take advantage of an unbalanced enemy through local and general exploitations and pursuit.

Terrain-Oriented

1-202. Leaders employ terrain-oriented attacks to seize control of terrain or facilities. Units conducting terrain-oriented attacks have less freedom of action to take advantage of a window of opportunity. The unit's first priority is the terrain or facility. Exploiting an enemy vulnerability can occur only when the security of the terrain or facility is no longer in question.

Tactical Enabling and Infantry Platoon Actions

1-203. Although friendly forces always remain enemy focused, there are many actions friendly forces conduct that are offensive in nature and are designed to shape or sustain other operations. Leaders employ tactical enabling operations to support the overall purpose of an operation.

Types of Offensive Operations

1-204. Types of offensive operations are described by the context surrounding an operation (terrain or force oriented). At the platoon and squad level, these offensive operations are basically planned, prepared for, and executed the same. The four types of offensive operations include:

(1) Movement to Contact – undertaken to gain or regain contact with the enemy (force-oriented).

(2) Attack – undertaken to achieve a decisive outcome (terrain-oriented or force-oriented).

(3) Exploitation – undertaken to take advantage of a successful attack (force-oriented).

(4) Pursuit – undertaken to destroy an escaping enemy (force-oriented).

1-205. This order of offensive operations is deliberate because they are listed in order of their normal occurrence. Generally, leaders conduct a movement to contact to find the enemy. When the leader has enough information about the enemy to be successful, he conducts an attack. Following a successful attack, the leader takes advantage of the enemy's disorganization and exploits the attack's success. After exploiting his success, the leader executes a pursuit to catch or cut off a fleeing enemy to complete its destruction. Although Infantry platoons and squads participate in exploit and pursuit operations, they do not plan them.

Defensive Operations

1-206. Defensive operations defeat an enemy attack, buy time, economize forces, or develop conditions favorable for offensive operations. Defensive operations alone normally cannot achieve a decision. Their overarching purpose is to create conditions for a counteroffensive that allows Army forces to regain the initiative (FM 1-02). Defensive operations do not exist in a vacuum—they exist side by side with offense, tactical enabling operations, and Infantry platoon actions. Leaders analyze the mission two levels up to determine how their unit's mission nests within the overall concept.

1-207. The principles of tactical maneuver also apply to the defense. To be decisive, defensive tactics must have both ingredients. Ensuring mobility remains a part of the defense is one of the leader's greatest challenges. While it is true that defending forces await the attacker's blow and defeat the attack by successfully deflecting it, this does not mean that defending is a passive activity. Leaders always look for ways to integrate movement into their defensive activities.

1-208. During the conduct of operations, regardless of type, friendly forces make many transitions requiring the unit to stop and restart movement. Infantry platoons and squads that are not moving are defending. Units that stop moving (attacking), immediately transition to defending. This transition is rapid and should be second nature to all Soldiers and their units. This is particularly relevant at the Infantry platoon and squad levels where the tactical situation can quickly shift to one where the unit is outnumbered and fighting for its survival.

Defensive Purposes

1-209. How a unit establishes its defenses is determined by the mission's purpose and intent. There are four general purposes for conducting a defense: defeat an attacking enemy; economize friendly forces in one area so they can be concentrated in another area; buy time; and develop conditions favorable for resuming offensive operations. In practice, each of these stated purposes for conducting a defense is considered in all defenses; the categories just describe the dominant purpose. Infantry platoons and squads can also be tasked to defend specific locations such as key terrain or facilities.

Defeat an Attacking Enemy and Develop Conditions for Offensive Operations

1-210. Defenses are designed to defeat enemy attack while preserving friendly forces. Defeating the enemy's attack requires him to transition to his own defensive actions. While this occurs, a window of opportunity for friendly forces may also occur. It is up to the leader to take advantage of an unbalanced enemy through local and general counterattacks.

Economy of Force to Concentrate in Another Area

1-211. Commanders seldom have all the combat forces they desire to conduct operations without accepting risk. Economy of force is defined as allocating minimum essential combat power to secondary efforts (FM 1-02). It requires accepting prudent risk in selected areas to achieve superiority—overwhelming effects—in the decisive operation. As a result, commanders arrange forces in space and time to create favorable conditions for a mobile defense and offensive operations in other areas.

Buy Time

1-212. Defenses to preserve friendly combat power are designed to protect the friendly force and prevent the destruction of key friendly assets. There are times when the unit establishes defenses to protect itself. Although friendly forces always remain enemy focused, there are many actions friendly forces conduct to sustain the unit. These sustaining actions typically require the unit to establish a defensive posture while the activity is conducted. Examples include: consolidation and reorganization, resupply/LOGPAC, pickup zone/landing zone, and CASEVAC/MEDEVAC. This type of defense can also be associated with assembly area activities, establishing lodgments for building up combat power, and facing a numerically-superior enemy force.

Develop Conditions Favorable for Resuming Offensive Operations

1-213. The enemy may have the advantage over friendly forces in areas such as combat power or position. This often occurs during forced entry operations where friendly forces defend in order to build up combat power.

Key Terrain or Facilities

1-214. Defenses for denying enemy access to an area are designed to protect specific location, key terrain, or facilities. Infantry platoons can be assigned missions to defend sites that range from hill tops—to key infrastructure—to religious sites. Because the defense is terrain oriented, leaders have less freedom of action when it comes to taking advantage of a window of opportunity. The unit's first priority is the terrain or facility. Exploiting an enemy vulnerability can occur only when the security of the terrain or facility is no longer in question.

Types of Defensive Operations

1-215. Defensive operations fall into one of the following three categories:

(1) Area defense – focuses on retaining terrain for a specified period of time (terrain-oriented).

(2) Mobile defense – stops an enemy attack with a fixing force and destroys it with a strike force (division level and higher operations [force-oriented]).

(3) Retrograde – a type of defensive operation that involves an organized movement away from the enemy. The three types of retrograde operations are: delay; withdrawal; and retirement.

Area Defense

1-216. The area defense is the most common defensive operation undertaken at the tactical level (brigade and below). This is discussed in Chapter 9.

Mobile Defense

1-217. The mobile defense is usually a corps-level operation. A mobile defense has three categories of forces: a fixing force, a strike force, or a reserve force. The decisive operation of a mobile defense is the strike force. Those units designated as the fixing force are essentially performing an area defense. Units designated as the strike force are essentially performing an attack. (For more information on the mobile defense, see FM 3-90, *Tactics*.)

Retrograde

1-218. The retrograde is a technique used by higher-level commanders to maintain or break contact with the enemy. This is done as part of a larger scheme of maneuver to create conditions to regain the initiative and defeat the enemy. Retrogrades improve the current situation or prevent a situation from deteriorating. These operations are a means to an end; not an end in itself. The Infantry platoon's fight in the higher commander's retrograde operation uses one of two techniques: fighting the enemy, or moving to the new location. Leaders must be aware of the potentially catastrophic impact a retrograde has on friendly troop's morale. The retrograde is the defensive counterpart to an offensive exploitation or pursuit. There are three techniques used to retrograde:

- Delay – trades space for time (attempting to slow the enemy's momentum).
- Withdrawal – trades time for space (breaking contact as far from the enemy as possible).
- Retirement – movement that is not in contact with the enemy.

STABILITY OPERATIONS

1-219. Stability operations encompass a range of actions that shape the political environment and respond to developing crises. This section provides an introductory discussion of stability operations (FM 3-0 and FM 3-07, *Stability Operations and Support Operations*).

1-220. Stability operations usually occur in conjunction with offensive and defensive operations. These operations are diverse, continuous, and often long-term. They may include both developmental and coercive actions. Developmental actions are aimed at enhancing a government's willingness and ability to care for its people, or simply providing humanitarian relief following a natural disaster. Coercive military actions involve the application of limited, carefully prescribed force, or the threat of force to achieve specific objectives. Stability operations are usually noncontiguous, and are often time and human intensive. Army elements might be tasked to conduct stability operations in a complex, dynamic, and often asymmetric environment to accomplish one or more of the following purposes:

- Deter or thwart aggression.
- Reassure allies, friendly governments, agencies, or groups.
- Provide encouragement and support for a weak or faltering government.
- Stabilize an area with a restless or openly hostile population.
- Maintain or restore order.
- Satisfy treaty obligations or enforce national or international agreements and policies.
- Provide humanitarian relief outside the continental United States and its territories.

CIVIL SUPPORT OPERATIONS

1-221. The overall purpose of civil support operations is to meet the immediate needs of designated groups, for a limited time, until civil authorities can accomplish these tasks without Army assistance. Civil support operations are a subset of Homeland Security. Operations support the nation's homeland defense (offensive and defensive), and are only conducted inside the U.S. and its territories.

1-222. During civil support operations, Infantry platoons and squads help provide essential services, assets, or specialized resources to help civil authorities deal with situations beyond their capabilities. The adversary is often disease, hunger, or the consequences of disaster. Civil support operations for the Infantry platoon and squad may include assisting civilians in extinguishing forest fires, in rescue and recovery efforts after floods or other natural disasters, or in supporting security operations before, during, or after terrorist attacks. Platoons and squads must maintain the capacity to conduct offensive, defensive, and tactical enabling operations during the conduct of civil support operations.

TACTICAL ENABLING OPERATIONS

1-223. Tactical enabling operations support the larger unit's effort to accomplish its mission. They always play a supporting role as part of one of the full spectrum operations. The effective planning, preparation, execution, and assessment of tactical enabling operations mirror that of traditional offense and defense operations.

1-224. There are six types of tactical enabling operations: reconnaissance; security; troop movement; relief in place; passage of lines; and combined arms breach.

Reconnaissance

1-225. Reconnaissance operations are undertaken to obtain (by visual observation or other detection methods) information about the activities and resources of an enemy or potential enemy. They are designed to secure data concerning the meteorological, hydrographical, or geographical characteristics and the indigenous population of a particular area (FM 1-02). The four forms of reconnaissance are route; zone; area; and reconnaissance in force.

1-226. Reconnaissance is performed before, during, and after other operations to provide information to the leader or higher commander for situational understanding. Reconnaissance identifies terrain characteristics, enemy and friendly obstacles to movement, and the disposition of enemy forces and civilian population; all of which enable the leader's movement and maneuver. Leaders also use reconnaissance prior to unit movements and occupation of assembly areas. It is critical to protect the force and preserve combat power. It also keeps the force free from contact as long as possible so it can concentrate on its decisive operation.

Security

1-227. Security operations are undertaken by the commander to provide early and accurate warning of enemy operations, to provide the force being protected with time and maneuver space within which to react to the enemy, and to develop the situation to allow the commander to effectively use the protected force. The five forms of security are cover, guard, screen, area, and local.

1-228. The ultimate goal of security operations is to protect the force from surprise and reduce the unknowns in any situation. Leaders employ security to the front, flanks, or rear of their force. The main difference between security and reconnaissance operations is that security operations orient on the force or facility being protected, while reconnaissance is enemy and terrain oriented. Security operations are shaping operations.

Troop Movement

1-229. Troop movement is the movement of troops from one place to another by any available means (FM 1-02). Troops move by foot, motor, rail, water, and air. There are three types of troop movement, with corresponding levels of security based on the presence of the enemy: administrative movement; road march; and approach march. (See Chapter 4.)

1-230. Successful movement places troops and equipment at their destination at the proper time, ready for combat. Commanders use various forms of troop movement to concentrate and disperse their forces for both decisive and shaping operations. Therefore, leaders and their Soldiers need to be familiar with all of the methods and types of troop movements and their roles within them.

Relief in Place

1-231. A relief in place (RIP) is an operation in which all or part of a unit is replaced in an area by the incoming unit. The responsibilities of the replaced elements for the mission and the assigned zone of operations are transferred to the incoming unit. The incoming unit continues the operation as ordered (FM 1-02).

Passage of Lines

1-232. A passage of lines is a tactical enabling operation in which one unit moves through another unit's positions with the intent of moving into or out of enemy contact (FM 1-02). Infantry platoons and squads perform roles as either the moving or stationary unit.

Combined Arms Breach

1-233. Combined arms breach operations are conducted to allow maneuver, despite the presence of obstacles. Breaching is a synchronized combined arms operation under the control of the maneuver commander. Breaching operations begin when friendly forces detect an obstacle and begin to apply the breaching fundamentals. However, they end when battle handover has occurred between follow-on forces and the unit conducting the breaching operation (FM 1-02).

SECTION VIII — SITUATION

1-234. Every military situation is unique and must be solved on its own merits. To better equip leaders to solve tactical problems, this section discusses some of the background issues that directly or indirectly affect Infantry platoons and squads. They are—

- The human dimension.
- The laws of war.
- The operational environment.

HUMAN DIMENSION

> "Were we able to examine all battles through a military microscope, it is probable that we would almost always find the small seed of victory sowed by a determined leader and a handful of determined men."
>
> *Infantry in Battle,* 1939

1-235. One of the toughest challenges faced by Infantry platoons is the need to reconcile the necessary orderliness of doctrine and training with a disorderly battlefield. The human dimension of "Army life" in garrison tends to be centralized and predictable. This is not true in combat, because operations usually do not proceed exactly as planned. For these reasons, leaders and their Soldiers must first understand that apparent contradiction between order and disorder is a normal aspect of combat. A working knowledge of the importance of will, skill, and the friction of combat is essential to fully comprehend the battlefield situation.

WILL

1-236. The human will is close combat's wild card. At times, human dynamics contribute more to victory in close combat than weapons and tactics. Close combat is messy, violent, and dirty. Although much of what happens in battle can be reduced to useful formulas (OPORDs, processes, drills, and methods), fighting and winning always includes the human dimension.

SKILL

1-237. Skill is tactical and technical competence. It is mastery of the generally-accepted tactics, techniques, and procedures used to carry out combat. Doctrine and training exist to promote the Soldier's skill to the highest level prior to combat, and to sustain it once in combat.

1-238. In close combat, commitment to winning and surviving the fight is the manifestation of human will. No other element has the potential of equalizing seemingly unequal opponents. Because the human will is difficult to measure, it is difficult to infuse into discussions of tactics. Concepts like tempo, initiative, flexibility, audacity, and momentum attempt to convey this critical aspect in doctrine. To win in combat, leaders and Soldiers must develop the will to adapt their training and doctrine to unique situations.

FRICTION

1-239. Friction is the resistance that comes from the environment that leaders and their units experience during the course of an operation. It is comprised of all the elements in the operational environment that come together to reduce the unit's ability to accomplish its mission. Some (but not all) factors that contribute to these incidents are—

- Danger.
- Unclear information or orders; misinterpreted orders.
- Rapidly-changing situations and continuous demands.
- Environmental factors such as noise, dirt, weather, and complex terrain.
- Physical factors such as hunger, fatigue, and lack of sleep.
- Fear.

1-240. Combat is where the positive aspects of will and skill battle with the negative aspects of friction. When will and skill are strong, no amount of friction can prevent a victory. Failure often results when the friction of close combat overcomes will and skill.

LAW OF WAR

1-241. The law of land warfare is an ever-present aspect of the operational environment. Leaders and their Soldiers have a legal and moral obligation to follow it. The law of war (LOW) explains rights afforded to everyone on the battlefield; both combatants and noncombatants.

WHY WE FOLLOW THE LAW OF WAR

1-242. U.S. Soldiers follow the LOW for five basic reasons. First, it is the law. Violations of the LOW are punishable under the UCMJ, the 1996 War Crimes Act, and international law. Second, following the LOW enhances public support for the military cause, contrasted by the lack of support displayed after incidents like the My Lai massacre and Abu Ghraib prisoner abuse case. Third, following the law of war may encourage some of our enemies to follow the law of war. Fourth, because they know American Soldiers will care for them, there is a greater chance our enemies will surrender rather than continue fighting. Fifth, it is morally right.

1-243. Although U.S. forces and their allies must respect the LOW, leaders remain aware that some of our enemies do not. In some cases, enemies seek an advantage by exploiting the LOW, after which some American Soldiers may have difficulty understanding why they should continue to follow the LOW. Leaders must set the example by adhering to the letter as well as the spirit of the LOW, even in the face of enemy violations.

PRINCIPLES

1-244. Under the LOW, leaders are legally accountable for the deadly force their units use during battle. Four principles exist to assist leaders in following the LOW: military necessity; distinction; avoiding unnecessary suffering; and proportionality. These principles guide the leader in making decisions that are consistent with international law:

(1) **Military Necessity.** The principle of military necessity states: "Soldiers may use force not forbidden by international law that is necessary to secure the proper submission of the enemy military force." In short, if you target someone or something with deadly force, doing so must offer a direct and concrete military advantage.

(2) **Distinction.** The principle of distinction states that combatants must distinguish combatants from noncombatants and military objects from civilian objects. On some contemporary battlefields, enemies may try to exploit this principle by fighting in civilian clothes and using civilian or protected structures.

(3) **Avoid Unnecessary Suffering.** The principle of avoiding unnecessary suffering allows you to cause only the amount of injury, destruction, and suffering that is necessary to accomplish your legitimate military purposes. Do not alter weapons to cause unnecessary suffering (such as making dumb-dumb rounds). Do not kill or destroy more than is necessary to win the fight or save another Soldier's life.

(4) **Proportionality.** The principle of proportionality states that "military forces may not cause suffering, injury, or destruction to noncombatants or civilian objects which would be excessive in relation to the concrete and direct military advantage anticipated." In other words, the military necessity of the target must outweigh the collateral damage caused by the commander's act.

1-245. Rules of engagement (ROE) are directives issued by competent military authority that delineate the circumstances and limitations under which United States forces will initiate or continue combat engagement with other forces encountered (FM 1-02). The ROE define the commander's rules for use of force and limit the commander's options to comply within the LOW. They take into account practical and political considerations and may limit the commander's use of force more than the LOW.

ENEMY PRISONERS OF WAR AND OTHER DETAINEES

1-246. The Geneva Convention acts as a shield to prevent the capturing force from prosecuting the captured force for lawful warlike acts. It requires all captured personnel to be treated humanely as enemy prisoner(s) of war (EPW) until a competent military tribunal determines that the captured personnel are not entitled to that status. AR 190-8, *Enemy Prisoners of War, Retained Personnel, Civilian Internees, and Other Detainees*, covers the proper treatment of EPWs and other detainees.

1-247. Injured enemy soldiers who are out of the fight and enemy soldiers making a clear attempt to surrender are protected under the LOW. However, because America's enemies know we follow the LOW, they may try to exploit the LOW to gain a tactical advantage. An enemy may not feign injury or surrender. For this reason, American Soldiers must maintain readiness to use deadly force when dealing with the injured or surrendering enemy until these individuals are in custody. Once American Soldiers determine that an enemy soldier is attempting to surrender or is injured so badly that he is out of the fight, that enemy soldier is protected unless he enters back into the fight.

1-248. At the Infantry platoon and squad levels, the six simple rules for EPWs are search, silence, segregate, safeguard, speed to the rear (the five S's), and tag. The tag includes the date of capture, location of capture (grid coordinate), capturing unit, and special circumstances of capture (how the person was captured). The five S's include:

(1) Search the EPW thoroughly and disarm him.
(2) Silence—require the EPW to be silent.
(3) Segregate the EPW from other EPWs (by sex and rank).
(4) Safeguard the EPW from harm while preventing him from escaping.
(5) Speed the EPW to the designated EPW collection point.

1-249. Once the enemy is under friendly control, they assume the protected status of detainee. This is an umbrella term that includes any person captured or otherwise detained by armed force. Under the LOW, leaders and Soldiers are personally responsible for detainees under their control. Mistreatment of EPWs is a criminal offense under the Geneva Convention, AR 190-8, and *The 1996 War Crimes Act* (18 U.S.C. § 2441). The War Crimes Act makes it a federal crime for any U.S. national, whether military or civilian, to violate the Geneva Convention by engaging in murder, torture, or inhuman treatment.

TEN SOLDIER RULES

1-250. The following 10 simple rules will assist Soldiers in living and enforcing the law of war (LOW) (use the mnemonic OBLIGATION):

(1) **O**nly fight individuals who are identified as uniformed combatants, terrorists, or insurgents committing hostile acts or demonstrating hostile intent.
(2) **B**ased on triage, medically care for all wounded, whether friend, foe, or noncombatant.
(3) **L**eave medical personnel, facilities, or equipment out of the fight unless they are being used by the enemy to attack U.S. forces.
(4) **I**njured or surrendering Soldiers who no longer have the means to fight are protected. Disarm them, treat their wounds, and speedily turn them over to the appropriate authorities.
(5) **G**uarantee humane treatment of noncombatants and enemy prisoners of war.
(6) **A**busing prisoners is never authorized. Do not kill, torture, or mistreat enemy prisoners of war or those being detained by U.S. forces.
(7) **T**aking private possessions is stealing. Respect private property.
(8) **I**ntervene, stop, or prevent violations of the law of war to the best of your ability.
(9) **O**nly use necessary force to eliminate the threat and accomplish the mission.
(10) **N**ever tolerate a LOW violation. Report all violations of the LOW to your superiors.

OPERATIONAL ENVIRONMENT

1-251. The operational environment is a composite of the conditions, circumstances, and influences that affect the employment of military forces and bear on the decisions of the unit leader (FM 1-02). In every day language, the operational environment is all of the variables that affect the leader's mission. It is essential for leaders to educate themselves on how to analyze and understand the variables within their operational environment.

1-252. Understanding the operational environment is perhaps the most difficult aspect of making decisions and conducting operations. The TTP for accomplishing tasks are fairly straightforward. This manual and many others contain numerous TTP for how to perform tasks and missions. Choosing and applying the appropriate TTP based on the specific conditions of a given operational environment, however, is never straightforward and always carries with it second and third order effects. Leaders must therefore educate themselves to understand their environment and the factors that affect their decisionmaking. This will contribute greatly to the development of their judgment in complex and uncertain situations.

1-253. Infantry platoon and squad leaders use the factors of METT-TC to understand and describe the operational environment. These six widely-known and used factors are categories for cataloging and analyzing information. Leaders and their Soldiers are constantly observing and assessing their environment.

Chapter 2
Employing Fires

Suppressing or destroying the enemy with direct and indirect fires is essential to success in close combat. Because fire and movement are complementary components of maneuver, the Infantry platoon leader must be able to effectively mass the fires of all available resources at critical points and times. Effective and efficient employment of fires is achieved when the platoon acquires the enemy rapidly and masses the effects of direct and indirect fires. When employed effectively the effects produce decisive results in the close fight.

SECTION I — CONSIDERATIONS FOR EMPLOYING AND CONTROLLING FIRE

2-1. When planning and executing fires, Infantry leaders must know how to apply several fundamental principles. The purpose of these principles is not to restrict the actions of subordinates. They are intended to help the platoon accomplish its primary goal in any engagement (acquire first, shoot first, and hit first) while giving subordinates the freedom to act quickly upon acquisition of the enemy. The principles of fire control are—

- Command and control.
- Mass the effects of fire.
- Destroy the greatest threat first.
- Avoid target overkill.
- Employ the best weapon for the target.
- Minimize friendly exposure (protection).
- Prevent fratricide.
- Plan for limited visibility conditions.
- Develop contingencies for diminished capabilities.

COMMAND AND CONTROL

2-2. Every time a Soldier fires a weapon or requests indirect fire, he does so with the intent to kill or destroy an enemy target. He may also affect an enemy target through nonlethal means such as smoke, illumination, or nonlethal fires. Platoon and squad leaders are the first leaders in the chain of command who are legally and morally responsible for the fires and effects produced by their subordinates.

2-3. Exercising control of the direct fires is founded upon the concept of authority. When given a mission, leaders are given the authority they need to accomplish the mission. This non-negotiable responsibility includes the need to fire weapons, move units, and conduct military actions. Leaders and their subordinates are accountable for carrying out these duties in a legal, moral, and competent manner.

2-4. Tactical reasons to exercise control include, combining weapons to achieve complementary and reinforcing effects, preventing fratricide on another unit, achieving a particular tempo, achieving surprise, and preventing detection. Technical reasons to exercise control include limited ammunition quantities, deconflicting fires, and managing surface danger zones (SDZs.)

2-5. Leaders must balance the need to personally control their subordinate's fires with the need for their units to be responsive to procedural control. The surest way for a leader to control his subordinate's fires is to withhold that authority to his level. The surest way to ensure his subordinates have maximum freedom of action is to provide them with rules and conditions to guide their personal fire decisions. These rules can be issued in the unit's TSOP, rules of engagement (ROE), and mission briefs.

MASS THE EFFECTS OF FIRE

2-6. Infantry units must mass the effects of fires to achieve decisive results. Leaders achieve fire superiority by concentrating all available fires. Massing involves focusing fires at critical points, distributing the effects, and shifting to new critical points as they appear. There are many ways to achieve fire superiority. They include:

- Using combinations of weapons and munitions.
- Applying the appropriate volume and accuracy of fire at enemy point and area targets.
- Establishing engagement criteria and engagement priorities.
- Assigning Soldiers mutually supporting positions and overlapping sectors of fire.
- Focusing fires on enemy vulnerabilities.

2-7. Concentration of fires, both preparatory and supporting, is necessary to gain and maintain fire superiority. Fires from weapons not organic to the platoon or squad are coordinated by the unit leader or his next higher headquarters. Artillery, tanks, and tactical air may be available to take part in the penetration and reduction of enemy prepared defenses. Fire superiority is particularly important while attacking when Infantry units begin breaching protective enemy obstacles and assaulting the enemy position itself. When defending, fire superiority defeats the enemy's attack, enabling the defender to transition to the offense by counterattacking.

2-8. Every tactical plan the leader develops (for both offense and defense) must have a concept of fires. (For example, how the platoon will gain and maintain fire superiority.) The plan to achieve fire superiority includes initiation, adjustments, and ceasing fire. Because the effects of fire tend to diminish as the enemy becomes accustomed to it, fires should initially be intense. Delivery of large volumes of concentrated fires into a specified area inflicts maximum damage and shock. Properly timed and delivered fires contribute to the achievement of surprise, and to the destruction of the enemy. Shifting and ceasing fires should be planned and executed with equal precision. If not, the complementary movement to positions of advantage is delayed, and the enemy could have an opportunity to recover and react.

2-9. Leaders concentrate the effects of combat power at the decisive place. First, leaders develop targets, target reference points (TRPs), and sectors of fire to integrate the effects of fires and maneuver with the terrain. Second, they select positions that maximize cover and concealment and emplace security elements to enhance protection. Third, they seek information from reconnaissance and surveillance elements to determine enemy dispositions and intentions. Finally, they exercise battlefield leadership before and after contact by making bold decisions and synchronizing other elements of combat power.

2-10. The fire plan is developed concurrently with the leader's scheme of maneuver, in as much detail as time will allow. When developing his fire plan, the leader considers—

- The use of all available assets.
- The enemy situation, disposition, and terrain.
- The nature of targets and the effects desired.
- The availability of ammunition and Soldier's combat load.
- Time of fire (initiation of fires, duration and rate, and cease fires).
- Scheduled and on-call fires.
- Use of smoke and illumination.
- Means of communication.

DESTROY THE GREATEST THREAT FIRST

2-11. The platoon engages targets in direct relation to the danger they present. If two or more targets of equal threat present themselves, the platoon should engage the closest target first. The platoon marks the defense engagement area (EA) so it can determine when to engage various targets, then plans these ranges on sketches and range cards. For example, the platoon should mark the EA at the Javelin maximum engagement distance (2,000 meters) to ensure gunners do not waste missiles.

AVOID TARGET OVERKILL

2-12. The Infantry platoon strives to avoid engaging a target with more than one weapon system at a time. To avoid target overkill, the platoon can divide EAs into sectors or quadrants of fire to better distribute direct fire among the platoon. The platoon can use many techniques to mark the EA. The platoon and company should develop a TSOP that divides the EA with both infrared and thermal TRPs to enable good distribution of fires within the EA. Squads and platoons should mark EAs with infrared devices for engagements during limited visibility. Thermal sights on the command launch unit (CLU) of the Javelin cannot detect infrared sources. Therefore, the EA must also be marked with thermal devices. The platoon can burn a mixture of rocks, sand, and diesel fuel inside a fuel drum, ammunition can, or bucket shortly before dusk to give off a heat source for most of the night.

2-13. The platoon leader may also designate rates of fire, by weapon system, to avoid target overkill. Predetermining the rates of fire and length of firing time allows the platoon leader to plan for sufficient ammunition needed for desired effect. The rates of fire are cyclic, rapid, and sustained.

2-14. In offensive operations, avoid overkill by—

- Establishing weapon system priorities to engage targets and distribute fires. The platoon leader may establish that a Javelin team engages a tank on the objective while the other Javelin team engages a bunker.
- Having the weapons squad leader control the support-by-fire element to prevent needless ammunition expenditure.
- Having the platoon leader use direct fire control measures as discussed in Section IV of this chapter.

EMPLOY THE BEST WEAPON FOR THE TARGET

2-15. Enemy target type, range, and exposure are key factors in determining the friendly weapon and munitions that should be employed for the desired target effects. Using the appropriate weapon against the enemy target increases the probability of its rapid destruction or suppression. The platoon leader task organizes and arrays his forces based on the terrain, enemy, and desired fires effects.

2-16. Weapons and munitions are designed with specifications that enable their effects to be forecasted with some degree of accuracy before being fired. They are also designed for a specific range versus specific targets. Platoon and squad Infantry leaders must have an intimate understanding of their organic and supporting weapons and munitions to include the following:

- Weapon characteristics, ranges, and optimal use.
- Munition characteristics, lethality, and optimal use (such as how to achieve intended effects and avoid unintended effects).
- Procedures to request, control, and adjust fires from other agencies.

2-17. Infantry platoon and squad leaders must ensure that they focus the fires of their weapons systems on targets their weapon systems are designed to engage (Figure 2-1). For example, CCMS are used against armored targets at ranges of up to 2,000 meters for stand-off protection. However, medium machine guns are used to destroy enemy unarmored vehicles and dismounted Infantry at ranges within 1,000 meters. Leaders plan and execute fires throughout the depth of the AO, engaging enemy targets early and continuously IAW weapon capabilities and standoff. The principle of depth enables Infantry units to achieve and maintain fire superiority. By engaging the enemy early, leaders disrupt the enemy's plans, forcing him to seek cover. To apply this principle, leaders are required to know weapon systems at their effective ranges as well as the movement rates of Soldiers and equipment. When moving, the friendly force echelons its fires in front of the friendly attacking force. This allows unhindered movement. When the friendly force defends, they echelon their forces against the approaching enemy force.

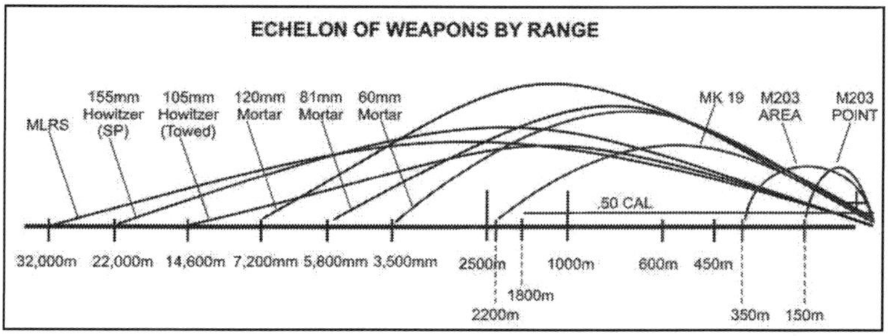

Figure 2-1. Weapon ranges.

MINIMIZE FRIENDLY EXPOSURE (PROTECTION)

2-18. Units increase their survivability by exposing themselves to the enemy only when necessary to engage him with effective fires. Natural or man-made defilade provides the best cover. Infantry units minimize their exposure by constantly seeking available cover, attempting to engage the enemy from the flank, remaining dispersed, firing from multiple positions, and limiting engagement exposure times.

PREVENT FRATRICIDE

2-19. Leaders must be proactive in reducing the risk of fratricide, especially when it concerns their Infantry platoon or squad on the multi-dimensional battlefield. There are numerous tools to assist them in fratricide avoidance. By monitoring unit locations, leaders at all levels can ensure that they know the precise locations of their own and other elements and can control their fires accordingly. Infantry leaders must know the location of each of the squads.

2-20. The platoon can use infrared and thermal marking techniques to ensure that adjacent units do not mistakenly fire at friendly forces during limited visibility. The assault element can use the infrared chemical lights, blacklight tube lights tied to poles, and many other methods to mark the assault element's progress. Leaders must ensure that the enemy does not have night vision capability before marking their Soldiers' progress with infrared marking devices. For a detailed discussion of fratricide avoidance, refer to Section III of this chapter.

PLAN FOR LIMITED VISIBILITY CONDITIONS

2-21. Dense fog, rain, heavy smoke, blowing sand, and the enemy's use of smoke may significantly reduce the leader's ability to control direct fires of the platoon. Therefore, Infantry units are equipped with thermal sights and night vision systems that allow squads to engage the enemy during limited visibility at nearly the same ranges normally engaged during the day.

DEVELOP CONTINGENCIES FOR DIMINISHED CAPABILITIES

2-22. A platoon leader usually develops a plan based on having all of his assets available and makes alternate plans to account for the loss of equipment or Soldiers. The platoon leader should develop a plan that maximizes his unit's capabilities while addressing the most probable occurrence. He should then factor in redundancy within the platoon. For example, he may designate alternate sectors of fire for the squads that provide him the means of shifting fires if one squad has been rendered ineffective. These contingencies may become items within a unit SOP.

SECTION II — WEAPON AND MUNITION CHARACTERISTICS

2-23. To better understand the science of employing fires, leaders should know the basic characteristics of weapons and munitions. This knowledge leads to an increased understanding of capabilities and the ability to achieve complementary, reinforcing effects.

COMMON WEAPONS AND MUNITION CHARACTERISTICS

2-24. There are five types of weapons used at the Infantry platoon level: small arms; machine guns; grenade launchers; shoulder-launched munitions (SLM)/Close Combat Missile System (CCMS); and mortars. These weapons are developed with emphasis on certain characteristics (Table 2-1).

Table 2-1. Common weapon characteristics.

	Small Arms	Machine Gun	Grenade Launcher	SLM/CCMS	Mortars
Lay	Direct fire	Direct fire	Direct fire	Direct fire	Indirect fire
Ammunition	Penetration	Penetration	HE	Penetration/ HE	HE WP ILLUM
Trajectory	Low trajectory	Low trajectory	High trajectory	Low trajectory	High trajectory
Point or Area Enemy Target	Point target	Point and area target	Point and area target	Point target	Area target
Organic Infantry Unit Weapons	M4	M249 MG M240 MG	M203	AT4 SMAW-D M72 Javelin	Organic to company/ battalion

LAY

2-25. The lay of a weapon is the characteristic that determines how a Soldier engages a target. A weapon's lay is either direct or indirect fire. Every weapon organic to the Infantry platoon or squad is direct fire, with the exception of company and battalion mortars. Infantry Soldiers armed with organic weapons engage the enemy with the weapon's own sight. The strength of a direct fire weapon is its responsiveness. The weapon does not need to be requested from higher, nor does higher have to "clear fires" before a round may be fired. Soldiers manning indirect fire weapons such as mortars engage the enemy by using a separate observer (Figure 2-2). Soldiers manning mortar weapon systems have the tactical advantage of avoiding direct contact with the enemy in the fight.

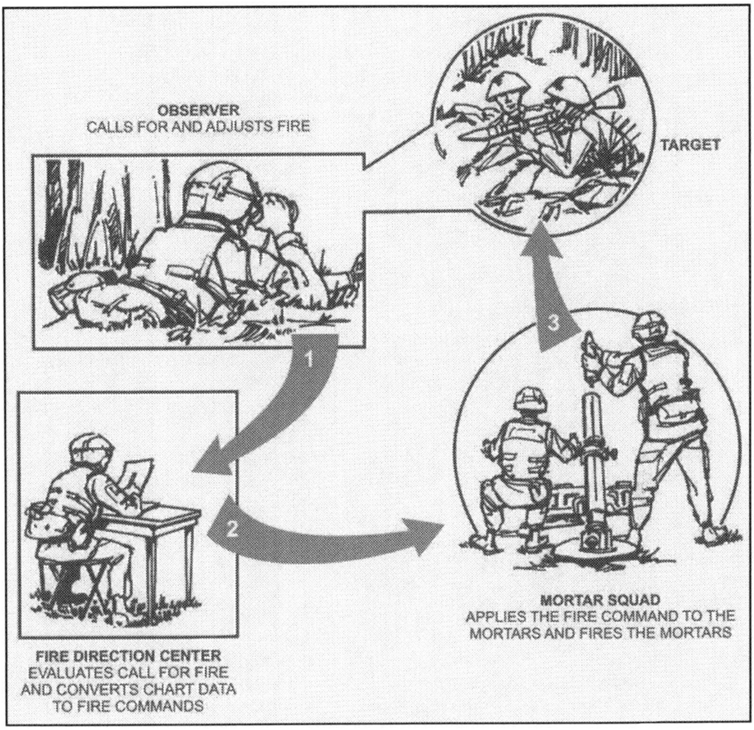

Figure 2-2. Indirect fire.

AMMUNITION

2-26. For the purpose of this manual, there are three categories of ammunition: high explosive (HE); penetration; and special purpose munitions. Only HE and penetration munitions are considered for achieving complementary and reinforcing effects. The leader is able to engage known enemy targets (those he can see and acquire) as well as likely enemy targets (those he cannot see and cannot clearly acquire). If the enemy remains hidden but suspected, the grenadier will engage him with high explosives. If the enemy attempts to move to a location that will protect him from HE munition, the automatic rifleman will engage him with a penetrating munition. Special purpose munitions are described for general information only.

High Explosive

2-27. HE munitions are used to kill enemy soldiers, force enemy soldiers to remain under protective field fortification cover, force an enemy vehicle to button up, or force an enemy vehicle into a less advantageous position. Only a direct hit will destroy or significantly damage an armored vehicle.

2-28. There are two noteworthy strengths of HE munitions. First, HE muntions do not have to score a direct hit to physically affect the target. This makes it possible to engage targets that are not clearly acquired, but are likely or suspected. Second, HE munitions are especially effective at destroying structures such as bunkers and vehicles.

Penetration

2-29. The effectiveness of penetration munitions is dependent on the weapon system's ability to generate velocity, and the ability of the munition's mass to punch a hole in the enemy target. It is fairly easy to gauge the effectiveness of penetration munitions. Soldiers can engage targets with confidence because of the known effect the round will have on a target. The three general categories of penetration munitions are ball and tracer, armor piercing, and high explosive antitank (HEAT).

Ball and Tracer

2-30. Ball and tracer rounds use high velocity to penetrate soft targets on impact. Penetration depends directly on the projectile's velocity, weight, and angle at which it hits. Ball and tracer rounds are usually small caliber (5.56 to 14.5 millimeters) and are fired from pistols, rifles, and machine guns.

Armor Piercing

2-31. Armor piercing rounds use shaped-charged or kinetic energy penetration warheads specially designed to penetrate armor plate and other types of homogeneous steel. They are used effectively against fuel supplies and storage areas.

HEAT (High Explosive Antitank)

2-32. HEAT rounds are designed to defeat armor through the use of shaped charge. A shaped charge is an explosive charge created so the force of the explosion is focused in a particular direction.

Special Purpose

2-33. There are many types of munitions that do not fit the profile of the two major categories (HE and penetration). These are called special purpose munitions. Examples are incendiary, obscuration, illumination, nuclear, and chemical rounds.

TRAJECTORY

2-34. Infantry Soldiers can more effectively engage moving enemy targets with low trajectory fire than high trajectory fire. Enemy reaction when engaged with friendly low trajectory fire is predictable: get down and seek frontal cover. When this happens, high trajectory fire can effectively engage enemy targets in fighting positions, holes, or deadspace where low trajectory fire cannot. Friendly high trajectory fire can also force the enemy to move out of the area and seek overhead cover, limiting their effectiveness.

2-35. Leaders create a dilemma for the enemy by combining low and high trajectory weapons. If the enemy gets up from his position and attempts to move, the automatic rifleman will engage him. If the enemy decides to stay in his position behind frontal cover, the grenadier will engage him. Either option results in the friendly force engaging the enemy. This united effect of the automatic rifleman and grenadier outweighs the effect either would have if they engaged the enemy without the other.

ENEMY TARGET TYPES

2-36. Weapons and munitions are designed for employment against the two general types of enemy targets: point, and area. A *point target* is located in a specific spot with a single aim point (enemy soldier, vehicle, piece of equipment). An *area target* is spread over an area with multiple aim points (formation of enemy soldiers, an enemy trench line). Some weapon systems such as machine guns and grenade launchers can effectively engage both point and area targets.

FIRE TEAM WEAPONS

2-37. The rate of fire is the number of rounds fired in a minute by a particular weapon system. The leader dictates the rate of fire for each weapon system under his control. There are two factors that contribute to leader decisions about rates of fire: achieving fire superiority; and ammunition constraints. For information on equipment in the weapons squad or other supporting weapons, see Appendix A and Appendix B.

RIFLE

2-38. Rifleman and Infantry leaders are currently armed with the M4 rifle. The M4 rifle is a direct fire weapon that fires ball and tracer 5.56-mm ammunition. The rifleman's primary role is to kill the enemy with precision fire. In this capacity, the rate of fire for the M4 rifle is not based on how fast the Soldier can pull the trigger. Rather, it is based on how fast the Soldier can accurately acquire and engage the enemy. The second role of the rifleman is to engage likely or suspected enemy targets with suppressive fire.

M249 MACHINE GUN

2-39. The automatic rifleman is currently armed with an M249 machine gun. The M249 is a direct-fire, low trajectory weapon that is primarily used to fire ball tracer 5.56-mm ammunition linked at area targets. The M249 also has the ability to fire unlinked 5.56-mm ammunition in 30-round magazines, but reliability is greatly reduced. Firing with a magazine should be limited to emergency situations.

M240B MACHINE GUN

2-40. Two medium machine guns (currently the M240B) and crews are found in the Infantry platoon's weapons squad. Machine gunners are a self-contained support by fire element or with a rifle squad to provide long range, accurate, sustained fires against enemy Infantry, apertures in fortifications, buildings, and lightly-armored vehicles. Machine gunners also provide a high volume of short-range fire in self defense against aircraft. THE M240B fires 7.62-mm ammunition. Refer to Appendix A for further information on machine guns.

GRENADE LAUNCHER

2-41. The grenadier is currently armed with the M203 40-mm grenade launcher. The M203 is a direct fire, high trajectory weapon that can be used for either point or area targets. The M203 fires several types of munitions including, HE, high explosive dual purpose (HEDP) (antipersonnel/antiarmor), riot control (CS), buckshot, and signaling. As with the rifleman, the grenadier's rate of fire is based on how quickly he can accurately acquire and engage the enemy.

SHOULDER-LAUNCHED MUNITIONS

2-42. Shoulder-launched munitions (SLM) are lightweight, self-contained, single-shot, disposable weapons that consist of unguided free flight, fin-stabilized, rocket-type cartridges packed in launchers. SLM provide the Soldier a direct fire capability to defeat enemy personnel within field fortifications, bunkers, caves, masonry structures, and lightly armored vehicles. Soldiers use SLM to engage enemy combatants at very close ranges—across the street or from one building to another. Likewise, SLM may be fired at long distances to suppress the enemy or kill him. Soldiers may employ the SLM as a member of a support-by-fire element to incapacitate enemy forces that threaten the friendly assault element. When the assault element clears a building, the leader may reposition the SLM gunner inside to engage a potential counterattack force. Refer to Appendix B for further information on SLM.

COMPLEMENTARY AND REINFORCING EFFECTS AT THE FIRE TEAM LEVEL

2-43. One of the leader's primary duties is to control the distribution of his unit's fires. An Infantry team leader tasked to establish support by fire uses the principles of complementary and reinforcing effects to guide his unit's actions. The goal of each weapon system combination is to create an effect that outweighs the effects that either weapon system would make acting alone. The primary combination team leaders strive to employ are the weapons systems of the automatic rifleman and the grenadier. This combination is the center around which the remainder of the fire team's functions revolves.

THE RIFLEMAN

2-44. The rifleman's role when the grenadier and automatic rifleman combine their fires is to perform one of three functions:

- Reinforce the automatic rifleman. If necessary, the rifleman can replace the automatic rifleman for a short time.
- Fix another target while the automatic rifleman and grenadier destroy the target they are engaging.
- Provide security and observation.

THE TEAM LEADER

2-45. In weapon employment, the team leader's role is to maximize the complementary effects of the combination of the grenadier and automatic rifleman. He does this through using proper fire commands and control measures. The team leader's second role is to assume the duties of the rifleman if necessary.

SECTION III — ENGAGING THE ENEMY WITHOUT ENDANGERING FRIENDLY TROOPS

2-46. In the offense, effective friendly supporting fires require firing on enemy targets that are close to assaulting friendly Infantry Soldiers. A safe integration of fires and maneuver this close demands careful planning, coordination, and knowledge of the supporting weapons. In the defense, the most common close support is the final protective fire (FPF), which is normally placed very close to friendly positions. When planning close supporting fires for the offense or defense, leaders consider the effect required, accuracy of the delivery system, protection of Soldiers, integration of assets, timings and control, echelonment of fires, and tactical risk from enemy forces.

2-47. Munition effects do not distinguish between friendly forces, noncombatants, and the enemy. To inflict maximum casualties on the enemy while minimizing effects to friendly Soldiers and noncombatants, leaders must have an understanding of weapon-munition effects, SDZ, minimum safe distances (MSD), risk estimate distances (RED), and the terrain's influence on projectiles. Failure to account for characteristics of direct and indirect weapon systems when considering tactics, techniques, and procedures can result in serious unintended consequences.

2-48. There are many variables that impact on the accuracy of the weapon. Artillery and mortars are referred to as area weapon systems because every round fired from the same tube impacts in an area around target aiming point. This dispersion is greater in length than in width. The weather conditions (wind, temperature, and humidity), the condition of the weapon, and the proficiency of the crew also affect accuracy.

SURFACE DANGER ZONE

2-49. The SDZ is the ground and airspace for vertical and lateral containment of projectiles, fragments, debris, and components resulting from the firing, launching, or detonation of weapon systems (including explosives and demolitions). Each weapon system or munition has its own unique SDZ. The critical components of the SDZ are the primary danger area and the buffer zone.

2-50. Understanding the components of the SDZ enable leaders and their Soldiers to make good decisions concerning how close they can get to the effects of friendly weapon fire. SDZs are developed using precise technical data without considering the effects of terrain. This data should be consulted whenever exact specifications are required. However, because the technical data can be confusing, it is useful to describe SDZs in a general manner. For exact weapon SDZs (see DA Pam 385-63, *Range Safety*).

2-51. The primary danger area consists of the dispersion and ricochet area along the gun-target line for the maximum range of the weapon. The dispersion area is a 5-degree angle to the right and left of the gun-target line that accounts for human error, gun or cannon tube wear, and propellant temperature. The ricochet area contains any projectiles that make contact with surrounding terrain following the munition's initial impact. It is located to the left and right of the dispersion area. The buffer zone is an area outside of the ricochet area allocated for additional safety measures. The buffer area exists to the sides of the gun-target line and at the far end of the weapon's maximum effect range.

DIRECT FIRE

2-52. For direct fire weapons, the risk of being hit by friendly munitions at the edge of the buffer zone is negligible. Based on the type of surface (earth, water, steel, or concrete), the risk increases significantly at the edge of the ricochet area. Risk is extremely high at the edge of the dispersion area. In accordance with DA PAM 385-63, the current level of acceptable risk in training is 1/1,000,000 (outside SDZ), but can be waived by the installation commander to 1/100,000 (outside ricochet area). In combat, most commanders use 1/100,000 (outside ricochet area) based upon METT-TC analysis and risk mitigation measures. Table 2-2 shows the probability of direct fire ricochets.

Table 2-2. Probability of ricochet.

Outside of Area	Probability
SDZ (Area A)	1/1,000,000
Ricochet area	1/100,000
Dispersion area	1/10,000

2-53. When Soldiers remain outside of the buffer zone, the probability of being hit by their own munitions is unlikely. This is true for training and combat. During training, units usually are not authorized to come any closer to the gun-target line than the buffer zone. However, there are many situations in combat that require Soldiers to get closer to the gun-target line of their supporting weapons than the buffer zone allows. In these situations, the leader must understand how to manage the risk to his unit. When assessing this risk the question for leaders to consider is: "Is the threat from the effects of my munitions greater than the threat from the enemy?"

REDUCING RISK

2-54. Given the uncertainty associated with combat and the threat of enemy action, leaders must understand how to reduce risks associated with fire and movement in proximity to direct and indirect fires. As a general rule, the dispersion and ricochet areas present an immediate danger to Soldiers. Observers and protective measures are therefore required.

2-55. The easiest way to protect friendly forces from unintended consequences of their own weapons is to always have an observer. Skilled observers can see the impact of the rounds and any maneuver elements near that area. In circumstances where assigning observers is not possible, leaders must take other measures to mitigate the risk of unintended consequences to friendly forces. Some of the most common include:

- Wearing and requiring Soldiers to wear protective equipment (body armor, Kevlar helmet, eye protection, hearing protection).
- Using terrain, natural or man-made, to mask effects of munitions.
- Adding a buffer zone of additional distance to the gun-target line.
- Using armored vehicles.
- Using graphic control measures.
- Ensuring a highly qualified Soldier is operating the weapons system.

MINIMUM SAFE DISTANCE AND RISK ESTIMATE DISTANCE

2-56. When determining risk with indirect fires, leaders use a combination of minimum safe distances (MSDs), and risk estimated distances (REDs). The MSD risk is designed for training and ensures that friendly Soldiers are far enough away from the effects of munitions so the risk to them is negligible. REDs refer to a safe distance away from a given type of friendly munitions and are only used in combat. REDs are divided into two categories based on the percent of incapacitation (PI) to friendly Soldiers, expressed as

.1 PI and 10 PI. The former (.1 PI) means that one in one thousand Soldiers will not be able to fight because of potential weapon munitions effects. The latter (10 PI) means that one in ten Soldiers will not be able to fight because of weapon effects. When MSDs and REDs are put together, the leader is able to manage his risk from negligible—to 10 PI—based on his distance from the impact of friendly supporting indirect fire. Table 2-3 contains a complete listing of MSDs and REDs for common fire support assets at maximum range of weapons systems. (At lesser ranges the RED decreases).

Table 2-3. MSDs and REDs for common fire support assets.

Weapon System	MSD (Training)	RED (Combat)	
		.1 PI	10 PI
60-mm Mortar (M224)	250m	175m	65m
81-mm Mortar (M252)	350m	230m	80m
120-mm Mortar (M120/M121)	600m	400m	100m
105-mm Artillery (M102/M119)	550m	275m	90m
155-mm Artillery (M109/M198)	725m	450m	125m
155-mm Artillery DPICM	725m	475m	200m

WARNING

REDs are for combat use and do not represent the maximum fragmentation envelopes of the weapons listed. REDs are not minimum safe distances for peacetime training use.

SECTION IV — EMPLOYING DIRECT FIRE

2-57. This section discusses direct fire control and employment rules of engagement, control measures, engagement techniques, fire commands, range cards, adjustments, and closure reports.

RULES OF ENGAGEMENT

2-58. The rules of engagement (ROE) specify the circumstances and limitations under which friendly forces may engage. They include definitions of combatant and noncombatant elements, and stipulate the treatment of noncombatants. Factors influencing ROE are national command policy, operational requirements, and the law of war. ROE always recognize a Soldier's right of self-defense while at the same time clearly defining circumstances in which he may fire.

CONTROL MEASURES

2-59. Direct fire control measures are the means by which the platoon leader or subordinate leaders control their unit's direct fires. Application of these concepts, procedures, and techniques assists the unit in acquiring the enemy, focusing fires on him, distributing the effects of the fires, effectively shifting fires, and preventing fratricide. No single measure is sufficient to effectively control fires. At the platoon level, fire control measures will be effective only if the entire unit has a common understanding of what the fire control measures mean and of how to employ them.

Leader Responsibilities

2-60. The Infantry platoon or squad leader communicates to his subordinates the manner, method, time to initiate, shift, mass fires, and when to disengage by using direct fire control measures. The leader should control his unit's fires so he can direct the engagement of enemy systems to gain the greatest effect. The commander uses the factors of METT-TC and reconnaissance to determine the most advantageous way to use direct fire control measures to mass the effects on the enemy and reduce fratricide from direct fire systems. He must understand the characteristics of weapon systems and available munitions (such as the danger to unprotected Soldiers when tanks fire, discarding sabot ammunition over Soldiers' heads or near them). The primary graphic direct fire control measures are—

- Unit boundary.
- Target reference point.
- Sector of fire.
- Engagement area (EA).

2-61. Other direct fire control measures include—

- Trigger line.
- Maximum engagement line (MEL).
- Final protective line (FPL).
- Principle direction of fire (PDF).
- Priority targets.

2-62. The noise and confusion of battle may limit the use of some of these methods. Therefore, the leader must select a method or combination of methods that will accomplish the mission. The leader should arrange to have a primary and secondary signaling method. The method may be positive (hands on) or procedural (prearranged). There are three types:

(1) Audio (radio, whistle, personal contact).
(2) Visual (hand-and-arm signals, pyrotechnics).
(3) Written (OPORD, range card, sector sketch).

Fire Control Process

2-63. To bring direct fires against an enemy force successfully, leaders must continuously apply the four steps of the fire control process. At the heart of this process are two critical actions intended to achieve decisive effects on the enemy: rapid, accurate target acquisition, and the massing of fires Target acquisition is the detection, identification, and location of a target in sufficient detail to permit the effective employment of the platoon's weapons. Massing of fires focuses direct fires at critical points, then distributes the fires for optimum effect. The four steps of the fire control process follow.

(1) Identify probable enemy locations and determine the enemy scheme of maneuver.
(2) Determine where and how to mass (focus and distribute) direct fires' effects.
(3) Orient forces to speed target acquisition.
(4) Shift direct fires to refocus or redistribute their effects.

Terrain-and Threat-Based Fire Control Measures

2-64. Table 2-4 lists the control measures by whether they are terrain or threat-based.

Table 2-4. Common fire control measures.

Terrain-Based Fire Control Measures	Threat-Based Fire Control Measures
Target reference point	Fire patterns
Engagement area	Engagement priorities
Sector of fire	Weapons ready posture
Maximum engagement line	Weapons control status
Final protective line	Trigger
Principal direction of fire	Weapons safety posture
Final protective fire	
Restrictive firing line	

Terrain-Based Fire Control Measures

2-65. The platoon leader uses terrain-based fire control measures to focus and control fires by directing the unit to engage a specific point or area rather than an enemy element. The following paragraphs describe the terrain-based fire associated with this type of control measure.

Target Reference Point

2-66. A TRP is a recognizable point on the ground that leaders use to orient friendly forces and to focus and control friendly direct and indirect fires. Soldiers use TRPs for target acquisition and range determination. Leaders designate TRPs to orient fires to a particular point, define sectors of fire and observation, and define the limits of an EA. A TRP can also designate the center of a sector or an area where the leader plans to distribute or converge with fires. In addition, when TRPs are designated as indirect fire targets, they can be used in calling for and adjusting indirect fires. Leaders designate TRPs at probable enemy locations and along likely avenues of approach. These points can be natural or man-made. A TRP can be an established site such as a hill or a building, or a feature designated as an impromptu TRP such as a burning enemy vehicle or smoke generated by an artillery round. Friendly units also can construct markers to serve as TRPs (Figure 2-3). TRPs include the following features and objects:

- Prominent hill mass.
- Distinctive building.
- Observable enemy position.
- Destroyed vehicle.
- Ground-burst illumination.
- Smoke round.
- Laser point.

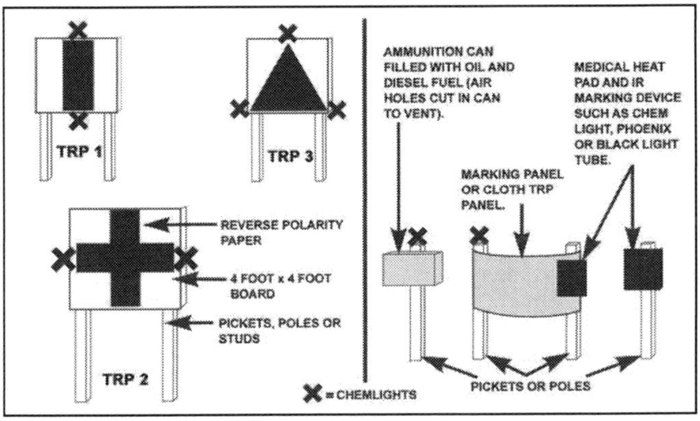

Figure 2-3. Example of constructed TRP markers.

2-67. Leaders designate natural terrain features, man-made terrain features, or any other visual means to be used as TRPs. While TRPs should be visible through all spectrums available to the unit, they should be visible in three observation modes (unaided, passive-infrared, and thermal). They must be easily identifiable to the defender during daylight, should be heated so they can be recognized with thermal sights, and should have an infrared signature so they can be recognized through night vision devices.

2-68. Leaders number TRPs for easy reference. For indirect fire systems, these numbers are assigned as targets (for example, AB1001). For direct fire systems, leaders use any system that is easy and recognizable to their subordinates. Figure 2-4 shows an example of a TRP numbering system when operating in a built-up area. The building and corner numbering system starts at the southwest corner of the objective area. Figure 2-5 shows an example of window and door numbering. In this technique, no distinction is made between windows and doors unless specified. The numbering and lettering always start at the bottom left of any completely visible structure. If a structure is obscured, an estimate is necessary until a more exact call can be made. Corrections to supporting fires are given like indirect fire corrections (for example, left 2, down 1).

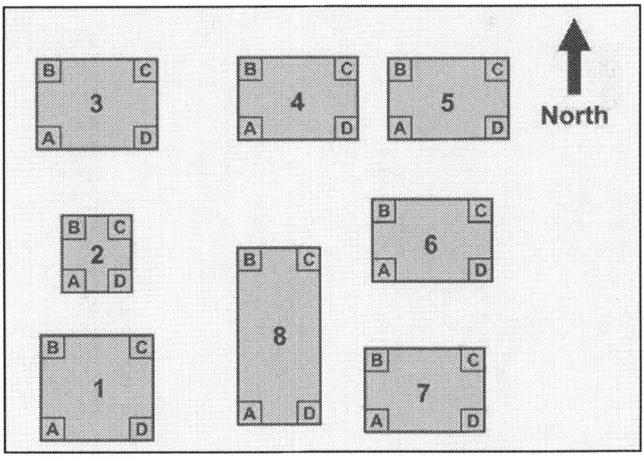

Figure 2-4. Example of TRP numbering system.

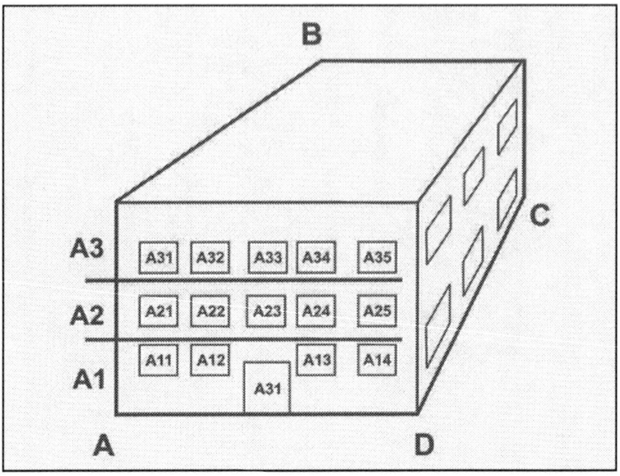

Figure 2-5. Example of window and door numbering.

Engagement Area

2-69. The engagement area (EA) is an area along a likely enemy avenue of approach where the platoon leader intends to mass the fires of all available weapons to destroy an enemy force. The size and shape of the EA are determined by the degree of relatively unobstructed visibility available to the friendly unit's weapons systems in their firing positions, and by the maximum range of those weapons. For an engagement area to be effective, the enemy must either choose to move through the area or be forced or channeled into the area by friendly action (obstacles, indirect fire). Typically, commanders delineate responsibility within the EA by assigning each platoon a sector of fire or direction of fire. These fire control measures are covered in the following paragraphs.

Sector of Fire

2-70. Leaders assign sectors of fire to Soldiers manning weapons or to a unit to cover a specific area of responsibility with observation and direct fire. In assigning sectors of fire, leaders consider the number and type of weapons available. The width of a sector of fire is defined by a right and left limit. Leaders may limit the assigned sector of fire to prevent accidental engagement of an adjacent friendly unit. The depth of a sector is usually the maximum range of the weapon system unless constrained by intervening terrain or by the leader (using a maximum engagement line [MEL]). At the platoon level, sectors of fire are assigned to each subordinate by the leader to ensure that the unit's area is completely covered by fire. Targets are engaged as they appear in accordance with established engagement priorities. Means of designating sectors of fire include:

- TRPs.
- Azimuth.
- Clock direction.
- Terrain-based quadrants.
- Friendly-based quadrants.

2-71. **Types of Sectors.** Leaders should assign a primary and a secondary sector of fire. The primary sector is the first priority; Soldiers and units are responsible for engaging and defeating the enemy here first. Fire then shifts to the secondary sector on order, when there are no targets in the primary sector, or when the leader needs to cover the movement of another friendly element. This secondary sector of fire can correspond to another friendly element's primary sector of fire to obtain overlapping fires and mutual support.

2-72. **Overlapping and Divided Sectors.** When assigning sectors, leaders attempt to build in mutual support and redundancy. By building redundancy into the observation and fire plan, leaders increase their probability of early detection of the enemy. Two common techniques are overlapping a sector and dividing a sector (Figure 2-6).

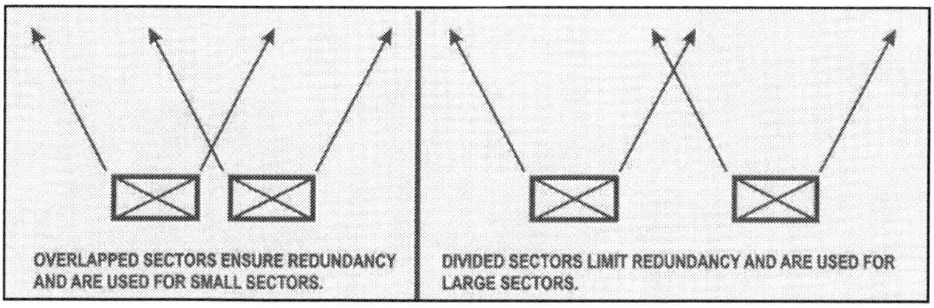

Figure 2-6. Overlapping and divided sectors.

2-73. **Dead Space.** It is important to identify dead space within a sector of fire. Dead space is any area that cannot be observed or covered by direct-fire weapons systems, including where the waist of a Soldier falls below a gunner or automatic rifleman's point of aim. When stationary, the most accurate method for determining dead space is to have one Soldier walk the weapon's line of sight and make a pace count of those areas where he encounters dead space. When the Soldier is not able to walk the line of fire, he can also determine dead space by observing the flight of tracer ammunition from a position behind and to the flank of the weapon.

2-74. All dead space within the sector must be identified to allow the leader and subordinate leaders to plan high trajectory fires (mortars, artillery, or M203) to cover that area.

2-75. **Searching the Sector.** Searching is the act of carefully watching the assigned sector. Individual and unit observation plans are inherent in all military operations. Individual Soldiers scan their sectors by conducting a rapid scan followed by a slow scan. When conducting a rapid scan, Soldiers make a quick overall search for obvious targets and unnatural colors, outlines, or movement. They follow the rapid scan with a slow deliberate scan, searching for signatures and indicators of common targets. Soldiers who use a more deliberate method to scan their sectors are generally more successful at detecting targets.

Maximum Engagement Line

2-76. The maximum engagement line (MEL) is the depth of the sector and is normally limited to the maximum effective engagement range of the weapons systems. However, it is also influenced by the enemy target description and the effects of terrain. Slope, vegetation, structures, and other features provide cover and concealment that may prevent the weapon from engaging out to the maximum effective range. To assist in determining the distance to each MEL, Soldiers should use a map to ensure the MELs are depicted accurately on the range card. Identifying the MEL prevents squads from engaging targets beyond the maximum effective ranges of their weapon systems and establishes criteria for triggers. This decreases needless and ineffective ammunition expenditure during an engagement.

Final Protective Line

2-77. If a final protective line (FPL) is assigned, a machine gun is sighted along it to employ grazing fire except when other targets are being engaged. An FPL becomes the machine gun's contribution to the unit's final protective fire (FPF). An FPL is fixed in direction and elevation. However, a small shift for search must be employed to prevent the enemy from crawling under the FPL. A small shift will also compensate for irregularities in the terrain or the sinking of the tripod legs into soft soil during firing.

Principal Direction of Fire

2-78. A PDF is generally assigned when the terrain does not lend itself to a FPL. A PDF is a direction of fire that is assigned priority to cover an area that has good fields of fire or has a likely dismounted avenue of approach. It also provides mutual support to the adjacent unit. Machine guns are sighted using a PDF if an FPL has not been assigned. If a PDF is assigned and other targets are not being engaged, machine guns remain on the PDF. The main difference between a PDF and an FPL is that the PDF is a sector, while the FPL is a fixed line. Means of designating a direction of fire include—

- Closest TRP.
- Clock direction.
- Cardinal direction and or magnetic azimuth.
- Tracer on target.
- Infrared laser pointer.

Final Protective Fire

2-79. The FPF is a line of fire established where an enemy assault is to be checked by the interlocking fires of all available friendly weapons, to include indirect fire. The FPF is reinforced with protective obstacles whenever possible. Initiation of the FPF is the signal for all squads, crews, and individual Soldiers to shift fires to their assigned portion of the FPL.

Restrictive Fire Line

2-80. An RFL is a linear fire control measure beyond which fires are prohibited without coordination. In the offense, the platoon leader may designate an RFL to prevent a base-of-fire squad(s) from firing into the area where an assaulting squad(s) is maneuvering. This technique is particularly important when mechanized vehicles directly support the maneuver of Infantry squads. In the defense, the platoon leader may establish an RFL to prevent squads from engaging one of the platoon's other rifle squads positioned in restricted terrain on the flank of an enemy avenue of approach. Figure 2-7 illustrates fire control measures on an example platoon sector sketch.

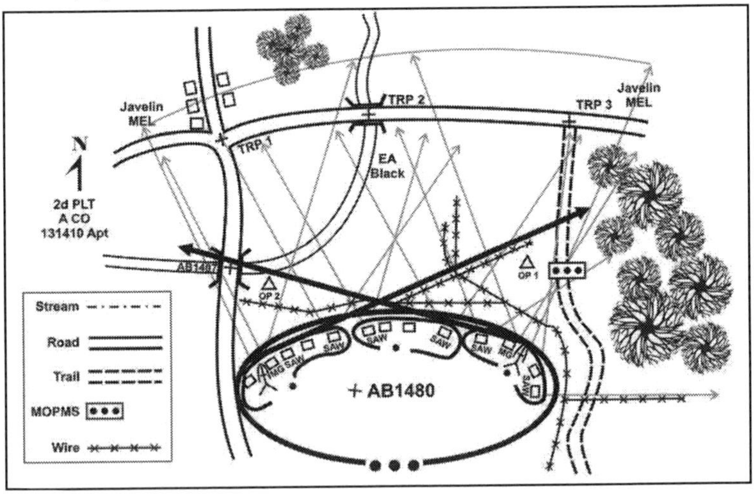

Figure 2-7. Example of fire control measures in platoon sector sketch.

Threat-Based Fire Control Measures

2-81. The platoon leader uses threat-based fire control measures to focus and control fires by directing the unit to engage a specific, templated enemy element rather than fire on a point or area. Threat-based fire control measures may be difficult to employ against an asymmetric threat. The following paragraphs describe the threat-based fire associated with this type of control measure.

Fire Patterns

2-82. Fire patterns are a threat-based measure designed to distribute the fires of a unit simultaneously among multiple, similar targets. Platoons most often use them to distribute fires across an enemy formation. Leaders designate and adjust fire patterns based on terrain and the anticipated enemy formation. The basic fire patterns are frontal fire, cross fire, and depth fire (Figure 2-8).

2-83. **Frontal Fire.** Leaders may initiate frontal fire when targets are arrayed in front of the unit in a lateral configuration. Weapons systems engage targets to their respective fronts. For example, the left flank weapon engages the left-most target; the right flank weapon engages the right-most target. As they destroy enemy targets, weapons shift fires toward the center of the enemy formation and from near to far.

2-84. **Cross Fire.** Leaders initiate cross fire when targets are arrayed laterally across the unit's front in a manner that permits diagonal fires at the enemy's flank, or when obstructions prevent unit weapons from firing frontally. Right flank weapons engage the left-most targets; left flank weapons engage the right-most targets. Firing diagonally across an EA provides more flank shots, increasing the chance of kills. It also reduces the possibility that friendly elements will be detected if the enemy continues to move forward. As they destroy enemy targets, weapons shift fires toward the center of the enemy formation.

2-85. **Depth Fire.** Leaders initiate depth fire when targets are dispersed in depth perpendicular to the unit. Center weapons engage the closest targets; flank weapons engage deeper targets. As they destroy targets, weapons shift fires toward the center of the enemy formation.

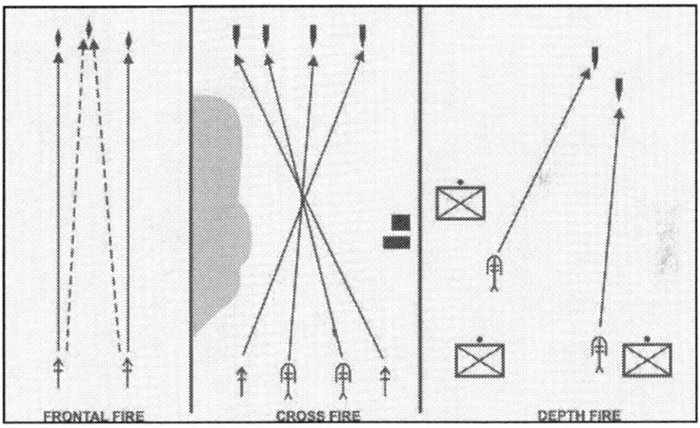

Figure 2-8. Fire patterns.

Engagement Priority

2-86. In concert with his concept of the operation, the company commander determines which target types provide the greatest payoff or present the greatest threat to his force. He then establishes these as a unit engagement priority. The platoon leader refines these priorities within his unit. Engagement priority specifies the order in which the unit engages enemy systems or functions. Engagement priorities are situational dependent. Subordinate elements can have different engagement priorities. For example, the leader establishes his engagement priorities so his medium machine guns engage enemy unarmored vehicles while his SLM and CCMS engage enemy tanks. Normally, units engage the most dangerous targets first, followed by targets in depth.

Weapons-Ready Posture

2-87. To determine the weapons-ready posture, leaders use their estimate of the situation to specify the ammunition and range for the engagement. Range selection is dependent on the anticipated engagement range. Terrain, visibility, weather, and light conditions affect range selection.

2-88. Within the platoon, weapons-ready posture affects the types and quantities of ammunition carried by the rifle and weapons squads.

2-89. For Infantry squads, weapons-ready posture is the selected ammunition and indexed range for individual and crew-served weapons. For example, an M203 grenadier whose likely engagement is to cover dead space at 200 meters from his position might load HEDP rounds. He will also set 200 meters on his quadrant sight for distance to the dead space. To prepare for an engagement in a wooded area where engagement ranges are extremely short, antiarmor specialists may be armed with SLM instead of CCMS.

Weapons Control Status

2-90. The three levels of weapons control status outline the conditions, based on target identification criteria, under which friendly elements may engage. The platoon leader sets and adjusts the weapons control status based on friendly and enemy disposition, and the clarity of the situation. In general, the higher the probability of fratricide, the more restrictive the weapons control status. The three levels are—

- **Weapons Hold.** Engage only if engaged or ordered to engage.
- **Weapons Tight.** Engage only targets that are positively identified as enemy.
- **Weapons Free.** Engage any targets that are not positively identified as friendly.

2-91. As an example, the platoon leader may establish the weapons control status as weapons hold when other friendly forces are passing friendly lines. Or the platoon leader may be able to set a weapons free status when he knows there are no friendly elements in the vicinity of the engagement. This permits his elements to engage targets at extended ranges even though it is difficult to distinguish targets accurately at ranges beyond 2,000 meters under battlefield conditions. The platoon leader may change the weapons control status for his elements based on situational updates. Weapons control status is extremely important for forces using combat identification systems. Establishing the weapons control status as weapons free permits leaders to engage an unknown target when they fail to get a friendly response.

Trigger

2-92. Triggers are an event or time-oriented criteria used to initiate planned actions to achieve surprise and inflict maximum destruction on the enemy. A designated point or points (selected along identifiable terrain) in an engagement area used to mass fires at a predetermined range (FM 1-02). Triggers can be a physical point on the ground (trigger line), a laser or lazed spot, or an action or event that causes friendly forces to do something. When using triggers to control fires, leaders ensure they have allocated them to start, shift, and cease fires. Leaders use triggers within the context of the ROE and the weapons control status. For example, a leader might say, WAIT UNTIL ENEMY SOLDIERS CROSS PL BLUE BEFORE ENGAGING.

2-93. A trigger line is a phase line used to mass fires at a predetermined range. The trigger line can be used when attacking or defending. In the offense, the trigger line is preferably perpendicular to the friendly axis of advance and is used to initiate or cease fires when reached by the unit. If defending, the leader initiates fire as the enemy reaches the trigger line.

Weapons Safety Posture

2-94. The weapons safety posture is an ammunition-handling command that allows leaders to control the safety status of their weapons. Soldier adherence to and leader supervision of the weapons safety posture prevents accidental discharge of weapons. Examples include:

- Handling live ammunition and weapons in peace time training in the same safe way during combat.
- Finger off the trigger and weapon on safe.
- Hand grenades attached correctly to the ammo pouches.
- Safety zones and back blast areas enforced.
- Strict enforcement of unit weapons and ammunition-handling SOPs at all times.

ENGAGEMENT TECHNIQUES

2-95. Engagement techniques are effects-oriented fire distribution measures. The most common engagement techniques in platoon operations are—

- Point fire.
- Area fire.
- Volley (or simultaneous) fire.
- Alternating fire.
- Sequential fire.
- Observed fire.
- Time of suppression.
- Reconnaissance by fire.

POINT FIRE

2-96. Point fire involves concentrating the effects of the platoon or squad's fire against a specific, identified target such as an enemy vehicle, machine gun bunker, or ATGM position. When leaders direct point fire, all the unit's weapons engage the target. They fire until they destroy it, or until the required time of suppression expires. Employing converging fires from dispersed positions makes point fire more effective because the target is engaged from multiple directions. The unit may initiate an engagement using point fire against the most dangerous threat, and then revert to area fire against other, less threatening point targets.

AREA FIRE

2-97. Area fire involves distributing the effects of a unit's fire over an area in which enemy positions are numerous or are not obvious. Typically, the primary purpose of area fire is suppression. However, sustaining effective suppression requires judicious control of the rate of fire.

VOLLEY FIRE

2-98. Volley fire is released when two or more firers engage a single target and the range is known. These firers engage the target at the same time on a prearranged signal such as a command, whistle, booby trap, mine, or TRP. This can be the most effective means of engagement as it places the most possible rounds on one enemy target at one time, thereby increasing the possibility of a kill.

2-99. Units employ simultaneous fire to rapidly mass the effects of their fires or to gain fire superiority. For example, a unit may initiate a support-by-fire operation with simultaneous fire, and then revert to alternating or sequential fire to maintain suppression. Volley fire is also employed to negate the chance that one of the Soldiers might miss his intended target with fire from his SLM. For example, a squad may employ volley fire with its SLM to ensure rapid destruction of an enemy vehicle that is engaging a friendly position.

ALTERNATING FIRE

2-100. During alternating fire, pairs of elements continuously engage the same point or area targets one at a time. For example, an Infantry platoon may alternate the fires of a pair of machine guns. Alternating fire permits the unit to maintain suppression for a longer duration than does volley fire. It also forces the enemy to acquire and engage alternating friendly points of fire.

SEQUENTIAL FIRE

2-101. In sequential fire, the subordinate elements of a unit engage the same point or area target one after another in an arranged sequence. Sequential fire can also help prevent the waste of ammunition, as when rifle squads wait to see the effects of the first CCMS before firing another. Additionally, sequential fire permits elements that have already fired to pass on information they have learned from the engagement. For example, an Infantryman who missed a BMP with SLM fires could pass range and lead information to the next Soldier preparing to engage the BMP with a SLM.

OBSERVED FIRE

2-102. Observed fire allows for mutual observation and assistance while protecting the location of the observing element and conserving ammunition. The company commander may employ observed fire between elements in the company. He may direct one platoon to observe while another platoon engages the enemy. The platoon may use observed fire when it is in protected defensive positions with engagement ranges of more than 800 meters. For example, the platoon leader may direct the weapons squad to engage an enemy at long range and the Infantry squads to observe the effects of the fires. The observing elements prepare to engage the enemy on order in case the weapons squad fails to effectively engage the enemy, encounters weapon malfunctions, or runs low on ammunition.

TIME OF SUPPRESSION

2-103. Time of suppression is the period, specified by the platoon leader, when an enemy position or force must be suppressed. Suppression time is typically dependent on the time it will take a supported element to maneuver, so suppression is generally more event- than time-driven. Normally, a friendly unit suppresses an enemy position using the sustained rate of fire of its automatic weapons. In planning for sustained suppression, leaders must consider several factors, including:

- The estimated time of suppression.
- The size of the area being suppressed.
- The type of enemy force to be suppressed.
- The range to the enemy target.
- The rates of fire.
- The available ammunition quantities.

RECONNAISSANCE BY FIRE

2-104. Reconnaissance by fire is the process of engaging possible enemy locations to elicit a tactical response from the enemy, such as return fire or movement. This response permits Infantry leaders to make accurate target acquisition and to mass fires against the enemy element. Typically, the platoon leader directs a subordinate squad to conduct the reconnaissance by fire. He may, for example, direct an overwatching squad to conduct the reconnaissance by fire against a probable enemy position before initiating movement by the bounding squad(s).

FIRE COMMANDS

2-105. Fire commands are the technical instructions used to initiate fires and can be used for individuals, crews, or units (but for simplicity, this section just refers to Soldiers). Fire commands are used to initiate, control, and synchronize fires. The fire command procedure takes the principles of direct fire employment and puts them into a coherent, usable format.

2-106. There are two types of commands: initial fire commands (issued to commence firing); and subsequent fire commands (issued to change firing data and to cease firing). The elements of both commands follow the same sequence. Subsequent commands include only such elements that are changed.

A correct fire command is brief, clear, and includes all the elements necessary for accomplishing the mission. Fire commands are sent to the firing unit or gunner by the best understood means (visually or vocally). To limit errors in transmission, the person receiving the commands repeats each element as it is received.

2-107. Fire commands for direct fire weapons consist of six elements: alert, location, target description, method of engagement, ammunition, and execution. When and how the leader issues a fire command is not as important as covering the information in the fire command with his subordinates. Frequently, especially at the fire team and crew-served weapon level, leaders use the elements of a fire command without adhering to a strict format. The point is not that the leader adheres to a format, but that he maintains positive control over his subordinates' fires. However, using a more formal approach to fire commands usually provides more clarity and certainty for Soldiers and crews.

ELEMENTS OF A FIRE COMMAND

2-108. Fire commands consist of—

- **Alert.** The leader designates which weapon(s) is to fire by weapon type, Soldier's position, or Soldier's name.
- **Location.** The leader guides the Soldier onto the target.
- **Target Description.** The leader identifies the target. For multiple targets, he also tells which target to engage first.
- **Method of Engagement.** The leader tells the Soldier how to deliver the fire onto the target.
- **Ammunition.** The leader tells the Soldier which ammunition to use if munitions are other than HE (this applies to M203 only).
- **Execution (Time).** The leader reconfirms that the target is hostile, then gives an execution command.

2-109. The full fire command is given when targets are not obvious and sufficient time is available to issue a full order.

2-110. Brief fire commands are given when the target is obvious and time is limited.

2-111. Delayed fire commands are used when the leader can anticipate what is going to happen. The Soldier or unit gets ready to fire but waits until the right moment before opening fire.

2-112. Subsequent fire commands are used to make adjustments in direction and elevation, change rates of fire after a fire mission is in progress, interrupt fires, or to terminate the alert.

TERMS AND TECHNIQUES

2-113. The following list of terms and techniques clarify the different elements of the fire command.

Location

2-114. Leaders can use one or more of the following methods to assist Soldiers in locating and distinguishing between targets (Table 2-5).

- **Use of Laser/Tracer.** ("On my laser/tracer.") To prevent loss of surprise when using tracer to designate targets, the leader's tracer fire becomes the last element of the fire command.
- **The Clock Method.** An imaginary clock face is superimposed on the landscape with 12 o'clock being the direction of travel.
- **The TRP Method.** The leader uses the closest, easily-recognizable point on the ground.
- **Cardinal Direction.** Uses general compass directions (N, NE, E, SE, S, SW, W, and NW).
- **Pointing.** The leader points his finger or weapon in the general direction of the target.
- **Orally.** The leader gives the direction to the target in relation to the Soldier's position (for example, front, left front, right front).

Table 2-5. Common means of identifying and marking target locations.

Terrain Features	Naked Eye (Day/Night)	Thermals (All Used at Night)
Hilltops	Azimuth (degree, mil) (D/N)	Burn barrels
Roads/streets	VS-17 panel (D)	BBQ grills
Streams	Engineer tape (D)	Reverse polarity paper
Road intersections	Chem light bundle (N)	Heated ammo can
Building corners	Strobe light (N)	IR (N)
Anything easily identifiable	Illumination (D/N)	Lasers (PAQ-4, PEQ-2, GCP, AIM 1)
	Pyrotechnics (D/N)	Beacon/firefly strobe
	Tracer fire (D/N)	Strobe light
	Destroyed vehicle (D/N)	

2-115. In defensive operations, the team leader and weapons squad leader use existing features as TRPs, or they can emplace specially-made markers. The Soldier captures these TRPs and sectors on a range card. In offensive operations, leaders normally predetermine location for TRPs and sectors based on the scheme of maneuver of the platoon leader or commander. These TRPs and sectors are useful for planning. However, the team leader/weapons squad leader must confirm them once they actually get on the ground.

Target Description

2-116. The most natural way for a leader to control his subordinates' fire when in contact is to simply describe the intended target(s). There are several terms used to shortcut the process, though leaders can use whatever means possible to ensure understanding. To shorten the target description, the team leader or weapons squad leader describes standard targets with standard procedure words (Table 2-6).

Table 2-6. Target descriptions and terms.

Target Type	Procedure Word
Tank or tank-like target	Tank
Personnel carrier	PC
Unarmored vehicle	Truck
Personnel	Troops
Helicopter	Chopper
Machine gun	Machine gun
Antitank gun or missile	AT weapon or RPG
Bunker	Bunker
Trench line	Trench
Urban structures	Door, window, room

Method of Engagement

2-117. The leader uses control to convey how he wants the target attacked. Common forms of this element of the fire command are—

- **Rates of Fire.** When changing rates, the leader needs only indicate rapid, sustained, or scan and shoot.
- **Machine Gun Manipulation.** Manipulation dictates the class of fire with respect to the weapon and is announced as FIXED, TRAVERSE, SEARCH, or TRAVERSE AND SEARCH.

Execution

2-118. The leader uses one of the following orders to initiate fires:

- **Fire.** The default rate of fire is at the sustained rate. The command to fire can occur in more than one form, including:
  - Pre-arranged visual signal.
  - Pre-arranged event.
  - Pre-arranged audio signal.
- **Rapid Fire.** Open fire at the rapid rate.
- **Scan and Shoot.** Fire when targets appear in the designated sector.
- **At My Command.** Be prepared to fire but do not initiate until the order to fire is given.

HAND-AND-ARM SIGNALS

2-119. Following are commonly used hand-and-arm signals for fire control (Figure 2-9).

- **Ready.** The Soldier indicates that he is ready to fire by yelling, UP or raising his hand above his head toward the leader.
- **Commence Firing or Change Rate of Firing.** The leader brings his hand (palm down) to the front of his body about waist level and moves it horizontally in front of his body. To signal an increase in the rate of fire, he increases the speed of the hand movement. To signal slower fire, he decreases the speed of the hand movement.
- **Change Direction or Elevation.** The leader extends his hand and arm in the new direction and indicates the amount of change necessary by the number of fingers extended. The fingers must be spread so the Soldier can easily see the number of fingers extended. Each finger indicates 1 meter of change for the weapon. If the desired change is more than 5 meters, the leader extends his hand the number of times necessary to indicate the total amount of change. For example, *right nine* would be indicated by extending the hand once with five fingers showing and a second time with four fingers showing for a total of nine fingers.
- **Interrupt or Cease Firing.** The leader raises his hand and arm (palm outward) in front of his forehead and brings it downward sharply.
- **Other Signals.** The leader can devise other signals to control his weapons. A detailed description of hand-and-arm signals is given in FM 21-60, *Visual Signals.*

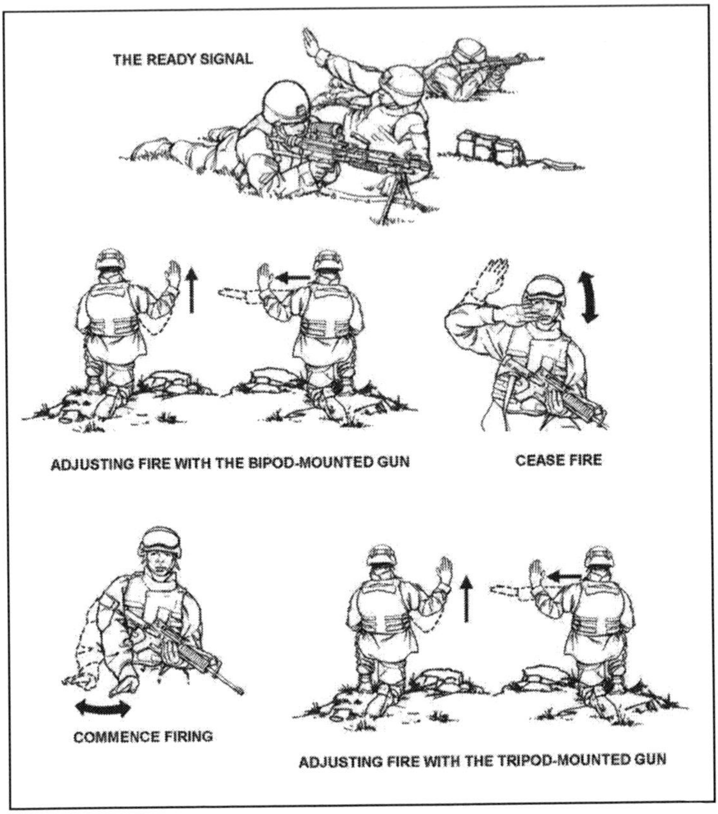

Figure 2-9. Hand-and-arm signals.

RANGE CARDS

2-120. A range card (DA Form 5517-R, *Standard Range Card*) is a sketch of the assigned sector for a direct fire weapon system on a given sector of fire (Figure 2-10). A range card aids in planning and controlling fires and aids the crews and squad gunners in acquiring targets during periods of limited visibility. Range cards show possible target areas and terrain features plotted in relation to a firing position. The process of walking and sketching the terrain to create a range card allows the individual Soldier or gunner to become more familiar with his sector. Range cards also aid replacement personnel in becoming oriented on the sector. Soldiers should continually assess the sector, and if necessary, update their range cards.

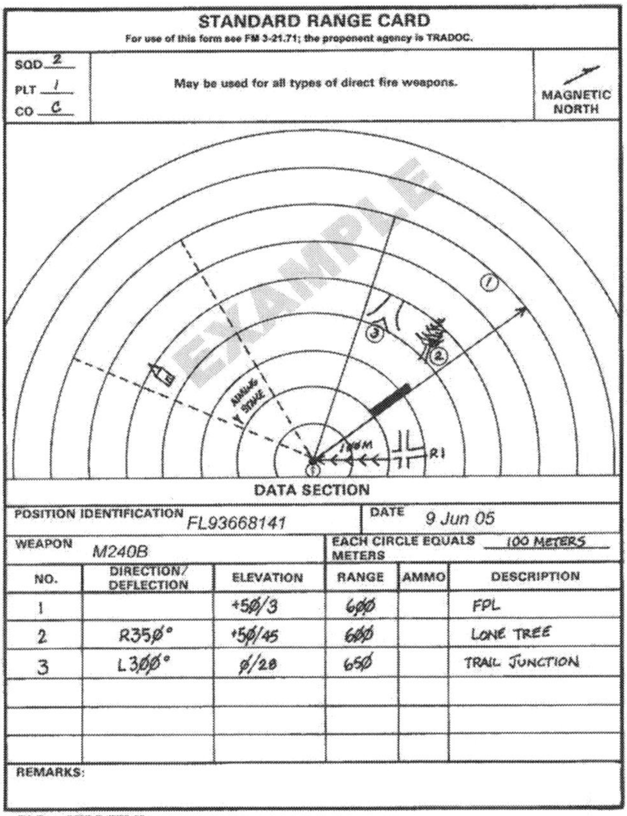

STANDARD RANGE CARD

For use of this form see FM 3-21.71; the proponent agency is TRADOC.

SQD 2

PLT 1

CO C

May be used for all types of direct fire weapons.

MAGNETIC NORTH

DATA SECTION

POSITION IDENTIFICATION FL93668141			DATE 9 Jun 05		
WEAPON M240B			EACH CIRCLE EQUALS 100 METERS METERS		
NO.	DIRECTION/ DEFLECTION	ELEVATION	RANGE	AMMO	DESCRIPTION
1		+5Ø/3	6ØØ		FPL
2	R35Ø°	+5Ø/45	6ØØ		LONE TREE
3	L3ØØ°	Ø/28	65Ø		TRAIL JUNCTION

REMARKS:

DA Form 5517-R, FEB 88

Figure 2-10. Example of completed DA Form 5517-R range card (primary sector with final protective line).

ADJUSTMENTS

2-121. Direct fire adjustments are fairly easy to make because the observer is also the shooter. However, when using an observer or spotter, direct fire adjustments are similar to those of indirect fire adjustments. This includes making deviation and range corrections. Deviation corrections move the round right or left toward the target, while range corrections add or drop the round toward the target with respect to the observer.

CLOSURE REPORT

2-122. The closure report completes the mission and provides a battle damage assessment. The report should go to both the FDC and the parent unit. Higher headquarters staff officers use battle damage assessment to update their running estimate and feed the common operating picture (COP).

SECTION V — EMPLOYING INDIRECT FIRES

2-123. The purpose of this section is to discuss techniques associated with calling for and adjusting indirect fires.

CALL FOR FIRE

2-124. The battalion fire support execution matrix may require the platoon to call for and adjust its own indirect fire support. Normally, the battalion fire support annex will designate company targets. However, the matrix also might designate platoon targets. The platoon uses these preplanned artillery targets to call for and adjust indirect fire. Either a Soldier or a forward observer (FO) can prepare and request a call for fire. To receive immediate indirect fire support, the observer must plan targets and follow proper call-for-fire procedures. If available, he should use a GPS and laser range finder.

2-125. The call for fire consists of required and optional elements. If the observer is untrained, FDC personnel are trained to assist him in the call-for-fire procedure and subsequent adjustments by asking leading questions to obtain the information needed. Optional elements, methods of engagement, and methods of fire and control require a relatively high level of experience, but are not necessary to get fire support.

REQUIRED ELEMENTS

2-126. Calls for fire must include the following three elements:

- Observer identification and warning order.
- Target location.
- Target description.

Observer Identification and Warning Order

2-127. Observer identification tells the fire direction center (FDC) who is calling. It also clears the net for the duration of the call. The WARNO tells the FDC the type of mission and the method of locating the target. The types of indirect fire missions are adjust fire, fire for effect (FFE), suppress, and immediate suppression.

Adjust Fire

2-128. Use this command when uncertain of target location. Calling an adjust fire mission means the observer knows he will need to make adjustments prior to calling a fire for effect.

Fire for Effect

2-129. Use this command for rounds on target, no adjustment. An example of this situation is if it is known that the target is in building X. Building X is easily identified on the map as Grid ML 12345678910.

Suppress

2-130. Use this command to obtain fire quickly. The suppression mission is used to initiate fire on a preplanned target (known to the FDC) and unplanned targets. An example is calling for fire to force the enemy to "get down and seek cover." This should enable friendly forces to close with and destroy the enemy with direct fire.

Immediate Suppression

2-131. Use this command to indicate the platoon is already being engaged by the enemy. Target identification is required. The term "immediate" tells the FDC that the friendly unit is in direct fire contact with the enemy target.

Target Location Methods

2-132. When locating a target for engagement, the observer must determine which of the target location methods he will use: grid, polar, or shift from a known point.

Grid Mission

2-133. The observer sends the enemy target location as an 8- or 10-digit grid coordinate. Before the first adjusting rounds are fired, the FDC must know the direction from the observer's location. The observer sends observer-target (OT) direction (to the nearest 10 mils) from his position to the target (Table 2-7).

Table 2-7. Example fire mission, grid.

Initial Fire Request From Observer to FDC	
Observer	***FDC***
Z57, THIS IS 271, ADJUST FIRE, OVER.	THIS IS Z57, ADJUST FIRE, OUT.
GRID NK180513, OVER.	GRID NK180513, OUT.
INFANTRY PLATOON IN THE OPEN, ICM IN EFFECT, OVER.	INFANTRY PLATOON IN THE OPEN, ICM IN EFFECT, OUT.
Message to Observer	
FDC	***Observer***
Z, 2 ROUNDS, TARGET, AF1027, OVER.	Z, 2 ROUNDS, TARGET IS AF1027, OUT.
For Subsequent Rounds (From Observer to FDC)	
Observer	***FDC***
DIRECTION 1680, OVER.	DIRECTION 1680, OUT.
Note: Send direction before or with the first subsequent correction.	

Polar Mission

2-134. The observer sends direction, distance, and an up or down measurement (if significant) from his location to the enemy target. The FDC must know the observer's location prior to initiating the call for fire. The word "polar" in the WARNO alerts the FDC that the target will be located with respect to the observer's position. The up or down correction is an estimated vertical shift from the observer's location to the target and is only significant if greater than or equal to 35 meters. If the target is higher, it is an up correction. If the target is lower, it is a down correction (Table 2-8 and Figure 2-11). Normally, inexperienced observers only send a direction and distance and ignore the up or down correction.

Table 2-8. Example fire mission, polar plot.

Initial Fire Request From Observer to FDC	
Observer	***FDC***
Z56, THIS IS Z31, FIRE FOR EFFECT, POLAR, OVER.	THIS IS Z56, FIRE FOR EFFECT, POLAR, OUT.
DIRECTION 4520, DISTANCE 2300, DOWN 35, OVER.	DIRECTION 4520, DISTANCE 2300, DOWN 35, OUT.
INFANTRY COMPANY IN OPEN, ICM, OVER.	INFANTRY COMPANY IN OPEN, ICM, OUT.
Message to Observer	
FDC	***Observer***
Y, VT, 3 ROUNDS, TARGET AF2036, OVER.	Y, VT, 3 ROUNDS, TARGET AF2036, OUT.

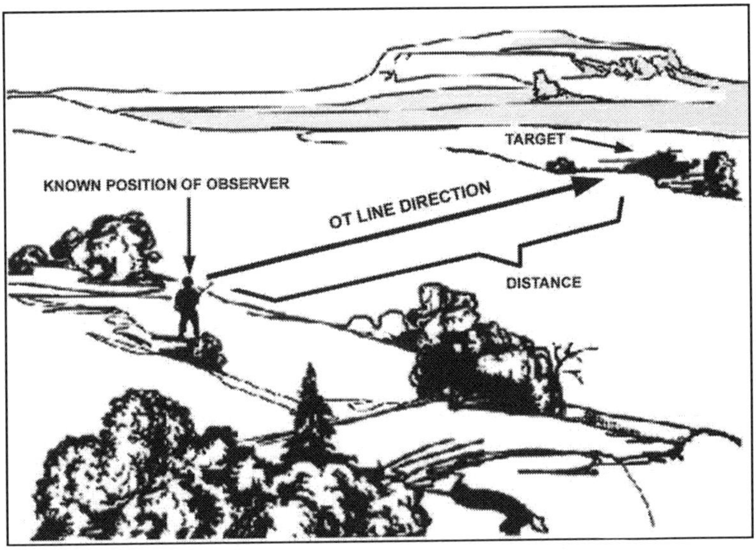

Figure 2-11. Polar plot method of target location.

Shift From a Known Point

2-135. Shift from a known point is performed when the observer and FDC have a common known point. The observer sends OT line and then determines the lateral and range shifts. The enemy target will be located in relation to a preexisting known point or recorded target. The point or target from which the shift is made is sent in the WARNO. (Both the observer and the FDC must know the location of the point or recorded target.) The observer sends a target/known point number, a direction, and left/right, add/drop, and up/down corrections as listed below (Table 2-9, and Figures 2-12 and 2-13):

- Direction from observer (grid azimuth in mils) to target.
- The lateral shift in meters (how far left or right the target is) from the known point (Figure 2-13).

- The range shift (how much farther [ADD] or closer [DROP] the target is in relation to the known point, to the nearest 100 meters) (Figure 2-13).
- The vertical shift (how much the altitude of the target is above [UP] or below [DOWN] the altitude of the known point, expressed to the nearest 5 meters). A vertical shift is usually only significant if it is greater than or equal to 35 meters.

Table 2-9. Example fire mission, shift from a known point.

Initial Fire Request From Observer to FDC	
Observer	***FDC***
H66 THIS IS H44, ADJUST FIRE, SHIFT AA7733, OVER.	THIS IS H66, ADJUST FIRE, SHIFT AA7733, OUT.
DIRECTION 5210, LEFT 380, ADD 400, DOWN 35, OVER.	DIRECTION 5210, LEFT 380, ADD 400, DOWN 35, OUT.
COMBAT OP IN OPEN, ICM IN EFFECT, OVER.	COMBAT OP IN OPEN, ICM IN EFFECT, OUT.
Message to Observer	
FDC	***Observer***
H, 1 ROUND, TARGET AA7742, OVER.	H, 1 ROUND, TARGET AA7742, OUT.

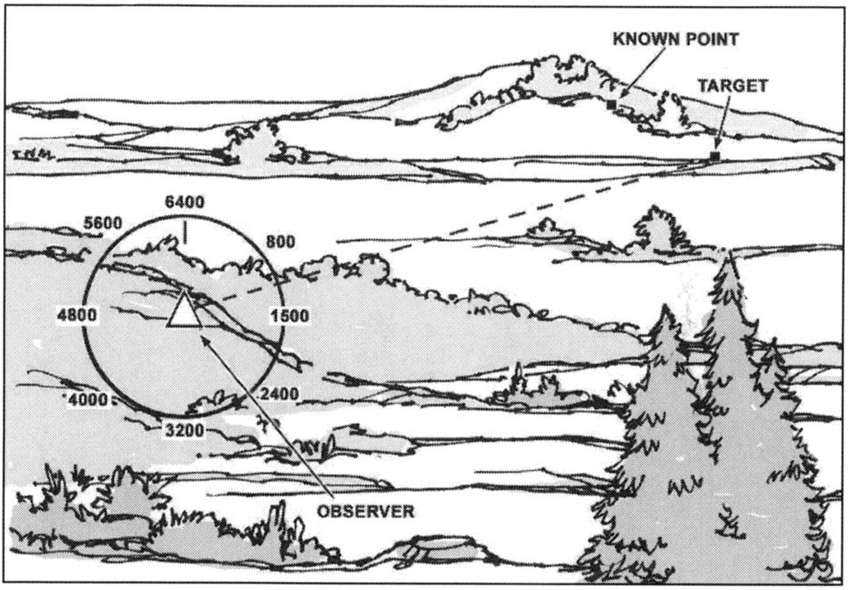

Figure 2-12. Shift from a known point method using direction (in mils).

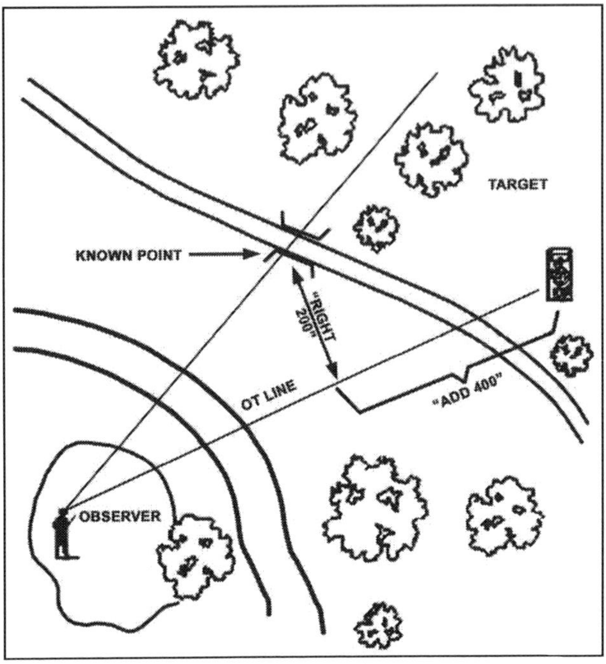

Figure 2-13. Lateral and range shifts from a known point.

> Sergeant Orest Bisko, a patrol leader from the 1st Marine Force Reconnaissance Company, knew how to use artillery. When occupying an observation post, Bisko fired his artillery at a set of known coordinates. This would enable him to later shift from that known point to the target with speed and precision. On 26 July 1966, while his four-man patrol was occupying an observation post, they spotted a large collection of enemy encamped in a small, wooded grove. The enemy force, he observed, apparently was in no hurry to move. Sergeant Bisko deliberately whispered fire commands over his radio to his direct support artillery. He ordered them to shift the distance from the known target and fire for effect. Three minutes later shells began crashing into the enemy perimeter. After approximately 30 minutes, 50 enemy were dead and the patrol had escaped in the confusion.
>
> *Shifting from a Known Point*
> *Small Unit Actions in Vietnam*
> *Francis J. West*

Target Description

2-136. The target description helps the FDC to select the type and amount of ammunition to best defeat the enemy target. Following is a brief description of the target using the mnemonic SNAP:

- Size and or shape ("one enemy soldier" or "platoon of enemy soldiers").
- Nature and or nomenclature ("T72," "sniper team," "machine gunner").

- Activity ("stationary" or "moving").
- Protection and or posture ("in the open," "dug in," or "on a rooftop").

Message to Observer

2-137. After the FDC receives the call for fire, it determines if and how the target will be attacked. That decision is announced to the observer in the form of a message to the observer (Tables 3-7, 3-8, and 3-9). The observer acknowledges the message to observer by reading it back in its entirety.

2-138. Additionally, the FDC will send the following transmissions:

- **Shot.** The term SHOT, OVER is transmitted by the FDC after each round fired in adjustment and after the initial round in the fire for effect (FFE) phase. The observer acknowledges with SHOT, OUT.
- **Splash.** The term SPLASH, OVER is transmitted by the FDC to inform the observer when his round is five seconds from detonation/impact. The observer responds with SPLASH, OUT.
- **Rounds Complete.** The term ROUNDS COMPLETE, OVER signifies that the number of rounds specified in the FFE have been fired. The observer responds with ROUNDS COMPLETE, OUT.

Optional Elements

2-139. A call for fire also might include the following information:

- Method of engagement.
- Danger close.
- Method of fire and control.
- Refinement and end of mission.

Method of Engagement

2-140. The observer uses the method of engagement portion of the call for fire to tell the FDC how to attack the enemy target. The method of engagement consists of the type of engagement, trajectory, danger close (if applicable), ammunition, and distribution.

Trajectory

2-141. A low-angle trajectory is standard without a request. A high-angle trajectory is at the request of the observer or when required due to masking terrain. An example of this terrain would be an enemy position in defilade on the backside of a mountain range. This allows the indirect fire munitions to successfully clear the top of the masking terrain and have more of a vertical descent, resulting in the munitions impacting directly on the enemy position.

Danger Close

2-142. Danger close is announced when applicable. Include the term *danger close* in the method-of-engagement portion of the call for fire when the target is within 600 meters of any friendly elements for both mortars and field artillery. When adjusting naval gunfire, announce DANGER CLOSE when the target is located within 750 meters and naval guns 5 inches or smaller are in use. For naval guns larger than 5 inches, announce DANGER CLOSE when the target is within 1,000 meters. The creeping method of adjustment will be used exclusively during danger close missions. The forward observer makes range changes by creeping the rounds to the target using corrections of less than 100 meters.

Ammunition

2-143. Ammunition is the type of projectile, the type of fuse action, and the volume of fire desired in the fire-for-effect phase stated in rounds per howitzer. The type of ammunition can be requested by the observer, but final determination is by the FDC based on Class V unit basic load and target description.

Method of Fire and Control

2-144. The method of fire and control indicates the desired manner of attacking the target, whether the observer wants to control the time or delivery of fire, and whether he can observe the target. The observer announces the appropriate method of fire and control.

Fire When Ready

2-145. FIRE WHEN READY is standard without request, and is not announced. The mission will be fired as soon as the data is processed, guns are laid on the target, and munitions are loaded.

At My Command

2-146. If the observer wishes to control the time of delivery of fire, he includes AT MY COMMAND in the method of control. When the pieces are ready to fire, the FDC announces PLATOON (or BATTERY or BATTALION) IS READY, OVER. (Call signs are used.) The observer announces FIRE when he is ready for the pieces to fire. In certain scenarios, the observer must consider the time of flight for the munitions to leave the indirect fire system and impact on the target. The "time of flight" data can be requested by the observer and determined by the FDC. This only applies to adjusting rounds and the first volley of an FFE. AT MY COMMAND remains in effect throughout the mission until the observer announces CANCEL AT MY COMMAND, OVER.

2-147. AT MY COMMAND can be further specified. BY ROUND AT MY COMMAND controls every round in adjustment and every volley in the FFE phase.

Time on Target

2-148. The observer may tell the FDC when he wants the rounds to impact by requesting, for example, TIME ON TARGET, 0859, OVER. The observer must ensure his time and the FDC's time are synchronized prior to the mission.

Time to Target

2-149. The observer may tell the FDC when he wants the rounds to impact by requesting TIME TO TARGET (so many) MINUTES AND SECONDS, OVER, STANDBY, READY, READY, HACK, OVER. Time to target is the time in minutes and seconds after the "hack" statement is delivered when rounds are expected to hit the target.

Check Firing

2-150. CHECK FIRING is used to cause an immediate halt in firing. Use this command only when necessary to *immediately* stop firing (for example, safety reasons) as it may result in cannons being out of action until any rammed/loaded rounds can be fired or cleared from the tubes.

Repeat

2-151. REPEAT can be given during adjustment or fire-for-effect missions. During adjustment, REPEAT means firing another round(s) with the last data and adjusting for any change in ammunition if necessary. REPEAT is not sent in the initial call for fire.

2-152. During fire for effect, REPEAT means fire the same number of rounds using the same method of fire for effect as last fired. Changes in the number of guns, the previous corrections, the interval, or the ammunition may be requested.

Request Splash

2-153. SPLASH can be sent at the observer's request. The FDC announces SPLASH to the observer 5 seconds prior to round impact. SPLASH must be sent to aerial observers and during high-angle fire missions.

Refinement and End of Mission

2-154. The observer should observe the results of the fire for effect and then take one of the following actions to complete the mission:

- Correct any adjustments.
- Record as target.
- Report battle damage assessment.
- Report end of mission.

ADJUST FIRE

2-155. If the rounds have accurately impacted the target after the initial call for fire, the observer requests fire for effect. If the rounds are not impacting the target, the observer adjusts the indirect fire onto the enemy target. Making adjustments to an indirect fire mission requires the observer to determine deviation and range corrections. Deviation corrections move the round right or left toward the target while range corrections add or drop the round toward the target with respect to the observer's position. If the observer cannot locate the target (due to deceptive terrain, lack of identifiable terrain features, poor visibility, or an inaccurate map), he adjusts the impact point of the rounds. The observer chooses an adjusting point. For a destruction mission (precision fire), the target is the adjusting point. For an area target (area fire), the observer picks a well defined adjusting point close to the center. The observer spots the first and each successive adjusting round and sends range and deviation corrections back to the FDC until rounds hit the target. The observer spots each round by relating the round's point of impact to the adjusting point. See FM 6-30, *Tactics, Techniques, and Procedures for Observed Fire,* for a more detailed discussion of adjusting mortar and artillery fire.

DEVIATION SPOTTING

2-156. Deviation spotting (left or right) involves measuring the horizontal angle (in mils) between the actual burst and the adjusting point (Figure 2-14). For example, a burst to the right of the target is spotted as "(so many) mils right." The observer uses an angle-measuring device to determine deviation. He might use the mil scale on his binoculars (Figure 2-15), or he might use his hand and fingers (Figure 2-16).

Figure 2-14. Deviation spotting.

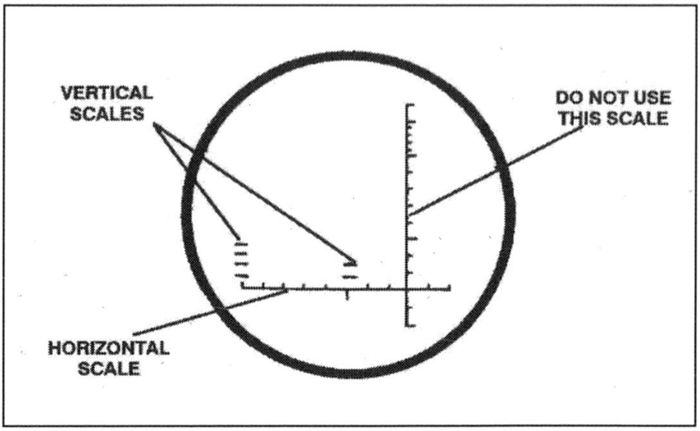

Figure 2-15. Binocular reticle with mil scale.

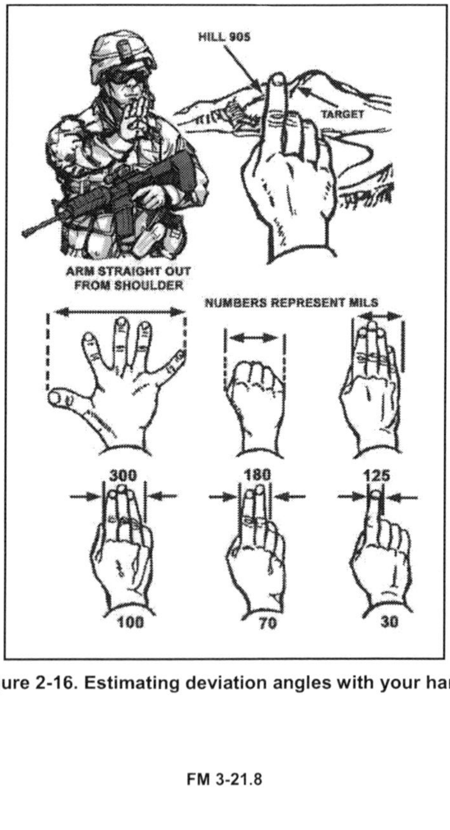

Figure 2-16. Estimating deviation angles with your hand.

2-157. On binoculars, the horizontal scale is divided into 10-mil increments and is used for measuring horizontal angles. The vertical scales in the center and on the left of the reticle (divided into 5-mil increments) are used for measuring vertical angles. The scale on the right, if present, is no longer used.

2-158. A burst on the OT line is spotted as "line." Deviation (left or right) should be measured to the nearest 5 mils for area targets, with measurements taken from the center of the burst. Deviation for a destruction mission (precision fire) is estimated to the nearest mil. Figure 2-17 shows the adjusting point at the center of the binocular horizontal scale.

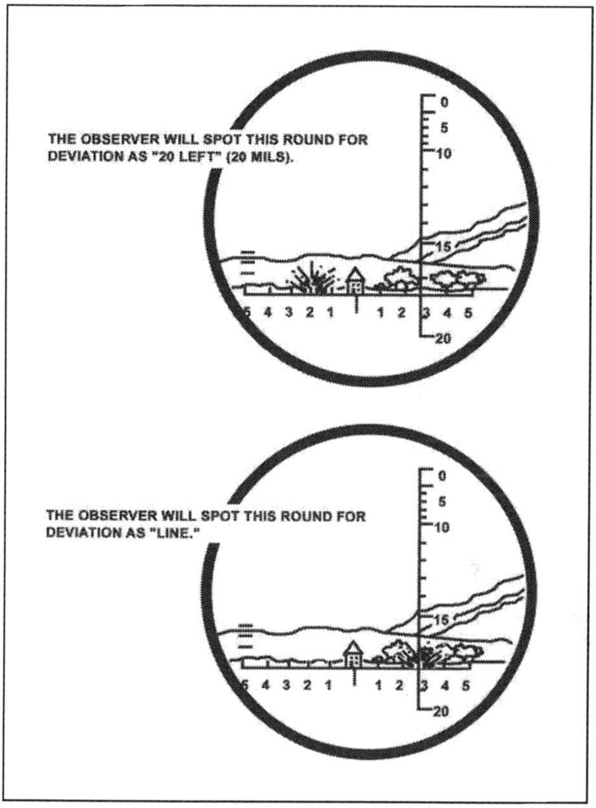

Figure 2-17. Deviation spotting with binoculars.

DEVIATION CORRECTION

2-159. Deviation correction is the distance (in meters) the burst must be moved left or right to be on line between the observer and the target. Once the mil deviation has been determined, the observer converts it into a deviation correction (in meters). The OT distance is converted to a number called the OT factor (see FM 6-30). The OT factor is used in adjusting fires after the initial call for fire. The OT direction is usually determined in mils but degree azimuths can be used if necessary. OT distance is determined through individual range estimation or through the use of specific technical laser range-finding equipment (such as

MELIOS). To determine the OT factor, take the range to the target, divide by 1,000, then round to the nearest even whole number.

2-160. The deviation correction is determined by multiplying the observed deviation in mils by the distance from the observer to the target in thousands of meters (the OT factor). The result is expressed to the nearest 10 meters (Figure 2-18 [Example 1]).

2-161. In adjustment of area fire, small deviation corrections (20 meters or less) can be ignored except when a small change determines a definite range spotting. Throughout the adjustment, the observer moves the adjusting rounds close enough to the OT line so range spotting is accurate. A minor deviation correction (10 to 20 meters) should be made in adjustment of precision fire.

2-162. If the OT distance is greater than 1,000 meters, round to the nearest thousand and express it in thousands of meters (Figure 2-18 [Example 2]). If the OT distance is less than 1,000 meters, round to the nearest 100 meters and express it as a decimal in thousands of meters (Figure 2-18 [Example 3]).

Deviation Correction

Example 1:
Observer deviation 20 mils
OT distance 2,000 meters
OT factor 2
Observer deviation x OT factor = deviation correction
20 x 2 = 40 meters

Example 2:
OT distance 4,200 meters – OT factor 4
OT distance 2,700 meters – OT factor 3

Example 3:
OT distance 800 meters – OT factor 0.8

Figure 2-18. Determing deviation correction.

ANGLE T

2-163. Angle T (Figure 2-19) is the angle formed by the intersection of the gun-target line and the OT line with its vertex at the target. If angle T is 500 mils or greater, the FDC should tell the observer. If this occurs, the observer continues to use the OT factor to make his deviation corrections. If he sees that he is getting more of a correction than he has asked for, the observer should consider cutting the corrections in half to better adjust rounds onto the target.

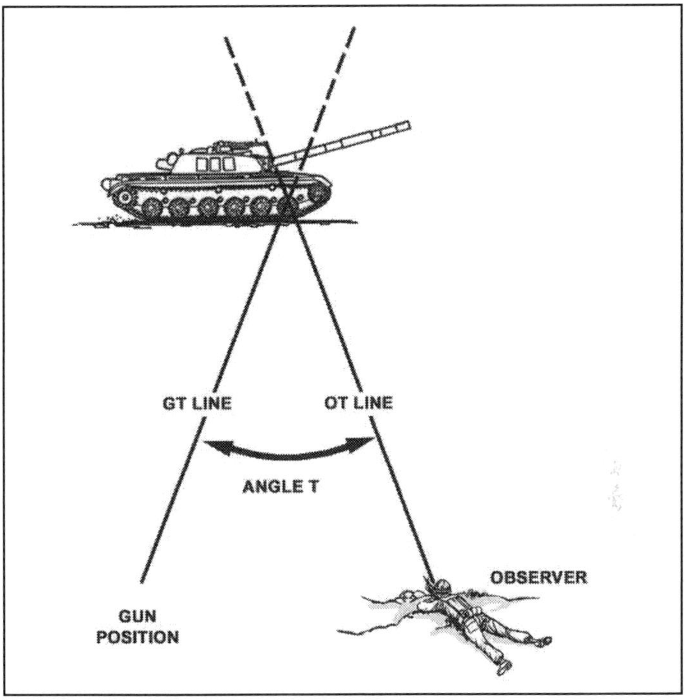

Figure 2-19. Angle T.

RANGE SPOTTING

2-164. Range spotting (short or over) requires adjusting the range to obtain fire on the target. An adjusting round's burst on or near the OT line gives a definite range spotting. If he cannot make a definite spotting, the observer announces a "lost" or "doubtful" spotting. In these situations only, he gives the deviation correction to the FDC. Deviation corrections include—

- **Over.** The observer sees the burst beyond the adjusting point.
- **Short.** The observer sees the burst between himself and the adjusting point.
- **Target.** The observer sees the burst hit the target. He uses this spotting only in precision fire (destruction missions).
- **Range Correct.** The observer believes that the burst occurred at the correct range.
- **Doubtful.** The observer sees the burst but cannot tell whether it occurred over, short, target, or range correct.
- **Lost,** The observer cannot see or hear the burst.

RANGE CORRECTION

2-165. With each successive correction, the adjusting round lands over or short of the adjusting point, but closes on the target. There are three methods of range corrections: successive bracketing; hasty bracketing; and the creeping method.

Successive Bracketing

2-166. In bracketing, the observer deliberately gives range corrections that land over or short of the target. After spotting the first round, the observer makes a drop/add correction which he believes will give him one round over and one round short of the target. For example, if the first round impacts over the target, the observer will give a drop correction which is large enough to cause the next round to impact short of the target. Once the observer meets the goal of one round over and one round shot, he cuts each correction in half and drops or adds as necessary. The observer continues bracketing until his correction is less than 50 meters. At this point his adjustment is finished and he transitions to a fire-for-effect mission. Using the above example, his final adjustment would be "add 50 meters fire for effect." This technique is called successive bracketing (Figure 2-20).

2-167. When bracketing, the observer uses the following guide to determine his first range correction:

- OT distance between 1,000 to 2,000 meters – initial add or drop at least 200 meters (+/- 200, +/- 100, +/- 50 fire for effect).
- OT distance greater than 2,000 meters – initial add or drop at least 400 meters (+/- 400, +/- 200, +/- 100, +/- 50 fire for effect).

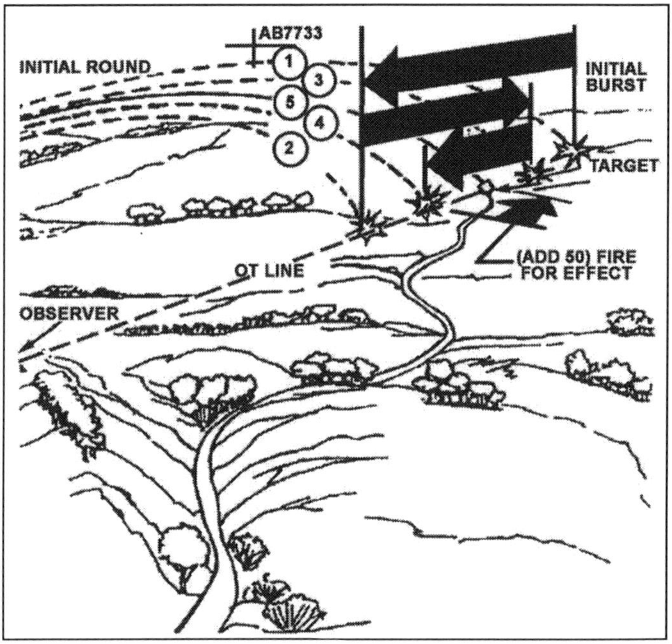

Figure 2-20. Successive bracketing technique.

Hasty Bracketing

2-168. An alternative to successive bracketing is hasty bracketing. Bracketing is an effective technique in that it is sure to bring fire on the target. However, bracketing is relatively time consuming. If the target is moving, bracketing may not be fast enough to engage the target.

2-169. A successful hasty bracket depends on a thorough terrain analysis, which gives the observer an accurate initial target location. For his first correction, the observer receives a bracket similar to that used for successive bracketing. Once the observer receives the initial bracket, he uses it like a yardstick to determine the subsequent correction. He then sends the FDC the correction to move the rounds to the target and to fire for effect (Figure 2-21). Hasty bracketing improves with observer experience and judgment.

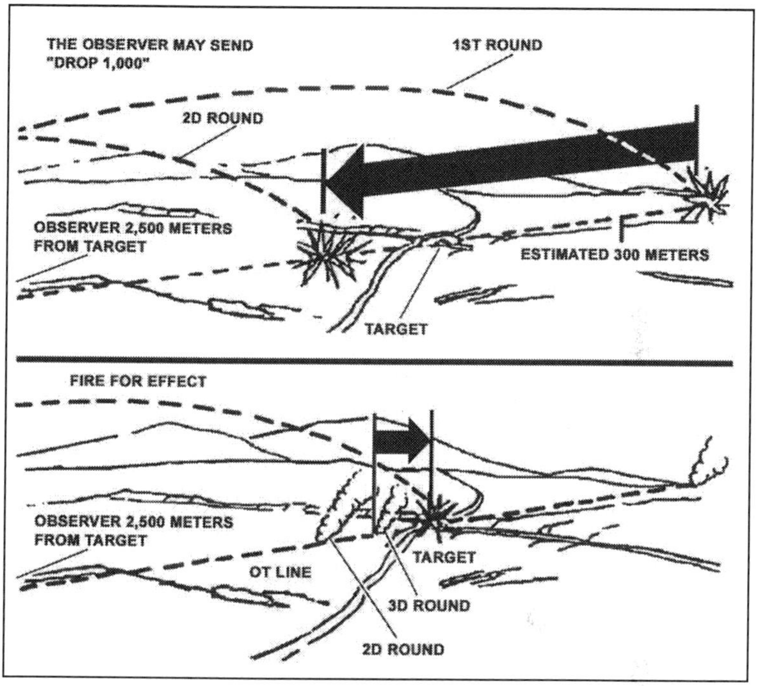

Figure 2-21. Hasty bracketing technique.

Creeping Method

2-170. In danger close situations, the observer uses the creeping method of adjustment. He calls for the first round and deliberately overshoots the target. He adjusts rounds in 100-meter increments or less until the fire hits the target (Figure 2-22). This method requires more time and ammunition than other methods. Therefore, the observer uses it only when he must consider safety first.

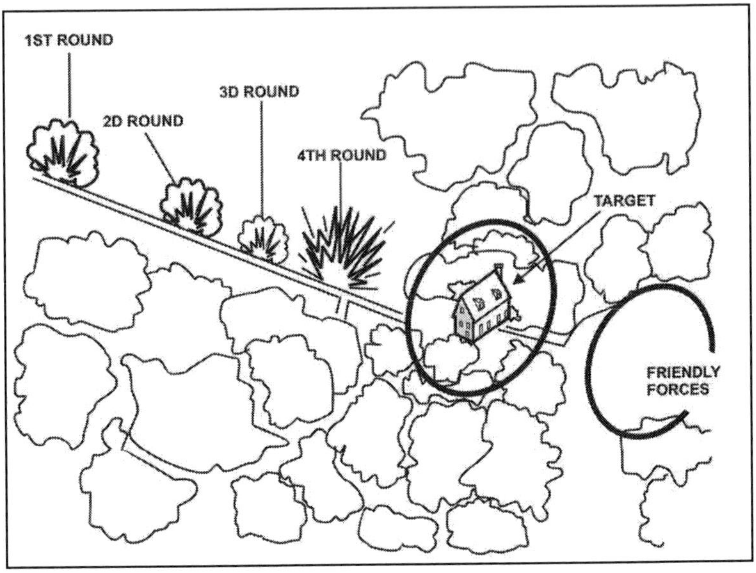

Figure 2-22. Creeping method of adjustment.

END OF MISSION

2-171. End of mission completes the mission and reports the battle damage assessment. The report should go to both the FDC and parent unit. Higher headquarters staff officers use battle damage assessment to update their staff estimates and feed the common operating picture. The proper report format for an indirect fire mission is END OF MISSION, TARGET# ____, BDA, OVER. An example of battle damage assessment is FOUR T72s DESTROYED, or ENEMY SNIPER TEAM SUPPRESSED.

FIRE SUPPORT COORDINATION MEASURES

2-172. Leaders use fire support coordination measures (FSCM) to facilitate both the engagement of targets and protection of friendly forces. Boundaries are the most basic FSCM. Boundaries are both permissive and restrictive FSCM. The fire support coordinator recommends FSCM to the leader based on the leader's guidance, location of friendly forces, scheme of maneuver, and anticipated enemy actions. Once the leader establishes FSCM, they are entered into or posted on all the unit's displays and databases (see FM 1-02).

PERMISSIVE FSCM

2-173. The primary purpose of permissive measures is to facilitate the attack of targets. Once they are established, further coordination is not required to engage targets affected by the measures. Permissive FSCM include a coordinated fire line, fire support coordination line, and free-fire area.

Coordinated Fire Line

2-174. A coordinated fire line (CFL) is a line beyond which conventional, direct, and indirect surface fire support means may fire at any time within the boundaries of the establishing headquarters. This is done without additional coordination. The purpose of the CFL is to expedite the surface-to-surface attack of targets beyond the CFL without coordination with the ground commander in whose area the targets are located (see JP 3-09, *Joint Fire Support*). Brigades or divisions usually establish a CFL, though a maneuver battalion may establish one. It is located as close as possible to the establishing unit without interfering with maneuver forces to open up the area beyond to fire support.

Fire Support Coordination Line

2-175. The fire support coordination line (FSCL) is an FSCM that facilitates the expeditious attack of surface targets of opportunity beyond the coordinating measure. The FSCL applies to all fires of air-, land-, and sea-based weapon systems using any type of ammunition. Forces attacking targets beyond an FSCL must inform all affected commanders in sufficient time to allow necessary reaction to avoid fratricide. Supporting elements attacking targets beyond the FSCL must ensure that the attack will not produce adverse effects on or to the rear of the line. Short of an FSCL, all air-to-ground and surface-to-surface attack operations are controlled by the appropriate leader responsible for that area.

Free-Fire Area

2-176. A free-fire area is a specific area into which any weapon system may fire without additional coordination with the establishing headquarters. Normally, division or higher headquarters establish a free-fire area on identifiable terrain.

Restrictive FSCM

2-177. A restrictive FSCM prevents fires into or beyond the control measure without detailed coordination. The primary purpose of restrictive measures is to provide safeguards for friendly forces. Restrictive FSCM include no-fire area, restrictive fire area, and restrictive fire line.

No-Fire Area

2-178. A no-fire area (NFA) is a land area designated by the appropriate commander into which fires or their effects are prohibited. Leaders use the NFA to protect independently-operating elements such as forward observers and special operating forces. They also use it for humanitarian reasons such as preventing the inadvertent engagement of displaced civilian concentrations, or to protect sensitive areas such as cultural monuments. There are two exceptions to this rule:

(1) The establishing headquarters may approve fires within the NFA on a case-by-case mission basis.

(2) When an enemy force within an NFA engages a friendly force, the friendly force may engage a positively-identified enemy force to defend itself.

Restrictive Fire Area

2-179. A restrictive fire area (RFA) is an area in which specific restrictions are imposed and into which fires that exceed those restrictions will not be delivered without coordination with the establishing headquarters. The purpose of the RFA is to regulate fires into an area according to the stated restrictions such as no unguided conventional or dud-producing munitions. For example, no DPICM rounds should be fired into an area of land that is later going to be occupied by friendly forces. These types of munition have a dud rate and could possibly result in friendly forces being incapacitated. Maneuver battalion or larger ground forces normally establish RFAs. On occasion, a company operating independently may establish an RFA. An RFA is usually located on identifiable terrain by grid or by a radius (in meters) from a center point. The restrictions on an RFA may be shown on a map or overlay, or reference can be made to an operation order that contains the restrictions.

Restrictive Fire Line

2-180. A restrictive fire line is a phase line established between converging friendly forces that prohibits fires or their effects across that line. The purpose of this phase line is to prevent fratricide between converging friendly forces. The next higher common commander of the converging forces establishes the restrictive fire line. Alternatively, the commander can use a restrictive fire line to protect sensitive areas such as cultural monuments. This control measure can also be used as a direct fire control measure.

SECTION VI — CLOSE AIR SUPPORT AND NAVAL GUNFIRE

2-181. Close air support (CAS) is defined in JP 3-09.3, *Joint Tactics, Techniques, and Procedures for Close Air Support (CAS),* as: air action by fixed- and rotary-wing aircraft against hostile targets that are in close proximity to friendly forces and that require detailed integration of each air mission with the fire and movement of these forces.

2-182. Very rarely will an Infantry platoon be directly supported by naval gunfire. But if they are, Navy liaison representatives located with supported ground forces coordinate the control of the fire. Naval gunfire can provide large volumes of immediately available, responsive fire support to land combat forces operating near coastal waters.

CLOSE AIR SUPPORT

2-183. The air liaison officer (ALO) is the battalion commander's advisor in planning, requesting, and executing CAS missions. The ALO serves as a link between the maneuver element and the attacking aircraft. The platoon may provide information that the ALO or tactical air control party (TACP) uses to target enemy forces. A joint terminal air controller (JTAC) may also be attached to the platoon to facilitate communication. The need for a JTAC should be identified during the planning phase of the mission.

2-184. Soldiers may provide emergency control if an ALO, FSO, FO, or JTAC is not available (ground force commander accepts responsibility for friendly casualties). This is possible only if the platoon has a UHF capable radio or if the aircraft is equipped with FM radios. Some U.S. Air Force, Navy, and Marine Corps fixed-wing aircrafts only have ultra high frequency (UHF) radios (AV-8B and F-14) (see FM 6-30). Others have FM capability (A/OA-10, F16, F/A-18, and AC-130). The platoon may also provide information on battle damage as observed. Figure 2-23 shows the format for assessing battle damage.

Battle Damage Assessment
Successful or unsuccessful
Target coordinates
Time on target
Number and type destroyed
Number and type damaged
Killed by air
Wounded by air
Dud bombs

Figure 2-23. Assessing battle damage.

AC-130 GUNSHIP

2-185. If the enemy air defense threat is low, the battalion requests CAS from an AC-130H or AC-130U gunship. The AC-130 provides effective fires night operations and flies CAS and special operations. The AC-130H aircraft contains one 40-mm gun and one 105-mm howitzer (the AC-130U has an additional 25-mm cannon). It is equipped with sensors and target acquisition systems that include forward-looking infrared radar and low-light television. It is effective in urban environments due to its advanced sensors.

Attack Helicopters and Close Combat Attack

2-186. The primary mission of attack helicopter units is to destroy enemy armor and mechanized forces or to provide precision fires. Employing attack helicopters increases the lethality of ground maneuver forces.

2-187. The close combat attack is a technique for using aviation direct fires closely integrated with close fight on the ground. It may be planned or unplanned, but works most effectively when the company integrates aviation assets into the planning process.

2-188. To request immediate close combat attack, if METT-TC permits, the ground unit in contact executes a face-to-face coordination or uses a radio transmission to provide a situation update to the attack aircraft. Figure 2-24 illustrates a close combat attack coordination checklist.

Close Combat Attack Coordination Checklist

1. Enemy situation – specific target identification.
2. Friendly situation – location and method of marking friendly positions.
3. Ground maneuver mission and scheme of maneuver.
4. Attack aircraft scheme of maneuver.
5. Planned EA and battle position/support-by-fire position.
6. Method of target marking.
7. Fire coordination and fire restrictions.
8. Map graphics update.
9. Request for immediate aviation close fight support – used for targets of opportunity or for ground-to-air target handoff.

Figure 2-24. Checking close combat coordination.

2-189. After receipt of a request for immediate close combat attack, the attack team leader informs the ground unit leader of the battle position, assault by fire position, or the series of positions his team will occupy. This information should provide the best observation and fields of fire into the engagement or target area. The attack team leader then provides the ground maneuver unit leader with his concept for the team's attack on the objective. Depending on SOP and tactical requirements, the flight lead may initially talk with the Infantry battalion, but will likely get pushed to the company net and may talk directly to the platoon leader.

2-190. Upon mission completion, the attack team leader provides the ground maneuver commander a battle damage assessment of the intended target.

Marking Friendly Positions

2-191. Whenever possible, friendly positions are marked to enhance safety, minimize the possibility of fratricide, and provide target area references. Methods of marking friendly positions are shown in Table 2-10.

Table 2-10. Methods of marking friendly positions.

METHOD	DAY/ NIGHT	ASSETS	FRIENDLY MARKS	TARGET MARKS	REMARKS
Smoke	D/N	All	Good	Good	Easily identifiable, may compromise friendly position, obscure target, or warn of fire support employment. Placement may be difficult due to structures.
Smoke (IR)	D/N	All/ NVD at night	Good	Good	Easily identifiable, may compromise friendly position, obscure target, or warn of fire support employment. Placement may be difficult due to structures. Night marking is greatly enhanced by the use of IR reflective smoke.
ILLUM, ground burst	D/N	All	N/A	Good	Easily identified, may wash out NVDs.

Table 2-10. Methods of marking friendly positions (continued).

METHOD	DAY/ NIGHT	ASSETS	FRIENDLY MARKS	TARGET MARKS	REMARKS
Signal mirror	D	All	Good	N/A	Avoids compromise of friendly location. Dependent on weather and available light and may be lost in reflections from other reflective surfaces such as windshields, windows, and water.
Spotlight	N	All	Good	Marginal	Highly visible to all. Compromises friendly position and warns of fire support employment. Effectiveness is dependent upon degree of urban lighting.
IR Spotlight	N	All NVD	Good	Marginal	Visible to all with NVDs. Less likely to compromise than overt light. Effectiveness dependent upon degree of urban lighting.
Visual laser	N	All	Good	Marginal	Highly visible to all. Risk of compromise is high. Effectiveness dependent upon degree of urban lighting.
Tracers	D/N	All	N/A	Marginal	May compromise position. May be difficult to distinguish mark from other gunfire. During daytime use, may be more effective to kick up dust surrounding target.
Electronic beacon	D/N	See remarks.	Excellent	Good	Ideal friendly marking device for AC-130 and some United States Air Force fixed wing (not compatible with Navy or Marine aircraft). Least impeded by urban terrain. Can be used as a TRP for target identification. Coordination with aircrews essential to ensure equipment and training compatibility.
Strobe (overt)	N	All	Marginal	N/A	Visible by all. Effectiveness dependent upon degree of urban lighting.
Strobe (IR)	N	All NVD	Good	N/A	Visible to all NVDs. Effectiveness dependent upon degree of urban lighting. Coded strobes aid in acquisition.
Flare (overt)	D/N	All	Good	N/A	Visible by all. Easily identified by aircrew.
Flare (IR)	N	All NVD	Good	N/A	Visible to all NVDs. Easily identified by aircrew.
Glint/IR panel	N	All NVD	Good	N/A	Not readily detectable by enemy. Very effective except in highly lit areas.
Combat identification panel	D/N	All FLIR	Good	N/A	Provides temperature contrast on vehicles or building. May be obscured by urban terrain.
VS-17 panel	D	All	Marginal	N/A	Only visible during daylight. Easily obscured by structures.
Chemical heat sources	D/N	All FLIR	Poor	N/A	Easily masked by urban structures and lost in thermal clutter. Difficult to acquire, can be effective when used to contrast cold background or when aircraft knows general location.
Spinning chem light (overt)	N	All	Marginal	N/A	Provides unique signature. May be obscured by structures. Effectiveness dependent upon degree of urban lighting.
Spinning chemlight (IR)	N	All NVD	Marginal	N/A	Provides unique signature. May be obscured by structures. Effectiveness dependent upon degree of urban lighting.

NAVAL GUNFIRE SUPPORT

2-192. Naval gunfire has a wide variety of weapons extending from light conventional armament to heavy missiles and nuclear weapons. It can play a vital role in reducing the enemy's capability of action by destroying enemy installations and fortifications before a ground assault, and by protecting and covering the supporting offensive operations of the land force after the assault.

Chapter 3

Tactical Movement

Tactical movement is the movement of a unit assigned a tactical mission under combat conditions when not in direct ground contact with the enemy. Tactical movement is based on the anticipation of early ground contact with the enemy, either en route or shortly after arrival at the destination. Movement ends when ground contact is made or the unit reaches its destination. Movement is not maneuver. Maneuver happens once a unit has made contact with the enemy. Because tactical movement shares many of the characteristics of an offensive action, the battlefield is organized in a manner similar to other offensive actions. This chapter discusses the basics and formations of tactical movement.

SECTION I — OVERVIEW

3-1. Movement refers to the shifting of forces on the battlefield. The key to moving successfully involves selecting the best combination of movement formations and movement techniques for each situation. Leaders consider the factors of METT-TC in selecting the best route and the appropriate formation and movement technique. The leader's selection must allow the moving platoon to—

- Maintain cohesion.
- Maintain communication.
- Maintain momentum.
- Provide maximum protection.
- Make enemy contact in a manner that allows them to transition smoothly to offensive or defensive action.

3-2. Careless movement usually results in contact with the enemy at a time and place of the enemy's choosing. To avoid this, leaders must understand the constantly-changing interrelationship between unit movement, terrain, and weapon systems within their area of operations. This understanding is the basis for employing movement formations, movement techniques, route selection and navigation, crossing danger areas, and security (Figure 3-1).

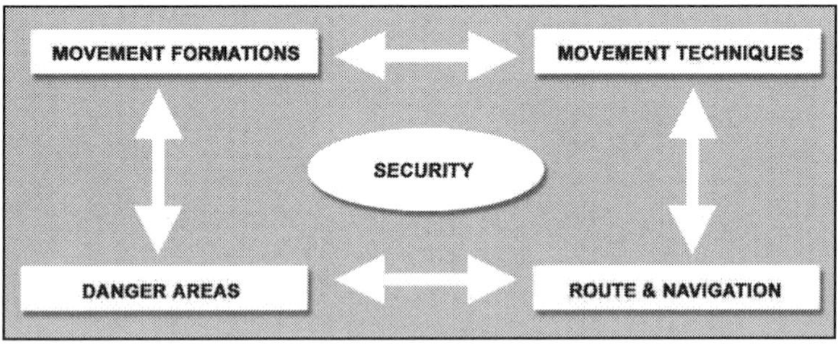

Figure 3-1. Basics of tactical movement.

3-3. Leaders executing tactical movement have three primary goals:

- Avoid surprise by the enemy.
- When necessary, transition quickly to maneuver while minimizing enemy effects.
- Get to the right place, at the right time, ready to fight.

3-4. Units moving behind enemy lines seek to avoid enemy contact. They choose the movement that allows them to retain security and control. To avoid loss of surprise and initiative, casualties, and mission failure, platoons normally—

- Avoid chance enemy contact, if possible.
- Move on covered and concealed routes.
- Avoid likely ambush sites and other danger areas.
- Practice camouflage, noise, and light discipline.
- Maintain 360-degree security.
- Make contact with the smallest element if enemy contact is unavoidable.
- Retain the initiative to attack at the time and place of the unit's choice.
- Take active countermeasures such as using smoke and direct and indirect fire to suppress or obscure suspected enemy positions.

3-5. Infantry platoons primarily move on foot. However, there are circumstances when they will move, and even fight, mounted. Because their units may operate with vehicle support, leaders must be comfortable employing tactical movement with a variety of vehicle platforms.

3-6. In selecting formations and movement techniques, leaders must consider other requirements such as speed and control as well as security. When conducting tactical movement, leaders must be prepared to quickly transition to maneuver and fight while minimizing the effects of the enemy. This requirement calls for the leader to determine which formation or combination of formations best suits the situation.

MOVEMENT FORMATIONS

3-7. Movement formations are the ordered arrangement of forces that describes the general configuration of a unit on the ground. They determine the distance between Soldiers, sectors of fire, and responsibilities for 360-degree security. Movement formations are used in combination with movement techniques (and other security measures), immediate action drills, and enabling tasks. Movement techniques define the level of security one subordinate provides another within a formation. Immediate action drills are those combat actions that enable the unit to quickly transition to maneuver during unexpected enemy contact. Enabling tasks facilitate transitions between other combat tasks. See Section II of this chapter for more on movement formations.

MOVEMENT TECHNIQUES

3-8. Movement techniques describe the position of squads and fire teams in relation to each other during movement. Platoons and squads use three movement techniques: traveling, traveling overwatch, and bounding overwatch.

3-9. Like formations, movement techniques provide varying degrees of control, security, and flexibility. Movement techniques differ from formations in two ways:

- Formations are relatively fixed; movement techniques are not. The distance between moving units or the distance that a squad bounds away from an overwatching squad varies based on factors of METT-TC.
- Formations allow the platoon to weight its maximum firepower in a desired direction; movement techniques allow squads to make contact with the enemy with the smallest element possible. This allows leaders to establish a base of fire, initiate suppressive fires, and attempt to maneuver without first having to disengage or be reinforced.

3-10. Leaders base their selection of a particular movement technique on the likelihood of enemy contact and the requirement for speed. See Section III of this chapter for more on movement techniques.

ROUTE AND NAVIGATION

3-11. Planning and selecting a route is a critical leader skill. One of the keys to successful tactical movement is the ability to develop routes that increase the unit's security, decrease the Soldier's effort, and get the unit to the objective on time in a manner prepared to fight. Good route selection begins with a thorough terrain analysis and ends with superior navigation. Planning and preparation are worthless if a unit cannot find its way to the objective, or worse, stumbles onto it because of poor navigation. See Section IV of this chapter for more on route and navigation.

DANGER AREAS

3-12. When analyzing the terrain (in the METT-TC analysis) during the troop-leading procedures (TLP), the platoon leader may identify danger areas. The term danger area refers to any area on the route where the terrain would expose the platoon to enemy observation, fire, or both. If possible, the platoon leader should plan to avoid danger areas. However, there are times when he cannot. When the unit must cross a danger area, it should do so as quickly and as carefully as possible. See Section V of this chapter for more information on danger areas.

SECURITY

3-13. Security during movement includes the actions that units take to secure themselves and the tasks given to units to provide security for a larger force. Platoons and squads enhance their own security during movement through the use of covered and concealed terrain; the use of the appropriate movement formation and technique; the actions taken to secure danger areas during crossing; the enforcement of noise, light, and radiotelephone discipline; and the use of proper individual camouflage techniques. See Section VII of this chapter for more on security.

3-14. Formations and movement techniques provide security by:

- Positioning each Soldier so he can observe and fire into a specific sector that overlaps with other sectors.
- Placing a small element forward to allow the platoon to make contact with only the lead element and give the remainder of the platoon freedom to maneuver.
- Providing overwatch for a portion of the platoon.

OTHER CONSIDERATIONS

3-15. In planning tactical movement, leaders should also consider the requirements for—

- Terrain.
- Planning.
- Direct fires.
- Fire support.
- Control.

TERRAIN

3-16. The formations and techniques shown in the illustrations in this chapter are examples only. They are generally depicted without terrain considerations (which are usually a critical concern in the selection and execution of a formation). Therefore, in both planning and executing tactical movement, leaders understand that combat formations and movement techniques require modification in execution. Spacing requirements and speed result from a continuous assessment of terrain. Leaders must stay ready to adjust the distance of individuals, fire teams, squads, and individual vehicles and vehicle sections based on terrain, visibility, and other mission requirements.

3-17. While moving, individual Soldiers and vehicles use the terrain to protect themselves during times when enemy contact is possible or expected. They use natural cover and concealment to avoid enemy fires. The following guidelines apply to Soldiers and vehicle crews using terrain for protection:

- Do not silhouette yourself against the skyline.
- Avoid possible kill zones because it is easier to cross difficult terrain than fight the enemy on unfavorable terms.
- Cross open areas quickly.
- Avoid large, open areas, especially when they are dominated by high ground or by terrain that can cover and conceal the enemy.
- Do not move directly forward from a concealed firing position.

PLANNING

3-18. One of the leader's primary duties is to develop a plan that links together route selection and navigation, combat formations, and appropriate security measures with enabling tasks that moves the unit from its current location to its destination. This plan must take into account the enemy situation and control during movement.

DIRECT FIRES

3-19. While moving or when stationary, each Soldier (or vehicle) has a sector to observe and engage enemy soldiers in accordance with the unit's engagement criteria (see Chapter 2). Individual and small unit sectors are the foundation of the unit's area of influence. Pre-assigned sectors are inherent in combat formations. When formations are modified, leaders must reconfirm their subordinates' sectors. Leaders have the added responsibility of ensuring their subordinates' sectors are mutually supporting and employing other security measures that identify the enemy early and allow the leader to shape the fight.

FIRE SUPPORT

3-20. Planning should always include arranging for fire support (mortars, artillery, CAS, attack helicopters, naval gunfire), even if the leader thinks it unnecessary. A fire plan can be a tool to help navigate and gives the leader the following options:

- Suppressing enemy observation posts or sensors.
- Creating a distraction.
- Achieving immediate suppression.
- Covering withdrawal off of an objective.
- Breaking contact.

CONTROL

3-21. Controlling tactical movement is challenging. The leader must be able to start, stop, shift left or right, and control the unit's direction and speed of movement while navigating, assessing the terrain, and preparing for enemy contact. Determining the proper movement formations and techniques during planning is important, but the leader must be able to assess his decision during execution and modify or change his actions based on the actual situation.

3-22. Without adequate procedural and positive control, it is difficult for the leader to make decisions and give orders, lead an effective response to enemy contact, or accurately navigate. Leaders exercise procedural control by unit training and rehearsals in the basics of tactical movement. The better trained and rehearsed subordinates are, the more freedom leaders have to concentrate on the situation, particularly the enemy and the terrain. Leaders exercise positive control by communicating to subordinates. They do so using hand-and-arm signals as a method of communication. They also use the other means of communication (messenger, visual, audio, radio, and digital) when appropriate.

3-23. All available communication is used (consistent with OPSEC and movement security) to assist in maintaining control during movement. March objectives, checkpoints, and phase lines may be used to aid in control. The number of reports is reduced as normally only exception reports are needed. The leader should be well forward in the formation but may move throughout as the situation demands. Communications with security elements are mandatory. Operations security often prevents the use of radios, so connecting files, runners, and visual signals can be used. Detailed planning, briefing, rehearsals, and control are valuable if there is enemy contact. Alternate plans are made to cover all possible situations.

SECTION II — MOVEMENT FORMATIONS

3-24. This section discusses movement formations of Infantry fire teams, squads, and platoons. The platoon leader uses formations for several purposes: to relate one squad to another on the ground; to position firepower to support the direct-fire plan; to establish responsibilities for sector security among squads; or to aid in the execution of battle drills. Just as they do with movement techniques, platoon leaders plan formations based on where they expect enemy contact, and on the company commander's plans to react to contact. The platoon leader evaluates the situation and decides which formation best suits the mission and situation.

3-25. Every squad and Soldier has a standard position. Soldiers can see their team leaders. Fire team leaders can see their squad leaders. Leaders control their units using hand-and-arm signals.

3-26. Formations also provide 360-degree security and allow units to give the weight of their firepower to the flanks or front in anticipation of enemy contact.

3-27. Formations do not demand parade ground precision. Platoons and squads must retain the flexibility needed to vary their formations to the situation. The use of formations allows Soldiers to execute battle drills more quickly and gives them the assurance that their leaders and buddy team members are in the expected positions and performing the right tasks.

3-28. Sometimes platoon and company formations differ due to METT-TC factors. For example, the platoons could move in wedge formations within a company vee. It is not necessary for the platoon formation to be the same as the company formation unless directed by the company commander. However, the platoon leader must coordinate his formation with other elements moving in the main body team's formation. Figure 3-2 illustrates platoon symbols.

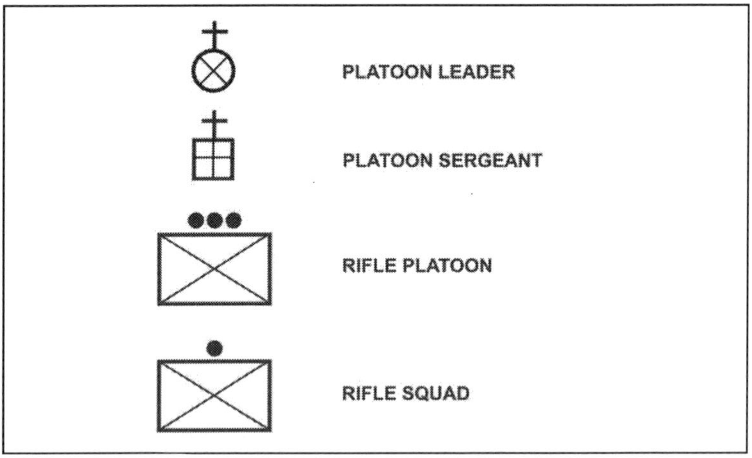

Figure 3-2. Legend of platoon symbols.

NOTE: The formations shown in the illustrations in this chapter are examples only. They generally are depicted without METT-TC considerations, which are always the most crucial element in the selection and execution of a formation. Leaders must be prepared to adapt their choice of formation to the specific situation.

PRIMARY FORMATIONS

3-29. Combat formations are composed of two variables: lateral frontage, represented by the line formation; and depth, represented by the column formation. The advantages attributed to any one of these variables are disadvantages to the other. Leaders combine the elements of lateral frontage and depth to determine the best formation for their situation. In addition to the line and column/file, the other five types of formations—box; vee; wedge; diamond; and echelon—combine these elements into varying degrees. Each does so with different degrees of emphasis that result in unique advantages and disadvantages (Table 3-1).

3-30. The seven formations can be grouped into two categories: formations with one lead element, and formations with more than one lead element. The formations with more than one lead element, as a general rule, are better for achieving fire superiority to the front, but are more difficult to control. Conversely, the formations with only one lead element are easier to control but are not as useful for achieving fire superiority to the front.

3-31. Leaders attempt to maintain flexibility in their formations. Doing so enables them to react when unexpected enemy actions occur. The line, echelon, and column formations are the least flexible of the seven formations. The line mass to the front has vulnerable flanks. The echelon is optimized for a flank threat—something that units want to avoid. The column has difficulty reinforcing an element in contact. Leaders using these formations should consider ways to reduce the risks associated with their general lack of flexibility.

Table 3-1. Primary formations.

Name/Formation/ Signal (if applicable)	Characteristics	Advantages	Disadvantages
Line Formation 	- All elements arranged in a row - Majority of observation and direct fires oriented forward; minimal to the flanks - Each subordinate unit on the line must clear its own path forward - One subordinate designated as the base on which the other subordinates cue their movement	Ability to: - Generate fire superiority to the front - Clear a large area - Disperse - Transition to bounding overwatch, base of fire, or assault	- Control difficulty increases during limited visibility and in restrictive or close terrain - Difficult to designate a maneuver element - Vulnerable assailable flanks - Potentially slow - Large signature
Column/File Formation 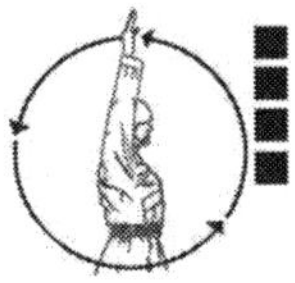	- One lead element - Majority of observation and direct fires oriented to the flanks; minimal to the front - One route means unit only influenced by obstacles on that one route	- Easiest formation to control (as long as leader can communicate with lead element) - Ability to generate a maneuver element - Secure flanks - Speed	- Reduced ability to achieve fire superiority to the front - Clears a limited area and concentrates the unit - Transitions poorly to bounding overwatch, base of fire, and assault - Column's depth makes it a good target for close air attacks and a machine gun beaten zone
Vee Formation 	- Two lead elements - Trail elements move between the two lead elements - Used when contact to the front is expected - "Reverse wedge" - Unit required to two lanes/routes forward	Ability to: - Generate fire superiority to the front - Generate a maneuver element - Secure flanks - Clear a large area - Disperse - Transition to bounding overwatch, base of fire, or assault	- Control difficulty increases during limited visibility and in restrictive or close terrain - Potentially slow
Box Formation 	- Two lead elements - Trail elements follow lead elements - All-around security	See vee formation advantages	See vee formation disadvantages
Wedge Formation 	- One lead element - Trail elements paired off abreast of each other on the flanks - Used when the situation is uncertain	Ability to: - Control, even during limited visibility, in restrictive terrain, or in close terrain - Transition trail elements to base of fire or assault - Secure the front and flanks - Transition the line and column	- Trail elements are required to clear their own path forward - Frequent need to transition to column in restrictive, close terrain
Diamond Formation 	- Similar to the wedge formation - Fourth element follows the lead element	See wedge formation advantages	See wedge formation disadvantages
Echelon Formation (Right) 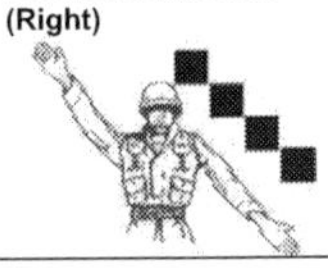	- Elements deployed diagonally left or right - Observation and fire to both the front and one flank - Each subordinate unit on the line clears its own path forward	- Ability to assign sectors that encompass both the front and flank	- Difficult to maintain proper relationship between subordinates - Vulnerable to the opposite flanks

FIRE TEAM FORMATIONS

3-32. The term fire team formation refers to the Soldiers' relative positions within the fire team. Fire team formations include the fire team wedge and the fire team file (Table 3-2). Both formations have advantages and disadvantages. Regardless of which formation the team employs, each Soldier must know his location in the formation relative to the other members of the fire team and the team leader. Each Soldier covers a set sector of responsibility for observation and direct fire as the team is moving. To provide the unit with all-round protection, these sectors must interlock. Team leaders must be constantly aware of their team's sectors and correct them as required.

Table 3-2. Comparison of fire team formations.

Movement Formation	When Most Often Used	CHARACTERISTICS			
		Control	Flexibility	Fire Capabilities and Restrictions	Security
Fire team wedge	Basic fire team formation	Easy	Good	Allows immediate fires in all directions	All-round
Fire team file	Close terrain, dense vegetation, limited visibility conditions	Easiest	Less flexible than wedge	Allows immediate fires to the flanks, masks most fires to the rear	Least

3-33. The team leader adjusts the team's formation as necessary while the team is moving. The distance between men will be determined by the mission, the nature of the threat, the closeness of the terrain, and by the visibility. As a general rule, the unit should be dispersed up to the limit of control. This allows for a wide area to be covered, makes the team's movement difficult to detect, and makes them less vulnerable to enemy ground and air attack. Fire teams rarely act independently. However, in the event that they do, when halted, they use a perimeter defense to ensure all-around security.

FIRE TEAM WEDGE

3-34. The wedge (Figure 3-3) is the basic formation for the fire team. The interval between Soldiers in the wedge formation is normally 10 meters. The wedge expands and contracts depending on the terrain. Fire teams modify the wedge when rough terrain, poor visibility, or other factors make control of the wedge difficult. The normal interval is reduced so all team members can still see their team leader and all team leaders can still see their squad leader. The sides of the wedge can contract to the point where the wedge resembles a single file. Soldiers expand or resume their original positions when moving in less rugged terrain where control is easier.

3-35. In this formation the fire team leader is in the lead position with his men echeloned to the right and left behind him. The positions for all but the leader may vary. This simple formation permits the fire team leader to lead by example. The leader's standing order to his Soldiers is: "Follow me and do as I do." When he moves to the right, his Soldiers should also move to the right. When he fires, his Soldiers also fire. When using the lead-by-example technique, it is essential for all Soldiers to maintain visual contact with the leader.

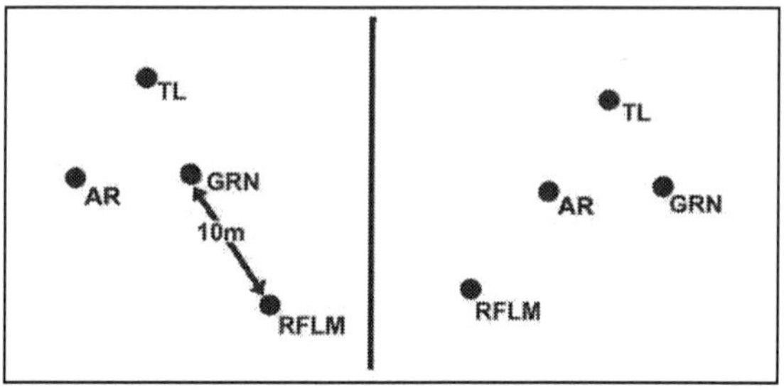

Figure 3-3. Fire team wedge.

FIRE TEAM FILE

3-36. Team leaders use the file when employing the wedge is impractical. This formation is most often used in severely restrictive terrain, like inside a building; dense vegetation; limited visibility; and so forth. The distance between Soldiers in the column changes due to constraints of the situation, particularly when in urban operations (Figure 3-4).

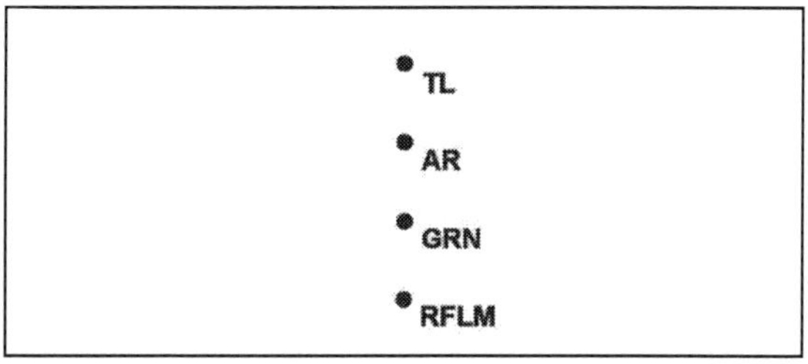

Figure 3-4. Fire team file.

SQUAD FORMATIONS

3-37. The term squad formation refers to the relative locations of the fire teams. Squad formations include the squad column, the squad line, and the squad file. Table 3-3 compares squad formations.

Table 3-3. Comparison of squad formations.

Movement Formation	When Most Often Used	CHARACTERISTICS			
		Control	Flexibility	Fire Capabilities and Restrictions	Security
Squad column	The main squad formation	Good	Aids maneuver, good dispersion laterally and in depth	Allows large volume of fire to the flanks but only limited volume to the front	All-around
Squad line	For maximum firepower to the front	Not as good as squad column	Limited maneuver capability (both fire teams committed)	Allows maximum immediate fire to the front	Good to the front, little to the flank and rear
Squad file	Close terrain, dense vegetation, limited visibility conditions	Easiest	Most difficult formation to maneuver from	Allows immediate fire to the flanks, masks most fire to the front and rear	Least

3-38. The squad leader adjusts the squad's formation as necessary while moving, primarily through the three movement techniques (see Section III). The squad leader exercises command and control primarily through the two team leaders and moves in the formation where he can best achieve this. The squad leader is responsible for 360-degree security, for ensuring the team's sectors of fire are mutually supporting, and for being able to rapidly transition the squad upon contact.

3-39. The squad leader designates one of the fire teams as the base fire team. The squad leader controls the squad's speed and direction of movement through the base fire team while the other team and any attachments cue their movement off of the base fire team. This concept applies when not in contact and when in contact with the enemy.

3-40. Weapons from the weapons squad (a machine gun or a Javelin) may be attached to the squad for the movement or throughout the operation. These high value assets need to be positioned so they are protected and can be quickly brought into the engagement when required. Ideally, these weapons should be positioned so they are between the two fire teams.

SQUAD COLUMN

3-41. The squad column is the squad's main formation for movement unless preparing for an assault (Figure 3-5). It provides good dispersion both laterally and in depth without sacrificing control. It also facilitates maneuver. The lead fire team is the base fire team. Squads can move in either a column wedge or a modified column wedge. Rough terrain, poor visibility, and other factors can require the squad to modify the wedge into a file for control purposes. As the terrain becomes less rugged and control becomes easier, the Soldiers assume their original positions.

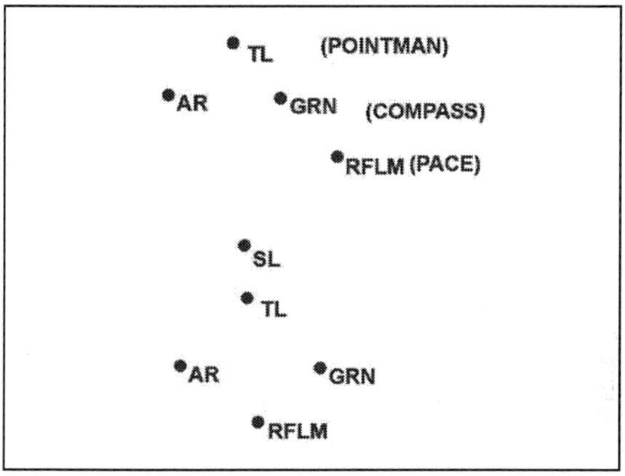

Figure 3-5. Squad column, fire teams in wedge.

SQUAD LINE

3-42. The squad line provides maximum firepower to the front and is used to assault or as a pre-assault formation (Figure 3-6). To execute the squad line, the squad leader designates one of the teams as the base team. The other team cues its movement off of the base team. This applies when the squad is in close combat as well. From this formation, the squad leader can employ any of the three movement techniques or conduct fire and movement (see Section III).

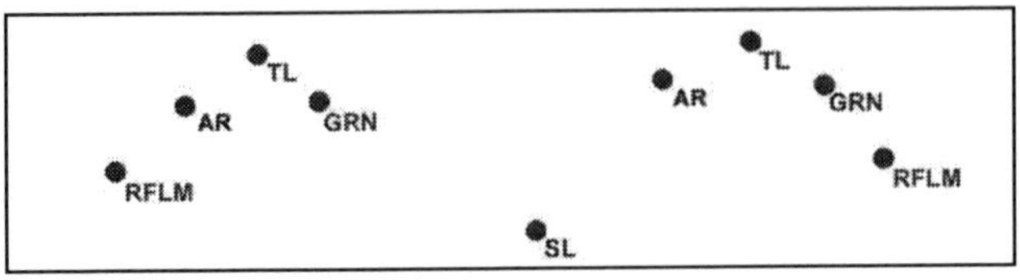

Figure 3-6. Squad line.

SQUAD FILE

3-43. The squad file has the same characteristics as the fire team file (Figure 3-7). In the event that the terrain is severely restrictive or extremely close, teams within the squad file may also be in file. This disposition is not optimal for enemy contact, but does provide the squad leader with maximum control. If the squad leader wishes to increase his control over the formation he moves forward to the first or second position. Moving forward also enables him to exert greater morale presence by leading from the front, and to be immediately available to make key decisions. Moving a team leader to the last position can provide additional control over the rear of the formation.

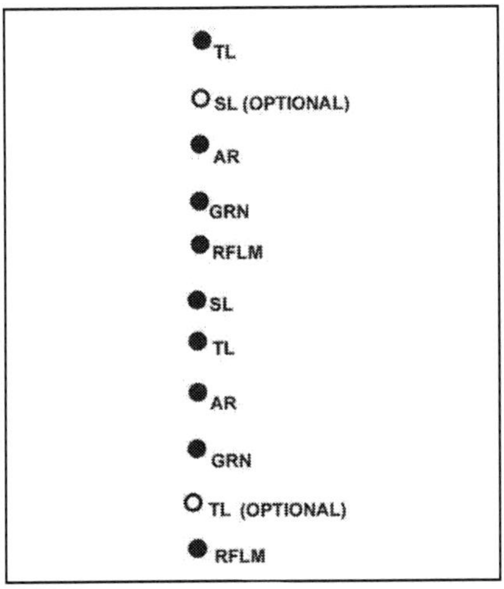

Figure 3-7. Squad file.

WEAPONS SQUAD MOVEMENT FORMATIONS

3-44. The weapons squad is not a rifle squad and should not be treated as such. During tactical movement the platoon leader has one of two options when it comes to positioning the weapons squad. The weapons squad can either travel together as a separate entity, or can be broken up and distributed throughout the formation. The advantage to keeping the weapons squad together is the ability to quickly generate a support by fire and gain fire superiority under the direction of the weapons squad leader. The disadvantage to this approach is the lack of redundancy throughout the formation. The advantage to distributing the weapons squad throughout the rifle squads is the coverage afforded to the entire formation. The disadvantage is losing the weapons squad leader as a single command and control element and the time required to reassemble the weapons squad if needed.

3-45. When the weapons squad travels dispersed, they can either be attached to squads or attached to the key leaders like the platoon leader, platoon sergeant, and weapons squad leader. There is no standard method for their employment. Rather, the platoon leader places the weapons using two criteria: ability to quickly generate fire superiority, and protection for these high value assets.

3-46. Like the rifle squad, the weapons squad, when traveling as a squad, uses either a column or line formation. Within these formations, the two sections can also be in column or line formation.

PLATOON FORMATIONS

3-47. The actual number of useful combinations of squad and fire team combat formations within the platoon combat formations is numerous, creating a significant training requirement for the unit. Add to that the requirement to modify formations with movement techniques, immediate action drills, and other techniques, and it is readily apparent that what the platoon leader needs is a couple of simple, effective strategies. These strategies should be detailed in the unit's SOPs. For a full description of each combat formation and advantages and disadvantages refer again to Table 3-1.

PLATOON LEADER RESPONSIBILITIES

3-48. Like the squad leader, the platoon leader exercises command and control primarily through his subordinates and moves in the formation where he can best achieve this. The squad and team leader execute the combat formations and movement techniques within their capabilities based on the platoon leader's guidance.

3-49. The platoon leader is responsible for 360-degree security, for ensuring that each subordinate unit's sectors of fire are mutually supporting, and for being able to rapidly transition the platoon upon contact. He adjusts the platoon's formation as necessary while moving, primarily through the three movement techniques (see Section III). Like the squad and team, this determination is a result of the task, the nature of the threat, the closeness of terrain, and the visibility.

3-50. The platoon leader is also responsible for ensuring his squads can perform their required actions. He does this through training before combat and rehearsals during combat. Well-trained squads are able to employ combat formations, movement techniques, actions on contact, and stationary formations.

PLATOON HEADQUARTERS

3-51. The platoon leader also has to decide how to disperse the platoon headquarters elements (himself, his RTO, his interpreter, the forward observer, the platoon sergeant, and the medic). These elements do not have a fixed position in the formations. Rather, they should be positioned where they can best accomplish their tasks. The platoon leader's element should be where he conducts actions on contact, where he can supervise navigation, and where he can communicate with higher. The FO's element should be where he can best see the battlefield and where he can communicate with the platoon leader and the battalion fire support officer (FSO). This is normally in close proximity to the platoon leader. The platoon sergeant's element should be wherever the platoon leader is not. Because of the platoon sergeant's experience, he should be given the freedom to assess the situation and advise the platoon leader accordingly. Typically, this means the platoon leader is more toward the front of the formation, while the platoon sergeant is more to the rear of the formation.

BASE SQUAD

3-52. The platoon leader designates one of the squads as the base squad. He controls the platoon's speed and direction of movement through the base squad, while the other squads and any attachments cue their movement off of the base squad.

MOVING AS PART OF A LARGER UNIT

3-53. Infantry platoons often move as part of a larger unit's movement. The next higher commander assigns the platoon a position within the formation. The platoon leader assigns his subordinates an appropriate formation based on the situation and uses the appropriate movement technique. Regardless of the platoon's position within the formation, it must be ready to make contact or to support the other elements by movement, by fire, or by both.

3-54. When moving in a company formation, the company commander normally designates a base platoon to facilitate control. The other platoons cue their speed and direction on the base platoon. This permits quick changes and lets the commander control the movement of the entire company by controlling only the base platoon. The company commander normally locates himself within the formation where he can best see and direct the movement of the base platoon. The base platoon's center squad is usually its base squad. When the platoon is not acting as the base platoon, its base squad is its flank squad nearest the base platoon.

PRIMARY FORMATIONS

3-55. Platoon formations include the column, the line (squads on line or in column), the vee, the wedge, and the file. The leader should weigh these carefully to select the best formation based on his mission and on METT-TC analysis. A comparison of the formations is in Table 3-4.

3-56. Within these platoon formations, the rifle squads are either in a column or a line. Within the rifle squad formations, the teams are in one of the six formations. Normally the platoon leader does not personally direct fire team formations, but he can do so if the situation dictates. He should at a minimum know the formation of the base fire team of the base squad. The weapons squad travels separately or attached to the rifle squads.

Table 3-4. Comparison of platoon formations.

Movement Formation	When Most Often Used	CHARACTERISTICS				
		Control	Flexibility	Fire Capability/ Restrictions	Security	Movement
Platoon column	Platoon primary movement formation	Good for maneuver (fire and movement)	Provides good dispersion laterally and in depth	Allows limited firepower to the front and rear, but high volume to the flanks	Extremely limited overall security	Good
Platoon line, squads on line	When the leader wants all Soldiers forward for maximum firepower to the front and the enemy situation is known	Difficult	Minimal	Allows maximum firepower to the front, little to flanks and rear	Less secure than other formations because of the lack of depth, but provides excellent security for the higher formation in the direction of the echelon	Slow
Platoon line, squads in column	May be used when the leader does not want everyone on line; but wants to be prepared for contact; when crossing the LD when LD is near the objective	Easier than platoon line, squads on line, but more difficult than platoon column	Greater than platoon column, squads on line, but less than platoon line, squads on line	Good firepower to the front and rear, minimum fires to the flanks; not as good as platoon column, better than platoon line	Good security all around	Slower than platoon column, faster than platoon line, squads on line
Platoon vee	When the enemy situation is vague, but contact is expected from the front	Difficult	Provides two squads up front for immediate firepower and one squad to the rear for movement (fire and movement) upon contact from the flank	Immediate heavy volume of firepower to the front or flanks, but minimum fires to the rear	Good security to the front	Slow
Platoon wedge	When the enemy situation is vague, but contact is not expected	Difficult but better than platoon vee and platoon line, squads on line	Enables leader to make contact with a small element and still have two squads to maneuver	Provides heavy volume of firepower to the front or flanks	Good security to the flanks	Slow, but faster than platoon vee
Platoon file	When visibility is poor due to terrain, vegetation, or light	Easiest	Most difficult formation from which to maneuver	Allows immediate fires to the flanks, masks most fires to front and rear	Extremely limited overall security	Fastest for dismounted movement

Platoon Column

3-57. In the platoon column formation, the lead squad is the base squad (Figure 3-8). It is normally used for traveling only.

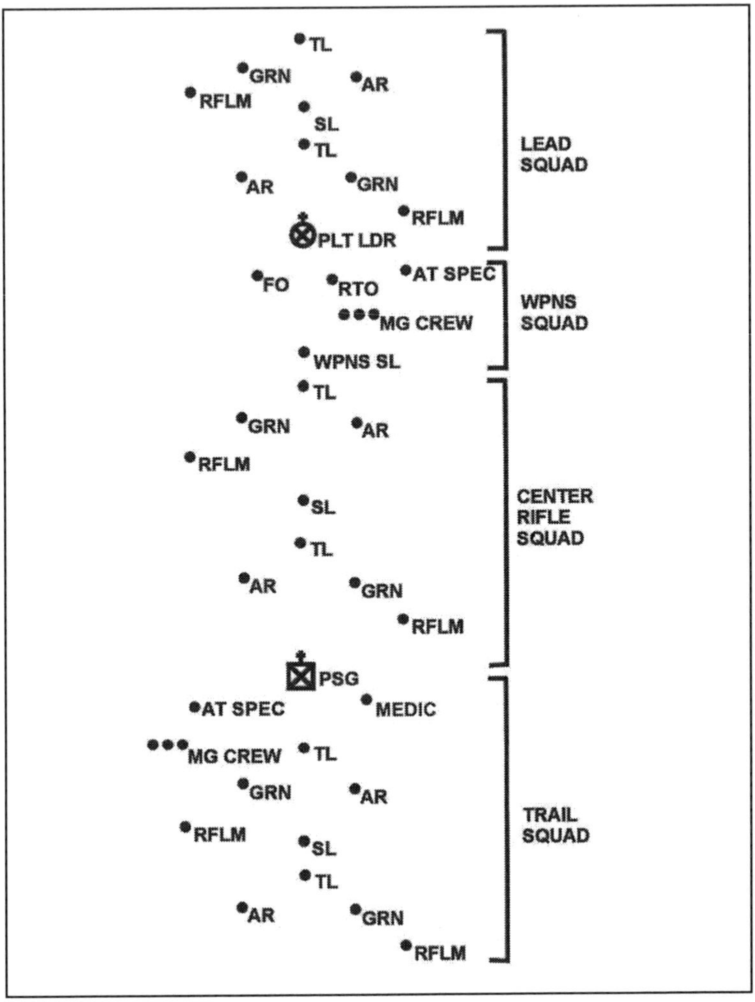

Figure 3-8. Platoon column.

NOTE: METT-TC considerations determine where the weapons squad or machine gun teams locate in the formation. They normally move with the platoon leader and /or PSG so he can establish a base of fire quickly.

Platoon Line, Squads on Line

3-58. In the platoon line, squads on line formation, when two or more platoons are attacking, the company commander chooses one of them as the base platoon. The base platoon's center squad is its base squad. When the platoon is not acting as the base platoon, its base squad is its flank squad nearest the base platoon. The weapons squad may move with the platoon, or it can provide the support-by-fire position. This is the basic platoon assault formation (Figure 3-9).

3-59. The platoon line with squads on line is the most difficult formation from which to make the transition to other formations.

3-60. It may be used in the assault to maximize the firepower and shock effect of the platoon. This normally is done when there is no more intervening terrain between the unit and the enemy, when antitank systems are suppressed, or when the unit is exposed to artillery fire and must move rapidly.

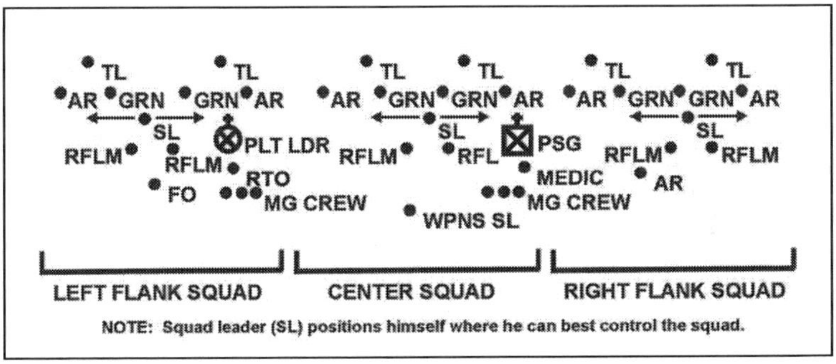

Figure 3-9. Platoon line, squads on line.

Platoon Line, Squads in Column

3-61. When two or more platoons are moving, the company commander chooses one of them as the base platoon. The base platoon's center squad is its base squad. When the platoon is not the base platoon, its base squad is its flank squad nearest the base platoon (Figure 3-10). The platoon line with squads in column formation is difficult to transition to other formations.

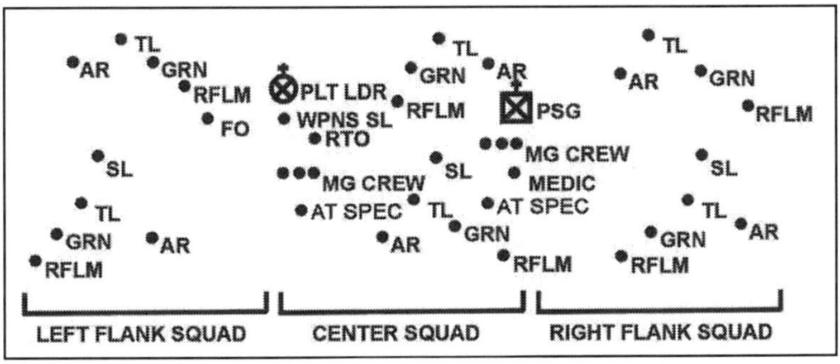

Figure 3-10. Platoon line, squads in column.

Platoon Vee

3-62. This formation has two squads up front to provide a heavy volume of fire on contact (Figure 3-11). It also has one squad in the rear that can either overwatch or trail the other squads. The platoon leader designates one of the front squads to be the platoon's base squad.

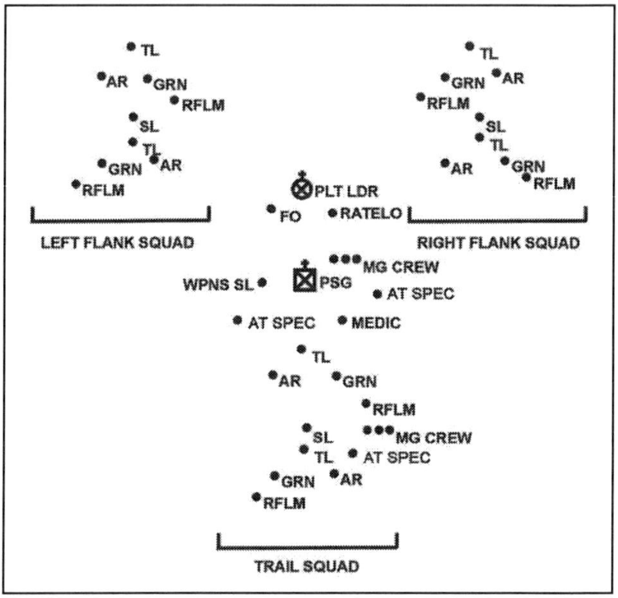

Figure 3-11. Platoon vee.

Platoon Wedge

3-63. This formation has two squads in the rear that can overwatch or trail the lead squad (Figure 3-12). The lead squad is the base squad. The wedge formation—

- Can be used with the traveling and traveling overwatch techniques.
- Allows rapid transition to bounding overwatch.

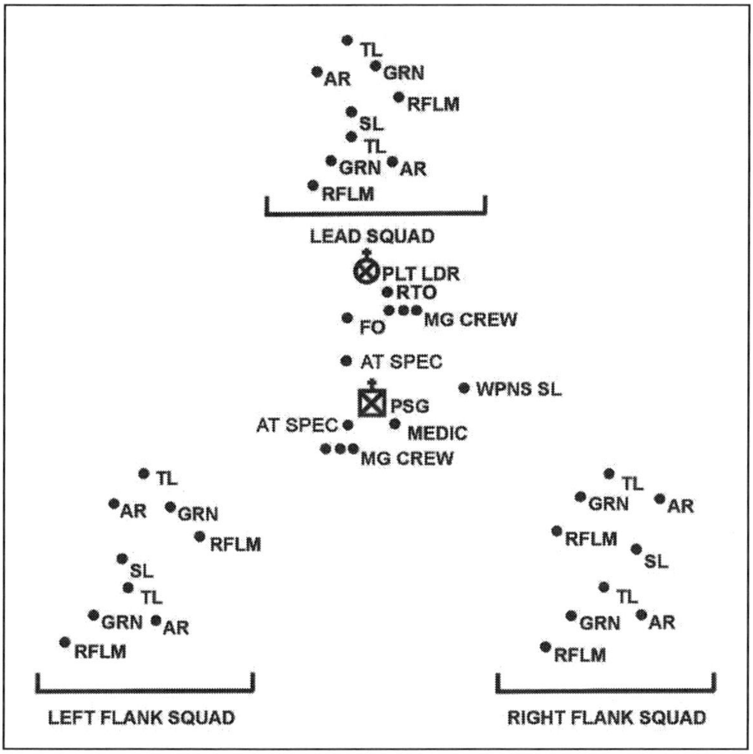

Figure 3-12. Platoon wedge.

Platoon File

3-64. This formation may be set up in several methods (Figure 3-13). One method is to have three-squad files follow one another using one of the movement techniques. Another method is to have a single platoon file with a front security element (point) and flank security elements. The distance between Soldiers is less than normal to allow communication by passing messages up and down the file. The platoon file has the same characteristics as the fire team and squad files. It is normally used for traveling only.

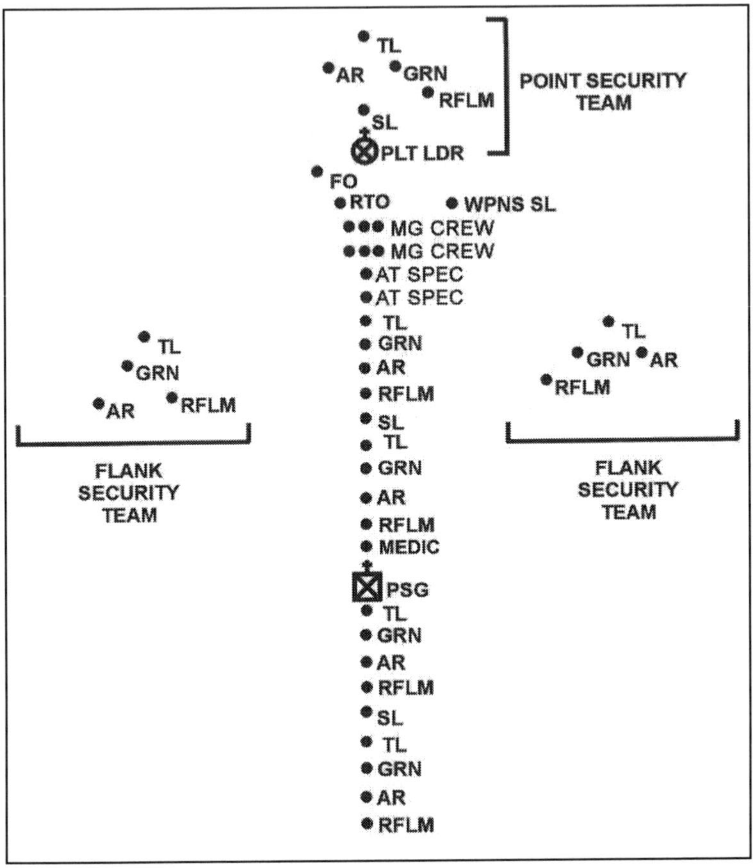

Figure 3-13. Platoon file.

SECTION III — MOVEMENT TECHNIQUES

3-65. Movement techniques are not fixed formations. They refer to the distances between Soldiers, teams, and squads that vary based on mission, enemy, terrain, visibility, and any other factor that affects control. There are three movement techniques: traveling; traveling overwatch; and bounding overwatch. The selection of a movement technique is based on the likelihood of enemy contact and the need for speed. Factors to consider for each technique are control, dispersion, speed, and security (Table 3-5). Individual movement techniques include high and low crawl, and three to five second rushes from one covered position to another (see FM 21-75, *Combat Skills of the Soldier*).

Table 3-5. Movement techniques and characteristics.

Movement Techniques	When Normally Used	CHARACTERISTICS			
		Control	Dispersion	Speed	Security
Traveling	Contact not likely	More	Less	Fastest	Least
Traveling overwatch	Contact possible	Less	More	Slower	More
Bounding overwatch	Contact expected	Most	Most	Slowest	Most

3-66. From these movement techniques, leaders are able to conduct actions on contact, making natural transitions to fire and movement as well as to conducting tactical mission tasks. When analyzing the situation, some enemy positions are known. However, most of the time enemy positions will only be likely (called templated positions). Templated positions are the leader's "best guess" based on analyzing the terrain and his knowledge of the enemy. Throughout the operation, leaders are continuously trying to confirm or deny both the known positions as well as the likely positions.

Methods of Maneuvering Subordinates

3-67. There are two methods of bounding the squads: successive; and alternate bounds. In successive bounds the lead element is always the same; in alternate bounds (called leapfrogging), the lead element changes each time (Figure 3-14).

Successive Bounds

3-68. If the platoon uses successive bounds, the lead squad, covered by the trail squad, advances and occupies a support-by-fire position. The trail squad advances to a support-by-fire position abreast of the lead squad and halts. The lead squad moves to the next position and the move continues. Only one squad moves at a time, and the trail squad avoids advancing beyond the lead squad.

Alternate Bounds

3-69. Covered by the rear squad, the lead squad moves forward, halts, and assumes overwatch positions. The rear squad advances past the lead squad and takes up overwatch positions. The initial lead squad then advances past the initial rear squad and takes up overwatch positions. Only one squad moves at a time. This method is usually more rapid than successive bounds.

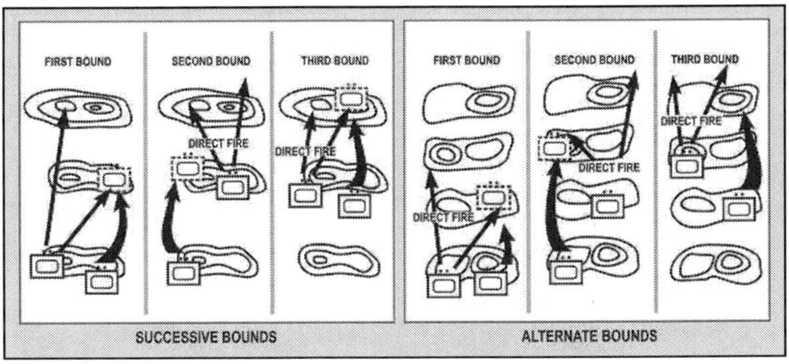

Figure 3-14. Successive and alternate bounds.

SQUAD MOVEMENT TECHNIQUES

3-70. The platoon leader determines and directs which movement technique the squad will use.

SQUAD TRAVELING

3-71. Traveling is used when contact with the enemy is not likely and speed is needed (Figure 3-15).

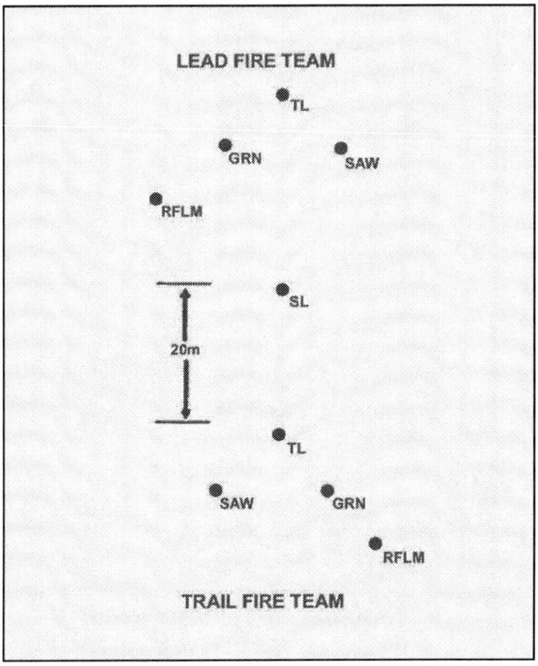

Figure 3-15. Squad traveling.

SQUAD TRAVELING OVERWATCH

3-72. Traveling overwatch is used when contact is possible. Attached weapons move near the squad leader and under his control so he can employ them quickly. Rifle squads normally move in column or wedge formation (Figure 3-16). Ideally, the lead team moves at least 50 meters in front of the rest of the element.

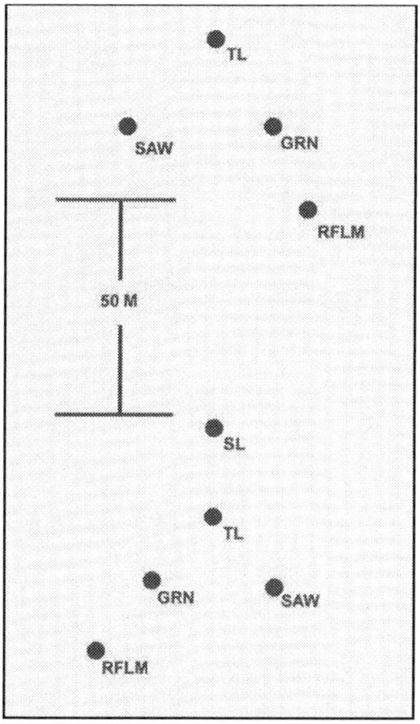

Figure 3-16. Squad traveling overwatch.

SQUAD BOUNDING OVERWATCH

3-73. Bounding overwatch is used when contact is expected, when the squad leader feels the enemy is near (based on movement, noise, reflection, trash, fresh tracks, or even a hunch), or when a large open danger area must be crossed. The lead fire team overwatches first. Soldiers in the overwatch team scan for enemy positions. The squad leader usually stays with the overwatch team. The trail fire team bounds and signals the squad leader when his team completes its bound and is prepared to overwatch the movement of the other team.

3-74. Both team leaders must know which team the squad leader will be with. The overwatching team leader must know the route and destination of the bounding team. The bounding team leader must know his team's destination and route, possible enemy locations, and actions to take when he arrives there. He must also know where the overwatching team will be and how he will receive his instructions (Figure 3-17). The cover and concealment on the bounding team's route dictates how its Soldiers move.

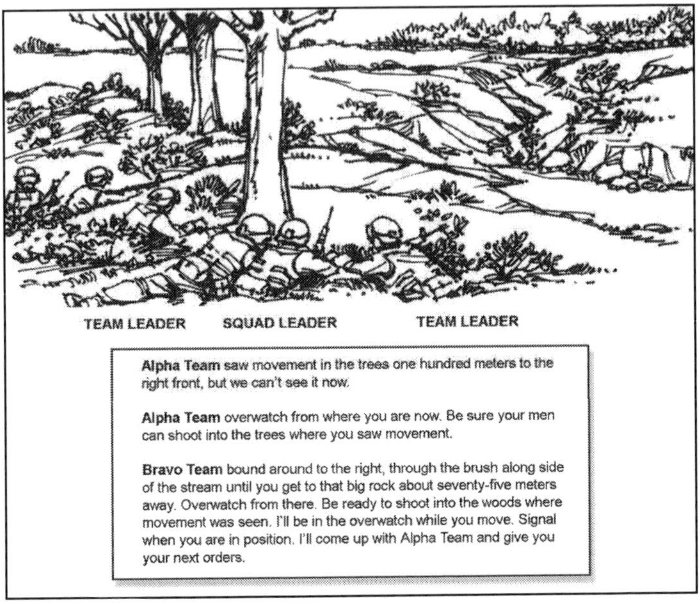

Figure 3-17. Squad bounding overwatch.

3-75. Teams can bound successively or alternately. Successive bounds are easier to control; alternate bounds can be faster (Figure 3-18).

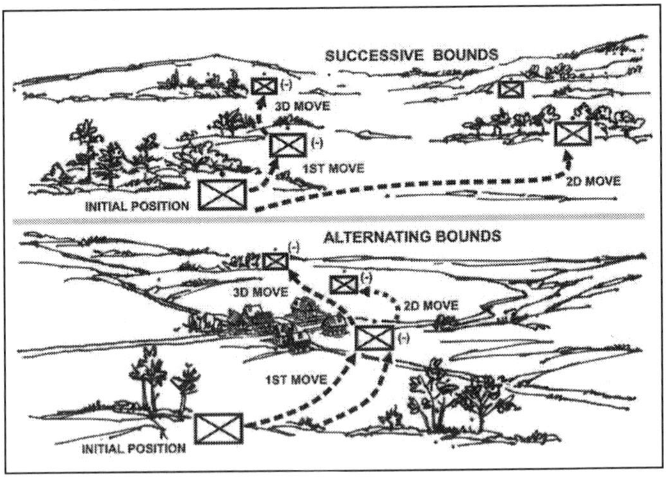

Figure 3-18. Squad successive and alternate bounds.

PLATOON MOVEMENT TECHNIQUES

3-76. The platoon leader determines and directs which movement technique the platoon will use. While moving, leaders typically separate their unit into two groups: a security element and the main body. In most scenarios, the Infantry platoon is not large enough to separate its forces into separate security forces and main body forces. However, it is able to accomplish these security functions by employing movement techniques. A movement technique is the manner a platoon uses to traverse terrain.

3-77. As the probability of enemy contact increases, the platoon leader adjusts the movement technique to provide greater security. The key factor to consider is the trail unit's ability to provide mutual support to the lead element. Soldiers must be able to see their fire team leader. The squad leader must be able to see his fire team leaders. The platoon leader should be able to see his lead squad leader.

TRAVELING

3-78. The platoon often uses the traveling technique when contact is unlikely and speed is needed (Figure 3-19). When using the traveling technique, all unit elements move continuously. In continuous movement, all Soldiers travel at a moderate rate of speed, with all personnel alert. During traveling, formations are essentially not altered except for the effects of terrain.

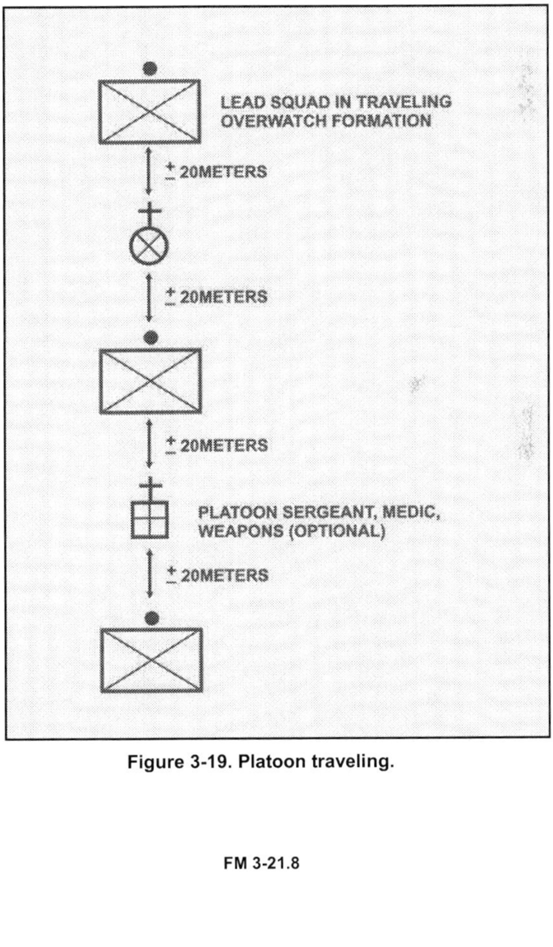

Figure 3-19. Platoon traveling.

TRAVELING OVERWATCH

3-79. Traveling overwatch is an extended form of traveling in which the lead element moves continuously but trailing elements move at varying speeds, sometimes pausing to overwatch movement of the lead element (Figure 3-20). Traveling overwatch is used when enemy contact is possible but not expected. Caution is justified but speed is desirable.

3-80. The trail element maintains dispersion based on its ability to provide immediate suppressive fires in support of the lead element. The intent is to maintain depth, provide flexibility, and sustain movement in case the lead element is engaged. The trailing elements cue their movement to the terrain, overwatching from a position where they can support the lead element if needed. Trailing elements overwatch from positions and at distances that will not prevent them from firing or moving to support the lead element. The idea is to put enough distance between the lead unit and the trail unit(s) so if the lead unit comes into contact, the trail unit(s) will be out of contact but have the ability to maneuver on the enemy.

3-81. Traveling overwatch requires the leader to control his subordinate's spacing to ensure mutual support. This involves a constant process of concentrating (close it up) and dispersion (spread it out). The primary factor is mutual support, with its two critical variables being weapon ranges and terrain. Infantry platoon's weapon range limitations dictate that units should not generally get separated by more than 300 meters. In compartmentalized terrain this distance is obviously closer while in open terrain this distance is greater.

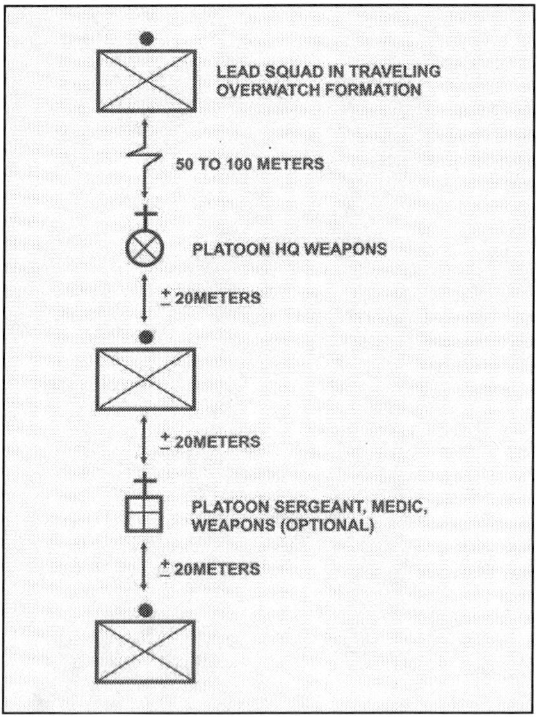

Figure 3-20. Platoon traveling overwatch.

BOUNDING OVERWATCH

3-82. Bounding overwatch is similar to fire and movement in which one unit overwatches the movement of another (Figure 3-21). The difference is there is no actual enemy contact. Bounding overwatch is used when the leader expects contact. The key to this technique is the proper use of terrain. Subordinate units fall into one of three categories: bounding, overwatching, or awaiting orders.

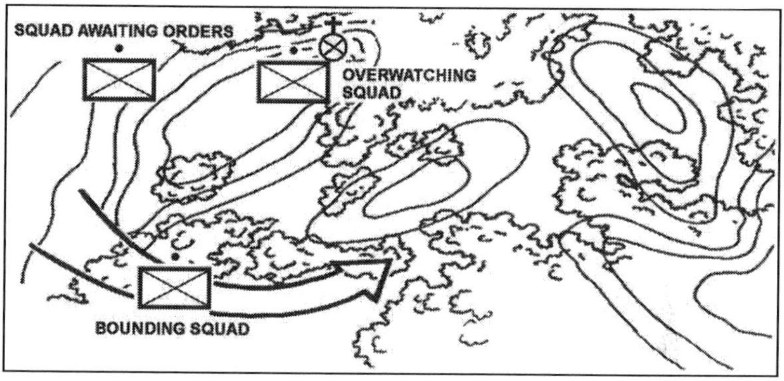

Figure 3-21. Platoon bounding overwatch.

One Squad Bounding

3-83. One squad bounds forward to a chosen position; it then becomes the overwatching element unless contact is made en route. The bounding squad can use traveling overwatch, bounding overwatch, or individual movement techniques (low and high crawl, and three to five second rushes by fire team or pairs).

3-84. Factors of METT-TC dictate the length of the bounds. However, the bounding squad(s) should never move beyond the range at which the base-of-fire squad(s) can effectively suppress known, likely, or suspected enemy positions. In severely restrictive terrain, the bounding squad(s) makes shorter bounds than it would in more open areas. The destination of the bounding element is based on the suitability of the next location as an overwatch position. When deciding where to send his bounding squad, a platoon leader considers—

- The requirements of the mission.
- Where the enemy is likely to be.
- The routes to the next overwatch position.
- The ability of an overwatching element's weapons to cover the bound.
- The responsiveness of the rest of the platoon.

One Squad Overwatching

3-85. One squad overwatches the bounding squad from covered positions and from where it can see and suppress likely enemy positions. The platoon leader remains with the overwatching squad. Normally the platoon's machine guns are located with the overwatching squad.

One Squad Awaiting Orders

3-86. Based on the situation, one squad is uncommitted and ready for employment as directed by the platoon leader. The platoon sergeant and the leader of the squad awaiting orders position themselves close to the platoon leader. On contact, this unit(s) should be prepared to support the overwatching element, move to assist the bounding squad, or move to another location based on the platoon leader's assessment.

Weapons Squad

3-87. Machine guns are normally employed in one of two ways:

- Attached to the overwatch squad or the weapons squad that supports the overwatch element.
- Awaiting orders to move (with the platoon sergeant [PSG]) or as part of a bounding element.

Command and Control of the Bounding Element

3-88. Ideally, the overwatch element maintains visual contact with the bounding element. However, the leader of the overwatch element may have the ability to digitally track the location of the bounding element without maintaining visual contact. This provides the bounding element more freedom in selecting covered and concealed routes to its next location. Before a bound, the platoon leader gives an order to his squad leaders from the overwatch position (Figure 3-22). He tells and shows them the following:

- The direction or location of the enemy (if known).
- The positions of the overwatching squad.
- The next overwatch position.
- The route of the bounding squad.
- What to do after the bounding squad reaches the next position.
- What signal the bounding squad will use to announce it is prepared to overwatch.
- How the squad will receive its next orders.

First Squad, overwatch from this position.

Second Squad, move through those trees to the left and clear that small hill one hundred meters to our front and set up an overwatch on the hill. Signal with your mirror when in position. I will move up with the first and third squads, and give you your next orders.

Third Squad, move up behind first squad and await orders.

Platoon Sergeant, position your machine guns and antitank weapon to the right of first squad. I will position mine to the left of first squad.

Figure 3-22. Example of platoon leader's orders for bounding overwatch.

SECTION IV — ROUTE SELECTION AND NAVIGATION

3-89. During planning and preparation for tactical movement, platoon leaders analyze the terrain from two perspectives. First, they analyze the terrain to see how it can provide tactical advantage, both to friendly and enemy forces. Second, they look at the terrain to determine how it can aid navigation. Leaders identify any areas or terrain features that dominate their avenue of approach. These areas are almost always considered key terrain and provide the unit possible intermediate and final objectives.

3-90. Ideally, the leader identifies along his route not only ground that is good for navigation, but also ground that facilitates destroying the enemy should contact occur. If the leader wants to avoid contact, he chooses terrain that will hide the unit. If he wants to make contact, he chooses terrain from where he can more easily scan and observe the enemy. On other occasions, the leader may require terrain that allows stealth or speed. Regardless of the requirement, the leader must ensure that most of the terrain along his route provides some tactical advantage.

3-91. Route Selection and Navigation are made easier with the aid of technology. Global Positioning System (GPS) devices or Force XXI Battle Command Brigade and Below Systems (FBCB2) enhance the Infantry platoon's ability to ensure they are in the right place at the right time and to determine the location of adjacent units.

NAVIGATION AIDS

3-92. There are two categories of navigational aids: linear; and point. Linear navigational aids are terrain features such as trails, streams, ridgelines, woodlines, power lines, streets, and contour lines. Point terrain features include hilltops, and prominent buildings. Navigation aids are usually assigned control measures to facilitate communication during the movement. Typically, linear features are labeled as phase lines while point features are labeled as checkpoints (or rally points). There are three primary categories of navigation aids: catching features; handrails; and navigational attack points.

CATCHING FEATURES

3-93. Catching features are obvious terrain features that go beyond a waypoint or control measure and can be either linear or point. The general idea is that if the unit moves past its objective, limit of advance, or checkpoint, the catching feature will alert them that they have traveled too far.

The Offset-Compass Method

3-94. If there is the possibility of missing a particular point along the route (such as the endpoint or a navigational attack point), it is sometimes preferable to deliberately aim the leg to the left or right of the end point toward a prominent catching feature. Once reached, the unit simply turns the appropriate direction and moves to the desired endpoint. This method is especially helpful when the catching feature is linear.

Boxing-In the Route

3-95. One of the techniques leaders can use to prevent themselves from making navigational errors is to "box in" the leg or the entire route. This method uses catching features, handrails, and navigational attack points to form boundaries. Creating a box around the leg or route assists in more easily recognizing and correcting deviation from the planned leg or route.

HANDRAILS

3-96. Handrails are linear features parallel to the proposed route. The general idea is to use the handrail to keep the unit oriented in the right direction. Guiding off of a handrail can increase the unit's speed while also acting as a catching feature.

NAVIGATIONAL ATTACK POINTS

3-97. Navigational attack points are an obvious landmark near the objective, limit of advance, or checkpoint that can be easily found. Upon arriving at the navigational attack point, the unit transitions from rough navigation (terrain association or general azimuth navigation) to point navigation (dead reckoning). Navigational attack points are typically labeled as checkpoints.

ROUTE PLANNING

3-98. Route planning must take into account enabling tasks specific to tactical movement. These tasks facilitate the overall operation. Tactical movement normally contains some or all of the following enabling tasks:

- Planning movement with GPS waypoints.
- Movement to and passage of friendly lines.
- Movement to an objective rally point (ORP).
- Movement to a phase line of deployment.
- Movement to a limit of advance.
- Linkup with another unit.
- Movement to a patrol base or assembly area.
- Movement back to and reentry of friendly lines.

3-99. Leaders first identify where they want to end up (the objective or limit of advance). Then, working back to their current location, they identify all of the critical information and actions required as they relate to the route. For example, navigational aids, tactical positions, known and templated enemy positions, and friendly control measures. Using this information, they break up their route in manageable parts called legs. Finally, they capture their information and draw a sketch on a route chart. There are three decisions that leaders make during route planning:

(1) The type of (or combination of) navigation to use.
(2) The type of route during each leg.
(3) The start point and end point of each leg.

3-100. The leader assesses the terrain in his proposed area of operation. In addition to the standard Army map, the leader may have aerial photographs and terrain analysis overlays from the parent unit, or he may talk with someone familiar with the area.

3-101. To control movement, leaders use axes of advance, directions of attack, infiltration lanes, phase lines, probable lines of deployment, checkpoints (waypoints), final coordination lines, rally points, assembly areas, and routes.

TYPES OF NAVIGATION

3-102. There are three types of navigation: terrain association; general azimuth method; and point navigation. Leaders use whichever type or combination best suits the situation.

TERRAIN ASSOCIATION

3-103. Terrain association is the ability to identify terrain features on the ground by the contour intervals depicted on the map. The leader analyzes the terrain using the factors of OAKOC and identifies major terrain features, contour changes, and man-made structures along his axis of advance. As the unit moves, he uses these features to orient the unit and to associate ground positions with map locations. The major advantage of terrain association is that it forces the leader to continually assess the terrain. This leads to identifying tactically-advantageous terrain and using terrain to the unit's advantage.

General Azimuth Method

3-104. For this method, the leader selects linear terrain features; then while maintaining map orientation and a general azimuth, he guides on the terrain feature. Advantages of the general azimuth method are that it speeds movement, avoids fatigue, and often simplifies navigation because the unit follows the terrain feature. The disadvantage is that it usually puts the unit on a natural line of drift. This method should end like terrain association, with the unit reaching a catching feature or a navigational attack point, then switching to point navigation.

Point Navigation

3-105. Point navigation, also called dead reckoning, is done by starting from a known point and then strictly following a predetermined azimuth and distance. This form of navigation requires a high level of leader control because even a slight deviation over the course of a movement can cause navigation errors. This method uses the dismounted compass and a distance from the pace man (or a vehicle's odometer when mounted) to follow a prescribed route. Point navigation requires the leader to follow these steps:

- Use the compass to maintain direction.
- Use the pace man's pace or a vehicle odometer to measure the distance traveled for each leg or part.
- Review the written description of the route plan to help prevent navigational errors.

3-106. When performed correctly, point navigation is very reliable, but time consuming. It is best used when the need for navigational accuracy outweighs the importance of using terrain. Point navigation is particularly useful when recognizable terrain features do not exist or are too far away to be helpful. For example, deserts, swamps, and thick forest make terrain association difficult. Using point navigation early on in a long movement can stress the compass man and it may be advisable to switch him. One of the problems with point navigation is negotiating severely restrictive terrain or danger areas.

Combinations

3-107. Leaders can benefit from combining the three types of navigation. Terrain association and the general azimuth method enable leaders to set a rough compass bearing and move as quickly as the situation allows toward a catching feature or a navigational attack point. Once reached, leaders switch to point navigation by paying extremely close attention to detail, taking as much time as necessary to analyze the situation and find their point. Terrain association and the general azimuth method allow for some flexibility in the movement, and therefore do not require the same level of control as point navigation. Point navigation, on the other hand, enables leaders to precisely locate their objective or point.

ROUTE TYPES

3-108. There are three types of routes leaders can choose from: those that follow linear terrain features; those that follow a designated contour interval; and those that go cross compartment. Terrain association can be used with all three route types. The general azimuth method is used with the contour and terrain feature method. Point navigation is used primarily with cross compartment.

Terrain Feature

3-109. Following a terrain feature is nothing more than moving along linear features such as ridges, valleys, and streets. The advantage of this method is that the unit is moving with the terrain. This is normally the least physically taxing of the methods. The disadvantage is that following terrain features also means following natural lines of drift, which leads to a higher probability of chance contact with the enemy.

Contouring

3-110. Contouring (remaining at the same height for the entire leg) follows the imaginary contour line around a hill or along a ridgeline. Contouring has two advantages. First, it prevents undue climbing or descending. Second, following the contour acts as handrail or catching feature. The disadvantage of contouring is that it can be physically taxing.

Cross Compartment

3-111. Cross compartment means following a predetermined azimuth and usually means moving against the terrain. The advantage of this method is that it provides the most direct route from the start point to the end point of the leg or route. There are two primary disadvantages to this type of route. First, this method can be physically taxing. Second, the unit might expose itself to enemy observation.

DEVELOP A LEG

3-112. The best way to manage a route is to divide it into segments called "legs." By breaking the overall route into several smaller segments, the leader is able to plan in detail. Legs typically have only one distance and direction. A change in direction usually ends the leg and begins a new one.

3-113. A leg must have a definite beginning and ending, marked with a control measure such as a checkpoint or phase line. (When using GPS, these are captured as waypoints.) When possible, the start point and end point should correspond to a navigational aid (catching feature or navigational attack point).

3-114. To develop a leg, leaders first determine the type of navigation and route that best suits the situation. Once these two decisions are made, the leader determines the distance and direction from the start point to the end point. He then identifies critical METT-TC information as it relates to that specific leg. Finally, leaders capture this information and draw a sketch on a route chart (Figure 3-23).

LEG	AZIMUTH/ DISTANCE	KEY INFORMATION
Leg 1: ATK PSN RED to CP 1 •CP 1 is a trail intersection. •Rally point #1 in effect.	150' / 800m	• O - Potential enemy EA vicinity Cp1. • C - Poor cover and concealment • O - Restricted movement throughout • K - N/A • A - 2 hard trails along leg
Leg 2: CP 1 to CP 2 •CP 2 is hill 213. •Rally point #2 in effect.	75' / 650m	• O - Potential enemy EA vicinity hill 213. • C - Good cover and concealment throughout • O - Restricted movement throughout; seasonal stream vicinity Nv 123094 • K - N/A • A - N/A
Leg 3: CP 2 to PL RED •PL Red 1 is a trail intersection. •Rally point #3 in effect.	90' / 900m	PL RED is PLD • O - Potential enemy EA vicinity PL BLUE. • C - Poor @ PL BLUE • O - Enemy obstacle NV 131950 • K - N/A • A - N/A
Leg 4: PL RED to OBJ PIG •OBJ PIG is a hill top. •ORP is rally point.	65' / 400m	ORP is Nv 134954 • O - Potential enemy EA vicinity PL BLUE. • C - Poor cover and concealment • O - Restricted movement throughout • K - OBJ PIG • A - 1 hardball road 100m from OBJ

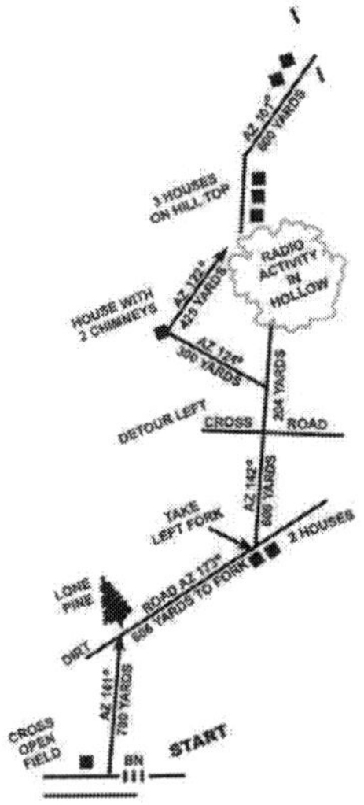

Figure 3-23. Sketch of legs example.

EXECUTE THE ROUTE

3-115. Using decisions about the route and navigation made during planning and preparation, leaders execute their route and direct their subordinates. In addition to executing the plan, leaders—

- Determine and maintain accurate location.
- Designate rally points.

DETERMINE LOCATION

3-116. Leaders must always know their units location during movement. Without accurate location, the unit cannot expect to receive help from supporting arms, integrate reserve forces, or accomplish their mission. To ensure accurate location, leaders use many techniques, including:

- Executing common skills.
- Designating a compass man and pace man.
- Using GPS / FBCB2.

Common Skills

3-117. All Infantrymen, particularly leaders, must be experts in land navigation. Important navigation tasks common to all include—

- Locating a point using grid coordinates. Using a compass (day/night).
- Determining location using resection, intersection, or modified resection.
- Interpreting terrain features.
- Measuring distance and elevation.
- Employing a GPS / FBCB2.

Compass Man

3-118. The compass man assists in navigation by ensuring the lead fire team leader remains on course at all times. The compass man should be thoroughly briefed. His instructions must include an initial azimuth with subsequent azimuths provided as necessary. The platoon or squad leader also should designate an alternate compass man. The leader should validate the patrol's navigation with GPS devices.

Pace Man

3-119. The pace man maintains an accurate pace at all times. The platoon or squad leader should designate how often the pace man is to report the pace. The pace man should also report the pace at the end of each leg. The platoon or squad leader should designate an alternate pace man.

Global Positioning Systems

3-120. GPSs receive signals from satellites or land-based transmitters. They calculate and display the position of the user in military grid coordinates as well as in degrees of latitude and longitude. During planning, leaders enter their waypoints into the GPS. Once entered, the GPS can display information such as distance and direction from waypoint to waypoint. During execution, leaders use the GPS to establish their exact location.

NOTE: Leaders need to remember that GPS and digital displays are not the only navigational tools they can use. The best use of GPS or digital displays is for confirming the unit's location during movement. Terrain association and map-reading skills are still necessary skills, especially for point navigation. Over reliance on GPS and digital displays can cause leaders to ignore the effects of terrain, travel faster than conditions allow, miss opportunities, or fail to modify routes when necessary.

DESIGNATE RALLY POINTS

3-121. A rally point is a place designated by the leader where the unit moves to reassemble and reorganize if it becomes dispersed. It can also be a place for a temporarily halt to reorganize and prepare for actions at the objective, to depart from friendly lines, or to reenter friendly lines (FM 1-02). Planned and unplanned rally points are common control measures used during tactical movement. Planned rally points include objective rally point(s) (ORP), initial rally point(s) (IRP), and reentry rally point(s) (RRP). Unplanned rally points are enroute rally points, near side rally points, and far side rally points. Despite the different types of rally points, the actions that occur there are generally the same.

3-122. Prior to departing, leaders designate tentative rally points and determine what actions will occur there. When occupying a rally point, leaders use a perimeter defense to ensure all-around security. Those rally points used to reassemble the unit after an event are likely to be chaotic scenes and will require immediate actions by whatever Soldiers happen to arrive. These actions and other considerations are listed in Table 3-6.

Table 3-6. Actions at rally point.

Rally Points	Soldier Actions at an RP	Other Considerations
Select a rally point that— • Is easily recognized. • Is large enough for the unit to assemble. • Offers cover and concealment. • Is defensible for a short time. • Is away from normal movement routes and natural lines of drift. Designate a rally point by one of the following three ways: • Physically occupy it for a short period. • Use hand-and-arm signals (either pass by at a distance or walk through). • Radio communication.	• Establish security. • Reestablish the chain of command. • Account for personnel and equipment status. • Determine how long to wait until continuing the unit's mission or linkup at a follow-on RP. • Complete last instructions.	• Travel time and distance. • Maneuver room needed. • Adjacent unit coordination requirements. • Line of sight and range requirements for communication equipment. • Trafficability and load bearing capacity of the soil (especially when mounted). • Ability to surprise the enemy. • Ability to prevent being surprised by the enemy. • Energy expenditure of Soldiers and condition they will be in at the end of the movement.

SECTION V — ACTIONS AT DANGER AREAS

3-123. When analyzing the terrain (in the METT-TC analysis) during the TLP, the platoon leader may identify danger areas. When planning the route, the platoon leader marks the danger areas on his overlay. The term *danger area* refers to any area on the route where the terrain could expose the platoon to enemy observation, fire, or both. If possible, the platoon leader plans to avoid danger areas, but sometimes he cannot. When the unit must cross a danger area, it does so as quickly and as carefully as possible. During planning, the leader designates near-side and far-side rally points. If the platoon encounters an unexpected danger area, it uses the en route rally points closest to the danger area as far-side and near-side rally points. Examples of danger areas include—

- **Open Areas.** Conceal the platoon on the near side and observe the area. Post security to give early warning. Send an element across to clear the far side. When cleared, cross the remainder of the platoon at the shortest exposed distance and as quickly as possible.
- **Roads and Trails.** Cross roads or trails at or near a bend, a narrow spot, or on low ground.

- **Villages.** Pass villages on the downwind side and well away from them. Avoid animals, especially dogs, which might reveal the presence of the platoon.
- **Enemy Positions.** Pass on the downwind side (the enemy might have scout dogs). Be alert for trip wires and warning devices.
- **Minefields.** Bypass minefields if at all possible, even if it requires changing the route by a great distance. Clear a path through minefields only if necessary.
- **Streams.** Select a narrow spot in the stream that offers concealment on both banks. Observe the far side carefully. Emplace near- and far-side security for early warning. Clear the far side and then cross rapidly but quietly.
- **Wire Obstacles.** Avoid wire obstacles (the enemy covers obstacles with observation and fire).

CROSSING OF DANGER AREAS

3-124. Regardless of the type of danger area, when the platoon must cross one independently, or as the lead element of a larger force, it must perform the following:

- When the lead team signals "danger area" (relayed throughout the platoon), the platoon halts.
- The platoon leader moves forward, confirms the danger area, and determines what technique the platoon will use to cross. The platoon sergeant also moves forward to the platoon leader.
- The platoon leader informs all squad leaders of the situation and the near-side and far-side rally points.
- The platoon sergeant directs positioning of the near-side security (usually conducted by the trail squad). These two security teams may follow him forward when the platoon halts and a danger area signal is passed back.
- The platoon leader reconnoiters the danger area and selects the crossing point that provides the best cover and concealment.
- Near-side security observes to the flanks and overmatches the crossing.
- When the near-side security is in place, the platoon leader directs the far-side security team to cross the danger area.
- The far-side security team clears the far side.
- The far-side security team leader establishes an observation post forward of the cleared area.
- The far-side security team signals to the squad leader that the area is clear. The squad leader relays the message to the platoon leader.
- The platoon leader selects the method the platoon will use to cross the danger area.
- The platoon quickly and quietly crosses the danger area.
- Once across the danger area, the main body begins moving slowly on the required azimuth.
- The near-side security element, controlled by the platoon sergeant, crosses the danger area where the platoon crossed. They may attempt to cover any tracks left by the platoon.
- The platoon sergeant ensures everyone crosses and sends up the report.
- The platoon leader ensures accountability and resumes movement at normal speed.

NOTE: The same principles stated above are used when crossing a smaller unit (such as a squad) across a danger area.

3-125. The platoon leader or squad leader decides how the unit will cross based on the time he has, size of the unit, size of the danger area, fields of fire into the area, and the amount of security he can post. An Infantry platoon or squad may cross all at once, in buddy teams, or one Soldier at a time. A large unit normally crosses its elements one at a time. As each element crosses, it moves to an overwatch position or to the far-side rally point until told to continue movement.

CROSSING OF LINEAR DANGER AREAS (PLATOON)

3-126. A linear danger area is an area where the platoon's flanks are exposed along a relatively narrow field of fire. Examples include streets, roads, trails, and streams. The platoon crosses a linear danger area in the formation and location specified by the platoon leader (Figure 3-24).

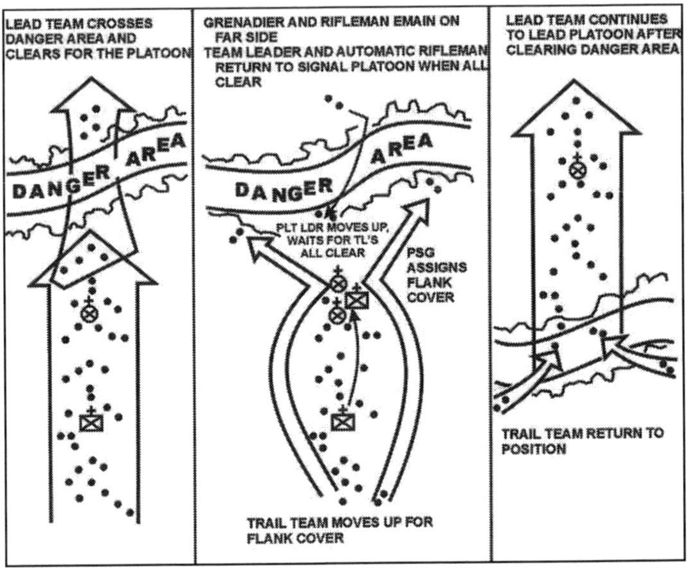

Figure 3-24. Crossing a linear danger area.

CROSSING OF LARGE OPEN AREAS

3-127. If the large open area is so large that the platoon cannot bypass it due to the time needed to accomplish the mission, a combination of traveling overwatch and bounding overwatch is used to cross the large open area (Figure 3-25). The traveling overwatch technique is used to save time. The squad or platoon moves using the bounding overwatch technique at any point in the open area where enemy contact may be expected. The technique may also be used once the squad or platoon comes within range of enemy small-arms fire from the far side (about 250 meters). Once beyond the open area, the squad or platoon re-forms and continues the mission.

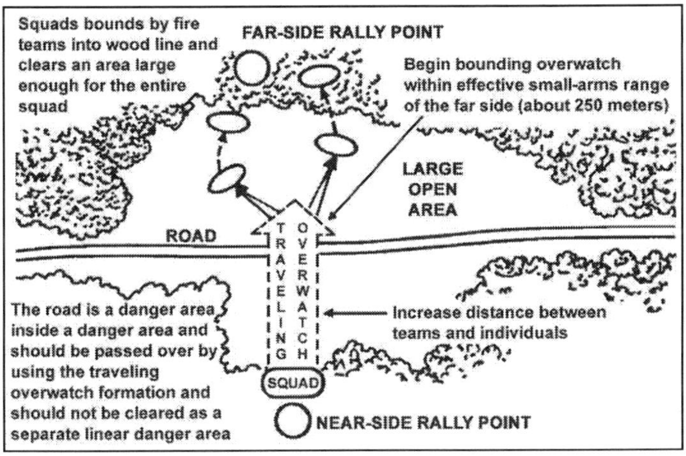

Figure 3-25. Crossing a large open area.

CROSSING OF SMALL OPEN AREAS

3-128. Small open areas are small enough to bypass in the time allowed for the mission. Two techniques can be used (Figure 3-26).

Contouring Around the Open Area

3-129. The leader designates a rally point on the far side with the movement azimuth. He then decides which side of the open area to contour around (after considering the distance, terrain, cover and concealment), and moves around the open area. He uses the wood line and vegetation for cover and concealment. When the squad or platoon arrives at the rally point on the far side, the leader reassumes the azimuth to the objective area and continues the mission (Figure 3-26).

Detour Bypass Method

3-130. The squad or platoon turns 90 degrees to the right or left around the open area and moves in the direction of travel. Once the squad or platoon has passed the danger area, the unit completes the box with another 90-degree turn and arrives at the far-side rally point, then continues the mission. The pace count of the offset and return legs is not added to the distance of the planned route (Figure 3-26).

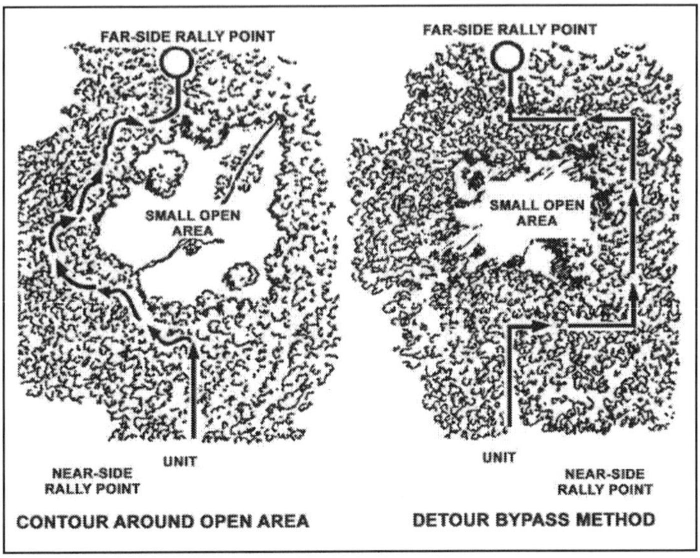

Figure 3-26. Crossing a small open area.

ENEMY CONTACT AT DANGER AREAS

3-131. An increased awareness of the situation helps the platoon leader control the platoon when it makes contact with the enemy. If the platoon makes contact in or near the danger area, it moves to the designated rally points. Based on the direction of enemy contact, the leader still designates the far- or near-side rally point. During limited visibility, he can also use his laser systems to point out the rally points at a distance. If the platoon has a difficult time linking up at the rally point, the first element to arrive should mark the rally point with an infrared light source. This will help direct the rest of the platoon to the location. During movement to the rally point, position updates allow separated elements to identify each other's locations. These updates help them link up at the rally point by identifying friends and foes.

SECTION VI — MOVEMENT WITH COMBAT VEHICLES

3-132. There are several options available to the platoon leader when augmented with vehicles. The platoon leader should employ the vehicles in conjunction with the rifle squads so each complements the other. Some options include—

- Employ them to support the Infantry rifle squads.
- Employ them separately to provide heavy direct fires or antiarmor fires.
- Leave in hide positions.
- Displace them to a secure location.

COMBAT VEHICLE AND INFANTRY SQUAD FORMATIONS

3-133. The principles of METT-TC guide the leader in selecting formations for combat vehicles and Infantry. The same principles for selecting combat formations with Infantrymen apply when selecting combat formations for combat vehicles moving with Infantrymen. The platoon leader can employ a variety of formations to meet the needs of his mission. The column, line, echelon, vee, and wedge are fundamental movement formations for combat vehicles.

3-134. After the leader combines the mounted and Infantry elements into one combat formation, it is his responsibility to ensure proper communication and fire control measures are implemented to maximize lethality and prevent fratricide.

3-135. After selecting the combat formations for the combat vehicles and Infantry, the leader can decide whether to lead with combat vehicles, Infantrymen, or a combination of the two. The default technique is to lead with Infantrymen.

LEAD WITH INFANTRY

3-136. Infantrymen are better suited for leading combat formations (Figure 3-27) when—

- A route leads through restrictive urban or rural terrain
- Stealth is desired.
- Enemy antitank minefields are templated.
- Enemy antitank teams are templated.

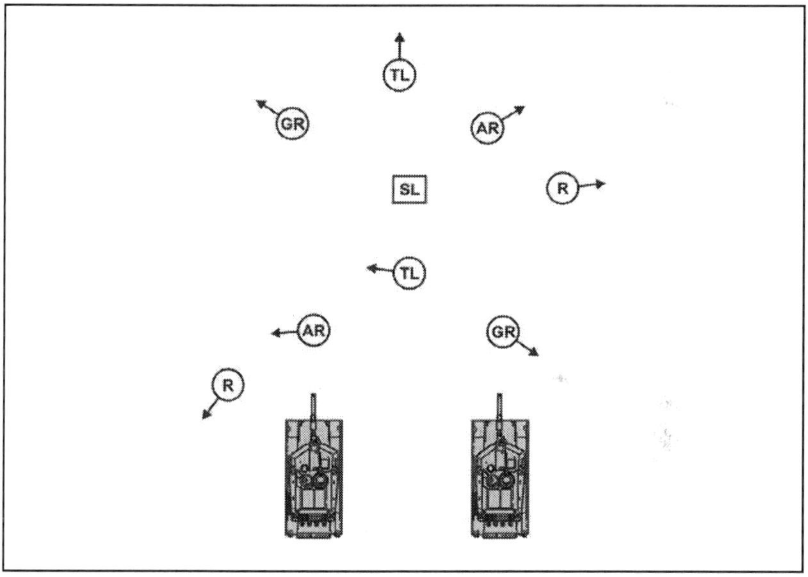

Figure 3-27. Lead with Infantry squad.

LEAD WITH COMBAT VEHICLES

3-137. Infantry leaders may choose to lead with combat vehicles (Figure 3-28) when—

- There is an armored or tank threat.
- Moving through open terrain with limited cover or concealment.
- There is a confirmed enemy location/direction.
- There are templated enemy antipersonnel minefields.

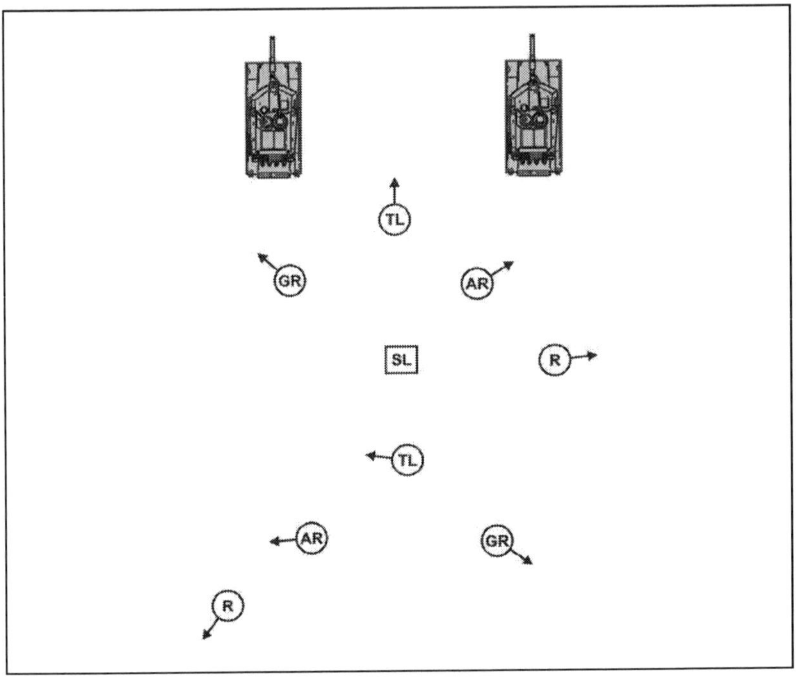

Figure 3-28. Lead with combat vehicles.

LEAD WITH BOTH COMBAT VEHICLES AND INFANTRY

3-138. Infantry leaders may choose to centrally locate the combat vehicles in their formation (Figure 3-29) when—

- Flexibility is desired.
- The enemy location is unknown.
- There is a high threat of dismounted enemy antitank teams.
- The ability to mass the fires of the combat vehicles quickly in all directions is desired.

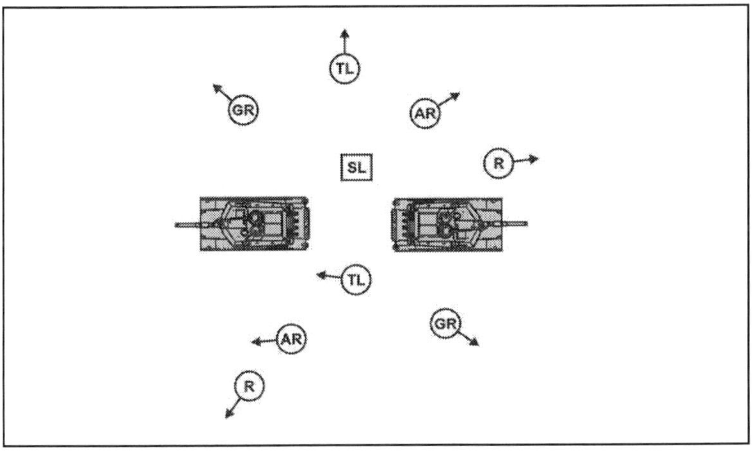

Figure 3-29. Lead with both combat vehicles and Infantry squad.

COMBAT VEHICLE AND INFANTRY PLATOON FORMATIONS

3-139. Infantry platoons can also incorporate their formations with those of combat vehicular units. The principles for choosing platoon combat formations are the same as squad combat formations. The Infantry platoon can conduct tactical movement with a platoon of combat vehicles (normally four) or a section of combat vehicles (normally two). Figures 3-30 and 3-31 detail some basic Infantry platoon formations with combat vehicle platoon formations.

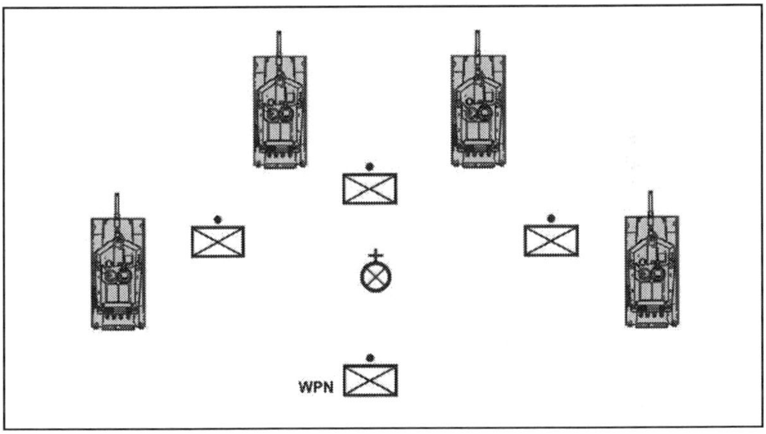

Figure 3-30. Combat vehicle wedge, Infantry platoon diamond.

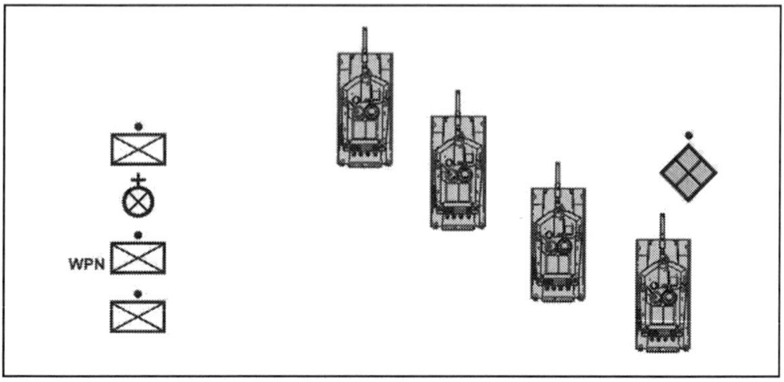

Figure 3-31. Combat vehicle echelon right, Infantry platoon column.

MOUNTED TACTICAL MOVEMENT

3-140. Mounted movement is very similar to dismounted movement. Depending on the vehicle type, a platoon may have a squad in one to four vehicles. Units with more than four vehicles should consider splitting the vehicles into two or more sections and control these sections much the same way squads control their teams.

3-141. Units augmented with four or more vehicles can use any of the seven formations. They use them within the context of the three movement techniques (see Section III) and should be prepared to execute immediate action drills when transitioning to maneuver. When the mounted unit stops, they use the coil and herringbone formations to ensure security.

3-142. In mounted successive bounds, vehicles keep their relative positions in the column. The first and second vehicles operate as a section in moving from one observation point to another. The second vehicle is placed in a concealed position, occupants dismounting if necessary, to cover movement of the first vehicle to an observation point. On reaching this point, occupants of the first vehicle observe and reconnoiter, dismounting if necessary. When the area is determined to be clear, the second vehicle is signaled forward to join the first vehicle. The commander of the first vehicle observes the terrain to the front for signs of enemy forces and selects the next stopping point. The first vehicle then moves out and the process is repeated. Movement distance of the lead vehicle does not exceed the limit of observation or the range of effective fire support from the second vehicle. The lead vehicle and personnel are replaced frequently to ensure constant alertness. The other vehicles in the column move by bounds from one concealed position to another. Each vehicle maintains visual contact with the vehicle ahead but avoids closing up (Figure 3-32). However, as a rule, vehicles always work in pairs and should never be placed in a situation where one vehicle is not able to be supported by the second.

3-143. In mounted alternate bounds, all except the first two vehicles keep their relative places in the column. The first two vehicles alternate as lead vehicles on each bound. Each covers the bound of the other. This method provides more rapid advance than movement by successive bounds, but is less secure. Security is obtained by the vehicle commander who assigns each Soldier a direction of observation (to the front, flank[s], or rear). This provides each vehicle with some security against surprise fire from every direction, and provides visual contact with vehicles to the front and rear.

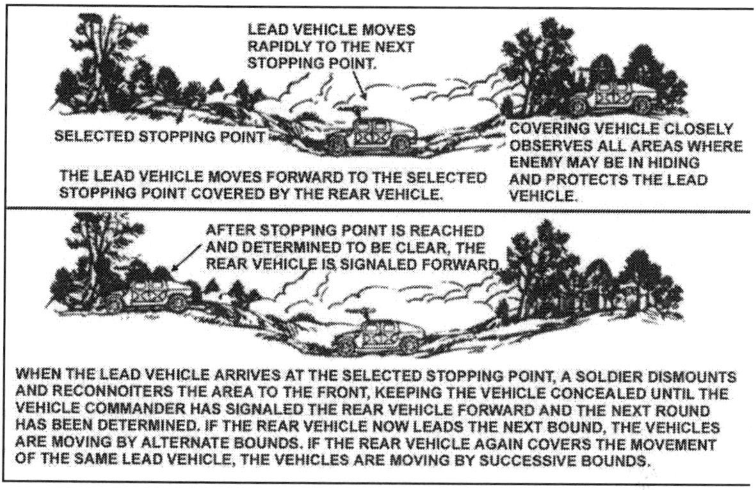

Figure 3-32. Lead vehicle moving by bounds.

CONVOYS

3-144. A convoy is a group of vehicles organized for the purpose of control and orderly movement with or without escort protection that moves over the same route at the same time under one commander (FM1-02).

3-145. The platoon conducts motor marches, usually in trucks. Some of the special considerations may include—

- **Protection.** Sandbag the bottom of the trucks to protect from mines. Ensure crew-served weapons are manned with qualified gunners.
- **Observation.** Ensure Soldiers sit facing outward and remove bows and canvas to allow 360-degree observation and rapid dismount.
- **Inspection.** Inspect vehicles and drivers to ensure they are ready. Perform before, during, and after preventive maintenance checks and services (PMCS). Ensure drivers' knowledge of the route, speed, and convoy distance.
- **Loading.** Keep fire team, squad, and platoon integrity when loading vehicles. Fire teams and squads are kept intact on the same vehicle. Platoon vehicles are together in the same march serial. Key weapons and equipment are crossloaded with platoon leaders and platoon sergeants in different vehicles.
- **Rehearsals.** Rehearse immediate action to enemy contact (near and far ambushes, air attack). Ensure drivers know what to do.
- **Air Guards.** Post air guards for each vehicle, with special consideration on the placement of crew served weapons.

ACTIONS AT DANGER AREAS (MOUNTED)

3-146. Infantry platoons must be prepared to negotiate danger areas when mounted. The discussion of leader and unit action are deliberately generic because of the wide variety of scenarios in which leaders might find themselves.

3-147. When moving mounted, units normally travel on roads, trails, and in unrestrictive terrain. Mounted units are typically vulnerable in the type of terrain favored by Infantry such as restrictive and close terrain. In addition, areas such as bridges, road junctions, defiles, and curves (that deny observation beyond the turn) are also considered danger areas. When leaders identify a danger area, they determine the appropriate movement technique to employ (traveling, traveling overwatch, or bounding overwatch). They then dismount their Infantry squads and clear the area or do a combination of both.

3-148. If time and terrain permit, the unit should either bypass a danger area or dismount Infantry to reconnoiter and clear it. However, the distances between covered and concealed positions may make this impractical. If time constraints prevent these options, the unit uses a combination of traveling overwatch and bounding overwatch to negotiate the danger area. As with dismounted actions at a danger area, the leader must be prepared to quickly transition to maneuver in case the unit makes contact with the enemy.

Mounted Traveling Overwatch

3-149. The lead element moves continuously along the covered and concealed routes that give it the best available protection from possible enemy observation and direct fire (Figure 3-33). The trail element moves at variable speeds providing continuous overwatch, keeping contact with the lead element, and stopping periodically to get a better look. The trail element stays close enough to ensure mutual support for the lead element. However, it must stay far enough to the rear to retain freedom of maneuver in case an enemy force engages the lead element.

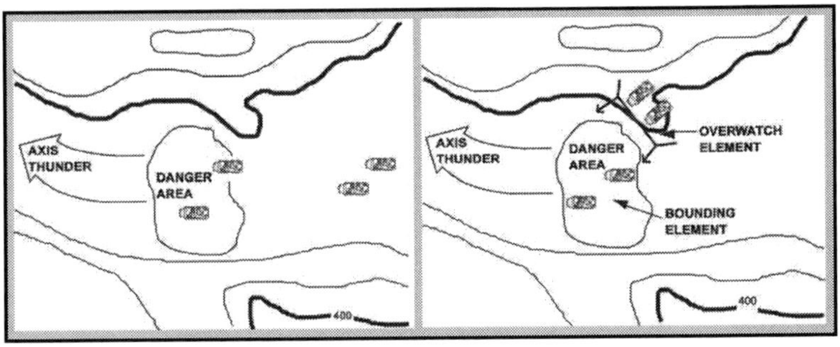

Figure 3-33. Mounted traveling overwatch.

Mounted Bounding Overwatch

3-150. With bounding overwatch, one section is always stopped to provide overwatching fire. The unit executing bounding overwatch uses either the successive or alternate bounding method.

Dismounting and Clearing the Area

3-151. The commander of the lead vehicle immediately notifies the platoon leader when he encounters an obstacle or other danger area. If needed, Soldiers dismount and take advantage of available cover and concealment to investigate these areas (Figure 3-34). If possible, the vehicle is moved off the road into a covered or concealed position. Weapons from the vehicle cover the advance of the dismounted element. Designated Soldiers reconnoiter these places under cover of the weapons in the vehicle. Obstacles are marked and bypassed, if possible. When they cannot be bypassed, they are cautiously removed.

3-152. Side roads intersecting the route of advance are investigated. Soldiers from one vehicle secure the road junction. One or two vehicles investigate the side road. The amount of reconnaissance on side roads is determined by the leader's knowledge of the situation. Soldiers investigating side roads do not move past supporting distance of the main body.

Figure 3-34. Dismounting and clearing the area.

SECTION VII — SECURITY

3-153. Maintaining security is a constant theme of tactical movement. Effective security can prevent enemy surprise. Security therefore requires everyone to concentrate on the enemy. Though this seems simple enough, in practice, it is not. This means that leaders and their Soldiers must be proficient in the basics of tactical movement. Failure to attain proficiency diverts attention away from the enemy, thereby directly reducing the unit's ability to fight.

3-154. Platoons and squads enhance their own security during movement through the use of covered and concealed terrain; the use of the appropriate movement formation and technique; the actions taken to secure danger areas during crossing; the enforcement of noise, light, and radiotelephone discipline; and the use of proper individual camouflage techniques.

3-155. During planning and preparation for movement, leaders analyze the enemy situation, determine known and likely enemy positions, and develop possible enemy courses of action. After first considering the enemy, leaders determine what security measures to emplace during tactical movement.

ENEMY

3-156. Leaders have to decide whether they are going to move aggressively to make contact, or stealthily to avoid contact. Either way, the leader has to anticipate enemy contact throughout. If possible, leaders should avoid routes with obvious danger areas such as built-up areas, roads, trails, and known enemy positions. If these places cannot be avoided, risk management should be conducted to develop ways to reduce danger to the unit. If stealth is desired, the route should avoid contact with local inhabitants, built-up areas, and natural lines of drift.

3-157. Movement techniques help the leader manage the amount of security his unit has during movement. Traveling is the least secure and used when contact is not likely. Traveling overwatch is used when contact is likely but not imminent. Bounding overwatch is used when contact is imminent. Leaders establish the probable line of deployment (PLD) to indicate where the transition from traveling overwatch to bounding overwatch should occur. When in contact with the enemy, the unit transitions from movement to maneuver (fire and movement) while the leader conducts actions on contact (Figure 3-35).

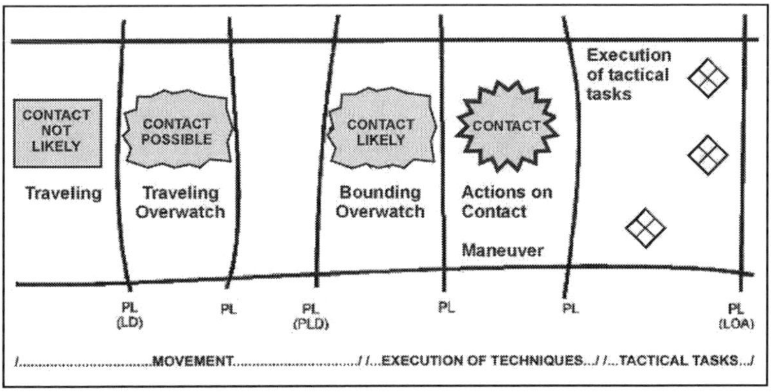

Figure 3-35. Movement to maneuver.

TERRAIN

3-158. When planning movements, the leader must consider how terrain affects security while simultaneously considering the other factors of METT-TC. Some missions may require the unit to move on other than covered and concealed routes. While leaders may not be able to prevent the unit's detection, they can ensure that they move on the battlefield in a time and place for which the enemy is unprepared. Particularly when moving in the open, leaders must avoid predictability and continue to use terrain to their advantage.

CAMOUFLAGE, NOISE, AND LIGHT DISCIPLINE

3-159. Leaders must ensure that camouflage used by their Soldiers is appropriate to the terrain and season. Platoon SOPs specify elements of noise and light discipline.

3-160. If Soldiers need more illumination than an image intensifier can provide in infrared mode during movement, they should use additional infrared light sources. The combination should provide the light needed with the least risk of enemy detection. When using infrared light, leaders must consider the enemy's night vision and infrared capabilities. For instance, an enemy with night vision capability can send infrared light signals, and he can concentrate direct and indirect fire on a platoon that is using infrared light.

SECURITY AT HALTS

3-161. Units conducting tactical movement frequently make temporary halts. These halts range from brief to extended periods of time. For short halts, platoons use a cigar-shaped perimeter intended to protect the force while maintaining the ability to continue movement. When the platoon leader decides not to immediately resume tactical movement, he transitions the platoon to a perimeter defense. The perimeter defense is used for longer halts or during lulls in combat.

Cigar-Shaped Perimeter

3-162. When the unit halts, if terrain permits, Soldiers should move off the route and face out to cover the same sectors of fire they were assigned while moving, allowing passage through the center of the formation. This results in a cigar-shaped perimeter. Actions by subordinate leaders and their Soldiers occur without an order from the leader. Soldiers are repositioned as necessary to take advantage of the best cover, concealment, and fields of fire.

Perimeter Defense

3-163. When operating independently, the platoon uses a perimeter defense during extended halts, resupply, and issuing platoon orders or lulls in combat. Normally the unit first occupies a short halt formation. Then after conducting a leader's reconnaissance of the position and establishing security, the unit moves into the perimeter defense.

Actions at Halts

3-164. Table 3-7 lists the standard actions taken at halts.

Table 3-7. Actions at halts.

Soldier (or Vehicle) Actions*	Squad Leader (or Section Leader) Actions	Platoon Leader Actions
• Moves to as much of a covered and concealed position as available. • Visually inspects and physically clears his immediate surroundings (a roughly 5-25m radius around his position). • Establishes a sector of fire for his assigned weapon (using 12 o'clock as the direction the Soldier is facing, the Soldier's sector of fire ranges from 10 o'clock to 2 o'clock). • Determines his observation and field of fire. Identifies dead space in his field of fire. • Identifies obstacles and determines enemy avenues of approach (both mounted and dismounted). • Identifies the dominant ground in his immediate surroundings. • Coordinates his actions with the Soldiers (or vehicles) on his left and right. (*These actions occur without leader prompting.)	• Adjusts his perimeter. ▪ If operating independently, the squad leader establishes 360-degree, three-dimensional security. ▪ Attempts to find terrain that anchors his position. ▪ If operating as part of a platoon, the squad leader arrays his teams to best fit into the platoon leader's defensive scheme, based on the platoon leader's guidance. • Visually inspects and physically clears (if required) the squad's immediate surrounding (about 35m, the distance within hand grenade range). • Ensures his squad's individual sectors of fire overlap with each other, creating a seamless perimeter with no gaps of fire coverage. • Identifies his dead space and adjusts his M203 grenadiers accordingly. • Identifies obstacles and the likely enemy avenue of approach (mounted and dismounted). • Identifies the dominant ground in his area of operation. • Coordinates responsibilities and sectors with the units on his left and right.	• Adjusts his perimeter. ▪ If operating independently, he establishes 360-degree, three-dimensional security. ▪ If operating as part of another organization, he arrays his squads to best fit into the controlling commander's defensive scheme. ▪ Supervises the emplacement of the weapons squad's weapon systems. • Dispatches an element (usually a fire team) to visually inspect and physically clear the platoon's immediate surrounding (an area out to small arms range, roughly 100-300m depending on terrain). • Ensures his squads' sectors of fire overlap with each other, creating a seamless perimeter with no gaps of fire coverage. • Identifies his dead space not covered and requests indirect fire support to overwatch dead space in the area of operation. • Identifies obstacles and the likely enemy avenue of approach (mounted and dismounted). • Identifies the dominant ground in his area of operation. • Coordinates with the units on his left and right.

SECTION VIII — OTHER MOVEMENT SITUATIONS

3-165. The platoon can use other formations for movement.

ADMINISTRATIVE MOVEMENT

3-166. Administrative movement is normally planned by the S4 as movements in which vehicles and Soldiers are arranged to expedite movement and conserve time and resources. No enemy interference is anticipated when planning administrative movement.

TACTICAL ROAD MARCHES

3-167. Infantry platoons participate in two types of tactical marches with the company: foot marches and motor marches. Tactical road marches are conducted to rapidly move units within an area of operations to prepare for combat operations. Commanders arrange troops and vehicles to expedite their movement, conserve time, energy, and unit integrity. They anticipate no interference except possible enemy air. For information on dismounted tactical road marches, see FM 21-18, *Foot Marches*.

MOVEMENT BY WATER

3-168. The platoon avoids crossing water obstacles when possible. Before crossing, however, leaders should identify weak or non-swimmers and pair them with a good swimmer in their squad.

3-169. When platoons or squads must move into, through, or out of rivers, lakes, streams, or other bodies of water, they treat the water obstacle as a danger area. While on the water, the platoon is exposed and vulnerable. To offset the disadvantages, the platoon—

- Moves during limited visibility.
- Disperses.
- Camouflages thoroughly.
- Moves near the shore to reduce the chances of detection.

3-170. When moving in more than one boat, the platoon—

- Maintains tactical integrity and self-sufficiency.
- Crossloads key Soldiers and equipment.
- Ensures that the radio is with the leader.

3-171. If boats are not available, several other techniques can be used such as—

- Swimming.
- Poncho rafts.
- Air mattresses.
- Waterproof bags.
- A 7/16-inch rope used as a semisubmersible, one-rope bridge or safety line.
- Water wings (made from a set of trousers).

MOVEMENT DURING LIMITED VISIBILITY CONDITIONS

3-172. At night or when visibility is poor, a platoon must be able to function in the same way as during daylight. It must be able to control, navigate, maintain security, move, and stalk at night or during limited visibility.

CONTROL

3-173. When visibility is poor, the following methods aid in control:

- Use of night vision devices.
- IR chemlights.
- Leaders move closer to the front.
- The platoon reduces speed.
- Each Soldier uses two small strips of luminous tape on the rear of his helmet to allow the Soldier behind him to see.
- Leaders reduce the interval between Soldiers and between units to make sure they can see each other.
- Leaders conduct headcounts at regular intervals and after each halt to ensure personnel accountability.

NAVIGATION

3-174. To assist in navigation during limited visibility, leaders use—

- Terrain association (general direction of travel coupled with recognition of prominent map and ground features).
- Dead reckoning, compass direction and specific distances or legs. (At the end of each leg, leaders should verify their location).
- Movement routes that parallel identifiable terrain features.
- Guides or marked routes. .
- GPS / FBCB2 devices.

SECURITY AT NIGHT

3-175. For stealth and security in night moves, squads and platoons—

- Designate a point man to maintain alertness, the lead team leader to navigate, and a pace man to count the distance traveled. Alternate compass and pace men are designated.
- Ensure good noise and light discipline.
- Use radio-listening silence.
- Camouflage Soldiers and equipment.
- Use terrain to avoid detection by enemy surveillance or night vision devices.
- Make frequent listening halts.
- Mask the sounds of movement with artillery fires.

Chapter 4

Protection

Protection is the preservation of the Infantry platoon and squad's fighting potential so leaders can apply maximum force at the decisive time and place. Protection is neither timidity nor an attempt to avoid all risk, because risk will always be present. Protection is a warfighting function (WFF) that encompasses the following areas: safety, fratricide avoidance, survivability, air and missile defense, antiterrorism, chemical biological radiological and nuclear (CBRN), defense information operations, and force health protection. This chapter covers the WWF areas of protection that are most relevant to the Infantry platoon and squad: risk management and fratricide avoidance, air defense, and CBRN.

SECTION I — RISK MANAGEMENT AND FRATRICIDE AVOIDANCE

4-1. Risk, or the potential for risk, is always present across full spectrum operations. The primary objective of risk management and fratricide avoidance is not to remove all risk, but to eliminate unnecessary risk. During peacetime leaders conduct tough, realistic training to help units protect their combat power through accident prevention. During combat operations units conduct risk management and fratricide avoidance to enable them to win the battle quickly and decisively with minimal losses. Risk management is an integral part of planning that takes place at all levels of the chain of command during each phase of every operation. This section outlines the process leaders use to identify hazards and implement plans to address each identified hazard. It also includes a detailed discussion of the responsibilities of the platoon's leaders and individual Soldiers in implementing a sound risk management program. For additional information on risk management, refer to FM 5-19, *Composite Risk Management.*.

RISK MANAGEMENT PROCEDURES

4-2. Risk management is the systematic process that identifies the relative risk of mission and training requirements. It weighs risk against training benefits and eliminates unnecessary risk that can lead to accidents. The platoon leader, his NCOs, and all other platoon Soldiers must know how to use risk management, coupled with fratricide reduction measures, to ensure that the mission is executed in the safest possible environment within mission constraints.

STEP 1 – IDENTIFY HAZARDS

4-3. A hazard is a source of danger. It is any existing or potential condition that can cause injury, illness, or death of personnel; damage to or loss of equipment and property; or some other sort of mission degradation. Tactical and training operations pose many types of hazards. The leader must identify the hazards associated with all aspects and phases of the Infantry platoon's mission, paying particular attention to the factors of METT-TC. Risk management must never be an afterthought; leaders must begin the process during their TLPs and continue it throughout the operation. Table 4-1 lists possible sources of battlefield hazards the Infantry platoon and squad might face during a typical tactical operation. The list is organized according to the factors of METT-TC.

Table 4-1. Potential hazards.

Potential Infantry Platoon and Squad Battlefield Hazards
Mission • Duration of the operation. • Mission complexity and difficulty/clarity of the plan. (Is the plan well-developed and easily understood?) • Proximity and number of maneuvering units.
Enemy • Knowledge of the enemy situation. • Enemy capabilities. • Availability of time and resources to conduct reconnaissance.
Terrain and Weather • Visibility conditions including light, dust, fog, and smoke. • Precipitation and its effect on mobility. Consider all aspects of the terrain as well as weather and trafficability. • Extreme heat or cold. • Additional natural hazards such as broken ground, steep inclines, or water obstacles.
Troops and Equipment • Experience the units conducting the operation have working together. • Danger areas associated with the platoon's weapons systems. • Soldier/leader proficiency. • Soldier/leader rest situation. • Degree of acclimatization to environment. • Impact of new leaders or crewmembers. • Friendly unit situation. • NATO or multinational military actions combined with U.S. forces.
Time Available • Time available for TLP and rehearsals by subordinates. • Time available for precombat checks and inspections.
Civil Considerations • Applicable ROE or ROI. • Potential operations that involve contact with civilians. • Potential for media contact and inquiries. • Interaction with host nation or other participating nation support.

Step 2 – Assess Hazards To Determine Risks

4-4. Hazard assessment is the process of determining the direct impact of each hazard on a training or operational mission. The following steps should be used when assessing hazards:

- Determine the hazards that can be eliminated or avoided.
- Assess each hazard that cannot be eliminated or avoided to determine the probability that the hazard will occur. A primary consideration is how likely the hazard is to cause injury, illness, loss, or damage.

- Assess the severity of hazards that cannot be eliminated or avoided. Severity is the result or outcome of a hazardous incident that is expressed by the degree of injury or illness (including death), loss of or damage to equipment or property, environmental damage, or other mission-impairing factors such as unfavorable publicity or loss of combat power.
- Accounting for both the probability and severity of a hazard, determine the associated risk level (extremely high, high, moderate, or low). Normally, the highest-level individual risk assessed is also the overall risk. Table 4-2 summarizes the four risk levels.
- Based on the factors of hazard assessment (probability, severity, and risk level, as well as the operational factors unique to the situation), complete the risk management worksheet. Figure 4-1 shows an example of a completed risk management worksheet.

Table 4-2. Risk levels and impact on mission execution.

Risk Level	Mission Effects
Extremely High (E)	Mission failure if hazardous incidents occur in execution.
High (H)	Significantly degraded mission capabilities in terms of required mission standards. Not accomplishing all parts of the mission or not completing the mission to standard (if hazards occur during mission).
Moderate (M)	Expected degraded mission capabilities in terms of required mission standards. Reduced mission capability (if hazards occur during the mission).
Low (L)	Expected losses have little or no impact on mission success.

A. Mission or Task:	B. Date/Time Group	C: Date Prepared:
Conduct a deliberate attack	Begin: 010035R May XX End: 010600R May XX	29 April XX

D. Prepared By: (Rank, Last Name, Duty Position) CPT Smith, Cdr

E. Task	F. Identify Hazard	G. Assess Hazard	H. Develop Controls	I. Determine Residual Risk	J. Implement Controls (How To)
Conduct obstacle breaching operations	Obstacles	High (H)	Develop and use obstacle reduction plan	Low (L)	Unit TSOP, OPORD, training handbook
	Inexperienced soldiers	High (H)	Additional training and supervision	Moderate (M)	Rehearsals, additional training
	Operating under limited visibility	Moderate (M)	Use NVDs, use IR markers on vehicles	Low (L)	Unit TSOP, OPORD
	Steep Cliffs	High (H)	Rehearse using climbing ropes	Moderate (M)	FM 3-97.6, Mountain Operations; FM 3-97.61 Military Mountaineering
	Insufficient planning time	High (H)	Plan and prepare concurrently	Moderate (M)	OPORD, Troop-leading procedures

K. Determine overall mission/task risk level after controls are implemented (circle one)

LOW (L) MODERATE (M) HIGH (H) EXTREMELY HIGH (E)

Figure 4-1. Example of completed risk management worksheet.

STEP 3 – DEVELOP CONTROLS AND MAKE RISK DECISIONS

4-5. This step consists of two substeps: develop controls and make risk decisions. These substeps are accomplished during the "make a tentative plan" step of the TLP.

Develop Controls

4-6. After assessing each hazard, develop one or more controls that will either eliminate the hazard or reduce the risk (probability, severity, or both) of potential hazardous incidents. Create as many control options as possible and then select those that best control risks without significantly impeding the training or operational mission objectives. When developing controls, consider the reason for the hazard, not just the hazard itself. For example, driving can be a hazard, but driving in inclement weather or with limited sleep may cause driving to be hazardous.

Make Risk Decisions

4-7. A key element in the process of making a risk decision is determining whether accepting the risk is justified or unnecessary. Risk decisionmaking should be made at the appropriate level—high enough to tap the experience and responsibility of those making the decision, and low enough to allow for the gaining of experience. As a guide, the leader responsible for executing the training or operational mission is authorized by the command or higher headquarters to make decisions at a specified risk level (extremely high, high, moderate, or low). When a leader is not authorized to make decisions for a risk level, the decision is referred to the next higher level of command. The decision maker must compare and balance the risk against mission expectations. If he determines the risk is unnecessary, he directs the development of additional controls or alternative controls; as another option, he can modify, change, or reject the selected COA for the operation.

STEP 4 – IMPLEMENT CONTROLS

4-8. Controls are the procedures and considerations the unit uses to eliminate hazards or reduce their risk. The implementation of controls is the most important part of the risk management process; it is the chain of command's contribution to the safety of the unit. Implementing controls includes coordination and communication with appropriate superior, adjacent, and subordinate units and with individuals executing the mission. The implementation of risk controls must be effectively communicated to all personnel, especially those responsible for the actual implementation of the controls. The platoon leader must ensure that specific controls are integrated into OPLANs, OPORDs, SOPs, and rehearsals. The critical check for this step is to ensure that controls are converted into clear, simple execution orders understood by all levels. Examples of risk management controls include:

- Thoroughly briefing all aspects of the mission, including related hazards and controls.
- Conducting thorough precombat checks and inspections.
- Allowing adequate time for rehearsals at all levels.
- Drinking plenty of water, eating well, and getting as much sleep as possible (at least 4 hours in any 24-hour period).
- Using buddy teams.
- Enforcing speed limits, using of seat belts, and driver safety.
- Establishing recognizable visual signals and markers to distinguish maneuvering units.
- Enforcing the use of ground guides in assembly areas and on dangerous terrain.
- Establishing marked and protected sleeping areas in assembly areas.
- Limiting single-vehicle movement.
- Establishing SOPs for the integration of new personnel.

Step 5 – Supervise And Evaluate

4-9. During mission execution, leaders must ensure that risk management controls are properly understood and executed. Leaders must continuously evaluate the unit's effectiveness in managing risks to gain insight into areas that need improvement.

Supervise

4-10. Leadership and unit discipline are the keys to ensuring that effective risk management controls are implemented. All leaders are responsible for supervising mission rehearsals and execution to ensure standards and controls are enforced. Effective supervision assures sustained effectiveness of risk controls. NCOs must enforce established safety policies as well as controls developed for a specific operation or task. Techniques include spot checks, inspections, SITREPs, confirmation briefs, buddy checks, and close supervision.

4-11. During mission execution, leaders must continuously monitor risk management controls to determine whether they are effective and to modify them as necessary. Leaders must also anticipate, identify, and assess new hazards. They ensure that imminent danger issues are addressed on the spot and that ongoing planning and execution reflect changes in hazard conditions.

Evaluate

4-12. Whenever possible, the risk management process should include an AAR to assess unit performance in identifying risks and preventing hazardous situations. During an AAR, leaders should assess whether the implemented controls were effective by specifically providing feedback on the effectiveness of risk controls. Following the AAR, leaders should incorporate lessons learned from the process into the Infantry platoon's SOPs and plans for future missions.

IMPLEMENTATION RESPONSIBILITIES

4-13. Leaders and individuals at all levels are responsible and accountable for managing risk. They must ensure that hazards and associated risks are identified and controlled during planning, preparation, and execution of operations. The platoon leader and his senior NCOs must look at both tactical risks and accident risks. The same risk management process is used to manage both types. In the Infantry platoon, the platoon leader alone determines how and where he is willing to take tactical risks. The platoon leader manages accident risks with the assistance of his PSG, NCOs, and individual Soldiers.

Breakdown of the Risk Management Process

4-14. If higher headquarters is not notified of a risk taken or about to be taken, the risk management process may break down. Such a failure can be the result of several reasons, but is usually one or more of the following factors:

- The risk denial syndrome in which leaders do not want to know about the risk.
- A Soldier who believes that the risk decision is part of his job and does not want to bother his leader.
- Outright failure to recognize a hazard or the level of risk involved.
- Overconfidence on the part of an individual or the unit in being able to avoid or recover from a hazardous incident.
- Subordinates who do not fully understand the higher commander's guidance regarding risk decisions.

FRATRICIDE AVOIDANCE

4-15. Fratricide is defined as the employment of friendly weapons with the intent of killing the enemy or destroying his equipment that results in the unforeseen and unintentional death or injury of friendly personnel. Fratricide prevention is the platoon leader's responsibility. Leaders across all WFF assist the platoon leader in accomplishing this mission. The following paragraphs focus on actions the platoon leader and his subordinate leaders can take with current resources to reduce the risk of fratricide.

4-16. In any tactical situation, it is critical that every Infantry platoon member know where he is and where other friendly elements are operating. With this knowledge, he must anticipate dangerous conditions and take steps to either to avoid or mitigate them. He must also ensure that all squad and team positions are constantly reported to higher headquarters so all other friendly elements are aware of where they are and what they are doing. When the platoon leader perceives a potential fratricide situation, he must personally use the higher net to coordinate directly with the friendly element involved.

EFFECTS

4-17. The effects of fratricide within a unit can be devastating to morale, good order, and discipline. Fratricide causes unacceptable losses and increases the risk of mission failure. It almost always affects the unit's ability to survive and function. Units experiencing fratricide suffer the following consequences:

- Loss of confidence in the unit's leadership.
- Self-doubt among leaders.
- Hesitancy in the employment of supporting combat systems.
- Over-supervision of units.
- Hesitancy in the conduct of limited visibility operations.
- Loss of aggressiveness in maneuver.
- Loss of initiative.
- Disrupted operations.
- General degradation of unit cohesiveness, morale, and combat power.

CAUSES

4-18. The lack of positive target identification and inability to maintain situational awareness during combat operations are major contributing factors to fratricide. The following paragraphs discuss the primary causes of fratricide. Leaders must identify any of the factors that may affect their units and then strive to eliminate or correct them.

Failures in the Direct Fire Control Plan

4-19. Failures in the direct fire control plan occur when units do not develop effective fire control plans, particularly in the offense. Units may fail to designate engagement areas, adhere to the direct fire plan, fail to understand surface danger areas, or position their weapons incorrectly. Under such conditions, fire discipline often breaks down upon contact. An area of particular concern is the additional planning that must go into operations requiring close coordination between mounted elements and dismounted elements.

Land Navigation Failures

4-20. Friendly units may stray out of assigned sectors, report wrong locations, and become disoriented. Much less frequently, they employ fire support weapons in the wrong location. In either type of situation, units that unexpectedly encounter another unit may fire their weapons at the friendly force.

Failures in Combat Identification

4-21. Vehicle commanders and machine gun crews cannot accurately identify the enemy near the maximum range of their weapons systems. During limited visibility, friendly units within that range may mistake each other as the enemy.

Inadequate Control Measures

4-22. Units may fail to disseminate the minimum necessary maneuver control measures and direct fire control measures. They may also fail to tie control measures to recognizable terrain or events.

Failures in Reporting and Communications

4-23. Units at all levels may fail to generate timely, accurate, and complete reports as locations and tactical situations change. This distorts the operating picture at all levels and can lead to erroneous clearance of fires.

Individual and Weapons Errors

4-24. Lapses in individual discipline can result in fratricide. Incidents such as these include negligent weapons discharges and mistakes with explosives and hand grenades.

Battlefield Hazards

4-25. A variety of explosive devices and materiel—unexploded ordnance, booby traps, and unmarked or unrecorded minefields, including scatterable mines—may create danger on the battlefield. Failures to mark, record, remove, or otherwise anticipate these threats lead to casualties.

Reliance on Instruments

4-26. A unit that relies too heavily on systems such as GPS devices, Force XXI Battle Command Brigade and Below System (FBCB2), or Land Warrior will find its capabilities severely degraded if these systems fail. The unit will be unable to maintain complete situational understanding because it will not have a common operations picture. To prevent potential dangers when system failure occurs, the platoon leader must ensure that he and his platoon balance technology with traditional basic Soldier skills in observation, navigation, and other critical activities.

PREVENTION

4-27. These guidelines are not intended to restrict initiative. Leaders must learn to apply them, as appropriate, based on the specific situation and the factors of METT-TC.

PRINCIPLES

4-28. At the heart of fratricide prevention are the following five key principles.

1 - Identify and Assess Potential Fratricide Risks During the TLP

4-29. Incorporate risk reduction control measures in WARNOs, the OPORD, and applicable FRAGOs.

2 - Maintain Situational Understanding

4-30. Focus on areas such as current intelligence, unit locations and dispositions, obstacles, CBRN contamination, SITREPs, and the factors of METT-TC. Leaders must accurately know their own location (and orientation) as well as the location of friendly, enemy, neutrals, and noncombatants.

3 - Ensure Positive Target Identification

4-31. Review vehicle and weapons ID cards. Become familiar with the characteristics of potential friendly and enemy vehicles, including their silhouettes and thermal signatures, combat identification panels, and thermal panels. This knowledge should include the conditions, including distance (range) and weather in which positive identification of various vehicles and weapons is possible. Enforce the use of challenge and password, especially during dismounted operations.

4 - Maintain Effective Fire Control

4-32. Ensure fire commands are accurate, concise, and clearly stated. Make it mandatory for Soldiers to ask for clarification of any portion of the fire command that they do not understand completely. Stress the importance of the chain of command in the fire control process and ensure Soldiers get in the habit of obtaining target confirmation and permission to fire from their leaders before engaging targets. Know who will be in and around the AO.

5 - Establish a Command Climate That Emphasizes Fratricide Prevention

4-33. Enforce fratricide prevention measures, placing special emphasis on the use of doctrinally-sound techniques and procedures. Ensure constant supervision in the execution of orders and in the performance of all tasks and missions to standard.

GUIDELINES AND CONSIDERATIONS

4-34. Additional guidelines and considerations for fratricide reduction and prevention include the following:

- Recognize the signs of battlefield stress. Maintain unit cohesion by taking quick, effective action to alleviate stress.
- Conduct individual, leader, and collective (unit) training covering fratricide awareness, target identification and recognition, and fire discipline.
- Develop a simple, executable plan.
- Give complete and concise orders. Include all appropriate recognition signals in paragraph 5 of the OPORD.
- To simplify OPORDs, use SOPs that are consistent with doctrine. Periodically review and update SOPs as needed.
- Strive to provide maximum planning time for leaders and subordinates.
- Use common language (vocabulary) and doctrinally-correct standard terminology and control measures.
- Ensure thorough coordination is conducted at all levels.
- Plan for and establish effective communications.
- Plan for collocation of command posts whenever it is appropriate to the mission such as during a passage of lines, or relief in place.
- Make sure ROE and ROI are clear.
- Conduct rehearsals whenever the situation allows adequate time to do so. Always conduct a rehearsal of actions on the objective.
- Be in the right place at the right time. Use position location and navigation devices (GPS or position navigation [POSNAV]), know your location and the locations of adjacent units (left, right, leading, and follow-on), and synchronize tactical movement. If the platoon or any element becomes lost, its leader must know how to contact higher headquarters immediately for instructions and assistance.
- Establish, execute, and enforce strict sleep and rest plans.

SECTION II — AIR AND MISSILE DEFENSE

4-35. Leaders must consider the use of air defense (AD) if evidence exists of enemy forces having the ability to employ fixed- or rotary-winged aircraft, or unmanned aircraft systems (UAS) against friendly forces. Operations in these situations require forces to be thoroughly trained on passive and active AD measures.

4-36. AD assets such as Stingers and Avengers may operate in and around the unit's AO, but the AD is not likely to be task-organized specifically to the Infantry platoon or squad. Therefore, the Infantry platoon and squad must conduct its own AD operations, relying on disciplined passive AD measures and the ability to actively engage aerial platforms with organic weapons systems.

EARLY WARNING PROCEDURES

4-37. Local AD warnings describe with certainty the air threat for a specific part of the battlefield. Air defense artillery (ADA) units use these local warnings to alert units to the state of the air threat in terms of "right here, right now." There are three local AD warning levels:

- **Dynamite.** Enemy aircraft are inbound or are attacking locally now.
- **Lookout.** Enemy aircraft are in the area of interest but are not threatening. They may be inbound, but there is time to react.
- **Snowman.** Enemy aircraft do not pose a threat at this time.

NOTE: The area ADA unit commander routinely issues AD warnings for dissemination throughout the theater of operations. These warnings describe the general state of the probable air threat and apply to the entire area.

PASSIVE AIR DEFENSE

4-38. Passive AD is the Infantry platoon and squad's primary method for avoiding enemy air attack. Passive AD consists of all measures taken to prevent the enemy from detecting or locating the unit, to minimize the target acquisition capability of enemy aircraft, and to limit damage to the unit if it comes under air attack. Target detection and acquisition are difficult for crews of high-performance aircraft, and the unit can exploit this advantage.

Guidelines

4-39. The Infantry platoon and squad should follow these guidelines to avoid detection or limit damage if detected:

- When stopped, occupy positions that offer cover and concealment and dig in and camouflage.
- When moving, use covered and concealed routes.
- Disperse as much as possible to make detection and attack more difficult.
- Eliminate or cover the spoil from dug-in positions.
- Do not fire on a hostile fixed-wing aircraft unless it is clear that the aircraft has identified friendly elements. Premature engagement compromises friendly positions.
- Designate air guards for every position; establish and maintain 360-degree security.
- Establish an air warning system in the unit SOP, including both visual and audible signals.

Procedures

4-40. When the Infantry platoon or squad observes enemy fixed-wing aircraft, helicopters, or unmanned aircraft systems (UAS) that could influence its mission, it initially takes passive AD measures unless the situation requires immediate active measures. Passive AD measures normally mean that friendly unit initiates its react-to-air-attack battle drill; however, the leader can initiate specific passive measures if necessary.

4-41. Passive AD involves these three steps:

- **Step 1.** Alert the friendly unit with a contact report.
- **Step 2.** Deploy or take the appropriate actions. If the Infantry platoon or squad is not in the direct path of an attacking aircraft, leaders have all friendly Soldiers seek cover and concealment and halt with as much dispersion as possible based on the terrain.
- **Step 3.** Prepare to engage the enemy aircraft.

Active Air Defense

4-42. Infantry platoons and squads avoid engaging enemy aircraft. If engagement is unavoidable, the friendly unit uses a technique known as volume of fire. This technique is based on the premise that the more bullets a unit can put in the sky, the greater the chance the enemy aircraft will fly into them. Even if these fires do not hit the enemy, a "wall of lead" in the sky can intimidate enemy pilots. This can cause them to break off their attack or distract them from taking proper aim. One of the most important points about volume of fire is that once the lead distance is estimated, Soldiers must aim at the estimated aiming point and fire at that single point until the aircraft has flown past it. Soldiers maintain the aiming point, not the lead distance. Once a Soldier starts firing, he does not adjust his weapon. Leaders establish the aiming point based on the type of aircraft that is attacking (Figure 4-2).

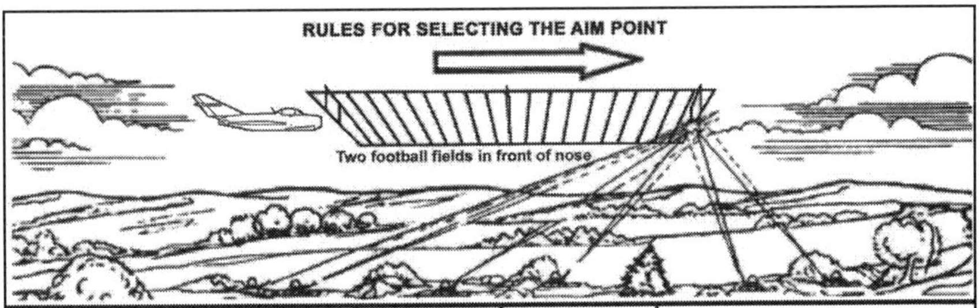

TYPE AERIAL PLATFORMS	COURSE	AIM POINT
JET/CRUISE MISSILE	CROSSING	TWO FOOTBALL FIELDS IN FRONT OF NOSE
JET/CRUISE MISSILE	OVERHEAD	TWO FOOTBALL FIELDS IN FRONT OF NOSE
JET/CRUISE MISSILE	DIRECTLY AT YOU	SLIGHTLY ABOVE AIRCRAFT NOSE
HELICOPTER/UAS	CROSSING	ONE-HALF FOOTBALL FIELD IN FRONT OF NOSE
HELICOPTER	HOVERING	SLIGHTLY ABOVE HELICOPTER BODY
HELICOPTER/UAS	DIRECTLY AT YOU	SLIGHTLY ABOVE HELICOPTER BODY

Figure 4-2. Volume of fire aim points.

SECTION III — CHEMICAL, BIOLOGICAL, RADIOLOGICAL, AND NUCLEAR DEFENSE

4-43. Chemical, biological, radiological, and nuclear (CBRN) weapons can cause casualties, destroy or disable equipment, restrict the use of terrain, and disrupt operations. They can be used separately or in combination to supplement conventional weapons. The Infantry platoon must be prepared to operate on a CBRN-contaminated battlefield without degradation of the unit's overall effectiveness.

4-44. CBRN defensive measures provide the capability to defend against enemy attack by chemical, biological, radiological, and chemical weapons and to survive and sustain combat operations in a CBRN environment. Survival and sustainment must use the following principles: avoidance of CBRN hazards, particularly contamination; protection of individuals and units from unavoidable CBRN hazards; and decontamination. An effective CBRN defense counters enemy threats and attacks by minimizing vulnerabilities, protecting friendly forces, and maintaining an operational tempo (OPTEMPO) that complicates targeting.

TENETS OF CBRN DEFENSE

4-45. Protection of the Infantry platoon and squad requires adherence to four rules of CBRN defense: contamination avoidance; reconnaissance; protection; and decontamination.

CONTAMINATION AVOIDANCE

4-46. Avoiding CBRN attacks and hazards is the first rule of CBRN defense. Avoidance allows leaders to shield Soldiers and units, and involves both active and passive measures. Passive measures include training, camouflage, concealment, hardening of positions, and dispersion. Active measures include employing detection equipment, reconnaissance, warnings and reports, markings, and contamination control.

RECONNAISSANCE

4-47. CBRN reconnaissance is detecting, identifying, reporting, and marking CBRN hazards. The process consists of search, survey, surveillance, and sampling operations. Due to the limited availability of the M93 Fox reconnaissance vehicle, commanders should consider as a minimum the following actions when planning and preparing for this type of reconnaissance:

- Use the intelligence preparation of the battlefield (IPB) process to orient on CBRN threat named areas of interest (NAIs).
- Pre-position reconnaissance assets to support requirements.
- Establish command and support relationships.
- Assess the time and distance factors for the conduct of CBRN reconnaissance.
- Report all information rapidly and accurately.
- Plan for resupply activities to sustain CBRN reconnaissance operations.
- Determine possible locations for post-mission decontamination.
- Plan fire support.
- Enact fratricide prevention measures.
- Establish MEDEVAC procedures.
- Identify CBRN warning and reporting procedures and frequencies.

PROTECTION

4-48. CBRN protection is an integral part of operations. Techniques that work for avoidance also work for protection (shielding Soldiers and units and shaping the battlefield). Other protection activities involve sealing or hardening positions, protecting Soldiers, assuming mission-oriented protective posture (MOPP) (Table 4-3), reacting to attack, and using collective protection. Individual protective items include the protective mask, joint service lightweight integrated suit technology (JSLIST) overgarments, multipurpose (rain/snow/chemical and biological) overboots (MULO), and gloves. The corps or higher-level commander establishes the minimum level of protection. Subordinate units may increase this level as necessary, but they may not decrease it.

Table 4-3. MOPP levels.

Equipment	MOPP Ready	MOPP0	MOPP1	MOPP2	MOPP3	MOPP4	Mask Only
Mask	Carried	Carried	Carried	Carried	Worn	Worn	Worn***
JSLIST	Ready*	Available**	Worn	Worn	Worn	Worn	NA
Overboots	Ready*	Available**	Available**	Worn	Worn	Worn	NA
Gloves	Ready*	Available**	Available**	Available**	Available**	Worn	NA
Helmet cover	Ready*	Available**	Available**	Worn	Worn	Worn	NA

*Items available to Soldier within two hours with replacement available within six hours.
**Items must be positioned within arm's reach of the Soldier.
***Never "mask only" if a nerve or blister agent has been used in the AO.

DECONTAMINATION

4-49. The use of CBRN weapons creates unique residual hazards that may force units into protective equipment. When the wearing of protective equipment is necessary, performance of individual and collective tasks can be degraded, and decontamination may be required. Decontamination is the removal or neutralization of CBRN contamination from personnel and equipment. It restores combat power and reduces casualties that may result from exposure, enabling commanders to sustain combat operations. In addition to the effects of CBRN weapons, contamination from collateral damage, natural disasters, and industrial emitters may also require decontamination. Use the four principles of decontamination when planning decontamination operations:

(1) Decontaminate as soon as possible.

(2) Decontaminate only what is necessary.

(3) Decontaminate as far forward as possible (METT-TC dependent).

(4) Decontaminate by priority.

Levels

4-50. The three levels of decontamination are immediate, operational, and thorough (Table 4-4).

Immediate Decontamination

4-51. Immediate decontamination requires minimal planning. It is a basic Soldier survival skill and is performed IAW STP 21-1-SMCT. The aim of immediate decontamination is to minimize casualties, save lives, and limit the spread of contamination. Personal wipedown with the M291 removes contamination from individual equipment.

Operational Decontamination

4-52. Operational decontamination reduces contact hazards and limits the spread of contamination through MOPP gear exchange and vehicle spraydown. It is done when a thorough decontamination cannot be performed. MOPP gear exchange should be performed within six hours of contamination, if possible.

Thorough Decontamination

4-53. Thorough decontamination involves detailed troop decontamination (DTD) and detailed equipment decontamination (DED). Thorough decontamination is normally conducted by company-size elements as part of restoration or during breaks in combat operations. These operations require support from a chemical decontamination platoon and a water source or supply.

Table 4-4. Decontamination levels and techniques.

Levels	Techniques[1]	Purpose	Best Start Time	Performed By
Immediate	Skin decontamination Personal wipe down Operator wipe down Spot decontamination	Saves lives Stops agent from penetrating Limits agent spread Limits agent spread	Before 1 minute Within 15 minutes Within 15 minutes Within 15 minutes	Individual Individual or buddy Individual or crew Individual or crew
Operational	MOPP gear exchange[2] Vehicle wash down	Provides temporary relief from MOPP4 Limits agent spread	Within 6 hours Within 1 hour (CARC) or within 6 hours (non-CARC)	Unit battalion crew or decontamination platoon
Thorough	DED and DAD DTD	Provides probability of long-term MOPP reduction	When mission allows reconstitution	Decontamination platoon Contaminated unit
[1] Techniques become less effective the longer they are delayed. 2 Performance degradation and risk assessment must be considered when exceeding 6 hours. See FM 3-11.5, *Multiservice Tactics, Techniques, and Procedures Chemical, Biological, Radiological, and Nuclear Contamination.*				

Planning Considerations

4-54. Leaders should include the following when planning for decontamination:

- Plan decontamination sites throughout the width and depth of the sector (identify water sources or supplies throughout the sector as well).
- Tie decontamination sites to the scheme of maneuver and templated CBRN strikes.
- Apply the principles of decontamination.
- Plan for contaminated routes.
- Plan for logistics and resupply of MOPP, mask parts, water, and decontamination supplies.
- Plan for medical concerns to include treatment and evacuation of contaminated casualties.
- Maintain site security.

Chapter 5

Command, Control, and Troop-Leading Procedures

The purpose of Command and Control (C2) is to implement the commander's will in pursuit of the unit's objective. C2 is both a system and a process. The essential component for both is leadership. This chapter provides techniques and procedures used by infantry platoons, squads, and sections for C2 and communications. It describes troop-leading procedures (TLP), communications in combat, and operation orders.

SECTION I — COMMAND AND CONTROL

5-1. C2 refers to the process of directing, coordinating, and controlling a unit to accomplish a mission. C2 implements the commander's will in pursuit of the unit's objective. The two components of C2 are the commander and the C2 system. At platoon level the C2 system consists of the personnel, information management, procedures, and equipment the platoon leader uses to carry out the operational process (plan, prepare, execute, and assess) within his platoon.

LEADERSHIP

5-2. Leadership means influencing people by providing purpose, direction, and motivation to accomplish a mission (Table 5-1). Leadership is the most vital component of C2.

Table 5-1. Elements of leadership.

Leadership: Influencing people to accomplish a mission by providing—	**PURPOSE**	The *reason* to accomplish the mission.
	DIRECTION	The *means* to accomplish the mission.
	MOTIVATION	The *will* to accomplish the mission.

MISSION-ORIENTED COMMAND AND CONTROL

5-3. Mission command is the conduct of military operations through decentralized execution based on mission orders for effective mission accomplishment. Successful mission command results from subordinate leaders at all echelons exercising disciplined initiative within the commander's intent to accomplish missions. It requires an environment of trust and mutual understanding. Successful mission command rests on the following four elements.

- ***Commander's Intent.*** The commander's intent is a clear, concise statement of what the force must do and the conditions the force must meet to succeed with respect to the enemy, terrain, and desired end state.
- ***Subordinates' Initiative.*** This is the assumption of responsibility for deciding and initiating independent actions when the concept of operations no longer applies or when an unanticipated opportunity leading to achieving the commander's intent presents itself.

- ***Mission Orders.*** Mission orders are a technique for completing combat orders. They allow subordinates maximum freedom of planning and action in accomplishing missions. They leave the "how" of mission accomplishment to subordinates.
- ***Resource Allocation.*** Commanders allocate enough resources for subordinates to accomplish their missions. Resources include Soldiers, material, and information.

MISSION COMMAND

5-4. Mission command concentrates on the objective of an operation, not on how to achieve it. It emphasizes timely decision-making. The platoon leader must understand the company commander's intent and his clear responsibility to act within that intent to achieve the desired end state. With the company commander's intent to provide unity of effort, mission command relies on decentralized execution and the platoon leader's initiative.

5-5. The company commander must create trust and mutual understanding between himself and his subordinates. This is more than just control. Commanders must encourage subordinates to exercise initiative. Mission command applies to all operations across the spectrum of conflict.

5-6. Mission command counters the uncertainty of war by reducing the amount of certainty needed to act. Commanders guide unity of effort through the commander's intent, mission orders, and the CCIR. Company commanders hold a "loose rein," allowing platoon leaders freedom of action and requiring them to exercise subordinates' initiative. Commanders make fewer decisions, but this allows them to focus on the most important ones. The command operates more on self-discipline than imposed discipline. Because mission command decentralizes decision-making authority and grants subordinates significant freedom of action, it demands more of commanders at all levels and requires rigorous training and education. If the platoon leader is new and has not reached the level of confidence or maturity of the commander, the commander may need to be more directive until the platoon leader is ready.

5-7. Mission command tends to be decentralized, informal, and flexible. Orders and plans are as brief and simple as possible, relying on implicit communication—subordinates' ability to coordinate and the human capacity to understand with minimal verbal information exchange. This can be a result of extended combat or training in which many actions and procedures have become standing operating procedure (SOP). By decentralizing decision-making authority, mission command increases tempo and improves the subordinates' ability to act in fluid and disorderly situations. Moreover, relying on implicit communication makes mission command less vulnerable to disruption of communications than detailed command.

5-8. Mission command is appropriate for operations in the often politically-charged atmosphere and complex conditions of stability operations. Company commanders must explain not only the tasks assigned and their immediate purpose, but also prescribe an atmosphere to achieve and maintain throughout the AO. They must explain what to achieve and communicate the rationale for military action throughout their commands. Doing this allows platoon leaders, squad leaders and their Soldiers to gain insight into what is expected of them, what constraints apply, and most important, why the mission is being undertaken.

5-9. Detailed command is ill-suited to the conditions of stability operations. Commanders using its techniques try to provide guidance or direction for all conceivable contingencies, which is impossible in dynamic and complex environments. Under detailed command, subordinates must refer to their headquarters when they encounter situations not covered by the commander's guidance. Doing this increases the time required for decisions and delays acting. In addition, success in interagency operations often requires unity of effort, even when there is not unity of command. In such an environment, detailed command is impossible. In contrast to the detailed instructions required by detailed command, mission command calls for a clear commander's intent. This commander's intent provides subordinates guidelines within which to obtain unity of effort with agencies not under military command. Subordinates then act within those guidelines to contribute to achieving the desired end state.

NOTE: The platoon leader must understand the situation and commander's intent one and two levels higher than his own. However, he must know the real-time battlefield situation in detail for his immediate higher level (company).

SECTION II — PLANS AND ORDERS

5-10. Plans are the basis for any mission. To develop his plan (concept of the operation), the platoon leader summarizes how best to accomplish his mission within the scope of the commander's intent one and two levels up. The platoon leader uses TLP to turn the concept into a fully developed plan and to prepare a concise, accurate operation order (OPORD). He assigns additional tasks (and outlines their purpose) for subordinate elements, allocates available resources, and establishes priorities to make the concept work. The following discussion covers important aspects of orders development and serves as an introduction to the discussion of the TLP. This section focuses on the mission statement and the commander's intent, which provide the doctrinal foundation for the OPORD. It also includes a basic discussion of the three types of orders (warning orders [WARNOs], OPORDs, and FRAGOs) used by the platoon leader. The platoon leader and his subordinates must have a thorough understanding of the building blocks for everything else that they do.

MISSION STATEMENT

5-11. The platoon leader uses the mission statement to summarize the upcoming operation. This brief paragraph (usually a single sentence) describes the type of operation, the unit's tactical task, and purpose. It is written based on the five Ws: who (unit), what (task[s]), when (date-time group), where (grid location or geographical reference for the AO or objective), and why (purpose). The platoon leader must ensure that the mission is thoroughly understood by all leaders and Soldiers one and two echelons down. The following considerations apply in development of the mission statement.

OPERATIONS

5-12. Full spectrum operations are groupings of related activities in four broad categories: offense, defense, stability, and civil support.

TASKS

5-13. Tactical tasks are specific activities performed by the unit while it is conducting a form of tactical operation or a choice of maneuver. The title of each task can also be used as an action verb in the unit's mission statement to describe actions during the operation. Tasks should be definable, attainable, and measurable. Tactical tasks that require specific tactics, techniques, and procedures (TTP) for the platoon are covered in detail throughout this manual. Figure 5-1 gives examples of tactical tasks the platoon and its subordinate elements may be called upon to conduct. Refer to FM 1-02 for definition of the tactical tasks listed in Figure 5-1.

Destroy Disrupt	Isolate Breach
Fix Suppress	Follow and Support
Block Support by Fire	Follow and Assume
Attack by Fire Interdict	Retain
Canalize Seize	Reduce
Secure Clear	

Figure 5-1. Examples of tactical tasks.

PURPOSE

5-14. A simple, clearly stated purpose tells subordinates the reason the platoon is conducting the mission.

PLACEMENT IN OPORD

5-15. The platoon leader has several options as to where in the OPORD he outlines his subordinates' tasks and purpose. His main concern is that placement of the mission statement should assist subordinate leaders in understanding the task and purpose and each of the five W elements exactly. Figure 5-2 shows an example of a mission statement the platoon leader might include in his order.

EXAMPLE:

3rd Platoon (Who performs the task?) Attacks to seize. (What is the task?) The bridge at (Nx330159). (Where do they perform the task?) At 040600Z FEB 01. (When do they perform the task?) To pass the 1st Platoon (company main effort) on to OBJ BOB. (Why must they perform the task?)

Or, broken out into the five W format:

Who? 3rd Platoon.
What? Seize.
Where? The bridge at (Nx330159).
When? At 040600Z FEB 01.
Why? To pass the 1st Platoon (company main effort) on to OBJ BOB.

Figure 5-2. Example mission statement.

COMBAT ORDERS

5-16. Combat orders are the means by which the platoon leader receives and transmits information from the earliest notification that an operation will occur through the final steps of execution. WARNOs, OPORDs, and FRAGOs are absolutely critical to mission success. In a tactical situation, the platoon leader and subordinate leaders work with combat orders on a daily basis, and they must have precise knowledge of the correct format for each type of order. At the same time, they must ensure that every Soldier in the platoon understands how to receive and respond to the various types of orders. The skills associated with orders are highly perishable. Therefore, the platoon leader must take every opportunity to train the platoon in the use of combat orders with realistic practice.

WARNING ORDER

5-17. Platoon leaders alert their platoons by using a WARNO during the planning for an operation. WARNOs also initiate the platoon leader's most valuable time management tool—the parallel planning process. The platoon leader may issue a series of warning orders to his subordinate leaders to help them prepare for new missions. The directions and guidelines in the WARNO allow subordinates to begin their own planning and preparation activities.

(1) The content of WARNOs is based on two major variables: information available about the upcoming operation and special instructions. The information usually comes from the company commander. The platoon leader wants his subordinates to take appropriate action, so he normally issues his WARNOs either as he receives additional orders from the company or as he completes his own analysis of the situation.

(2) In addition to alerting the unit to the upcoming operation, WARNOs allow the platoon leader to issue tactical information incrementally and, ultimately, to shorten the length of the actual OPORD. WARNOs do not have a specific format, but one technique to follow is the five-

paragraph OPORD format. Table 5-2 shows an example of how the platoon leader might use WARNOs to alert the platoon and provide initial planning guidance.

Table 5-2. Example of multiple warning orders.

PLATOON LEADER'S ACTION	POSSIBLE CONTENT OF WARNING ORDER	PLATOON LEADER'S PURPOSE
Receive the company warning order	Warning order #1 covers: Type of mission and tentative task organization. Movement plan. Tentative timeline. Standard drills to be rehearsed.	Prepare squads for movement to the tactical assembly area. Obtain map sheets.
Conduct METT-TC analysis	Warning order #2 covers: Friendly situation. Enemy situation. Security plan. Terrain analysis. Platoon mission.	Initiate squad-level mission analysis. Initiate generic rehearsals (drill- and task-related). Prepare for combat.
Develop a plan	Warning order #3 covers: Concept of the operation. Concept of fires. Subordinate unit tasks and purposes. Updated graphics.	Identify platoon-level reconnaissance requirements. Direct leader's reconnaissance. Prepare for combat.

OPERATIONS ORDER

5-18. The OPORD is the five-paragraph directive issued by a leader to subordinates for the purpose of implementing the coordinated execution of an operation. When time and information are available, the platoon leader will normally issue a complete OPORD as part of his TLP. However, after issuing a series of WARNOs, he does not need to repeat information previously covered. He can simply review previously issued information or brief the changes or earlier omissions. He then will have more time to concentrate on visualizing his concept of the fight for his subordinates. As noted in his WARNOs, the platoon leader also may issue an execution matrix either to supplement the OPORD or as a tool to aid in the execution of the mission. However, the matrix order technique does not replace a five-paragraph OPORD.

FRAGMENTARY ORDER

5-19. A FRAGO is an abbreviated form of an OPORD (verbal, written, or digital) that normally follows the five-paragraph format. It is usually issued on a day-to-day basis that eliminates the need for restating information contained in a basic OPORD. It may be issued in sections. It is issued after an OPORD to change or modify that order and is normally focused on the next mission. The platoon leader uses a FRAGO to—

- Communicate changes in the enemy or friendly situation.
- Task subordinate elements based on changes in the situation.
- Implement timely changes to existing orders.
- Provide pertinent extracts from more detailed orders.
- Provide interim instructions until he can develop a detailed order.
- Specify instructions for subordinates who do not need a complete order.

SECTION III — TROOP-LEADING PROCEDURES

5-20. The TLP begin when the platoon leader receives the first indication of an upcoming mission. They continue throughout the operational process (plan, prepare, execute, and assess). The TLP comprise a sequence of actions that help platoon leaders use available time effectively and efficiently to issue orders and execute tactical operations. TLP are not a hard and fast set of rules. Some actions may be performed simultaneously or in an order different than shown in Figure 5-3. They are a guide that must be applied consistent with the situation and the experience of the platoon leader and his subordinate leaders. The tasks involved in some actions (such as initiate movement, issue the WARNO, and conduct reconnaissance) may recur several times during the process. The last action (activities associated with supervising and refining the plan) occurs continuously throughout TLP and execution of the operation. The following information concerning the TLP assumes that the platoon leader will plan in a time-constrained environment. All steps should be done, even if done in abbreviated fashion. As such, the suggested techniques are oriented to help a platoon leader quickly develop and issue a combat order.

RECEIVE THE MISSION
ISSUE A WARNING ORDER
MAKE A TENTATIVE PLAN
INITIATE MOVEMENT
CONDUCT RECONNAISSANCE
COMPLETE THE PLAN
ISSUE THE OPERATIONS ORDER
SUPERVISE AND REFINE

Figure 5-3. Troop-leading procedures.

RECEIVE THE MISSION

5-21. This step begins with the receipt of an initial WARNO from the company. It also may begin when the platoon leader receives the commander's OPORD, or it may result from a change in the overall situation. Receipt of mission initiates the planning and preparation process so the platoon leader can prepare an initial WARNO as quickly as possible. At this stage of the TLP, mission analysis should focus on determining the unit's mission and the amount of available time. For the platoon leader, mission analysis is essentially the analysis of the factors of METT-TC, but he must not become involved in a detailed METT-TC analysis. This will occur after issuing the initial WARNO. The platoon leader should use METT-TC from the enemy's perspective to develop the details of possible enemy courses of action (COA). The following can assist in this process.

- Understand the enemy's mission. Will the enemy's likely mission be based on his doctrine, knowledge of the situation, and capabilities? This may be difficult to determine if the enemy has no established order of battle. Enemy analysis must consider situational reports of enemy patterns. When does the enemy strike, and where? Where does the enemy get logistical support and fire support? What cultural or religious factors are involved?
  - Why is the enemy conducting this operation?
  - What are the enemy's goals and are they tied to specific events or times?
  - What are the enemy's capabilities?
  - What are the enemy's objectives? Based on the situation template (SITEMP) and the projected enemy mission, what are the enemy's march objectives (offense) or the terrain or force he intends to protect (defense)? The commander normally provides this information.
- If the enemy is attacking, which avenues will he use to reach his objectives in executing his COAs and why?
- How will terrain affect his speed and formations?
- How will he use key terrain and locations with clear observation and fields of fire?

- How will terrain affect his speed and formations?
- How will he use key terrain and locations with clear observation and fields of fire?
- Does the weather aid or hinder the enemy in accomplishing his mission or does the weather degrade the enemy's weapons or equipment effectiveness?
- Enemy obstacles are locations provided by the company commander, platoon leader's assessment, or obtained from reconnaissance that give the platoon leader insight into how the enemy is trying to accomplish his mission.
- Perhaps the most critical aspect of mission analysis is determining the combat power potential of one's force. The platoon leader must realistically and unemotionally determine what tasks his Soldiers are capable of performing. This analysis includes the troops attached to or in direct support of the platoon. The platoon leader must know the status of his Soldiers' experience and training level, and the strengths and weaknesses of his subordinate leaders. His assessment includes knowing the status of his Soldiers and their equipment, and it includes understanding the full array of assets that are in support of the platoon such as additional AT weapons, snipers, and engineers. For example, how much indirect fire is available and when is it available?

5-22. As addressed in the "receive the mission" TLP, time analysis is a critical aspect to planning, preparation, and execution. Not only must the platoon leader appreciate how much time is available, he must be able to appreciate the time-space aspects of preparing, moving, fighting, and sustaining. He must be able to see his own tasks and enemy actions in relation to time. The platoon leader should conduct backward planning and observe the "1/3 – 2/3 rule" to allow subordinates their own planning time. Examples of time analysis are as follows.

(1) He must be able to assess the impact of limited visibility conditions on the TLP.

(2) He must know how long it takes to conduct certain tasks such as order preparation, rehearsals, back-briefs, and other time-sensitive preparations for subordinate elements.

(3) He must understand how long it takes to deploy a support by fire (SBF) element, probably the weapons squad, and determine the amount of ammunition needed to sustain the support for a specific period of time.

(4) He must know how long it takes to assemble a bangalore torpedo and to breach a wire obstacle.

(5) Most importantly, as events occur, the platoon leader must adjust his analysis of time available to him and assess the impact on what he wants to accomplish.

(6) Finally, he must update previous timelines for his subordinates, listing all events that affect the platoon.

5-23. The commander will provide the platoon leader with civil considerations that may affect the company and platoon missions. The platoon leader also must identify any civil considerations that may affect only his platoon's mission. Platoons are likely to conduct missions in areas where there are numerous non-combatants and civilians on the battlefield. Some considerations may include refugee movement, humanitarian assistance requirements, or specific requirements related to the rules of engagement (ROE) or rules of interaction (ROI).

ISSUE A WARNING ORDER

5-24. After the platoon leader determines the platoon's mission and gauges the time available for planning, preparation, and execution, he immediately issues an oral WARNO to his subordinates. In addition to telling his subordinates of the platoon's new mission, the WARNO also gives them the platoon leader's planning timeline. The platoon leader relays all other instructions or information that he thinks will assist the platoon in preparing for the new mission. Such information includes information about the enemy, the nature of the overall plan, and specific instructions for preparation. Most importantly, by issuing the initial WARNO as quickly as possible, the platoon leader enables his subordinates to begin their own planning and preparation while he begins to develop the platoon operation order. An example may include the squads rehearsing designated battle drills. This is called parallel planning.

MAKE A TENTATIVE PLAN

5-25. After receiving the company OPORD (or FRAGO), the platoon leader develops a tentative plan. The process of developing this plan in a time-constrained environment usually has six steps: receipt of the mission, mission analysis, COA development, COA analysis, COA selection, and issue the order. The platoon leader relies heavily on the company commander's METT-TC analysis. This allows the platoon leader to save time by focusing his analysis effort on areas that affect his plan. Typically, a platoon leader will develop one COA. If more time is available, he may develop more than one, in which case he will need to compare these COAs and select the best one.

MISSION ANALYSIS

5-26. This is a continuous process during the course of the operation. It requires the platoon leader to analyze all the factors of METT-TC in as much depth as time and quality of information will allow. The factors of METT-TC are not always analyzed sequentially. How and when the platoon leader analyzes each factor depends on when information is made available to him. One technique for the analysis is based on the sequence of products that the company commander receives and produces: mission, enemy, terrain and weather, troops, time, civil considerations. The platoon leader must develop significant conclusions about how each element will affect mission accomplishment and then account for it in his plan.

MISSION

5-27. Leaders at every echelon must have a clear understanding of the mission, intent, and concept of the operation of the commanders one and two levels higher. Without this understanding, it would be difficult to exercise disciplined initiative. One technique to quickly understand the operation is to draw a simple sketch of the battalion and company's concepts of the operation (if not provided by the commander). The platoon leader now can understand how the platoon mission relates to the missions of other units and how his mission fits into the overall plan, and he can capture this understanding of the purpose (why) in his restated mission statement. The platoon leader will write a restated mission statement using his analysis of these areas: the battalion mission, intent, and concept; the company mission, intent, and concept; identification of specified, implied, and essential tasks; identification of risks; and any constraints.

- *Battalion Mission, Intent, and Concept.* The platoon leader must understand the battalion commander's concept of the operation. He identifies the battalion's task and purpose, and how his company is contributing to the battalion's fight. The platoon leader also must understand the battalion commander's intent found in the friendly forces paragraph (paragraph 1b) of the company order.
- *Company Mission, Intent, and Concept.* The platoon leader must understand the company's concept of the operation. He identifies the company's task and purpose, as well as his contribution to the company's fight. The platoon leader must clearly understand the commander's intent from the order (paragraph 3a). Additionally, the platoon leader identifies the task, purpose, and disposition for all adjacent maneuver elements under company control.
- *Platoon Mission.* The platoon leader finds his platoon's mission in the company's concept of the operation paragraph. The purpose of the main effort platoon usually matches the purpose of the company. Similarly, shaping operation platoons' purposes must relate to the purpose of the main effort platoon. The platoon leader must understand how his purpose relates to the other platoons in the company. He determines the platoon's essential tactical task to successfully accomplish his given purpose. Finally, he must understand why the commander gave his platoon a particular tactical task and how it fits into the company's concept of the operation.
- *Constraints.* Constraints are restrictions placed on the platoon leader by the commander to dictate action or inaction, thus restricting the freedom of action the platoon leader has for planning by stating the things that must or must not be done. The two types of constraints are: requirements for action (for example, maintain a squad in reserve); and prohibitions of action (for example, do not cross phaseline [PL] BULL until authorized).
- *Identification of Tasks.* The platoon leader must identify and understand the tasks required to accomplish the mission. There are three types of tasks: specified; implied; and essential.

  - Specified Tasks. These are tasks specifically assigned to a platoon by the commander. Paragraphs 2 and 3 from the company OPORD state specified tasks. Specified tasks may also be found in annexes and overlays (see p. 5-21 for OPORD example).
  - Implied Tasks. These are tasks that must be performed to accomplish a specified task, but which are not stated in the OPORD. Implied tasks are derived from a detailed analysis of the OPORD, the enemy situation, the COAs, and the terrain. Analysis of the platoon's current location in relation to future areas of operation as well as the doctrinal requirements for each specified task also might provide implied tasks. SOP tasks are not considered implied tasks.
  - Essential Tasks. An essential task is one that must be executed to accomplish the mission derived from a review of the specified and implied tasks. This is normally the task found in the mission statement
- *Identification of Risks.* Risk is the chance of injury or death to individuals and damage to or loss of vehicles and equipment. Risk, or the potential for risk, is always present in every combat and training situation the platoon faces. Risk management must take place at all levels of the chain of command during every operation. It is an integral part of tactical planning. The platoon leader, his NCOs, and all other platoon Soldiers must know how to use risk management, coupled with fratricide avoidance measures, to ensure that the mission is executed in the safest possible environment within mission constraints. The platoon leader should review risk from a tactical perspective (how can they best accomplish the mission with the least damage to their unit?) and an individual perspective (how do I minimize the chances of my Soldiers getting hurt and keep my equipment from being damaged?). Refer to Chapter 4 for a detailed discussion of risk management and fratricide avoidance.
- *Restated Platoon Mission Statement.* The platoon leader restates his mission statement using the five Ws: who, what, when, where, and why. The "who" is the platoon. The "what" is the type of operation and the platoon's essential tactical task. The "when" is the date-time group (DTG) given in the OPORD. The "where" is the objective or location taken from the OPORD. The "why" is the purpose for the platoon's essential tactical task taken from the commander's paragraph 3.

Analysis of Terrain and Weather

5-28. The platoon leader must conduct a detailed analysis of the terrain to determine how it will uniquely affect his unit and the enemy he anticipates fighting. The platoon leader must gain an appreciation of the terrain before attempting to develop either enemy or friendly COA. He must exceed merely making observations (for example, this is high ground, this is an avenue of approach). He must arrive at significant conclusions concerning how the ground will affect the enemy and his unit. Because of limited planning time, the platoon leader normally prioritizes his terrain analysis. For example, in the conduct of an assault, his priority may be the area around the objective followed by the platoon's specific axis leading to the objective.

5-29. Terrain mobility is classified in one of three categories:

(1) Unrestricted. This is terrain free of any movement restrictions. No actions are required to enhance mobility. For mechanized forces, unrestricted terrain is typically flat or moderately sloped, with scattered or widely spaced obstacles such as trees or rocks. Unrestricted terrain generally allows wide maneuver and offers unlimited travel over well-developed road networks. Unrestricted terrain is an advantage in situations requiring rapid movement for mechanized forces.

(2) Restricted. This terrain hinders movement to some degree, and units may need to detour frequently. Restricted terrain may cause difficulty in maintaining optimal speed, moving in some types of combat formations, or transitioning from one formation to another. This terrain typically encompasses moderate to steep slopes or moderate to dense spacing of obstacles such as trees, rocks, or buildings. The terrain may not require additional assets or time to traverse, but it may

hinder movement to some degree due to increased security requirements. In instances when security is the paramount concern, both friendly and enemy elements may move in more restricted terrain that may provide more cover and concealment.

(3) Severely Restricted. This terrain severely hinders or slows movement in combat formations unless some effort is made to enhance mobility. It may require a commitment of engineer forces to improve mobility or a deviation from doctrinal tactics, such as using a column rather than a wedge formation or moving at speeds much slower than otherwise preferred. Severely restricted terrain includes any terrain that requires equipment not organic to the unit to cross (for example, a large body of water and slopes requiring mountaineering equipment).

5-30. The military aspects of terrain observation (Figure 5-4) are used to analyze the ground. The sequence used to analyze the military aspects of terrain can vary. The platoon leader may prefer to determine obstacles first, avenues of approach second, key terrain third, observation and fields of fire fourth, and cover and concealment last. For each aspect of terrain, the platoon leader determines its effect on both friendly and enemy forces. The following are OAKOC aspects of terrain.

OAKOC

OBSERVATION AND FIELDS OF FIRE.
AVENUES OF APPROACH.
KEY AND DECISIVE TERRAIN.
OBSTACLES.
COVER AND CONCEALMENT.

Figure 5-4. Military aspects of terrain.

OBSTACLES

5-31. The platoon leader first identifies existing and reinforcing obstacles in his AO that limit his mobility with regards to the mission. Existing obstacles are typically natural terrain features present on the battlefield. These may include ravines, gaps, or ditches over 3-meters wide; tree stumps and large rocks over 18-inches high; forests with trees 8 inches or greater in diameter and with less than 4 meters between trees; and manmade obstacles such as towns or cities. Reinforcing obstacles are typically manmade obstacles that augment existing obstacles. These may include minefields, AT ditches, road craters, abatis and log cribs, wire obstacles, and infantry strongpoints. Figure 5-5 lists several offensive and defensive considerations the platoon leader can include in his analysis of obstacles and restricted terrain.

OFFENSIVE CONSIDERATIONS

- How is the enemy using obstacles and restricted terrain features?
- What is the composition of the enemy's reinforcing obstacles?
- How will obstacles and terrain affect my movement and or maneuver?
- If necessary, how can the company avoid such features?
- How do we detect and, if desired, bypass the obstacles?
- Where has the enemy positioned weapons to cover the obstacles, and what type of weapons is he using?
- If I must support or execute a breach, where is the expected breach site?

DEFENSIVE CONSIDERATIONS

- Where do I want to kill the enemy? Where do I want him to go?
- How will existing obstacles and restricted terrain affect the enemy?
- Where does the enemy want to go?
- How can I use these features to force the enemy into my engagement area, deny him an avenue, or disrupt his movement?

Figure 5-5. Considerations in obstacle and terrain analysis.

AVENUES OF APPROACH

5-32. An avenue of approach is an air or ground route of an attacking force leading to its objective or key terrain. For each avenue of approach, the platoon leader determines the type (mounted, dismounted, air, or subterranean), size, and formation and speed of the largest unit that can travel along it. The commander may give him this information. Mounted forces may move on avenues along unrestricted or restricted terrain (or both). Dismounted avenues and avenues used by reconnaissance elements and infantry platoons normally include terrain that is restricted and at times severely restricted to mounted forces. The terrain analysis also must identify avenues of approach for both friendly and enemy units. Figure 5-6 lists several considerations for avenue of approach analysis.

OFFENSIVE CONSIDERATIONS

- How can I use each avenue of approach to support my movement and maneuver?
- How will each avenue support movement techniques, formations, and (once we make enemy contact) maneuver?
- Will variations in trafficability or lane width force changes in formations or movement techniques or require defile drills?
- What are the advantages and disadvantages of each avenue?
- What are the enemy's likely counterattack routes?
- Do lateral routes exist that we can use to shift to other axes or that the enemy can use to threaten our flanks?

DEFENSIVE CONSIDERATIONS

- What are all likely enemy avenues into my sector?
- How can the enemy use each avenue of approach?
- Do lateral routes exist that the enemy can use to threaten our flanks?
- What avenues would support a friendly counterattack?

Figure 5-6. Considerations for avenue of approach analysis.

KEY TERRAIN

5-33. Key terrain affords a marked advantage to the combatant who seizes, retains, or controls it. The platoon leader identifies key terrain starting at the objective or main battle area and working backwards to his current position. It is a conclusion rather than an observation. The platoon leader must assess what terrain is key to accomplishing his mission. Key terrain may allow the platoon leader to apply direct fire or achieve observation of the objective (or avenue of approach). Key terrain may also be enemy oriented, meaning that if the enemy controls the terrain it could prevent the platoon from accomplishing its mission.

- An example of key terrain for a platoon could be a tree line on a hillside that provides overwatch of a high-speed avenue of approach. Controlling this tree line may be critical in passing follow-on forces (main effort) to their objective. High ground is not necessarily key terrain. A prominent hilltop that overlooks an avenue of approach and offers clear observation and fields of fire, if it is easily bypassed, is not key terrain.
- Although unlikely, the platoon leader may identify decisive terrain—key terrain that holds such importance that the seizure, retention, and control of it will be necessary for mission accomplishment and may decide the outcome of the battle. Use the following two military aspects of terrain (observation and fields of fire, and cover and concealment) to analyze each piece of key terrain. Figure 5-7 depicts operational considerations to use when analyzing key terrain.

OPERATIONAL CONSIDERATIONS

- What terrain is key to the company and to the battalion and why?
- Is the enemy controlling this key terrain?
- What terrain is key to the enemy and why?
- How do I gain or maintain control of key terrain?
- What terrain is key for friendly observation, both for command and control and for calling for fires?

Figure 5-7. Considerations in key terrain analysis.

OBSERVATION AND FIELDS OF FIRE

5-34. The platoon leader analyzes areas surrounding key terrain, objectives, avenues of approach, and obstacles to determine if they provide clear observation and fields of fire for both friendly and enemy forces. He locates intervisibility lines (terrain that inhibits observation from one point to another) that have not been identified by the commander and determines where visual contact between the two forces occurs. When analyzing fields of fire, the platoon leader focuses on both friendly and enemy direct fire capabilities. Additionally, he identifies positions that enable artillery observers to call for indirect fires and permit snipers to engage targets. Figure 5-8 provides considerations for analysis of observation and fields of fire. Whenever possible, the platoon leader conducts a ground reconnaissance from both the friendly and enemy perspective.

OFFENSIVE CONSIDERATIONS
• Are clear observation and fields of fire available on or near the objective for enemy observers and weapon systems? • Where can the enemy concentrate fires? • Where is he vulnerable? • Where are possible SBF or assault-by-fire positions for friendly forces? • Where are the natural target reference points (TRPs)? • Where do I position indirect fire observers?
DEFENSIVE CONSIDERATIONS
• What locations afford clear observation and fields of fire along enemy avenues of approach? • Where will the enemy set firing lines and or antitank weapons? • Where will I be unable to mass fires? • Where is the dead space in my sector? Where am I vulnerable? • Where are the natural TRPs? • Where do I position indirect fire observers?

Figure 5-8. Considerations for analysis of observation and fields of fire.

COVER AND CONCEALMENT

5-35. Cover is protection from the effects of fires. Concealment is protection from observation but not direct fire or indirect fires. Figure 5-9 provides considerations for analysis of cover and concealment. Consideration of these elements can lead the platoon leader to identify areas that can, at best, achieve both facets. The platoon leader looks at the terrain, foliage, structures, and other features on the key terrain, objective, and avenues of approach to identify sites that offer cover and concealment.

OFFENSIVE CONSIDERATIONS
• What axes afford both clear fields of fire and effective cover and concealment? • Which terrain provides bounding elements with cover and concealment while facilitating lethality?

Figure 5-9. Considerations in analysis of cover and concealment.

Five military aspects of weather

(1) Visibility.
(2) Winds.
(3) Precipitation.
(4) Cloud cover.
(5) Temperature/humidity.

5-36. The platoon leader must go beyond merely making observations. He must arrive at significant conclusions about how the weather will affect his platoon and the enemy. He receives conclusions from the commander and identifies his own critical conclusions about the weather. Most importantly, the platoon leader must apply these conclusions when he develops friendly and enemy COAs. The five military aspects of weather are—

- Visibility. The platoon leader identifies critical conclusions about visibility factors (such as fog, smog, and humidity) and battlefield obscurants (such as smoke and dust). Some visibility considerations are—

- Will the current weather favor the use of smoke to obscure during breaching?
- Will fog affect friendly and enemy target acquisition?

- Light Data. The platoon leader identifies critical conclusions about beginning morning nautical twilight (BMNT), sunrise (SR), sunset (SS), end of evening nautical twilight (EENT), moonrise (MR), moonset (MS), and percentage of illumination. Some light data considerations are—
  - Will the sun rise behind my attack?
  - How can I take advantage of the limited illumination?
  - How will limited illumination affect friendly and enemy target acquisition?
- Temperature. The platoon leader identifies critical conclusions about temperature factors (such as high and low temperatures and infrared crossover times) and battlefield factors (such as use of smoke or chemicals). Some temperature considerations are—
  - How will temperature (hot or cold) affect rate of foot march for the platoon?
  - How will temperature (hot or cold) affect the Soldiers and equipment?
  - Will temperatures favor the use of nonpersistent chemicals?
- Precipitation. The platoon leader identifies critical conclusions about precipitation factors (such as type, amount, and duration). Some precipitation considerations are—
  - How will precipitation affect mobility?
  - How can precipitation add to the platoon achieving surprise?
- Winds. The platoon leader identifies critical conclusions about wind factors (such as direction and speed). Some wind considerations are—
  - Will wind speed cause smoke to dissipate quickly?
  - Will wind speed and direction favor enemy use of smoke?

5-37. The platoon leader identifies critical conclusions about cloud cover (such as target acquisition degradation, aircraft approach, and radar effectiveness). Some cloud cover considerations are—

- Will heavy cloud cover limit illumination and solar heating of targets?
- Will heavy cloud cover degrade the use of infrared-guided artillery?
- Will cloud cover cause glare, a condition that attacking aircraft might use to conceal their approach?
- Will the cloud cover affect ground surveillance radar (GSR) coverage of the AO?

ANALYSIS OF ENEMY

5-38. This step allows the platoon leader to identify the enemy's strength and potential weaknesses or vulnerabilities so he can exploit them to generate overwhelming combat power in achieving his mission. The platoon leader must understand the assumptions the commander used to portray the enemy's COAs covered in the company's plan. Furthermore, the platoon leader's assumptions about the enemy must be consistent with those of the company commander. To effectively analyze the enemy, the platoon leader must know how the enemy may fight. It is equally important for the platoon leader to understand what is actually known about the enemy as opposed to what is only assumed or templated.

5-39. During doctrinal analysis, it is not enough only to know the number and types of vehicles, soldiers, and weapons the enemy has. The platoon leader's analysis must extend down to the individual key weapon system. During stability operations or small-scale contingency (SSC) operations in an underdeveloped area where little is known about the combatants, it may be difficult to portray or template the enemy doctrinally. In this case, the platoon leader must rely on brigade and battalion analyses funneled through the company commander as well as his own knowledge of recent enemy activities. The platoon leader should consider the following areas as he analyzes the enemy.

- *Composition.* The platoon leader's analysis must determine the number and types of enemy vehicles, soldiers, and equipment that could be used against his platoon. He gets this information from paragraph 1a of the company OPORD. His analysis also must examine how the enemy organizes for combat to include the possible use of a reserve.

- *Disposition.* From the commander's information, the platoon leader identifies how the enemy that his platoon will fight is arrayed.
- *Strength.* The platoon leader identifies the strength of the enemy. It is imperative that the platoon leader determines the actual numbers of equipment and personnel that his platoon is expected to fight or that may affect his platoon. Again, much of this information is gained through the detailed OPORD.
- *Capabilities.* Based on the commander's assessment and the enemy's doctrine and current location, the platoon leader must determine what the enemy is capable of doing against his platoon during the mission. Such an analysis must include the planning ranges for each enemy weapons system that the platoon may encounter.
- *Anticipated Enemy Courses of Action.* To identify potential enemy COAs, the platoon leader weighs the result of his initial analysis of terrain and weather against the enemy's composition, capabilities, and doctrinal objectives. He then develops an enemy SITEMP for his portion of the company plan. The end product is a platoon SITEMP, a graphic overlay depiction of how he believes the enemy will fight under the specific conditions expected on the battlefield. The commander's analysis and understanding of the current enemy and friendly situation will provide the platoon leader with most of this information. Included in the SITEMP is the range fan of the enemy's weapons and any tactical and protective obstacles, either identified or merely templated. Once the SITEMP has been developed it should be transferred to a large-scale sketch to enable subordinates to see the details of the anticipated enemy COA. After the platoon leader briefs the enemy analysis to his subordinates, he must ensure they understand what is known, what is suspected, and what merely templated (educated guess) is. The platoon's SITEMP should depict individual Soldier and weapons positions and is a refinement of the commander's SITEMP.

Summary of Mission Analysis

5-40. The end result of mission analysis, as done during the formulation of a tentative plan, is a number of insights and conclusions regarding how the factors of METT-TC affect accomplishment of the platoon's mission. The platoon leader must determine how he can apply his strengths against enemy weakness, while protecting his weaknesses from enemy strength. From these the platoon leader will develop a COA.

Course of Action Development

5-41. The purpose of COA development is to determine one (or more) way(s) to achieve the mission by applying the overwhelming effects of combat power at the decisive place or time with the least cost in friendly casualties. If time permits, the platoon leader may develop several COAs. The platoon leader makes each COA as detailed as possible to describe clearly how he plans to use his forces to achieve the unit's purpose and mission-essential task(s) consistent with the commander's intent. He focuses on the actions the unit must take at the decisive point and works backward to his start point. A COA should satisfy the criteria listed in Table 5-3.

NOTE: The platoon leader should consider (METT-TC dependent) incorporating his squad leaders and platoon sergeant in COA development. Incorporating the squad leaders and platoon sergeant in the process may add time to the initial COA development process, but it will save time by increasing their understanding of the platoon's plan.

Table 5-3. Course of action criteria.

Suitable	If the COA were successfully executed, would the unit accomplish the mission consistent with the battalion and company commander's concept and intent?
Feasible	The platoon must have the technical and tactical skill and resources to successfully accomplish the COA. In short, given the enemy situation and terrain, the unit must have the training, equipment, leadership, and rehearsal time necessary to successfully execute the mission.
Distinguishable	If more than one COA is developed, then each COA must be sufficiently different from the others to justify full development and consideration. At platoon level, this is very difficult to accomplish, particularly if the platoon has limited freedom of action or time to plan and prepare.
Complete	The COA must include the operational factors of who, what, when, where, and how. The COA must address the doctrinal aspects of the operation. For example, in the attack against a defending enemy, the COA must cover movement to, deployment against, assault of, and consolidation upon the objective.

(1) ***COA Development Step 1: Analyze Relative Combat Power.*** This step compares combat power strengths and weaknesses of both friendly and enemy forces. At the platoon level this should not be a complex process. However, if the platoon is attacking or defending against a force that has no order of battle but has exhibited guerrilla- or terrorist-type tactics, it could be difficult. For the platoon leader, it starts by returning to the conclusions the commander arrived at during mission analysis, specifically the conclusions about the enemy's strength, weakness, and vulnerabilities. In short, the platoon leader is trying to ascertain where, when, and how the platoon's combat power (Intelligence, Movement and Maneuver, Fire Support, Protection, Sustainment, and Command and Control) can be superior to the enemy's while achieving the mission. This analysis should lead to techniques, procedures, and a potential decisive point that will focus the COA development. See FM 1-02 for the definition of a decisive point.

- ***COA Development Step 2: Generate Options.*** The platoon leader must first identify the objectives or times at which the unit will mass overwhelming firepower to achieve a specific result (with respect to terrain, enemy, and or time) that will accomplish the platoon's mission. He should take the following action.
- *Determine the Doctrinal Requirements.* As the platoon leader begins to develop a COA he should consider, if he has not done so in mission analysis, what doctrine suggests in terms of accomplishing the mission. For example, in an attack of a strongpoint, doctrine outlines several steps: isolate the objective area and the selected breach site, attack to penetrate and seize a foothold in the strongpoint, exploit the penetration, and clear the objective. In this case, doctrine gives the platoon leader a framework to begin developing a way to accomplish the mission.
- *Determine the Decisive Point.* The next and most important action is to identify a decisive point in order to progress with COA development. The decisive point may be given to the platoon leader by the company commander or be determined by the platoon leader through his relative combat power analysis.
- *Determine the Purpose of Each Element.* Determine the purpose of the subordinate elements starting with the main effort. The main effort's purpose is nested to the platoon's purpose and is achieved at the platoon leader's decisive point. The platoon leader next identifies the purposes of shaping efforts. These purposes are nested to the main effort's purpose by setting the conditions for success of the main effort.
- *Determine Tasks of Subordinate Elements.* Starting with the main effort, the platoon leader specifies the essential tactical tasks that will enable the main and shaping efforts to achieve their purpose.

(2) ***COA Development Step 3: Array Initial Forces.*** The platoon leader next must determine the specific number of squads and weapons necessary to accomplish the mission and provide a basis for

development of a scheme of maneuver. He will consider the platoon's restated mission statement, the commander's intent, and the enemy's most probable COA. He should allocate resources to the main effort (at the decisive point) and continue with shaping efforts in descending order of importance to accomplish the tasks and purposes he assigned during Step 2. For example, the main effort in an attack of a strong point may require a rifle squad and an engineer squad to secure a foothold, whereas an SBF force may require the entire weapons squad.

(3) ***COA Development Step 4: Develop Schemes of Maneuver.*** The scheme of maneuver is a description of how the platoon leader envisions his subordinates will accomplish the mission from the start of the operation until its completion. He does this by determining how the achievement of one task will lead to the execution of the next. He clarifies in his mind the best ways to use the available terrain as well as how best to employ the platoon's strengths against the enemy's weaknesses (gained from his relative combat power analysis). This includes the requirements of indirect fire to support the maneuver. The platoon leader then develops the maneuver control measures necessary to enhance understanding of the scheme of maneuver, ensure fratricide avoidance, and to clarify the task and purpose of the main and shaping efforts. (Refer to Chapter 4 for a detailed discussion of fratricide avoidance.) He also determines the supply and medical evacuation aspects of the COA.

(4) ***COA Development Step 5: Assign Headquarters.*** The platoon leader assigns specific elements (for example, squads) as the main and shaping efforts. The platoon leader ensures that he has employed every element of the unit and has C2 for each element.

(5) ***COA Development Step 6: Prepare COA Statements and Sketches.*** The platoon leader's ability to prepare COA sketches and statements will depend on the amount of time available and his skill and experience as a platoon leader. Whenever possible, the platoon leader should prepare a sketch showing the COA. The COA statement is based on the scheme of maneuver the commander has already developed and the platoon leader's situational analysis. It focuses on all significant actions from the start of the COA to its finish. The company commander should provide the platoon and squad leaders his COA analysis when time is a limiting factor. Particularly if the order is verbal, it is extremely useful to have one or more sketches of critical events within the plan that require coordinated movement of two or more subordinate units.

- **Wargaming of COA.** After developing a COA, the platoon leader wargames it to determine its advantages and disadvantages, to visualize the flow of the battle, and to identify requirements to synchronize actual execution. This is typically done during a discussion with the squad leaders, platoon sergeant, or other key personnel. This technique is not complicated, and it facilitates a total understanding of the plan. This is not a rehearsal. The wargame is designed to synchronize all platoon actions, whereas during COA development the leader is focused on simply integrating all platoon assets into the fight.
- **COA Comparison and Selection.** If the platoon leader develops more than one COA, he must compare them by weighing the specific advantages, disadvantages, strengths, and weaknesses of each. These attributes may pertain to the accomplishment of the platoon purpose, the use of terrain, the destruction of the enemy, or any other aspect of the operation that the platoon leader believes is important. The platoon leader uses these factors as his frame of reference in tentatively selecting the best COA. He makes the final selection of a COA based on his own analysis.

INITIATE MOVEMENT

5-42. The platoon leader initiates any movement that is necessary to continue preparations or to posture the unit for the operation. This may include movement to an assembly area (AA), battle position, perimeter defense, or attack position; movement of reconnaissance elements; or movement to compute time-distance factors for the unit's mission.

OFFENSIVE CONSIDERATIONS
• Are clear observation and fields of fire available on or near the objective for enemy observers and weapon systems? • Where can the enemy concentrate fires? • Where is he vulnerable? • Where are possible SBF or assault-by-fire positions for friendly forces? • Where are the natural target reference points (TRPs)? • Where do I position indirect fire observers?
DEFENSIVE CONSIDERATIONS
• What locations afford clear observation and fields of fire along enemy avenues of approach? • Where will the enemy set firing lines and or antitank weapons? • Where will I be unable to mass fires? • Where is the dead space in my sector? Where am I vulnerable? • Where are the natural TRPs? • Where do I position indirect fire observers?

Figure 5-8. Considerations for analysis of observation and fields of fire.

COVER AND CONCEALMENT

5-35. Cover is protection from the effects of fires. Concealment is protection from observation but not direct fire or indirect fires. Figure 5-9 provides considerations for analysis of cover and concealment. Consideration of these elements can lead the platoon leader to identify areas that can, at best, achieve both facets. The platoon leader looks at the terrain, foliage, structures, and other features on the key terrain, objective, and avenues of approach to identify sites that offer cover and concealment.

OFFENSIVE CONSIDERATIONS
• What axes afford both clear fields of fire and effective cover and concealment? • Which terrain provides bounding elements with cover and concealment while facilitating lethality?

Figure 5-9. Considerations in analysis of cover and concealment.

Five military aspects of weather

(1) Visibility.
(2) Winds.
(3) Precipitation.
(4) Cloud cover.
(5) Temperature/humidity.

5-36. The platoon leader must go beyond merely making observations. He must arrive at significant conclusions about how the weather will affect his platoon and the enemy. He receives conclusions from the commander and identifies his own critical conclusions about the weather. Most importantly, the platoon leader must apply these conclusions when he develops friendly and enemy COAs. The five military aspects of weather are—

- Visibility. The platoon leader identifies critical conclusions about visibility factors (such as fog, smog, and humidity) and battlefield obscurants (such as smoke and dust). Some visibility considerations are—

- The format of the five-paragraph OPORD helps the platoon leader paint a complete picture of all aspects of the operation: terrain, enemy, higher and adjacent friendly units, platoon mission, execution, support, and command. The format also helps him address all relevant details of the operation. Finally, it provides subordinates with a predictable, smooth flow of information from beginning to end.

SUPERVISE AND REFINE

5-46. The platoon leader supervises the unit's preparation for combat by conducting confirmation briefs, rehearsals, and inspections. Table 5-4 lists the items the unit should have.

Table 5-4. Precombat checklist.

Precombat Checklist		
ID card	Pintels	Grappling hook
ID tags	T&E mechanisms	Sling sets
Ammunition	Spare barrels	PZ marking kit
Weapons	Spare barrel bags	ANCD
Protective mask	Extraction tools	Plugger/GPS
Knives	Asbestos gloves	Handheld microphones
Flashlights	Barrel changing handles	NVDs
Radios and backup communication	Headspace and timing gauges	Batteries and spare batteries
Communication cards	M249 tools	Picket pounder
9-line MEDEVAC procedures	BII	Engineer stakes
	Oil & transmission fluids	Pickets
OPORD	Anti-freeze coolant	Concertina wire
FRAGOs	5-gallon water jugs	TCP signs
Maps	MREs	IR lights
Graphics, routes, OBJs, LZs, and PZs	Load plans	Glint tape
Protractors	Fuel cans	Chemical lights
Alcohol pens	Fuel spout	Spare hand sets
Alcohol erasers	Tow bars	Pencil with eraser
Pen and paper	Slave cables	Weapon tie downs
Tripods	Concertina wire gloves	

5-47. Platoon leaders should conduct a confirmation brief after issuing the oral OPORD to ensure subordinates know the mission, the commander's intent, the concept of the operation, and their assigned tasks. Confirmation briefs can be conducted face to face or by radio, depending on the situation. Face to face is the desired method, because all section and squad leaders are together to resolve questions, and it ensures that each leader knows what the adjacent squad is doing.

5-48. The platoon conducts rehearsals. During the rehearsals, leaders practice sending tactical reports IAW the unit's SOPs. Reporting before, during, and after contact with the enemy is rehearsed in detail starting with actions on the objective. Rehearsals are not intended to analyze a COA.

(1) The platoon leader uses well-planned, efficiently run rehearsals to accomplish the following:

- Reinforce training and increase proficiency in critical tasks.
- Reveal weaknesses or problems in the plan.
- Integrate and synchronize the actions of attached elements.
- Confirm coordination requirements between the platoon and adjacent units.
- Confirm each Soldier's understanding of the mission, concept of the operation, the direct fire plan, anticipated contingencies, and possible actions and reactions for various situations that may arise during the operation.

(2) Rehearsal techniques include the following:

- *Map Rehearsal.* A map rehearsal is usually conducted as part of a confirmation brief involving subordinate leaders or portions of their elements. The leader uses the map and overlay to guide participants as they brief their role in the operation. If necessary, he can use a sketch map. A sketch map provides the same information as a terrain model and can be used at any time.
- *Sand Table or Terrain Model.* This reduced-force or full-force technique employs a small-scale sand table or model that depicts graphic control measures and important terrain features for reference and orientation. Participants walk around the sand table or model to practice the actions of their own elements or vehicles (if working with mechanized units) in relation to other members of the platoon.
- *Radio Rehearsal.* This is a reduced-force or full-force rehearsal conducted when the situation does not allow the platoon to gather at one location. Subordinate elements check their communications systems and rehearse key elements of the platoon plan.
- *Reduced-Force Rehearsal.* In this rehearsal, leaders discuss the mission while moving over key terrain or similar terrain.
- *Full-Force Rehearsal.* This technique is used during a full-force rehearsal. Rehearsals begin in good visibility over open terrain and become increasingly realistic until conditions approximate those expected in the AO.

NOTE: If time permits, the platoon should conduct a full-force rehearsal of the plan.

SQUAD ORDERS

5-49. The squad leader follows the same format as in Figure 5-10 and issues his five-paragraph format OPORD to his squad. Because the squad is the smallest maneuver element, he does not develop COAs. He must, however, assign specific tasks and purposes to his team leaders to ensure his squad mission is accomplished.

EXAMPLE

1. **Situation.**
   a. **Enemy Forces.** The enemy the squad is expected to encounter or who could affect the squad mission. Include the composition, disposition, and capabilities of the enemy.
   b. **Friendly Forces.** The squad leader states the company's mission and the commander's intent along with the platoon leader's mission. Also states the missions of sister squads or other units on his flanks and rear, and how they influence his squad.
   c. **Attachments and Detachments.** Squads normally fight as a unit without attachments.

2. **Mission.** The squad's mission statement includes the 5 W's: Who, What (task), Where, When, and Why (purpose).

3. **Execution.**
   a. **Concept of the Operation.** The squad leader describes in one or more paragraphs the employment of the other squads and platoon attachments for the platoon mission. The platoon main effort, shaping effort, and general plan for fire support. Any engineer, reconnaissance, or security operations are also included.
   b. **Maneuver.** The squad leader describes in detail the task and purpose for each of his fire teams. Movement formations, actions on the objective, and engagement/disengagement criteria from start to mission completion are included.
   C. **Fires.** The squad leader explains the concept of indirect and direct fires to include priority of fires for artillery and mortars, when fires will begin, cease or shift, and target integration.

4. **Service Support.** Here the squad leader explains logistical items needed for the mission that are not covered in the unit SOPs. Additional ammunition, location of casualty collection points (CCPs), caches, enemy prisoner of war (EPW) collection points, and any other items the squad members may need that are unique to the mission are included.

5. **Command and Signal.**
   a. **Command.** Location of the platoon and company command posts or key leaders during the mission.
   b. **Signal.** Unit SOPs normally cover most of the signal instructions. Unique areas for the mission may be methods of communication, pyrotechnics to be used to signal, code words, running password, challenge and passwords (used when behind friendly lines), and other special signal instructions.

Figure 5-10. Five-paragraph format OPORD example.

Chapter 6
Sustainment

Sustainment facilitates uninterrupted operations by means of logistical support. It is accomplished through supply systems, maintenance, and other services that ensure continuous support throughout combat operations. The platoon leader is responsible for planning sustainment. The platoon sergeant is the platoon's main sustainment operator. The platoon sergeant works closely with the company executive officer (XO) and first sergeant (1SG) to ensure the platoon receives the required support for its assigned missions. Sustainment responsibilities and procedures in the platoon are the same as those that are habitually associated with Infantry units. The platoon and company rely heavily upon their higher headquarters for their sustainment needs. The company normally forecasts supplies with input from each platoon. in the process.

SECTION I — INDIVIDUAL RESPONSIBILITIES

6-1. Sustainment is an ever present requirement in all operations. All Soldiers, leaders, and units have sustainment responsibilities. At the tactical level there are two main categories of personnel: sustainment providers; and sustainment users. Both the sustainment provider and the sustainment user have responsibilities for making the system work. The sustainment provider brings the sustainment user the supplies needed to fight. An Infantry platoon is normally a sustainment user only, having no organic sustainment assets. This section focuses on specific individual responsibilities within the platoon's sustainment chain.

PLATOON SERGEANT

6-2. As the platoon's main sustainment operator, the platoon sergeant executes the platoon's logistical plan based on mission requirements, and platoon and company SOPs. The platoon sergeant's sustainment duties include—

- Participating in sustainment rehearsals at the company level and integrating sustainment into the platoon's maneuver rehearsals.
- Receiving, consolidating, and forwarding all administrative, personnel, and casualty reports to the 1SG as directed or IAW unit SOP.
- Obtaining supplies, equipment (except Class VIII), and mail from the supply sergeant and ensuring proper distribution.
- Supervising evacuation of casualties, KIAs, EPWs, and damaged equipment.
- Maintaining the platoon's manning roster.
- Cross-leveling supplies and equipment throughout the platoon.
- Coordinating logistics/personnel requirements with attached or OPCON units.

SQUAD LEADER

6-3. Each squad leader's sustainment duties include:

- Ensuring Soldiers perform proper maintenance on all assigned equipment.
- Ensuring Soldiers maintain personal hygiene.
- Compiling personnel and logistics reports for the platoon and submitting them to the platoon sergeant as directed or IAW unit SOP.

- Obtaining supplies, equipment (except Class VIII), and mail from the platoon sergeant and ensuring proper distribution.
- Cross-leveling supplies and equipment throughout the squad.

TRAUMA SPECIALIST/PLATOON MEDIC

6-4. The trauma specialist/platoon medic is attached from the battalion medical platoon to provide emergency medical treatment for sick, injured, or wounded platoon personnel. Emergency medical treatment procedures performed by the trauma specialist may include opening an airway, starting intravenous fluids, controlling hemorrhage, preventing or treating for shock, splinting fractures or suspected fractures, and providing relief for pain. The trauma specialist is trained under the supervision of the battalion surgeon or physician's assistant (PA) and medical platoon leader. The trauma specialist is also responsible for—

- Triaging injured, wounded, or ill friendly and enemy personnel for priority of treatment.
- Conducting sick call screening for the platoon.
- Assisting in the evacuation of sick, injured, or wounded personnel under the direction of the platoon sergeant.
- Assisting in the training of the platoon's combat lifesavers in enhanced first-aid procedures.
- Requisitioning Class VIII supplies from the battalion aid station (BAS) for the platoon according to the tactical standing operating procedure (TSOP).
- Recommending locations for platoon casualty collection point(s) (CCP).
- Providing guidance to the platoon's combat lifesavers as required.

COMBAT LIFESAVER

6-5. The combat lifesaver (CLS) is a nonmedical Soldier trained to provide advanced first aid/lifesaving procedures beyond the level of self-aid or buddy aid. The CLS is not intended to take the place of medical personnel. His specialized training can slow deterioration of a wounded Soldier's condition until treatment by medical personnel is possible. Each certified combat lifesaver is issued a CLS aid bag. Whenever possible, the platoon leader ensures there is at least one CLS in each fire team.

TRAINING

6-6. Because combat lifesaving is an organic capability, the platoon should make it a training priority. An emerging "first responder" program is now expanding CLS trauma treatment with increased emphasis on combat and training injuries.

DUTIES

6-7. The combat lifesaver ensures that the squad CLS bag, litters, and IVs are properly packed. He also identifies any Class VIII shortages to the platoon medic, and participates in all casualty treatment and litter-carry drills. His advanced first-aid skills are called upon in the field until casualties can be evacuated. The combat lifesaver must know the location of the CCP and the SOP for establishing it. The CLS has a laminated quick reference nine-line MEDEVAC card.

SECTION II — PLANNING CONSIDERATIONS

6-8. Planning sustainment operations is primarily a company- and battalion-level operation. While the company commander and XO plan the operation, the platoon leader is responsible for execution at platoon level.

PLANNING

6-9. The platoon sergeant executes the plan at squad level. Sustainment at the Infantry platoon level is characterized by the following: responsiveness, economy, flexibility, integration, and survivability

RESPONSIVENESS

6-10. To be effective sustainment needs to be responsive. This requires users to provide timely requests for supplies and support while requiring providers to anticipate user needs in advance.

ECONOMY

6-11. To be efficient, sustainment providers and users exercise conservation. Because resources are always limited, it is in the best interest of everyone to use only what is needed. The principle of economy necessitates that Soldiers, leaders, and their units conserve resources whenever possible. This also ensures other Soldiers and units will have the supplies they need.

FLEXIBILITY

6-12. The principle of flexibility embodies the chaotic nature of combat. Providers and users alike remain aware that, despite the best efforts of all involved, things seldom go as planned; shipments are delayed, convoys are attacked, and supplies are destroyed. To support the needs of both the individual unit and the rest of the units on the battlefield requires both the user and provider to know what they need, when they need it and possible substitutes.

INTEGRATION

6-13. To function properly, sustainment considerations must be integrated into every aspect of an operation. Sustainment is not branch or rank specific – it is an essential part of all operations at all levels by all Soldiers. Again, without sustainment units can not accomplish their mission

SURVIVABILITY

6-14. On the whole, sustainment assets are necessary yet finite resources that are easily destroyed. Units without their classes of supply can not fight. Accordingly, survivability of sustainment assets is a high priority for everyone. This affects the platoon in two ways. First units may be required to conduct security missions for sustainment assets, such as convoy security, base security, and response force activities. Second, units must ensure the survivability of their own supplies and any asset that might be under their charge by properly safeguarding them

DEVELOPMENT OF THE SUSTAINMENT PLAN

6-15. The platoon leader develops his sustainment plan by determining exactly what he has on hand to accurately predict his support requirements. This process is important not only in confirming the validity of the sustainment plan, but also in ensuring the platoon submits support requests as early as possible. The platoon leader formulates his sustainment execution plan and submits support requests to the company based on his maneuver plan. It is critical for the company to know what the platoon has on hand for designated critical supplies.

OPERATIONAL QUESTIONS

6-16. The sustainment plan should provide answers to the following types of operational questions:

TYPES OF SUPPORT

- Based on the nature of the operation and specific tactical factors, what types of support will the platoon need?

Quantities

- In what quantities will this support be required?
- If occupying a SBF position, how long is the platoon likely to fire, and at what rate of fire? This drives the estimate for required Class V.
- Will emergency resupply be required during the battle? Potentially when and where?
- Does this operation require prestocked supplies (cache points)?

Threat

- What are the composition, disposition, and capabilities of the expected enemy threat?
- How will these affect sustainment operations during the battle?
- Where and when will the expected contact occur?
- What are the platoon's expected casualties and equipment losses based on the nature and location of expected contact?
- What impact will the enemy's special weapons capabilities (such as CBRN) have on the battle and on expected sustainment requirements?
- How many EPWs are expected, and where?

Terrain and Weather

- What ground will provide the best security for CCPs?
- What are the platoon's casualty evacuation routes?
- What are the company's dirty routes for evacuating contaminated personnel and equipment?

Time and Location

- When and where will the platoon need sustainment?
- Based on the nature and location of expected contact, what are the best sites for the CCP?
- Where will the EPW collection points be located? Who secures them, when does the platoon turn them over, and to whom?

Requirements

6-17. Determine support requirements by asking the following questions:

- What are the support requirements by element and type of support?
- Which squad has priority for emergency Class V resupply?

Risk Factor

- Will lulls in the battle permit support elements to conduct resupply operations in relative safety?
- If no lulls are expected, how can the platoon best minimize the danger to the sustainment vehicles providing the required support?

Resupply Technique

6-18. Resupply techniques the platoon use will be based on information developed during the sustainment planning process.

CLASSES OF SUPPLY CONSIDERATIONS

6-19. The platoon sergeant obtains supplies and delivers them to the platoon. The platoon leader establishes priorities for delivery, but combat demands that Class I, V, and IX supplies and equipment take priority because they are the most critical to successful operations.

CLASS I

6-20. This class includes rations, water, and ice. It also includes gratuitous issue of items related to health, morale, and welfare. The Daily Strength Report triggers an automatic request for Class I supplies. Personnel in the field trains prepare rations and deliver them with the LOGPAC. If the unit has special food requests, they must request them (for example, if a mission calls for MREs in lieu of planned hot rations).

CLASS II

6-21. This class includes clothing, individual equipment, mission-oriented protective posture (MOPP) suits, tentage, tool sets, and administrative and housekeeping supplies. The platoon sergeant normally distributes expendable items such as soap, toilet tissue, and insecticide based on battalion and company LOGPAC schedules.

CLASS III

6-22. This class includes bulk and packaged petroleum, oil, and lubricants (POL) products, which Infantry platoons do not normally require. Unusual Class III requests are coordinated by the company and then delivered to the battalion combat trains.

CLASS IV

6-23. This class includes construction materials, pickets, sandbags, and concertina wire.

CLASS V

6-24. This class covers all types of ammunition and mines including, C4, and other explosives.

CLASS VI

6-25. This class includes personal-demand items including, candy, soaps, cameras, film, and sundry packets that are normally sold through the exchange system.

CLASS VII

6-26. Infantry platoons do not normally have vehicles. However, this class includes major end items such as major equipment and vehicles. Battle loss reports trigger the issuance of Class VII items.

CLASS VIII

6-27. This class covers medical supplies. The BAS replaces combat lifesaver bags and first-aid kits on a one-for-one basis.

CLASS IX

6-28. This class includes repair parts and documents required for equipment maintenance operations. Repair parts are issued in response to a specific request or are obtained by direct exchange of repairable parts. The latter can include batteries for NVDs, and man-portable radios. In combat situations, exchange and cannibalization are normal ways to obtain Class IX items.

CLASS X

6-29. This class includes materials to support nonmilitary programs such as agricultural and economic development. Division level or higher will provide the platoon with instructions for requesting and issuing Class X supplies.

MISCELLANEOUS

6-30. This category covers anything that does not fall under one of the existing classes of supply.

MAINTENANCE

6-31. Proper maintenance is the key to keeping equipment and other materials in serviceable condition. It is a continuous process, starting with preventive measures taken by each Soldier responsible for a piece of equipment, and continuing on through repair and recovery efforts. Equipment services include inspecting, cleaning, testing, servicing, repairing, requisitioning, recovering, and evacuating damaged equipment for repair.

SOLDIER'S LOAD

6-32. The Soldier's load is a main concern of the leader. How much is carried, how far, and in what configuration are important mission considerations. Leaders must learn to prepare for the most likely contingencies based on available information, because they cannot be prepared for all possible operations. See FM 21-18, *Foot Marches*, and FM 3-21.10, *The Infantry Rifle Company,* for detailed discussions on load planning, calculating, and management techniques used to assist leaders and Soldiers in organizing tactical loads to ensure safety and combat effectiveness.

COMBAT LOAD AND BASIC LOAD

6-33. The platoon's combat load varies by mission and includes the supplies physically carried into the fight. The company commander may direct minimum requirements or be very specific for the composition of the combat load. Often, the unit SOP or the platoon leader specifies most items. The basic load includes supplies kept by the platoon for use in combat. The quantity of most basic load supply items depends on how many days in combat the platoon might have to sustain itself without resupply. For Class V ammunition, the higher commander or SOP specifies the platoon's basic load.

TRANSPORTATION

6-34. Because the Infantry platoon leader has no organic transportation, they request transportation support through the 1SG or company XO. They, in turn, request it from the battalion S4 for ground transportation or S3 air operations if the transportation is for helicopters. Whenever possible, unless there is a specific reason not to, rucksacks and excess equipment should be transported by vehicle.

SECTION III — RESUPPLY OPERATIONS

6-35. Resupply operations fall into one of three classifications: routine, emergency, or prestock. The platoon SOP specifies cues for each method. The platoon should rehearse or conduct resupply operations every time they conduct field training. The actual method selected for resupply in the field depends on METT-TC factors.

ROUTINE RESUPPLY

6-36. Routine resupply operations primarily include Classes I, V, and IX; mail; and other items requested by the platoon. When possible, the platoon should conduct routine resupply daily. Ideally, it does so during periods of limited visibility.

6-37. The LOGPAC technique offers a simple, efficient way to accomplish routine resupply operations. The key feature of LOGPAC, a centrally organized resupply convoy, originates at the battalion trains. The convoy carries all items needed to sustain the platoon for a specific period (usually 24 hours) or until the next scheduled LOGPAC. The battalion SOP will specify the LOGPAC's exact composition and march order.

6-38. As directed by the commander or XO, the 1SG establishes the company resupply point. He uses either the service station method (Figure 6-1), the tailgate method (Figure 6-2), or the in-position method (Figure 6-3). He briefs each LOGPAC driver on which method to use. When he has the resupply point ready, the 1SG informs the commander. The company commander then directs each platoon or element to conduct resupply based on the tactical situation.

6-39. The service station method allows the squads to move individually to a centrally located resupply point. This method requires the Soldiers to leave their fighting positions. Depending on the tactical situation, a squad moves out of its position, conducts resupply operations, and moves back into position. The squads rotate individually to eat; pick up mail, Class IX supplies, and other supplies and sundries; and refill or exchange water. This process continues until the entire platoon has received its supplies. The technique is used when contact is not likely and for the resupply of one or several classes of supplies.

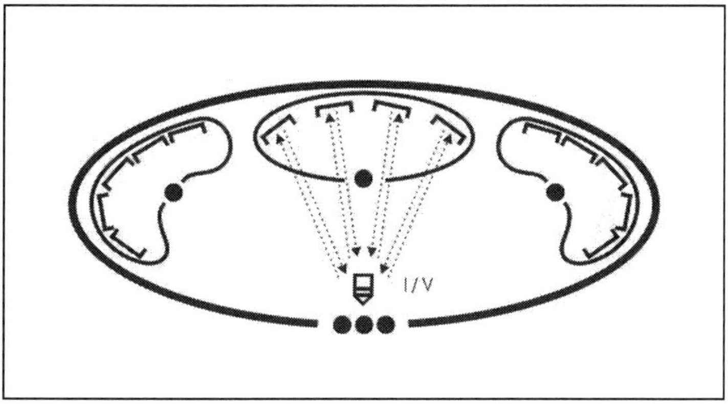

Figure 6-1. Service station resupply method.

NOTE: The platoon order should state the sequence for moving squads or portions of squads out of position. Companies may vary the technique by establishing a resupply point for each platoon and moving the supplies to that point.

6-40. In AAs, the 1SG normally uses the tailgate method (Figure 6-2). Individual Soldiers rotate through the feeding area. While there, they pick up mail and sundries and refill or exchange water cans. They centralize and guard any EPW. They take Soldiers killed in action and their personal effects to the holding area (normally a location downwind and out of sight of the platoon/company), where the 1SG assumes responsibility for them.

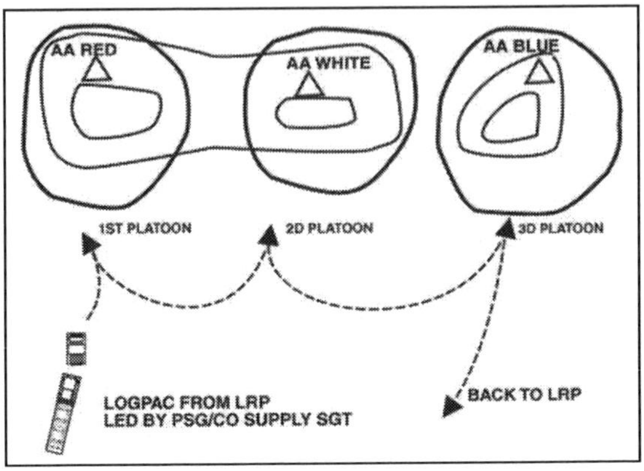

Figure 6-2. Tailgate resupply method.

6-41. During operations when contact with the enemy is imminent, the in-position resupply method (Figure 6-3) may be required to ensure adequate supplies are available to the squads. This method requires the company to bring forward supplies, equipment, or both to individual fighting positions. The platoon normally provides a guide to ensure the supplies are distributed to the most critical position first. This method—

- Is used when an immediate need exists.
- Is used to resupply single classes of supply.
- Enables leaders to keep squad members in their fighting positions.

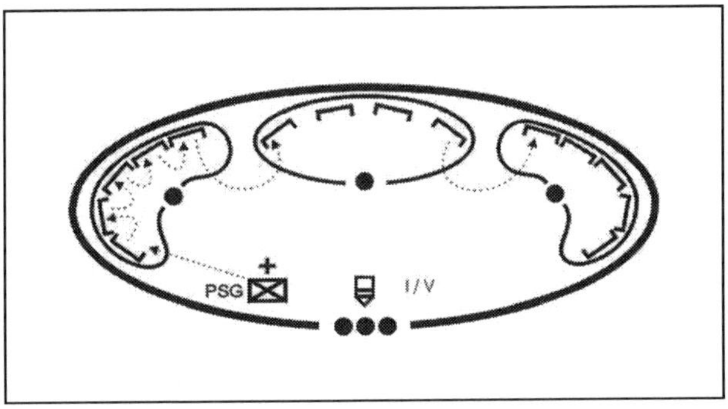

Figure 6-3. In-position resupply method.

NOTE: If resupply vehicles cannot move near platoon positions, platoon members may need to help the resupply personnel move supplies and equipment forward.

EMERGENCY RESUPPLY

6-42. Occasionally during combat operations, the platoon may have such an urgent need for resupply that it cannot wait for a routine LOGPAC. Emergency resupply may involve Classes I (usually water), V, VII,

VIII, and CBRN equipment. Emergency resupply can be conducted using either the service station or tailgate method, but more often uses the in-position method. The fastest appropriate means is normally used, although procedures may have to be adjusted when the company is in contact with the enemy. In the service station method, individual squads may pull back during a lull in combat to conduct resupply and then return to the fight. With tailgate resupply, the company brings limited supplies forward to the closest concealed position behind each element.

PRESTOCK RESUPPLY

6-43. In defensive or stay-behind operations and at some other times, the platoon may need prestocked supplies (also known as prepositioned or cached resupply). Normally, the platoon only prepositions items directed by the company.

6-44. All levels must carefully plan and execute prestock operations. All leaders, down to squad leader level, must know the exact locations of prestock sites. They verify these locations during reconnaissance or rehearsals. The platoon takes steps to ensure the survivability of the prestocked supplies. These measures include selecting covered and concealed positions and digging in the prestock positions. The platoon leader must have a removal and destruction plan to prevent the enemy from capturing prepositioned supplies.

6-45. During offensive operations, the company can preposition supplies on trucks well forward on the battlefield. This works well if the company expects to use a large volume of fire, with corresponding ammunition requirements. It allows the platoons to quickly resupply during consolidation or during lulls.

AERIAL SUSTAINMENT

6-46. Aerial sustainment is an aviation mission that consists of moving personnel, equipment, materiel, and supplies by utility, cargo, and fixed-wing assets for use in operations. Overland resupply might not work due to terrain, distance, or the existing enemy threat. The platoon must initiate a request for resupply and must push it through company to battalion. The platoon must prepare to receive the supplies at the specified time and location.

6-47. A aerial sustainment with speed balls is a technique with preconfigured loads to resupply Infantry platoons in urban areas (Figure 6-4). Sustainment personnel prepackage supplies in aviation kit bags, duffle bags, or other suitable containers. Helicopters fly as close to the drop point as possible, reduce speed, drop supplies, and leave the area quickly. Supplies should be packaged in bubble wrap or other shock-absorbing material to minimize damage.

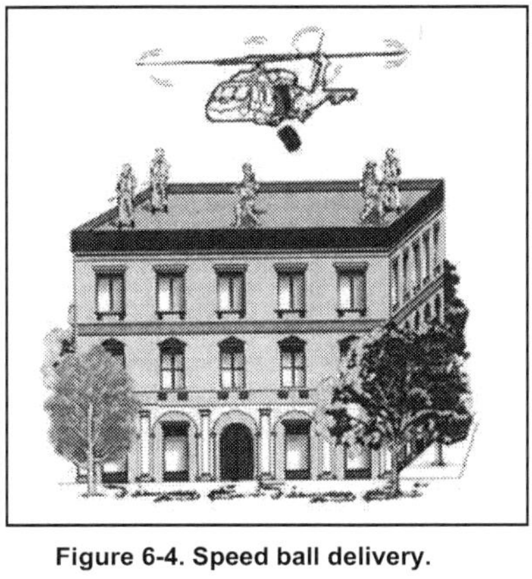

Figure 6-4. Speed ball delivery.

SECTION IV — CASUALTY PROCEDURES

6-48. Following are the procedures that should be followed in the treatment, evacuation, and reporting of combat casualties.

INITIAL CARE

6-49. When combat begins and casualties occur, the platoon first must provide initial care to those wounded in action (WIA).

6-50. Effective casualty evacuation provides a major increase in the morale of a unit. This is accomplished through the administration of first aid (self-aid/buddy aid), enhanced first aid (by the combat lifesaver), and emergency medical treatment (by the trauma specialist/platoon medic). Casualties are cared for at the point of injury or under nearby cover and concealment.

6-51. During the fight, casualties should remain under cover where they received initial treatment. As soon as the situation allows, squad leaders arrange for casualty evacuation to the platoon CCP. The platoon normally sets up the CCP in a covered and concealed location to the rear of the platoon position. At the CCP, the platoon medic conducts triage on all casualties, takes steps to stabilize their conditions, and starts the process of moving them to the rear for advanced treatment. Before the platoon evacuates casualties to the CCP or beyond, leaders should remove all key operational items and equipment from each person. Removal should include automated network control devices (ANCD), GPS maps, position-locating devices, and laser pointers. Every unit should establish an SOP for handling the weapons and ammunition of its WIA.

6-52. The tactical situation will determine how quickly fellow Soldiers can treat wounded Soldiers. Understandably, fewer casualties occur if Soldiers focus on destroying or neutralizing the enemy that caused the casualties. This is a critical situation that should be discussed and rehearsed by the squads and platoons prior to executing a mission.

MOVEMENT

6-53. Timely movement of casualties from the battlefield is important not only for safety and care for the wounded, but also for troop morale.

6-54. Squad leaders are responsible for casualty evacuation from the battlefield to the platoon CCP. At the CCP, the senior trauma specialist assists the platoon sergeant and 1SG in arranging evacuation by ground or air ambulance or by non-standard means. Leaders must minimize the number of Soldiers required to evacuate casualties. Casualties with minor wounds can walk or even assist with carrying the more seriously wounded. Soldiers can make field-expedient litters by cutting small trees and putting the poles through the sleeves of zippered Army combat uniform (ACU) blouses or ponchos. A travois, or skid, may be used for casualty evacuation. This is a type of litter on which wounded can be strapped; it can be pulled by one person. It can be locally fabricated from durable, rollable plastic on which tie-down straps are fastened. In rough terrain (or on patrols), casualties may be evacuated all the way to the BAS by litter teams. From there they can be carried with the unit until transportation can reach them, or left at a position and picked up later.

6-55. From the platoon area, casualties are normally evacuated to the company CCP and then back to the BAS. The company 1SG, with the assistance of the platoon sergeant, is normally responsible for movement of the casualties from the platoon to the company CCP. The unit SOP should address this activity, including the marking of casualties during limited visibility operations. Small, standard, or infrared chemical lights work well for this purpose. Once the casualties are collected, evaluated, and treated, they are sent to company CCP. Once they arrive, the above process is repeated while awaiting their evacuation back to the BAS.

6-56. When the company is widely dispersed, the casualties may be evacuated directly from the platoon CCP by vehicle or helicopter. Helicopter evacuation may be restricted due to the enemy air defense artillery (ADA) or small arms/RPG threat. In some cases, casualties must be moved to the company CCP or battalion combat trains before helicopter evacuation. When there are not enough battalion organic ambulances to move the wounded, unit leaders may direct supply vehicles to "backhaul" casualties to the BAS after supplies are delivered. Normally, urgent casualties will move by ambulance. Less seriously hurt Soldiers are moved through other means. If no ambulance is available, the most critical casualties must get to the BAS as quickly as possible. In some cases, the platoon sergeant may direct platoon litter teams to carry casualties to the rear.

6-57. The senior military person present determines whether to request medical evacuation and assigns precedence. These decisions are based on the advice of the senior medical person at the scene, the patient's condition, and the tactical situation. Casualties will be picked up as soon as possible, consistent with available resources and pending missions. Following are priority categories of precedence and the criteria used in their assignment.

PRIORITY I-URGENT

6-58. Assigned to emergency cases that should be evacuated as soon as possible and within a maximum of two hours in order to save life, limb, or eyesight; to prevent complications of serious illness; or to avoid permanent disability.

PRIORITY IA-URGENT-SURG

6-59. Assigned to patients who must receive far forward surgical intervention to save their lives and stabilize them for further evacuation.

PRIORITY II-PRIORITY

6-60. Assigned to sick and wounded personnel requiring prompt medical care. The precedence is used when special treatment is not available locally and the individual will suffer unnecessary pain or disability (becoming URGENT precedence) if not evacuated within four hours.

PRIORITY III-ROUTINE

6-61. Assigned to sick and wounded personnel requiring evacuation but whose condition is not expected to deteriorate significantly. The sick and wounded in this category should be evacuated within 24 hours.

PRIORITY IV-CONVENIENCE

6-62. Assigned to patients for whom evacuation by medical vehicle is a matter of medical convenience rather than necessity.

CASEVAC

6-63. Casualty evacuation (CASEVAC) is the term used to refer to the movement of casualties by air or ground on nonmedical vehicles or aircraft. CASEVAC operations normally involve the initial movement of wounded or injured Soldiers to the nearest medical treatment facility. Casualty evacuation operations may also be employed in support of mass casualty operations. Medical evacuation (MEDEVAC) includes the provision of en route medical care, whereas CASEVAC does not provide any medical care during movement. For definitive information on CASEVAC, see FM 8-10-6, *Medical Evacuation in a Theater of Operations, Tactics, Techniques, and Procedures*, FM 8-10-26, *Employment of the Medical Company (Air Ambulance)*, and Table 6-1.

6-64. When possible, medical platoon ambulances provide evacuation and en route care from the Soldier's point of injury or the platoon's or company's CCP to the BAS. The ambulance team supporting the company works in coordination with the senior trauma specialist supporting the platoons. In mass casualty situations, non-medical vehicles may be used to assist in casualty evacuation as directed by the Infantry company commander. However, plans for the use of non-medical vehicles to perform casualty evacuation should be included in the unit SOP.

Table 6-1. Procedures to Request Medical Evacuation (MEDEVAC).

Line/ Item	Explanation	Where/How Obtained	Who Normally Provides	Reason
1/Location of pickup site by grid coordinates with grid zone letters	Encrypt the grid coordinates of the pickup site. When using the DRYAD Numeral Cipher, the same "SET" line will be used to encrypt the grid zone letters and the coordinates. To prevent misunderstanding, it is stated that grid zone letters are included in the message (unless SOP specifies its use at all times).	From map	Unit leader(s)	Required so evacuation vehicle knows where to pick up patient, and, unit coordinating the evacuation mission can plan the route for the evacuation vehicle (if evacuation vehicle must pick up from more than one location).
2/Requesting unit radio frequency, call signal, and suffix	Encrypt the frequency of the radio at the pickup site, not a relay frequency. The call sign (and suffix if used) of the person to be contacted at the pickup site may be transmitted in the clear.	From SOI	RATELO	Required so evacuation vehicle can contact requesting unit while en route (or obtain additional information and change in situation or directions).
3/Number of patients by precedence. Note the brevity codes used.	Report only applicable information and encrypt the brevity codes. A-Urgent B-Urgent-Surgical C-Priority D-Routine E-Convenience If two or more categories must be reported in the same request, insert the word *BREAK* between each category.	From evaluation of patient(s)	Medic or senior person present	Required by unit controlling the evacuation vehicles to assist in prioritizing missions.
4/Special equipment required	Encrypt the applicable brevity codes. A-None B-Hoist C-Extraction equipment D-Ventilator	From evaluation of patient/ situation	Medic or senior person present	Required so equipment can be placed on board the evacuation vehicle prior to the start of the mission.
5/Number of patients	Report only applicable information and encrypt the brevity code. If requesting MEDEVAC for both types, insert the word *BREAK* between the litter entry and ambulatory entry. For example: L + # of PNT-litter A + # of PNT-ambulatory	From evaluation of patient	Medic or senior person present	Required so appropriate number of evacuation vehicles may be dispatched to the pickup site. They should be configured to carry the patients requiring evacuation.

Table 6-1. Procedures to Request Medical Evacuation (MEDEVAC) (continued).

Line/ Item	Explanation	Where/How Obtained	Who Normally Provides	Reason
6/Security of pickup site (wartime)	N-No enemy troops in area P-Possibly enemy troops in area (approach with caution) E-Enemy troops in area (approach with caution) X-Enemy troops in area (armed escort required)	From evaluation of situation	Unit leader	Required to assist the evacuation crew in assessing the situation and determining if assistance is required. More definitive guidance (such as specific location of enemy to assist an aircraft in planning its approach) can be furnished by the evacuation aircraft while it is en route.
7/Number and type of wound, injury, or illness (peacetime)	Specific information regarding patient wounds by type (gunshot or shrapnel). Report serious bleeding and patient blood type (if known).	From evaluation of patient	Medic or senior person present	Required to assist evacuation personnel in determining treatment and special equipment needed.
8/Method of marking pickup site	Encrypt the brevity codes. A-Panels B-Pyrotechnic signal C-Smoke signal D-None E-Other	Based on situation and availability of materials	Medic or senior person present	Required to assist the evacuation crew in identifying the specific location of the pick up. Note that the color of the panels or smoke should not be transmitted until the evacuation vehicle contacts the unit just prior to its arrival. For security, the crew should identify the color. The unit should verify it.
9/Patient nationality and status	The number of patients in each category does not need to be transmitted. Encrypt only the applicable brevity codes. A-US military B-US civilian C-Non-US military D-Non-US civilian E-EPW	From evaluation of patient	Medic or senior person present	Required to assist in planning for destination facilities and need for guards. Unit requesting support should ensure that there is an English-speaking representative at the pickup site.
10/CBRN contamin-ation (wartime)	Include this line only when applicable. Encrypt the applicable brevity codes. N-Nuclear B-Biological C-Chemical	From situation	Medic or senior person present	Required to assist in planning for the mission. Determine which evacuation vehicle will accomplish the mission and when it will be accomplished.
11/Terrain description (peacetime)	Include details of terrain features in and around proposed landing site. If possible, describe relationship of site to prominent terrain feature (lake, mountain, and tower).	From area survey	Personnel at site	Required to allow evacuation personnel to assist route/avenue of approach into area. Of particular importance if hoist operation is required.

UNIT SOPS

6-65. Unit SOPs and OPORDs must address casualty treatment and evacuation in detail. They should cover the duties and responsibilities of key personnel, the evacuation of chemically contaminated casualties (on routes separate from noncontaminated casualties), and the priority for manning key weapons and positions. They should specify preferred and alternate methods of evacuation and make provisions for retrieving and safeguarding the weapons, ammunition, and equipment of casualties. Slightly wounded personnel are treated and returned to duty by the lowest echelon possible. Platoon medic evaluate sick Soldiers and either treat or evacuate them as necessary. Casualty evacuation should be rehearsed like any other critical part of an operation.

CASUALTY REPORT

6-66. A casualty report is filled out when a casualty occurs, or as soon as the tactical situation permits. This is usually done by the Soldier's squad leader and turned in to the platoon sergeant, who forwards it to the 1SG. A brief description of how the casualty occurred (including the place, time, and activity being performed) and who or what inflicted the wound is included. If the squad leader does not have personal knowledge of how the casualty occurred, he gets this information from any Soldier who does have the knowledge. Department of the Army (DA) Form 1156, *Casualty Feeder Card* (Figure 6-5A and B), is used to report those Soldiers who have been killed and recovered, and those who have been wounded. This form is also used to report KIA Soldiers who are missing, captured, or not recovered. The Soldier with the most knowledge of the incident should complete the witness statement. This information is used to inform the Soldier's next of kin and to provide a statistical base for analysis of friendly or enemy tactics. Once the casualty's medical condition has stabilized, the company commander may write a letter to the Soldier's next of kin. During lulls in the battle, the platoon forwards casualty information to the company headquarters.

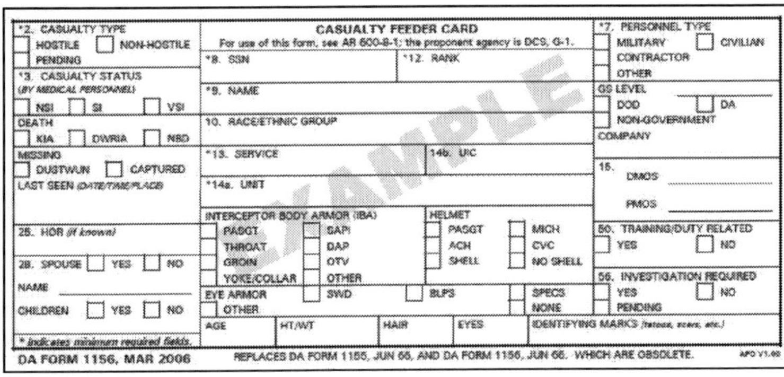
*2. CASUALTY TYPE
HOSTILE NON-HOSTILE
PENDING
*3. CASUALTY STATUS (BY MEDICAL PERSONNEL)
NSI SI VSI
DEATH
KIA DWRIA NBD
MISSING
DUSTWUN CAPTURED
LAST SEEN (DATE/TIME/PLACE)
25. HOR (if known)
28. SPOUSE YES NO
NAME
CHILDREN YES NO
* indicates minimum required fields.

CASUALTY FEEDER CARD
For use of this form, see AR 600-8-1; the proponent agency is DCS, G-1.
*8. SSN
*12. RANK
*9. NAME
10. RACE/ETHNIC GROUP
*13. SERVICE
14b. UIC
*14a. UNIT
INTERCEPTOR BODY ARMOR (IBA)
PASGT SAPI
THROAT DAP
GROIN OTV
YOKE/COLLAR OTHER
HELMET
PASGT MICH
ACH CVC
SHELL NO SHELL
EYE ARMOR
SWD BLPS SPECS
OTHER NONE
AGE
HT/WT
HAIR
EYES
IDENTIFYING MARKS (tattoos, scars, etc.)

*7. PERSONNEL TYPE
MILITARY CIVILIAN
CONTRACTOR
OTHER
GS LEVEL
DOD DA
NON-GOVERNMENT
COMPANY
15. DMOS
PMOS
50. TRAINING/DUTY RELATED
YES NO
56. INVESTIGATION REQUIRED
YES NO
PENDING

DA FORM 1156, MAR 2006
REPLACES DA FORM 1155, JUN 66, AND DA FORM 1156, JUN 66, WHICH ARE OBSOLETE.
APD V1.00

Figure 6-5A. DA Form 1156, casualty feeder card (report front).

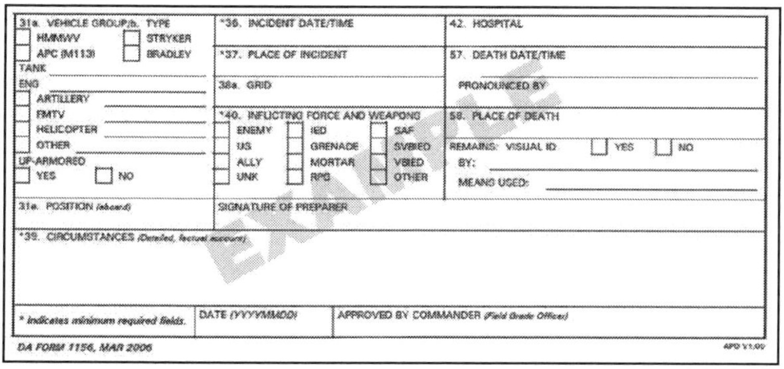

31a. VEHICLE GROUP/b. TYPE
HMMWV
STRYKER
APC (M113)
BRADLEY
TANK
ENG
ARTILLERY
FMTV
HELICOPTER
OTHER
UP-ARMORED
YES
NO
31a. POSITION (aboard)

*36. INCIDENT DATE/TIME
*37. PLACE OF INCIDENT
38a. GRID
*40. INFLICTING FORCE AND WEAPONS
ENEMY
IED
SAF
US
GRENADE
SVBIED
ALLY
MORTAR
VBIED
UNK
RPG
OTHER
SIGNATURE OF PREPARER

42. HOSPITAL
57. DEATH DATE/TIME
PRONOUNCED BY
58. PLACE OF DEATH
REMAINS: VISUAL ID
YES
NO
BY:
MEANS USED:

*39. CIRCUMSTANCES (Detailed, factual account)

* indicates minimum required fields.
DATE (YYYYMMDD)
APPROVED BY COMMANDER (Field Grade Officer)

DA FORM 1156, MAR 2006
APD V1.00

Figure 6-5B. DA Form 1156, casualty feeder card (report back).

KILLED IN ACTION

6-67. The platoon leader designates a location for the collection of KIAs. All personal effects remain with the body. However, squad leaders remove and safeguard any equipment and issue items. He keeps these until he can turn the equipment and issue items over to the platoon sergeant. The platoon sergeant turns over the KIA to the 1SG. As a rule, the platoon should not transport KIA remains on the same vehicle as wounded Soldiers. KIAs are normally transported to the rear on empty resupply trucks, but this depends on unit SOP.

SECTION V — ENEMY PRISONERS OF WAR AND RETAINED/DETAINED PERSONS

6-68. Enemy prisoners of war (EPW) and captured enemy equipment or materiel often provide excellent combat information. This information is of tactical value only if the platoon processes and evacuates prisoners and materiel to the rear quickly. In any tactical situation, the platoon will have specific procedures and guidelines for handling prisoners and captured materiel.

6-69. All persons captured, personnel detained or retained by U.S. Armed Forces during the course of military operations, are considered "detained" persons until their status is determined by higher military and civilian authorities. The BCT has an organic military police platoon organic to the BSTB to take control of and evacuate detainees (Figure 6-6). However, as a practical matter, when Infantry squads, platoons, companies, and battalions capture enemy personnel, they must provide the initial processing and holding for detainees. Detainee handling is a resource intensive and politically sensitive operation that requires detailed training, guidance, and supervision.

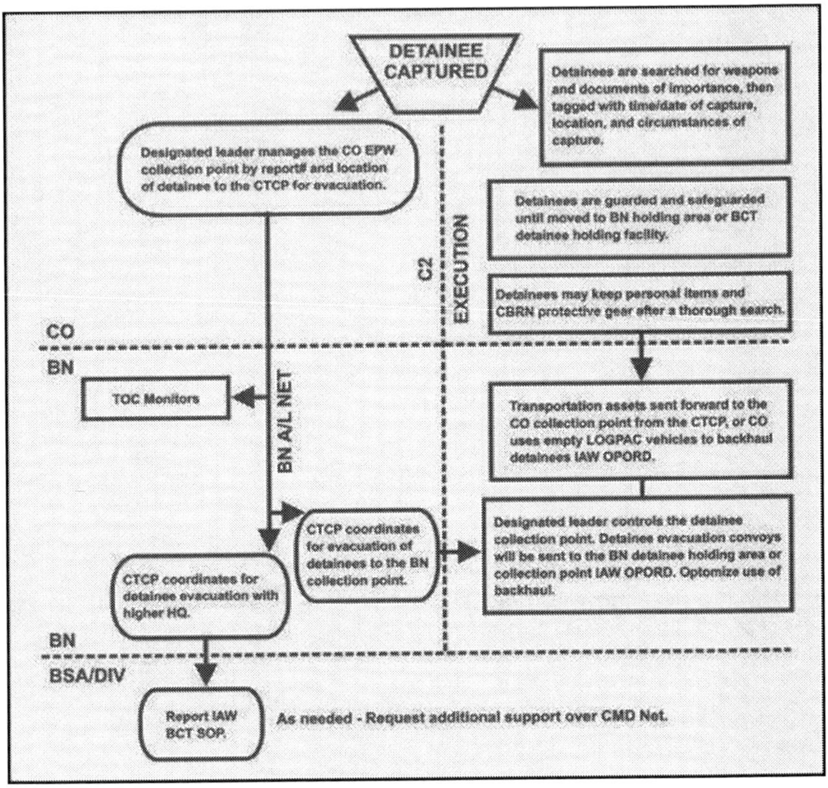

Figure 6-6. Detainee handling.

6-70. All detained persons shall be immediately given humanitarian care and treatment. U.S. Armed Forces will never torture, maltreat, or purposely place detained persons in positions of danger. There is never a military necessity exception to violate these principles.

6-71. Field processing of detainees is always handled IAW the 5 Ss and T method:

- *Search:* Confiscate weapons and items of intelligence value or items that might assist the detainee to escape. Let the detainee keep protective clothing, equipment, identification and personal items. All confiscated items must be tagged.
- *Silence:* Direct the detainees not to talk, or make facial or hand gestures. They may be gagged.
- *Segregate:* Leaders are separated from the rest of the population. Separate hostile elements such as religious, political, or ethnic groups. Separate women and minors from adult male detainees.
- *Safeguard:* Ensure detainees are provided adequate food, potable water, clothing, shelter, medical attention, and that they not exposed to unnecessary danger. Do not use coercion to obtain information. Immediately report allegations of abuse through command channels.
- *Speed to a safe area/rear*: Evacuate detainees from the battlefield to a holding area or facility as soon as possible. Transfer captured documents and other property to the forces assuming responsibility for the detainees.

- *Tag*: Before evacuating an EPW detainee, he must be tagged with Department of the Defense (DD) Form 2745, *Enemy Prisoner of War (EPW) Capture Tag* (Part A) (Figure 6-7), or by field expedient means. Field expedient means should include tagging with date and time of capture, location of capture, capturing unit, and circumstances of capture. DD Form 2745, *Unit Record Card* (Part B), is the unit record copy (Figure 6-8). DD Form 2745, *Document/Special Equipment Weapons Card* (Part C), is for the detainee's confiscated property (Figure 6-9). Tagging is critical. If it does not happen the ability of higher headquarters to quickly obtain pertinent tactical information is greatly reduced.

6-72. Detainees should be evacuated as soon as is practical to the BCT detainee collection point. Tactical questioning of detainees is allowed relative to collection of CCIR. However, detainees must always be treated IAW the U.S. Law of War Policy as set forth in the Department of Defense Directive 2311.01E, *DoD Law of War Program*.

6-73. Soldiers capturing equipment, documents, and detainees should tag them using DD Form 2745, *Enemy Prisoner of War (EPW) Capture Tag*, take digital pictures, and report the capture immediately. Detainees are allowed to keep protective equipment such as protective masks. Other captured military equipment and detainee personal effects are inventoried on DA Form 4137, *Evidence/Property Custody Document*. Soldiers then coordinate with the platoon and company headquarters to link up and turn the documents and prisoners over to designated individuals.

6-74. In addition to initial processing, the capturing element provides guards and transportation to move prisoners to the designated EPW collection points. The capturing element normally carries prisoners on vehicles already heading toward the rear, such as tactical vehicles returning from LOGPAC operations. The capturing element must also feed, provide medical treatment, and safeguard EPWs until they reach the collection point.

6-75. Once EPWs arrive at the collection point, the platoon sergeant assumes responsibility for them. He provides for security and transports them to the company EPW collection point. He uses available personnel as guards, including walking wounded or Soldiers moving to the rear for reassignment.

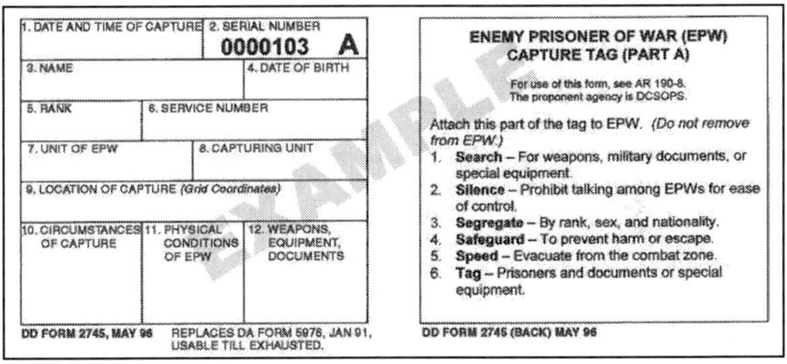
1. DATE AND TIME OF CAPTURE
2. SERIAL NUMBER 0000103 A
3. NAME
4. DATE OF BIRTH
5. RANK
6. SERVICE NUMBER
7. UNIT OF EPW
8. CAPTURING UNIT
9. LOCATION OF CAPTURE *(Grid Coordinates)*
10. CIRCUMSTANCES OF CAPTURE
11. PHYSICAL CONDITIONS OF EPW
12. WEAPONS, EQUIPMENT, DOCUMENTS

DD FORM 2745, MAY 96 REPLACES DA FORM 5976, JAN 91, USABLE TILL EXHAUSTED.

ENEMY PRISONER OF WAR (EPW) CAPTURE TAG (PART A)

For use of this form, see AR 190-8. The proponent agency is DCSOPS.

Attach this part of the tag to EPW. *(Do not remove from EPW.)*

1. **Search** – For weapons, military documents, or special equipment.
2. **Silence** – Prohibit talking among EPWs for ease of control.
3. **Segregate** – By rank, sex, and nationality.
4. **Safeguard** – To prevent harm or escape.
5. **Speed** – Evacuate from the combat zone.
6. **Tag** – Prisoners and documents or special equipment.

DD FORM 2745 (BACK) MAY 96

Figure 6-7. DA Form 2745, enemy prisoner of war (EPW) capture tag (part A).

1. DATE AND TIME OF CAPTURE | 2. SERIAL NUMBER **0000103 B**
3. NAME | 4. DATE OF BIRTH
5. RANK | 6. SERVICE NUMBER
7. UNIT OF EPW | 8. CAPTURING UNIT
9. LOCATION OF CAPTURE *(Grid Coordinates)*
10. CIRCUMSTANCES OF CAPTURE | 11. PHYSICAL CONDITION OF EPW | 12. WEAPONS, EQUIPMENT, DOCUMENTS

DD FORM 2745, MAY 96 REPLACES DA FORM 5976, JAN, 91 USABLE TILL EXHAUSTED

UNIT RECORD CARD *(PART B)*

Forward to Unit.

(Capturing unit retains for records.)

Use string, wire, or other durable material to attach the appropriate section of this form to the EPW's equipment or property.

DD FORM 2745 (BACK), MAY 96

EXAMPLE

Figure 6-8. DD 2745, unit record card (part B).

1. DATE AND TIME OF CAPTURE | 2. SERIAL NUMBER **0000103 C**
3. NAME | 4. DATE OF BIRTH
5. RANK | 6. SERVICE NUMBER
7. UNIT OF EPW | 8. CAPTURING UNIT
9. LOCATION OF CAPTURE *(Grid Coordinates)*
10. DESCRIPTION OF WEAPONS, SPECIAL EQUIPMENT, DOCUMENTS

DD FORM 2745, MAY 96 REPLACES DA FORM 5976, JAN 91 USABLE TILL EXHAUSTED.

DOCUMENT/SPECIAL EQUIPMENT WEAPONS CARD (PART C)

Attach this part of the tag to property taken. *(Do not remove from property.)*

As a minimum, the tag must include the following information:

Item 1. Date and time of capture *(YYYYMMDD)*.
Item 8. Capturing unit.
Item 9. Place of capture *(grid coordinates)*.
Item 10. Circumstances of *capture (how the EPW was captured)*.

DD FORM 2745 (BACK), MAY 96

EXAMPLE

Figure 6-9. DD 2745, document/special equipment weapons card (part C).

Chapter 7

OFFENSIVE OPERATIONS

Platoon and squad leaders must understand the principles, tactics, techniques, and procedures associated with the offense. They must comprehend their role when operating within a larger organization's operations, and when operating independently. They must recognize the complementary and reinforcing effects of other maneuver elements and supporting elements with their own capabilities. They must also understand the impact of terrain, open or restrictive, on their operations. This chapter discusses offensive operations and the elements that affect tactical success.

SECTION I — INTRODUCTION TO OFFENSIVE OPERATIONS

7-1. Infantry platoon offensive actions can occur during all types of full spectrum operations. The enemy situation affects the type of operation conducted. METT-TC influences the actions of leaders and options available to them.

7-2. The outcome of decisive combat derives from offensive operations. The platoon can best close with the enemy by means of fire and maneuver to destroy or capture him, repel his assault by fire, engage in close combat, or counterattack through offensive operations. While tactical considerations call for the platoon to execute defensive operations for a period of time, defeating the enemy requires a shift to offensive operations. This is also true in stability operations in which transitions to the offense can occur suddenly and unexpectedly. To ensure the success of the attack, the platoon leader must understand the following fundamentals of offensive operations and apply the TLP during the operations process. (For a discussion on the TLP operations process, refer to Chapter 5.) A sound doctrinal foundation during offensive planning assists the platoon leader in capitalizing on the tactical employment of the Infantry platoon.

CHARACTERISTICS OF OFFENSIVE OPERATIONS

7-3. Surprise, concentration, tempo, and audacity characterize all offensive operations. To maximize the value of these characteristics, platoon leaders must apply the following considerations.

SURPRISE

7-4. Platoons achieve surprise by attacking the enemy at a time or place he does not expect or in a manner for which he is unprepared. Unpredictability and boldness, within the scope of the commander's intent, help the platoon gain surprise. Total surprise is rarely essential; simply delaying or disrupting the enemy's reaction is usually effective. Surprise also stresses the enemy's command and control and induces psychological shock in his Soldiers and leaders. The platoon's ability to infiltrate during limited visibility and to attack are often key to achieving surprise.

CONCENTRATION

7-5. Platoons achieve concentration by massing the effects of their weapons systems and rifle squads to achieve a single purpose. Massing effects does not require all elements of the platoon to be co-located; it simply requires the effects of the weapons systems to be applied at the right place and time. Because the attacker moves across terrain the enemy has prepared, he may expose himself to the enemy's fires. By concentrating combat power, the attacker can reduce the effectiveness of enemy fires and the amount of time he is exposed to those fires. Modern navigation tools such as global positioning systems (GPSs) allow the platoon leader to disperse, while retaining the ability to quickly mass the effects of the platoon's weapons systems whenever necessary.

TEMPO

7-6. Tempo is the rate of speed of military action. Controlling or altering that rate is essential for maintaining the initiative. While a fast tempo is preferred, the platoon leader must remember that synchronization sets the stage for successful accomplishment of the platoon's mission. To support the commander's intent, the platoon leader must ensure his platoon's movement is synchronized with the company's movement and with the other platoons. If the platoon is forced to slow down because of terrain or enemy resistance, the commander can alter the tempo of company movement to maintain synchronization. The tempo may change many times during an offensive operation. The platoon leader must remember that it is more important to move using covered and concealed routes (from which he can mass the effects of direct fires), than it is to maintain precise formations and predetermined speeds.

AUDACITY

7-7. Audacity is a simple plan of action, boldly executed. It is the willingness to risk bold action to achieve positive results. Knowledge of the commander's intent one and two levels up allows the platoon leader to take advantage of battlefield opportunities whenever they present themselves. Audacity enhances the effectiveness of the platoon's support for the entire offensive operation. Marked by disciplined initiative, audacity also inspires Soldiers to overcome adversity and danger.

TYPES OF OFFENSIVE OPERATIONS

7-8. The four types of offensive operations, described in FM 3-90, are movement to contact, attack, exploitation, and pursuit. Companies can execute movements to contact and attacks. Platoons generally conduct these forms of the offense as part of a company. Companies and platoons participate in an exploitation or pursuit as part of a larger force. The nature of these operations depends largely on the amount of time and enemy information available during the planning and preparation for the operation phases. All involve designating decisive points, maintaining mutual support, gaining fire superiority over the enemy, and seizing positions of advantage without prohibitive interference by the enemy.

MOVEMENT TO CONTACT

7-9. Movement to contact is a type of offensive operation designed to develop the situation and establish or regain contact. The platoon will likely conduct a movement to contact as part of a company when the enemy situation is vague or not specific enough to conduct an attack. For a detailed discussion of movement to contact, refer to Section V.

ATTACK

7-10. An attack is an offensive operation that destroys enemy forces, seizes, or secures terrain. An attack differs from a movement to contact because the enemy disposition is at least partially known. Movement supported by fires characterizes an attack. The platoon will likely participate in a synchronized company attack. However, the platoon may conduct a special purpose attack as part of or separate from a company offensive or defensive operation. Special purpose attacks consist of ambush, spoiling attack, counterattack, raid, feint, and demonstration. For a detailed discussion of attack and special purpose attacks, refer to Section VI.

EXPLOITATION

7-11. All commanders are expected to exploit successful attacks. In the exploitation, the attacker extends the destruction of the defending force by maintaining constant offensive pressure. Exploitations are conducted at all command levels, but divisions and brigades are the echelons that conduct major exploitation operations. The objective of exploitation is to disintegrate the enemy to the point where they have no alternative but surrender or fight following a successful attack. Indicators such as increased enemy prisoners of war (EPW), lack of organized defense, loss of enemy unit cohesion upon contact, and capture of enemy leaders indicate the opportunity to shift to an exploitation. Companies and platoons may conduct movements to contact or attacks as part of a higher unit's exploitation.

PURSUIT

7-12. Pursuits are conducted at the company level and higher. A pursuit typically follows a successful exploitation. The pursuit is designed to prevent a fleeing enemy from escaping and to destroy him. Companies and platoons may conduct pursuits as part of a higher unit's exploitation.

FORMS OF MANEUVER

7-13. In the typical offensive operations sequence (see Section II), the platoon maneuvers against the enemy in an area of operation. Maneuver places the enemy at a disadvantage through the application of friendly fires and movement. The five forms of maneuver are—

(1) Envelopment.
(2) Turning movement.
(3) Infiltration.
(4) Penetration.
(5) Frontal attack.

ENVELOPMENT

7-14. Envelopment (Figure 7-1) is a form of maneuver in which an attacking element seeks to avoid the principal enemy defenses by seizing objectives to the enemy flank or rear in order to destroy him in his current positions. Flank attacks are a variant of envelopment in which access to the enemy's flank and rear results in enemy destruction or encirclement. A successful envelopment requires discovery or creation of an assailable flank. The envelopment is the preferred form of maneuver because the attacking element tends to suffer fewer casualties while having the most opportunities to destroy the enemy. A platoon may conduct the envelopment by itself or as part of the company's attack. Envelopments focus on—

- Seizing terrain.
- Destroying specific enemy forces.
- Interdicting enemy withdrawal routes.

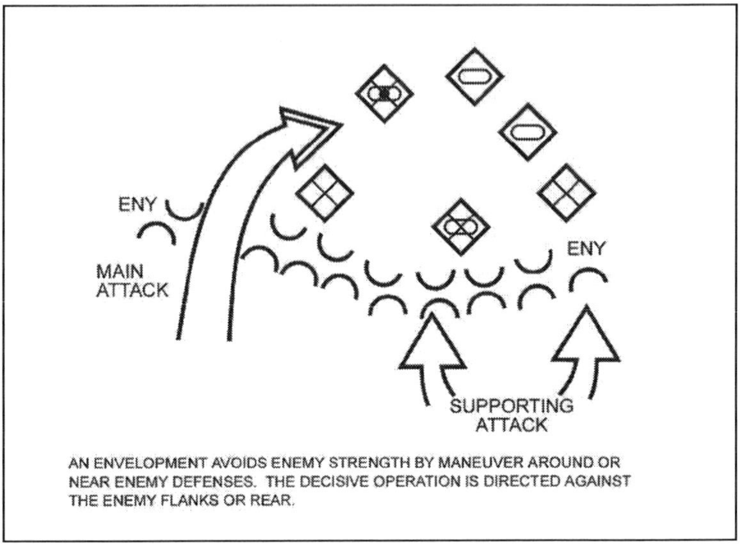

Figure 7-1. Envelopment.

TURNING MOVEMENT

7-15. The turning movement (Figure 7-2) is a form of maneuver in which the attacking element seeks to avoid the enemy's principal defensive positions by seizing objectives to the enemy's rear. This causes the enemy to move out of his current positions or to divert major forces to meet the threat. For a turning movement to be successful, the unit trying to turn the enemy must attack something the enemy will fight to save or that will cause him to move to avoid destruction. This may be a supply route, an artillery emplacement, or a headquarters. In addition to attacking a target that the enemy will fight to save, the attacking unit should be strong enough to pose a real threat. A platoon will likely conduct a turning movement as part of a company supporting a battalion attack.

NOTE: The turning movement is different from envelopment because the element conducting the turning movement seeks to make the enemy displace from his current location. An enveloping element seeks to engage the enemy in his current location from an unexpected direction.

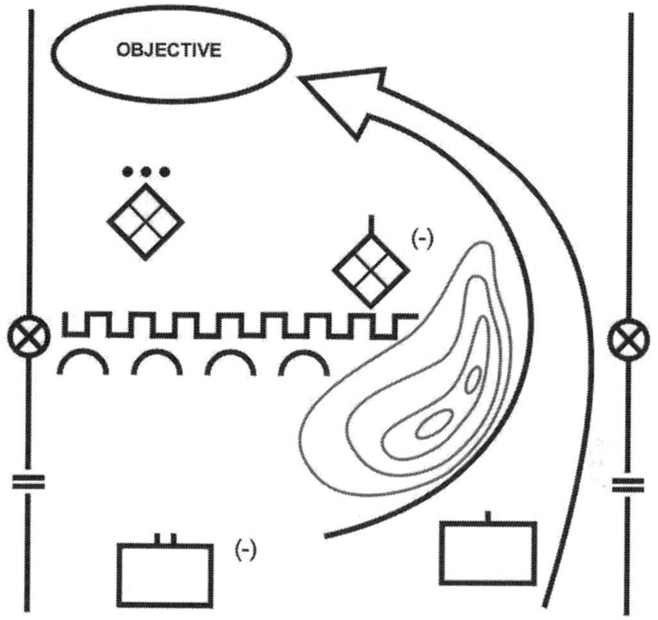

A TURNING MOVEMENT AVOIDS THE ENEMY'S PRINCIPLE DEFENSIVE POSTIONS BY SEIZING OBJECTIVES TO THE ENEMY REAR AND CAUSING THE ENEMY TO MOVE OUT OF HIS CURRENT POSITION.

Figure 7-2. Turning movement.

INFILTRATION

7-16. Infiltration (Figure 7-3) is a form of maneuver in which an attacking element conducts undetected movement through or into an area occupied by enemy forces to gain a position of advantage in the enemy rear. When conducted efficiently only small elements will be exposed to enemy defensive fires. Moving and assembling forces covertly through enemy positions takes a considerable amount of time. A successful infiltration reaches the enemy's rear without fighting through prepared positions. An infiltration is normally used in conjunction with and in support of a unit conducting another form of maneuver. A platoon may conduct an infiltration as part of a larger unit's attack with the company employing another form of maneuver. A platoon may conduct an infiltration to—

- Attack enemy-held positions from an unexpected direction.
- Occupy a support-by-fire position to support an attack.
- Secure key terrain.
- Conduct ambushes and raids.
- Conduct a covert breach of an obstacle.

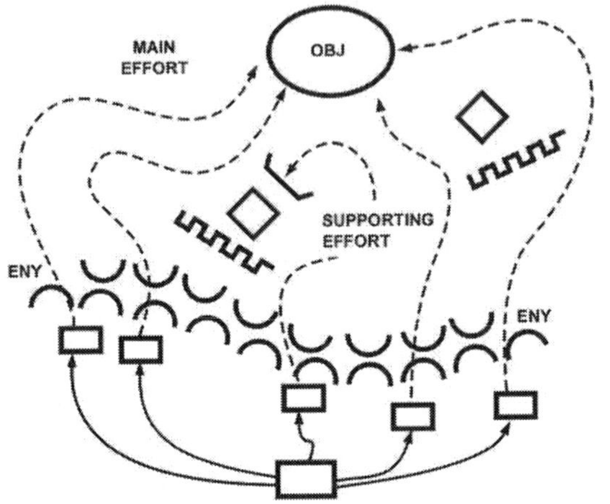

AN INFILTRATION USES COVERT MOVEMENT OF FORCES THROUGH ENEMY LINES TO ATTACK POSITIONS IN THE ENEMY REAR.

Figure 7-3. Infiltration.

Penetration

7-17. Penetration (Figure 7-4) is a form of maneuver in which an attacking element seeks to rupture enemy defenses on a narrow front to create both assailable flanks and access to the enemy's rear. Penetration is used when enemy flanks are not assailable; when enemy defenses are overextended; when weak spots in the enemy defense are identified; and when time does not permit some other form of maneuver. A penetration normally consists of three steps: breach the enemy's main defense positions, widen the gap created to secure flanks by enveloping one or both of the newly exposed flanks, and seize the objective. As part of a larger force penetration the platoon will normally isolate, suppress, fix, or destroy enemy forces; breach tactical or protective obstacles in the enemy's main defense; secure the shoulders of the penetration; or seize key terrain. Similar to breaching obstacles, the platoon will be designated as a breach, support, or assault element. A company may also use the penetration to secure a foothold within a built-up area.

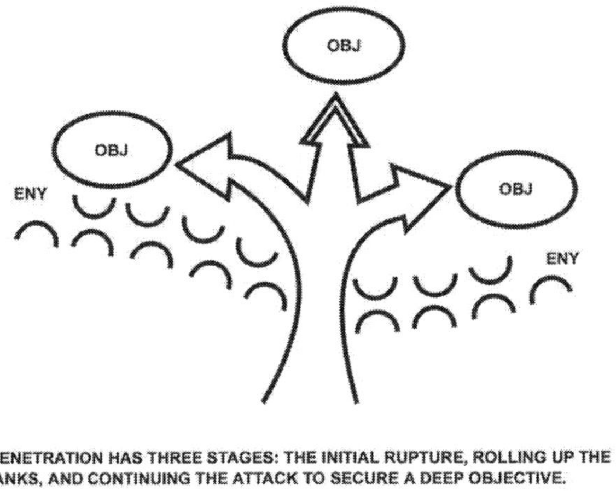

Figure 7-4. Penetration.

FRONTAL ATTACK

7-18. Frontal attack (Figure 7-5) is a form of maneuver in which an attacking element seeks to destroy a weaker enemy force or fix a larger enemy force along a broad front. It is the least desirable form of maneuver because it exposes the attacker to the concentrated fire of the defender and limits the effectiveness of the attacker's own fires. However, the frontal attack is often the best form of maneuver for an attack in which speed and simplicity are key. It is useful in overwhelming weak defenses, security outposts, or disorganized enemy forces, and is also often used when a unit conducts a reconnaissance in force.

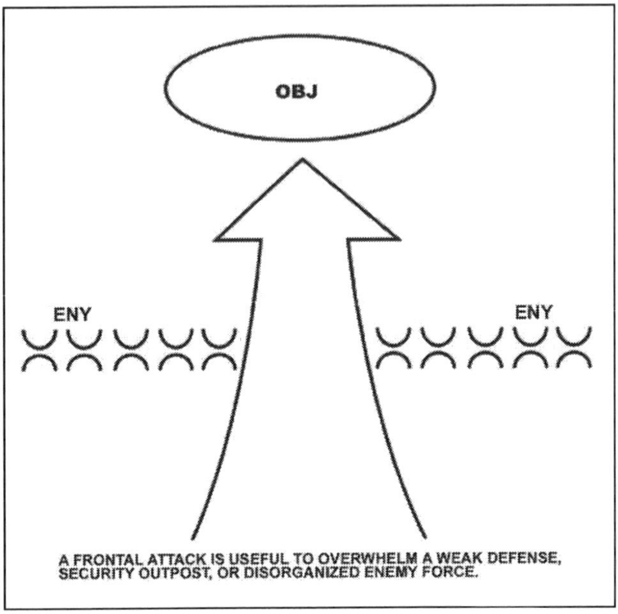

Figure 7-5. Frontal attack.

SECTION II — SEQUENCE OF OFFENSIVE OPERATIONS

7-19. As the platoon leader plans for an offensive mission, he generally considers the following, which apply to many, but not all, offensive operations:

- Assembly area.
- Reconnaissance.
- Movement to the line of departure.
- Maneuver.
- Deployment.
- Assault.
- Consolidation and reorganization.

PREPARATION IN THE OFFENSE

7-20. The friendly Infantry attacker has the advantage of choosing the time, place, and method of the engagement. Infantry units should maximize this advantage by engaging the enemy defender in a way that the he is unprepared for. Preparations for offensive operations include planning and rehearsals enabled by friendly reconnaissance operations that determine the enemy defender's disposition, composition, strength, capabilities, and possible courses of action. Friendly Infantry units then use this knowledge to develop their own courses of action.

ASSEMBLY AREA

7-21. The assembly area (AA) is the area a unit occupies to prepare for an operation. To prepare the platoon for the upcoming battle, the platoon leader plans, directs, and supervises mission preparations in the assembly area. This time allows the platoon to conduct precombat checks and inspections, rehearsals, and sustainment activities. The platoon will typically conduct these preparations within a company assembly area; it will rarely occupy its own assembly area.

RECONNAISSANCE

7-22. All leaders should aggressively seek information about the terrain and the enemy. Because the enemy situation and available planning time may limit a unit's reconnaissance, the platoon will likely conduct reconnaissance to answer the company commander's priority intelligence requirements (PIR). An example is reconnoitering and timing routes from the assembly area to the line of departure. The platoon may also augment the efforts of the battalion reconnaissance platoon to answer the commander's PIR. Other forms of reconnaissance include maps, and if available, terrain software/databases. Updates from reconnaissance can occur at any time while the platoon is planning for, preparing for, or executing the mission. As a result, the leader must always be prepared to adjust his plans.

MOVEMENT TO THE LINE OF DEPARTURE

7-23. The platoon will typically move from the AA to the line of departure as part of the company movement plan. This movement plan may direct the platoon to move to an attack position to await orders to cross the line of departure. If so, the platoon leader must reconnoiter, time, and rehearse the route to the attack position. Section and squad leaders must know where they are to locate within the assigned attack position, which is the last position an attacking element occupies or passes through before crossing the line of departure. The company commander may order all of the platoons to move within a company formation from the assembly area directly to the point of departure at the line of departure. The point of departure is the point where the unit crosses the line of departure and begins moving along a direction of attack or axis of advance. If one point of departure is used, it is important that both the lead platoon and trail platoons reconnoiter, time, and rehearse the route to it. This allows the company commander to maintain synchronization. To maintain flexibility and to further maintain synchronization, he may also designate a point of departure along the line of departure for each platoon.

MANEUVER

7-24. The company commander will plan the approach of all platoons to the objective to ensure synchronization, security, speed, and flexibility. He will select the platoons' routes, movement techniques, formations, and methods of movement to best support his intent for actions on the objective. The platoon leader must recognize this portion of the battle as a fight, not as a movement. He must be prepared to make contact with the enemy. (For a detailed discussion of actions on contact, refer to Chapter 1, paragraph 1-160, and Chapter 3, Section V.) He must plan accordingly to reinforce the commander's needs for synchronization, security, speed, and flexibility. During execution, he may display disciplined initiative and alter his platoon's formation, technique, or speed to maintain synchronization with the other platoons and flexibility for the company commander.

DEPLOYMENT

7-25. As the platoon deploys and moves toward the assault position, it minimizes delay and confusion by beginning the final positioning of the squads as directed by the company commander. An assault position is the last covered and concealed position short of the objective from which final preparations are made to assault the objective. This tactical positioning allows the platoon to move in the best tactical posture through the assault position into the attack. Movement should be as rapid as the terrain, unit mobility, and enemy situation permit. A common control measure used in or just beyond the assault position is the probable line of deployment (PLD), which is used most often under conditions of limited visibility. The probable line of deployment is a phase line the company commander designates as a location where he intends to completely deploy his unit into the assault formation before beginning the assault.

ASSAULT

7-26. During an offensive operation, the platoon's objective may be terrain-oriented or force-oriented. Terrain-oriented objectives may require the platoon to seize a designated area and often require fighting through enemy forces. If the objective is force-oriented, an objective may be assigned for orientation, while the platoon's efforts are focused on the enemy's actual location. Actions on the objective begin when the company or platoon begins placing direct and indirect fires on the objective. This may occur while the platoon is still moving toward the objective from the assault position or probable line of deployment.

CONSOLIDATION AND REORGANIZATION

7-27. The platoon consolidates and reorganizes as required by the situation and mission. Consolidation is the process of organizing and strengthening a newly captured position so it can be defended. Reorganization is the actions taken to shift internal resources within a degraded unit to increase its level of combat effectiveness. Reorganization actions can include, cross-leveling ammunition, ensuring key weapons systems are manned, and ensuring key leadership positions are filled if the operators/crew become casualties. The platoon executes follow-on missions as directed by the company commander. A likely mission may be to continue the attack against the enemy within the area of operations. Regardless of the situation, the platoon must posture itself and prepare for continued offensive operations. Table 7-1 contains common consolidation and reorganization activities.

Table 7-1. Consolidation and reorganization activities.

Consolidation Activities	Reorganization Activities
• **S**ecurity measures include— ▪ Establishing 360-degree local security. ▪ Using security patrols. ▪ Using observation posts/outposts. ▪ Emplacing early warning devices. ▪ Establishing and registering final protective fires. ▪ Seeking out and eliminating enemy resistance (on and off the objective). • **A**utomatic weapons (man, position, and assign principal directions of fire [PDFs] to Soldiers manning automatic weapons). • **F**ields of fire (establish sectors of fire and other direct fire control measures for each subunit/Soldier). • **E**ntrenchment (provide guidance on protection requirements such as digging/building fighting positions).	• Reestablishing the chain of command. • Manning key weapon systems. • Maintaining communications and reports, to include— ▪ Restoring communication with any unit temporarily out of communication. ▪ Sending unit situation report. ▪ Sending SITREPs (at a minimum, subordinates report status of mission accomplishment). ▪ Identifying and requesting resupply of critical shortages. • Resupplying and redistributing ammunition and other critical supplies. • Performing special team actions such as— ▪ Consolidating and evacuating casualties, EPWs, enemy weapons, noncombatants/ refugees, and damaged equipment (not necessarily in the same location). ▪ Treating and evacuating wounded personnel. ▪ Evacuating friendly KIA. ▪ Treating and processing EPWs. ▪ Segregating and safeguarding noncombatants/ refugees. ▪ Searching and marking positions to indicate to other friendly forces that they have been cleared.

SECTION III — PLANNING CONSIDERATIONS

7-28. The warfighting functions are a group of tasks and systems united by a common purpose that Infantry leaders use to accomplish missions and training objectives. Planning, synchronization and coordination among the warfighting functions are critical for success. The warfighting functions are addressed in this section.

INTELLIGENCE

The Intelligence warfighting function consists of the related tasks and systems that facilitate understanding of the enemy, terrain, weather, and civil considerations. In offensive operations the Infantry platoon leader uses his intelligence, surveillance, reconnaissance (ISR) assets to study the terrain and confirm or deny the enemy's strengths, dispositions, and likely intentions, especially where and in what strength the enemy will defend. These assets also gather information concerning the civilian population within the AO to confirm or deny their numbers, locations, and likely intentions.

MOVEMENT AND MANEUVER

7-29. The movement and maneuver warfighting function consists of the related tasks and systems that move forces to achieve a position of advantage in relation to the enemy. The purpose of maneuver is to close with and destroy the defending enemy. Maneuver requires a base-of-fire element to suppress or destroy enemy forces with accurate direct fires and bounding elements to gain positional advantage over the enemy. When effectively executed, maneuver leaves enemy elements vulnerable by forcing them to fight in at least two directions, robbing them of initiative, and ultimately limiting their tactical options. Movement and maneuver are the means by which Infantry leaders mass the effects of combat power to achieve surprise, shock, momentum, and dominance.

7-30. The platoon will likely focus on mobility during the movement phase of offensive operations and may be required to breach obstacles as part of an offensive operation. These obstacles may be protective (employed to assist units in their close-in protection), which the platoon is expected to breach without additional assets. Tactical obstacles, however, which block, disrupt, turn, or fix unit formations, normally require engineer assets to breach. Refer to FM 3-34.2, *Combined-Arms Breaching Operations,* for a more detailed discussion of breaching.

FIRE SUPPORT

7-31. The fire support warfighting function consists of the related tasks and systems that provide collective and coordinated use of Army indirect fires, joint fires, and offensive information operations. The platoon may be able to employ indirect fires from field artillery or company and or battalion mortars to isolate a small part of the enemy defense or to suppress the enemy on the objective. The platoon leader must always keep in mind the potential danger to friendly elements created by indirect fires used in support of the assault. He must ensure that the indirect fire assets always know the position and direction of movement of his platoon.

PROTECTION

7-32. The protection warfighting function consists of the related tasks and systems that preserve the force, so the Infantry leader can apply maximum combat power. Preserving the force includes protecting personnel, physical assets, and information of the Infantry platoon. Areas included in protection at the Infantry platoon level are:

- Safety
- Fratricide avoidance
- Survivability
- Air and missile defense
- Force health protection

SUSTAINMENT

7-33. The sustainment warfighting function includes related tasks and systems that provide support and services to ensure freedom of action, extend operational reach, and prolong endurance. The primary purpose of sustainment in the offense is to assist the platoon and company in maintaining momentum during the attack. Key sustainment planning considerations for the platoon leader during the offense include:

- High expenditure of ammunition for selected tactical tasks.
- Friendly casualty rate and how to evacuate the casualties to what locations.
- Availability of water and other mission-essential supplies before, during, and after actions on the objective.

COMMAND AND CONTROL

7-34. The command and control warfighting function consists of the related tasks and systems that support Infantry leaders in exercising authority and direction. At the Infantry platoon and squad level, command and control refers to the process of directing, coordinating, and controlling a unit to accomplish a mission. During offensive operations Infantry leaders must establish control measures to provide a way to direct and coordinate the platoon or squad's movement.

7-35. Control measures are directives given graphically or orally by a commander to subordinate commands to assign responsibilities, coordinate fires and maneuver, and control combat operations. Each control measure can be portrayed graphically. In general, all control measures should be easily identifiable on the ground. Leaders organize the battlefield by establishing control measures that dictate responsibility, control movement, and manage fires.

AREA OF OPERATION

7-36. The area of operation (AO) is the basic control measure for assigning responsibility and conducting operations. An AO is a clearly defined geographical area with associated airspace where leaders conduct operations within the limits of their authority. Within an AO leaders are responsible for accomplishing their mission and are accountable for their unit's actions. Units acting as part of a larger unit operate within the AO of the next higher commander. When assigned their own AO, leaders of Infantry platoons or squads usually have expanded planning, preparation, and execution responsibilities. At lower levels, the term AO is often synonymous with a unit's current location and any associated operational environment, usually without formal boundaries.

7-37. Boundaries control the maneuver and fire of adjacent units. They are normally drawn along recognizable terrain features and are situated so key terrain features and avenues of approach are inclusive to one unit.

7-38. Leaders use boundaries as their basic control measure to divide up the battlefield and assign responsibilities. When given a boundary, the owning unit may employ any direct or indirect fire in accordance with previously-issued orders and ROE without receiving further clearance from the controlling headquarters. The following exceptions apply:

- Munitions that produce effects outside of the boundary must be authorized by higher.
- Munitions that are restricted must authorized.

Direct Fire and Boundaries

7-39. Direct fire may be used across a unit boundary without prior coordination if the enemy target is clearly identified. When possible, direct fire boundries should be coordinated with adjacent units. Unless the target poses an imminent threat, the leader authorizing the fire should attempt prior coordination before engaging targets across his boundary. Indirect fire will not be used across a unit's boundary unless prior coordination is made.

Basic Control Measures

7-40. Leaders use the boundary to divide up their AOs for subordinates. An AO normally contains one or more engagement areas (defense operations) and or objectives (offensive operations). Leaders use additional control measures to specify responsibilities, control movement and fires (direct and indirect), sequence subordinate activities, and synchronize other resources.

Types of AOs

7-41. The type of AO is defined by whether a unit shares a boundary with an adjacent unit. If it does, it is a contiguous AO. If a boundary is not shared with another unit, it is a noncontiguous AO (Figure 7-6). The higher headquarters is responsible for the area between noncontiguous AOs.

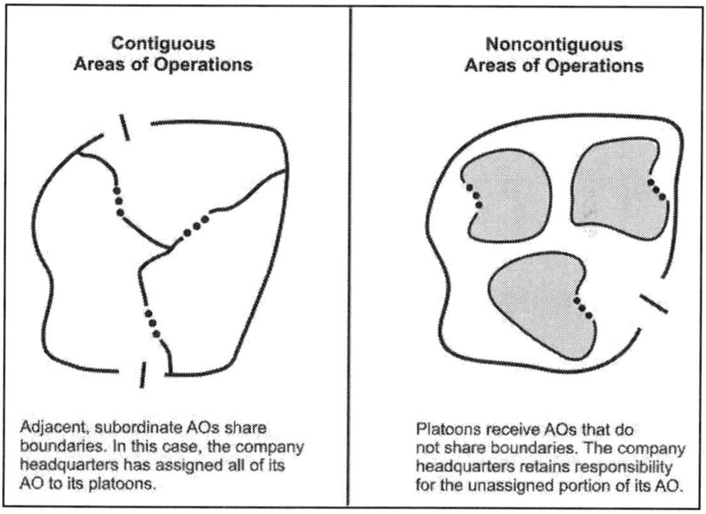

Figure 7-6. Types of AOs.

Mutually Supporting Units

7-42. Regardless of whether a unit shares a common boundary, leaders must determine if they have mutually-supporting adjacent units. The presence of a mutually-supporting unit indicates an increased requirement for coordination. A position without mutually-supporting adjacent units indicates an increased requirement for security—360-degree security.

INTEGRATING CONTROL MEASURES WITH TERRAIN

7-43. When looking for terrain features to use as control measures, leaders consider three types: linear; point; and area. Linear features follow major natural and man-made features such as ridgelines, valleys, trails, streams, power lines, and streets. Point features can be identified by a specific feature or a grid coordinate including, hilltops, and prominent buildings. Area features are significantly larger than point features and require a combination of grid coordinates and terrain orientation. Table 7-2 lists common uses of terrain features for control measures.

Table 7-2. Terrain feature control measures.

Use of Linear Terrain Features in the Offense	Use of Point Terrain Features in the Offense	Use of Area Terrain Features in the Offense
Axis of advance Direction of attack Infiltration lane Limit of advance Line of contact Line of departure Phase line Probable line of deployment Route	Check point Coordination point Linkup point Point of departure Rally point Target reference point	Assembly area Assault position Battle position Objective Named area of interest Targeted area of interest
Use of Linear Terrain Features in the Defense	**Use of Point Terrain Features in the Defense**	**Use of Area Terrain Features in the Defense**
Battle handover line Final protective line Forward edge of battle area (FEBA) Forward line of own troops (FLOT) Screen line Guard line	Observation post Target reference point	Battle position Main battle area Security zone

SECTION IV — PLATOON ATTACKS

7-44. Platoons and squads normally conduct an attack as part of the Infantry company. An attack requires detailed planning, synchronization, and rehearsals to be successful. The company commander designates platoon objectives with a specific mission for his assault, support, and breach elements. To ensure synchronization, all leaders must clearly understand the mission, with emphasis on the purpose, of peer and subordinate elements. Leaders must also know the location of their subordinates and adjacent units during the attack. In addition to having different forms based on their purposes (refer to Section VII), attacks are characterized as hasty, or deliberate. The primary difference between the hasty and deliberate attack is the planning and coordination time available to allow the full integration and synchronization of all available combined arms assets. Attacks may take the form of one of the following:

- Enemy-oriented attacks against a stationary force.
- Enemy-oriented attacks against a moving force.
- Terrain-oriented attacks.

7-45. Additionally, some attacks may be significantly focused on executing a select task by a certain date/time group. Attacks will either be daylight attacks or limited visibility attacks. Limited visibility attacks are further divided into illuminated and nonilluminated attacks. Leaders must always plan on nonilluminated attacks becoming illuminated at some point, whether due to friendly or enemy efforts.

DELIBERATE ATTACK

7-46. A deliberate attack is a type of offensive action characterized by preplanned coordinated employment of firepower and maneuver to close with and destroy the enemy. The deliberate attack is a fully coordinated operation that is usually reserved for those situations in which the enemy defense cannot be overcome by a hasty attack. Commanders may order a deliberate attack when the deployment of the enemy shows no identifiable exposed flank or physical weakness, or when a delay will not significantly improve the enemy's defenses. The deliberate attack is characterized by detailed intelligence concerning a situation that allows the leader to develop and coordinate detailed plans. The leader task-organizes his forces specifically for the operation to provide a fully synchronized combined arms team. Time taken to prepare a deliberate attack is also time in which the enemy can continue defensive improvements, disengage, or launch a spoiling attack. The phases of the deliberate attack are reconnaissance, move to the objective, isolate the objective, seize a foothold and exploit the penetration (actions on the objective), and consolidate and reorganize (Figure 7-7).

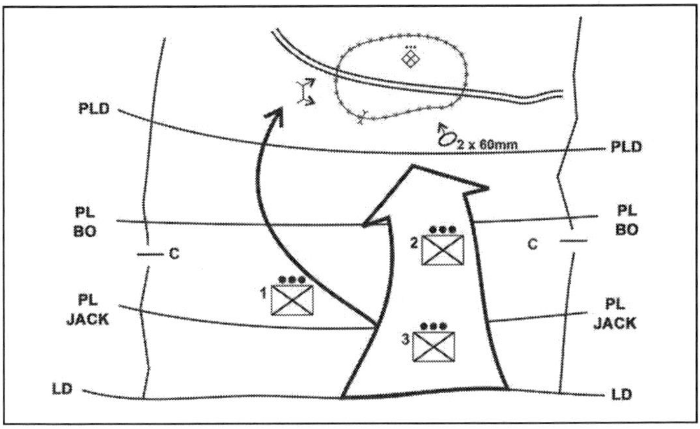

Figure 7-7. Company deliberate attack.

Reconnaissance

7-47. Before a deliberate attack, the platoon and company should gain enemy, terrain, and friendly information from the reconnaissance conducted by the battalion reconnaissance platoon. However, this may not always occur. The platoon and company should be prepared to conduct their own reconnaissance of the objective to confirm, modify, or deny their tentative plan.

7-48. Platoons should not conduct reconnaissance unless specifically tasked to do so in a consolidated reconnaissance plan. If possible, the company should determine the enemy's size, location, disposition, most vulnerable point, and most probable course of action. At this point, and with permission from battalion, the company should direct the platoon to conduct a reconnaissance patrol. This element conducts a reconnaissance of the terrain along the axis of advance and on the objective. It determines where the enemy is most vulnerable to attack and where the support element can best place fires on the objective.

7-49. The tentative plan may change as a result of the reconnaissance if the platoon or squad discovers that terrain or enemy dispositions are different than determined earlier in the TLP. The platoon or squad leader may modify control measures based on the results of the reconnaissance, and must send these adjustments to their leader as soon as possible. For example, the platoon may discover the weapons squad cannot suppress the enemy from the north side of the objective as originally planned because of terrain limitations. Therefore, the platoon leader moves the support-by-fire positions to the south side of the objective, adjusts the tentative plan's control measures, and radios the control measures to his commander for approval. The graphics are subsequently disseminated throughout the company and to adjacent units as needed.

Advance to the Objective

7-50. The attacking element advances to within assault distance of the enemy position under supporting fires using a combination of traveling, traveling overwatch, or bounding overwatch. Platoons advance to successive positions using available cover and concealment. The company commander may designate support-by-fire positions to protect friendly elements with suppressive direct fires. As the company maneuvers in zone, it employs fires to suppress, neutralize, and obscure the enemy positions. The support-by-fire elements may need to occasionally change locations to maintain the ability to support the advancing assault element.

Assembly Area to the Line of Departure

7-51. The line of departure is normally a phase line where elements of the attacking element transition to secure movement techniques in preparation for contact with the enemy. Platoons may maneuver from the line of departure to designated support-by-fire positions, assault positions, and breach or bypass sites. Before leaving the assembly area, the platoon leader should receive an update of the location of forward and adjacent friendly elements. He should also receive updated enemy locations. The platoon leader then disseminates these reports to each squad leader.

7-52. The platoon moves forward from the assembly area to the line of departure, usually as part of a company formation along a planned route. The platoon leader should have reconnoitered the route to the line of departure and specifically to the crossing point. During the planning stage, he plots a waypoint on the line of departure at the point he intends to cross. The platoon navigates to the waypoint during movement. The move from the assembly area is timed during the reconnaissance so the lead section crosses the line of departure at the time of attack without halting in the attack position. If the platoon must halt in the attack position, the squads establish security and take care of last minute coordination.

Line of Departure to Assault Position

7-53. The platoon moves from the line of departure to the assault position. The platoon leader plots waypoints to coincide with checkpoints along the route. During movement, he ensures the platoon navigates from checkpoint to checkpoint or phase line by using basic land navigation skills supplemented by precision navigation.

Assault Position to the Objective

7-54. The assault position is the last covered and concealed position before reaching the objective. Ideally, the platoon occupies the assault position without the enemy detecting any of the platoon's elements. Preparations in the assault position may include preparing bangalores, other breaching equipment or demolitions, fixing bayonets, ceasing or shifting fires, or preparing smoke pots. The platoon may halt in the assault position if necessary to ensure it is synchronized with friendly forces. Once the assault element moves forward of the assault position, the assault must continue. If stopped or turned back, the assault element could sustain unnecessary casualties.

7-55. Supporting fire from the weapons squad must continue to suppress the enemy and must be closely controlled to prevent fratricide. At times, the assault element may mark each Soldier or just the team on the flank nearest the support element. The key is to ensure the support-by-fire element knows the location of the assault element at all times. The assaulting Soldiers and the support element sustain a high rate of fire to suppress the enemy.

7-56. When the assault element moves to the breach point, the base-of-fire leader verifies the assault element is at the right location. The base-of-fire leader is responsible for tracking the assault element as it assaults the objective. The company commander shifts or ceases indirect fire when it endangers the advancing Soldiers and coordinates this with the platoon's assault. As the fire of the platoon's support is masked, the platoon leader shifts or ceases it or displaces the weapons squad to a position where continuous fire can be maintained.

Isolate the Objective

7-57. The goals of isolation are to prevent the enemy from reinforcing the objective and to prevent enemy forces on the objective from leaving. Infantry platoons will probably be an isolating element within a company.

SEIZE A FOOTHOLD AND EXPLOIT THE PENETRATION (ACTIONS ON THE OBJECTIVE)

7-58. The platoon leader often designates assault, support, and breach elements within his platoon to conduct a deliberate attack. One technique is to designate the weapons squad as the support element, an Infantry squad as the breach element, and the remainder of the platoon as the assault element.

7-59. The supporting elements assist the breach element's initial breach of the objective by placing suppressive fires on the most dangerous enemy positions. As the breach is being established, the weapons squad shifts fires (or local self-defense weapons) to allow the breach element to penetrate the objective and avoid fratricide. Visual observation and information provided through the radio are vital to maintain suppressive fires just forward of the breach and assault elements.

7-60. The supporting elements monitor the forward progress of the assault element and keep shifting suppressive fire at a safe distance in front of them. The weapons squad positions itself to provide continual close-in suppressive fire to aid the actions of the assault squad(s) as it moves across the objective.

7-61. Once the breach element has seized the initial foothold on the objective, the assault element may then move through the breach lane to assault the objective. As this occurs, the platoon leader closely observes the progress of the breach and assault elements to ensure there is no loss in momentum, and that assault and breach elements do not cross in front of the supporting elements.

7-62. All communication from the support element to the breach, assault, and weapons support is by frequency modulated (FM) radio or signals. If the platoon sergeant or squad leader observes problems, they radio the platoon leader. The platoon leader uses this information and what he personally sees on the objective to control the assault.

CONSOLIDATE AND REORGANIZE

7-63. Once enemy resistance on the objective has ceased, the platoon quickly consolidates to defend against a possible counterattack and prepares for follow-on missions.

7-64. Consolidation consists of actions taken to secure the objective and defend against an enemy counterattack.

7-65. Reorganization, normally conducted concurrently with consolidation, consists of preparing for follow-on operations. As with consolidation, the platoon leader must plan and prepare for reorganization as he conducts his TLP.

SITE EXPLOITATION

7-66. Once the sensitive site is secure, enemy resistance eliminated, and safe access established, exploitation of the site begins. Subject matter experts and teams carefully enter and exploit every structure, facility, and vehicle on the site and determine its value and its hazard to the platoon. The security force continues to secure the site. Leaders may elect to rotate the assault, support, and security forces if the site exploitation lasts for a prolonged period of time. .

HASTY ATTACK

7-67. The platoon normally participates in a hasty attack as part of a larger unit, during movement to contact, as part of a defense, or whenever the commander determines that the enemy is vulnerable. A hasty attack is used to—

- Exploit a tactical opportunity.
- Maintain the momentum.
- Regain the initiative.
- Prevent the enemy from regaining organization or balance.
- Gain a favorable position that may be lost with time.

7-68. Because its primary purpose is to maintain momentum or take advantage of the enemy situation, the hasty attack is normally conducted with only the resources that are immediately available. Maintaining constant pressure through hasty attacks keeps the enemy off balance and makes it difficult for him to react effectively. Rapidly attacking before the enemy can act often results in success even when the combat power ratio is not as favorable as desired. With its emphasis on agility and surprise, however, this type of attack may cause the attacking element to lose a degree of synchronization. To minimize this risk, the commander should maximize use of standard formations, well-rehearsed, thoroughly-understood battle and crew drills, and SOPs. The hasty attack is often the preferred option during continuous operations. It allows the commander to maintain the momentum of friendly operations while denying the enemy the time needed to prepare his defenses and to recover from losses suffered during previous action. Hasty attacks normally result from a movement to contact, successful defense, or continuation of a previous attack.

TASK ORGANIZATION

7-69. The hasty attack is conducted using the principles of fire and movement. The controlling headquarters normally designates a base-of-fire element and a maneuver element.

CONDUCT OF THE HASTY ATTACK

7-70. By necessity, hasty attacks are simple and require a minimum of coordination with higher and adjacent leaders. Leaders, however, still take the necessary measures to assess the situation, decide on an appropriate course of action, and direct their subordinates in setting conditions and execution.

7-71. Execution begins with establishment of a base of fire, which then suppresses the enemy force. The maneuver element uses a combination of techniques to maintain its security as it advances in contact to a position of advantage. These techniques include:

- Use of internal base-of-fire and bounding elements.
- Use of covered and concealed routes.
- Use of indirect fires and smoke grenades or pots to suppress or obscure the enemy or to screen friendly movement.
- Execution of bold maneuver that initially takes the maneuver element out of enemy direct fire range.

SECTION V — OTHER OFFENSIVE OPERATIONS

7-72. This section focuses on offensive operations of movement to contact, exploitation, and pursuit the platoon normally conducts as part of an Infantry company or larger element:

MOVEMENT TO CONTACT

7-73. Platoons and squads participate in a movement to contact as part of a company using movement formations and techniques explained in Chapter 4. A company generally conducts a movement to contact when it must gain or maintain contact with the enemy, or when it lacks sufficient time to gain intelligence or make extensive plans to defeat the enemy (Figure 7-8). Higher intelligence assets should attempt to find the enemy through reconnaissance and surveillance. Battalions may task or allow companies to gather intelligence through reconnaissance and surveillance if the company commander needs to further develop the intelligence picture. In this case, the company tasks a platoon or squad to conduct reconnaissance, surveillance, or both.

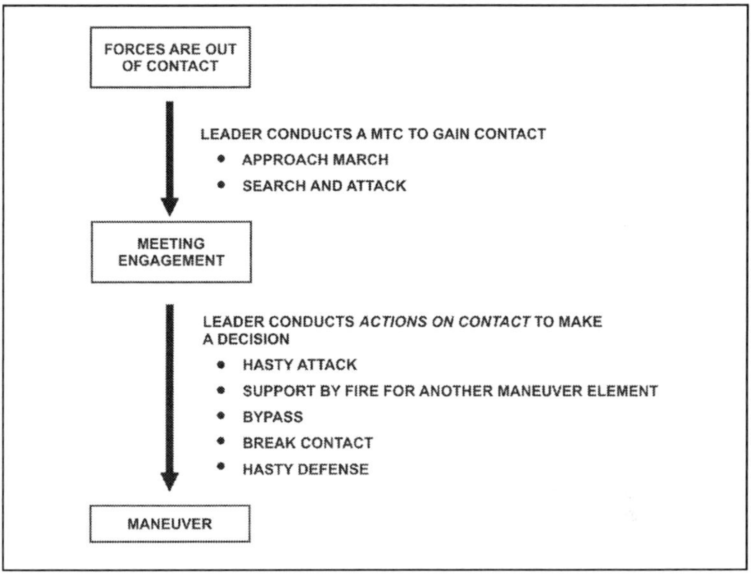

Figure 7-8. Movement to contact framework.

7-74. The movement to contact results in a meeting engagement. A meeting engagement is the combat action that occurs when a moving element engages a stationary or moving enemy at an unexpected time and place. Meeting engagements are characterized by—

- Limited knowledge of the enemy.
- Minimum time available for the leader to conduct actions on contact.
- Rapidly changing situation.
- Rapid execution of battle and crew drills.

PLANNING CONSIDERATIONS

7-75. The company commander will not have a complete visualization of the situation. The leader's role is to gain as much firsthand information as possible. Combined with information on the enemy and the terrain, firsthand information provides knowledge and understanding necessary to respond to the enemy. However, if the enemy situation remains vague, the platoon must be prepared to act in any situation. This is accomplished through proper planning, appropriate movement formations and techniques, fire control measures, platoon SOPs, engagement criteria, and studying the terrain before and during movement to anticipate likely enemy locations. While moving, all leaders study the terrain and anticipate enemy contact. Based on these terrain studies, leaders should avoid likely areas of enemy ambush or areas that expose their platoons to long-range observation and fires. If the enemy is a conventional force, his units may use a doctrinal approach to their disposition, making it easier to find them. If faced with an asymmetric threat, there may be no doctrinal template for the enemy. In this instance, the leader must look for historical patterns in the enemy's operations. In both cases, the leader has to analyze how the enemy fights, how he uses terrain, and what he hopes to accomplish against friendly elements.

TECHNIQUES

7-76. A movement to contact is conducted using one of two techniques: approach march, or search and attack (Table 7-3). The approach march technique is used when the enemy is expected to deploy using relatively fixed offensive or defensive formations, and the situation remains vague. The search and attack technique is used when the enemy is dispersed, when he is expected to avoid contact or quickly disengage and withdraw, or when the higher unit needs to deny him movement in an area of operation.

Table 7-3. The two types of movement to contact.

Approach march is best used when the—	Search and attack is best used when the—
• Enemy force is more conventional in nature. • Enemy force follows a more structured order of battle and is more predictable. • Enemy force is more centrally located. • Enemy conducts more centralized operations.	• Enemy conducts operations over a very large area in a dispersed manner, forcing friendly units to disperse to locate him. • Enemy forces and operations are unconventional or guerilla in nature. • Enemy typically operates in small teams and only makes contact when he feels he has the advantage.

COMMAND AND CONTROL

7-77. The company commander will dictate a number of command and control techniques for the unit to employ. The platoon leader, within the scope of the commander's intent and guidance and the factors of METT-TC, may modify these techniques to better control his sections and squads. The platoon leader will tell the company commander of any additional graphic control measures he builds into his plan. Some examples of command and control techniques are discussed below.

Graphic Control Measures

7-78. The company commander will normally assign lines of departure, phase lines, checkpoints, and GPS waypoints to control the forward movement of the platoon. The platoon does not stop at a phase line unless told to do so. If necessary, the platoon leader designates additional phase lines, checkpoints, or waypoints for use within the platoon to reduce the number and length of radio transmissions used to control movement.

Fire Control Measures

7-79. The platoon uses boundaries, direct fire plans, pyrotechnics, signals, and FRAGOs for direct fire control and distribution. (For a detailed discussion of direct fire control and distribution, refer to Chapter 2.) The variety of weapons in the Infantry platoon makes it critical for all squads to understand the observation plan and the designated sectors of fire during a movement to contact. This takes on importance because of the scarcity of information about the enemy.

Indirect Fire Plan

7-80. The platoon leader must have a good indirect fire plan for his route to cover anticipated places of contact. These targets are a product of the platoon leader's analysis of the factors of METT-TC and must be incorporated into the company indirect fire plan.

DEVELOPING THE SITUATION

7-81. Once the platoon makes contact with the enemy, it maintains contact until the commander orders otherwise. The platoon leader develops the situation based on the effectiveness of enemy fire, friendly casualties, size of enemy force, and freedom to maneuver. He gathers and reports critical information about the enemy and recommends a course of action. The platoon can bypass the enemy with permission from the commander, conduct an attack, fix the enemy so another platoon can conduct the assault, conduct a defense, establish an ambush, or break contact. The following guidelines apply for the platoon to develop the situation after making contact.

DEFENSIVE CONSIDERATIONS

7-82. In some situations, a platoon conducting a movement to contact makes contact with a much larger and more powerful enemy force. If the platoon encounters a larger enemy force where the terrain gives the platoon an advantage, it should attempt to fix the enemy force. This allows the rest of the company to maneuver against the force. If the platoon cannot fix the enemy, it may have to assume a defensive posture (see Chapter 8) or break contact, but it should do so only if it is in danger of being overwhelmed. Surrendering the initiative to the enemy means the enemy has fixed the platoon in place. Exposed rifle squads are vulnerable to enemy indirect fires. If the platoon receives indirect fire during movement, it should attempt to move out of the area or find a covered position for the rifle squads. Once the indirect fires cease, the platoon prepares for an enemy assault. In the defense, the platoon leader—

- Keeps the company commander informed and continues to report on enemy strength, dispositions, and activities.
- Positions squads to cover dismounted avenues of approach in preparation for the enemy's attack.
- Orients the weapons squad and their Javelins along mounted avenues of approach and establishes positions for the M240B machine guns.
- Establishes direct fire control and distribution measures.
- Calls for and adjusts indirect fires.

APPROACH MARCH TECHNIQUE

7-83. The approach march advances a combat unit when direct contact with the enemy is intended. It can be performed dismounted, mounted, or a combination of the two. The concept behind the approach march as a technique for movement to contact is to make contact with the smallest enemy element. When executed effectively it allows the commander the flexibility of maneuvering or bypassing the enemy force. During an approach march, the company commander will organize his unit into two elements (advance guard, and main body). As part of a company using the approach march technique, platoons may act as the advance guard, the flank or rear guard, or may receive on-order missions as part of the main body.

Advance Guard

7-84. The advance guard operates forward of the main body to ensure its uninterrupted advance. It protects the main body from surprise attack and fixes the enemy to protect the deployment of the main body. As the advance guard, the platoon finds the enemy and locates gaps, flanks, and weaknesses in his defense. The advance guard attempts to make contact on ground of its own choosing, to gain the advantage of surprise, and to develop the situation (either fight through or support the assault of all or part of the main body). The advance guard operates within the range of indirect fire support weapons. The platoon uses appropriate formations and movement techniques based on the factors of METT-TC.

7-85. The advance guard is normally the most robust of the security elements. In addition to the general security measures described above, the advance guard and its sub elements—

- Preserve the main body's freedom of maneuver.
- Prevent unnecessary delay in movement of the main body.
- Learn the whereabouts of the enemy.
- Develop intelligence about the terrain and the environment.
- Detect and overcome enemy security measures.
- Identify and disrupt enemy attempts to ambush the main body.
- Must be ready to gain fire superiority and fight any enemy forces encountered.
- Watch the enemy (if direct fire contact is not pending).
- Delay any enemy attacks to gain time for the main body to deploy.

Main Body

7-86. When moving as part of the main body platoons may be tasked to assault, bypass, or fix an enemy force; or to seize, secure, or clear an assigned area. The platoon also may be detailed to provide squads as flank or rear guards, stay-behind ambushes, or additional security to the front. Platoons and squads use appropriate formations and movement, assault, and ambush techniques.

7-87. The main body moves to reinforce any success achieved by the advance guard, flank the enemy position, or apply overwhelming combat power to seize the contested area. During the attack, the leader on the field takes care to isolate the objective by positioning the flank guard to prevent interdiction of enemy reinforcements into the engagement. The positioning of flank guard blocking positions must be far enough away from the area that no enemy weapons can bring fires to effect the attack by the main body.

Flank or Rear Guard

7-88. The platoon will have the responsibilities of flank or rear guard when moving within the company main body. However, the platoon may act as the flank or rear guard for a battalion conducting a movement to contact using approach march technique. In either situation, the platoon—

- Moves using the appropriate formation and movement technique. (It must maintain the same momentum as the main body.)
- Provides early warning.
- Destroys enemy reconnaissance units.
- Prevents direct fires or observation of the main body.

Actions on Contact

7-89. Once the advance guard makes contact, the main body's leader conducts actions on contact to determine how the main body will fight the enemy. To facilitate this, the advance guard reports enemy contact or disruption. It also deploys and attempts to overcome enemy based on information from point patrol. If the advance guard is not able to overcome the enemy, it assumes a support-by-fire position to support maneuver of the remainder of the advance guard. The remainder of the advance guard attempts a close envelopment to defeat the enemy unless the enemy force is overwhelmingly superior. If successful, the advance guard reforms and resumes march or initiates pursuit. If unsuccessful, the advance guard holds its positions, blocks the enemy, and continues supporting the subsequent maneuver and attack of the main body.

Additional Approach March COAs

7-90. There are several courses of action available to the leader when the advance guard comes into contact with a force that it cannot overcome with its organic forces. These courses of action include—

- Frontal attack.
- Fix and bypass.
- Fix, isolate, and attack.
- Oblique attack.
- Withdrawal.

Search and Attack Technique

7-91. The search and attack is a technique conducted when the enemy is operating as small, dispersed elements, or when the task is to deny the enemy the ability to move within a given area of operations. The platoon will participate as part of company or battalion search and attack. A unit conducts a search and attack for one or more of the following reasons:

- Render the enemy in the area of operations combat-ineffective.
- Prevent the enemy from operating unhindered in a given area of operations.

- Prevent the enemy from massing to disrupt or destroy friendly military or civilian operations, equipment, or facilities.
- Gain information about the enemy and the terrain.

Organization of Elements

7-92. The higher commander will task-organize the subordinate units into reconnaissance (finding, fixing, and finishing) elements. He will assign specific tasks and purposes to his search and attack elements. It is important to note that within the concept of find, fix, and finish, all platoons could be the reconnaissance element. Depending on the size of the enemy they find, they could end up executing a reconnaissance mission, become the fixing element, or find that they are able to finish the enemy. Planning considerations for organizing include—

- The factors of METT-TC.
- The requirement for decentralized execution.
- The requirement for mutual support. (The platoon leader must be able to respond to contact with his rifle squads or to mutually support another platoon within the company.)
- The Soldier's load. (The leader should ask, "Does the Soldier carry his rucksack, cache it, or leave it at a central point? How will the rucksacks be linked up with the Soldier?")
- Resupply and CASEVAC.
- The employment of key weapons.
- The requirement for patrol bases.

Find (Reconnaissance Element)

7-93. The size and composition of the reconnaissance element is based on the available information on the size and activity of the enemy operating in the designated area of operations. The reconnaissance element typically consists of the battalion reconnaissance platoon plus other battalion and higher level assets. Reconnaissance operations are used to answer information requirements used for leader decisionmaking and are not normally followed immediately by a hasty attack. The find action of a search and attack is used to locate the enemy with the expressed intent of making a hasty attack as soon as possible with the main body. The platoon will reconnoiter named area(s) of interest (NAI) and other areas as designated. The platoon may find the enemy through zone reconnaissance, patrolling, and establishing observation posts.

7-94. The task of the search element is to locate the enemy or information leading to the enemy. The techniques used to search are unique to the area of operations and should be developed and adapted to the specifics of the particular environment. What works in one location may not work in another.

7-95. The security element has two tasks: early warning of approaching enemy and providing support forces to the search elements if in contact with the enemy. The purpose of the security element is to protect the search element allowing them to search. Security elements tasked to provide early warning must be able to observe avenues of approach into and out of the objective area. If the search element is compromised, the security element must be able to quickly support them. These positions must also be able to facilitate communication to higher as well as any supporting assets.

Fixing Element

7-96. The fixing element must have sufficient combat power to isolate the enemy and develop the situation once the reconnaissance element finds him. When developing the situation, the fixing element either continues to maintain visual contact with the enemy until the finishing element arrives, or conducts an attack to physically fix the enemy until the finishing element arrives. The goal is to keep the enemy in a position in which he can be destroyed by the finishing element. Sometimes the fixing element may have sufficient combat power to destroy the enemy themselves. The platoon maintains visual contact to allow the reconnaissance element to continue to other NAIs and isolates the immediate area. The fixing element makes physical contact only if the enemy attempts to leave the area or other enemy elements enter the area. At all times after contact is made, the platoon integrates as many combat multipliers into the fight as possible. Examples include indirect fire support, attack aviation, close air support (CAS), and antiarmor sections or platoons, if they are available.

7-97. The fix element can consist of maneuver elements, fire support assets, and aviation elements. To isolate the enemy, fixing elements normally establish a cordon of blocking positions on possible avenues of approach out of the engagement area. The fix element is also responsible for ensuring its own internal security, conducting link ups with the find/finish elements as required and coordinating fire support assets.

Finishing Element

7-98. The finishing element must have sufficient combat power to destroy enemy forces located within the area of operations. The finishing element must be responsive enough to engage the enemy before he can break contact, yet patient enough not to rush to failure. A platoon, as the finishing element, may be tasked to—

- Destroy the enemy with an attack.
- Block enemy escape routes while another unit conducts the attack.
- Destroy the enemy with an ambush while the reconnaissance or fixing elements drive the enemy toward the ambush location.
- Not allow the enemy to break contact.

Control Measures

7-99. The higher commander will define commander's intent and establish control measures that allow for decentralized execution and platoon leader initiative to the greatest extent possible. The minimum control measures for a search and attack include—

- Areas of operation.
- Named areas of interest.
- Phase lines.
- TRPs.
- Objectives.
- Checkpoints.
- Contact points.
- GPS waypoints.

7-100. An area of operation defines the location in which the subordinate units will conduct their searches. A technique called the "horse blanket" breaks the battalion and company area of operation into many named smaller areas of operation. Units remain in designated areas of operation as they conduct their missions. Battalion and higher reconnaissance assets might be used to observe areas of operation with no platoons in them, while platoons or companies provide their own reconnaissance in the AO. This command and control technique, along with TRPs, assists in avoiding fratricide in a noncontiguous environment. A TRP facilitates the responsiveness of the fixing and finishing elements once the reconnaissance element detects the enemy. Objectives and checkpoints guide the movement of subordinates and help leaders control their organizations. Contact points aid coordination among the units operating in adjacent areas.

EXPLOITATION

7-101. A platoon normally takes part in exploitations as part of a larger force. However, the platoon should exploit tactical success at the local level within the higher commanders' concept of the operation and intent.

PURSUIT

7-102. The objective of the pursuit is the total destruction of the enemy force. Forces equally as or more mobile than the enemy normally conduct the pursuit. The platoon may take part in a pursuit after a successful hasty attack, as part of a company mission, or as part of a task-organized company acting as a designated pursuit element.

ATTACKS DURING LIMITED VISIBILITY

7-103. Effective use of night vision device(s) (NVD) and thermal weapons site(s) (TWS) during limited visibility attacks enhance squad and platoon abilities to achieve surprise and cause panic in a lesser-equipped enemy. NVD enhancements allow the Infantry Soldier to see farther and with greater clarity and provide a marked advantage over the enemy.

7-104. Leaders have an increased ability to control fires during limited visibility. The platoon has three types of enhancements for use in fire control: target designators (GCP-1 and AIM-1); aiming lights (AIM-1 and AN/PAQ-4B/C); and target illuminators designed for use with NVDs. These include infrared parachute flares, infrared trip flares, infrared 40-mm rounds, infrared mortar rounds, infrared bike lights, and remote black lights. These assets greatly aid in target acquisition and fire control. If the engagement becomes illuminated, there are a variety of target illuminators for the unaided eye.

7-105. Soldiers carrying weapons with NVD enhancements have greater accuracy of fires during limited visibility. Each Soldier in the platoon is equipped with an AN/PAQ-4B/C aiming light for his individual weapon. The AN/PAQ-4B/C enables the rifleman to put infrared light on the target at the point of aim.

7-106. Leaders can designate targets with greater precision using the PEQ-2. The PEQ-2 is an infrared laser pointer that uses an infrared light to designate targets and sectors of fire and to concentrate fire. The leader lazes a target and directs his Soldiers to place their fires on the target. Soldiers then use the aiming lights on their AN/PAQ-4B/Cs to engage the target.

7-107. Leaders also can designate larger targets using target illuminators. Target illuminators are essentially infrared light sources that light the target, making it easier to acquire effectively. Target illuminators consist of infrared illumination rounds, infrared M203 40-mm rounds, infrared trip flares, and infrared parachute flares. Leaders and Soldiers use the infrared devices to identify enemy or friendly personnel and then engage targets using their aiming lights.

7-108. The platoon leader and squad leaders follow tactical standing operating procedures (TSOP) and sound courses of action to synchronize the employment of infrared illumination devices, target designators, and aiming lights. This is done during their assault on the objective, while remaining prepared for a noninfrared illuminated attack.

7-109. Leaders use luminous tape or chemical lights to mark assault personnel to prevent fratricide. The enemy must not be able to see the marking. Two techniques are to place tape on the back of the helmet or to use small infrared chemical lights (if the enemy has no NVDs). Supporting elements must know the location of the lead assault element.

7-110. To reduce the risk to the assault element, the platoon leader may assign weapons control restrictions. For example, the squad on the right in the assault might be assigned weapons free to the right flank because no friendly Soldiers are there. The squad on the left may be assigned weapons tight or weapons hold, which means that another friendly unit is located there.

7-111. The platoon leader may do the following to increase control during the assault:

- Avoid use of flares, grenades, or smoke on the objective.
- Allow only certain personnel with NVDs to engage targets on the objective.
- Use a magnetic azimuth for maintaining direction.
- Use mortar or artillery rounds to orient attacking units.
- Assign a base squad or fire team to pace and guide others.
- Reduce intervals between Soldiers and squads.

7-112. As in daylight, mortar, artillery, and antiarmor fires are planned, but are not fired unless the platoon is detected or is ready to assault. Some weapons may fire before the attack and maintain a pattern to deceive the enemy or to help cover noise made by the platoon's movement. This is not done if it will disclose the attack.

7-113. Indirect fire is hard to adjust when visibility is poor. If the exact location of friendly units is not clearly known, indirect fire is directed first at enemy positions beyond the objective, then moved onto the objective.

7-114. Illuminating rounds that are fired to burn on the ground can be used to mark objectives. This helps the platoon orient on the objective but may adversely affect NVDs.

7-115. Smoke is planned to further reduce the enemy's visibility, particularly if he has NVDs. The smoke is laid close to or on enemy positions so it does not restrict friendly movement or hinder the breaching of obstacles. Employing smoke on the objective during the assault may make it hard for assaulting Soldiers to find enemy fighting positions. If enough thermal sights are available, smoke on the objective may provide a decisive advantage for a well-trained platoon.

7-116. Illumination is always planned for limited visibility attacks, giving the leader the option of calling for it. Battalion commanders normally control the use of conventional illumination, but may authorize the company commander to do so. If the commander decides to use conventional illumination, he should not call for it until the assault is initiated or the attack is detected. It should be placed on several locations over a wide area to confuse the enemy as to the exact place of the attack. Also, it should be placed beyond the objective to help assaulting Soldiers see and fire at withdrawing or counterattacking enemy Soldiers.

NOTE: If the enemy is equipped with NVDs, leaders must evaluate the risk of using each technique and ensure the mission is not compromised because the enemy can detect infrared light sources.

SECTION VI — SPECIAL PURPOSE ATTACKS

7-117. When the company commander directs it, the platoon conducts a special attack. The commander bases his decision on the factors of METT-TC. Special purpose attacks are subordinate forms of an attack and they include—

- Ambush.
- Raid.
- Counterattack.
- Spoiling attack.
- Feint.
- Demonstration.

7-118. As forms of the attack, they share many of the same planning, preparation, and execution considerations of the offense. Feints and demonstrations are also associated with military deception operations.

AMBUSH

7-119. An ambush is a form of attack by fire or other destructive means from concealed positions on a moving or temporarily halted enemy. It may take the form of an assault to close with and destroy the enemy, or be an attack by fire only. An ambush does not require ground to be seized or held. Ambushes are generally executed to reduce the enemy force's overall combat effectiveness. Destruction is the primary reason for conducting an ambush. Other reasons to conduct ambushes are to harass the enemy, capture the enemy, destroy or capture enemy equipment, and gain information about the enemy. Ambushes are classified by category (deliberate or hasty), formation (linear or L-shaped), and type (point, area, or antiarmor). The platoon leader uses a combination of category, type, and formation for developing his ambush plan. See Chapter 9 for greater detail on ambushes.

OPERATIONAL CONSIDERATIONS

7-120. The execution of an ambush is offensive in nature. However, the platoon may be directed to conduct an ambush during offensive or defensive operations. The platoon must take all necessary precautions to ensure that it is not detected during movement to or preparation of the ambush site. The platoon also must have a secure route of withdrawal following the ambush. An ambush normally consists of the following actions:

- Tactical movement to the objective rally point (ORP).
- Reconnaissance of the ambush site.
- Establishment of the ambush security site.
- Preparation of the ambush site.
- Execution of the ambush.
- Withdrawal.

TASK ORGANIZATION

7-121. The Infantry platoon is normally task-organized into assault, support, and security elements for execution of the ambush.

Assault Element

7-122. The assault element executes the ambush. It may employ an attack by fire, an assault, or a combination of those techniques to destroy the ambushed enemy force. The assault element generally consists of a rifle squad. The platoon leader is normally located with the assault element.

Support Element

7-123. The support element fixes the enemy force to prevent it from moving out of the kill zone, which allows the assault element to conduct the ambush. The support element generally uses direct fires in this role, but it may be responsible for calling indirect fires to further fix the ambushed enemy force. The support element generally consists of the weapons squad. The platoon sergeant is normally located with the support element.

Security Element

7-124. The security element provides protection and early warning to the assault and support elements, and secures the objective rally point. It isolates the ambush site both to prevent the ambushed enemy force from moving out of the ambush site and to prevent enemy rescue elements from reaching the ambush site. The security element may also be responsible for securing the platoon's withdrawal route. The security element generally consists of a rifle squad.

PLANNING

7-125. The platoon leader's key planning considerations for any ambush include the following:

- Cover the entire kill zone (engagement area) by fire.
- Use existing terrain features (rocks or fallen trees, for example) or reinforcing obstacles (Claymores or other mines) orienting into the kill zone to keep the enemy in the kill zone.
- Determine how to emplace reinforcing obstacles on the far side of the kill zone.
- Protect the assault and support elements with mines, Claymores, or explosives.
- Use the security element to isolate the kill zone.
- Establish rear security behind the assault element.
- Assault into the kill zone to search dead and wounded, to assemble prisoners, and to collect equipment. The assault element must be able to move quickly on its own through the ambush site protective obstacles.
- Time the actions of all elements of the platoon to prevent the loss of surprise.

NOTE: When manning an ambush for long periods of time, the platoon leader may use only one squad to conduct the entire ambush, rotating squads over time. The platoon leader must consider the factors of METT-TC and must especially consider the company commander's intent and guidance.

CATEGORY

7-126. The leader determines the category of ambush through an analysis of the factors of METT-TC. Typically, the two most important factors are time and enemy.

Deliberate

7-127. A deliberate ambush is a planned offensive action conducted against a specific target for a specific purpose at a predetermined location. When planning a deliberate ambush, the leader requires detailed information on the—

- Size and composition of the targeted enemy unit.
- Weapons and equipment available to the enemy.
- Enemy's route and direction of movement.
- Times that the targeted enemy unit will reach or pass specified points along the route.

Hasty

7-128. The platoon (or squad) conducts a hasty ambush when it makes visual contact with an enemy force and has time to establish an ambush without being detected. The conduct of the hasty ambush should represent the execution of disciplined initiative within the parameters of the commander's intent. The actions for a hasty ambush should be established in a unit SOP and rehearsed so Soldiers know what to do on the leader's signal.

FORMATIONS

7-129. The platoon leader considers the factors of METT-TC to determine the required formation.

Linear

7-130. In an ambush using a linear formation, the assault and support elements deploy parallel to the enemy's route. This position forces the enemy on the long axis of the kill zone, and subjects the enemy to flanking fire. The linear formation can be used in close terrain that restricts the enemy's ability to maneuver against the platoon, or in open terrain (provided a means of keeping the enemy in the kill zone can be effected).

L-Shaped

7-131. In an L-shaped ambush the assault element forms the long leg parallel to the enemy's direction of movement along the kill zone. The support element forms the short leg at one end of and at a right angle to the assault element. This provides both flanking (long leg) and enfilading (short leg) fires against the enemy. The L-shaped ambush can be used at a sharp bend in a road, trail, or stream. It should not be used where the short leg would have to cross a straight road or trail. The platoon leader must consider the other factors of METT-TC before opting for the L-shaped formation. Special attention must be placed on sectors of fire and SDZ of weapons because of the risk of fratricide when conducting an L-shaped ambush.

V-Shaped Ambush

7-132. The V-shaped ambush assault elements are placed along both sides of the enemy route so they form a V. Take extreme care to ensure neither group fires into the other. This formation subjects the enemy to both enfilading and interlocking fire.

TYPE

7-133. The company commander, following an analysis of the factors of METT-TC, determines the type of ambush that the platoon will employ.

CONDUCTING AN AREA AMBUSH

7-134. An area ambush (more than one point ambush) is not conducted by a unit smaller than a platoon. This ambush works best where enemy movement is restricted. Once the platoon is prepared, the area ambush is conducted the same as a point ambush. The dominating feature of an area ambush is the amount of synchronization between the separate point ambushes.

7-135. Area ambushes require more planning and control to execute successfully. Surprise is more difficult to achieve simply because of the unit's dispersion in the AO. Having more than one ambush site increases the likelihood of being detected by the enemy or civilians. This major disadvantage is offset by the increased flexibility and sophistication available to the leader.

CONDUCTING A POINT AMBUSH

7-136. Point ambushes are set at the most ideal location to inflict damage on the enemy. Such ambushes must be able to handle being hit by the enemy force from more than one direction. The ambush site should enable the unit to execute an ambush in two or three main directions. The other directions must be covered by security that gives early warning of enemy attack.

RAID

7-137. A raid is a limited-objective form of an attack, usually small-scale, involving swift penetration of hostile territory to secure information, confuse the enemy, or destroy installations. A raid always ends with a planned withdrawal to a friendly location upon completion of the mission. The platoon can conduct an independent raid in support of the battalion or higher headquarters operation, or it can participate as part of the company in a series of raids. Rifle squads do not execute raids; they participate in a platoon raids.

OPERATIONAL CONSIDERATIONS

7-138. The platoon may conduct a raid to accomplish a number of missions, including the following:

- Capture prisoners.
- Destroy specific command, control, and or communications locations.
- Destroy logistical areas.
- Obtain information concerning enemy locations, dispositions, strengths, intentions, or methods of operation.
- Confuse the enemy or disrupt his plans.
- Seize contraband.

TASK ORGANIZATION

7-139. The task organization of the raiding element is determined by the purpose of the operation. However, the raiding force normally consists of the following elements:

- Support element (support by fire).
- Assault element (with the essential task of the mission).
- Breach element (if required to reduce enemy obstacles).
- Isolation/security element.

CONDUCT OF THE RAID

7-140. The main differences between a raid and other special purpose attacks are the limited objectives of the raid and the associated withdrawal following completion. However, the sequence of platoon actions for a raid is very similar to those for an ambush. Additionally, the assault element of the platoon may have to conduct a breach of a protective obstacle (if a breach element has not been designated). Raids may be conducted in daylight or darkness, within or beyond the supporting distances of the parent unit. When the enemy location to be raided is beyond supporting distances of friendly lines, the raiding party operates as a separate element. An objective, usually very specific in nature, is assigned to orient the raiding unit (Figure 7-9). During the withdrawal, the attacking element should use a route different from that used to conduct the raid itself.

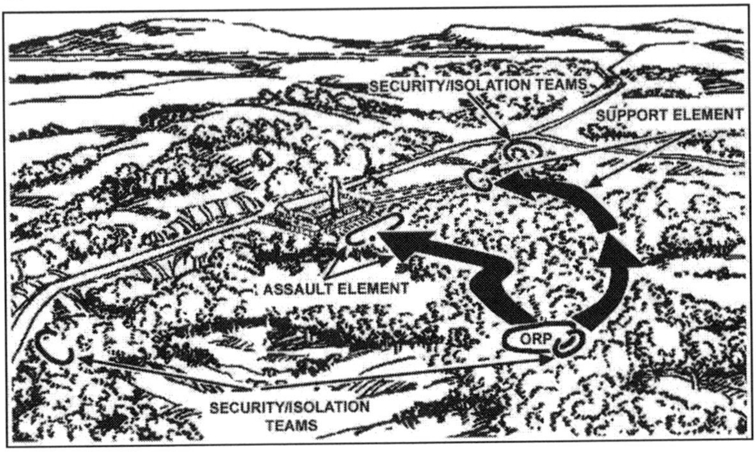

Figure 7-9. Platoon raid.

COUNTERATTACK

7-141. The counterattack is a form of attack by part or all of a friendly defending element against an enemy attacking force. The general objective of a counterattack is to deny the enemy his goal of attacking. This attack by defensive elements regains the initiative or denies the enemy success with his attack. The platoon may conduct a counterattack as a lightly committed element within a company or as the battalion reserve. Counterattacks afford the friendly defender the opportunity to create favorable conditions for the commitment of combat power. The platoon counterattacks after the enemy begins his attack, reveals his main effort, or creates an assailable flank. As part of a higher headquarters, the platoon conducts the counterattack much like other attacks. However, the platoon leader must synchronize the execution of his counterattack within the overall defensive effort. The platoon should rehearse the counterattack and prepare the ground to be traversed, paying close attention to friendly unit locations, obstacles, and engagement areas.

SPOILING ATTACK

7-142. A spoiling attack is a form of attack that preempts or seriously impairs an enemy attack while the enemy is in the process of planning or preparing to attack. The purpose of a spoiling attack is to disrupt the enemy's offensive capabilities and timelines, destroy his personnel and equipment, and gain additional time for the defending element to prepare positions. The purpose is not to secure terrain or other physical objectives. A commander (company or battalion) may direct a platoon to conduct a spoiling attack during friendly defensive preparations to strike the enemy while he is in assembly areas or attack positions where he is preparing offensive operations. The platoon leader plans for a spoiling attack as he does for other attacks.

FEINT

7-143. A feint is a form of attack used to deceive the enemy as to the location and time of the actual operation. Feints attempt to induce the enemy to move reserves and shift his fire support to locations where they cannot immediately impact the actual operation. When directed to conduct a feint, the platoon seeks direct fire or contact with the enemy, but avoids decisive engagement. The commander (company or battalion) will assign the platoon an objective limited in size or scope. The planning, preparation, and execution considerations are the same as for other forms of attack. The enemy must be convinced that the feint is the actual attack.

DEMONSTRATION

7-144. A demonstration is a form of attack designed to deceive the enemy as to the location or time of the actual operation by a display of force. Demonstrations attempt to deceive the enemy and induce him to move reserves and shift his fire support to locations where they cannot immediately impact the actual operation. When directed to conduct a demonstration, the platoon does not seek to make contact with the enemy. The planning, preparation, and execution considerations are the same as for other forms of attack. It must appear to be an actual impending attack.

SECTION VII — OFFENSIVE TACTICAL TASKS

7-145. Tactical tasks are specific activities performed by units as they conduct tactical operations or maneuver. At the platoon level, these tasks are the warfighting actions the platoon may be called on to perform in battle. This section provides discussion and examples of some common actions and tasks the platoon may perform during a movement to contact, a hasty attack, or a deliberate attack. It is extremely important to fully understand the purpose behind a task (what) because the purpose (why) defines what the platoon must achieve as a result of executing its mission. A task can be fully accomplished, but if battlefield conditions change and the platoon is unable to achieve the purpose, the mission is a failure.

NOTE: The situations used in this section to describe the platoon leader's role in the conduct of tactical tasks are examples only. They are not applicable in every tactical operation, nor are they intended to prescribe any specific method or technique the platoon must use in achieving the purpose of the operation. Ultimately, it is up to the commander or leader on the ground to apply both the principles discussed here, and his knowledge of the situation. An understanding of his unit's capabilities, the enemy he is fighting, and the ground on which the battle is taking place are critical when developing a successful tactical solution.

SEIZE

7-146. Seizing involves gaining possession of a designated objective by overwhelming force. Seizing an objective is complex. It involves closure with the enemy, under fire of the enemy's weapons to the point that the friendly assaulting element gains positional advantage over, destroys, or forces the withdrawal of the enemy.

7-147. A platoon may seize prepared or unprepared enemy positions from either an offensive or defensive posture. Examples include the following:

- A platoon seizes the far side of an obstacle as part of a company breach or seizes a building to establish a foothold in an urban environment.
- A platoon seizes a portion of an enemy defense as part of a company deliberate attack.
- A platoon seizes key terrain to prevent its use by the enemy.

7-148. There are many inherent dangers in seizing an objective. They include the requirement to execute an assault, prepared enemy fires, a rapidly changing tactical environment, and the possibility of fratricide when friendly elements converge. These factors require the platoon leader and subordinate leaders to understand the following planning considerations.

7-149. Developing a clear and current picture of the enemy situation is very important. The platoon may seize an objective in a variety of situations, and the platoon leader will often face unique challenges in collecting and disseminating information on the situation. For example, if the platoon is the seizing element during a company deliberate attack, the platoon leader should be able to develop an accurate picture of the enemy situation during the planning and preparation for the operation. He must be prepared to issue modifications to the platoon as new intelligence comes in or as problems are identified in rehearsals.

7-150. In another scenario, the platoon leader may have to develop his picture of the enemy situation during execution. He must rely more heavily on reports from units in contact with the enemy and on his own development of the situation. In this type of situation, such as when the platoon is seizing an enemy combat security outpost during a movement to contact, the platoon leader must plan on relaying information as it develops. He uses clear, concise FRAGOs to explain the enemy situation, and give clear directives to subordinates.

CLEAR

7-151. Clearing requires the platoon to remove all enemy forces and eliminate organized resistance within an assigned area. The platoon may be tasked with clearing an objective area during an attack to facilitate the movement of the remainder of the company, or may be assigned clearance of a specific part of a larger objective area. Infantry platoons are normally best suited to conduct clearance operations, which in many cases will involve working in restrictive terrain. Situations in which the platoon may conduct the clearance tactical task include clearing a—

- Defile, including choke points in the defile and high ground surrounding it.
- Heavily wooded area.
- Built-up or strip area. Refer to FM 3-06, *Urban Operations*, and FM 3-06.11, *Combined Operations in Urban Terrain,* for a detailed discussion of urban combat.
- Road, trail, or other narrow corridor, which may include obstacles or other obstructions on the actual roadway and in surrounding wooded and built-up areas.

General Terrain Considerations

7-152. The platoon leader must consider several important terrain factors when planning and executing the clearance task. Observation and fields of fire may favor the enemy. To be successful, the friendly attacking element must neutralize this advantage by identifying dead spaces where the enemy cannot see or engage friendly elements. It should also identify multiple friendly support-by-fire positions that are necessary to support a complex scheme of maneuver which cover the platoon's approach, the actual clearance task, and friendly maneuver beyond the restrictive terrain.

7-153. When clearing in support of tactical vehicles, cover and concealment are normally abundant for Infantry elements, but scarce for trail-bound vehicles. Lack of cover leaves vehicles vulnerable to enemy antiarmor fires. While clearing in support of mechanized vehicles, obstacles influence the maneuver of vehicles entering the objective area. The narrow corridors, trails, or roads associated with restrictive terrain can be easily obstructed with wire, mines, and log cribs.

7-154. Key terrain may include areas dominating the objective area, approaches, or exits, and any terrain dominating the area inside the defile, wooded area, or built-up area. Avenues of approach will be limited. The platoon must consider the impact of canalization and estimate how much time will be required to clear the objective area.

Restrictive Terrain Considerations

7-155. Conducting clearance in restrictive terrain is both time consuming and resource intensive. During the planning process, the platoon leader evaluates the tactical requirements, resources, and other considerations for each operation.

7-156. During the approach, the platoon leader focuses on moving combat power into the restrictive terrain and posturing it to start clearing the terrain. The approach ends when the rifle squads complete their preparations to conduct an attack. The platoon leader—

- Establishes support-by-fire positions.
- Destroys or suppresses any known enemy positions to allow elements to approach the restrictive terrain.
- Provides more security by incorporating suppressive indirect fires and obscuring or screening smoke.

7-157. The platoon leader provides support by fire for the rifle squads. He prepares to support the rifle squads where they enter the restrictive terrain by using—

- High ground on either side of a defile.
- Wooded areas on either side of a trail or road.
- Buildings on either side of a road in a built-up area.
- Movement of rifle squads along axes to provide cover and concealment.

7-158. Clearance begins as the rifle squads begin their attack in and around the restrictive terrain. Examples of where this maneuver may take place include—

- Both sides of a defile, either along the ridgelines or high along the walls of the defile.
- Along the wood lines parallel to a road or trail.
- Around and between buildings on either side of the roadway in a built-up area.

7-159. The following apply during clearance:

- The squads provide a base of fire to allow the weapons squad or support-by-fire element to bound to a new support-by-fire position. This cycle continues until the entire area is cleared.
- Direct-fire plans should cover responsibility for horizontal and vertical observation, and direct fire.
- Squads should clear a defile from the top down and should be oriented on objectives on the far side of the defile.
- Engineers with manual breaching capability should move with the rifle squads. Engineers may also be needed in the overwatching element to reduce obstacles.

7-160. At times, the unit may encounter terrain that restricts or severely restricts movement. Movement through these areas is vulnerable to ambush and road blocks. Clearance techniques can also be loosely applied to other terrain features. Bridges, city streets, road bends, corridors, thickly wooded areas, and any other area where a narrow passage wall has severely restrictive terrain on both sides may need clearing when advancing in the fight.

7-161. The platoon must secure the far side of the defile, built-up area, or wooded area until the company moves forward to pick up the fight beyond the restrictive terrain. If the restrictive area is large, the platoon may be directed to assist the passage of another element forward to continue the clearance operation. The platoon must be prepared to—

- Destroy enemy forces.
- Secure the far side of the restrictive terrain.
- Maneuver squads to establish support-by-fire positions on the far side of the restrictive terrain.
- Support by fire to protect the deployment of the follow-on force assuming the fight.
- Suppress any enemy elements that threaten the company while it exits the restrictive terrain.
- Disrupt enemy counterattacks.
- Protect the obstacle reduction effort.
- Maintain observation beyond the restrictive terrain.
- Integrate indirect fires as necessary.

Enemy Analysis

7-162. Careful analysis of the enemy situation is necessary to ensure the success of clearing. The enemy evaluation should include the following:

- Enemy vehicle location, key weapons, and Infantry elements in the area of operations.
- Type and locations of enemy reserve forces.
- Type and locations of enemy OPs.
- The impact of the enemy's CBRN and or artillery capabilities.

Belowground Operations

7-163. Belowground operations involve clearing enemy trenches, tunnels, caves, basements, and bunker complexes. The platoon's base-of-fire element and maneuvering squads must maintain close coordination. The weapons squad or support-by-fire element focuses on protecting the squads as they clear the trench line, or maneuver to destroy individual or vehicle positions. The base-of-fire element normally concentrates on destroying key surface structures (especially command posts and crew-served weapons bunkers) and the suppression and destruction of enemy vehicles.

7-164. The platoon must establish a base of fire to allow the rifle squads to then maneuver or enter the trench line, tunnel, basement, or bunker. The direct-fire plan must be thoroughly developed and rehearsed to ensure it will facilitate effective protection for the Infantry while preventing fratricide.

7-165. The platoon leader must also consider specific hazards associated with the platoon or supporting weapons systems. An example is the downrange hazard for the rifle squads created by the CCMS.

7-166. The platoon should consider using restrictive fire measures to protect converging friendly elements. It must also use other direct-fire control measures such as visual signals to trigger the requirement to lift, shift, or cease direct fires. Techniques for controlling direct fires during trench, tunnel, basement, and bunker clearance may include the following: attaching a flag to a pole carried by the Soldier who follows immediately behind the lead clearing team; using panels to mark cleared bunkers, tunnels, and basements; using visual signals to indicate when to lift, shift, or cease fires.

7-167. Once the rifle squads enter the belowground area, the combined effects of the platoon's assets place the enemy in a dilemma. Every action the enemy takes to avoid direct fire from the support-by-fire element, such as maintaining defilade positions or abandoning bunker complexes, leaves him vulnerable to attack from the rifle squads maneuvering down the trench. Every time the enemy moves his vehicles to avoid attacking squads, or when his Infantry elements stay in bunkers or command posts, he exposes himself to support fires.

7-168. Consolidation consists of securing the objective and defending against an enemy counterattack.

7-169. Reorganization, normally conducted concurrently with consolidation, consists of preparing for follow-on operations. As with consolidation, the platoon leader must plan and prepare for reorganization as he conducts his TLP. He ensures the platoon is prepared to—

- Provide essential medical treatment and evacuate casualties as necessary.
- Cross-level personnel and adjust task organization as required.
- Conduct resupply operations, including rearming and refueling.
- Redistribute ammunition.
- Conduct required maintenance.

SUPPRESS

7-170. The platoon maneuvers to a position on the battlefield where it can observe the enemy and engage him with direct and indirect fires. The purpose of suppressing is to prevent the enemy from effectively engaging friendly elements with direct or indirect fires. To accomplish this, the platoon must maintain orientation both on the enemy force and on the friendly maneuver element it is supporting. During planning and preparation, the platoon leader should consider—

- Conducting a line-of-sight analysis during his terrain analysis to identify the most advantageous positions from which to suppress the enemy.
- Planning and integrating direct and indirect fires.
- Determining control measures (triggers) for lifting, shifting, or ceasing direct fires (see Chapter 2).
- Determining control measures for shifting or ceasing indirect fires.
- Planning and rehearsing actions on contact.
- Planning for large Class V expenditures. (The company commander and the platoon leader must consider a number of factors in assessing Class V requirements including the desired effects of the platoon direct fires; the composition, disposition, and strength of the enemy force; and the time required to suppress the enemy.)
- Determining when and how the platoon will reload ammunition during the fight while still maintaining suppression for the assaulting element.

SUPPORT BY FIRE

7-171. The platoon maneuvers to a position on the battlefield from where it can observe the enemy and engage him with direct and indirect fires. The purpose of support by fire is to prevent the enemy from engaging friendly elements.

7-172. To accomplish this task, the platoon must maintain orientation both on the enemy force and on the friendly maneuver element it is supporting. The platoon leader should plan and prepare by—

- Conducting line-of-sight analysis to identify the most advantageous support-by-fire positions.
- Conducting planning and integration for direct and indirect fires.
- Determining triggers for lifting, shifting, or ceasing direct and indirect fires.
- Planning and rehearsing actions on contact.
- Planning for large Class V expenditures, especially for the weapons squad and support elements, because they must calculate rounds per minute. (The platoon leader and weapons squad leader must consider a number of factors in assessing Class V requirements, including the desired effects of platoon fires; the time required for suppressing the enemy; and the composition, disposition, and strength of the enemy force.)

7-173. A comprehensive understanding of the battlefield and enemy and friendly disposition is a crucial factor in all support-by-fire operations. The platoon leader uses all available intelligence and information resources to stay abreast of events on the battlefield. Additional considerations may apply. The platoon may have to execute an attack to secure the terrain from where it will conduct the support by fire. The initial support-by-fire position may not afford adequate security or may not allow the platoon to achieve its intended purpose. This could force the platoon to reposition to maintain the desired weapons effects on the enemy. The platoon leader must ensure the platoon adheres to these guidelines:

- Maintain communication with the moving element.
- Be prepared to support the moving element with both direct and indirect fires.
- Be ready to lift, shift, or cease fires when masked by the moving element.
- Scan the area of operations and prepare to acquire and destroy any enemy element that threatens the moving element.
- Maintain 360-degree security.
- Use Javelins to destroy any exposed enemy vehicles.
- Employ squads to lay a base of sustained fire to keep the enemy fixed or suppressed in his fighting positions.
- Prevent the enemy from employing accurate direct fires against the protected force.

ATTACK BY FIRE

7-174. The platoon maneuvers to a position on the battlefield from where it can observe the enemy and engage him with direct and indirect fires at a distance to destroy or weaken his maneuvers. The platoon destroys the enemy or prevents him from repositioning. The platoon employs long-range fires from dominating terrain. It also uses flanking fires or takes advantage of the standoff range of the unit's weapons systems. The company commander may designate an attack-by-fire position from where the platoon will fix the enemy. An attack-by-fire position is most commonly employed when the mission or tactical situation focuses on destruction or prevention of enemy movement. In the offense, it is usually executed by supporting elements. During defensive operations, it is often a counterattack option for the reserve element.

7-175. When the platoon is assigned an attack-by-fire position, the platoon leader obtains the most current intelligence update on the enemy and applies his analysis to the information. During planning and preparation, the platoon leader should consider—

- Conducting a line-of-sight analysis during terrain analysis to identify the most favorable locations to destroy or fix the enemy.
- Conducting direct and indirect fire planning and integration.
- Determining control measures (triggers) for lifting, shifting, or ceasing direct fires.
- Determining control measures for shifting or ceasing indirect fires.
- Planning and rehearsing actions on contact.

7-176. Several other considerations may affect the successful execution of an attack by fire. The platoon may be required to conduct an attack against enemy security forces to seize the ground from where it will establish the attack-by-fire position. The initial attack-by-fire position may afford inadequate security or may not allow the platoon to achieve its task or purpose. This could force the platoon to reposition to maintain the desired weapons effects on the enemy force. Because an attack by fire may be conducted well beyond the direct fire range of other platoons, it may not allow the platoon to destroy the targeted enemy force from its initial positions. The platoon may begin to fix the enemy at extended ranges. Additional maneuver would then be required to close with the enemy force and complete its destruction. Throughout an attack by fire, the platoon should reposition or maneuver to maintain flexibility, increase survivability, and maintain desired weapons effects on the enemy. Rifle squad support functions may include:

- Seizing the attack-by-fire position before occupation by mounted sections.
- Providing local security for the attack-by-fire position.
- Executing timely, decisive actions on contact.
- Using maneuver to move to and occupy attack-by-fire positions.
- Destroying enemy security elements protecting the targeted force.
- Employing effective direct and indirect fires to disrupt, fix, or destroy the enemy force.

SECTION VIII — URBAN AREAS

7-177. Infantry platoons conduct operations in urban areas using the same principles applicable to other offensive operations. This section explains the general tactics, techniques, and procedures used for a limited attack in an urban area. Depending on the scale of the operation, Infantry platoons or squads may be required to conduct any or all of the find, fix, fight, and follow-through functions. Leaders should expect trouble in the process of determining the exact location of the enemy and should anticipate enemy knowledge of their movements prior to arriving in the objective area. For a more detailed discussion on urban operations see FM 3-06.11.

CRITICAL TASKS

7-178. There are a number of critical tasks that need emphasis for Infantry platoons assaulting a building:

- Isolate the building.
- Gain and maintain fire superiority inside and outside the building.
- Gain access to the inside of the building.

- Move inside the building.
- Seize positions of advantage.
- Control the tempo.

FIND

7-179. The compartmentalized nature of urban terrain, limited observation and fields of fire, and the vast amounts of potential cover and concealment mean that defenders can disperse and remain undetected. The origin of enemy gunfire can be difficult to detect, because distance and direction become distorted by structures. The nature of urban conflicts makes it more difficult for leaders to exercise command and control verbally, and for Soldiers to pass and receive information. Situational understanding is normally limited to the platoon's immediate area.

ISOLATE THE BUILDING

7-180. The fix function has two aspects: isolating the objective to prevent interference from the outside (while preventing enemy from exiting), and separating forces on the objective from each other (denying mutual support and repositioning). This is accomplished by achieving fire superiority and seizing positions of advantage. If the platoon is conducting a semi-independent assault, it should be organized to accomplish both the fix and finish function.

7-181. A cordon is a line of troops or military posts that enclose an area to prevent passage. The Infantry platoon normally conducts a cordon as part of a larger unit. It is established by positioning one or more security elements on key terrain that dominates avenues of approach in and out of the objective area. The overall goal is the protection of the maneuver element, and to completely dominate what exits or enters the objective area. This requires a detailed understanding of avenues of approach in the area. There are many techniques used to facilitate isolation including, blocking positions, direct fire (precision and area), indirect fire, roadblocks, checkpoints, and observation posts. The same techniques can be used to cordon and search a small urban area (such as a village) surrounded by other terrain (Figure 7-10).

7-182. Ideally these positions are occupied simultaneously, but a sequential approach can also be useful. Limited visibility aids can be used in the establishment and security of the cordon. The security element can either surround the area while the maneuver element simultaneously moves in, or it can use a sequential technique in which they use stealth to get into position before the actual assault.

7-183. Plans should be developed to handle detained personnel. Infantrymen will normally provide security and accompany police and intelligence forces who will identify, question, and detain suspects. Infantry may also conduct searches and assist in detaining suspects, but their principal role is to reduce any resistance that may develop and to provide security for the operation. Use of force is kept to a minimum unless otherwise directed.

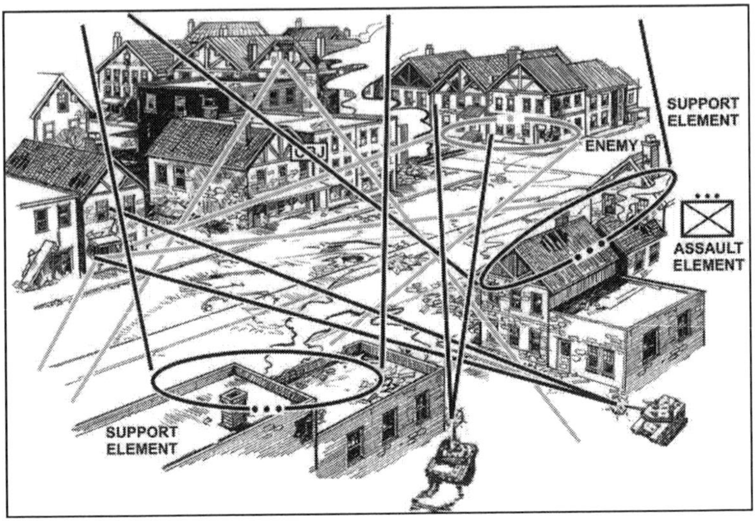

Figure 7-10. Isolate the building.

ASSAULT A BUILDING

7-184. Squads and platoons, particularly when augmented with engineers, are the best organized and equipped units in the Army for breaching protective obstacles; gaining access to buildings; and assaulting rooms, hallways, and stairways. Although there are specific drills associated with fighting in buildings, the overall assault is an operation, not a drill. During planning, the leader's level of detail should identify each window (aperture, opening, or firing port) in his sector fortifications. He should then consider assigning these as a specific TRP when planning fires.

On 21 July 2003, the 3rd battalion 327 Infantry Regiment and an assault team of elite Special Operations Soldiers from Task Force 20 conducted an assault on a building as part of a raid to kill or capture high value targets.

Surprise, created by leveraging the aspect of time, enabled the leader to control the tempo by creating an initial advantage for the attackers. Instead of sequential actions, the leaders began actions on the objective with the near simultaneous arrival of the assault, support element, and security elements. The assault element arrived at an assault position right outside of the building. The support element occupied three separate support-by-fire positions. Two security elements were organized to establish inner and outer cordons. The first security element isolated the objective by establishing six blocking positions that denied enemy escape and blocked local counterattacks. The second security element formed the outer cordon to prevent a general counterattack and protect the population. With the enemy force found and fixed, an interpreter using a bullhorn requested their quiet surrender. This request was met by gunfire from the objective.

The fight began with the enemy concentrated on the building's second floor. The support element easily achieved enough fire superiority to enable movement outside the building. With those conditions created, the assault element moved from the assault position to objective and without difficulty seized the building's first floor. However, once the assault element attempted to go to the second floor, the support element was not able to maintain the fire superiority necessary to facilitate the move. The assault element was also unable to achieve fire superiority. Undeterred, the assault element chanced a move up the stairway only to be beaten back by effective enemy fires resulting in casualties. They attempted several times to gain fire superiority but the defender's position of advantage gave them the firepower advantage and the assault was halted.

When the leader realized the team would not be able to gain fire superiority from their current locations, he slowed the operation's tempo down by ordering another element to seize a position of advantage on top of the objective building. Several members of the follow-on force gained access to the roof tops of neighboring houses. From there, they were able to use a supersurface avenue of approach to seize the objective building's rooftop. From this position of advantage, the Soldiers were able to communicate target locations and monitor munition effects, increasing the support element's ability to destroy or suppress the enemy.

With fire superiority completely gained outside and inside the building, the assault team successfully renewed its attempt to move to the second floor. Once up the stairs, they quickly eliminated remaining resistance and cleared the remainder of the objective, finishing the fight and accomplishing the mission.

ENTERING THE BUILDING

7-185. After establishing suppression and obscuration, leaders deploy their subordinates to secure the near side and then, after gaining access, secure the far side. Gaining access to the inside of the building normally requires reducing protective obstacles. Reducing obstacles is discussed at length in Appendix F.

7-186. Units gain access by using either a top or bottom entry. The entry point is the same thing as a point of penetration for an obstacle breach and as such is a danger area. The entry point will become the focus of fires for any enemy in a position to fire at it. It is commonly referred to as the "fatal funnel." Leaders ensure they have established measures to ensure the assault team has fire superiority when moving through the fatal funnel. Grenades (ROE determines fragmentation or concussion) are used to gain enough of a window of opportunity until the assault element can employ its small arms fire.

Top Entry

7-187. The top of a building is ordinarily considered a position of advantage. Entering at the top and fighting downward is the preferred method of gaining access to a building for a number of reasons. First, just as in operations on other types of terrain, it is easier to own the high ground and work your way down than it is to fight your way up when the enemy owns the high ground. Second, an enemy forced down to ground level may be tempted to withdraw from the building and expose himself to the fire of covering units or weapons. Third, the ground floor and basements are normally more heavily defended. Finally, the roof of a building is ordinarily weaker than the walls (and therefore easier to penetrate).

7-188. Top entry is only feasible when the unit can gain access to an upper floor or rooftop. Rooftops are danger areas when surrounding buildings are higher and forces can be exposed to fire from those buildings. Soldiers should consider the use of devices and other techniques that allow them upper level access without using interior stairways. Those devices and techniques include, but are not limited to, adjacent rooftops, fire escapes, portable ladders, and various Soldier-assisted lifts. For more information on top entry breaching, see FM 3-06.11.

Bottom Entry

7-189. Entry at the bottom is common and may be the only option available. When entering from the bottom, breaching a wall to create a "mousehole" is the preferred method because doors and windows may be booby-trapped and covered by fire from inside the structure. There are many ways to accomplish this, including employing CCMS, SLM, demolitions, hand tools, machine guns, artillery fire, and tank fire. The actual technique used depends on the ROE, assets available, building structure, and the enemy situation. If the assault element must enter through a door or window, it should enter from a rear or flank position after ensuring the entry point is clear of obstacles.

Secure the Near and Far Side of the Point of Penetration

7-190. Infantry platoons use the following drill for gaining access to the building. The steps of this drill are very similar to those drills described in Section IX to secure the near and far side of the point of penetration—

- The squad leader and the assault fire team move to the last covered and concealed position near the entry point.
- The squad leader confirms the entry point.
- The platoon leader or squad leader shifts the support fire away from the entry point.
- The support-by-fire element continues to suppress building and adjacent enemy positions as required.
- Buddy team #1 (team leader and automatic rifleman) remain in a position short of the entry point to add suppressive fires for the initial entry.
- Buddy team #2 (grenadier and rifleman) and the squad leader move to the entry point. They move in rushes or by crawling.
- The squad leader positions himself where he can best control his teams.
- Buddy team #2 position themselves against the wall to the right or left of the entry point.
- On the squad leader command of COOK OFF GRENADES (2 seconds maximum), the Soldiers employing the grenades shout, FRAG OUT, and throw the grenades into the building. (If the squad leader decides not to use grenades, he commands, PREPARE TO ENTER—GO!)
- Upon detonation of both grenades (or command GO), the buddy team flows into the room/hallway and moves to points of domination engaging all identified or likely enemy positions.
- Both Soldiers halt and take up positions to block any enemy movement toward the entry point.
- Simultaneously, buddy team #1 moves to and enters the building, joins buddy team #2, and announces, CLEAR.
- The squad leader remains at the entry point and marks it IAW unit SOP. He calls forward the next fire team with, NEXT TEAM IN.
- Once the squad has secured a foothold, the squad leader reports to the platoon leader, FOOTHOLD SECURE. The platoon follows the success of the seizure of the foothold with the remainder of the platoon.

7-191. When using a doorway as the point of entry, the path of least resistance is initially determined on the way the door opens. If the door opens inward, the Soldier plans to move away from the hinged side. If the door opens outward, he plans to move toward the hinged side. Upon entering, the size of the room, enemy situation, and obstacles in the room (furniture and other items) that hinder or channel movement become factors that influence the number one man's direction of movement.

CLEAR A ROOM

7-192. The term *room* in this FM means any enclosed space or partition within a building. Although rooms come in all shapes and sizes, there are some general principles that apply to most room clearing tasks. For clearing large open buildings such as hangars or warehouses, it may be necessary to use subordinate units using a line formation while employing traveling or bounding overwatch. These methods can effectively clear the entire structure while ensuring security.

7-193. Room clearing techniques differ based on METT-TC, ROE, and probability of noncombatants inside the building. If there are known or suspected enemy forces, but no noncombatants inside the building, the platoon may conduct *high intensity* room clearings. If there are known or suspected noncombatants within the building, the platoon may conduct *precision* room clearings. High intensity room clearing may consist of fragmentation grenade employment and an immediate and high volume of small arms fire placed into the room, precision room clearing will not.

7-194. Room clearing techniques are described using the standard four-man fire team. This does not mean that all four members must enter a room, nor does it mean that more than four men cannot enter. The fire team organization is the baseline from where units adapt to the specific situation. This is because the compartmentalized nature typical of buildings and rooms makes units larger than squads awkward and unmanageable.

7-195. For this battle drill to be effectively employed, each member of the team must know his sector of fire and how his sector overlaps and links with the sectors of the other team members. No movement should mask the fire of any of the other team members.

7-196. On the signal, the team enters through the entry point (or breach). As the team members move to their points of domination, they engage all threats or hostile targets in sequence in their sector. The direction each man moves should not be preplanned unless the exact room layout is known. Each man should, however, go in a direction opposite the man in front of him (Figure 7-11). For example:

- **#1 Man.** The #1 man enters the room and eliminates any immediate threat. He can move left or right, moving along the path of least resistance to a point of domination—one of the two corners and continues down the room to gain depth.
- **#2 Man.** The #2 man enters almost simultaneously with the first and moves in the opposite direction, following the wall. The #2 man must clear the entry point, clear the immediate threat area, and move to his point of domination.
- **#3 Man.** The #3 man simply moves in the opposite direction of the #2 man inside the room, moves at least 1 meter from the entry point, and takes a position that dominates his sector.
- **#4 Man.** The #4 man moves in the opposite direction of the #3 man, clears the doorway by at least 1 meter, and moves to a position that dominates his sector.

7-197. Once the room is cleared, the team leader may order some team members to move deeper into the room overwatched by the other team members. The team leader must control this action. In addition to dominating the room, all team members are responsible for identifying possible loopholes and mouseholes in the ceiling, walls, and floor. Cleared rooms should be marked IAW unit SOP.

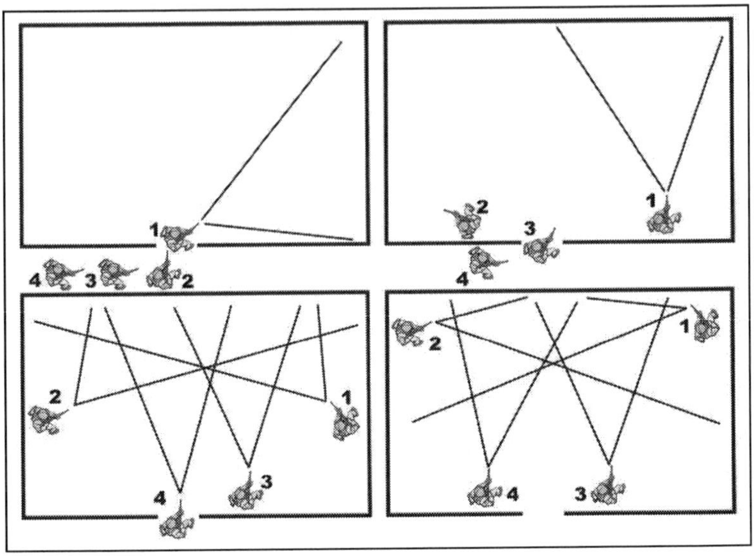

Figure 7-11. Clearing a room.

MOVING IN THE BUILDING

7-198. Movement techniques used inside a building are employed by teams to negotiate hallways and other avenues of approach. They are similar to movement techniques employed when clearing enemy trenches, which is discussed in Section IX.

Diamond Formation (Serpentine Technique)

7-199. The serpentine technique is a variation of a diamond formation that is used in a narrow hallway. The #1 man provides security to the front. His sector of fire includes any enemy Soldiers who appear at the far end or along the hallway. The #2 and #3 men cover the left and right sides of the #1 man. Their sectors of fire include any enemy combatants who appear suddenly from either side of the hall. The #4 man (normally carrying the M249 machine gun) provides rear protection against any enemy Soldiers suddenly appearing behind the team.

Vee Formation (Rolling-T Technique)

7-200. The rolling-T technique is a variation of the Vee formation and is used in wide hallways (Figure 7-12). The #1 and #2 men move abreast, covering the opposite side of the hallway from the one they are walking on. The #3 man covers the far end of the hallway from a position behind the #1 and #2 men, firing between them. The #4 man provides rear security.

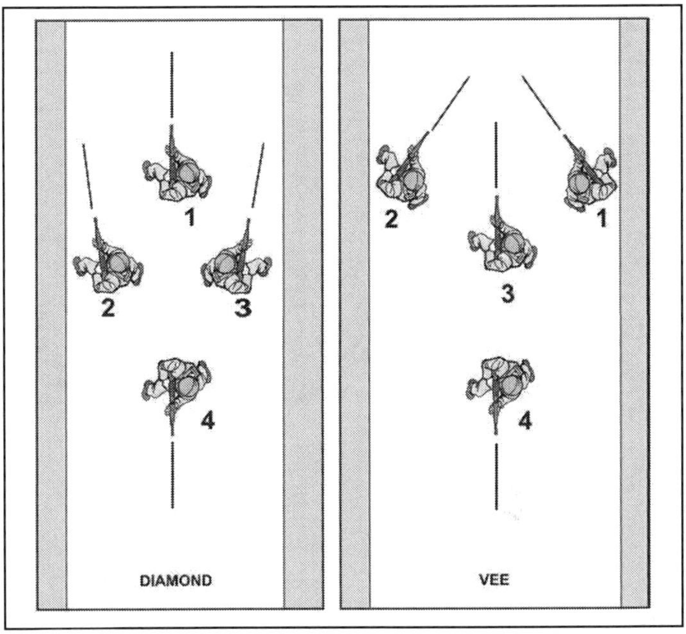

Figure 7-12. Diamond and vee formation techniques.

Clearing Hallway Junctions

7-201. Hallway intersections are danger areas and should be approached cautiously. Figure 7-13 depicts the fire team's actions upon reaching a "T" intersection when approaching along the "cross" of the "T". The unit is using the diamond (serpentine) formation for movement (Figure 7-13 A). To clear a hallway—

- The team configures into a modified 2-by-2 (box) formation with the #1 and #3 men abreast and toward the right side of the hall. The #2 man moves to the left side of the hall and orients to the front, and the #4 man shifts to the right side (his left) and maintains rear security. (When clearing a right-hand corner, use the left-handed firing method to minimize exposure [Figure 7-13 B]).
- The #1 and #3 men move to the edge of the corner. The #3 man assumes a low crouch or kneeling position. On signal, the #3 man, keeping low, turns right around the corner and the #1 man, staying high, steps forward while turning to the right. (Sectors of fire interlock and the low/high positions prevent Soldiers from firing at one another [Figure 7-13 C]).
- The #2 and #4 men continue to move in the direction of travel. As the #2 man passes behind the #1 man, the #1 man shifts laterally to his left until he reaches the far corner (Figure 7-13 D).
- The #2 and #4 men continue to move in the direction of travel. As the #4 man passes behind the #3 man, the #3 man shifts laterally to his left until he reaches the far corner. As the #3 man begins to shift across the hall, the #1 man turns into the direction of travel and moves to his original position in the diamond (serpentine) formation (Figure 7-13 E).
- As the #3 and #4 men reach the far side of the hallway, they, too, assume their original positions in the serpentine formation, and the fire team continues to move (Figure 7-13 F).

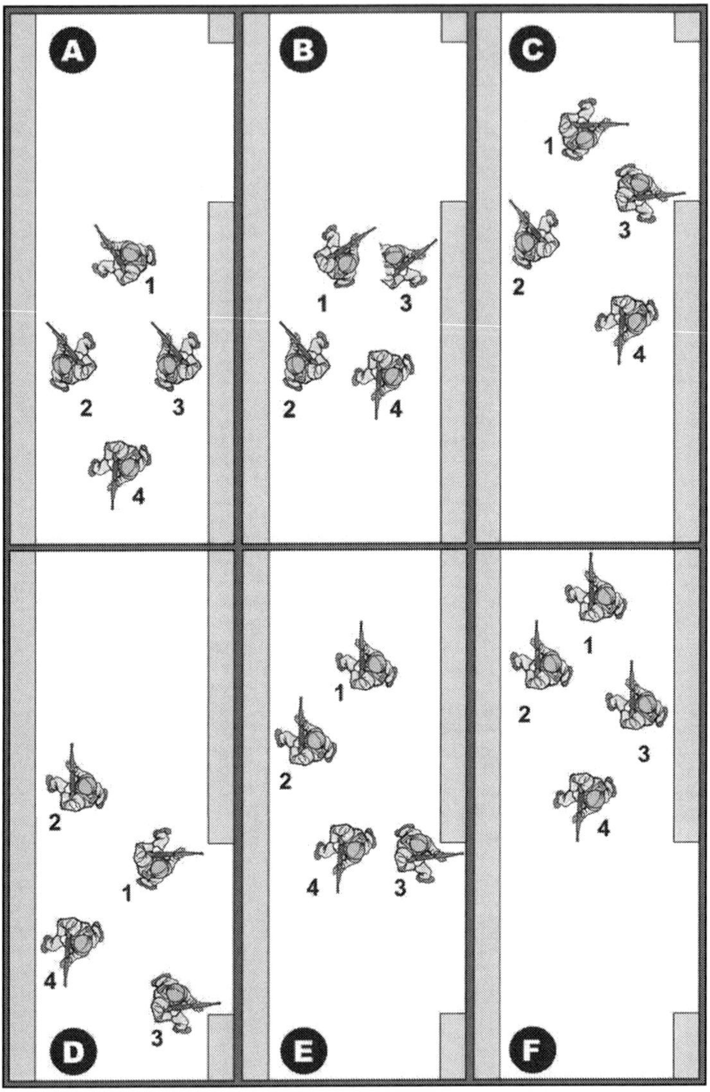

Figure 7-13. Clearing hallway junctions.

Clearing a "T" Intersection

7-202. Figure 7-14 depicts the fire team's actions upon reaching a "T" intersection when approaching from the base of the "T". The fire team is using the diamond (serpentine) formation for movement (Figure 7-14 A). To clear a "T" intersection—

- The team configures into a 2-by-2 (box) formation with the #1 and #2 men left and the #3 and #4 men right. (When clearing a right-hand corner, use the left-handed firing method to minimize exposure [Figure 7-14 B]).
- The #1 and #3 men move to the edge of the corner and assume a low crouch or kneeling position. On signal, the #1 and #3 men simultaneously turn left and right respectively (Figure 7-14 C).
- At the same time, the #2 and #4 men step forward and turn left and right respectively while maintaining their (high) position. (Sectors of fire interlock and the low/high positions prevent Soldiers from firing at another [Figure 7-14 D]).
- Once the left and right portions of the hallway are clear, the fire team resumes the movement formation (Figure 7-14 E). Unless security is left behind, the hallway will no longer remain clear once the fire team leaves the immediate area.

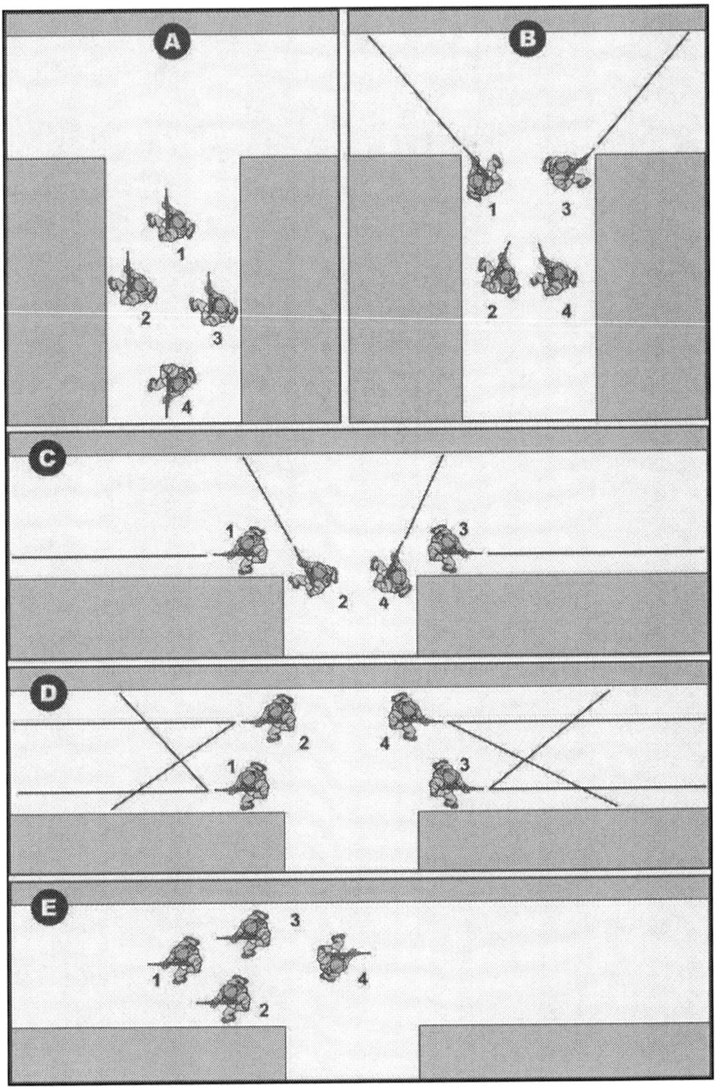

Figure 7-14. Clearing a "T" intersection.

Clearing Stairwells and Staircases

7-203. Stairwells and staircases are comparable to doorways because they create a fatal funnel. The danger is intensified by the three-dimensional aspect of additional landings. The ability of units to conduct the movement depends upon which direction they are traveling and the layout of the stairs. Regardless, the clearing technique follows a basic format:

- The leader designates an assault element to clear the stairs.
- The unit maintains 360-degree, three-dimensional security in the vicinity of the stairs.
- The leader then directs the assault element to locate, mark, bypass, and or clear any obstacles or booby traps that may be blocking access to the stairs.
- The assault element moves up (or down) the stairway by using either the two-, three-, or four-man flow technique, providing overwatch up and down the stairs while moving. The three-man variation is preferred (Figure 7-15).

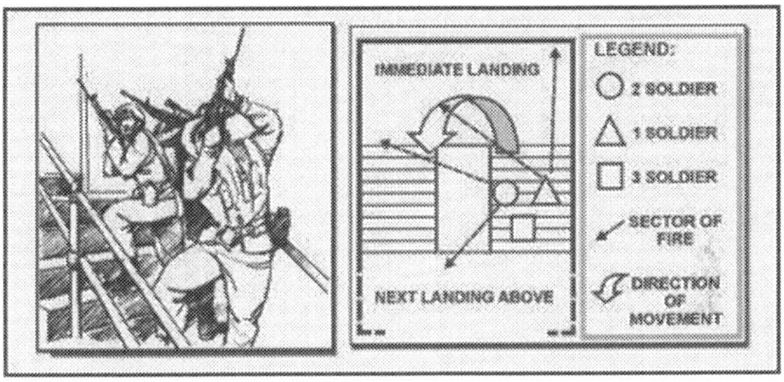

Figure 7-15. Three-man-flow clearing technique.

FOLLOW THROUGH

7-204. After securing a floor (bottom, middle, or top), selected members of the unit are assigned to cover potential enemy counterattack routes to the building. Priority must be given initially to securing the direction of attack. Security elements alert the unit and place a heavy volume of fire on enemy forces approaching the unit.

7-205. Units must guard all avenues of approach leading into their area. These may include—

- Enemy mouseholes between adjacent buildings.
- Covered routes to the building.
- Underground routes into the basement.
- Approaches over adjoining roofs or from window to window.

7-206. Units that performed missions as assault elements should be prepared to assume an overwatch mission and to support another assault element.

7-207. To continue the mission—

- Momentum must be maintained. This is a critical factor in clearing operations. The enemy cannot be allowed to move to its next set of prepared positions or to prepare new positions.
- The support element pushes replacements, ammunition, and supplies forward to the assault element.
- Casualties must be evacuated and replaced.
- Security for cleared areas must be established IAW the OPORD or TSOP.
- All cleared areas and rooms must be marked IAW unit SOP.
- The support element must displace forward to ensure that it is in place to provide support (such as isolation of the new objective) to the assault element.

SECTION IX — ATTACKING FORTIFIED POSITIONS

7-208. Fortifications are works emplaced to defend and reinforce a position. Time permitting, enemy defenders build bunkers and trenches, emplace protective obstacles, and position mutually supporting fortifications when fortifying their positions. Soldiers who attack prepared positions should expect to encounter a range of planned enemy fires to include small arms fire, mortars, artillery, antitank missiles, antitank guns, tanks, attack aviation, and close air support. Attacking forces should also expect a range of offensive type maneuver options to include spoiling attacks, internal repositioning, counterattacks, and withdrawing to subsequent defensive positions. Spoiling attacks will attempt to disrupt the attacker's momentum and possibly seize key terrain. If driven out of their prepared positions, enemy troops may try to win them back by hasty local counterattacks or through deliberate, planned combined arms counterattacks. If forced to withdraw, the enemy forces may use obstacles, ambushes, and other delaying tactics to slow down pursuing attackers.

7-209. The attack of a fortified position follows the basic principles of tactical maneuver. However, greater emphasis is placed upon detailed planning, special training and rehearsals, increased fire support, and the use of special equipment. The degree of special preparation depends upon the character, and extent of the defense.

7-210. The deliberate nature of defenses requires a deliberate approach to the attack. These types of operations are time consuming. Leaders must develop schemes of maneuver that systematically reduce the area. Initially, these attacks should be limited in scope, focusing on individual positions and intermediate terrain objectives. Leaders must establish clear bypass criteria and position destruction criteria as well as allocate forces to secure cleared enemy positions. Failure in this will likely result in enemy reoccupying the positions, isolating lead elements, and ambushing follow-on units.

7-211. The intense, close combat prevalent in trench clearing is remarkably similar to fighting in built up areas. Comparable characteristics include:

- **Restricted Observation and Fields of Fire.** Once the trench is entered, visibilities may be limited to a few meters in either direction. This compartmentalization necessarily decentralizes the engagement to the lowest level.
- **Cover and Concealment.** The nature of a trench system allows covered movement of both friendly and enemy forces. To prevent being flanked or counterattacked, junctions, possible entry points, and corners should be secured.
- **Difficulty in Locating the Enemy.** The assault element may come under fire from multiple mutually supporting positions in the trench or a nearby position. The exact location of the fire may be difficult to determine. Supporting elements outside the trench should be capable of locating, suppressing, or destroying such threats.
- **Close Quarters Fighting.** Because of the close nature of the trench system, Soldiers should be prepared to use close quarters marksmanship, bayonet, and hand-to-hand fight techniques.
- **Restricted Movement.** Trench width and height will severely restrict movement inside the system. This will ordinarily require the assault element to move at a low crouch or even a crawl. Sustainment functions such as ammunition resupply, EPW evacuation, casualty evacuation, and reinforcement will also be hampered.
- **Sustainment.** The intensity of close combat in the trench undoubtedly results in increased resource requirements.

FIND

7-212. Finding the enemy's fortified positions relates back to the position's purpose. There are two general reasons to create fortified positions. The first includes defending key terrain and using the position as a base camp, shelter, or sanctuary for critical personnel or activities. This type of position is typically camouflaged and difficult to locate. When U.S. forces have air superiority and robust reconnaissance abilities, enemy forces will go to great lengths to conceal these positions. Sometimes the only way to find these enemy positions is by movement to contact. When Infantry platoons or squads encounter a previously unidentified prepared enemy position, they should not, as a general rule, conduct a hasty attack until they have set conditions for success.

7-213. The second general purpose for fortified positions is to create a situation in which the attacker is required to mass and present a profitable target. This type of position normally occurs in more conventional battles. These positions can be relatively easy to find because they occupy key terrain, establish identifiable patterns, and generally lack mobility.

7-214. Attacking fortified positions requires thorough planning and preparation based on extensive reconnaissance.

FIX

7-215. An enemy in fortified defenses has already partially fixed himself. This does not mean he will not be able to maneuver or that the fight will be easy. It does mean that the objective is probably more defined than with an enemy with complete freedom of movement. Fixing the enemy will still require measures to prevent repositioning to alternate, supplementary, and subsequent positions on the objective and measures to block enemy counterattack elements.

FINISH — FIGHTING ENEMIES IN FORTIFICATIONS

7-216. Finishing an enemy in prepared positions requires the attacker to follow the fundamentals of the offense-surprise, concentration, tempo, and audacity to be successful.

7-217. The actual fighting of enemy fortifications is clearly an Infantry platoon unit function because squads and platoons, particularly when augmented with engineers, are the best organized and equipped units in the Army for breaching protective obstacles. They are also best prepared to assault prepared positions such as bunkers and trench lines. Infantry platoons are capable of conducting these skills with organic, supplementary, and supporting weapons in any environment.

7-218. Leaders develop detailed plans for each fortification, using the SOSRA technique to integrate and synchronize fire support and maneuver assets. Although there are specific drills associated with the types of fortifications, the assault of a fortified area is an operation, not a drill. During planning, the leader's level of detail should identify each aperture (opening or firing port) of his assigned fortification(s) and consider assigning these as a specific target when planning fires. Contingency plans are made for the possibility of encountering previously undetected fortifications along the route to the objective, and for neutralizing underground defenses when encountered.

SECURING THE NEAR AND FAR SIDE—BREACHING PROTECTIVE OBSTACLES

7-219. To fight the enemy almost always requires penetrating extensive protective obstacles, both antipersonnel and antivehicle. Of particular concern to the Infantrymen are antipersonnel obstacles. Antipersonnel obstacles (both explosive and nonexplosive) include, wire entanglements; trip flares; antipersonnel mines; field expedient devices (booby traps, nonexplosive traps, punji sticks); flame devices; rubble; warning devices; CBRN; and any other type of obstacle created to prevent troops from entering a position. Antipersonnel obstacles are usually integrated with enemy fires close enough to the fortification for adequate enemy surveillance by day or night, but beyond effective hand grenade range. Obstacles are also used within the enemy position to compartmentalize the area in the event outer protective barriers are breached. See Appendix F for more information on obstacles.

7-220. The following steps are an example platoon breach:

- The squad leader and the breaching fire team move to the last covered and concealed position near the breach point (point of penetration).
- The squad leader confirms the breach point.
- The platoon leader or squad leader shifts the suppressing element away from the entry point.
- The fire element continues to suppress enemy positions as required.

- Buddy team #1 (team leader and the automatic rifleman) remains in a position short of the obstacle to provide local security for buddy team #2.
- The squad leader and breaching fire team leader employ smoke grenades to obscure the breach point.
- Buddy team #2 (grenadier and rifleman) moves to the breach point. They move in rushes or by crawling.
- The squad leader positions himself where he can best control his teams.
- Buddy team #2 positions themselves to the right and left of the breach point near the protective obstacle.
- Buddy team #2 probes for mines and creates a breach, marking their path as they proceed.
- Once breached, buddy team #1 and buddy team #2 move to the far side of the obstacle and take up covered and concealed positions to block any enemy movement toward the breach point. They engage all identified or likely enemy positions.
- The squad leader remains at the entry point and marks it. He calls forward the next fire team with, "Next team in."
- Once the squad has secured a foothold, the squad leader reports to the platoon leader, "Foothold secure." The platoon follows the success of the seizure of the foothold with the remainder of the platoon.

KNOCKING OUT BUNKERS

7-221. The term *bunker* in this FM covers all emplacements having overhead cover and containing apertures (embrasures) through which weapons are fired. The two primary types are reinforced concrete pillboxes, and log bunkers. There are two notable exploitable weaknesses of bunkers.

7-222. First, bunkers are permanent, their location and orientation fixed. Bunkers cannot be relocated or adjusted to meet a changing situation. They are optimized for a particular direction and function. The worst thing an Infantry platoon or squad can do is to approach the position in the manner it was designed to fight. Instead, the unit should approach the position from the direction it is least able to defend against—the flank or rear.

7-223. Second, bunkers must have openings (doors, windows, apertures, or air vents). There are two disadvantages to be exploited here. First, structurally, the opening is the weakest part of the position and will be the first part of the structure to collapse if engaged. Second, a single opening can only cover a finite sector, creating blind spots. To cover these blind spots, the defender has to either rely on mutually supporting positions or build an additional opening. Mutual support may be disrupted, thereby enabling the attacker to exploit the blind spot. Adding additional openings correspondingly weakens the position's structural soundness, in which case the attacker targets the opening to collapse the position.

7-224. Ideally the team is able to destroy the bunker with standoff weapons and HE munitions. However, when required, the fire team can assault the bunker with small arms and grenades. A fire team (two to four men) with HE and smoke grenades move forward under cover of the suppression and obscuration fires from the squad and other elements of the base of fire. When they reach a vulnerable point of the bunker, they destroy it or personnel inside with grenades or other hand-held demolitions. All unsecured bunkers must be treated as if they contain live enemy, even if no activity has been detected from them. The clearing of bunkers must be systematic or the enemy will come up behind assault groups. To clear a bunker—

- The squad leader and the assault fire team move to the last covered and concealed position near the position's vulnerable point.
- The squad leader confirms the vulnerable point
- The platoon leader/squad leader shifts the base of fire away from the vulnerable point.
- The base of fire continues to suppress the position and adjacent enemy positions as required.
- Buddy team #1 (team leader and the automatic rifleman) remain in a position short of the position to add suppressive fires for buddy team #2 (grenadier and rifleman).
- Buddy team #2 moves to the vulnerable point. They move in rushes or by crawling.

- One Soldier takes up a covered position near the exit.
- The other Soldier cooks off a grenade (2 seconds maximum), shouts, FRAG OUT, and throws it through an aperture.
- After the grenade detonates, the Soldier covering the exit enters and clears the bunker.
- Simultaneously, the second Soldier moves into the bunker to assist Soldier #1.
- Both Soldiers halt at a point of domination and take up positions to block any enemy movement toward their position.
- Buddy team #1 moves to join buddy team #2.
- The team leader inspects the bunker, marks the bunker, and signals the squad leader.
- The assault squad leader consolidates, reorganizes, and prepares to continue the mission.

ASSAULTING TRENCH SYSTEMS

7-225. Trenches are dug to connect fighting positions. They are typically dug in a zigzagged fashion to prevent the attacker from firing down a long section if he gets into the trench, and to reduce the effectiveness of high explosive munitions. Trenches may also have shallow turns, intersections with other trenches, firing ports, overhead cover, and bunkers. Bunkers will usually be oriented outside the trench, but may also have the ability to provide protective fire into the trench.

7-226. The trench provides defenders with a route that has frontal cover, enabling them to reposition without the threat of low trajectory fires. However, unless overhead cover is built, trenches are subject to the effects of high trajectory munitions like the grenade, grenade launcher, plunging machine gun fire, mortars, and artillery. These types of weapon systems should be used to gain and maintain fire superiority on defenders in the trench.

7-227. The trench is the enemy's home, so there is no easy way to clear it. Their confined nature, extensive enemy preparations, and the limited ability to integrate combined arms fires makes trench clearing hazardous for even the best trained Infantry. If possible, a bulldozer or plow tank can be used to fill in the trench and bury the defenders. However, since this is not always feasible, Infantry units must move in and clear trenches. Although obscuration is necessarily outside the trench, it can be more of hindrance to the attacker inside the trench. Use of night vision equipment also requires special considerations.

Entering the Trenchline

7-228. To enter the enemy trench the platoon takes the following steps:

- The squad leader and the assault fire team move to the last covered and concealed position near the entry point.
- The squad leader confirms the entry point.
- The platoon leader or squad leader shifts the base of fire away from the entry point.
- The base of fire continues to suppress trench and adjacent enemy positions as required.
- Buddy team #1 (team leader and automatic rifleman) remains in a position short of the trench to add suppressive fires for the initial entry.
- Buddy team #2 (grenadier and rifleman) and squad leader move to the entry point. They move in rushes or by crawling (squad leader positions himself where he can best control his teams).
- Buddy team #2 positions itself parallel to the edge of the trench. Team members get on their backs.
- On the squad leader command of COOK OFF GRENADES (2 seconds maximum), they shout, FRAG OUT, and throw the grenades into the trench.
- Upon detonation of both grenades, the Soldiers roll into the trench, landing on their feet and back-to-back. They engage all known, likely or suspected enemy positions.
- Both Soldiers immediately move in opposite directions down the trench, continuing until they reach the first corner or intersection.

- Both Soldiers halt and take up positions to block any enemy movement toward the entry point.
- Simultaneously, buddy team #1 moves to and enters the trench, joining buddy team #2. The squad leader directs them to one of the secured corners or intersections to relieve the Soldier who then rejoins his buddy at the opposite end of the foothold.
- At the same time, the squad leader rolls into the trench and secures the entry point.
- The squad leader remains at the entry point and marks it. He calls forward the next fire team with, NEXT TEAM IN.
- Once the squad has secured a foothold, the squad leader reports to the platoon leader, FOOTHOLD SECURE. The platoon follows the success of the seizure of the foothold with the remainder of the platoon.

7-229. The leader or a designated subordinate must move into the trench as soon as possible to control the tempo, specifically the movement of the lead assault element and the movement of follow-on forces. He must resist the temptation to move the entire unit into the trench as this will unduly concentrate the unit in a small area. Instead,, he should ensure the outside of the trench remains isolated as he maintains fire superiority inside the trench. This may require a more deliberate approach. When subordinates have reached their objectives or have exhausted their resources, the leader commits follow-on forces or requests support from higher. Once stopped, the leader consolidates and reorganizes.

7-230. The assault element is organized into a series of three-man teams. The team members are simply referred to as number 1 man, number 2 man, and number 3 man. Each team is armed with at least one M249 and one grenade launcher. All men are armed with multiple hand grenades.

7-231. The positioning within the three-man team is rotational, so the men in the team must be rehearsed in each position. The number 1 man is responsible for assaulting down the trench using well aimed effective fire and throwing grenades around pivot points in the trenchline or into weapons emplacements. The number 2 man follows the number 1 man closely enough to support him but not so closely that both would be suppressed if the enemy gained local fire superiority. The number 3 man follows the number 2 man and prepares to move forward when positions rotate.

7-232. While the initial three-man assault team rotates by event, the squad leader directs the rotation of the three-man teams within the squad as ammunition becomes low in the leading team, casualties occur, or as the situation dictates. Since this three-man drill is standardized, three-man teams may be reconstituted as needed from the remaining members of the squad. The platoon leader controls the rotation between squads using the same considerations as the squad leaders.

Clearing the Trenchline

7-233. Once the squad has secured the entry point and expanded it to accommodate the squad, the rest of the platoon enters and begins to clear the designated section of the enemy position. The platoon may be tasked to clear in two directions if the objective is small. Otherwise, it will only clear in one direction as another platoon enters alongside and clears in the opposite direction.

7-234. The lead three-man team of the initial assault squad moves out past the security of the support element and executes the trench clearing drill. The number 1 man, followed by number 2 man and number 3 man, maintains his advance until arriving at a pivot, junction point, or weapons emplacement in the trench. He alerts the rest of the team by yelling out, POSITION or, JUNCTION, and begins to prepare a grenade. The number 2 man immediately moves forward near the lead man and takes up the fire to cover until the grenade can be thrown around the corner of the pivot point. The number 3 man moves forward to the point previously occupied by number 2 and prepares for commitment.

7-235. If the lead man encounters a junction in the trench, the platoon leader should move forward, make a quick estimate, and indicate the direction the team should continue to clear. This will normally be toward the bulk of the fortification or toward command post emplacements. He should place a marker (normally specified in the unit TSOP) pointing toward the direction of the cleared path. After employing a grenade, the number 2 man moves out in the direction indicated by the platoon leader and assumes the duties of the number 1 man. Anytime the number 1 man runs out of ammunition, he shouts, MAGAZINE, and immediately moves against the wall of the trench to allow the number 2 man to take up the fire. Squad leaders continue to push uncommitted three-man teams forward, securing bypassed trenches and rotating fresh teams to the front. It is important to note that trenches are cleared in sequence not simultaneously.

Moving in a Trench

7-236. Once inside, the trench teams use variations of the combat formations described in Chapter 3 to move. These formations are used as appropriate inside buildings as well. The terms *hallway* and *trench* are used interchangeably. The column (file) and box formations are self explanatory. The line and echelon formations are generally infeasible.

FOLLOW-THROUGH

7-237. The factors for consolidation and reorganization of fortified positions are the same as consolidation and reorganization of other attacks. If a fortification is not destroyed sufficiently to prevent its reuse by the enemy, it must be guarded until means can be brought forward to complete the job. The number of positions the unit can assault is impacted by the—

- Length of time the bunkers must be guarded to prevent reoccupation by the enemy.
- Ability of the higher headquarters to resupply the unit.
- Availability of special equipment in sufficient quantities.
- Ability of the unit to sustain casualties and remain effective.

7-238. As part of consolidation, the leader orders a systematic search of the secured positions for booby traps and spider holes. He may also make a detailed sketch of his area and the surrounding dispositions if time allows. This information will be helpful for the higher headquarters intelligence officer or if the unit occupies the position for an extended length of time.

Chapter 8

Defensive Operations

Though the outcome of decisive combat derives from offensive actions, leaders often find it is necessary, even advisable, to defend. The general task and purpose of all defensive operations is to defeat an enemy attack and gain the initiative for offensive operations. It is important to set conditions of the defense so friendly forces can destroy or fix the enemy while preparing to seize the initiative and return to the offense. The platoon may conduct the defense to gain time, retain key terrain, facilitate other operations, preoccupy the enemy in one area while friendly forces attack him in another, or erode enemy forces. A well coordinated defense can also set the conditions for follow-on forces and follow-on operations.

SECTION I — CHARACTERISTICS OF THE DEFENSE

8-1. Following are the characteristics of the defense that constitute the planning fundamentals for the Infantry platoon:

- Preparation
- Security
- Disruption
- Massing effects
- Flexibility

8-2. To ensure the success of the defense, the platoon leader must understand the characteristics of the defense and apply TLP during planning, preparation, and execution of the operation.

PREPARATION

8-3. The friendly defender arrives in the battle area before the enemy attacker. As the defender, the platoon must take advantage of this by making the most of preparations for combat in the time available. By thoroughly analyzing the factors of METT-TC, the platoon leader gains an understanding of the tactical situation and identifies potential friendly and enemy weaknesses.

8-4. By arriving in the battle area first, the Infantry platoon has the advantage of preparing the terrain before the engagement. Through the proper selection of terrain and reinforcing obstacles, friendly forces can direct the energy of the enemy's attack into terrain of their choosing. Friendly forces must take advantage of this by making the most thorough preparations that time allows while always continuing to improve their defenses—security measures, engagement areas, and survivability positions. Preparation of the ground consists of plans for fires and movement; counterattack plans; and preparation of positions, routes, obstacles, logistics, and command and control (C2) facilities.

8-5. The Infantry platoon must exploit every aspect of terrain and weather to its advantage. In the defense, as in the attack, terrain is valuable only if the friendly force gains advantage from its possession or control. In developing a defensive plan, the friendly force takes account of key terrain and attempts to visualize and cover with fire all possible enemy avenues of approach into their sector. The friendly defense seeks to defend on terrain that maximizes effective fire, cover, concealment, movement, and surprise.

8-6. Friendly forces must assume that their defensive preparations are being observed. To hinder the enemy's intelligence effort, leaders establish security forces to conduct counter reconnaissance and deceive the enemy as to the exact location of the main defenses.

SECURITY

8-7. The goals of the platoon's security efforts are normally tied to the company efforts. These efforts include providing early warning, destroying enemy reconnaissance units, and impeding and harassing elements of the enemy main body. The platoon will typically continue its security mission until directed to displace.

DISRUPTION

8-8. Defensive plans vary with the circumstances, but all defensive concepts of the operation aim at disrupting the enemy attacker's synchronization. Counterattacks, indirect fires, obstacles, and the retention of key terrain prevent the enemy from concentrating his strength against selected portions of the platoon's defense. Destroying enemy command and control vehicles disrupts the enemy synchronization and flexibility. Separating enemy units from one another allows them to be defeated piecemeal.

MASSING EFFECTS

8-9. The platoon must mass the overwhelming effects of combat power at the decisive place and time if it is to succeed. It must obtain a local advantage at points of decision. Offensive action may be a means of gaining this advantage. The platoon leader must remember that this massing refers to combat power and its effects—not just numbers of Soldiers and weapons systems.

FLEXIBILITY

8-10. Flexibility is derived from sound preparation and effective command and control and results from a detailed analysis of the factors of METT-TC, an understanding of the unit's purpose, and aggressive reconnaissance and surveillance. The platoon must be agile enough to counter or avoid the enemy attacker's blows and then strike back effectively. For example, supplementary positions on a secondary avenue of approach may provide additional flexibility to the platoon. Immediate transitions from defense to offense are difficult. To ease this transition, the platoon leader must think through and plan for actions his platoon may need to take, and then rehearse them in a prioritized sequence based on time available.

SECTION II — SEQUENCE OF THE DEFENSE

8-11. As part of a larger element, the platoon conducts defensive operations in a sequence of integrated and overlapping phases. This section focuses on the following phases within the sequence of the defense:

- Reconnaissance, security operations, and enemy preparatory fires.
- Occupation.
- Approach of the enemy main attack.
- Enemy assault.
- Counterattack.
- Consolidation and reorganization.

RECONNAISSANCE, SECURITY OPERATIONS, AND ENEMY PREPARATORY FIRES

8-12. Security forces must protect friendly forces in the main battle area (MBA) and allow them to prepare for the defense. The goals of a security force include providing early warning, destroying enemy reconnaissance elements (within its capability), and disrupting enemy forward detachments or advance guard elements. The platoon may be attached to a larger element or remain with the parent company to conduct counter-reconnaissance. Additionally, the platoon may conduct security operations as part of the company defensive plan by conducting patrols or manning observation post(s) (OP) to observe named area(s) of interest (NAI).

8-13. The platoon may also be required to provide guides to the passing friendly security force and may be tasked to close the passage lanes. The passage could be for friendly forces entering or departing the security zone, and may include logistics units supporting the security forces. The platoon, as part of a larger force, may also play a role in shaping the battlefield. The battalion or brigade combat team commander may position the company to deny likely enemy attack corridors. This will enhance flexibility and force enemy elements into friendly engagement areas.

8-14. When not conducting security or preparation tasks, the Infantry platoon normally occupies dug-in positions with overhead cover to avoid possible enemy artillery preparatory fires.

OCCUPATION

8-15. The occupation phase of the defense includes moving from one location to the defensive location. A quartering party under company control normally leads this movement to clear the defensive position and prepares it for occupation. The platoon plans, reconnoiters, and then occupies the defensive position. The battalion establishes security forces. The remaining forces prepare the defense. To facilitate maximum time for planning, occupying, and preparing the defense, leaders and Soldiers at all levels must understand their duties and responsibilities, including priorities of work (covered in the WARNO or by a unit TSOP).

8-16. Occupation and preparation of the defense site (see Section V of this chapter) is conducted concurrently with the TLP and the development of the engagement area (if required). The platoon occupies defensive positions IAW the company commander's plan and the results of the platoon's reconnaissance. To ensure an effective and efficient occupation, the reconnaissance element marks the friendly positions. These tentative positions are then entered on the operational graphics. Each squad moves in or is led in by a guide to its marker. Once in position, each squad leader checks his position location. As the platoon occupies its positions, the platoon leader manages the positioning of each squad to ensure they locate IAW the tentative plan. If the platoon leader notes discrepancies between actual positioning of the squads and his plan, he makes the corrections. Security is placed out in front of the platoon. The platoon leader must personally walk the fighting positions to ensure that everyone understands the plan and that the following are IAW the plan:

- Weapons orientation and general sectors of fire.
- Crew served weapons positions.
- Rifle squads' positions in relation to each other.

8-17. Each squad leader ensures he knows the location of the platoon leader and platoon sergeant for command and control purposes, and where the casualty collection point is located. The platoon may be required to assist engineers in the construction of tactical obstacles in their sector. All leaders must know where these obstacles are so they can tie them into their fire plan.

8-18. When the occupation is complete, subordinate leaders can begin to develop their sector sketches (paragraph 8-100) based on the basic fire plan developed during the leader's reconnaissance. Positions are improved when the direct fire plan is finalized and proofed. In addition to establishing the platoon's primary positions, the platoon leader and subordinate leaders normally plan for preparation and occupation of alternate, supplementary, and subsequent positions. This is done IAW the company order. The platoon and/or company reserve need to know the location of these positions. The following are tactical considerations for these positions.

ALTERNATE POSITIONS

8-19. The following characteristics and considerations apply to an alternate position:

- Covers the same avenue of approach or sector of fire as the primary position.
- Located slightly to the front, flank, or rear of the primary position.
- Positioned forward of the primary defensive positions during limited visibility operations.
- Normally employed to supplement or support positions with weapons of limited range, such as Infantry squad positions. They are also used as an alternate position to fall back to if the original position is rendered ineffective or as a position for Soldiers to rest or perform maintenance.

SUPPLEMENTARY POSITIONS

8-20. The following characteristics and considerations apply to a supplementary position:

- Covers an avenue of approach or sector of fire different from those covered by the primary position.
- Occupied based on specific enemy actions.

SUBSEQUENT POSITIONS

8-21. The following characteristics and considerations apply to a subsequent position:

- Covers the same avenue of approach and or sector of fire as the primary position.
- Located in depth through the defensive area.
- Occupied based on specific enemy actions or conducted as part of the higher headquarters' scheme of maneuver.

APPROACH OF THE ENEMY MAIN ATTACK

8-22. As approach of the enemy main attack begins, brigade combat team and higher headquarters engage the enemy at long range using indirect fires, electronic warfare, Army attack aviation, and close air support (CAS). The goal is to use these assets and disrupting obstacles to shape the battlefield and or to slow the enemy's advance and break up his formations, leaving him more susceptible to the effects of crew served weapons. As the enemy's main body echelon approaches the battalion engagement area, the battalion may initiate indirect fires and CAS to weaken the enemy through attrition. At the same time, the brigade combat team's effort shifts to second-echelon forces, depending on the commander's plan. Based on an event stated in the company commander's order, Infantry platoons cease security patrols and bring OPs back into the defense at a predetermined time. Positions may be shifted in response to enemy actions or other tactical factors.

ENEMY ASSAULT

8-23. During an enemy assault attacking enemy forces attempt to fix and finish friendly forces. Their mission will be similar to those in friendly offensive operations: destroy forces, seize terrain, and conduct a penetration to pass follow-on forces through. During execution of the defense, friendly forces will mass the effects of fires to destroy the assaulting enemy. The platoon leader must determine if the platoon can destroy the enemy from its assigned positions.

FIGHTING FROM ASSIGNED POSITIONS

8-24. If the platoon can destroy the enemy from its assigned positions, the platoon continues to fight the defense.

8-25. The platoon leader continues to call for indirect fires as the enemy approaches. The platoon begins to engage the enemy at their weapon systems' maximum effective range. They attempt to mass fires and initiate them simultaneously to achieve maximum weapons effects. Indirect fires and obstacles integrated with direct fires should disrupt the enemy's formations, channel him toward EAs, prevent or severely limit his ability to observe the location of friendly positions, and destroy him as he attempts to breach tactical and or protective obstacles. If there is no enlisted tactical air controller (ETAC) available, the forward observer or platoon leader will be prepared to give terminal guidance to attack aviation if available and committed into his area of operations.

8-26. Leaders control fires using standard commands, pyrotechnics, and other prearranged signals. (See Chapter 2, Employing Fires, for more information.) The Infantry platoon increases the intensity of fires as the enemy closes within range of additional friendly weapons. Squad leaders and team leaders work to achieve a sustained rate of fire from their positions by having buddy teams engage the enemy so both

Soldiers are not reloading their weapons at the same time. To control and distribute fires, leaders consider—

- Range to the enemy.
- Engagement criteria (what to fire at, when to fire [triggers], and why).
- Most dangerous or closest enemy targets.
- Shifting to concentrate direct fires either independently or as directed by higher headquarters.
- Ability of the platoon to engage dismounted enemy with enfilading, grazing fires.
- Ability of the platoon's SLM and CCMS to achieve flank shots against enemy vehicles.

8-27. When the enemy closes on the platoon's protective wire, machine guns fire along interlocking principal direction(s) of fire (PDF) or final protective line(s) (FPL) as previously planned and designated. Other weapons fire at their designated PDFs. Grenadiers engage the enemy with grenade launchers in dead space or as the enemy attempts to breach protective wire. The platoon leader requests final protective fire (FPF) if it is assigned in support of his positions.

8-28. The platoon continues to defend until it repels the enemy or is ordered to disengage.

FIGHTING FROM OTHER THAN ASSIGNED POSITIONS

8-29. If the platoon cannot destroy the enemy from its assigned positions, the platoon leader reports the situation to the company commander and continues to engage the enemy. He repositions the platoon (or squads of the platoon) when directed by the commander in order to—

- Continue fires into the platoon engagement area.
- Occupy supplementary or alternate positions.
- Reinforce other parts of the company.
- Counterattack locally to retake lost fighting positions.
- Withdraw from an indefensible position using fire and movement to break contact.

NOTE: The platoon leader does not move his platoon out of position if it will destroy the integrity of the company defense. All movements and actions to reposition squads and the platoon must be thoroughly rehearsed.

COUNTERATTACK

8-30. As the enemy's momentum is slowed or stopped, friendly forces may counterattack. The counterattack may be launched to seize the initiative from the enemy or to completely halt his attack. In some cases, the purpose of the counterattack will be mainly defensive (for example, to reestablish the forward edge of the battle area [FEBA] or to restore control of the area). The Infantry platoon may participate in the counterattack as a base-of-fire element or as the counterattack force. This counterattack could be planned or conducted during the battle when opportunities to seize the initiative present themselves.

CONSOLIDATION AND REORGANIZATION

8-31. The platoon secures its sector and reestablishes the defense by repositioning friendly forces, destroying enemy elements, treating and evacuating casualties, processing EPWs, and reestablishing obstacles. The platoon conducts all necessary sustainment functions, such as cross-leveling ammunition and weapons, as it prepares to continue defending. Squad and team leaders provide liquid, ammunition, casualty, and equipment (LACE) reports to the platoon leader. The platoon leader reestablishes the platoon chain of command. He consolidates squad LACE reports and provides the platoon report to the company commander. The platoon sergeant coordinates for resupply and supervises the execution of the casualty and

EPW evacuation plan. The platoon continues to repair or improve positions, quickly reestablishes observation posts, and resumes security patrolling as directed.

8-32. Consolidation includes organizing and strengthening a position so it can continue to be used against the enemy. Platoon consolidation requirements include:

- Adjusting other positions to maintain mutual support.
- Reoccupying and repairing positions and preparing for renewed enemy attack.
- Relocating selected weapons to alternate positions if leaders believe the enemy may have pinpointed them during the initial attack.
- Repairing any damaged obstacles and replacing any Claymore mines.
- Reestablishing security and communications.

8-33. Reorganization includes shifting internal resources within a degraded friendly unit to increase its level of combat effectiveness. Platoon consolidation requirements include:

- Manning key weapons as necessary.
- Providing first aid and preparing wounded Soldiers for CASEVAC.
- Redistributing ammunition and supplies.
- Processing and evacuating EPWs.

SECTION III — PLANNING CONSIDERATIONS

8-34. The Army warfighting functions incorporate a list of critical tactical activities that provide a structure for leaders to prepare and execute the defense. Synchronization and coordination among the warfighting functions are critical for success.

MOVEMENT AND MANEUVER

8-35. Effective weapons positioning enables the platoon to mass fires at critical points on the battlefield to effectively engage the enemy in the engagement area. (See Section IV for more information on engagement area development.) The platoon leader must maximize the strengths of the platoon's weapons systems while minimizing its exposure to enemy observation and fires.

8-36. Mobility focuses on the ability to reposition friendly forces, including unit displacement and the commitment of reserve forces. The company commander's priorities may specify that some routes be improved to support such operations. Countermobility channels the enemy into the engagement area as it limits the maneuver of enemy forces and enhances the effectiveness of the defender's direct and indirect fires.

DEPTH AND DISPERSION

8-37. Dispersing positions laterally and in depth helps protect the force from enemy observation and fires. Platoon positions are established to allow sufficient maneuver space within each position for in-depth placement of crew-served weapons systems and Infantry squads. Infantry fighting positions are positioned to allow massing of direct fires at critical points on the battlefield, as well as to provide overlapping fire in front of other fighting positions. Although the factors of METT-TC ultimately determine the placement of weapons systems and unit positions, the following also apply:

- Infantry squads can conduct antiarmor fires in depth with CCMS, which have a maximum range of 2,000 meters.
- Infantry squads can retain or deny key terrain if employed in strongpoints or protected positions.
- Infantry squads can protect obstacles or flank positions that are tied into severely restrictive terrain.

FLANK POSITIONS

8-38. Flank positions enable a defending friendly force to bring direct fires to bear on an attacking force. An effective flank position provides the friendly defender with a larger, more vulnerable enemy target while leaving the attacker unsure of the location of the defender. Major considerations for successful employment of a flank position are the friendly defender's ability to secure the flank, and his ability to achieve surprise by remaining undetected. Effective direct fire control (see Chapter 2, Employing Fires) and fratricide avoidance measures (see Chapter 5, Command, Control, and Troop-Leading Procedures) are critical considerations when employing flank positions.

MOBILITY

8-39. During defensive preparations, mobility focuses initially on the ability to resupply, CASEVAC, reposition, and the rearward and forward passage of forces, supplies, and equipment. Once defensive preparations are complete, the mobility focus shifts to routes to alternate, supplementary, or subsequent positions. The company commander will establish the priority of mobility effort within the company.

COUNTERMOBILITY

8-40. To be successful in the defense, the platoon leader must integrate obstacles into both the direct and indirect fire plans. (Refer to FM 90-7, *Combined Arms Obstacle Integration,* for additional information on obstacle planning, siting, and turnover.) A tactical obstacle is designed or employed to disrupt, fix, turn, or block the movement of the enemy. Platoons construct tactical obstacles when directed by the company commander.

Disrupting Effects

8-41. Disrupting effects focus a combination of fires and obstacles to impede the enemy's attack in several ways, including breaking up his formations, interrupting his tempo, and causing early commitment of breaching assets. These effects are often the product of situational obstacles such as scatterable mines, and are normally used forward within engagement areas or in support of forward positions within a defensive sector. Normally, only indirect fires and long-range direct fires are planned in support of disrupting obstacles (Figure 8-1).

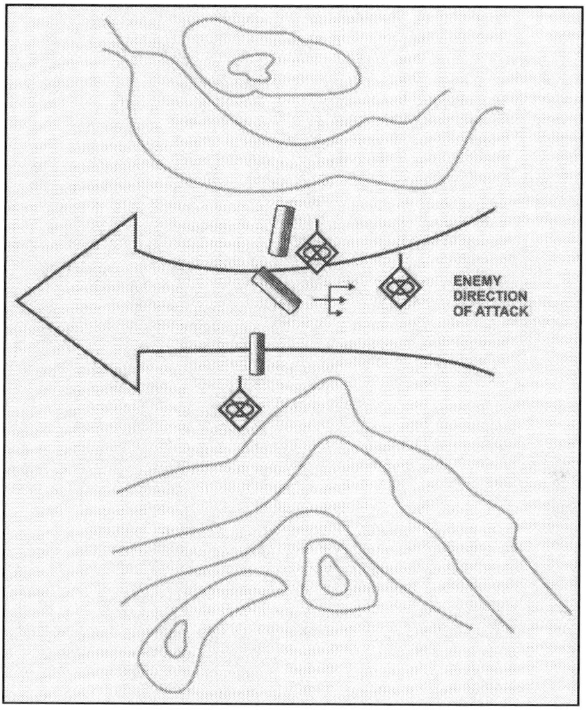

Figure 8-1. Disrupt obstacle effect.

Fixing Effects

8-42. Fixing effects use a combination of fires and obstacles to slow or temporarily stop an attacker within a specified area, normally an engagement area (Figure 8-2). The defending unit can then focus on defeating the enemy by using indirect fires to fix him in the engagement area while direct fires inflict maximum casualties and damage. If necessary, the defender can reposition his forces using the additional time gained as a result of fixing the enemy. To fully achieve the fixing effect, direct and or indirect fires must be integrated with the obstacles. The company commander must specify the size of the enemy unit to be fixed.

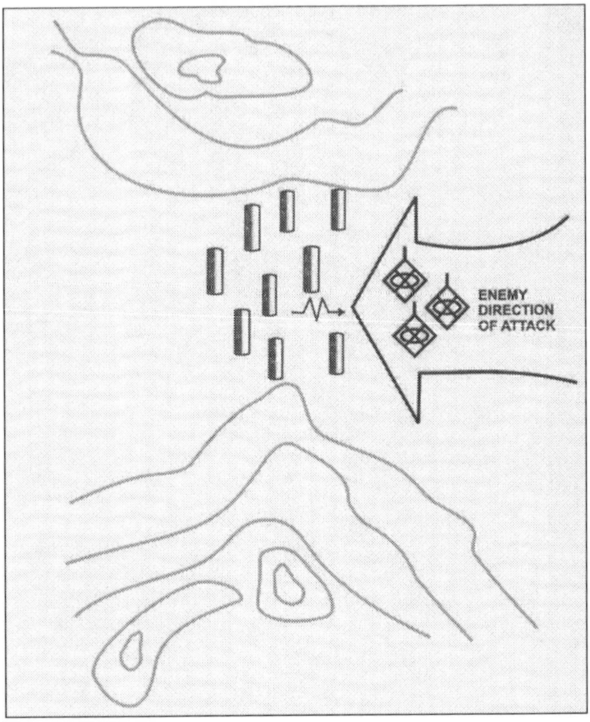

Figure 8-2. Fix obstacle effect.

Turning Effects

8-43. Turning effects (Figure 8-3) use the combination of direct and indirect fires and obstacles to support the company commander's scheme of maneuver in several ways, including the following:

- Diverting the enemy into an engagement area and exposing his flanks when he makes the turn.
- Diverting an enemy formation from one avenue of approach to another.
- Denying the enemy the ability to mass his forces on a flank of the friendly force.

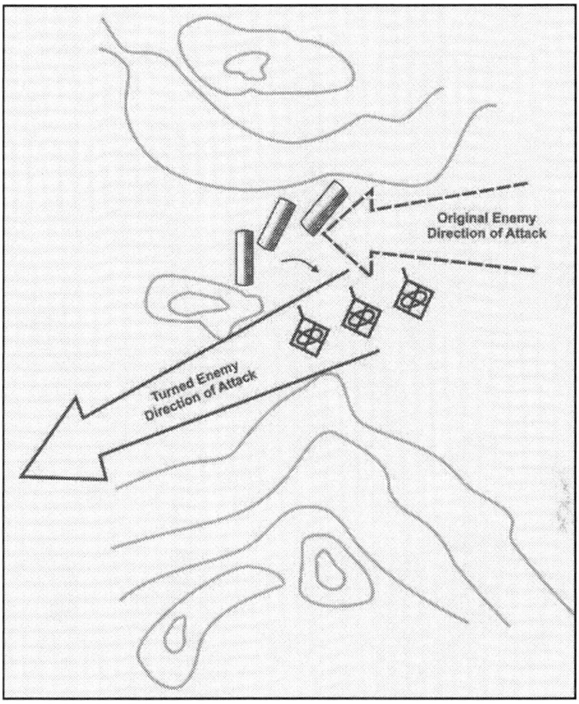

Figure 8-3. Turn obstacle effect.

Blocking Effects

8-44. Blocking effects use the combination of direct and indirect fires and obstacles to stop an attacker along a specific avenue of approach (Figure 8-4). Fires employed to achieve blocking effects are primarily oriented on preventing the enemy from maneuvering. Because they require the most extensive engineer effort of any type of obstacle, blocking effects are employed only at critical choke points on the battlefield. Blocking obstacles must be anchored on both sides by existing obstacles (severely restrictive terrain). Direct and or indirect fires must cover the obstacles to achieve the full blocking effect. The company commander must clearly specify the size of enemy force that he intends to block.

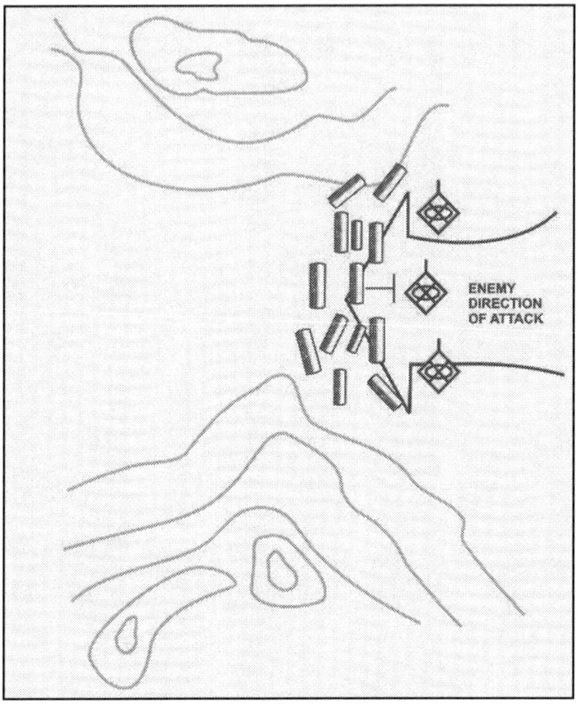

Figure 8-4. Block obstacle effect.

DISPLACEMENT AND DISENGAGEMENT PLANNING

8-45. Displacement and disengagement are key control measures that allow the platoon to retain its operational flexibility and tactical agility. The ultimate goals of displacement and disengagement are to enable the platoon to maintain standoff range of the CCMS and to avoid being fixed or decisively engaged by the enemy.

Considerations

8-46. While displacement and disengagement are valuable tactical tools, they can be extremely difficult to execute in the face of a rapidly advancing enemy force. In fact, displacement in contact poses great problems. The platoon leader must therefore plan for it thoroughly before the operation and rehearse moving to alternate and supplementary positions if time permits. Even then, he must carefully evaluate the situation whenever displacement in contact becomes necessary to ensure it is feasible, and that it will not result in unacceptable personnel or equipment losses. The platoon leader must consider several important factors in displacement planning:

- The enemy situation (for example, an enemy attack with battalion-sized element may prevent the platoon from disengaging).
- Higher headquarters' disengagement criteria.
- Availability of friendly direct fire to facilitate disengagement by suppressing or disrupting the enemy.

- Availability of cover and concealment, indirect fires, and smoke to assist disengagement.
- Obstacle integration, including situational obstacles.
- Positioning of forces on terrain (such as reverse slopes or natural obstacles) that provides an advantage to the disengaging elements.
- Identification of displacement routes and times that disengagement and or displacement will take place.
- The size of the friendly force available to engage the enemy in support of the displacing unit.

Disengagement Criteria

8-47. Disengagement criteria dictate to subordinate elements the circumstances under which they will displace to alternate, supplementary, or subsequent defensive positions. The criteria are tied to an enemy action (such as one motorized rifle platoon advancing past Phase Line Delta) and are linked to the friendly situation. For example, they may depend on whether a friendly overwatch element or artillery unit can engage the enemy. Disengagement criteria are developed during the planning process based on the unique conditions of a specific situation. They should not be part of the unit's SOP.

Direct Fire Suppression

8-48. The attacking enemy force must not be allowed to bring effective fires to bear on a disengaging force. Direct fires from the base-of-fire element, employed to suppress or disrupt the enemy, are the most effective way to facilitate disengagement. The platoon may also receive base-of-fire support from another element in the company, but in most cases the platoon will establish its own base of fire. Employing an internal base of fire requires the platoon leader to carefully sequence the displacement of his elements.

Cover and Concealment

8-49. Ideally, the platoon and subordinate elements should use covered and concealed routes when moving to alternate, supplementary, or subsequent defensive positions. Regardless of the degree of protection the route itself affords, the platoon should rehearse the movement. By rehearsing, the platoon can increase the speed at which it moves and provide an added measure of security. The platoon leader must make a concerted effort whenever time is available to rehearse movement in limited visibility and degraded conditions.

Indirect Fires and Smoke

8-50. Artillery or mortar fires can be employed to assist the platoon during disengagement. Suppressive fires, placed on an enemy force as it is closing inside the defender's standoff range, will disrupt his formations, slow his progress, and if the enemy is a mechanized force, cause him to button up. The defending force engages the enemy with long-range direct fires, then disengages and moves to new positions. Smoke may be employed to obscure the enemy's vision, slow his progress, or screen the defender's movement out of the defensive positions or along his displacement route.

Obstacle Integration

8-51. Obstacles should be integrated with direct and indirect fires to assist disengagement. By slowing and disrupting enemy movement, obstacles provide the defender the time necessary for displacement. Obstacles also allow friendly forces to employ direct and indirect fires against the enemy. The modular pack mine system (MOPMS) can be employed in support of the disengagement to either block a key displacement route once the displacing unit has passed through it, or to close a lane through a tactical obstacle. The location of obstacle emplacement depends in large measure on METT-TC factors. An obstacle should be positioned far enough away from the defender so enemy elements can be effectively engaged on the far side of the obstacle while the defender remains out of range of the enemy's massed direct fires.

FIRE SUPPORT

8-52. For the indirect fire plan to be effective in the defense, the unit must plan and execute indirect fires in a manner that achieves the intended task and purpose of each target. Indirect fires serve a variety of purposes in the defense, including:

- Slowing and disrupting enemy movement.
- Preventing the enemy from executing breaching operations at turning or blocking obstacles.
- Destroying or delaying enemy forces at obstacles using massed indirect fires or precision munitions (such as Copperhead rounds).
- Defeating attacks along dismounted avenues of approach using FPF.
- Disrupting the enemy to allow friendly elements to disengage or conduct counterattacks.
- Obscuring enemy observation or screening friendly movement during disengagement and counterattacks.
- Based on the appropriate level of approval, delivering scatterable mines to close lanes and gaps in obstacles, disrupting or preventing enemy breaching operations, disrupting enemy movement at choke points, or separating or isolating enemy echelons.

PROTECTION

8-53. Platoons are responsible for coordinating and employing their own protective obstacles to protect their defensive positions. To be most effective, these obstacles should be tied into existing obstacles and FPFs. The platoon may use mines and wire from its basic load or pick up additional assets (including MOPMS, if available) from the engineer Class IV or V supply point. (See Appendix F for details on MOPMS and mines.) The platoon, through the company, also may be responsible for any other required coordination (such as that needed in a relief in place) for recovery of the obstacle or for its destruction (as in the case of MOPMS). A detail discussion of Protection can be found in Chapter 4.

8-54. In planning for protective obstacles, the platoon leader must evaluate the potential threat to the platoon position and employ the appropriate asset. For example, MOPMS is predominately an antitank system best used on mounted avenues of approach, but it does have some antipersonnel applications. Wire obstacles may be most effective when employed on dismounted avenues of approach. FM 90-7 provides detailed planning guidance for the emplacement of protective obstacles.

8-55. Protective obstacles are usually located beyond hand grenade range (40 to 100 meters) from a Soldier's fighting position. They may extend out 300 to 500 meters to tie into tactical obstacles and existing restrictive or severely restrictive terrain. The platoon leader should therefore plan protective obstacles in depth and attempt to maximize the effective range of his weapons.

8-56. When planning protective obstacles, the platoon leader should consider the amount of time required to prepare them, the resources available after constructing necessary tactical obstacles, and the priorities of work for the Soldiers in the platoon.

WIRE OBSTACLES

8-57. There are three types of wire obstacles: protective wire; tactical wire; and supplementary wire (Figure 8-5).

Protective Wire

8-58. Protective wire may be a complex obstacle providing all-round protection of a platoon perimeter, or it may be a simple wire obstacle on the likely dismounted avenue of approach toward a squad position (Figure 8-6). Command-detonated M18 Claymore mines may be integrated into the protective wire or used separately.

Tactical Wire

8-59. Tactical wire is positioned to increase the effectiveness of the platoon's direct fires. It is usually positioned along the friendly side of a machine gun FPL. Tactical minefields may also be integrated into these wire obstacles or be employed separately.

Supplementary Wire

8-60. Supplementary wire obstacles are employed to break up the line of tactical wire to prevent the enemy from locating platoon weapons (particularly CCMS and machine guns) by following the tactical wire.

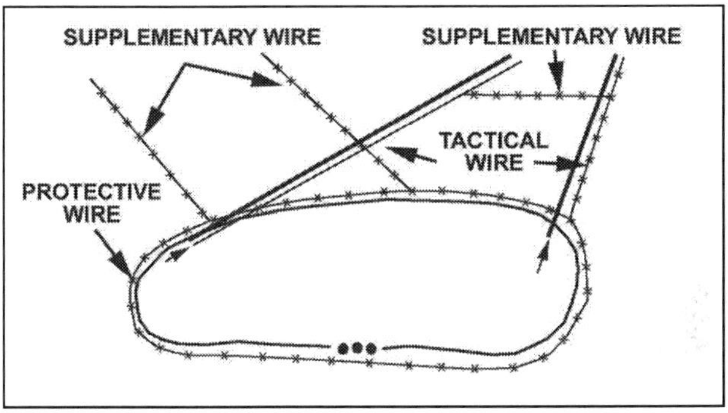

Figure 8-5. Three types of protective wire obstacles.

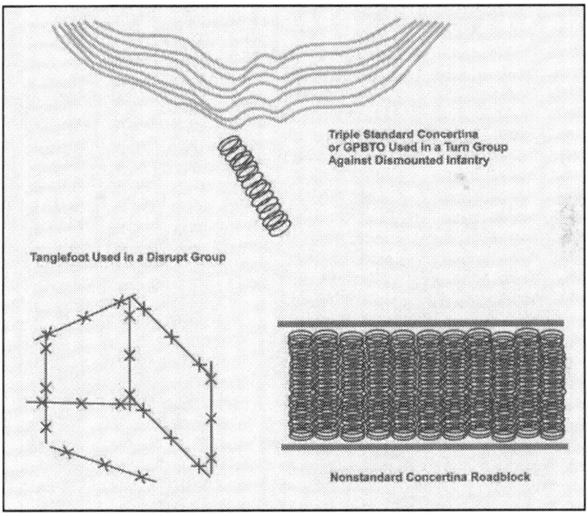

Figure 8-6. Protective wire groups.

Obstacle Lanes

8-61. The platoon may be responsible for actions related to lanes through obstacles. These duties may include overwatching lanes in the obstacle, marking lanes in an obstacle, reporting the locations of the entry and exit points of each lane, manning contact points, providing guides for elements passing through the obstacle, and closing lanes when directed.

Survivability

8-62. Survivability focuses on protecting friendly forces from the effect of enemy weapons systems. Survivability positions are prepared in defensive positions or strongpoints to protect weapons systems and rifle squads. Positions can be dug in and reinforced with overhead cover to provide rifle squads and crew-served weapons with protection against shrapnel from air bursts. The company may dig in ammunition prestocks at platoon alternate, supplementary, or subsequent defensive positions. The platoon leader may have time only to dig in positions that have the least amount of natural cover and concealment. Soil composition should also be a consideration in the selection of defensive positions. Sites to be avoided include those where the soil is overly soft, hard, wet, or rocky.

Air and Missile Defense

8-63. The focus of an air and missile defense plan is on likely air avenues of approach for enemy fixed-wing, helicopters, and unmanned aircraft systems that may not correspond with the enemy's ground avenues of approach. A platoon leader is not likely to emplace air defense assets, but he must be aware that higher headquarters may employ air defense assets near his defensive position. For a detailed discussion of air defense, see Section II, Chapter 4.

SUSTAINMENT

8-64. In addition to the sustainment function required for all operations, the platoon leader should consider prestocking (also known as pre-positioning or caches). The platoon leader's mission analysis (or guidance from the company commander) may reveal that the platoon's ammunition needs during an operation may exceed its basic load. This requires the platoon to establish ammunition caches. The caches, which may be positioned at an alternate or subsequent position, should be dug in. Security should be provided by active or passive means (guarded or observed) to indicate when and if the cache is tampered with.

8-65. The platoon must have a plan to recover their assets when quickly transitioning to the offense or counterattack or when disengaging.

INTELLIGENCE

8-66. The intelligence warfighting function consists of the related tasks and systems that facilitate understanding of the enemy, terrain, weather, and civil considerations. It includes tasks associated with ISR. It is a flexible, adjustable architecture of procedures, personnel, organizations, and equipment. These provide relevant information and products relating to the threat, civil populace, and environment to commanders. Intelligence warfighting function focuses on four primary tasks:

(1) Support to situational understanding.
(2) Support to strategic responsiveness.
(3) Conduct ISR.
(4) Provide intelligence support to targeting.

COMMAND AND CONTROL

8-67. The command and control warfighting function consists of the related tasks and systems that support commanders in exercising authority and direction. It includes those tasks associated with acquiring friendly information, managing all relevant information, and directing and leading subordinates.

SECTION IV — ENGAGEMENT AREA DEVELOPMENT

8-68. The engagement area is the place where the platoon leader intends to destroy an enemy force using the massed fires of all available weapons. The success of any engagement depends on how effectively the platoon leader can integrate the obstacle and indirect fire plans with his direct fire plan in the engagement area to achieve the platoon's purpose. At the platoon level, engagement area development remains a complex function that requires parallel planning and preparation if the platoon is to accomplish its assigned tasks. Despite this complexity, engagement area development resembles a drill. The platoon leader and his subordinate leaders use a standardized set of procedures. Beginning with an evaluation of the factors of METT-TC, the development process covers these steps:

- Identify likely enemy avenues of approach.
- Identify the enemy scheme of maneuver.
- Determine where to kill the enemy.
- Plan and integrate obstacles.
- Emplace weapons systems.
- Plan and integrate indirect fires.
- Conduct an engagement area rehearsal.

IDENTIFY LIKELY ENEMY AVENUES OF APPROACH

8-69. The platoon leader conducts an initial reconnaissance from the enemy's perspective along each avenue of approach into the sector or engagement area. During his reconnaissance, he confirms key terrain identified by the company commander, including locations that afford positional advantage over the enemy and natural obstacles and choke points that restrict forward movement. The platoon leader determines which avenues will afford cover and concealment for the enemy while allowing him to maintain his tempo. The platoon leader also evaluates lateral mobility corridors (routes) that adjoin each avenue of approach.

IDENTIFY ENEMY SCHEME OF MANEUVER

8-70. The platoon leader greatly enhances this step of the engagement area development process by gaining information early. He receives answers to the following questions from the company commander:

- Where does the enemy want to go?
- Where will the enemy go based on terrain?
- What is the enemy's mission (or anticipated mission)?
- What are the enemy's objectives?
- How will the enemy structure his attack?
- How will the enemy employ his reconnaissance assets?
- What are the enemy's expected rates of movement?
- How will the enemy respond to friendly actions?

DETERMINE WHERE TO KILL THE ENEMY

8-71. As part of his TLP, the platoon leader must determine where he will mass combat power on the enemy to accomplish his purpose. This decision is tied to his assessment of how the enemy will fight into the platoon's engagement area. Normally this entry point is marked by a prominent TRP that all platoon elements can engage with their direct fire weapons. This allows the commander to identify where the platoon will engage enemy forces through the depth of the company engagement area. In addition, the leader—

- Identifies TRPs that match the enemy's scheme of maneuver, allowing the platoon (or company) to identify where it will engage the enemy through the depth of the engagement area.
- Identifies and records the exact location of each TRP.

- Determines how many weapons systems can focus fires on each TRP to achieve the desired purpose.
- Determines which squad(s) can mass fires on each TRP.
- Begins development of a direct fire plan that focuses at each TRP.

NOTE: In marking TRPs, use thermal sights to ensure visibility at the appropriate range under varying conditions, including daylight and limited visibility.

PLAN AND INTEGRATE OBSTACLES

8-72. To be successful in the defense, the platoon leader must integrate tactical obstacles with the direct fire plan, taking into account the intent of each obstacle. At the company level, obstacle intent consists of the target of the obstacle, the desired effect on the target, and the relative location of the group. A platoon must have a clear task and purpose to properly emplace a tactical obstacle. The company or battalion will normally designate the purpose of the tactical obstacle. The purpose will influence many aspects of the operation, from selection and design of obstacle sites, to actual conduct of the defense. Once the tactical obstacle has been emplaced, the platoon leader must report its location and the gaps in the obstacle to the company commander. This ensures that the company commander can integrate obstacles with his direct and indirect fire plans, refining his engagement area development.

EMPLACE WEAPONS SYSTEMS

8-73. To position weapons effectively, leaders must know the characteristics, capabilities, and limitations of the weapons as well as the effects of terrain and the tactics used by the enemy. Platoon leaders should position weapons where they have protection, where they can avoid detection, and where they can surprise the enemy with accurate, lethal fires. In order to position the weapons, the platoon leader must know where he wants to destroy the enemy and what effect he wants the weapon to achieve. He should also consider—

- Selecting tentative squad defensive positions.
- Conducting a leader's reconnaissance of the tentative defensive positions.
- Walking the engagement area to confirm that the selected positions are tactically advantageous.
- Confirming and marking the selected defensive positions.
- Developing a direct fire plan that accomplishes the platoon's purpose.
- Ensuring the defensive positions do not conflict with those of adjacent units and is effectively tied in with adjacent positions.
- Selecting primary, alternate, and supplementary fighting positions to achieve the desired effect for each TRP.
- Ensuring the squad leaders position weapons systems so the required numbers of weapons or squads effectively cover each TRP.
- Inspecting all positions.

NOTE: When possible, select fighting and crew-served weapon positions while moving in the engagement area. Using the enemy's perspective enables the platoon leader to assess survivability of the positions.

PLAN AND INTEGRATE INDIRECT FIRES

8-74. In planning and integrating indirect fires, the platoon leader must accomplish the following:

- Determine the purpose of fires if the company commander has not already done so.
- Determine where that purpose will best be achieved if the company commander has not done so.
- Establish the observation plan with redundancy for each target. Observers include the platoon leader as well as members of subordinate elements (such as team leaders) with fire support responsibilities.

- Establish triggers based on enemy movement rates.
- Obtain accurate target locations using survey and navigational equipment.
- Refine target locations to ensure coverage of obstacles.
- Register artillery and mortars.
- Plan FPF.

CONDUCT AN ENGAGEMENT AREA REHEARSAL

8-75. The purpose of rehearsal is to ensure that every leader and every Soldier understands the plan (Figure 8-7), and is prepared to cover his assigned areas with direct and indirect fires.

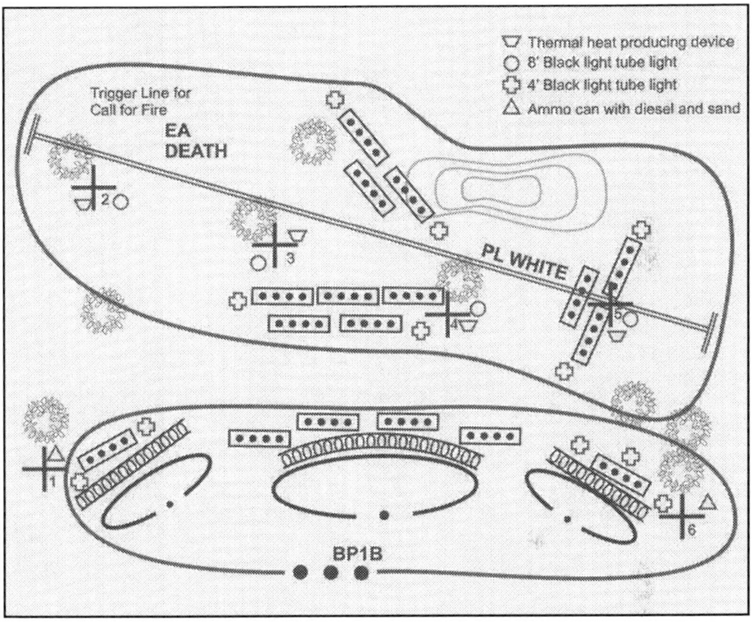

Figure 8-7. Integrated engagement area plan.

8-76. The platoon will probably participate in a company-level engagement area rehearsal. The company commander has several options for conducting a rehearsal, but the combined arms rehearsal produces the most detailed understanding of the plan. One technique the platoon leader may use for his rehearsal is the full dress rehearsal. In the defense, the platoon leader may have the platoon sergeant and squads conduct a movement through the engagement area to depict the attacking enemy force, while the platoon leader and squad leaders rehearse the battle from the platoon defensive positions. The rehearsal should cover—

- Rearward passage of security forces (as required).
- Closure of lanes (as required).
- Use of fire commands, triggers, and or maximum engagement lines (MELs) to initiate direct and indirect fires.
- Shifting of fires to refocus and redistribute fire effects.
- Disengagement criteria.

- Identification of displacement routes and times.
- Preparation and transmission of critical reports.
- Assessment of the effects of enemy weapons systems.
- Displacement to alternate, supplementary, or subsequent defensive positions.
- Cross-leveling or resupply of Class V items.
- Evacuation of casualties.

NOTE: When conducting his rehearsal, the platoon leader should coordinate the platoon rehearsal with the company to ensure other units' rehearsals are not planned for the same time and location. Coordination will lead to more efficient use of planning and preparation time for all company units. It will also eliminate the danger of misidentification of friendly forces in the rehearsal area.

SECTION V — OCCUPATION AND PREPARATION OF DEFENSIVE POSITIONS

8-77. Occupation and preparation of defensive positions is conducted concurrently with the TLP and engagement area development. The process is not sequential. The potential problem associated with this process is the lack of adequate preparation time if the platoon has several other defensive positions (alternate, supplementary, and subsequent) and engagement areas to develop.

OCCUPATION OF THE DEFENSE

8-78. The platoon occupies defensive positions IAW the platoon leader's plan and the results of the reconnaissance.

8-79. To ensure an effective and efficient occupation, rifle squads move to the locations marked previously by the reconnaissance element. These positions may also be on the operational graphics. Once in position, each squad leader checks his location on the map to ensure he is complying with the platoon leader's graphics. As the platoon occupies its positions, the platoon leader ensures that each squad locates IAW his plan. If the platoon leader notes discrepancies between actual positioning of the squads and his plan, he corrects it immediately.

8-80. Once each rifle squad has occupied its position, the platoon leader must walk the positions to ensure that weapons orientation, positioning of the rifle squads, and understanding of the plan are IAW the pre-established plan. The platoon leader should not rely on updates from his subordinates. He should always walk his defensive perimeter. For command and control purposes, each squad leader must know the location of the platoon leader and the platoon sergeant.

8-81. Night vision equipment enhances the occupation process under limited visibility conditions. For instance, the platoon leader can mark his position with an infrared light source and squad leaders can move to premarked positions with infrared light sources showing them where to locate. Additionally, the squad leaders can use AN/PAQ-4B/Cs or AN/PEQ-2As to point out sectors of fire and TRPs to their Soldiers, using infrared light sources to keep the occupation clandestine.

8-82. The platoon may conduct a hasty occupation in the defense during a counterattack or after disengagement and movement to alternate, supplementary, or subsequent defensive positions.

8-83. The platoon leader issues a FRAGO covering the following minimum information:

- Changes in the enemy or friendly situation.
- The platoon task and purpose (what the platoon must accomplish and why).
- The task and purpose for each subordinate element.
- The scheme of fires.
- Coordinating instructions.

8-84. At a minimum, the following actions must be taken:

- The platoon approaches the defensive positions from the rear or flank.
- The platoon establishes direct fire control measures or, if these are preplanned, reviews the plan.
- The platoon leader reports, "Occupied" to the company commander.

8-85. The platoon conducts deliberate occupation of defensive positions when time is available, when enemy contact is not expected, and when friendly elements are positioned forward in the sector to provide security for forces in the main battle area. Actually establishing defensive positions is accomplished concurrently with the development of the engagement area. The platoon leader directs the initial reconnaissance from the engagement area and then tentatively emplaces crew-served weapon systems.

8-86. Once the defensive positions are established, subordinate leaders can begin to develop their sector sketches and fire plans based on the basic fire plan developed during the leader's reconnaissance. Fighting positions are improved while the direct fire plan is finalized and proofed. The platoon leader, with guidance from the company commander, designates the level of preparation for each defensive position based on the time available and other tactical considerations for the mission. The three levels of defensive position preparation (occupy, prepare, and reconnoiter) are listed here in descending order of thoroughness and time required.

OCCUPY

8-87. Complete the preparation of the position from where the platoon will initially defend. The position is fully reconnoitered, prepared, and occupied prior to the "defend not later than (NLT)" time specified in the company order. The platoon must rehearse the occupation, and the platoon leader must establish a trigger for occupation of the position.

PREPARE

8-88. The position and the corresponding engagement area will be fully reconnoitered. Squad positions in the defensive positions and direct fire control measures in the engagement area should be marked. Survivability positions may be dug, ammunition caches pre-positioned, and protective obstacles emplaced.

RECONNOITER

8-89. Both the engagement area and defensive positions will be fully reconnoitered. Tentative weapon positions should be planned in the defensive positions, and direct fire control measures should be established in the engagement area.

8-90. In addition to establishing the platoon's primary defensive positions, the platoon leader and subordinate leaders normally plan for preparation and occupation of alternate, supplementary, and subsequent defensive positions. This is done IAW the company order. See Section II for characteristics of alternate, supplementary, and subsequent defensive positions.

PRIORITY OF WORK

8-91. Leaders must ensure that Soldiers prepare for the defense quickly and efficiently. Work must be done in order of priority to accomplish the most in the least amount of time while maintaining security and the ability to respond to enemy action. Below are basic considerations for priorities of work.

- Emplace local security (all leaders).
- Position and assign sectors of fire for each squad (platoon leader).
- Position and assign sectors of fire for the CCMS and medium machine gun teams (platoon leader).
- Position and assign sectors of fire for M249 MG, grenadiers, and riflemen (squad leaders).
- Establish command post and wire communications.
- Designate FPLs and FPFs.
- Clear fields of fire and prepare range cards.

- Prepare sector sketches (leaders).
- Dig fighting positions (stage 1 [see Section VII]).
- Establish communication and coordination with the company and adjacent units.
- Coordinate with adjacent units. Review sector sketches.
- Emplace antitank and Claymore mines, then wire and other obstacles.
- Mark or improve marking for TRPs and other fire control measures.
- Improve primary fighting positions and add overhead cover (stage 2).
- Prepare supplementary and then alternate positions (same procedure as the primary position).
- Establish sleep and rest plans.
- Distribute and stockpile ammunition, food, and water.
- Dig trenches to connect positions.
- Continue to improve positions—construct revetments, replace camouflage, and add to overhead cover.

8-92. Unit priorities of work are normally found in SOPs. However, the commander will dictate the priorities of work for the company based on the factors of METT-TC. Several actions may be accomplished at the same time. Leaders must constantly supervise the preparation of fighting positions, both for tactical usefulness and proper construction.

SECURITY IN THE DEFENSE

8-93. Security in the defense includes all active and passive measures taken to avoid detection by the enemy, deceive the enemy, and deny enemy reconnaissance elements accurate information on friendly positions. The two primary tools available to the platoon leader are observation posts and patrols. In planning for the security in the defense, the platoon leader considers the terrain in terms of OAKOC. He uses his map to identify terrain that will protect the platoon from enemy observation and fires while providing observation and fires into the engagement area. Additionally, he uses intelligence updates to increase his situational understanding, reducing the possibility of the enemy striking at a time or in a place for which the platoon is unprepared.

OBSERVATION POSTS

8-94. An observation post gives the platoon its first echelon of security in the defense. The observation post provides early warning of impending enemy contact by reporting direction, distance, and size. It detects the enemy early and sends accurate reports to the platoon. The platoon leader establishes observation posts along the most likely enemy avenues of approach into the position or into the area of operations. Leaders ensure that observation posts have communication with the platoon.

8-95. Early detection reduces the risk of the enemy overrunning the observation post. Observation posts may also be equipped with a Javelin CLU to increase the ability to detect the enemy. They may receive infrared trip flares, infrared parachute flares, infrared M203 rounds, and even infrared mortar round support to illuminate the enemy. The platoon leader weighs the advantages and disadvantages of using infrared illumination when the enemy is known to have night vision devices that detect infrared light. Although infrared and thermal equipment within the platoon enables the platoon to see the observation post at a greater distance, the observation post should not be positioned outside the range of the platoon's small-arms weapons.

8-96. To further reduce the risk of fratricide, observation posts use GPS, if available, to navigate to the exit and entry point in the platoon's position. The platoon leader submits an observation post location to the company commander to ensure a no-fire area (NFA) is established around each observation post position. The commander sends his operational overlay with observation post positions to the battalion and adjacent units. He receives the same type overlay from adjacent units to assist in better command and control and fratricide avoidance. The platoon leader confirms that the company fire support element (FSE) has forwarded these locations to the battalion FSO and has received the appropriate NFAs on the fire support graphics.

Patrols

8-97. Platoons actively patrol in the defense. Patrols enhance the platoon's ability to fill gaps in security between observation posts (see Chapter 9). The platoon leader forwards his tentative patrol route to the commander to ensure they do not conflict with other elements within the company. The commander forwards the entire company's patrol routes to the battalion. This allows the battalion S3 and S2 to ensure all routes are coordinated for fratricide prevention, and that the company and platoons are conforming to the battalion intelligence, surveillance, and reconnaissance (ISR) plan. The patrol leader may use a GPS to enhance his basic land navigational skills as he tracks his patrol's location on a map, compass, and pace count or odometer reading.

ESTABLISHMENT OF DEFENSIVE POSITIONS

8-98. Platoons establish defensive positions IAW the platoon leader and commander's plan. They mark engagement areas using marking techniques prescribed by unit SOP. The platoon physically marks obstacles, TRPs, targets, and trigger lines in the engagement area. During limited visibility, the platoon can use infrared light sources to mark TRPs for the rifle squads. When possible, platoons should mark TRPs with both a thermal and an infrared source so the rifle squads can use the TRP.

Range Card

8-99. A range card is a sketch of a sector that a direct fire weapons system is assigned to cover. Range cards aid in planning and controlling fires. They also assist crews in acquiring targets during limited visibility, and orient replacement personnel, platoons, or squads that are moving into position. During good visibility, the gunner should have no problems maintaining orientation in his sector. During poor visibility, he may not be able to detect lateral limits. If the gunner becomes disoriented and cannot find or locate reference points or sector limit markers, he can use the range card to locate the limits. The gunner should make the range card so he becomes more familiar with the terrain in his sector. He should continually assess the sector and, if necessary, update his range card.

Sector Sketches

8-100. Detailed sketches aid in the planning, distribution, and control of the platoon fires. Gunners prepare the range cards. Squad leaders prepare squad sector sketches, section leaders prepare section sketches, and the platoon leader prepares the platoon sketch.

WEAPONS PLACEMENT

8-101. To position weapons effectively, leaders must know the characteristics, capabilities, and limitations of the weapons; the effects of terrain; and the tactics used by the enemy. Additionally, the platoon leader must consider whether his primary threat will be vehicles or Infantry. His plan should address both mounted and dismounted threats. Also, the platoon leader may have an antitank section attached.

Close combat missile systems Employment

8-102. The primary role of Close Combat Missile Systems (CCMS) is to destroy enemy armored vehicles. When there is no armored vehicle enemy, CCMS can be employed in a secondary role of providing fire support against point targets such as crew-served weapons positions. CCMS optics (such as the Javelin's command launch unit [CLU]) can be used alone or as an aided vision device for reconnaissance, security operations, and surveillance. Reduced or limited visibility will not degrade the effectiveness of the CCMS. This fact allows the antiarmor specialist to continue to cover his sector without having to reposition closer to the avenue of approach. The platoon leader's assessment of the factors of METT-TC will determine the employment of CCMS. (For a detailed discussion on the employment of the Javelin, refer to Appendix B.) Based on the situation, the platoon leader may employ all or some of the CCMS. He may use centralized control or decentralized control.

Centralized Control

8-103. The platoon leader controls the fires of his CCMS gunners by both physically locating the weapons in his vicinity and personally directing their fires, or by grouping them together under the control of the platoon sergeant or weapons squad leader.

Decentralized Control

8-104. CCMS gunners operate with and are controlled by their weapons squad leader. A rifle squad leader may need to employ one fire team with a CCMS. The platoon leader normally gives the command to fire.

Medium Machine Gun Employment

8-105. Medium machine guns are the platoon's primary crew-served weapons that are positioned first if the enemy is a dismounted force. (For a detailed discussion on the employment of the M240B and the M249, refer to Appendix A.) Once these guns are sited, the leader positions riflemen to protect them. The guns are positioned to place direct fire on locations where the platoon leader wants to concentrate combat power to destroy the enemy.

M203 Employment

8-106. The M203 grenade launcher is the squad leader's indirect fire weapon. The platoon leader positions the grenadier to cover dead space in the squad's sector, especially the dead space for the medium machine guns. The grenadier is also assigned a sector of fire overlapping the riflemen's sectors of fire. The high-explosive dual purpose (HEDP) round is effective against lightly armored vehicles.

Employment of Riflemen

8-107. The platoon and squad leaders assign positions and sectors of fire to each rifleman in the platoon. Normally, they position the riflemen to support and protect machine guns and antiarmor weapons. Riflemen are also positioned to cover obstacles, provide security, cover gaps between platoons and companies, or provide observation.

COORDINATION

8-108. Coordination is important in every operation. In the defense, coordination ensures that units provide mutual support and interlocking fires. In most circumstances, the platoon leader conducts face-to-face coordination to facilitate understanding and resolve issues effectively. The platoon leader should send and receive the following information prior to conducting face-to-face coordination:

- Location of leaders.
- Location of fighting positions.
- Location of observation posts and withdrawal routes.
- Location and types of obstacles, including Claymores.
- Location, activities, and passage plan for reconnaissance platoon and other units forward of the platoon's position.
- Location of all Soldiers and units operating in and around the platoon's area of operations.

SECTION VI — DEFENSIVE TECHNIQUES

8-109. The platoon will normally defend IAW command orders using one of these basic techniques:

- Defend an area.
- Defend a battle position.
- Defend a strongpoint.
- Defend a perimeter.
- Defend a reverse slope.

DEFEND AN AREA

8-110. Defending an area sector allows a unit to maintain flank contact and security while ensuring unity of effort in the scheme of maneuver. Areas afford depth in the platoon defense. They allow the platoon to achieve the platoon leader's desired end state while facilitating clearance of fires at the appropriate level of responsibility. The company commander normally orders a platoon to defend an area (Figure 8-8) when flexibility is desired, when retention of specific terrain features is not necessary, or when the unit cannot concentrate fires because of any of the following factors:

- Extended frontages.
- Intervening, or cross-compartmented, terrain features.
- Multiple avenues of approach.

8-111. The platoon is assigned an area defense mission to prevent a specific amount of enemy forces from penetrating the area of operations. To maintain the integrity of the area defense, the platoon must remain tied to adjacent units on the flanks. The platoon may be directed to conduct the defense in one of two ways.

8-112. He may specify a series of subsequent defensive positions within the area from where the platoon will defend to ensure that the fires of two platoons can be massed.

8-113. He may assign an area to the platoon. The platoon leader assumes responsibility for most tactical decisions and controlling maneuvers of his subordinate squads by assigning them a series of subsequent defensive positions. This is done IAW guidance from the company commander in the form of intent, specified tasks, and the concept of the operation. The company commander normally assigns an area to a platoon only when it is fighting in isolation.

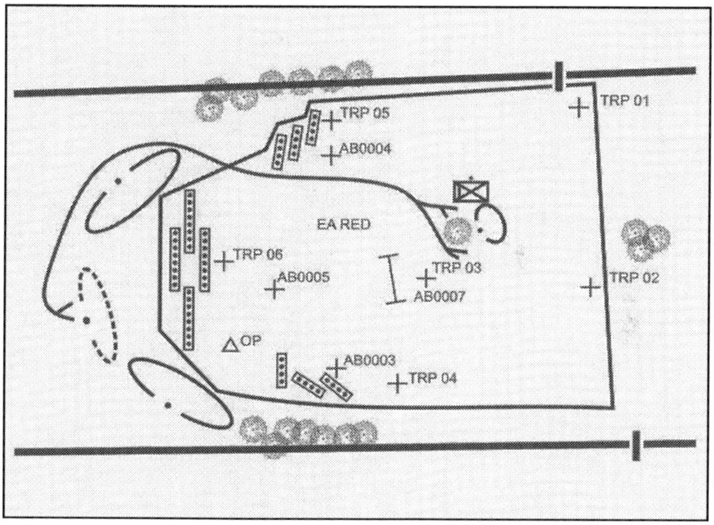

Figure 8-8. Concept of the operation for defending an area.

DEFEND A BATTLE POSITION

8-114. The company commander assigns the defensive technique of defending a battle position to his platoons when he wants to mass the fires of two or more platoons in a company engagement area, or to position a platoon to execute a counterattack. A unit defends from a battle position to—

- Destroy an enemy force in the engagement area.
- Block an enemy avenue of approach.
- Control key or decisive terrain.
- Fix the enemy force to allow another friendly unit to maneuver.

8-115. The company commander designates engagement areas to allow each platoon to concentrate its fires or to place it in an advantageous position for the counterattack. Battle positions are developed in such a manner to provide the platoon the ability to place direct fire throughout the engagement area. The size of the platoon battle position can vary, but it should provide enough depth and maneuver space for subordinate squads to maneuver into alternate or supplementary positions and to counterattack. The battle position is a general position on the ground. The platoon leader places his squads on the most favorable terrain in the battle position based on the higher unit mission and commander's intent. The platoon then fights to retain the position unless ordered by the company commander to counterattack or displace. The following are basic methods of employing a platoon in a battle position:

- Same battle position, same avenue of approach.
- Same battle position, multiple avenues of approach.
- Different battle positions, same avenue of approach.
- Different battle positions, multiple avenues of approach.

SAME BATTLE POSITION, SAME AVENUE OF APPROACH

8-116. Rifle squads are on the same battle position covering the same avenue of approach (Figure 8-9). The platoon can defend against mounted and dismounted attacks and move rapidly to another position.

8-117. All squads are in the same battle position when the terrain provides good observation, fields of fire, and cover and concealment.

8-118. Employing all the squads of the platoon on the same battle position covering the same avenue of approach is the most conservative use of the platoon. Its primary advantages are that it facilitates command and control functions because of the proximity of squad elements on the same approach and it provides increased security.

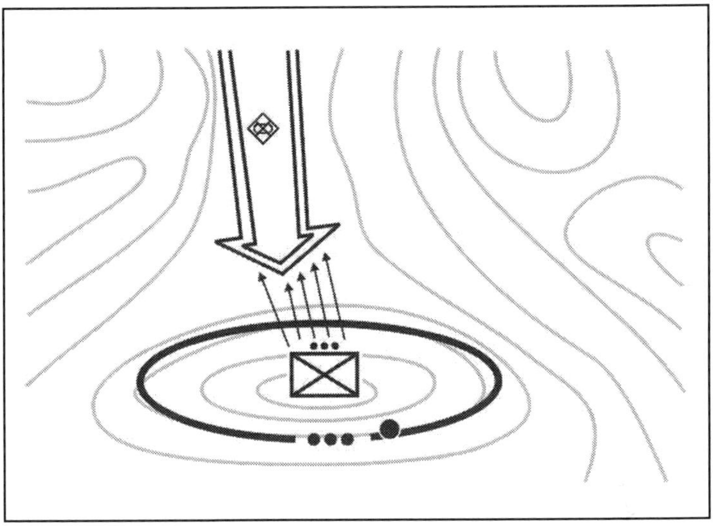

Figure 8-9. Same battle position, same avenue of approach.

SAME BATTLE POSITION, MULTIPLE AVENUES OF APPROACH

8-119. Rifle squads occupy the same battle position but cover multiple enemy avenues of approach (Figure 8-10).

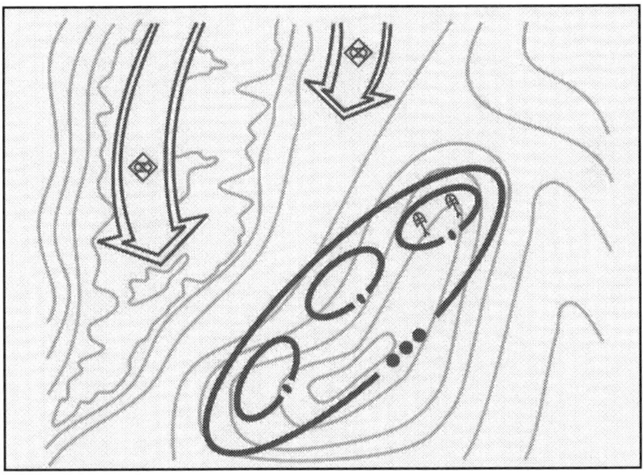

Figure 8-10. Same battle position, multiple avenues of approach.

DIFFERENT BATTLE POSITIONS, SAME AVENUE OF APPROACH

8-120. Rifle squads are on different battle positions covering the same avenue of approach (Figure 8-11). If positioned on separate battle positions, rifle squads must fight in relation to each other when covering the same avenues of approach. A weapons squad can provide supporting fires for the rifle squads from their primary, alternate, or supplementary positions. All squads are positioned to engage enemy forces on the same avenue of approach, but at different ranges.

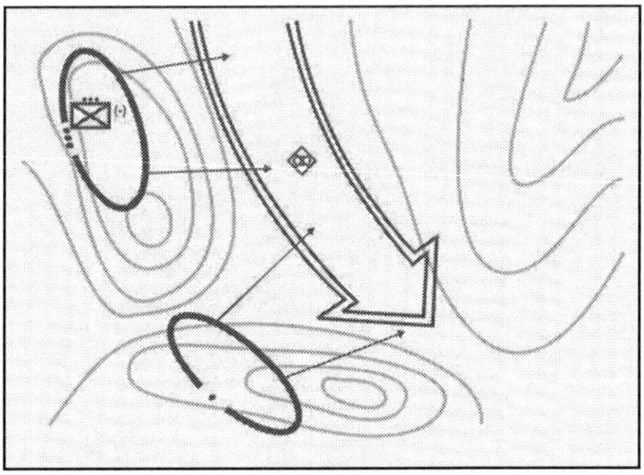

Figure 8-11. Different battle positions, same avenue of approach.

DIFFERENT BATTLE POSITIONS, MULTIPLE AVENUES OF APPROACH

8-121. Squads may be employed on different battle positions and multiple avenues of approach (Figure 8-12) to ensure that the squad battle positions cannot be fixed, isolated, or defeated by the enemy.

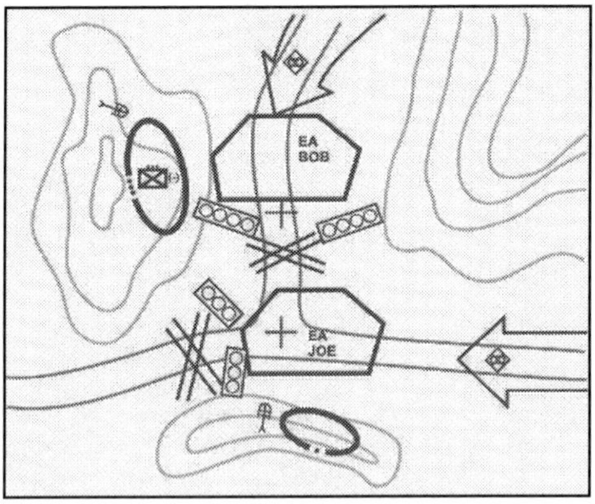

Figure 8-12. Different battle positions, multiple avenues of approach.

DEFEND A STRONGPOINT

8-122. Defending a strongpoint (Figure 8-13) is not a common mission for an Infantry platoon. A strongpoint defense requires extensive engineer support (expertise, materials, and equipment), and takes a long time to complete. When the platoon is directed to defend a strongpoint, it must retain the position until ordered to withdraw. The success of the strong-point defense depends on how well the position is tied into the existing terrain. This defense is most effective when it is employed in terrain that provides cover and concealment to both the strongpoint and its supporting obstacles. Mountainous, forested, or urban terrain can be adapted easily to a strongpoint defense. Strongpoints placed in more open terrain require the use of reverse slopes or of extensive camouflage and deception efforts. This defensive mission may require the platoon to—

- Hold key or decisive terrain critical to the company or battalion scheme of maneuver.
- Provide a pivot to maneuver friendly forces.
- Block an avenue of approach.
- Canalize the enemy into one or more engagement areas.

CHARACTERISTICS OF THE STRONGPOINT DEFENSE

8-123. The prime characteristic of an effective strongpoint is that it cannot be easily overrun or bypassed. It must be positioned and constructed so the enemy knows he can reduce it only at the risk of heavy casualties and significant loss of materiel. He must be forced to employ massive artillery concentrations and dismounted Infantry assaults in his attack, so the strongpoint must be tied in with existing obstacles and positioned to afford 360-degree security in observation and fighting positions.

TECHNIQUES AND CONSIDERATIONS

8-124. A variety of techniques and considerations are involved in establishing and executing the strongpoint defense, including considerations for displacement and withdrawal from the strongpoint.

8-125. The platoon leader begins by determining the projected size of the strongpoint. He does this through assessing the number of weapons systems and individual Soldiers available to conduct the assigned mission, and by assessing the terrain on which the platoon will fight. He must remember that although a strongpoint is usually tied into a company defense and flanked by other defensive positions, it must afford 360-degree observation and firing capability.

8-126. The platoon leader must ensure that the layout and organization of the strongpoint maximizes the capabilities of the platoon's personnel strength and weapons systems without sacrificing the security of the position. Platoon options range from positioning CCMS outside the strongpoint (with the rifle squads occupying fighting positions inside it), to placing all assets within the position. From the standpoint of planning and terrain management, placing everything in the strongpoint is the most difficult option and potentially the most dangerous because of the danger of enemy encirclement.

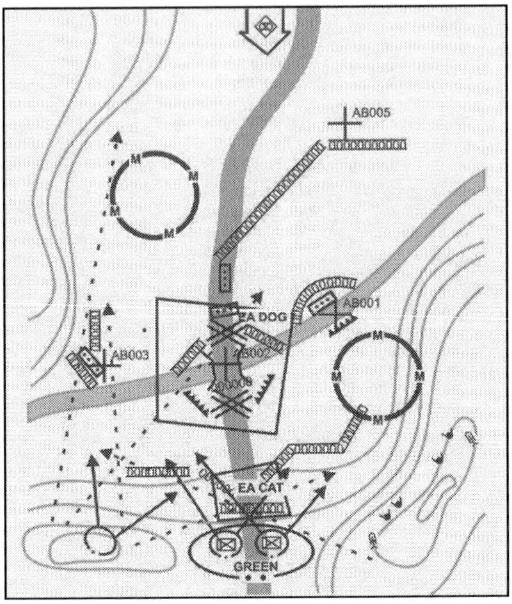

Figure 8-13. Defending a strongpoint.

8-127. In laying out the strongpoint, the platoon leader designates weapon positions that support the company defensive plan. Once these primary positions have been identified, he continues around the strongpoint, siting weapons on other possible enemy avenues of approach and engagement areas until he has the ability to orient effectively in any direction. The fighting positions facing the company engagement area may be along one line of defense or staggered in depth along multiple lines of defense (if the terrain supports positions in depth).

8-128. The platoon's reserve may be comprised of a fire team, squad, or combination of the two. The platoon leader must know how to influence the strongpoint battle by employing his reserve. He has several employment options including reinforcing a portion of the defensive line or counterattacking along a portion of the perimeter against an identified enemy main effort.

8-129. The platoon leader should identify routes or axes that will allow the reserve to move to any area of the strongpoint. He should then designate positions the reserve can occupy once they arrive. These routes and positions should afford sufficient cover to allow the reserve to reach its destination without enemy interdiction. The platoon leader should give special consideration to developing a direct fire plan for each contingency involving the reserve. The key area of focus may be a plan for isolating an enemy penetration of the perimeter. Rehearsals cover actions the platoon takes if it has to fall back to a second defensive perimeter, including direct fire control measures necessary to accomplish the maneuver. FPF may be employed to assist in the displacement.

8-130. Engineers support strongpoint defense by reinforcing the existing obstacles. Priorities of work will vary depending on the factors of METT-TC, especially the enemy situation and time available. For example, the first 12 hours of the strongpoint construction effort may be critical for emplacing countermobility obstacles and survivability positions, and command and control bunkers. If the focus of engineer support is to make the terrain approaching the strongpoint impassable, the battalion engineer effort must be adjusted accordingly.

8-131. The battalion obstacle plan provides the foundation for the company strongpoint obstacle plan. The commander or platoon leader determines how he can integrate protective obstacles (designed to defeat dismounted enemy Infantry assaults) into the overall countermobility plan. If adequate time and resources are available, he should plan to reinforce existing obstacles using field-expedient demolitions.

8-132. Once the enemy has identified the strongpoint, he will mass all the fires he can spare against the position. To safeguard his rifle squads, the platoon leader must arrange for construction of overhead cover for individual fighting positions. If the strongpoint is in a more open position (such as on a reverse slope), he may also plan for interconnecting trenchlines. This will allow Soldiers to move between positions without exposure to direct and indirect fires. If time permits, these crawl trenches can be improved to fighting trenches or standard trenches.

DEFEND A PERIMETER

8-133. A perimeter defense allows the defending force to orient in all directions. In terms of weapons emplacement, direct and indirect fire integration, and reserve employment, a platoon leader conducting a perimeter defense should consider the same factors as for a strongpoint operation.

8-134. The perimeter defense allows only limited maneuver and limited depth. Therefore, the platoon may be called on to execute a perimeter defense under the following conditions:

- Holding critical terrain in areas where the defense is not tied in with adjacent units.
- Defending in place when it has been bypassed and isolated by the enemy.
- Conducting occupation of an independent assembly area or reserve position.
- Preparing a strongpoint.
- Concentrating fires in two or more adjacent avenues of approach.
- Defending fire support or engineer assets.
- Occupying a patrol base.

8-135. The major advantage of the perimeter defense (Figure 8-14) is the platoon's ability to defend against an enemy avenue of approach. A perimeter defense differs from other defenses in that—

- The trace of the platoon is circular or triangular rather than linear.
- Unoccupied areas between squads are smaller.
- Flanks of squads are bent back to conform to the plan.
- The bulk of combat power is on the perimeter.
- The reserve is centrally located.

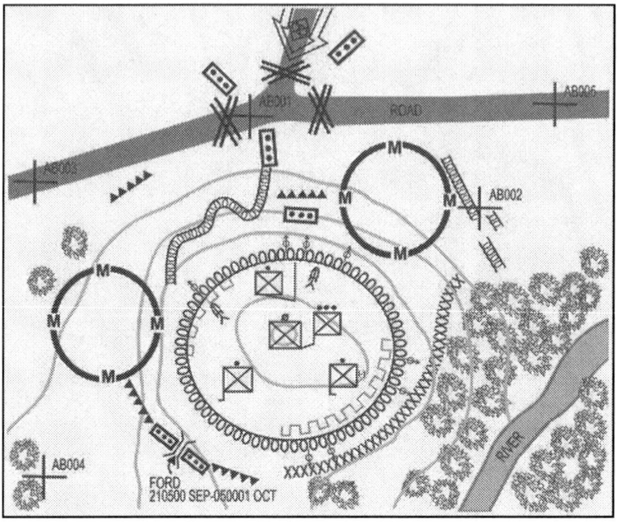

Figure 8-14. Perimeter defense with rifle team in reserve.

NOTE: A variant of the perimeter defense is the use of the shaped defense, which allows two of the platoon's squads to orient at any particular time on any of three engagement areas.

DEFEND A REVERSE SLOPE

8-136. The platoon leader's analysis of the factors of METT-TC often leads him to employ his forces on the reverse slope (Figure 8-15). If the rifle squads are on a mounted avenue of approach, they must be concealed from enemy direct fire systems. This means rifle squads should be protected from enemy tanks and observed artillery fire.

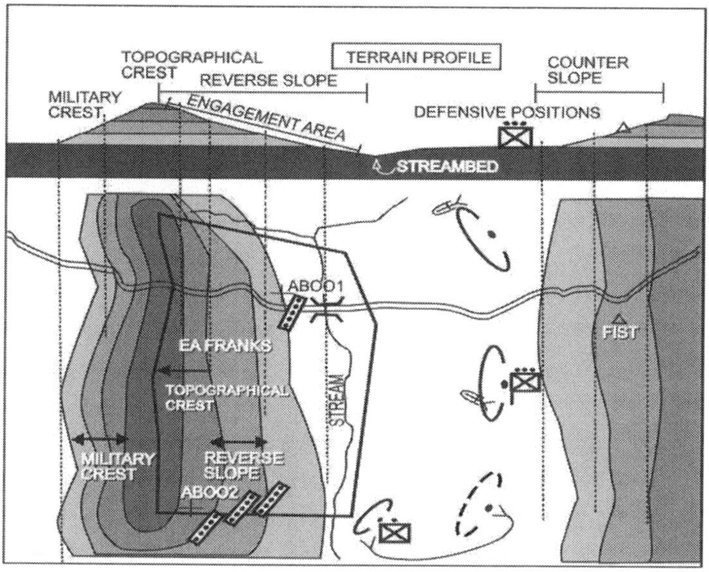

Figure 8-15. Reverse-slope defense options.

8-137. The majority of a rifle squad's weapons are not effective beyond 600 meters. To reduce or prevent destruction from enemy direct and indirect fires beyond that range, a reverse-slope defense should be considered. Using this defense conflicts to some extent with the need for maximum observation forward to adjust fire on the enemy, and the need for long-range fields of fire for CCMS. In some cases it may be necessary for these weapons systems to be deployed forward while the rifle squads remain on the reverse slope. CCMS gunners withdraw from their forward positions as the battle closes. Their new positions should be selected to take advantage of their long-range fires, and to get enfilade shots from the depth and flanks of the reverse slope.

8-138. The nature of the enemy may change at night, and the rifle squads may occupy the forward slope or crest to deny it to the enemy. In these circumstances, it is feasible for a rifle squad to have an alternate night position forward. The area forward of the topographical crest must be controlled by friendly forces through aggressive patrolling and both active and passive reconnaissance measures. The platoon should use all of its night vision devices to deny the enemy undetected entry into the platoon's defensive area. CCMS are key parts of the platoon's surveillance plan and should be positioned to take advantage of their thermal sights. The enemy must not be allowed to take advantage of reduced visibility to advance to a position of advantage without being taken under fire.

8-139. The company commander normally makes the decision to position platoons on a reverse slope. He does so when—

- He wishes to surprise or deceive the enemy about the location of his defensive position.
- Forward slope positions might be made weak by direct enemy fire.
- Occupation of the forward slope is not essential to achieve depth and mutual support.
- Fields of fire on the reverse slope are better or at least sufficient to accomplish the mission.
- Forward slope positions are likely to be the target of concentrated enemy artillery fires.

8-140. The following are advantages of a reverse-slope defense:

- Enemy observation of the position, including the use of surveillance devices and radar, is masked.
- Enemy cannot engage the position with direct fire without coming within range of the defender's weapons.
- Enemy indirect fire will be less effective because of the lack of observation.
- Enemy may be deceived about the strength and location of positions.
- Defenders have more freedom of movement out of sight of the enemy.

8-141. Disadvantages of a reverse-slope defense include the following:

- Observation to the front is limited.
- Fields of fire to the front are reduced.
- Enemy can begin his assault from a closer range.

8-142. Obstacles are necessary in a reverse-slope defense. Because the enemy will be engaged at close range, obstacles should prevent the enemy from closing too quickly and overrunning the positions. Obstacles on the reverse slope can halt, disrupt, and expose enemy vehicles to flank antitank fires. Obstacles should also block the enemy to facilitate the platoon's disengagement.

SECTION VII — FIGHTING AND SURVIVABILITY POSITIONS

8-143. The defensive plan normally requires building fighting positions. Fighting positions protect Soldiers by providing cover from direct and indirect fires and by providing concealment through positioning and proper camouflage. Because the battlefield conditions confronting Infantrymen are never standard, there is no single standard fighting position design that fits all tactical situations.

8-144. Soldiers prepare fighting positions even when there is little or no time before contact with the enemy is expected (Figure 8-16). They locate them behind whatever cover is available and where they can engage the enemy. The position should give frontal protection from direct fire while allowing fire to the front and oblique. Occupying a position quickly does not mean there is no digging. Soldiers can dig initial positions in only a few minutes. A fighting position just 18 inches deep will provide a significant amount of protection from direct fire and even fragmentation. All positions are built by stages. The initial fighting position construction can be improved over time to a more elaborate position.

Figure 8-16. Initial fighting position.

PRINCIPLES

8-145. Leaders follow three basic principles to effectively and efficiently prepare fighting positions: site positions to best engage the enemy, prepare positions by stages, and inspect all positions. The leader's responsibilities include the following:

- Protect troops.
- Plan and select fighting position sites.
- Supervise construction.
- Inspect periodically.
- Depending on assets, request technical advice from engineers as required.
- Improve and maintain unit survivability continuously.
- Determine if there is a need to build the overhead cover up or down.

SITE POSITIONS TO BEST ENGAGE THE ENEMY

8-146. The most important aspect of a fighting position is that it must be tactically well positioned. Leaders must be able to look at the terrain and quickly identify the best location for fighting positions. Good positions allow—

- Soldiers to engage the intended enemy element within their assigned sectors of fire.
- Soldiers to fire out to the maximum effective range of their weapons with maximum grazing fire and minimal dead space.
- Grenadiers to be placed in positions to cover dead space.

8-147. Leaders must ensure fighting positions provide mutually supporting, interlocking fires. This allows them to cover the platoon's sector from multiple positions. When possible, they site positions behind natural cover and in easily camouflaged locations. The enemy must not be able to identify the position until it is too late and he has been effectively engaged.

PREPARE POSITIONS BY STAGES

8-148. Leaders must ensure their Soldiers understand when and how to prepare fighting positions based on the situation. Soldiers prepare fighting positions every time the platoon makes an extended halt. Half of the platoon digs in while the other half maintains security. Soldiers prepare positions in stages and a leader inspects the position at each stage before the Soldiers move to the next stage. When expecting an immediate enemy attack, Infantrymen dig stage 1 fighting positions. As time becomes available, these defensive positions are continually improved, enlarged, and strengthened.

Stage 1

8-149. The platoon leader checks fields of fire from the prone position. For a stage 1 position (Figure 8-17) the Soldiers—

- Emplace sector stakes.
- Stake the primary sector.
- Position grazing fire log or sandbag between the sector stakes.
- Place the aiming stake(s) to allow limited visibility engagement of a specific target.
- Trace the outline of the position on the ground.
- Clear the fields of fire for both the primary and secondary sectors of fire.
- Ensure the leader inspects the position before they move to stage 2.

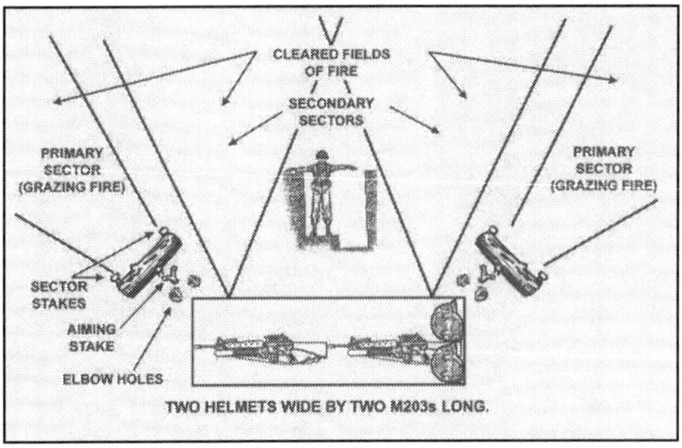

Figure 8-17. Stage 1, preparation of a fighting position.

Stage 2

8-150. Soldiers prepare retaining walls (Figure 8-18) for the parapets. They ensure that—

- There is a minimum distance (equal to the width of one helmet) from the edge of the hole to the beginning of the front, flank, and rear cover.
- The cover to the front consists of sandbags (or logs), two to three high, and for a two-Soldier position, about the length of two M302 rifles (about 7 feet).
- The cover to the flanks is the same height, but only one M203 rifle length (about 3.5 feet).
- The cover to the rear is one sandbag high and one M203 long (about 3.5 feet).
- If logs are used, they must be held firmly in place with strong stakes.
- The leader inspects the retaining wall before they begin stage 3.

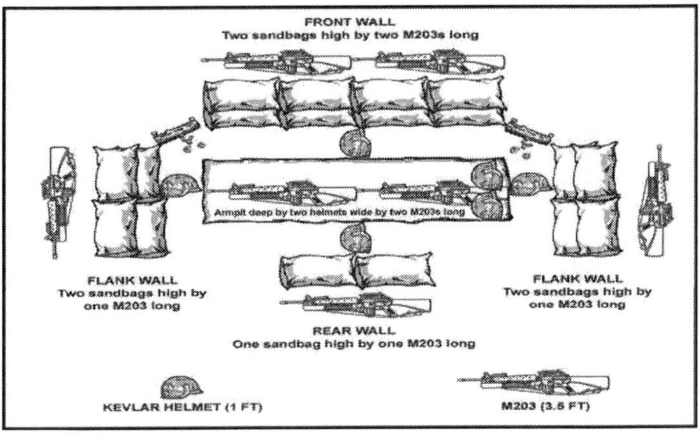

Figure 8-18. Stage 2, preparation of a fighting position.

Stage 3

8-151. Soldiers dig the position and throw dirt forward of the parapet retaining walls and pack it down hard (Figure 8-19). They—

- Dig the position armpit (of the tallest Soldier) deep.
- Fill the parapets in order of front, flanks, and rear.
- Camouflage the parapets and the entire position.
- Dig grenade sumps and slope the floor toward them.
- Dig storage areas for two rucksacks into the rear wall if needed.
- Ensure the leader inspects the work.

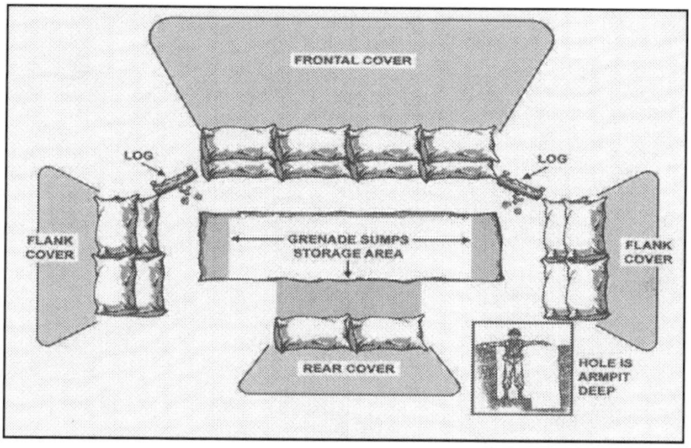

Figure 8-19. Stage 3, preparation of a fighting position.

Stage 4

8-152. In stage 4, Soldiers prepare the overhead cover (Figure 8-20). At times, the terrain will accommodate the construction of a position with overhead cover that protects Soldiers from indirect fire fragmentation while allowing them to return fire. Sometimes, especially on open terrain, this is not possible, and the entire position must be built below ground level. Although this type of position offers excellent protection and concealment to Soldiers, it limits their ability to return fire from within a protected area. To prepare overhead cover, Soldiers—

- Always provide solid lateral support. They build the support with 4- to 6-inch logs on top of each other running the full length of the front and rear cover.
- Place five or six logs 4 to 6 inches in diameter and two M203s long (about 7 feet) over the center of the position, resting them on the overhead cover support, not on the sandbags.
- Place waterproofing (plastic bags, ponchos) on top of these logs.
- Put a minimum of 18 inches of packed dirt or sandbags on top of the logs.
- Camouflage the overhead cover and the bottom of the position.
- Ensure the leader inspects the position.

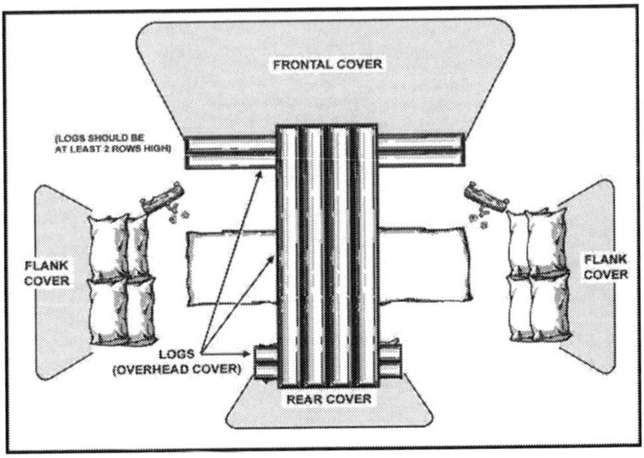

Figure 8-20. Stage 4, preparation of a fighting position.

INSPECT ALL POSITIONS

8-153. Leaders must ensure their Soldiers build fighting positions that are both effective and safe. An improperly sited position cannot be used and an improperly constructed position is a danger to its occupants. Leaders should inspect the progress of the fighting position at each stage in its preparation.

FIGHTING POSITION MATERIALS

8-154. Sometimes Soldiers must construct fighting positions using only the basic tools and materials they can carry or find in the local area such as entrenching tools, sandbags, and locally cut timber. At other times, significant amounts of Class IV construction materials and heavier digging tools may be available (Table 8-1).

Table 8-1. Examples of field-expedient fighting position materials.

Wall Revetment	Stand Alone Positions
• Sheet metal • Corrugated sheet metal • Plastic sheeting • Plywood • Air mat panels • Air Force air load pallets	• Prefabricated concrete catch basins • Military vans • Shipping containers • Large diameter pipe/culvert • Steel water tanks • Vehicle hulks
Overhead Cover Stringers	**Wall Construction (Building Up)**
• Single pickets • Double pickets • Railroad rails • "I" beams • 2-inch diameter pipe • Timbers (2"x 4", 4"x 4", and larger) • Reinforced concrete beams • 55-gallon drums cut in half • Culverts cut in half • Pre-cast concrete panels 6-8 inches thick • Airfield panels	• 55-gallon drums filled with sand • Shipping boxes/packing material • Expended artillery shells filled with sand • Prefabricated concrete panels • Prefabricated concrete traffic barriers • Sand grid material
Aiming Stakes	**Limiting Stakes**
• 2-foot pickets • Wooden tent poles	• 2-foot pickets • Wooden tent poles • Filled sandbags

NOTE: Regardless of the position design, the type of construction materials, the tools available, or the terrain, all fighting positions must incorporate sound engineering construction principles. Unless it is constructed properly, a fighting position can easily collapse and crush or bury the Soldiers within. FM 5-103, *Survivability*, and FM 5-34, *Engineer Field Data*, provide excellent information on these principles. Additionally, GTA 05-08-001, *Survivability Positions, and* GTA 07-06-001, *Fighting Position Construction--Infantry Leader's Reference Card,* contain detailed information in easy-to-use formats.

TYPES OF FIGHTING POSITIONS

8-155. There are many different types of fighting positions. The number of occupants; types of weapons; tools, materials, and time available; and terrain dictate the type of position.

8-156. The do's and don'ts of fighting position construction are listed in Table 8-2.

Table 8-2. Do's and don'ts of fighting position construction.

DO...	DON'T...
• Construct to standard. • Ensure adequate material is available. • Dig down as much as possible. • Maintain, repair, and improve positions continuously. • Inspect and test position safety daily, after heavy rain, and after receiving direct and indirect fire. • Revet walls in unstable and sandy soil. • Interlock sandbags for double wall construction and corners. • Check stabilization of wall bases. • Fill sandbags about 75% full. • Use common sense. • Use soil to fill sandbags, fill in any cavities in overhead cover, or spread to blend with surroundings.	• Fail to supervise. • Use sandbags for structural support. • Put Soldiers in marginally safe positions. • Take short cuts. • Build above ground unless absolutely necessary. • Forget lateral bracing on stringers. • Forget to camouflage. • Drive vehicles within 6 feet of a fighting position.

8-157. Infantry fighting positions are normally are constructed to hold one, two, or three Soldiers. There are special designs adapted for use by machine gun (M240B) and antiarmor (Javelin) teams.

One-Soldier Fighting Position

8-158. Positions that contain a single Soldier are the least desirable, but they are useful in some situations. One-Soldier positions may be required to cover exceptionally wide frontages. They should never be positioned out of sight of adjacent positions. The one-Soldier fighting position (Figure 8-21) should allow the Soldier to fire to the front or to the oblique from behind frontal cover. Advantages and disadvantages to consider when choosing a one-Soldier fighting position include:

- The one-Soldier position allows choices in the use of cover.
- The hole only needs to be large enough for one Soldier and his gear.
- It does not have the security of a two-Soldier position.

Figure 8-21. One-soldier fighting position.

Two-Soldier Fighting Position

8-159. A two-Soldier fighting position (Figure 8-22) is normally more effective than a one-Soldier fighting position. It can be used to provide mutual support to adjacent positions on both flanks and to cover dead space immediately in front of the position. One or both ends of the hole may extend around the sides of the frontal cover. Modifying a position in this way allows both Soldiers to have better observation and greater fields of fire to the front. Also, during rest or eating periods, one Soldier can watch the entire sector while the other sleeps or eats. If they receive fire from their front, they can move back to gain the protection of the frontal cover. By moving about one meter, the Soldiers can continue to find and hit targets to the front during lulls in enemy fire. This type of position—

- Requires more digging.
- Is more difficult to camouflage.
- Provides a better target for enemy hand grenades.

Figure 8-22. Two-soldier fighting position.

Three-Soldier Fighting Position

8-160. A three-Soldier position has several advantages. A leader can be in each position, making command and control easier. It supports continuous security operations better than other positions. One Soldier can provide security; one can do priority work; and one can rest, eat, or perform maintenance. This allows the priority of work to be completed more quickly than in a one- or two-Soldier position. This position allows the platoon to maintain combat power and security without shifting personnel or leaving positions unmanned. It provides 360-degree observation and fire, and is more difficult for the enemy to destroy because he must kill or suppress three Soldiers.

8-161. When using three-Soldier positions, the leader must consider several things. Either the distance between positions must be increased, or the size of the squad's sector must be reduced. The choice depends mainly on visibility and fields of fire. Because the squad leader is in a fighting position that will most likely be engaged during the battle, he cannot exert personal control over the other two positions. The squad leader controls the battle by—

- Communicating his plans and intent to his squad, including control measures and fire plans.
- Using prearranged signals like flares, whistles, or tracers.
- Positioning key weapons in his fighting position.
- Placing his fighting position so it covers key or decisive terrain.
- Placing his fighting position where his team might be able to act as a reserve.

8-162. The three-Soldier emplacement is a T-position (Figure 8-23). This basic design can be changed by adding or deleting berms, changing the orientation of the T, or shifting the position of the third Soldier to form an L instead of a T. The layout of the position can be oriented to fire on expected enemy avenues of approach from any direction. Berms must not block observation or fire into assigned primary or alternate sectors. Care must be taken to properly support the overhead cover.

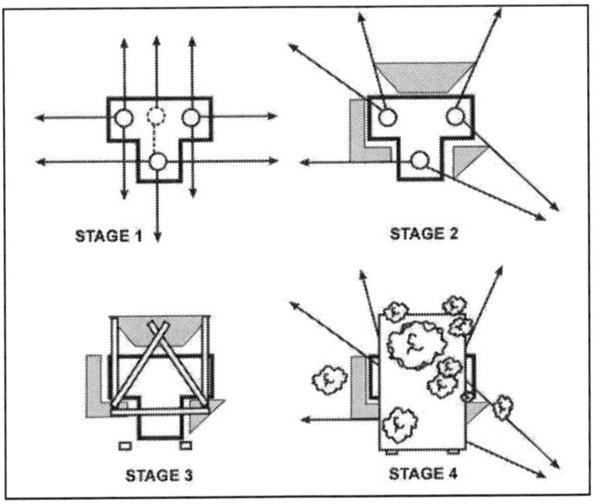

Figure 8-23. Three-soldier T-position.

Machine Gun Position

8-163. The primary sector of fire is usually to the oblique so a machine gun can fire across the platoon's front. The tripod is used on the side covering the primary sector of fire. The bipod legs are used on the side covering the secondary sector of fire. When changing from primary to secondary sectors, the gunner moves only the machine gun. Occasionally a sector of fire that allows firing directly to the front is assigned, but this can reduce the frontal cover for the crew when firing to the oblique (Figure 8-24). For a detailed discussion on the employment of the M240B, refer to Appendix A.

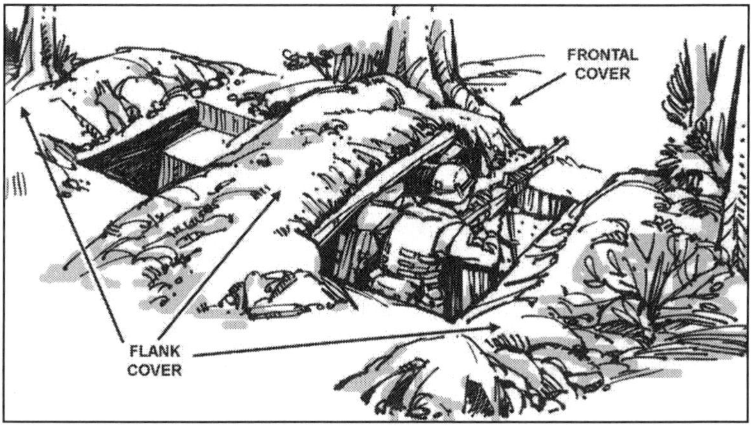

Figure 8-24. Machine gun position.

8-164. After the platoon leader positions the machine gun, he marks the position of the tripod legs and the limits of his sectors of fire. The crew then traces the outline of the hole and the frontal cover (if it must be improved).

8-165. The crew digs firing platforms first to lessen their exposure in case they must fire before completing the position. The platforms must not be so low that the gun cannot be traversed across its entire sector of fire, reducing the profile of the gunner when firing and reducing the frontal cover height.

8-166. After digging the firing platforms, the crew digs the hole. They first place the dirt where frontal cover is needed, digging the hole deep enough (usually armpit deep) to protect them while allowing the gunner to fire with comfort. When the frontal cover is high enough and thick enough, the crew uses the rest of the dirt to build flank and rear cover. Trench-shaped grenade sumps are dug at various points so either Soldier can kick a grenade into one if needed. Overhead cover for a machine gun position is constructed following the steps of stage 4, preparation of a fighting position (see paragraph 8-152f and Figure 8-20).

NOTE: In some positions, a machine gun might not have a secondary sector of fire. In this case, dig only half the position.

8-167. For a three-Soldier crew for a machine gun, the ammunition bearer digs a one-Soldier fighting position to the flank that is connected with the gun position by a crawl trench. From this position, the ammunition bearer can see and fire to the front and to the oblique. Usually the ammunition bearer is on the same side as the FPL or PDF. This allows him to see and fire his rifle into the machine gun's secondary sector and to see the gunner and assistant gunner.

Javelin Position

8-168. The Javelin can be employed from initial or completed positions (Figure 8-25). However, some changes are required. For a detailed discussion on the employment of the Javelin, refer to Appendix B.

Figure 8-25. Javelin position.

8-169. The gunner must keep the weapon at least 6 inches above the ground to allow room for the stabilizing fins to unfold. The hole is only waist deep to allow the gunner to move while tracking to acquire a target. Because the Javelin gunner must be above ground level, the frontal cover should be high enough to hide his head, and, if possible, the backblast of the Javelin. A hole is dug in front of the position for the bipod legs.

8-170. When the Javelin can be fired in one direction only, the position is adjusted to provide cover and concealment from all other directions, and the Javelin should be fired to the oblique. This protects the position from frontal fire and allows engagement of the target from the flank. Both ends of the launcher must extend out over the edges of the hole.

8-171. Overhead cover must be built on the flanks. Cover must be large enough for the gunner, the tracker, and the missiles. Overhead cover that allows fire from underneath can be built if the backblast area is clear. Overhead cover must be well camouflaged.

8-172. The Javelin is an important weapon and is easy to detect. Therefore, selection and preparation of alternate positions have high priority. When preparing an alternate position, the gunner should select and improve a covered route to it so he can move to the position under fire.

SLM Position

8-173. The AT4 can be fired from Infantry fighting positions. If the AT4 is to be fired from a two-Soldier position, the gunner must ensure the other Soldier is not in the backblast area. Assume the basic standing position, but instead of stepping forward, lean against the back wall of the fighting position. Ensure that the rear of the weapon extends beyond the rear of the fighting position.

NOTE: Leaders must ensure that light antiarmor weapons are positioned so the backblast misses other fighting positions.

TRENCHES

8-174. When there is time and help available, trenches should be dug to connect fighting positions so Soldiers can move by covered routes. The depth of a trench depends on the type of help and equipment available. Without engineer help, platoons dig crawl trenches (about 3 feet deep by 2 feet wide) (Figure 8-26). With engineer help, they dig standard trenches. The trench should zigzag so the enemy cannot fire down a long section. Platoons normally dig crawl trenches because engineer assets are usually limited. Platoons use crawl trenches to conceal their movement into and within positions. Spoil is placed on parapets, normally on each side of the trench. If the trench runs across a forward slope, all the spoil is placed on the enemy side to make the forward parapet higher. All spoil needs careful concealment from enemy direct observation.

Figure 8-26. Crawl trenches.

SECTION VIII — RETROGRADE

8-175. The retrograde is a type of defensive operation that involves organized movement away from the enemy. The enemy may force these operations, or a commander may execute them voluntarily. Retrograde operations are transitional and are not considered in isolation. There are three forms of retrograde: withdrawal; delay; and retirement. Platoons may participate in stay-behind missions as part of a withdrawal or delay.

WITHDRAWAL

8-176. A withdrawal occurs when an element disengages from enemy contact to reposition itself for another mission. A platoon usually conducts a withdrawal as part of a larger force. As part of a company, a platoon may withdraw with the main element (under pressure) or may be used as the detachment left in contact (DLIC) in a withdrawal not under pressure. This information applies whether or not the platoon is under pressure from the enemy. Regardless of employment, the platoon leader conducts his withdrawal IAW his higher commander's guidance. On receipt of the order to conduct a withdrawal, the platoon leader begins preparing his order based on his higher unit's FRAGO. He identifies possible key terrain and routes based on the higher unit's graphics and his map. He formulates and briefs his FRAGO to his squad leaders. When the withdrawal is executed, squad leaders ensure they are moving IAW the platoon leader's plan by monitoring position locations.

WITHDRAWAL NOT UNDER PRESSURE

8-177. In a withdrawal not under pressure, platoons may serve as or as part of the DLIC. A DLIC is used to deceive the enemy into thinking that the entire force is still in position (Figure 8-27). As the DLIC, the platoon—

- Repositions squads and weapons to cover the company's withdrawal.
- Repositions a squad in each of the other platoon positions to cover the most dangerous avenue of approach into the position.
- Continues the normal operating patterns of the company and simulates company radio traffic.
- Covers the company withdrawal with planned direct and indirect fires if the company is attacked during withdrawal.
- Withdraws by echelon once the company is at its next position.

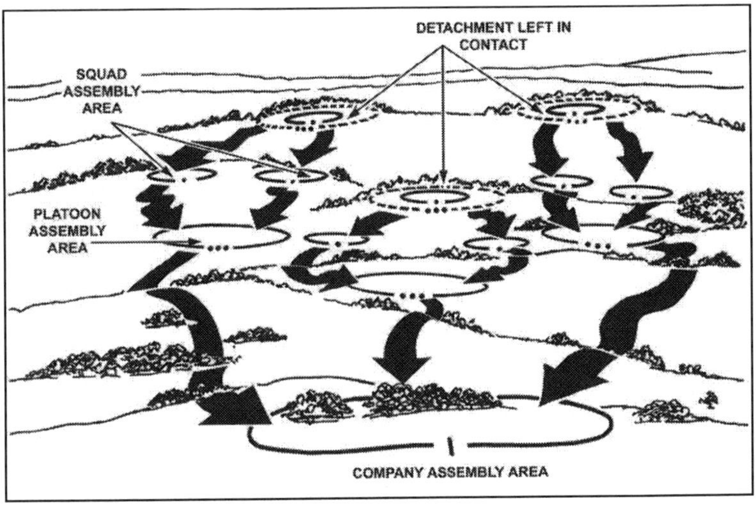

Figure 8-27. Withdrawal not under pressure.

WITHDRAWAL UNDER PRESSURE

8-178. If the platoon cannot prepare and position the security force, it conducts a fighting withdrawal. The platoon disengages from the enemy by maneuvering to the rear. Soldiers and squads not in contact are withdrawn first to provide suppressive fire and to allow Soldiers and squads in contact to withdraw.

DISENGAGEMENT

8-179. Based on orders from the battalion commander, the company commander determines how long to retain defensive positions. The company may be required to remain and fight for a certain amount of time, or it may be required to disengage and displace to subsequent positions. A platoon, as part of a company, may disengage to defend from another battle position, prepare for a counterattack, delay, withdraw, or prepare for another mission.

8-180. Fire and movement to the rear is the basic tactic for disengaging. All available fires are used to slow the enemy and allow platoons to move away. The company commander may move his platoons and mass fires to stop or slow the enemy advance before beginning the movement away from the enemy.

8-181. Using bounding overwatch, a base of fire is formed to cover platoons or squads moving away from the enemy. One platoon or squad acts as the base of fire, delaying the enemy with fire or retaining terrain blocking his advance, while other platoons or squads disengage.

8-182. Moving platoons or squads get to their next position and provide a base of fire to cover the rearward movement of forward platoons and squads.

8-183. Fire and movement is repeated until contact with the enemy is broken, the platoons pass through a different base-of-fire force, or the platoons are in position to resume their defense (Figure 8-28).

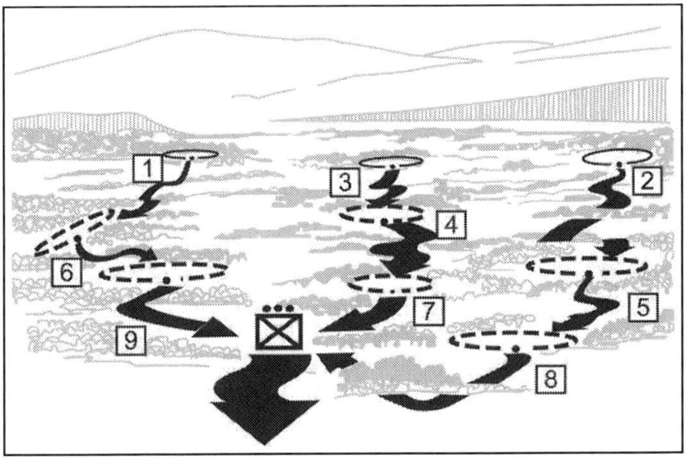

Figure 8-28. Bounding overwatch to the rear.

8-184. Tactics used by the platoon to disengage from the enemy differ according to the company commander's plan for disengagement, how the platoon is deployed, and other factors. The following actions apply in all cases:

- Maximum use is made of the terrain to cover rearward movement. Squads back out of position and move, attempting to keep a terrain feature between them and the enemy.
- Rapid movement and effective base of fire enhance mobility and are key to a successful disengagement.

8-185. Plans for disengagement may be part of any defensive plan. When squads are separated, there are three ways they can disengage: by teams; by thinning the lines when they must cover their own movement; or simultaneously when they are covered by another force.

Teams

8-186. When the rifle platoon must cover their own movement, two squads stay in position as a base of fire (Figure 8-29). The third squad and weapons squad move to the rear (crew served weapons move based on the platoon leader's assessment of when they could best move). The squads left in position must fire into the entire element's sector to cover the movement of the other squad(s). Sectors of fire are adjusted for better coverage of the element's sector. The moving squad may displace by fire teams or as squads because there are two squads covering their movement. The squads left in position sequentially disengage. Movement to the rear by alternating squads continues until contact is broken.

Figure 8-29. Disengagement by squads.

Thinning the Lines

8-187. When disengaging by thinning the lines, selected Soldiers from each fire team (usually one Soldier from each fighting position) disengage and move to the rear (Figure 8-30). The Soldiers still in position become the base of fire to cover the movement.

Figure 8-30. Disengagement by thinning the lines.

Simultaneous

8-188. Squads disengage simultaneously when they are covered by another force. Simultaneous disengagement is favored when rapid movement is critical; when the disengaging element is adequately covered by overwatching fires; when the enemy has not closed on the rifle squad or cannot fire effectively at it; and when there are obstacles to delay the enemy. Simultaneous disengagement is used when rifle squads are able to move before the enemy can close on their position. Other platoons of the company or battalion cover the disengagement with supporting fires.

DELAY

8-189. In a delay, the enemy slows its forward momentum when the platoon forces him to repeatedly deploy for the attack. After causing the enemy to deploy, the delaying force withdraws to new positions, trading space for time. A delay is typically done to buy time for friendly forces to regain the offensive. It is also done to buy time so friendly forces can establish an effective defense, or to determine enemy intentions. Inflicting casualties on the enemy is normally secondary to slowing the enemy approach. As part of a company or larger operation, the platoon can expect to be tasked as a reserve, security force, or part of the main body. The squads or sections and platoons disengage from the enemy as described in a withdrawal under pressure (see paragraph 8-176) and move directly to their next position and defend again. The squads and platoons slow the advance of the enemy by causing casualties and equipment losses by employing—

- Ambushes.
- Snipers.
- Obstacles.
- Minefields (including phony minefields).
- Artillery and mortar fire.

8-190. A common control measure used in these missions is the delay line, which is a phase line the enemy is not allowed to cross until a specified date and time. Infantry must carefully consider the mobility difference between themselves and the attacking force, maximizing the use of both terrain and counter-mobility obstacles. A delay operation terminates when the delaying force conducts a rearward passage of lines through a defending force, the delaying force reaches defensible terrain and transitions to the defense, the advancing enemy force reaches a culminating point and can no longer continue to advance, or the delaying force goes on the offensive.

STAY-BEHIND OPERATIONS

8-191. Stay-behind operations can be used as part of defensive or retrograde operations. In these operations, the commander leaves a unit in position to conduct a specified mission while the remainder of his forces withdraw or retire from an enemy. Stay-behind is inherently risky, and resupply and casualty evacuation are difficult. Conducting stay-behind operations places a premium on Infantry leadership and initiative, and ultimately terminates when the unit conducts a linkup with attacking friendly forces or reenters friendly lines.

TYPES

8-192. The two types of stay-behind operations are unplanned; and deliberate.

Unplanned

8-193. An unplanned stay-behind operation is one in which a unit finds itself cut off from other friendly elements for an indefinite time. In this kind of operation the unit has no specific planning or targets, and must rely on its organic assets.

Deliberate

8-194. A deliberate stay-behind operation is one in which a unit plans to operate in an enemy-controlled area as a separate yet cohesive element for a certain amount of time or until a specified event occurs. A deliberate stay-behind operation requires extensive planning. Squads, sections, and platoons conduct this type of operation as part of larger units.

PLANNING

8-195. Troop-leading procedures (TLP) apply to stay-behind operations. Planners must pay strict attention to task organization, reconnaissance, and sustainment.

Task Organization

8-196. A stay-behind unit includes only the Soldiers and equipment needed for the mission. It provides its own logistics support and security, and must be able to hide easily and move through restrictive terrain.

Reconnaissance

8-197. Reconnaissance is most important in a stay-behind operation. Reporting tasks and information requirements can include suitable sites for patrol bases, hide positions, observation posts, caches, water sources, dismounted and mounted avenues of approach, kill zones, engagement areas, and covered and concealed approach routes. The unit may be required to collect intelligence on enemy forces around them.

Logistics

8-198. Because the stay-behind unit will not be in physical contact with its supporting unit, supplies of rations, ammunition, radio batteries, water, and medical supplies are cached. Provisions for casualty and EPW evacuation depend on company and battalion plans.

RETIREMENT

8-199. Retirement is a form of retrograde in which a force not in contact with the enemy, moves away from the enemy. Retiring units organize to fight but do so only in self defense. Retirements are usually not as risky as delays or withdrawals. Retiring units normally road march away from the enemy. Infantry platoons participate in retirements as part of their company and higher headquarters.

Chapter 9

Patrols and Patrolling

A patrol is a detachment sent out by a larger unit to conduct a specific mission. Patrols operate semi-independently and return to the main body upon completion of their mission. Patrolling fulfills the Infantry's primary function of finding the enemy to either engage him or report his disposition, location, and actions. Patrols act as both the eyes and ears of the larger unit and as a fist to deliver a sharp devastating jab and then withdraw before the enemy can recover.

SECTION I — OVERVIEW

PATROLS AND PATROLLING

9-1. A patrol is sent out by a larger unit to conduct a specific combat, reconnaissance, or security mission. A patrol's organization is temporary and specifically matched to the immediate task. Because a patrol is an organization, not a mission, it is not correct to speak of giving a unit a mission to "Patrol."

9-2. The terms "patrolling" or "conducting a patrol" are used to refer to the semi-independent operation conducted to accomplish the patrol's mission. Patrols require a specific task and purpose.

9-3. A commander sends a patrol out from the main body to conduct a specific tactical task with an associated purpose. Upon completion of that task, the patrol leader returns to the main body, reports to the commander and describes the events that took place, the status of the patrol's members and equipment, and any observations.

9-4. If a patrol is made up of an organic unit, such as a rifle squad, the squad leader is responsible. If a patrol is made up of mixed elements from several units, an officer or NCO is designated as the patrol leader. This temporary title defines his role and responsibilities for that mission. The patrol leader may designate an assistant, normally the next senior man in the patrol, and any subordinate element leaders he requires.

9-5. A patrol can consist of a unit as small as a fire team. Squad- and platoon-size patrols are normal. Sometimes, for combat tasks such as a raid, the patrol can consist of most of the combat elements of a rifle company. Unlike operations in which the Infantry platoon or squad is integrated into a larger organization, the patrol is semi-independent and relies on itself for security.

PATROL LEADERS

9-6. The leader of every patrol, regardless of the type or the tactical task assigned, has an inherent responsibility to prepare and plan for possible enemy contact while on the mission. Patrols are never administrative. They are always assigned a tactical mission. On his return to the main body, the patrol leader must always report to the commander. He then describes the patrol's actions, observations, and condition.

PURPOSE OF PATROLLING

9-7. There are several specific purposes that can be accomplished by patrolling:

- Gathering information on the enemy, on the terrain, or on the populace.
- Regaining contact with the enemy or with adjacent friendly forces

- Engaging the enemy in combat to destroy him or inflict losses.
- Reassuring or gaining the trust of a local population.
- Preventing public disorder.
- Deterring and disrupting insurgent or criminal activity.
- Providing unit security.
- Protecting key infrastructure or bases.

TYPES OF PATROLS

9-8. Patrol missions can range from security patrols in the close vicinity of the main body, to raids deep into enemy territory. Successful patrolling requires detailed contingency planning and well-rehearsed small unit tactics. The planned action determines the type of patrol.

COMBAT AND RECONNAISSANCE PATROLS

9-9. The two categories of patrols are combat and reconnaissance. Regardless of the type of patrol being sent out, the commander must provide a clear task and purpose to the patrol leader. Any time a patrol leaves the main body of the unit there is a possibility that it may become engaged in close combat.

9-10. Patrols that depart the main body with the clear intent to make direct contact with the enemy are called *combat patrols*. The three types of combat patrols are raid patrols, ambush patrols (both of which are sent out to conduct special purpose attacks), and security patrols.

9-11. Patrols that depart the main body with the intention of avoiding direct combat with the enemy while seeing out information or confirming the accuracy of previously-gathered information are called *reconnaissance patrols*. The most common types reconnaissance patrols are area, route, zone, and point. Leaders also dispatch reconnaissance patrols to track the enemy, and to establish contact with other friendly forces. Contact patrols make physical contact with adjacent units and report their location, status, and intentions. Tracking patrols follow the trail and movements of a specific enemy unit. Presence patrols conduct a special form of reconnaissance, normally during stability or civil support operations.

ORGANIZATION OF PATROLS

9-12. A patrol is organized to perform specific tasks. It must be prepared to secure itself, navigate accurately, identify and cross danger areas, and reconnoiter the patrol objective. If it is a combat patrol, it must be prepared to breach obstacles, assault the objective, and support those assaults by fire. Additionally, a patrol must be able to conduct detailed searches as well as deal with casualties and prisoners or detainees.

9-13. The leader identifies those tasks the patrol must perform and decides which elements will implement them. Where possible, he should maintain squad and fire team integrity.

9-14. Squads and fire teams may perform more than one task during the time a patrol is away from the main body or it may be responsible for only one task. The leader must plan carefully to ensure that he has identified and assigned all required tasks in the most efficient way.

9-15. Elements and teams for platoons conducing patrols include the common and specific elements for each type of patrol. The following elements are common to all patrols.

HEADQUARTERS ELEMENT

9-16. The headquarters element normally consists of the patrol leader and his radio operator. The platoon sergeant may be designated as the assistant patrol leader. Combat patrols may include a forward observer and perhaps his radio operator. Any attachments the platoon leader decides that he or the platoon sergeant must control directly are also part of the headquarters element.

AID AND LITTER TEAM(S)

9-17. Aid and litter teams are responsible for locating, treating, and evacuating casualties.

ENEMY PRISONER OF WAR/DETAINEE TEAM(S)

9-18. EPW teams are responsible for controlling enemy prisoners IAW the five S's and the leader's guidance. These teams may also be responsible for accounting for and controlling detainees or recovered personnel.

SURVEILLANCE TEAM(S)

9-19. Surveillance teams are used to establish and maintain covert observation of an objective for as long as it takes to complete the patrol's mission.

EN ROUTE RECORDER

9-20. An en route recorder can be designated to record all information collected during the mission.

COMPASS AND PACE MAN

9-21. If the patrol does not have access to global positioning systems, or if it is operating in a location where there is no satellite reception, it may be necessary to navigate by dead reckoning. This is done with a compass man and a pace man.

ASSAULT TEAM(S)

9-22. Combat patrols designate assault teams to close with the enemy on the objective or to clear the ambush kill zone.

SUPPORT TEAM(S)

9-23. Combat patrols designate teams to provide direct fire in support of the breach and assault teams.

BREACH TEAM(S) AND SEARCH TEAM(S)

9-24. Combat patrols have breach teams to assist the assault team in getting to the objective. Search teams are designated to conduct a cursory or detailed search of the objective area.

INITIAL PLANNING AND COORDINATION FOR PATROLS

9-25. Leaders plan and prepare for patrols using troop-leading procedures and an estimate of the situation. They must identify required actions on the objective, plan backward to the departure from friendly lines, then forward to the reentry of friendly lines.

9-26. The patrol leader will normally receive the OPORD in the battalion or company CP where communications are good and key personnel are available for coordination. Because patrols act semi-independently, move beyond the direct-fire support of the parent unit, and often operate forward of friendly units, coordination must be thorough and detailed.

9-27. Patrol leaders may routinely coordinate with elements of the battalion staff directly. Unit leaders should develop tactical SOPs with detailed checklists to preclude omitting any items vital to the accomplishment of the mission.

9-28. Items coordinated between the leader and the battalion staff or company commander include:

- Changes or updates in the enemy situation.
- Best use of terrain for routes, rally points, and patrol bases.
- Light and weather data.
- Changes in the friendly situation.
- The attachment of Soldiers with special skills or equipment (engineers, sniper teams, scout dog teams, FOs, or interpreters).

- Use and location of landing or pickup zones.
- Departure and reentry of friendly lines.
- Fire support on the objective and along the planned routes, including alternate routes.
- Rehearsal areas and times. The terrain for the rehearsal should be similar to that at the objective, to include buildings and fortifications if necessary. Coordination for rehearsals includes security of the area, use of blanks, pyrotechnics, and live ammunition.
- Special equipment and ammunition requirements.
- Transportation support, including transportation to and from the rehearsal site.
- Signal plan—call signs frequencies, code words, pyrotechnics, and challenge and password.

9-29. The leader coordinates with the unit through which his platoon or squad will conduct its forward and rearward passage of lines.

9-30. The platoon leader also coordinates patrol activities with the leaders of other units that will be patrolling in adjacent areas at the same time.

COMPLETION OF THE PATROL PLAN

9-31. As the platoon leader completes his plan, he considers the following elements.

ESSENTIAL AND SUPPORTING TASKS

9-32. The leader ensures that he has assigned all essential tasks to be performed on the objective, at rally points, at danger areas, at security or surveillance locations, along the route(s), and at passage lanes.

KEY TRAVEL AND EXECUTION TIMES

9-33. The leader estimates time requirements for movement to the objective, leader's reconnaissance of the objective, establishment of security and surveillance, compaction of all assigned tasks on the objective, movement to an objective rally point to debrief the platoon, and return through friendly lines.

PRIMARY AND ALTERNATE ROUTES

9-34. The leader selects primary and alternate routes to and from the objective (Figure 9-1). Return routes should differ from routes to the objective.

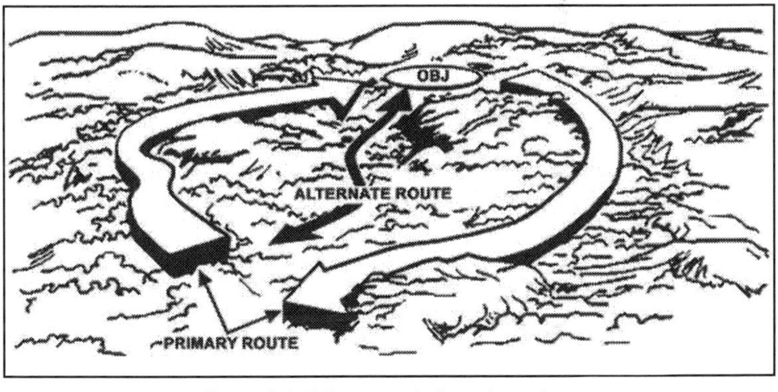

Figure 9-1. Primary and alternate routes.

SIGNALS

9-35. The leader should consider the use of special signals. These include arm-and-hand signals, flares, voice, whistles, radios, visible and nonvisible lasers. All signals must be rehearsed to ensure all Soldiers know what they mean.

CHALLENGE AND PASSWORD OUTSIDE OF FRIENDLY LINES

9-36. The challenge and password from the SOI must not be used when the patrol is outside friendly lines. The unit's tactical SOP should state the procedure for establishing a patrol challenge and password as well as other combat identification features and patrol markings.

LOCATION OF LEADERS

9-37. The leader considers where he, the platoon sergeant, and other key leaders should be located for each phase of the patrol mission. The platoon sergeant is normally with the following elements for each type of patrol:

- On a raid or ambush, he normally controls the support element.
- On an area reconnaissance, he normally supervises security in the objective rally point (ORP).
- On a zone reconnaissance, he normally moves with the reconnaissance element that sets up the link-up point.

ACTIONS ON ENEMY CONTACT

9-38. The leader's plan must address actions on chance contact at each phase of the patrol mission.

- The plan must address the handling of seriously wounded and KIAs.
- The plan must address the handling of prisoners captured as a result of chance contact who are not part of the planned mission.

DEPARTURE FROM FRIENDLY LINES OR FIXED BASE

9-39. The departure from friendly lines, or from a fixed base, must be thoroughly planned and coordinated.

COORDINATION

9-40. The platoon leader must coordinate with the commander of the forward unit and leaders of other units that will be patrolling in the same or adjacent areas. The coordination includes SOI information, signal plan, fire plan, running passwords, procedures for departure and reentry of lines, planned dismount points, initial rally points, actions at departure and reentry points, and information about the enemy.

(1) The platoon leader provides the forward unit leader with the unit identification, size of the patrol, departure and return times, and area of operation.

(2) The forward unit leader provides the platoon leader with the following:

- Additional information on terrain just outside the friendly unit lines.
- Known or suspected enemy positions in the near vicinity.
- Likely enemy ambush sites.
- Latest enemy activity.
- Detailed information on friendly positions, obstacles, and OPs.
- Friendly unit fire plan.
- Support the unit can provide (fire support, litter teams, guides, communications, and reaction force).

PLANNING

9-41. In his plan for the departure of friendly lines, the leader should consider the following sequence of actions:

- Making contact with friendly guides at the contact point.
- Moving to a coordinated initial rally point just inside friendly lines.
- Completing final coordination.
- Moving to and through the passage point.
- Establishing a security-listening halt beyond the friendly unit's final protective fires.

RALLY POINTS

9-42. The leader considers the use and locations of rally points. A rally point is a place designated by the leader where the platoon moves to reassemble and reorganize if it becomes dispersed.

SELECTION OF RALLY POINTS

9-43. The leader physically reconnoiters routes to select rally points whenever possible. He selects tentative points if he can only conduct a map reconnaissance. Routes are confirmed by the leader through actual inspection as the platoon moves through them. Rally points must—

- Be easy to recognize on the ground.
- Have cover and concealment.
- Be away from natural lines of drift.
- Be defendable for short periods.

TYPES OF RALLY POINTS

9-44. The most common types of rally points are initial, en route, objective, reentry, near- and far-side. Soldiers must know which rally point to move to at each phase of the patrol mission. They should know what actions are required there and how long they are to wait at each rally point before moving to another. Following are descriptions of these five rally points.

(1) ***Initial rally point.*** An initial rally point is a place inside of friendly lines where a unit may assemble and reorganize if it makes enemy contact during the departure of friendly lines or before reaching the first en route rally point. It is normally selected by the commander of the friendly unit.

(2) ***En route rally point.*** The leader designates en route rally points based on the terrain, vegetation, and visibility.

(3) ***Objective rally point.*** The objective rally point (ORP) is a point out of sight, sound, and small-arms range of the objective area. It is normally located in the direction that the platoon plans to move after completing its actions on the objective. The ORP is tentative until the objective is pinpointed (Figure 9-2). Actions at or from the ORP include—

- Issuing a final FRAGO.
- Disseminating information from reconnaissance if contact was not made.
- Making final preparations before continuing operations.
- Accounting for Soldiers and equipment after actions at the objective are complete.
- Reestablishing the chain of command after actions at the objective are complete.

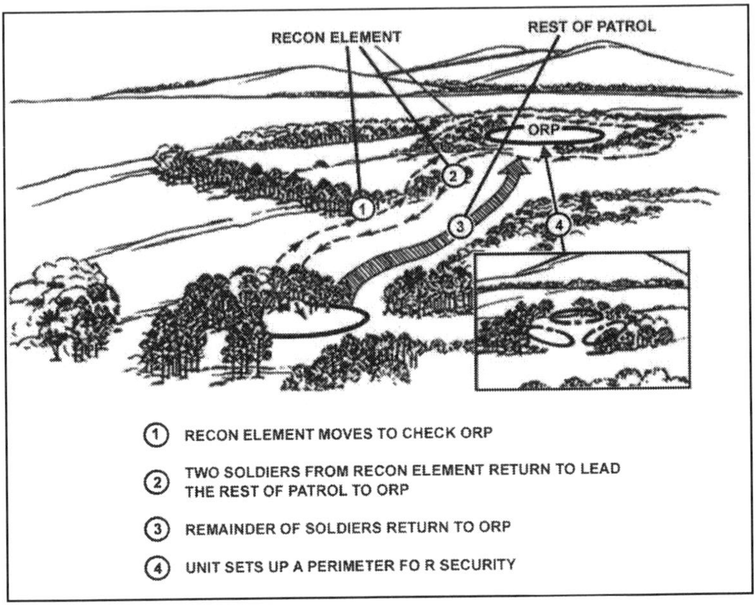

Figure 9-2. Objective rally point.

(4) ***Reentry rally point.*** The reentry rally point is located out of sight, sound, and small-arms weapons range of the friendly unit through which the platoon will return. This also means that the RRP should be outside the final protective fires of the friendly unit. The platoon occupies the RRP as a security perimeter.
(5) ***Near-and far-side rally points.*** These rally points are on the near and far side of danger areas. If the platoon makes contact while crossing the danger area and control is lost, Soldiers on either side move to the rally point nearest them. They establish security, reestablish the chain of command, determine their personnel and equipment status, continue the patrol mission, and link up at the OR.

SECTION II — COMBAT PATROLS

9-45. A combat patrol provides security and harasses, destroys, or captures enemy troops, equipment, or installations. When the commander gives a unit the mission to send out a combat patrol, he intends for the patrol to make contact with the enemy and engage in close combat. A combat patrol always attempts to remain undetected while moving, but of course it ultimately discloses its location to the enemy in a sudden, violent surprise attack. For this reason, the patrol normally carries a significant amount of weapons and ammunition. It may carry specialized munitions. A combat patrol collects and reports any information gathered during the mission, whether related to the combat task or not. The three types of combat patrols are raid, ambush, and security.

RAID

9-46. A raid is a surprise attack against a position or installation for a specific purpose *other than* seizing and holding the terrain. It is conducted to destroy a position or installation, to destroy or capture enemy soldiers or equipment, or to free prisoners. A raid patrol retains terrain just long enough to accomplish the intent of the raid. A raid always ends with a planned withdrawal off the objective and a return to the main body.

AMBUSH

9-47. An ambush is a surprise attack from a concealed position on a moving or temporarily halted target. An ambush patrol does not need to seize or hold any terrain. It can include an assault to close with and destroy the target, or an attack by fire only.

SECURITY

9-48. A security patrol is sent out from a unit location when the unit is stationary or during a halt to search the local area, detect any enemy forces near the main body, and to engage and destroy the enemy within the capability of the patrol. This type of combat patrol is normally sent out by units operating in close terrain with limited fields of observation and fire. Although this type of combat patrol seeks to make direct enemy contact and to destroy enemy forces within its capability, it should try to avoid decisive engagement. A security patrol detects and disrupts enemy forces that are conducting reconnaissance of the main body or that are massing to conduct an attack. Security patrols are normally away from the main body of the unit for limited periods, returning frequently to coordinate and rest. They do not operate beyond the range of communications and supporting fires from the main body, especially mortar fires.

COMBAT PATROL PLANNING

9-49. There are three essential elements for a combat patrol: security; support; and assault (Figure 9-3). Assault elements accomplish the mission during actions on the objective. Support elements suppress or destroy enemy on the objective in support of the assault element. Security elements assist in isolating the objective by preventing enemy from entering and leaving the objective area as well as by ensuring the patrol's withdrawal route remains open. The size of each element is based on the situation and the leader's analysis of METT-TC.

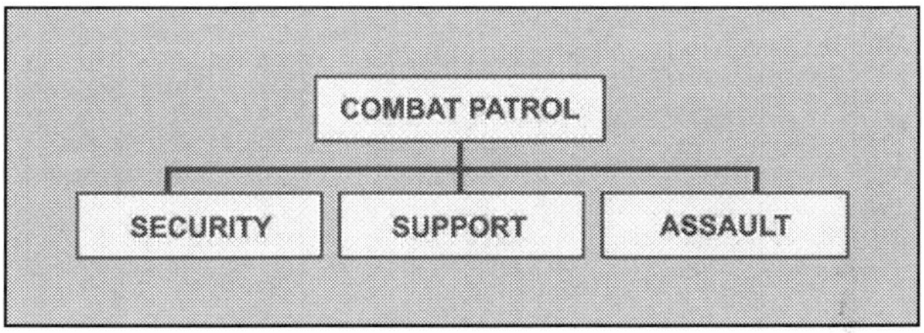

Figure 9-3. Organization of forces.

ASSAULT ELEMENT

9-50. The assault element is the combat patrol's decisive effort. Its task is to conduct actions on the objective. The assault element is responsible for accomplishing the unit's task and purpose. This element must be capable (through inherent capabilities or positioning relative to the enemy) of destroying or seizing the target of the combat patrol. Tasks typically associated with the assault element include:

- Conduct of assault across the objective to destroy enemy equipment, capture or kill enemy, and clearing of key terrain and enemy positions.
- Deployment close enough to the objective to conduct an immediate assault if detected.
- Being prepared to support itself if the support element cannot suppress the enemy.
- Providing support to a breach element in reduction of obstacles (if required).
- Planning detailed fire control and distribution.
- Conducting controlled withdrawal from the objective.

9-51. Analysis of METT-TC, particularly for a raid, may result in the requirement to organize a separate breach force. At times this may include breaching an obstacle.

9-52. Additional tasks/special purpose teams assigned may include:

- Search teams – to find and collect documents, equipment and information that can be used as intelligence.
- Prisoner teams – to capture, secure, and account for prisoners and detainees.
- Demolition teams – to plan and execute the destruction of obstacles and enemy equipment.
- Breach team – to create small-scale breaches in protective obstacles to facilitate the completion of the patrol's primary task
- Aid and litter teams – to identify, collect, render immediate aid and coordinate medical evacuation for casualties

SUPPORT ELEMENT

9-53. The support element suppresses the enemy on the objective using direct and indirect fires. The support element is a shaping effort that sets conditions for the mission's decisive effort. This element must be capable, through inherent means or positioning relative to the enemy, of supporting the assault element. The support force can be divided up into two or more elements if required.

9-54. The support element is organized to address a secondary threat of enemy interference with the assault element(s). The support force suppresses, fixes, or destroys elements on the objective. The support force's primary responsibility is to suppress enemy to prevent reposition against decisive effort. The support force—

- Initiates fires and gains fire superiority with crew-served weapons and indirect fires.
- Controls rates and distribution of fires.
- Shifts/ceases fire on signal.
- Supports the withdrawal of the assault element.

SECURITY ELEMENT

9-55. The security element(s) is a shaping force that has three roles. The first role is to isolate the objective from enemy personnel and vehicles attempting to enter the objective area. Their actions range from simply providing early warning, to blocking enemy movement. This element may require several different forces located in various positions. The patrol leader is careful to consider enemy reserves or response forces that, once the engagement begins, will be alerted. The second role of the security element is to prevent enemy from escaping the objective area. The third role is to secure the patrol's withdrawal route.

9-56. There is a subtle yet important distinction for the security element. All elements of the patrol are responsible for their own local security. What distinguishes the security element is that they are protecting the entire patrol. Their positions must be such that they can, in accordance with their engagement criteria, provide early warning of approaching enemy.

9-57. The security element is organized to address the primary threat to the patrol—being discovered and defeated by security forces prior to execution of actions on the objective. To facilitate the success of the assault element, the security element must fix or block (or at a minimum screen) all enemy security or response forces located on parts of the battlefield away from the raid.

LEADER LOCATIONS

9-58. Leaders locate where they can best influence the situation, which is usually with either the support element or assault element. The second in charge normally locates at the opposite location of the leader.

ACTIONS ON THE OBJECTIVE – RAID

9-59. A raid is a surprise attack against a position or installation for a specific purpose *other than* seizing and holding the terrain. It is conducted to destroy a position or installation, destroy or capture enemy soldiers or equipment, or free prisoners. A raid patrol retains terrain just long enough to accomplish the intent of the raid. A raid always ends with a withdrawal off the objective and a return to the main body.

9-60. Raids are characterized by the following:

- Destruction of key systems or facilities (C2 nodes, logistical areas, other high value areas).
- Provide or deny critical information.
- Securing of hostages or prisoners.
- Confusing the enemy or disrupting his plans.
- Detailed intelligence (significant ISR assets committed).
- Command and control from the higher HQ to synchronize the operation.
- Creating a window of opportunity for the raiding force.

9-61. Raids are normally conducted in five phases (Figure 9-4):

- Approach the objective.
- Isolate the objective area.
- Set conditions for the assault element.
- Assault the objective.
- Tactical movement away from the objective area.

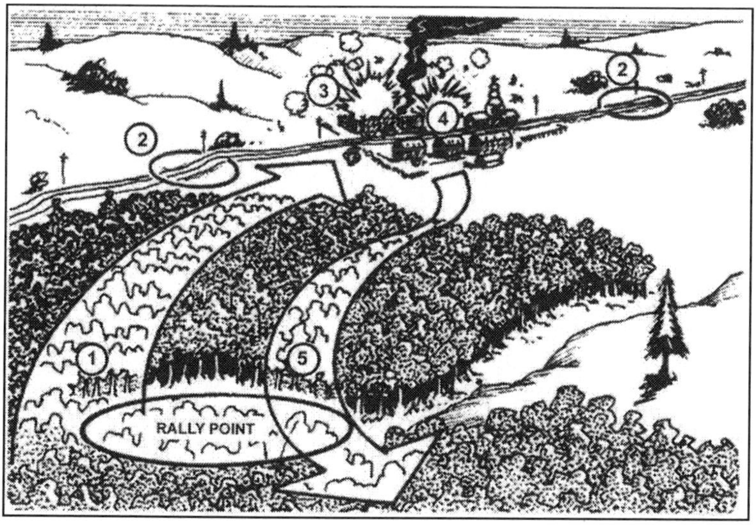

Figure 9-4. The five phases of a raid.

ACTIONS ON THE OBJECTIVE – AMBUSH

9-62. An ambush is a surprise attack from a concealed position on a moving or temporarily halted target. It can include an assault to close with and destroy the target, or only an attack by fire. An ambush need not seize or hold ground. The purpose of an ambush is to destroy or harass enemy forces. The ambush combines the advantages of the defense with the advantages of the offense, allowing a smaller force with limited means the ability to destroy a much larger force. Ambushes are enemy-oriented. Terrain is only held long enough to conduct the ambush and then the force withdraws. Ambushes range from very simple to complex and synchronized; short duration of minutes to long duration of hours; and within hand grenade range, to maximum standoff. Ambushes employ direct fire systems as well as other destructive means, such as command-detonated mines and explosives, and indirect fires on the enemy force. The attack may include an assault to close with and destroy the enemy or may just be a harassing attack by fire. Ambushes may be conducted as independent operations or as part of a larger operation.

9-63. There are countless ways for leaders to develop an ambush. To assist the leader clarify what he wants, he develops the ambush based on its purpose, type, time, and formation.

9-64. The purpose of an ambush is either harassment or destruction. A harassing ambush is one in which attack is by fire only (meaning there is no assault element). A destruction ambush includes assault to close with and destroy the enemy.

9-65. The two types of ambushes are point, and area. In a *point ambush*, Soldiers deploy to attack a single kill zone. In an *area ambush*, Soldiers deploy as two or more related point ambushes. These ambushes at separate sites are related by their purpose (Figure 9-5).

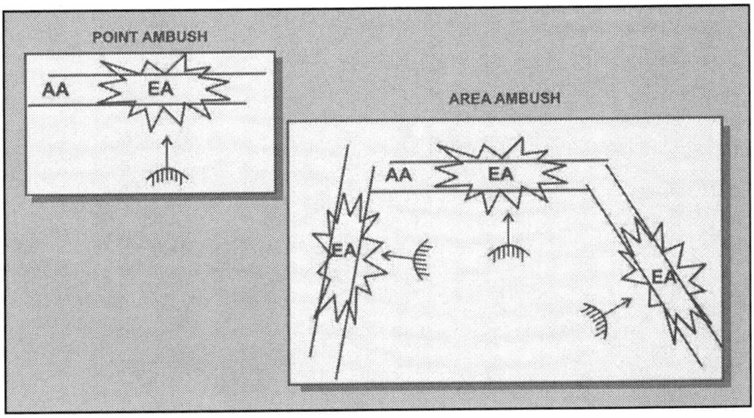

Figure 9-5. Point and area ambush.

9-66. Based on the amount of time available to set an ambush, ambushes are hasty and deliberate.

9-67. A *hasty ambush* is conducted based on an unanticipated opportunity. It is used when a patrol sees the enemy before the enemy sees them, and the patrol has time to act. The leader gives the prearranged signal to start the action and all Soldiers move to concealed firing positions, prepared to engage the enemy. Depending on the mission, the patrol may allow the enemy to pass if the enemy does not detect the patrol.

9-68. A *deliberate ambush* is conducted against a specific target at a location chosen based on intelligence. With a deliberate ambush, leaders plan and prepare based on detailed information that allows them to anticipate enemy actions and enemy locations. Detailed information includes: type and size of target, organization or formation, routes and direction of movement, time the force will reach or pass certain points on its route, and weapons and equipment carried.

TERMINOLOGY

9-69. During terrain analysis, leaders identify at least four different locations: the kill zone, the ambush site, security positions, and rally points. As far as possible, so-called "ideal" ambush sites should be avoided because alert enemies avoid them if possible and increase their vigilance and security when they must be entered. Therefore, surprise is difficult to achieve. Instead, unlikely sites should be chosen when possible. Following are characteristics of these four ideal positions.

Ambush Site

9-70. The ambush site is the terrain on which a point ambush is established. The ambush site consists of a support-by-fire position for the support element and an assault position for the assault element. An ideal ambush site—

- Has good fields of fire into the kill zone.
- Has good cover and concealment.
- Has a protective obstacle.
- Has a covered and concealed withdrawal route.
- Makes it difficult for the enemy to conduct a flank attack.

Kill Zone

9-71. The kill zone is the part of an ambush site where fire is concentrated to isolate or destroy the enemy. An ideal kill zone has these characteristics:

- Enemy forces are likely to enter it.
- It has natural tactical obstacles.
- Large enough to observe and engage the anticipated enemy force.

Near Ambush

9-72. A near ambush is a point ambush with the assault element within reasonable assaulting distance of the kill zone (less than 50 meters). Close terrain, such as an urban area or heavy woods, may require this positioning. It may also be appropriate in open terrain in a "rise from the ground" ambush.

Far Ambush

9-73. A far ambush is a point ambush with the assault element beyond reasonable assaulting distance of the kill zone (beyond 50 meters). This location may be appropriate in open terrain offering good fields of fire or when attack is by fire for a harassing ambush.

Security Positions

9-74. An ideal security position —

- Does not mask fires of the main body.
- Provides timely information for the main body (gives the leader enough time to act on information provided).
- Can provide a support by fire position.

Rally Points

9-75. The platoon leader considers the use and locations of rally points (see paragraph 9-42). The rally point is a place designated by the leader where the platoon moves to reassemble and reorganize if it becomes dispersed.

9-76. The leader physically reconnoiters routes to select rally points whenever possible. He selects tentative points if he can only conduct a map reconnaissance. He confirms them by actual inspection as the platoon moves through them. Rally points must—

- Be easy to find
- Have cover and concealment
- Be away from natural lines of drift
- Be defendable for short periods

FORMATIONS

9-77. Many ambush formations exist. This FM only discusses the linear, L-shaped, and V-shaped (Figures 9-6 through 9-8). All of these formations require leaders to exercise strict direct fire control. Leaders need to understand the strengths and weaknesses of their units and plan accordingly.

9-78. The formation selected is based on the following:

- Terrain.
- Visibility.
- Soldiers available.
- Weapons and equipment.
- Ease of control.
- Target to be attacked.

Linear Ambush

9-79. In an ambush using a linear formation, the assault and support elements parallel the target's route. This positions the assault and support elements on the long axis of the kill zone and subjects the target to flanking fire (Figure 9-6). Only a target that can be covered with a full volume of fire can be successfully engaged in the kill zone. A dispersed target might be too large for the kill zone. This is the disadvantage of linear formations.

9-80. The linear formation is good in close terrain restricting the target's maneuver, and in open terrain where one flank is blocked by natural obstacles or can be blocked by other means such as claymore mines. Claymore mines or explosives can be placed between the assault and support elements and the kill zone to protect the unit from counter-ambush actions.

9-81. When the ambushing unit deploys this way, it leaves access lanes through the obstacles so it can assault the target. An advantage of the linear formation is the relative ease by which it can be controlled under all visibility conditions.

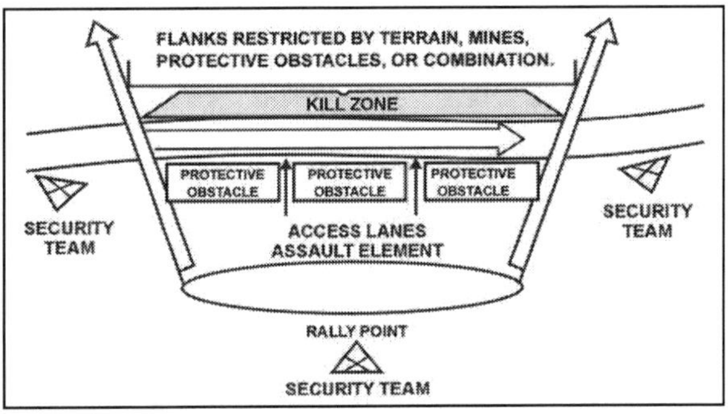

Figure 9-6. Linear ambush.

L-Shaped Ambush

9-82. An ambush in the L-shaped formation (Figure 9-7) is a variation of the linear formation. The long leg of the L (assault element) is parallel to the kill zone. This leg provides flanking fire. The short leg (support element) is at the end of and at a right angle to the kill zone. This leg provides enfilade fire that works with fire from the other leg. The L-shaped formation can be used at a sharp bend in a trail, road, or stream.

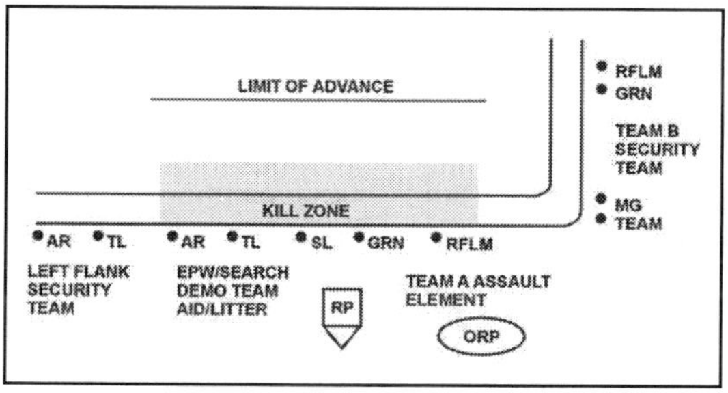

Figure 9-7. L-shaped ambush.

V-Shaped Ambush

9-83. The V-shaped ambush assault elements (Figure 9-8) are placed along both sides of the enemy route so they form a V. Take extreme care to ensure neither group fires into the other. This formation subjects the enemy to both enfilading and interlocking fire.

9-84. When performed in dense terrain, the legs of the V close in as the lead elements of the enemy force approach the point of the V. The legs then open fire from close range. Here, even more than in open terrain, all movement and fire is carefully coordinated and controlled to avoid fratricide.

9-85. Wider separation of the elements makes this formation difficult to control, and there are fewer sites which favor its use. Its main advantage is it is difficult for the enemy to detect the ambush until well into the kill zone.

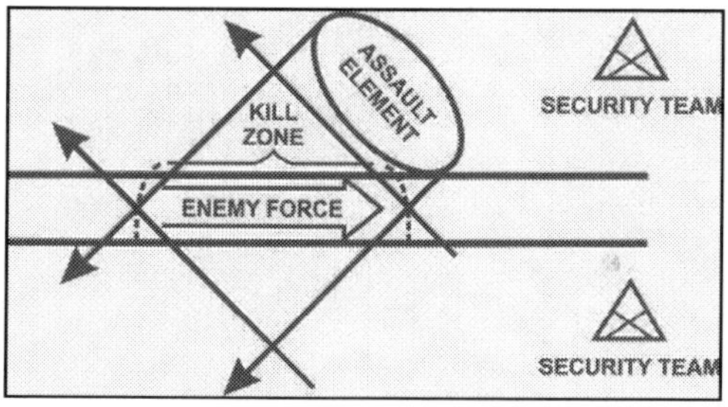

Figure 9-8. V-shaped ambush.

Final Preparations

9-86. Final preparations begin with the unit occupying an ORP and end with the main body prepared to depart for the ambush site. The unit halts at the ORP and establishes security. When ready, the leader conducts his reconnaissance to confirm the plan, positions the security element, and returns to the ORP. The security element leaves the ORP first. Teams of the security element move to positions from which they can secure the ORP and the flanks of the ambush site (Figure 9-9).

NOTE: The security elements should use a release point if there is a great distance between the ORP and objective.

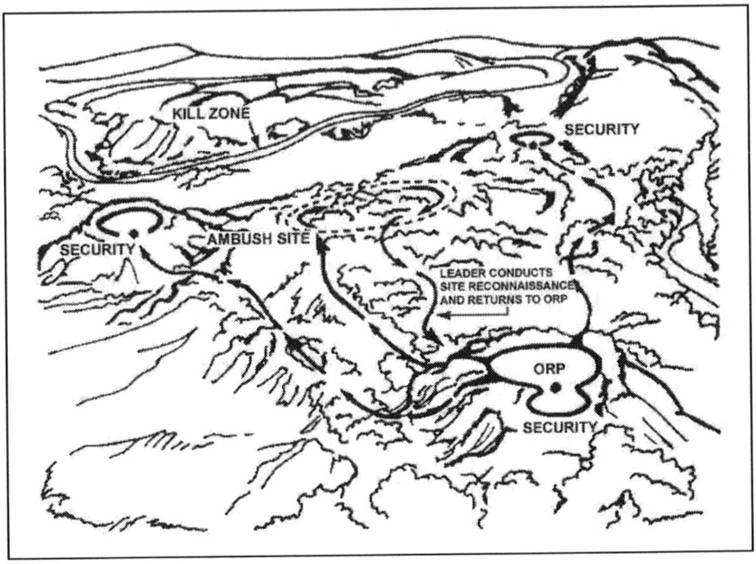

Figure 9-9. Security teams in position.

Occupy the Site and Conduct Ambush

9-87. Occupying the site and conducting the ambush begins with main body movement out of the ORP, and ends when the leader initiates a withdrawal. Common control measures include—

- Kill Zone.
- Limit of advance.
- ABF/SBF position.
- Assault position.
- Target reference point (TRP).
- Phase line.

Time of Occupation

9-88. As a rule, the ambush force occupies the ambush site at the latest possible time permitted by the tactical situation and the amount of site preparation required. This reduces the risk of discovery and the time that Soldiers must remain still and quiet in position.

Occupying the Site

9-89. Security elements are positioned first to prevent surprise while the ambush is being established. When the security teams are in position, the support and assault elements leave the ORP and occupy their positions. If there is a suitable position, the support element can overwatch the assault element's move to the ambush site. If not, both elements leave the ORP at the same time (Figure 9-10).

9-90. The main body moves into the ambush site from the rear. Ideally, leaders emplace the most casualty-producing weapons first, ensuring they have line of sight along the entire kill zone. Once positioned, the leader emplaces his subordinate units to complement and reinforce the key positions. The leader selects his location where he can best initiate and control the action. Once on the objective, movement is kept to a minimum and the number of men moving at a time is closely controlled. Leaders emplace and enforce local security measures.

Positions

9-91. Each Soldier must be hidden from the target and have line of sight into the kill zone. At the ambush site, positions are prepared with minimal change in the natural appearance of the site. Soldiers conceal debris resulting from preparation of positions.

Confirming the Direct Fire Plan

9-92. Claymore mines, explosives, and grenade launchers may be used to cover any dead space left by automatic weapons. All weapons are assigned sectors of fire to provide mutual support. The unit leader sets a time by which positions must be prepared.

Movement in the Kill Zone

9-93. The kill zone is not entered if entry can be avoided. When emplacing tactical obstacles, care is taken to remove any tracks or signs that might alert the enemy and compromise the ambush. If claymore mines or explosives are placed on the far side, or if the appearance of the site might cause the enemy to check it, a wide detour around the kill zone should be made. Here, too, care is taken to remove any traces which might reveal the ambush. An alternate route from the ambush site is also planned.

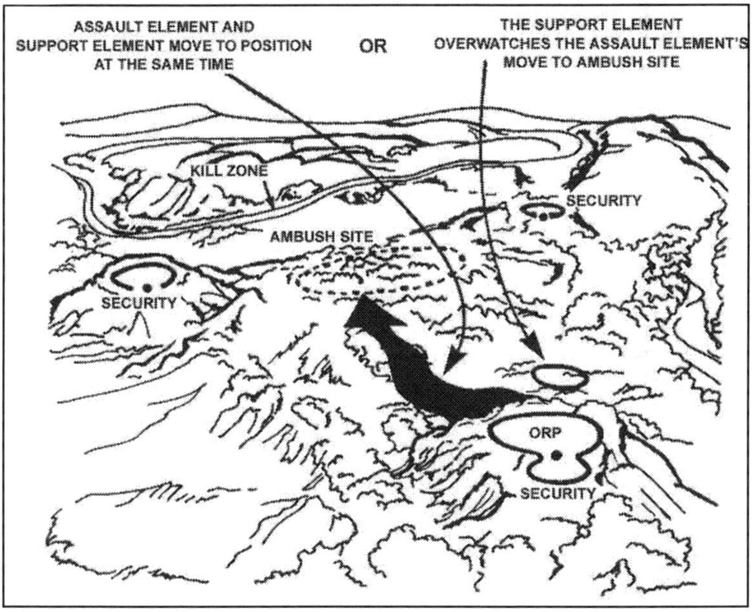

Figure 9-10. Assault element moving to the ambush site.

Initiating the Ambush

9-94. Once all friendly elements are in position, the unit waits for the enemy target. When the target approaches, the security team that spots it alerts the ambush leader. The security team reports the target's direction of movement, size, and any special weapons or equipment. Upon receipt of the report, the leader alerts the other elements.

9-95. When most of the enemy force is in the kill zone, the leader initiates the ambush with the most casualty-producing weapon, machine gun fire, or the detonation of mines or explosives. The detonation of explosives can cause a pause in the initiation of fires due to the obscuration created by the explosion. Once conditions are set, cease or shift fires. The assault element may conduct an assault through the kill zone to the limit of advance (LOA). If the assault element must assault the kill zone, the leader signals to cease or shift fire. This also signals the assault to start. Besides destruction of the enemy force, other kill zone tasks can include searching for items of intelligence value, capturing prisoners, and completing the destruction of enemy equipment. When the assault element has finished its mission in the kill zone, the leader gives the signal to withdraw to the ORP.

9-96. Fire discipline is critical during an ambush. Soldiers do not fire until the signal is given. Then it must be delivered at once in the heaviest, most accurate volume possible. Well-trained gunners and well-aimed fire help achieve surprise and destruction of the target. When the target is to be assaulted, the ceasing or shifting of fire must also be precise. If it is not, the assault is delayed, and the target has a chance to react. Sector stakes should be used if possible.

Withdrawal

9-97. The withdrawal begins once the assault element completes its actions on the objective and ends with consolidation/reorganization at a designated rally point. On signal, the unit withdraws to the ORP, reorganizes, and continues its mission. At a set terrain feature the unit halts and disseminates information. If the ambush fails and the enemy pursues, the unit withdraws by bounds. Units should use smoke to help conceal the withdrawal. Obstacles already set along the withdrawal routes can help stop the pursuit.

Conducting an Area Ambush

Area

9-98. In an area ambush, Soldiers deploy in two or more related point ambushes. The platoon may conduct an area ambush as part of a company offensive or defensive plan, or it may conduct a point ambush as part of a company area ambush.

9-99. The platoon is the smallest level to conduct an area ambush). Platoons conduct area ambushes (Figure 9-11) where enemy movement is largely restricted to trails or streams.

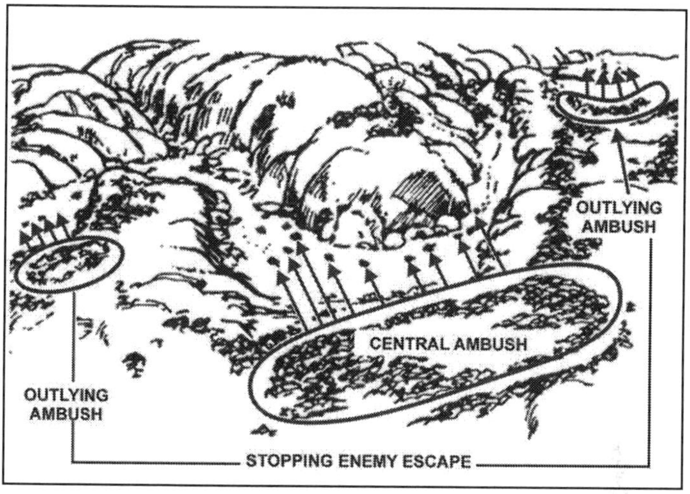

Figure 9-11. Area ambush.

9-100. The platoon leader (or company commander) selects one principal ambush site around which he organizes outlying ambushes. These secondary sites are located along the enemy's most likely avenue of approach and escape routes from the principal ambush site. Squads are normally responsible for each ambush site.

9-101. The platoon leader considers the factors of METT-TC to determine the best employment of the weapons squad. He normally locates the medium machine guns with the support element in the principal ambush site.

9-102. Squads (or sections) responsible for outlying ambushes do not initiate their ambushes until the principal one has been initiated. They then engage to prevent enemy forces from escaping the principal ambush or reinforcing the ambushed force.

9-103. Smaller ambushes can be used to isolate the main ambush kill zone (Figure 9-12).

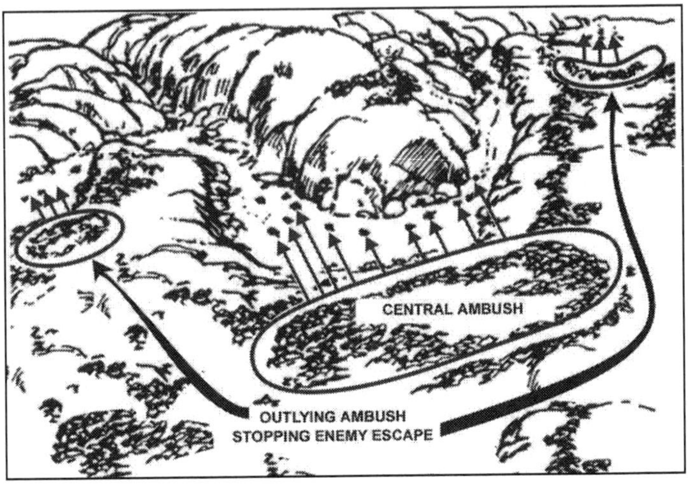

Figure 9-12. Use of smaller ambushes to isolate the main ambush kill zone.

ANTIARMOR AMBUSH

9-104. Platoons and squads conduct antiarmor ambushes (Figure 9-13) to destroy armored vehicles. The antiarmor ambush may be part of an area ambush. The antiarmor ambush consists of the assault element (armor-killer element) and the support-security element.

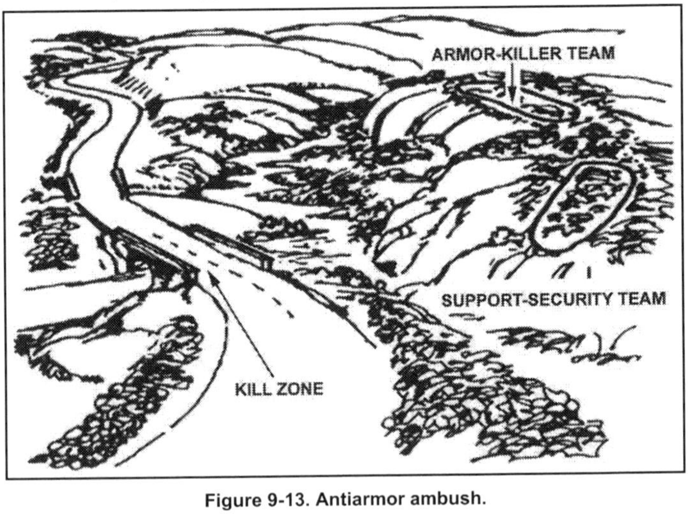

Figure 9-13. Antiarmor ambush.

9-105. The armor-killer element is built around the close combat missile systems. (Refer to Appendix B for information about employment of the Javelin.) The leader should consider additional shoulder-launched munitions available to supplement the CCMS fires. The leader considers the factors of METT-TC to position all antiarmor weapons to ensure the best engagement (rear, flank, or top). The remainder of the platoon must function as support and security elements in the same manner as the other types of ambushes.

9-106. In a platoon antiarmor ambush, the company commander selects the general site for the ambush. The platoon leader must find a specific site that restricts the movement of enemy armored vehicles out of the designated kill zone. The platoon leader should emplace his weapons so an obstacle is between the platoon and the kill zone. In a squad antiarmor ambush, the platoon leader selects the general site for the ambush. The squad leader must then find a site that restricts the movement of enemy armored vehicles out of the kill zone.

9-107. The support-security elements are emplaced to cover dismounted enemy avenues of approach into the ambush site.

9-108. The leader should consider the method for initiating the antiarmor ambush. The preferred method is to use a command-detonated AT mine placed in the kill zone. The Javelin can be used to initiate the ambush, but even with its limited signature, it may be less desirable than an AT mine.

9-109. The armor-killer team destroys the first and last vehicle in the enemy formation, if possible. All other weapons begin firing once the ambush has been initiated.

9-110. The leader must determine how the presence of dismounted enemy soldiers with armored vehicles will affect the success of the ambush. The leader's choices include:

- Initiate the ambush as planned.
- Withdraw without initiating the ambush.
- Initiate the ambush with machine guns without firing antiarmor weapons.

9-111. Because of the speed enemy armored forces can reinforce the ambushed enemy with, the leader should plan to keep the engagement short and have a quick withdrawal planned. The platoon, based on the factors of METT-TC, may not clear the kill zone as in other types of ambushes.

CONDUCTING A POINT AMBUSH

Point

9-112. In a point ambush, Soldiers deploy to attack an enemy in a single kill zone. The platoon leader is the leader of the assault element. The platoon sergeant will probably locate with the platoon leader in the assault element.

9-113. The security or surveillance team(s) should be positioned first. The support element should then be emplaced before the assault element moves forward. The support element must overwatch the movement of the assault element into position.

9-114. The platoon leader must check each Soldier once he emplaces. The platoon leader signals the surveillance team to rejoin the assault element if it is positioned away from the assault location. Actions of the assault element, support element, and security element are shown in Table 9-1.

Table 9-1. Actions by ambush elements.

Assault Element	Support Element	Security Element
Identify individual sectors of fire assigned by the platoon leader; emplace aiming stakes. Emplace Claymores and other protective obstacles. Emplace Claymores, mines, or other explosives in dead space within the kill zone. Camouflage positions. Take weapons off safe when directed by the platoon leader.	Identify sectors of fire for all weapons, especially machine guns. Emplace limiting stakes to prevent friendly fires from hitting the assault element in an L-shaped ambush. Emplace Claymores and other protective bstacles. Camouflage positions.	Identify sectors of fire for all weapons; emplace aiming stakes. Emplace Claymores and other protective obstacles. Camouflage positions. Secure the ORP. Secure a route to the ORP, as required.

9-115. The platoon leader instructs the security element (or teams) to notify him of the enemy's approach into the kill zone using the size, activity, location, unit, time, and equipment (SALUTE) reporting format. The security element must also keep the platoon leader informed if any additional enemy forces are following the lead enemy force. This will allow the platoon leader to know if the enemy force meets the engagement criteria directed by the company commander. The platoon leader must be prepared to give free passage to enemy forces that are too large or that do not meet the engagement criteria. He must report to the company commander any enemy forces that pass through the ambush unengaged.

9-116. The platoon leader initiates the ambush with the greatest casualty-producing weapon, typically a command-detonated Claymore. He must also plan a back-up method, typically a machine gun, to initiate the ambush should the primary means fail. All Soldiers in the ambush must know the primary and back-up methods. The platoon should rehearse with both methods to avoid confusion and the loss of surprise during execution of the ambush.

9-117. The platoon leader must include a plan for engaging the enemy during limited visibility. Based on the company commander's guidance, the platoon leader should consider the use and mix of tracers and the employment of illumination, NVDs, and TWSs. For example, if Javelins are not used during the ambush, the platoon leader may still employ the command launch unit with its thermal sights in the security or support element to observe enemy forces.

9-118. The platoon leader also may include the employment of indirect fire support in his plan. Based on the company commander's guidance, the platoon leader may employ indirect fires to cover flanks of the kill zone to isolate an enemy force or to assist the platoon's disengagement if the ambush is compromised or if the platoon must depart the ambush site under pressure.

9-119. The platoon leader must have a good plan (day and night) to signal the advance of the assault element into the kill zone to begin its search and collection activities. He should take into consideration the existing environmental factors. For example, smoke may not be visible to the support element because of limited visibility or the lay of the terrain. All Soldiers must know and practice relaying the signal during rehearsals to avoid the potential of fratricide.

9-120. The assault element must be prepared to move across the kill zone using individual movement techniques if there is any return fire once they begin to search. Otherwise, the assault element moves across by bounding fire teams.

9-121. The assault element collects and secures all EPWs and moves them out of the kill zone to an established location before searching dead enemy bodies. The EPW collection point should provide cover and should not be easily found by enemy forces following the ambush. The friendly assault element searches from the far side of the kill zone to the near side.

9-122. Once the bodies have been thoroughly searched, search teams continue in this manner until all enemy personnel in and near the kill zone have been searched. Enemy bodies should be marked once searched (for example, folded arms over the chest and legs crossed) to ensure thoroughness and speed and to avoid duplication of effort.

9-123. The platoon identifies and collects equipment to be carried back and prepares it for transport. Enemy weapon chambers are cleared and put on safe. The platoon also identifies and collects at a central point the enemy equipment to be destroyed. The demolition team prepares the fuse and awaits the signal to initiate. This is normally the last action performed before departing the ambush site. The flank security element returns to the ORP after the demolition team completes its task. The platoon will treat friendly wounded first and then enemy wounded.

9-124. The flank security teams may also emplace antiarmor mines after the ambush has been initiated if the enemy is known to have armored vehicles that can quickly reinforce the ambushed enemy force. If a flank security team makes enemy contact, it fights as long as possible without becoming decisively engaged. It uses prearranged signals to inform the platoon leader it is breaking contact. The platoon leader may direct a portion of the support element to assist the security element in breaking contact.

9-125. The platoon leader must plan the withdrawal of the platoon from the ambush site. The planning process should include the following:

- Elements are normally withdrawn in the reverse order that they established their positions.
- Elements may return to the release point, then to the objective rally point, depending on the distance between the elements.
- The security element at the objective rally point must be alert to assist the platoon's return. It maintains security for the ORP while the remainder of the platoon prepares to depart.

9-126. Actions back at the ORP include, but are not limited to, accounting for personnel and equipment, stowing captured equipment, and first aid (as necessary).

SECURITY PATROLS

9-127. Security patrols prevent surprise of the main body by screening to the front, flank, and rear of the main body and detecting and destroying enemy forces in the local area. Security patrols do not operate beyond the range of communication and supporting fires from the main body; especially mortar fires, because they normally operate for limited periods of time, and are combat oriented.

9-128. Security patrols are employed both when the main body is stationary and when it is moving. When the main body is stationary, the security patrol prevents enemy infiltration, reconnaissance, or attacks. When the main body is moving, the security patrol prevents the unit from being ambushed or coming into surprise chance contact.

SECTION III — RECONNAISSANCE PATROLS

9-129. A reconnaissance patrol collects information to confirm or disprove the accuracy of information previously gained. The intent for this type of patrol is to move stealthily, avoid enemy contact, and accomplish its tactical task without engaging in close combat. With one exception (presence patrols), reconnaissance patrols always try to accomplish their mission without being detected or observed. Because detection cannot always be avoided, a reconnaissance patrol carries the necessary arms and equipment to protect itself and break contact with the enemy. A reconnaissance patrol normally travels light, with as few personnel, arms, ammunition, and equipment as possible. This increases stealth and cross-country mobility in close terrain. Regardless of how the patrol is armed and equipped, the leader always plans for the worst case: direct-fire contact with a hostile force. Leaders must anticipate where they may possibly be observed and control the hazard by emplacing measures to lessen their risk. If detected or unanticipated opportunities arise, reconnaissance patrols must be able to rapidly transition to combat. Types of reconnaissance patrols follow

Area Reconnaissance Patrol

9-130. The area reconnaissance patrol focuses only on obtaining detailed information about the terrain or enemy activity within a prescribed area. See Section IV for further details.

Route Reconnaissance Patrol

9-131. The route reconnaissance patrol obtains detailed information about a specified route and any terrain where the enemy could influence movement along that route. See Section V for further details.

Zone Reconnaissance Patrol

9-132. Zone reconnaissance patrols involve a directed effort to obtain detailed information on all routes, obstacles, terrain, and enemy forces within a zone defined by boundaries. See Section VI for further details.

Point Reconnaissance Patrol

9-133. The point reconnaissance patrol goes straight to a specific location and determines the situation there. As soon as it does so, it either reports the information by radio or returns to the larger unit to report. This patrol can obtain, verify, confirm, or deny extremely specific information for the commander. These patrols are often used in stability or civil support operations. Normally, the patrol leader is the individual responsible for making the assigned assessment. This may involve interacting with the local populace. To allow this, interpreters or local civil leaders might accompany the patrol. The patrol leader may be required to participate in lengthy discussions or inspections with individuals at the site. During that time he is vulnerable to attack. The assistant patrol leader should not become involved in these talks, but should remain focused on external security to prevent attack from outside and on the personal security of the patrol leader. One or two specially-designated members of the patrol may be needed to protect the patrol leader while his attention is focused on discussions.

Leader's Reconnaissance Patrol

9-134. The leader's reconnaissance patrol reconnoiters the objective just before an attack or prior to sending elements forward to locations where they will support by fire. It confirms the condition of the objective, gives each subordinate leader a clear picture of the terrain where he will move, and identifies any part of the objective he must seize or suppress. The leader's reconnaissance patrol can consist of the unit commander or representative, the leaders of major subordinate elements, and (sometimes) security personnel and unit guides. It gets back to the main body as quickly as possible. The commander can use the aid in Figure 9-14 to help in remembering a five-point contingency:

G	Going—where is the leader going?
O	Others—what others are going with him?
T	Time (duration)—how long will the leader be gone?
W	What do we do if the leader fails to return?
A	Actions—what actions do the departing reconnaissance element and main body plan to take on contact?

Figure 9-14. Reconnaisance patrol five-point congengency

Contact Patrol

9-135. A contact patrol is a special reconnaissance patrol sent from one unit to physically contact and coordinate with another. Modern technology has reduced, but not eliminated, the need for contact patrols. They are most often used today when a U.S. force must contact a non-U.S. coalition partner who lacks compatible communications or position-reporting equipment. Contact patrols may either go to the other unit's position, or the units can meet at a designated contact point. The leader of a contact patrol provides the other unit with information about the location, situation, and intentions of his own unit. He obtains and reports the same information about the contacted unit back to his own unit. The contact patrol also observes and reports pertinent information about the area between the two units.

Presence Patrols

9-136. A presence patrol is used in stability or civil support operations. It has many purposes, but should always see and be seen, but seen in a specific manner determined by the commander. Its primary goal is to gather information about the conditions in the unit's AO. To do this, the patrol gathers critical (as determined by the commander) information, both specific and general. The patrol seeks out this information, and then observes and reports. Its secondary role is to be seen as a tangible representation of the U.S. military force, projecting an image that furthers the accomplishment of the commander's intent.

9-137. In addition to reconnaissance tasks, presence patrols demonstrate to the local populace the presence and intent of the U.S. forces. Presence patrols are intended to clearly demonstrate the determination, competency, confidence, concern, and when appropriate, the overwhelming power of the force to all who observe it, including local and national media.

9-138. The commander always plans for the possibility that a presence patrol may make enemy contact, even though that is not his intent. Rarely should a commander use a presence patrol where enemy contact is likely. Presence patrols work best for some types of stability operations such as peace operations, humanitarian and civic assistance, non-combatant evacuations, or shows of force. Before sending out a presence patrol, the commander should carefully consider what message he wants to convey, and then clearly describe his intent to the patrol leader.

9-139. To accomplish the "to be seen" part of its purpose, a presence patrol reconnoiters overtly. It takes deliberate steps to visibly reinforce the *impression* the commander wants to convey to the populace. Where the patrol goes, what it does there, how it handles its weapons, what equipment and vehicles it uses, and how it interacts with the populace are all part of that impression. When the presence patrol returns to the main body, the commander thoroughly debriefs it; not only for hard information, but also for the patrol leader's impressions of the effects of the patrol on the populace. This allows the commander to see to modify the actions of subsequent patrols.

Tracking Patrol

9-140. A tracking patrol is normally a squad-size, possibly smaller, element. It is tasked to follow the trail of a specific enemy unit in order to determine its composition, final destination, and actions en route. Members of the patrol look for subtle signs left by the enemy as he moves. As they track, they gather information about the enemy unit, the route it took, and the surrounding terrain. Normally, a tracking patrol avoids direct fire contact with the tracked unit, but not always. Tracking patrols often use tracker dog teams to help them maintain the track.

CONTROL MEASURES

9-141. Control measures help leaders anticipate being detected. They include:

- **Rendezvous point**: a location designated for an arranged meeting from which to begin an action or phase of an operation or to return to after an operation. This term is generally synonymous with linkup point.
- **Release point**: a location on a route where marching elements are released from centralized control (FM1-02). The release point is also used after departing the ORP.
- **Linkup point**: a point where two infiltrating elements in the same or different infiltration lanes are scheduled to consolidate before proceeding with their missions (FM 1-02).

9-142. Leaders use the three fundamentals of reconnaissance to organize their patrols into two forces: a reconnaissance element, and a security element. The first fundamental of reconnaissance (gain the information required), is the patrol's decisive action. Using the second principle (avoid detection), leaders organize this element accordingly. The remainder of the patrol is organized as a security element designed according to the third principle (employ security measures).

RECONNAISSANCE ELEMENTS

9-143. The task of the reconnaissance element is to obtain the information requirements for the purpose of facilitating tactical decision making. The primary means is reconnaissance (or surveillance) enabled by tactical movement and continuous, accurate reporting. The reconnaissance patrol leader decides how in depth the reconnaissance will be. A thorough and accurate reconnaissance is important. However, avoiding detection is equally important.

9-144. Below are some of the additional tasks normally associated with a reconnaissance element:

- Reconnoiter all terrain within the assigned area, route, or zone.
- Determine trafficability routes or potential avenues of approach (based on the personnel or vehicles to be used on the route).
  - Inspect and classify all bridges, overpasses, underpasses, and culverts on the route.*
  - Locate fords or crossing sites near bridges on the route.
- Determine the time it takes to traverse the route.
- Reconnoiter to the limit of direct fire range.
  - Terrain that influences the area, route, or zone.
  - Built-up areas.
  - Lateral routes.
- Within capabilities, reconnoiter natural and man-made obstacles to ensure mobility along the route. Locate a bypass or reduce/breach, clear, and mark—
  - Lanes.
  - Defiles and other restrictive/severely restrictive terrain.
  - Minefields.
  - Contaminated areas.
  - Log obstacles such as abatis, log cribs, stumps, and posts.
  - AT ditches.
  - Wire entanglements.
  - Fills, such as a raised railroad track.
  - Other obstacles along the route.
- Determine the size, location, and composition of society/human demographics.
- Identify key infrastructure that could influence military operations, including the following:
  - Political, government, and religious organizations and agencies.
  - Physical facilities and utilities (such as power generation, transportation, and communications networks).
- Find all threat forces that influence movement along the area, route, or zone.
- Report information.

***NOTE:** Infantry platoons typically do not have the expertise to complete a full technical inspection of bridges, roads, and culverts; this task normally requires augmentation. Infantry platoons do, however, have the ability to conduct a general assessment.

SECURITY ELEMENTS

9-145. The security element has two tasks: provide early warning of approaching enemy; and provide support by fire to the reconnaissance elements if they come in contact with the enemy. The purpose of the security element is to protect the reconnaissance element, thereby allowing them to obtain the IR. Security elements tasked to provide early warning must be able to observe avenues of approach into and out of the objective area. If the reconnaissance element is compromised, the security element must be able to quickly support them. They do so by occupying positions that enable them to observe the objective as well as cover the reconnaissance element. Soldiers in these positions must be able to engage the enemy with direct and indirect fire. They must also be able to facilitate communication to higher as well as any supporting assets. This worst-case scenario must be well rehearsed and well thought out.

Organizing the Reconnaissance Patrol

9-146. Regardless of how the reconnaissance and security elements are organized, each element always maintains responsibility for its own local security. In a small reconnaissance patrol, the patrol headquarters may form a part of one of the subordinate elements rather than being a separate element. The number and size of the various teams and elements must be determined through the leader's METT-TC analysis. There are three ways to organize the reconnaissance and security elements (Figure 9-15).

9-147. The first technique is to organize the reconnaissance elements separate from security elements. This technique is used when the security element is able to support the reconnaissance element from one location. This requires the reconnaissance objective to be clearly defined and the area to be fairly open.

9-148. The second technique is to organize the reconnaissance elements and security elements together into R&S teams. This technique is used when the reconnaissance objective is not clearly defined or the teams are not mutually supporting and each reconnaissance potentially needs its own security force. Within the R&S team, the reconnaissance can be done by one or two individuals while the rest of the element provides security. The number of Soldiers in an R&S team may vary depending on the mission. Usually a fire team (three to four Soldiers) is required for an adequate reconnaissance and still provide local security for the team.

9-149. The third technique is to establish R&S teams with an additional, separate security element. The separate security element can also act as a reserve or as a quick reaction force.

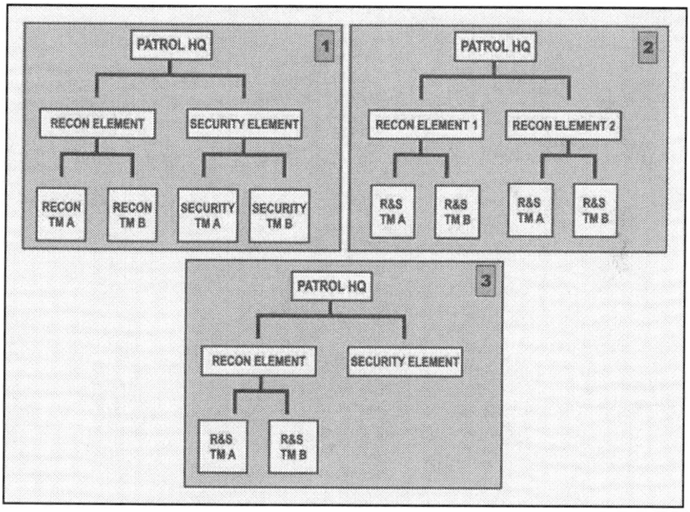

Figure 9-15. Organization of reconnaissance patrol.

ACTIONS ON THE RECONNAISSANCE OBJECTIVE

9-150. The actual reconnaissance begins at the designated transition point and ends with a follow-on transition to tactical movement away from the reconnaissance objective. Leaders mark the follow-on transition point with a control measure similar to the first transition point, using a linkup point, rendezvous point, a limit of advance, or a phase line. During this phase, leaders execute one of the three types of reconnaissance (area, zone, and route). These types of reconnaissance are distinguished by the scope of the reconnaissance objective. The types of reconnaissance patrols Infantry units conduct are area, zone, and route (Figure 9-16).

9-151. An *area reconnaissance* is conducted to obtain information about a certain location and the area around it such as road junctions, hills, bridges, or enemy positions. The location of the objective is shown by either grid coordinates or a map overlay. A boundary line encircles the area.

9-152. A *zone reconnaissance* is conducted to obtain information on all the enemy, terrain, and routes within a specific zone. The zone is defined by boundaries.

9-153. A *route reconnaissance* can orient on a road, a narrow axis such as an infiltration lane, or a general direction of attack. A platoon conducts a hasty route reconnaissance when there is too little time for a detailed route reconnaissance or when the mission requires less detailed information. Information sought in a hasty route reconnaissance is restricted to the type of route (limited or unlimited), obstacle limitations (maximum weight, height, and width), and observed enemy.

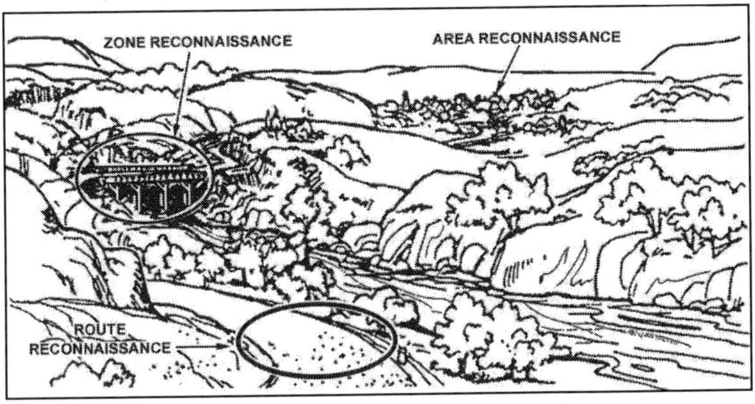

Figure 9-16. Types of reconnaissance patrols.

9-154. To plan for a reconnaissance, use the reverse planning process. The leader first determines the reconnaissance objective, an information requirement (IR) that corresponds to the terrain and or enemy in a specific area, route, or zone; it may be designated by a control measure such as an NAI, checkpoint, objective, route, phase lines, or boundaries. Once the leader has clarified the reconnaissance objective, he determines the observation plan that will enable the patrol to obtain the IR. After determining the observation plan, the leader determines the tactical movement necessary to position the patrol to achieve his observation plan.

INFORMATION REQUIREMENTS

9-155. Information requirements (IR) are the basis for the commander's critical information requirements (CCIR) needed to make tactical decisions. It is the responsibility of the controlling headquarters to clearly define the IR they want the patrol to determine. It is the responsibility of the patrol leader to clarify these IR prior to conducting the mission. Table 9-2 illustrates an example matrix that can be used to capture the IR for the controlling headquarters' collection plan.

Table 9-2. Example IR collection matrix.

Information Requirement	Location/ Description	Time	Purpose
1. Enemy forces within small arms range of intersection	NV12349875 road intersection	From: 201700Nov To: 210600Nov	Facilitate the company's passage through the area

9-156. IR can be enemy oriented, terrain oriented, civil oriented, or a combination. It is important for the leader to clarify the requirement prior to conducting the reconnaissance. Knowing this orientation enables the leader to demonstrate the initiative required to meet the higher leader's IR.

9-157. *Terrain-oriented IR* focus on determining information on the terrain of a particular area, route, or zone. While the unit will certainly look for the enemy presence, the overall intent is to determine the terrain's usefulness for friendly purposes. For example, the company commander may send out a squad-sized reconnaissance patrol to identify a location for the company's future assembly area. The patrol leader may send out a squad-sized reconnaissance patrol to obtain information about a bridge on a proposed infiltration route.

9-158. *Enemy-oriented IR* focus on finding a particular enemy force. The purpose of enemy-oriented reconnaissance is to confirm or deny planning assumptions. While the unit may be given a terrain feature as a reference point, the overall intent is to find the enemy. This means that if the enemy is not in the location referenced, it is usually necessary for the leader to demonstrate the initiative to find the enemy force within his given parameters.

9-159. *Civil-oriented IR* focus on determining information on the human environment in a particular area, route, or zone. Civil-oriented IR is a larger, vaguer category that requires more clarification than the other two categories. Examples of IR that are civil-oriented are the physical infrastructure; service infrastructures such as sewer, water, electric, and trash; the political situation; demographics; and refugees.

OBSERVATION PLAN

9-160. Once the patrol leader understands the IR, he then determines how it is that he will obtain it by developing an observation plan. The leader captures the observation plan as part of the patrol leader's course of action sketch. This is done through asking two basic questions:

(1) What is the best location(s) to obtain the information required?

(2) How can I best obtain the information without compromising the patrol?

9-161. The answer to the first question is: all vantage points and observation posts from which the patrol can best obtain the IR. A vantage point is a temporary position that enables observation of the enemy. It is meant to be occupied only until the IR is confirmed or denied. The answer to the second question is: use the routes and number of teams necessary to occupy the vantage points and OPs. An OP is a position from where military observations can be made and fire can be directed and adjusted. OPs must possess appropriate communications. The OP can either be short term (12 hours or less) or long term, depending on guidance from higher. Unlike a vantage point, the OP is normally occupied and surveillance is conducted for a specified period of time. The patrol views the reconnaissance objective from as many perspectives as possible, using whatever combinations of OPs and vantage points are necessary. The leader selects the tentative locations for the patrol's vantage points, OPs, and movement after analyzing METT-TC factors. These locations are proposed and must be confirmed and adjusted as necessary by the actual leader on the ground. From his analysis, he determines how many vantage points and OPs he must establish and where to position them. Once he decides on these general locations, he designs the routes for necessary movement between these and other control measures (such as the release points and linkup points). Positions should have the following characteristics:

- Covered and concealed routes to and from each position.
- Unobstructed observation of the assigned area, route, or zone. Ideally, the fields of observation of adjacent positions overlap to ensure full coverage.

- Effective cover and concealment. Leaders select positions with cover and concealment to reduce their vulnerability on the battlefield. Leaders may need to pass up a position with favorable observation capability but no cover and concealment to select a position that affords better survivability.
- A location that will not attract attention. Positions should not be sited in such locations as a water tower, an isolated grove of trees, or a lone building or tree. These positions draw enemy attention and may be used as enemy artillery TRPs.
- A location that does not skyline the observers. Avoid hilltops. Locate positions farther down the slope of the hill or on the side, provided there are covered and concealed routes into and out of the position.

9-162. The locations selected by the patrol are either long range or short range. Long-range positions must be far enough from the objective to be outside enemy's small-arms weapons, sensors, and other local security measures. Long-range positions are the most desirable method for executing a reconnaissance because the patrol does not come in close enough to be detected. If detected, the patrol is able to employ direct and indirect fires. Therefore, it is used whenever METT-TC permits the required information to be gathered from a distance. Security must be maintained by:

- Selecting covered and concealed OPs.
- Using covered and concealed routes in and around the objective area.
- Deploying security elements, including sensors, to give early warning, and providing covering fire if required.

9-163. Short-range positions are within the range of enemy local security measures and small-arms fire. When information required cannot be obtained by a long-range position, reconnaissance elements move closer to the objective. The vantage points and routes used during short-range observation should be carefully planned out and verified prior to using them. Doing so prevents detection by the enemy or friendly units from stumbling into one another or covering ground already passed over by another element.

SECTION IV — AREA RECONNAISSANCE

9-164. Area reconnaissance is a directed effort to obtain detailed information concerning the terrain or enemy activity within a prescribed area (FM 1-02). That area may be given as a grid coordinate, an objective, on an overlay. In an area reconnaissance, the patrol uses surveillance points, vantage points, or OPs around the objective to observe it and the surrounding area.

9-165. Actions at the objective for an area reconnaissance begin with the patrol in the ORP, and end with a dissemination of information after a linkup of the patrol's subordinate units. The critical actions include:

- Actions from the ORP.
- Execute the observation plan.
- Link up and continue the mission.

ACTIONS FROM THE OBJECTIVE RALLY POINT

9-166. The patrol occupies the ORP and conducts associated priorities of work. While the patrol establishes security and prepares for the mission, the patrol leader and selected personnel conduct a leader's reconnaissance. The leader must accomplish three things during this reconnaissance: pin point the objective and establish surveillance, identify a release point and follow-on linkup point (if required), and confirm the observation plan.

OBSERVATION PLAN FOR AN AREA RECONNAISSANCE

9-167. Upon returning from the leader's reconnaissance, the patrol leader disseminates information and FRAGOs as required. Once ready, the patrol departs. The leader first establishes security. Once security is in position, the reconnaissance element moves along the specified routes to the observation posts and vantage points in accordance with the observation plan.

SHORT RANGE

9-168. On nearing the objective, the patrol commander should establish a forward release point. It should be sited so it is well hidden, no closer than 200 meters from known enemy patrol routes, OPs, or sentry positions. The forward RP provides the patrol leader with a temporary location close to the objective from which he can operate. While the close reconnaissance is in progress, it should be manned by the patrol second in charge and the radio operator. Only vital transmissions should be made while in the forward release point. The volume setting should be as low as possible on the radio, and if available, the operator should use an earphone.

9-169. The close reconnaissance team should make its final preparation in the forward release point. Movement from the forward release point must be very slow and deliberate. Leaders should allow sufficient time for the team to obtain the information. If time is limited, the team should only be required to obtain essential information. If the enemy position is large, or time is limited, the leader may employ more than one close reconnaissance team. If this occurs, each patrol must have clearly defined routes for movement to and from the forward release point. They must also have clearly defined areas in which to conduct their reconnaissance in order to avoid clashes.

9-170. The close reconnaissance team normally consists of one to two observers and two security men. The security men should be sufficiently close to provide protection to the observer, but far enough away so his position is not compromised. When moving in areas close to the enemy position, only one man should move at any one time. Accordingly, bounds should be very short.

9-171. Once in position, the patrol observes and listens to acquire the needed information. No eating, no talking, and no unnecessary movement occurs at this time. If the reconnaissance element cannot acquire the information needed from its initial position, it retraces the route and repeats the process. This method of reconnaissance is extremely risky. The reconnaissance element must remember that the closer it moves to an objective, the greater the risk of being detected.

MULTIPLE RECONNAISSANCE AND SURVEILLANCE TEAMS

9-172. When information cannot be gathered from just one OP/vantage point, successive points may be used. Once determined, the leader must decide how his patrol will actually occupy them. The critical decision is to determine the number of teams in the reconnaissance element. The advantages of a single team in the reconnaissance element are the leader's ability to control the team, and the decreased probability of enemy detection. The disadvantages of the single team are the lack of redundancy and the fact that the objective area is only observed by one team. The advantages of using multiple teams include, affording the leader redundancy in accomplishing his mission, and the ability to look at the objective area from more than one perspective. The disadvantages include, the increased probability of being detected by the enemy, and increased difficulty of controlling the teams.

9-173. The leader may include a surveillance team in his reconnaissance of the objective from the ORP. He positions these surveillance teams while on the reconnaissance. He may move them on one route, posting them as they move, or he may direct them to move on separate routes to their assigned locations.

SECURITY ELEMENT

9-174. The subordinate leader responsible for security establishes security at the ORP and positions other security teams as required on likely enemy avenues of approach into the objective area.

SURVEILLANCE TEAMS

9-175. The platoon and squad use the surveillance/vantage point method that utilizes a series of surveillance or vantage points around the objective to observe it and the surrounding areas.

9-176. The unit halts in the ORP and establishes security while they confirm the location. The platoon leader conducts a leader's reconnaissance of the objective area to confirm the plan, and then returns to the ORP.

9-177. Once the security teams are in position, the reconnaissance element leaves the ORP. The element moves to several surveillance or vantage points around the objective. Instead of having the entire element move as a unit from point to point, the element leader might decide to have only a small reconnaissance team move to each surveillance or vantage point. After reconnoitering the objective, elements return to the ORP and disseminate information.

SECTION V — ROUTE RECONNAISSANCE

CONDUCT

9-178. A route reconnaissance is conducted to obtain detailed information about one route and all its adjacent terrain, or to locate sites for emplacing obstacles. Route reconnaissance is oriented on a road, a narrow axis such as an infiltration lane, or on a general direction of attack. Patrols conducting route reconnaissance operations attempt to view the route from both the friendly and enemy perspective. Infantry platoons require augmentation with technical expertise for a complete detailed route reconnaissance. However, platoons are capable of conducting hasty route reconnaissance or area reconnaissance of selected route areas.

9-179. Route reconnaissance is conducted to obtain and locate the following:

- Detailed information about trafficability on the route and all adjacent terrain.
- Detailed information about an enemy activity or enemy force moving along a route.
- Sites for emplacing hasty obstacles to slow enemy movement.
- Obstacles, CBRN contamination, and so forth.

9-180. The Infantry platoon unit can also be tasked to survey a route in a planned infiltration lane. After being briefed on the proposed infiltration, the patrol leader conducts a thorough map reconnaissance and plans a series of fans along the route (Figure 9-17). The coverage must reconnoiter all intersecting routes for a distance greater than the range at which enemy direct-fire weapons could influence the infiltrating forces.

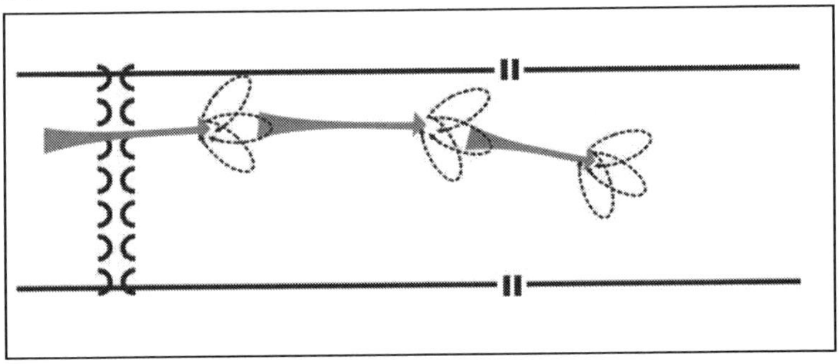

Figure 9-17. Route reconnaissance using fans.

9-181. The platoon reports conditions likely to affect friendly movement. These conditions include:

- Presence of the enemy.
- Terrain information.
- Location and condition of bypasses, fords, and obstacles.
- Choke points.
- Route and bridge conditions.

9-182. If all or part of the proposed route is a road, the leader must treat the road as a danger area. The platoon moves parallel to the road, using a covered and concealed route. When required, reconnaissance and security teams move close to the road to reconnoiter key areas. The platoon plans a different route for its return.

9-183. The leader should submit the patrol report in an overlay format (Figure 9-18) that includes—

- Two grid references (required).
- Magnetic north arrow (required).
- Route drawn to scale (required).
- Title block (required).
- Route classification formula (required).
- Road curves with a radius of less than 45 degrees.
- Steep grades and their maximum gradients.
- Road width of constrictions such as bridges and tunnels, with the widths and lengths of the traveled ways (in meters).
- Underpass limitations with limiting heights and widths.
- Bridge bypasses classified as easy, hard, or impossible.
- Civil or military road numbers or other designations.
- Locations of fords, ferries, and tunnels with limiting information.
- Causeways, snow sheds, or galleries if they are in the way. Data about clearance and load-carrying capacity should be included to permit an evaluation to decide whether to strengthen or remove them.

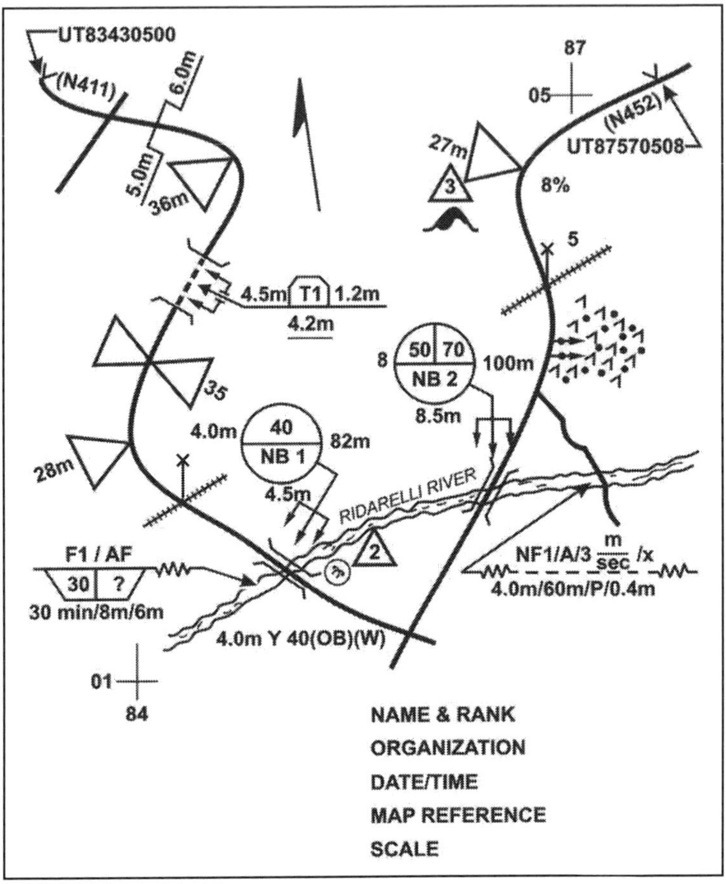

Figure 9-18. Route reconnaissance overlay.

SECTION VI — ZONE RECONNAISSANCE

9-184. A zone reconnaissance is conducted to obtain information on enemy, terrain, and routes within a specified zone. Zone reconnaissance techniques include the use of moving elements, stationary teams, or multiple area reconnaissance actions.

MOVING ELEMENT TECHNIQUES

9-185. When moving elements are used, the elements (squads or fire teams) move along multiple routes to cover the whole zone. When the mission requires a unit to saturate an area, the unit uses one of the following techniques: the fan; the box; converging routes; or successive sectors.

FAN METHOD

9-186. When using the fan method, the leader first selects a series of ORPs throughout the zone to operate from. The patrol establishes security at the first ORP. Upon confirming the ORP location, the leader confirms reconnaissance routes out from and back to the ORP. These routes form a fan-shaped pattern around the ORP. The routes must overlap to ensure the entire area is reconnoitered. Once the routes are confirmed, the leader sends out R&S teams along the routes. When all R&S teams have returned to the ORP, the platoon collects and disseminates all information to every Soldier before moving on to the next ORP.

9-187. Each R&S team moves from the ORP along a different fan-shaped route that overlaps with others to ensure reconnaissance of the entire area (Figure 9-19). These routes should be adjacent to each other. Adjacent routes prevent the patrol from potentially making contact in two different directions. The leader maintains a reserve at the ORP.

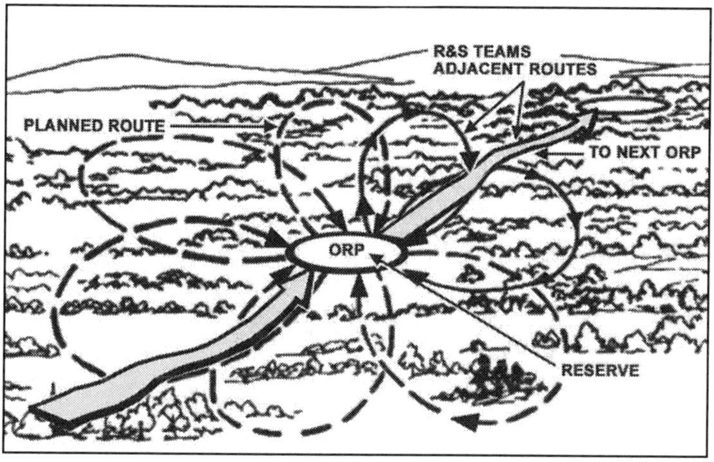

Figure 9-19. Fan method.

BOX METHOD

9-188. When using the box method, the leader sends his R&S teams from the ORP along routes that form a boxed-in area. He sends other teams along routes through the area within the box (Figure 9-20). All teams meet at a link-up point at the far side of the box from the ORP.

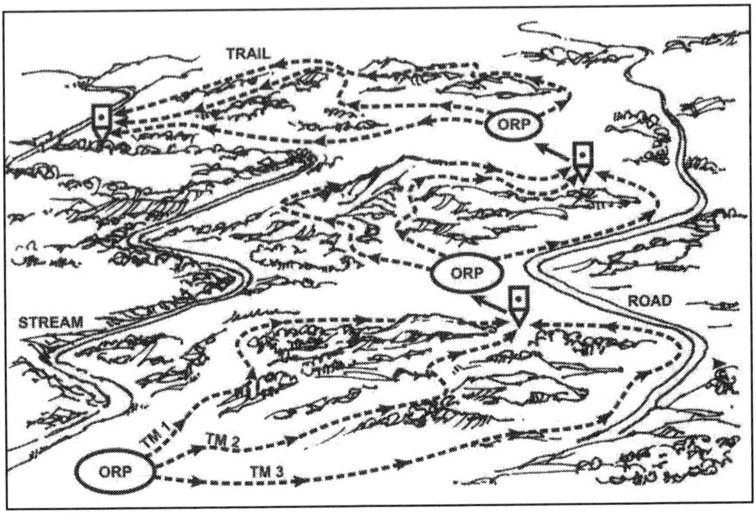

Figure 9-20. Box method.

Converging Routes Method

9-189. When using the converging routes method, the leader selects routes from the ORP through the zone to a rendezvous point at the far side of the zone from the ORP. Each R&S team moves along a specified route and uses the fan method to reconnoiter the area between routes (Figure 9-21). The leader designates a time for all teams to link up. Once the unit arrives at the rendezvous point, it halts and establishes security.

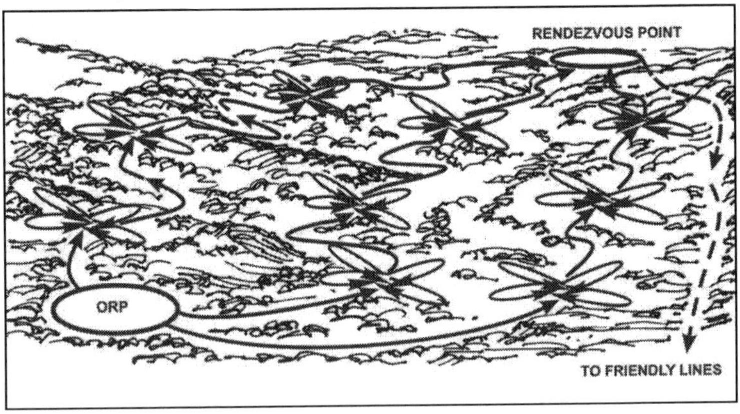

Figure 9-21. Converging routes method.

SUCCESSIVE SECTOR METHOD

9-190. The successive sector method is a continuation of the converging routes method (Figure 9-22). The leader divides the zone into a series of sectors. The platoon uses the converging routes within each sector to reconnoiter to an intermediate link-up point where it collects and disseminates the information gathered to that point. It then reconnoiters to the next sector. Using this method, the leader selects an ORP, a series of reconnaissance routes, and linkup points. The actions from each ORP to each linkup point are the same as in the converging routes method. Each linkup point becomes the ORP for the next phase. Upon linkup at a linkup point, the leader again confirms or selects reconnaissance routes, a linkup time, and the next linkup point. This action continues until the entire zone has been reconnoitered. Once the reconnaissance is completed, the unit returns to friendly lines.

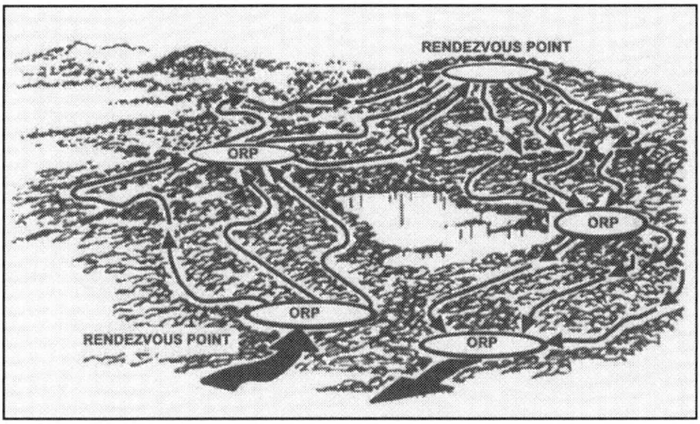

Figure 9-22. Successive sector method.

STATIONARY ELEMENT TECHNIQUES

9-191. Using the stationary element technique, the leader positions surveillance teams in locations where they can collectively observe the entire zone for long-term, continuous information gathering (Figure 9-23). The leader must consider sustainment requirements when developing his Soldiers' load plan.

Figure 9-23. Zone reconnaissance using the stationary element technique.

MULTIPLE AREA RECONNAISSANCE

9-192. When using multiple area reconnaissance the leader tasks each of his subordinate units to conduct a series of area reconnaissance actions within the zone (Figure 9-24).

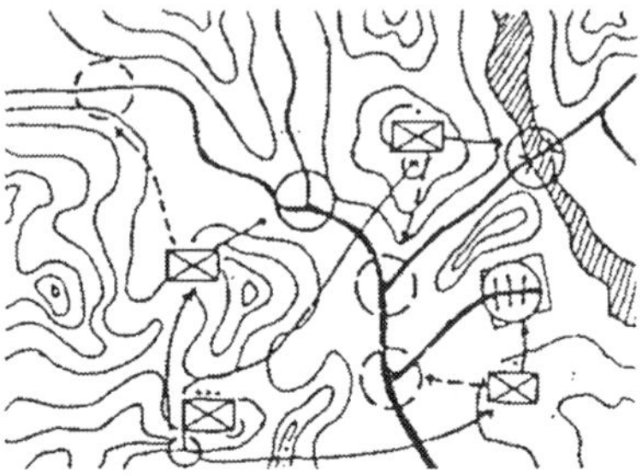

Figure 9-24. Zone reconnaissance using multiple area reconnaissance.

SECTION VII — PATROL PREPARATIONS

PREPARATIONS

9-193. Units send out patrols under many and varied conditions on the battlefield. Patrols are often used during high-intensity combat. They are also sent out during stability operations, and when the unit is providing support to civil authorities. The specific actions taken in preparing for a patrol, while conducting the mission, and after returning to the main body will vary depending on the tactical situation. The principles, however, will remain the same. During high-intensity combat, some of the actions described below may be abbreviated. Those same actions may be executed in much greater detail and specificity during stability operations or during support to civil authority. In general, patrol activities are much more closely documented during operations in other than high-intensity combat. Successful patrol operations require considerable preparation before a patrol departs. The commander or platoon leader should brief the patrol leader and give him clear orders before sending him away from the main body. Patrol members should depart on patrol confident of the patrol's capabilities. This can be understood through detailed knowledge of the mission's task and purpose, the threats that may be encountered during the patrol, and good situational awareness.

BRIEFINGS AND ORDERS

9-194. Patrol orders, pre-patrol briefings, and rehearsals should cover the following subjects:

- *Environment, local situation and possible threats.* The patrol leader should coordinate an intelligence briefing that covers the operating environment, local civil situation, terrain and weather that might affect the patrol's mission, general and specific threats to the patrol, suspect persons, and vehicles and locations known to be in the patrol's area.
- *Mine and IED threat.* The patrol leader should make a mine and IED risk assessment based on the latest information available. This will determine many of the actions of the patrol. Patrol members must be informed of the latest mine and IED threats and the restrictions to the unit's tactical SOPs that result.
- *Operations Update.* The patrol leader should coordinate for an up-to-date briefing on the location and intentions of other friendly patrols and units in the patrol's area. This briefing should include the existing fire and maneuver control measures in effect, any no-go or restricted

areas, any special instructions in effect for the patrol's area, and all other operational issues that may affect the patrol and its mission.

- *Mission and Tasks.* Every patrol leader should be given a specific task and purpose to accomplish with his patrol. Accordingly, each patrol member must know the mission and be aware of their responsibilities.
- *Locations and Route.* The patrol leader must brief his patrol on all pertinent locations and routes. Locations and routes may include drop-off points, pick-up points, planned routes; rally points, exit and re-entry points, and alternates for each should be covered in detail.
- *Posture.* This is a key consideration during a presence patrol. The patrol leader should not depart until he is sure that he completely understands what posture or attitude the commander wishes the patrol to present to the populace it encounters. The posture may be soft or hard depending on the situation, and the environment. The patrol posture may have to change several times during a patrol.
- *Actions on Contact and Actions at the Scene of an Incident.* These are likely to be part of the unit's tactical SOPs but should be covered especially if there are local variations or new members in the patrol.
- Rules of Engagement, Rules of Interaction and Rules for Escalation of Force: Each member of the patrol must know and understand these rules.
- *Communications Plan/Lost Communications Plan.* Every patrol member should know the means in which the patrol plans to communicate, to whom, how, and when it should report. The patrol leader must ensure that he has considered what actions the patrol will take in the event it loses communications. The unit may have established these actions in its tactical SOP, but all patrol members should be briefed on the communication plan and be given the appropriate frequencies, contact numbers, and passwords that are in effect.
- *Electronic Countermeasures Plan.* This is especially important if the IED threat level is high. The patrol leader should clearly explain to all members of the patrol which ECM devices are being employed, and their significant characteristics. These issues may be covered by the unit's tactical SOP but all patrol members should be briefed on the ECM plan that is in effect during the patrol.
- *Standard and Special Uniforms and Equipment:* Equipment should be distributed evenly among the patrol members. The location of key or unique equipment should be known by all members of the patrol. SOPs should be developed to stipulate what dress is to be worn for the various types of patrol. The dress state will be linked to threats and posture of the patrol, so patrol members should be briefed in sufficient time to enable proper preparations. All patrols must have a day and night capability regardless of the expected duration of the patrol.
- *Medical.* Every Soldier should carry his own first aid dressing per the unit tactical SOP. If possible, every patrol should have at least one combat lifesaver with a CLS bag. All patrol members must know who is responsible for carrying the pack and know how to use its contents.
- *Attachments.* The patrol leader must ensure that all personnel attached to the patrol are introduced to the other patrol members and briefed thoroughly on the tactical SOP; all patrol special orders; and the existing chain of command. The following type personnel may be attached to a unit going out to patrol:
- Interpreters.
  - Police (either military police or local security forces).
  - Specialists in search or explosive demolitions.
  - Female Soldiers specifically designated and trained to search local women.
  - Dog and dog handlers.

EQUIPMENT

9-195. Equipment carried by the patrol will be environment and task specific.

- *Radios and* electronic countermeasures *(ECM) Equipment.* Radios and ECM equipment should be checked prior to every patrol to ensure that it is serviceable and operates correctly.

batteries must be taken for the expected duration of the patrol plus some extra as backup. Patrol members must be trained in the operation of all ECM and radio equipment. It is the patrol leader's responsibility to ensure that radios and ECM equipment are switched on and working and communication checks are conducted prior to leaving the base location.

- *Weapons.* All weapons must be prepared for firing prior to departure from the larger unit. Slings should be used to ensure weapons do not become separated from any Soldier who becomes incapacitated. This also ensures that a weapon cannot be snatched away from a distracted Soldier while he is speaking with locals and used against him.
- *Ammunition.* Sufficient ammunition, signal pyrotechnics, smoke, and non-lethal munitions must be carried to enable the patrol to conduct its mission. The amount of each a patrol carries may be established by the unit's tactical SOP or by the patrol leader based on his evaluation of the situation the patrol will face.
- *Load-carrying Equipment.* Patrol members should carry sufficient team and personal equipment to enable them to accomplish other missions (such as reassignment to a cordon position before returning to the larger unit for resupply). The unit's tactical SOP should establish the standard amount of equipment and supplies to be carried. The commander must consider carefully the burden he places on his Soldiers going on a foot patrol, especially in extreme weather conditions or rugged terrain.
- *Documentation.* Team leaders are responsible to the patrol leader for ensuring that appropriate documentation is carried by individuals for the conduct of the mission. Under normal circumstances, Soldiers should carry just their identification card and tags. The unit tactical SOP may prohibit or require the carrying of other appropriate theatre specific documentation such as cards with rules on escalation of force, rules of engagement, or rules of interaction.

Equipment Checks

9-196. A number of equipment checks should be conducted prior to the patrol departing:

- *Individual Equipment Check.* It is the responsibility of every patrol member to check his or her individual equipment. Soldiers should ensure any loose items of equipment carried are secured.
- *Team Leader's Equipment Check.* Leaders must ensure that individual team members limit what they carry to that which is required for the patrol. Team equipment must be checked for serviceability.
- *Patrol Leader's Equipment Check.* Patrol leaders should check individual and team equipment from each team prior to deploying, paying particular attention to the serviceability of mission specific equipment.

Rehearsals

9-197. Patrols should rehearse any specific tactical actions or drills for situations the patrol leader anticipates they might encounter.

Communications Checks

9-198. Communications checks should be conducted with the unit headquarters or the tactical operations center before every patrol. Patrols should not leave the vicinity of the main body until all communication systems are operating correctly.

Patrol Manifest

9-199. When the situation allows, the patrol leader should submit a written patrol manifest to the commander or to Tactical Operations Center personnel prior to departing the main body. Regardless of the situation, whenever the unit sends out a patrol there should be a specific list of the patrol members made before it departs. The unit tactical SOP may establish a specific format for this manifest, but generally it should contain the following information:

- Patrol number or call sign designation.

before it departs. The unit tactical SOP may establish a specific format for this manifest, but generally it should contain the following information:

- Patrol number or call sign designation.
- Unit designation of unit sending the patrol out.
- Patrol task and purpose (mission).
- Names and rank of patrol leader and all subordinate leaders.
- Estimated DTG Out.
- Estimated DTG In.
- Brief description of the patrol's intended route.
- Complete names, rank, and unit of all members of the patrol, including attachments.
- Number, nomenclature, and serial number of all weapons with the patrol.
- Number, nomenclature, and serial number of all ECM devices, radios, and any other special or sensitive equipment with the patrol.
- Vehicle type and registration number (if appropriate)

9-200. The purpose of the manifest is to allow the higher headquarters to keep track of all the patrols that are out and those that have returned. If the patrol engages the enemy or fails to return on time without reporting, the headquarters has information on the size, capability and intentions of the patrol that it may need. If the patrol suffers casualties or has a vehicle disabled, this manifest can be used to check that all personnel, weapons and sensitive items were recovered.

DEPARTURE REPORT

9-201. The patrol leader should render a departure report just as the patrol departs the main body location or the base. Depending on the procedure established by the unit's tactical SOP, this might include a detailed listing of the patrol's composition. It may also simply state the patrol's call sign or patrol number and report its departure.

WEAPONS STATUS

9-202. Immediately upon leaving an established base or the main body position, the patrol leader and team leaders should ensure that all the patrol weapons are loaded and prepared for immediate action. Electronic countermeasures should be checked to ensure they are turned on if appropriate and all radio frequency settings should be confirmed.

9-203. When the patrol returns to the base, each Soldier should clear his weapon immediately after entering the protected area. The unit's tactical SOP will normally establish precise procedures for this clearing. Patrol leaders should ensure that all individual and crew-served weapons are unloaded.

EXITING AND ENTERING A FIXED BASE

9-204. Exiting and entering a fixed operating base is a high risk activity due to the way troops are channeled through narrow entry or exit points. Insurgents are known to monitor patrols leaving and entering base locations to identify patterns and areas of weakness that they can exploit. Patrols leaving and entering a base can reduce the risks of attack by varying the points used to exit and enter the base, and any routes used to transit the immediate area around the base. If this is not possible, extreme caution should be used in the vicinity of the exit and entry points. Patrol leaders must ensure their patrols do no become complacent. Units should ensure close coordination between patrol leaders and guards at the entry point while the patrol is transiting the gate.

SECURITY CHECKS WHILE ON PATROL

9-205. Patrol members must assist their patrol leader by applying basic patrolling techniques consistently. This gives the team leader more time to concentrate on assisting the patrol leader in the conduct of the patrol. Team members should concentrate on maintaining spacing, formation, alertness, conducting 5 and 20 meter checks and taking up effective fire positions without supervision.

5 and 20 Meter Checks

9-206. Every time a patrol stops, it should use a fundamental security technique known as the 5 and 20 meter check. The technique involves every patrol member requiring him to make detailed, focused examinations of the area immediately around him, and looking for anything out of the ordinary that might be dangerous or significant. Five meter checks should be conducted every time a patrol member stops. Twenty meter checks should be conducted when a patrol halts for more than a few minutes.

9-207. Soldiers should conduct a visual check using their unaided vision, and by using the optics on their weapons and binoculars. They should check for anything suspicious, and anything out of the ordinary. This might be as minor as bricks missing from walls, new string or wire run across a path, mounds of fresh soil dirt, or any other suspicious signs. Check the area at ground level through to above head height.

9-208. When the patrol makes a planned halt, the patrol leader identifies an area for occupation and stops 50 meters short of it. While the remainder of the patrol provides security, the patrol leader carries out a visual check using binoculars. He then moves the patrol forward to 20 meters from the position and conducts a visual check using optics on his weapon or with unaided vision.

9-209. Before actually occupying the position, each Soldier carries out a thorough visual and physical check for a radius of 5 meters. They must be systematic, take time and show curiosity. Use touch and, at night, white light if appropriate.

9-210. Any obstacles must be physically checked for command wires. Fences, walls, wires, posts and the ground immediately underneath must be carefully felt by hand, without gloves.

SECTION VIII — POST PATROL ACTIVITIES

ACCOUNTING FOR PATROL MEMBERS

9-211. Immediately on re-entering the secure base or rejoining the unit, the patrol leader should positively verify that all members of the patrol and any included attachments, prisoners, or detainees are accounted for.

Checking In

9-212. The patrol leader should check in with the company command post or the battalion tactical operations center as soon as possible after entering the base location or rejoining the unit.

Accounting for Weapons and Equipment

9-213. The patrol leader is responsible for verifying that all the patrol's weapons, ammunition, munitions and equipment are properly accounted for and reporting that status to the commander or the operations center. Lost or missing equipment must be reported immediately. The patrol may be ordered to return to the area where it was lost, if it is assessed safe to do so, and look for the item.

Hot debrief

9-214. The patrol leader should conduct a "hot debrief" with the entire patrol as soon as possible after entering the base or rejoining the main body. This allows him to capture low level information while the Soldiers' memories are fresh and the information relevant. Every member of the patrol should participate. If there was an interpreter or other attachments with the patrol, they too should be de-briefed as a source of human intelligence (HUMINT) by allowing them to pass on any information they obtained during the patrol. The patrol leader includes the significant information that he gleans during the hot debrief in his patrol report to the commander.

PATROL REPORT

9-215. Immediately after the hot debrief, the patrol leader should render his patrol report to the commander. This report may be verbal or written, simple, or elaborate depending on the situation and the commander's requirements. The commander may have the patrol leader render his report to the battalion intelligence officer or to the duty officer at the battalion tactical operations center, especially during stability or civil support operations. The patrol commander is responsible for the patrol report. He may be assisted by his assistant patrol leaders and any specialist personnel that were attached to the patrol.

ACTUAL PATROL ROUTE

9-216. The patrol report (Figure 9-25) should include a description of the actual route taken by the patrol (as opposed to the planned route), including any halt locations. If the unit uses digital command and control systems that automatically track and display the patrol's route, the information is already known. If not, the patrol leader must report it. When global positioning devices are used by the patrol, gathering route information is easier and faster. The actual route the patrol took is important for planning future patrol routes and actions. Enemy intelligence operations will attempt to identify any pattern setting by U.S. and coalition patrols, including the locations of halts. This may result in attack against locations regularly used by security forces.

Patrol Report (Example)

To: (Commander of unit ordering the patrol)
From: (Rank an name of the patrol leader)
Title: PATROL SITREP for Patrol # (Patrol designation or number per unit tactical SOP)
DTG Patrol Departed and DTG Patrol Returned: (All dates and times per the unit tactical SOP)

Mission: (Restatement of original mission, noting any modifications or FRAGOs received during the patrol's duration.)

Friendly forces (Only specify details on patrol composition that have changed.)

Situation: (The patrol leader's evaluation of mission accomplishment with a general description of any significant patrol sightings.)

Specific Incidents

-Time of incident
-Location of incident (grid/name)
-Type/description of incident
-Persons involved or witnesses to the incident
-Number and types of casualties
-Location of casualties
-Actions taken by friendly forces
-Details of hostile persons/terrorists/insurgents
-General comments/additional info

Figure 9-25. Patrol report example.

Appendix A

Machine Gun Employment

Whether organic to the unit or attached, machine guns provide the heavy volume of close and continuous fire needed to achieve fire superiority. They are the Infantry platoon's most effective weapons against a dismounted enemy force. These formidable weapons can engage enemy targets beyond the capability of individual weapons with controlled and accurate fire. This appendix addresses the capabilities, limitations, and fundamental techniques of fire common to machine guns.

SECTION I — TECHNICAL DATA AND CONSIDERATIONS

A-1. Leaders must know the technical characteristics of their assigned weapon systems and associated ammunition to maximize their killing and suppressive fires while minimizing the risk to friendly forces. Table A-1 lists machine gun specifications and technical data. Read the FMs specific to the machine guns listed in Table A-1 for complete information regarding their technical specifications.

Table A-1. Machine gun specifications.

WEAPON	M249	M240B	M2	MK 19
FIELD MANUAL	FM 3-22.68	FM 3-22.68	FM 3-22.65	FM 3-22.27
TM	9-1005-201-10	9-1005-313-10	9-1005-213-10	9-1010-230-10
DESCRIPTION	5.56-mm gas-operated automatic weapon	7.62-mm gas-operated medium machine gun	.50-caliber recoil-operated heavy machine gun	40-mm air-cooled, blowback-operated automatic grenade launcher
WEIGHT	16.41 lbs (gun with barrel) 16 lbs (tripod)	27.6 lbs (gun with barrel) 20 lbs (tripod)	128 lbs (gun with barrel and tripod)	140.6 lbs (gun with barrel and tripod)
LENGTH	104 cm	110.5 cm	156 cm	109.5 cm

Table A-1. Machine gun specifications (continued).

WEAPON	M249	M240B	M2	MK 19
SUSTAINED RATE OF FIRE Rounds/burst Interval Minutes to barrel change	50 RPM 6-9 rounds 4-5 seconds 10 minutes	100 RPM 6-9 rounds 4-5 seconds 10 minutes	40 RPM 6-9 rounds 10-15 seconds Change barrel end of day or if damaged	40 RPM
RAPID RATE OF FIRE Rounds/burst Interval Minutes to barrel change	100 RPM 6-9 rounds 2-3 seconds 2 minutes	200 RPM 10-13 rounds 2-3 seconds 2 minutes	40 RPM 6-9 rounds 5-10 seconds Change barrel end of day or if damaged	60 RPM
CYCLIC RATE OF FIRE	850 RPM in continuous burst Barrel change every 1 minute	650-950 RPM in continuous burst Barrel change every 1 minute	450-550 RPM in continuous burst	325-375 RPM in continuous burst
MAXIMUM EFFECTIVE RANGES	Bipod/point: 600 m Bipod/area: 800 m Tripod/area: 1,000 m Grazing: 600 m	Bipod/point: 600 m Tripod/point: 800 m Bipod/area: 800 m Tripod/area: 1,100 m Suppression: 1,800 m Grazing: 600 m	Point: 1,500 m (single shot) Area: 1,830 m Grazing: 700 m	Point: 1,500 m Area: 2,212 m
MAXIMUM RANGE	3,600 m	3,725 m	6,764 m	2,212 m

A-2. Machine gun fire has different effects on enemy targets depending on the type of ammunition used, the range to target, and the nature of the target. It is important that gunners and leaders understand the technical aspects of the different ammunition available to ensure the machine guns and automatic weapons are employed in accordance with their capabilities. Machine guns and automatic weapons use several different types of standard military ammunition. Soldiers should use only authorized ammunition that is manufactured to U.S. and NATO specifications.

M249 MACHINE GUN

A-3. The M249 machine gun is organic to the Infantry platoon and provides rifle squads with a light automatic weapon for employment during assault (Figure A-1). The M249 can also be used in the machine gun role in the defense or support-by-fire position. The M249 fires from the bipod, the hip, or from the underarm position. The hip and underarm positions are normally used for close-in fire during an assault when the M249 gunner is on the move and does not have time to set the gun in the bipod position. It is best used when a high rate of fire is needed immediately. Accuracy of fire is decreased when firing from either the hip or shoulder.

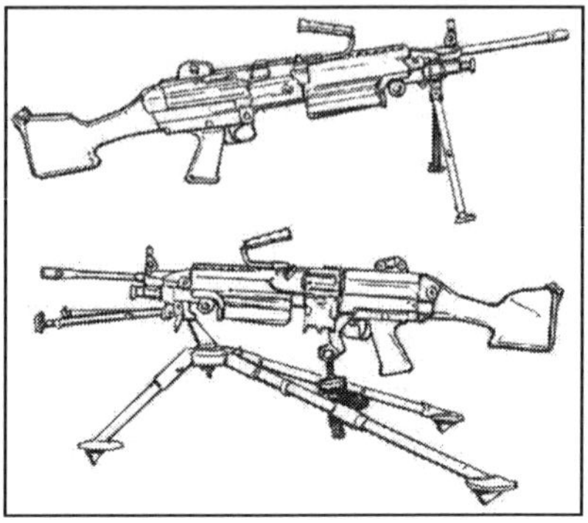

Figure A-1. M249 machine gun, bipod and tripod mounted.

A-4. Available M249 ammunition is classified as follows (Table A-2).

- **M855 5.56-mm Ball.** For use against light materials and personnel, but not vehicles.
- **M856 5.56-mm Tracer.** Generally used for adjustments after observation, incendiary effects, and signaling. When tracer rounds are fired, they are normally mixed with ball ammunition in a ratio of four ball rounds to one tracer round.
- **M193 5.56-mm Ball.** M193 ball ammunition can be fired with the M249, but accuracy is degraded. It should therefore only be used in emergency situations when M855 ball is not available.
- **M196 5.56-mm Tracer.** M196 tracer ammunition can be fired with the M249, but accuracy is degraded. It should therefore only be used in emergency situations when M856 ammunition is not available.

Table A-2. M249 ballistic data.

AVAILABLE M249 CARTRIDGES	MAXIMUM RANGE (meters)	TRACER BURNOUT (meters)	USES
Ball, M855	3,600	——	Light materials, personnel
Tracer, M856	3,600	900	Observation and adjustment of fire, incendiary effects, signaling

M240B MACHINE GUN

A-5. The M240B is organic to the Infantry platoon. Two machine guns and crews are found in the weapons squad (Figure A-2). The M240B can be fired in the assault mode in emergencies, but is normally fired from the bipod or tripod platform. It can also be vehicle mounted. The platoon leader (through his weapons squad leader) employs his M240B machine guns with a rifle squad to provide long range, accurate, sustained fires against dismounted infantry, apertures in fortifications, buildings, and lightly-armored vehicles. The M240B also provides a high volume of short-range fire in self defense against aircraft. Machine gunners use point, traversing, searching, or searching and traversing fire to kill or suppress targets.

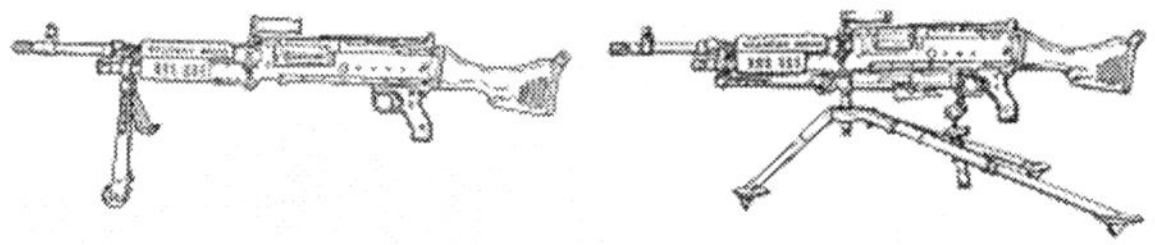

Figure A-2. M240B machine gun, bipod and tripod mounted.

A-6. Available M240B machine gun ammunition is classified as follows (Table A-3).

- **M80 7.62-mm Ball.** For use against light materials and personnel.
- **M61 7.62-mm Armor Piercing.** For use against lightly-armored targets.
- **M62 7.62-mm Tracer.** For observation of fire, incendiary effects, signaling, and for training. When tracer rounds are fired, they are normally mixed with ball ammunition in a ratio of four ball rounds to one tracer round.

Table A-3. M240B ballistic data.

AVAILABLE M240B CARTRIDGES	MAXIMUM RANGE (meters)	TRACER BURNOUT (meters)	USES
Ball, M80	3,725	——	Light materials, personnel
Armor Piercing, M61	3,725	——	Lightly-armored targets
Tracer, M62	3,725	900	Observation and adjustment of fire, incendiary effects, signaling

MK 19 40-MM MACHINE GUN, MOD 3

A-7. The MK 19 is not organic to the weapons company, not the Infantry platoon, but because there are many times when Infantrymen use it, it is described in this appendix. The MK 19 supports the Soldier in both the offense and defense. It gives the unit the capability of laying down a heavy volume of close, accurate, and continuous fire (Figure A-3). The MK 19 can also—

- Protect motor movements, assembly areas, and supply trains in a bivouac.
- Defend against hovering rotary aircraft.
- Destroy lightly-armored vehicles.
- Fire on enemy prepared positions.
- Provide high volumes of fire into an engagement area (EA).
- Cover obstacles.
- Provide indirect fires from defilade positions.

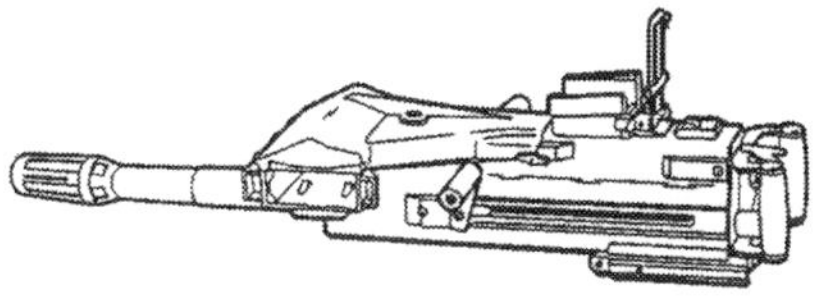

Figure A-3. MK 19, 40-mm grenade machine gun, MOD 3.

A-8. The MK 19 is normally vehicle mounted on a pedestal, ring, or weapon platform, but can also be fired from the M3 tripod. It fires high explosive (HE) and high explosive, dual purpose (HEDP) rounds. The HE round is effective against unarmored vehicles and personnel.

A-9. Available MK 19 machine gun ammunition is classified as follows (Table A-4).

- **M430 40-mm HEDP.** This is the standard round for the MK 19 and comes packed in either 48- or 32- round ammunition containers. It can penetrate 2 inches of steel armor at zero-degree obliquity and inflict casualties out to 15 meters from impact. It arms within 18 to 30 meters of the gun muzzle.
- **M383 40-mm HE.** Comes packed in a 48-round container. It has a wound radius of 15 meters, but lacks the armor piercing capabilities of the HEDP round. It arms 18 to 36 meters from the muzzle.

Table A-4. MK 19 ballistic data.

AVAILABLE MK 19 CARTRIDGES	MAXIMUM RANGE (meters)	PENETRATION/ CASUALTY RADIUS	USES
HEDP, M430	2,212	2-inch armor/ 15-meter casualty radius	Lightly-armored targets, light material targets, personnel.
HE, M383	2,212	15-meter casualty radius	Unarmored vehicles, light material targets, personnel

M2 .50 CALIBER MACHINE GUN

A-10. The M2 .50 caliber machine gun is not organic to the Infantry platoon, but as there are many times when Infantrymen use it, it is described in this appendix (Figure A-4).

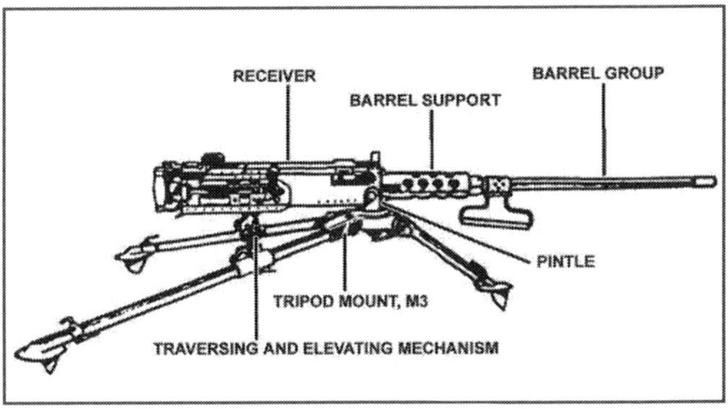

Figure A-4. M2 .50 caliber machine gun.

A-11. The available M2 .50 caliber machine gun ammunition is classified as follows (Table A-5).

- **M2 .50-Caliber Ball.** For use against enemy personnel and light material targets.
- **M1/M17 .50-Caliber Tracer.** Aids in observing fire. Secondary purposes are for incendiary effect and for signaling.
- **M1 .50-Caliber Incendiary.** For incendiary effect, especially against aircraft.

- **M2 .50-Caliber AP.** For use against armored aircraft and lightly-armored vehicles, concrete shelters, and other bullet-resisting targets.
- **M8 .50-Caliber API.** For combined armor-piercing and incendiary effect.
- **M20 .50-Caliber API Tracer.** For combined armor-piercing and incendiary effect, with the additional tracer feature.

Table A-5. M2 Ballistic data.

AVAILABLE M2 CARTRIDGES	MAXIMUM RANGE (meters)	TRACER BURNOUT (meters)	AVERAGE MUZZLE VELOCITY (feet per second)
Ball, M2	7,400	——	2,930
Tracer, M1 (with gilding metal jacket)	5,575	1,800	2,860
Tracer, M1 (with clad steel jacket)	5,450	1,800	3,030
Tracer, M17	5,450	2,450	3,030
Incendiary, M1	6,050	——	3,090
Armor-piercing, M2	7,400	——	2,930
Armor-piercing incendiary, M8	6,470	——	3,050
Armor-piercing incendiary tracer, M20	6,470	*300-1,750	3,050
* This tracer is dim at near ranges but increases in brightness as it moves farther from the gun.			

SECTION II — COMBAT TECHNIQUES OF FIRE

A-12. This section is designed to illustrate the characteristics of machine gun fire, the types of enemy targets that might be engaged, and how to successfully apply machine gun fire on those enemy targets.

A-13. Read the appropriate FM (as shown in Table A-1) for more weapon-specific information on engaging enemy targets with a particular machine gun.

CHARACTERISTICS OF FIRE

A-14. The gunner's or leader's knowledge of the machine gun is not complete until he learns about the action and effect of the projectiles when fired. The following definitions will help the leader, gunner, and automatic rifleman understand the characteristics of fire for the platoon's machine guns.

LINE OF SIGHT

A-15. Line of sight is an imaginary line drawn from the firer's eye through the sights to the point of aim.

BURST OF FIRE

A-16. A burst of fire is a number of successive rounds fired with the same elevation and point of aim when the trigger is held to the rear. The number of rounds in a burst can vary depending on the type of fire employed.

TRAJECTORY

A-17. Trajectory is the curved path of the projectile in its flight from the muzzle of the weapon to its impact. The major factors that influence trajectory are the velocity of the round, gravity, rotation of the round, and resistance of the air. As the range to the target increases, so does the curve of trajectory (Figure A-5).

MAXIMUM ORDINATE

A-18. Maximum ordinate is the highest point above the line of sight the trajectory reaches between the muzzle of the weapon and the base of the target. It always occurs at a point about two-thirds of the distance from weapon to target and increases with range. Like trajectory, maximum ordinate increases as the range increases (Figure A-5).

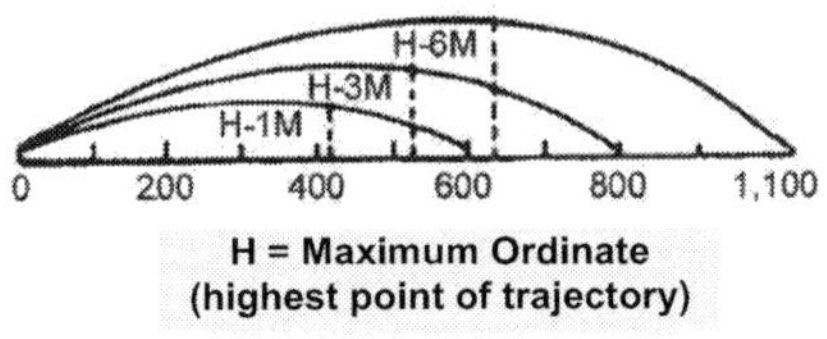

Figure A-5. Trajectory and maximum ordinate.

CONE OF FIRE

A-19. The cone of fire is the pattern formed by the different trajectories in each burst as they travel downrange. Vibration of the weapon and variations in ammunition and atmospheric conditions all contribute to the trajectories that make up the cone of fire (Figure A-6).

BEATEN ZONE

A-20. The beaten zone is the elliptical pattern formed when the rounds within the cone of fire strike the ground or target. The size and shape of the beaten zone change as a function of the range to and slope of the target, but is normally oval or cigar shaped and the density of the rounds decreases toward the edges. Gunners and automatic riflemen should engage targets to take maximum effect of the beaten zone. The simplest way to do this is to aim at the center base of the target. Most rounds will not fall over the target, and any that fall short will create ricochets into the target (Figure A-6).

Effective Beaten Zone

A-21. Because of dispersion, only that part of the beaten zone in which 85 percent of the rounds fall is considered the effective beaten zone.

Effect of Range on the Beaten Zone

A-22. As the range to the target increases, the beaten zone becomes shorter and wider. Conversely, as the range to the target decreases, the beaten zone becomes longer and narrower (Table A-6).

Effect of Slope on the Beaten Zone

A-23. The length of the beaten zone for any given range will vary according to the slope of the ground. On rising ground, the beaten zone becomes shorter but remains the same width. On ground that slopes away from the gun, the beaten zone becomes longer but remains the same width.

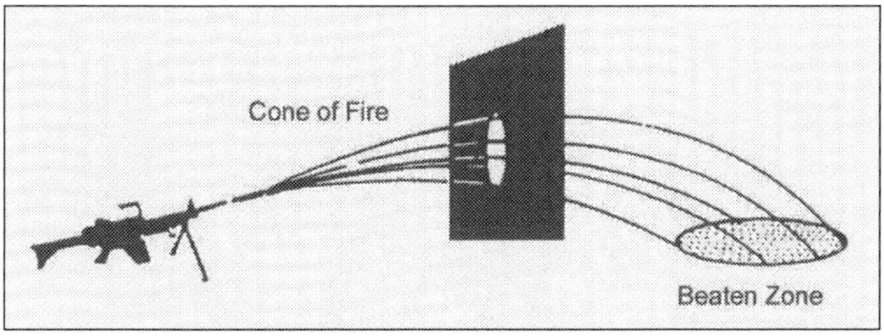

Figure A-6. Cone of fire and beaten zone.

Table A-6. Beaten zones of M240B.

M240B
Range: 500m (1m wide x 110m long)
Range: 1,000m (2m wide x 75m long)
Range: 1,500m (3m wide x 55m long)
Range: 2,000m (4m wide x 50m long)

DANGER SPACE

A-24. This is the space between the muzzle of the weapon and the target where the trajectory does not rise above 1.8 meters (the average height of a standing Soldier) that includes the beaten zone. Gunners should consider the danger space of their weapons when planning overhead fires.

SURFACE DANGER ZONE

A-25. Surface danger zones (SDZs) were developed for each weapon and are defined as the area in front, back, or side of the muzzle of the weapon that provides a danger to friendly forces when the weapon is fired. The SDZ is not just the area that comprises the cone of fire as it moves downrange. It also involves the possible impact area on both sides of the gun target line and the possible dispersion of material caused by the strike of the rounds, the possible ricochet area, and any area to the rear that is adversely affected by the effects of firing the weapon (Figure A-7).

A-26. SDZs were developed primarily for ranges and must be complied with when training, but they should also be complied with in combat when possible to minimize risk to friendly forces.

A-27. Refer to DA PAM 385-63 for a more detailed discussion of the SDZs for machine guns.

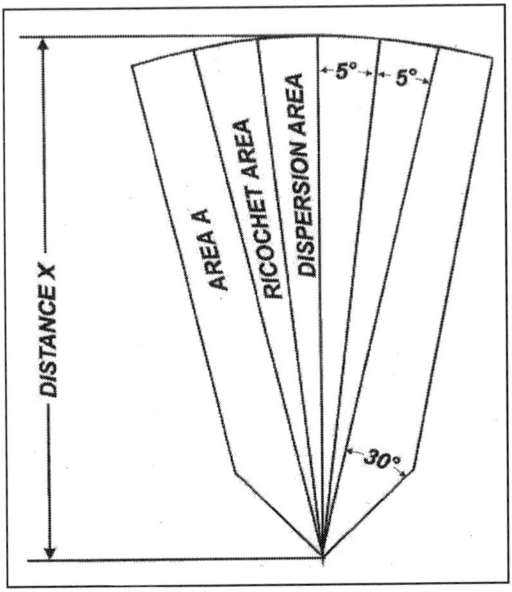

Figure A-7. Example surface danger zone for machine guns.

CLASSIFICATIONS OF AUTOMATIC WEAPONS FIRE

A-28. The U.S. Army classifies automatic weapons fires with respect to the ground, target, and weapon.

CLASSIFICATION OF FIRES WITH RESPECT TO THE GROUND

A-29. Fires with respect to the ground include grazing and plunging fire.

Dead Space

A-30. Any fold or depression in the ground that prevents a target from being engaged from a fixed position is termed *dead space*. Paragraph A-81 discusses methods of determining dead space.

Grazing Fires

A-31. Automatic weapons achieve grazing fire when the center of the cone of fire does not rise more than 1 meter above the ground. Grazing fire is employed in the final protective line (FPL) in defense and is only possible when the terrain is level or uniformly sloping. Any dead space encountered along the FPL must be covered by indirect fire, such as from an M203. When firing over level or uniformly sloping terrain, the machine gun M240B and M249 can attain a maximum of 600 meters of grazing fire. The M2 can attain a maximum of 700 meters. Paragraphs A-78 and A-79 discuss the FPL.

Plunging Fires

A-32. Plunging fire occurs when there is little or no danger space from the muzzle of the weapon to the beaten zone. It occurs when weapons fire at long range, when firing from high ground to low ground, when firing into abruptly rising ground, or when firing across uneven terrain, resulting in a loss of grazing fire at any point along the trajectory (Figure A-8).

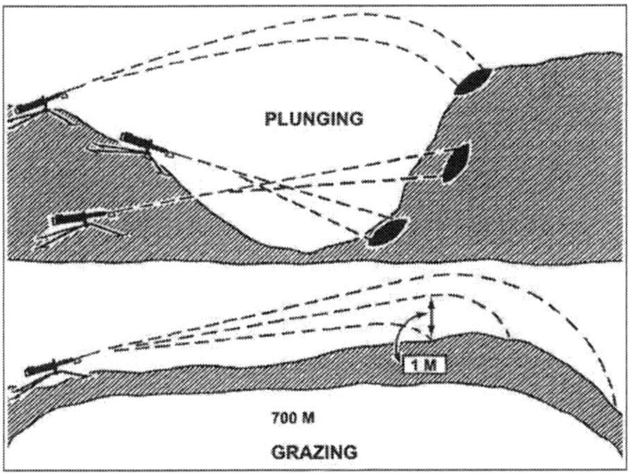

Figure A-8. Classes of fire with respect to the ground.

Classification of Fires With Respect to the Target

A-33. Fires with respect to the target include enfilade, frontal, flanking, and oblique fire (Figures A-9, A-10, and A-11). These targets are normally presented to gun teams by the enemy and must be engaged as they are presented. For instance, if the enemy presents its flank to the gun crew as it moves past their position from the left or right, the gun crew will have no choice but to employ flanking fire on the enemy.

A-34. Leaders and gunners should strive at all times to position their gun teams where they can best take advantage of the machine gun's beaten zone with respect to an enemy target. Channeling the enemy by use of terrain or obstacles so they approach a friendly machine gun position from the front in a column formation is one example. In this situation, the machine gun would employ enfilade fire on the enemy column, and the effects of the machine gun's beaten zone would be much greater than if it engaged that same enemy column from the flank.

Enfilade Fire

A-35. Enfilade fire occurs when the long axis of the beaten zone coincides or nearly coincides with the long axis of the target. It can be frontal fire on an enemy column formation or flanking fire on an enemy line formation. *This is the most desirable class of fire with respect to the target because it makes maximum use of the beaten zone.* Leaders and gunners should always strive to position the guns to the extent possible that they can engage enemy targets with enfilade fire (Figures A-9 and A-11).

Frontal Fire

A-36. Frontal fire occurs when the long axis of the beaten zone is at a right angle to the front of the target. This type of fire is highly desirable when engaging a column formation. It then becomes enfilade fire as the beaten zone coincides with the long axis of the target (Figures A-9 and A-10). Frontal fire is not as desirable when engaging a line formation because the majority of the beaten zone normally falls below or after the enemy target.

Flanking Fire

A-37. Flanking fire is delivered directly against the flank of the target. Flanking fire is highly desirable when engaging an enemy line formation. It then becomes enfilade fire as the beaten zone will coincide with the long axis of the target (Figures A-9 and A-10). Flanking fire against an enemy column formation is least desirable because the majority of the beaten zone normally falls before or after the enemy target.

Oblique Fire

A-38. Gunners and automatic riflemen achieve oblique fire when the long axis of the beaten zone is at an angle other than a right angle to the front of the target (Figures A-9 and A-11).

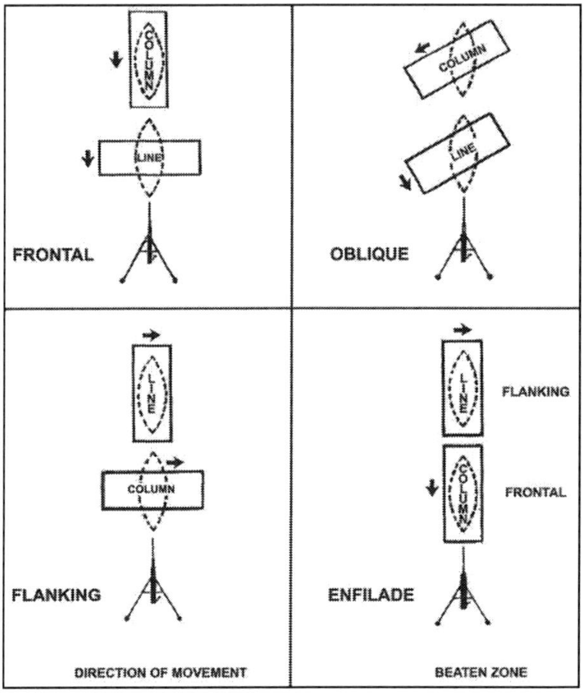

Figure A-9. Classes of fire with respect to the target.

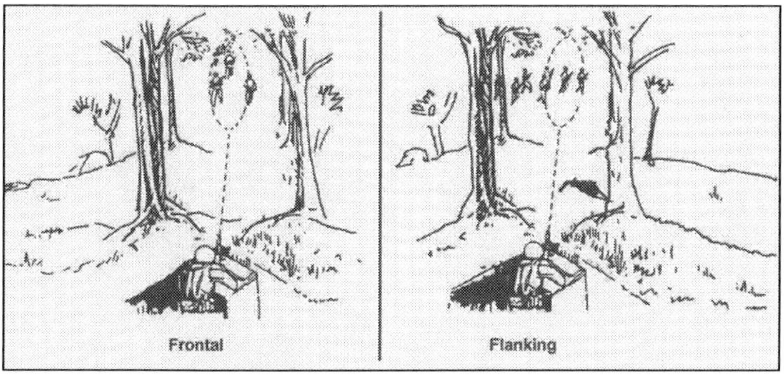

Figure A-10. Frontal fire and flanking fire.

Figure A-11. Oblique fire and enfilade fire.

Classification of Fires With Respect to the Machine Gun

A-39. Fires with respect to the weapon include fixed, traversing, searching, traversing and searching, swinging traverse, and free gun fires (Figure A-12).

Fixed Fire

A-40. Fixed fire is delivered against a stationary point target when the depth and width of the beaten zone will cover the target with little or no manipulation needed. After the initial burst, the gunners will follow any change or movement of the target without command.

Traversing Fire

A-41. Traversing disperses fires in width by successive changes in direction, but not elevation. It is delivered against a wide target with minimal depth. When engaging a wide target requiring traversing fire, the gunner should select successive aiming points throughout the target area. These aiming points should be close enough together to ensure adequate target coverage. However, they do not need to be so close that they waste ammunition by concentrating a heavy volume of fire in a small area.

Searching Fire

A-42. Searching distributes fires in depth by successive changes in elevation. It is employed against a deep target or a target that has depth and minimal width, requiring changes in only the elevation of the gun. The amount of elevation change depends upon the range and slope of the ground.

Traversing and Searching Fire

A-43. This class of fire is a combination in which successive changes in direction *and* elevation result in the distribution of fires both in width and depth. It is employed against a target whose long axis is oblique to the direction of fire.

Swinging Traverse

A-44. Swinging traverse fire is employed against targets that require major changes in direction but little or no change in elevation. Targets may be dense, wide, in close formations moving slowly toward or away from the gun, or vehicles or mounted troops moving across the front. If tripod mounted, the traversing slide lock lever is loosened enough to permit the gunner to swing the gun laterally. When firing swinging traverse, the weapon is normally fired at the cyclic rate of fire. Swinging traverse consumes a lot of ammunition and does not have a beaten zone because each round seeks its own area of impact.

Free Gun

A-45. Free gun fire is delivered against moving targets that must be rapidly engaged with fast changes in both direction and elevation. Examples are aerial targets, vehicles, mounted troops, or infantry in relatively close formations moving rapidly toward or away from the gun position. When firing free gun, the weapon is normally fired at the cyclic rate of fire. Free gun fire consumes a lot of ammunition and does not have a beaten zone because each round seeks its own area of impact.

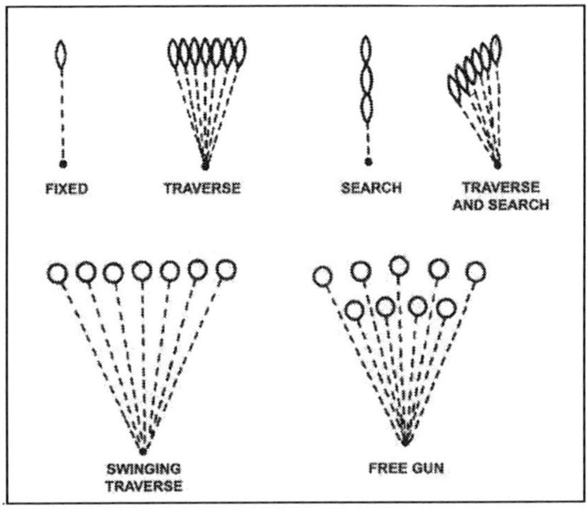

Figure A-12. Classes of fire with respect to the gun.

APPLICATION OF FIRE

A-46. Application of fire consists of the methods the gunner uses to effectively cover an enemy target area. Training these methods of applying fire can be accomplished only after the weapons squad leader and the gunners have learned how to recognize the different types of targets they may find in combat. They must also know how to distribute and concentrate their fire, and how to maintain the proper rate of fire. Normally, the gunner is exposed to two types of targets in the squad or platoon sector: enemy soldiers, and supporting automatic weapons. Leaders must ensure that these targets have priority and that they are engaged immediately.

A-47. To be effective, machine gun fire must be distributed over the entire target area. Improper distribution of fire results in gaps that allow the enemy to escape or use their weapons against friendly positions without effective opposition.

A-48. The method of applying fire to a target is generally the same for either a single gun or a pair of guns. Direct lay is pointing the gun for direction and elevation so the sights are aligned directly on the target. Fire is delivered in width, depth, or in a combination of the two. To distribute fire properly, gunners must know where to aim, how to adjust their fire, and the direction to manipulate the gun. The gunner must aim, fire, and adjust on a certain point of the target. Binoculars may be used by the leader to facilitate fire adjustment.

SIGHT PICTURE

A-49. A correct sight picture has the target, front sight post, and rear sight aligned. The sight picture has sight alignment and placement of the aiming point on the target. The gunner aligns the front sight post in the center of the rear sight and then aligns the sights with the target. *The top of the front sight post is aligned on the center base of the target.*

BEATEN ZONE

A-50. The gunner ensures throughout his firing that the center of the beaten zone is maintained at the center base of the target for maximum effect from each burst of fire. When this is done, projectiles in the upper half of the cone of fire will pass through the target if it has height, and the projectiles in the lower half of the beaten zone may ricochet into the target (Figure A-13).

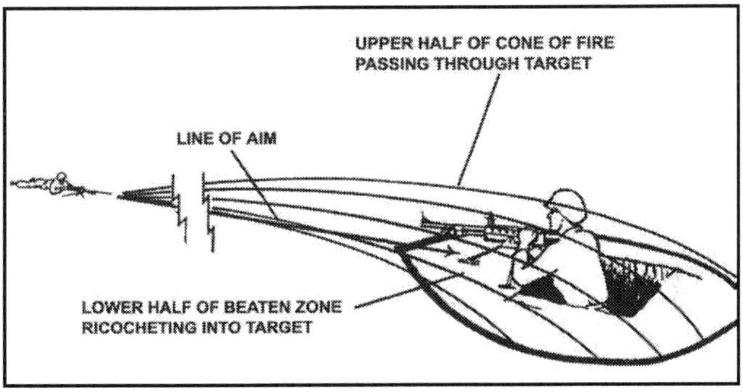

Figure A-13. Line of aim and placement of center of beaten zone on target.

A-51. The gunner must move his beaten zone in a certain direction over the target. The direction depends on the type of target and whether the target is engaged with a pair of guns or a single gun. When engaging targets other than point targets with a pair of guns, the targets are divided so fire is evenly distributed throughout the target area. Fire delivered on point targets or a specific area of other target configurations is called concentrated fire.

TARGET ENGAGEMENTS BY TYPES OF TARGETS

A-52. Gunners engage targets throughout their respective sectors. They must know how to effectively engage all types of targets, either individually or with other gunners.

A-53. Gunner's targets in combat are normally enemy troops in various formations or displacements, which require distribution and concentration of fire. These targets often have both width and depth, and the application of machine gun fire is designed to completely cover the area in which the enemy is known or suspected to be. These targets may be easy to see or may be indistinct and difficult to locate. The size of the target, stated in terms of the number of aiming points required to engage it completely, determines its type.

A-54. When a single gunner is assigned any target he is responsible for covering the entire target.

A-55. When a pair of gunners engage an enemy target, each gunner is normally responsible for effectively covering one half of the target. The gunners must be prepared to engage the entire target should the other gun go down.

A-56. The machine gun can provide units with a self-defense capability against hostile low-flying, low-performance aircraft. These guns are employed in the air defense role as part of the unit's local defense. The machine guns are not components of an integrated and coordinated air defense system. Unless otherwise directed, hostile aircraft within range of the gun (about 800 meters maximum effective range) should be engaged. The decision will be made by the commander. Typical targets are surveillance, reconnaissance, and liaison aircraft; troop carriers; helicopters; and drones.

Engagement and Employment

A-57. The mission is to impose maximum attrition upon the attacking enemy such as low-flying, low-performance aircraft. Employment of machine guns used for air defense is guided by the following defense design factors:

- Defense design should produce an equally balanced defense that is effective in all directions, unless a forced route of approach exists.
- Machine guns should be sited so the maximum number of targets can be engaged, continuous fire can be delivered, and the most likely routes of approach are covered.
- Machine guns used to defend march columns should be interspersed in the convoy, with emphasis on the lead and rear elements (Figure A-14).

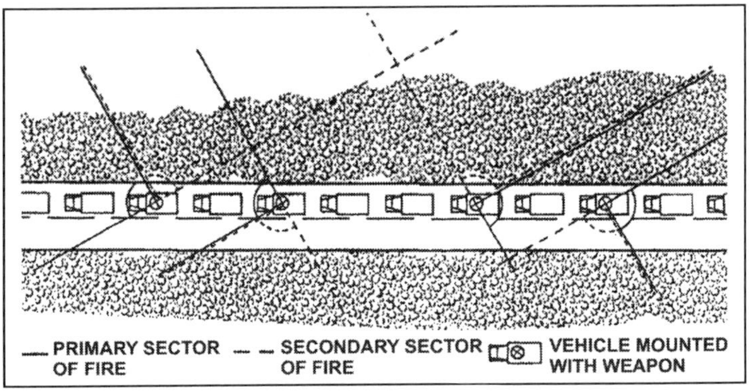

Figure A-14. March column with four machine guns.

Target Selection and Engagement Control

A-58. These actions depend upon visual means. The sites selected for the guns must provide maximum observation and unobstructed sectors of fire. Units furnished machine guns in sufficient numbers should site them within mutual support distances of 90 to 360 meters. Each gun is assigned a primary and secondary sector of fire. Weapon crews maintain constant vigilance in their primary sectors of fire, regardless of the sector in which the guns are actually engaged.

DISTRIBUTION, CONCENTRATION, AND RATE OF FIRE

A-59. The size and nature of the enemy target determines how machine gun fire is applied. Automatic weapons fire in one of three rates: rapid, sustained, or cyclic. The rates of fire for each machine gun are shown in Table A-1. The situation normally dictates the rate used, but the availability of ammunition and need for barrel changes play important roles as well. The rate of fire must be controlled to adequately cover the target, but not waste ammunition or destroy the barrel.

DISTRIBUTED AND CONCENTRATED FIRE

A-60. Distributed fire is delivered in width and depth such as at an enemy formation. Concentrated fire is delivered at a point target such as an automatic weapon or an enemy fighting position.

RAPID FIRE

A-61. Rapid rate of fire places an exceptionally high volume of fire on an enemy position. Machine gunners normally engage targets at the rapid rate to suppress the enemy quickly. Rapid fire requires much more ammunition than sustained fire and requires frequent barrel changes.

SUSTAINED FIRE

A-62. Once the enemy has been suppressed, machine gunners fire at the sustained rate. Sustained fire conserves ammunition and requires only infrequent barrel changes, but it might not be enough volume of fire to effectively suppress or destroy.

CYCLIC RATE OF FIRE

A-63. To fire the cyclic rate, the gunner holds the trigger to the rear while the assistant gunner feeds ammunition into the weapon. This is normally only used to engage aerial targets in self-defense or to fire the final protective fire in the defense to protect the perimeter. This produces the highest volume of fire the machine gun can fire, but can permanently damage the machine gun and barrel and should be used only in case of emergency.

TARGET ENGAGEMENT DURING LIMITED VISIBILITY

A-64. Gunners have difficulty detecting and identifying targets during limited visibility. The leader's ability to control the fires of his weapons is also reduced; therefore, he may instruct the gunners to fire without command when targets present themselves.

A-65. Gunners should engage targets only when they can identify the targets, unless ordered to do otherwise. For example, if one gunner detects a target and engages it, the other gunner observes the area fired upon and adds his fire only if he can identify the target or if ordered to fire.

A-66. Tracer ammunition helps a gunner engage targets during limited visibility and should be used if possible. It is important to note that in certain circumstances the enemy will have an easy time identifying the machine gun's position if the gunner uses tracer ammunition. The need to effectively engage targets must be balanced with the need to keep the guns safe before deciding to employ tracers. If firing unaided, gunners must be trained to fire low at first and adjust upward. This overcomes the tendency to fire high.

A-67. When two or more gunners are engaging linear targets, linear targets with depth, or deep targets, they do not engage these targets as they would when visibility is good. With limited visibility, the center and flanks of these targets may not be clearly defined. Therefore, each gunner observes his tracers and covers what he believes to be the entire target.

TECHNIQUES

A-68. Techniques of fire include assault fire; overhead fire; and fire from a defilade position. Only automatic rifles use assault fire.

ASSAULT FIRE

A-69. Automatic riflemen use assault fire when in close combat. Assault fire involves firing without the aid of sights using the hip, shoulder, and underarm positions. The underarm position is best when rapid movement is required. In all three positions, automatic riflemen adjust their fire by observing the tracer and the impact of the bullets in the target area. Additional considerations for automatic riflemen using assault fire include—

- Maintaining alignment with the rest of the assault element.
- Reloading rapidly.
- Aiming low and adjusting the aim upward toward the target.
- Distributing fires across the objective when not engaging enemy automatic weapons.

OVERHEAD FIRE

A-70. Gunners can use overhead fire when there is sufficient low ground between the machine gun and the target area for the maneuver of friendly forces. A machine gun on a tripod is capable of delivering this type of fire because of the small and uniform dispersion of the cone of fire. Gunners must accurately estimate range to the target and establish a safety limit that is an imaginary line parallel to the target where fire would cause casualties to friendly Soldiers. Gun crews and leaders must be aware of this safety limit. Leaders must designate signals for lifting or shifting fires. Gunners should not attempt overhead fires if the terrain is level or slopes uniformly, if the barrel is badly worn, or if visibility is poor.

Gunner's Rule

A-71. The gunner's rule can be applied when the friendly troops are at least 350 meters in front of the gun position and the range to the target is 850 meters or less (Figure A-15). The rule follows:

- Lay the gun on the target with the correct sight setting to hit the target.
- Without disturbing the lay of the gun, set the rear sight at a range of 1,600 meters.
- Look through the sights and notice where the new line of aim strikes the ground. This is the limit of troop safety. When the feet of the friendly troops reach this point, fire must be lifted or shifted.

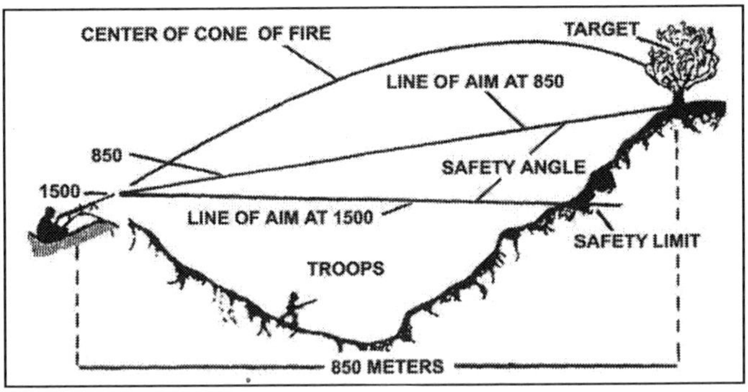

Figure A-15. Application of gunner's rule.

Leader's Rule

A-72. When the range to the target is greater than 850 meters, overhead fire should be delivered only in an emergency. Even then, fire should only extend to a range at which the tracers or strike of the bullets can be seen by the gunner. In this situation the leader's rule applies (Figure A-16). The platoon or section leader uses the leader's rule only when the target is greater than 850 meters. The rule follows:

- Select a point on the ground where it is believed friendly troops can advance with safety.
- Determine the range to this point by the most accurate means available.
- Lay the gun on the target with the correct sight setting to hit the target.
- Without disturbing the lay of the gun, set the rear sight to 1,600 meters or the range to the target plus 500 meters, whichever is the greater of the two ranges. Under no conditions should the sight setting be less than 1,500 meters.
- Note the point where the new line of aim strikes the ground.
  - If it strikes at the selected point, that point marks the limit of safety.
  - If it strikes short of the selected point, it is safe for troops to advance to the point where the line of aim strikes the ground and to an unknown point beyond. If fire is called for after friendly troops advance farther than the point where the line of aim strikes the ground, this farther point is determined by testing new selected points until the line of aim and the selected point coincide.
  - If it clears the selected point, it is safe for the troops to advance to the selected point and to an unknown point beyond. If it is advantageous to have troops advance beyond the selected point, this farther point must be determined by testing new selected points until the line of aim and the selected point coincide. This point marks the line of safety.

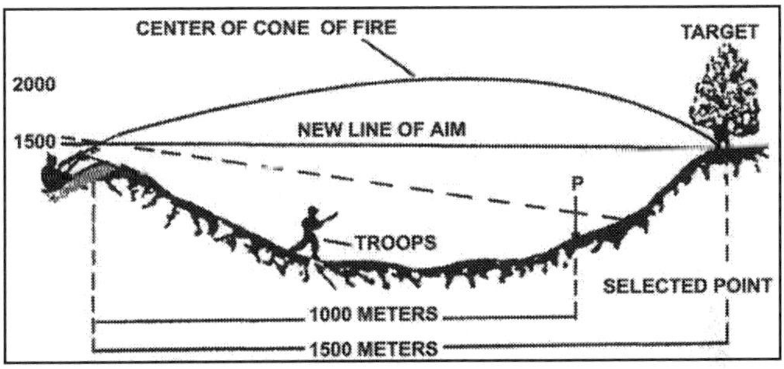

Figure A-16. Application of leader's rule.

Fire From a Defilade Position

A-73. Defilade positions protect gunners from frontal or enfilading fires (Figure A-17). Cover and concealment may not provide the gunner a view of some or all of the target area. In this instance, some other member of the platoon must observe the impact of the rounds and communicate adjustments to the gunner (Figure A-18). Gunners and leaders must consider the complexity of laying on the target. They must also take into account the gunner's inability to make rapid adjustments to engage moving targets, the ease with which targets are masked, and the difficulty in achieving grazing fires for an FPL.

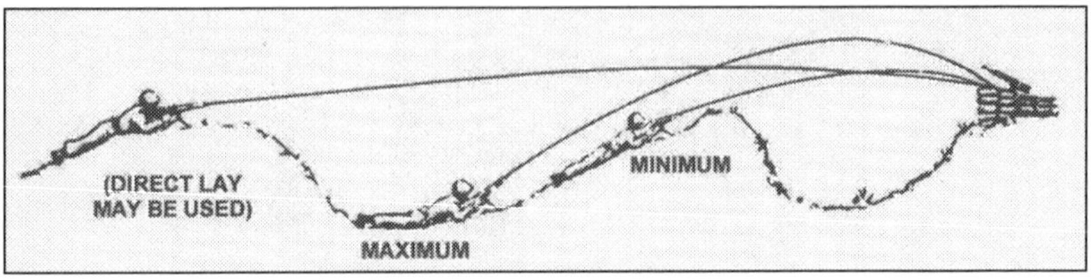

Figure A-17. Defilade positions.

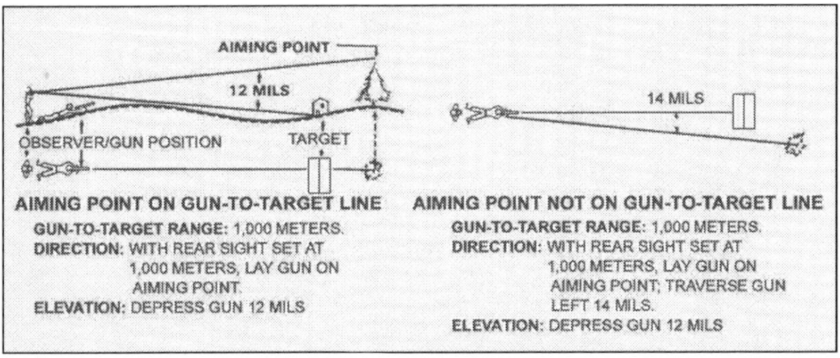

Figure A-18. Observer adjusting fire.

SECTION III — PREDETERMINED FIRES

A-74. Predetermined fires organize the battlefield for the gunners. They allow the leader and gunner to select potential targets or target areas that will most likely be engaged or that have tactical significance. This includes dismounted enemy avenues of approach, likely positions for automatic weapons, and probable enemy assault positions. The gunners do this by using sectors of fire, final protective lines, or a principal direction of fire and selected target areas. This preparation maximizes the effectiveness of the machine gun during good as well as limited visibility. It enhances fire control by reducing the time required to identify targets, determine range, and manipulate the weapon onto the target. Abbreviated fire commands and previously-recorded data enable the gunner to aim or adjust fire on the target quickly and accurately. Selected targets should be fired on in daylight whenever practical to confirm data. The range card identifies the targets and provides a record of firing data.

TERMINOLOGY

A-75. Gunners need to know several terms associated with predetermined fire.

SECTOR OF FIRE

A-76. A sector of fire is an area to be covered by fire that is assigned to an individual, a weapon, or a unit. Gunners are normally assigned a primary and a secondary sector of fire.

FINAL PROTECTIVE FIRE

A-77. A final protective fire (FPF) is an immediately-available, prearranged barrier of fire to stop enemy movement across defensive lines or areas.

Final Protective Line

A-78. An FPL is a predetermined line along which grazing fire is placed to stop an enemy assault. If an FPL is assigned, the machine gun is sighted along it except when other targets are being engaged. An FPL becomes the machine gun's part of the unit's final protective fires. An FPL is fixed in direction and elevation. However, a small shift for search must be employed to prevent the enemy from crawling under the FPL and to compensate for irregularities in the terrain or the sinking of the tripod legs into soft soil during firing. Fire must be delivered during all conditions of visibility.

A-79. A good FPL covers the maximum area with grazing fire. Grazing fire can be obtained over various types of terrain out to a maximum of 600 meters. To obtain the maximum extent of grazing fire over level or uniformly sloping terrain, the gunner sets the rear sight at 600 meters. He then selects a point on the ground that he estimates to be 600 meters from the machine gun, and he aims, fires, and adjusts on that point. To prevent enemy soldiers from crawling under grazing fire, he searches (downward) by lowering the muzzle of the weapon.

Principal Direction of Fire

A-80. A principal direction of fire (PDF) is assigned to a gunner to cover an area that has good fields of fire or has a likely dismounted avenue of approach. It also provides mutual support to an adjacent unit. Machine guns are sighted using the PDF if an FPL has not been assigned. If a PDF is assigned and other targets are not being engaged, machine guns remain on the PDF. A PDF has the following characteristics:

- It is used only if an FPL is not assigned; it then becomes the machine gun's part of the unit's final protective fires.
- When the target has width, direction is determined by aiming on one edge of the target area and noting the amount of traverse necessary to cover the entire target.
- The gunner is responsible for the entire wedge-shaped area from the muzzle of the weapon to the target, but elevation may be fixed for a priority portion of the target.

Dead Space and Grazing Fire

A-81. The extent of grazing fire and the extent of dead space may be determined in two ways. In the preferred method, the machine gun is adjusted for elevation and direction. A member of the squad then walks along the FPL while the gunner aims through the sights. In places where the Soldier's waist (midsection) falls below the gunner's point of aim, dead space exists. Arm-and-hand signals must be used to control the Soldier who is walking and to obtain an accurate account of the dead space and its location. Another method is to observe the flight of tracer ammunition from a position behind and to the flank of the weapon.

Primary Sector of Fire

A-82. The primary sector of fire is assigned to the gun team to cover the most likely avenue of enemy approach from all types of defensive positions.

Secondary Sector of Fire

A-83. The secondary sector of fire is assigned to the gun team to cover the second most likely avenue of enemy approach. It is fired from the same gun position as the primary sector of fire.

RANGE CARD

A-84. DA Form 5517-R, *Standard Range Card*, provides a record of firing data and aids defensive fire planning.

FIELD EXPEDIENTS

A-85. When laying the machine gun for predetermined targets, the gunner can use field expedients as a means of engaging targets when other sources are not available.

BASE STAKE TECHNIQUE

A-86. A base stake is used to define sector limits and may provide the lay for the FPL or predetermined targets along a primary or secondary sector limit. This technique is effective in all visibility conditions. The gunner uses the following steps:

- Defines the sector limits by laying the gun for direction along one sector limit and by emplacing a stake along the outer edge of the folded bipod legs. Rotates the legs slightly on the receiver, so the gunner takes up the "play." Uses the same procedure for placing a stake along the opposite sector limit.
- Lays the machine gun along the FPL by moving the muzzle of the machine gun to a sector limit. Adjusts for elevation by driving a stake into the ground so the top of the stake is under the gas cylinder extension. This allows a few mils of depression to cover irregularities in the terrain.
- Lays the machine gun to engage other targets within a sector limit. Done in a primary sector by using the procedure described previously, except he keeps the elevation fixed.

NOTCHED-STAKE OR TREE-CROTCH TECHNIQUE

A-87. The gunner uses the notched-stake or tree-crotch technique with the bipod mount to engage predetermined targets within a sector or to define sector limits. This technique is effective during all conditions of visibility and requires little additional material. The gunner uses the following steps:

- Drives either a notched stake or tree crotch into the ground where selected targets are anticipated. Places the stock of the machine gun in the nest of the stake or crotch and adjusts the weapon to hit the selected targets and to define his sector limits.
- Digs shallow, curved trenches or grooves for the bipod feet. (These trenches allow for rotation of the bipod feet as the gunner moves the stock from one crotch or stake to another.)

HORIZONTAL LOG OR BOARD TECHNIQUE

A-88. This technique is used with the bipod or tripod mount to mark sector limits and engage wide targets. It is good for all visibility conditions and is best suited for flat, level terrain. The gunner uses the following steps.

Bipod-Mounted Machine Gun

A-89. Using a bipod-mounted machine gun, the gunner places a log or board beneath the stock of the weapon so the stock can slide across it freely. He digs shallow, curved trenches or grooves for the bipod feet to allow rotation of the feet as he moves the stock along the log or board. (The gunner may mark the sector limits by notching or placing stops on the log or board. The gunner uses the bipod firing position and grip.)

Tripod-Mounted Machine Gun

A-90. Using a tripod-mounted machine gun, the gunner places a log or board beneath the barrel, positioning it so the barrel, when resting on the log or board, is at the proper elevation to obtain grazing fire. When appropriate, he marks the sector limits as described for the bipod in the preceding paragraph. (This technique is used only if a T&E mechanism is not available.)

SECTION IV — FIRE CONTROL

A-91. Fire control includes all actions of the leader and Soldiers in planning, preparing, and applying fire on a target. The leader selects and designates targets. He also designates the midpoint and flanks or ends of a target, unless they are obvious to the gunner. The gunner fires at the instant desired. He then adjusts fire, regulates the rate of fire, shifts from one target to another, and ceases fire. When firing, the gunner should continue to fire until the target is neutralized or until signaled to do otherwise by the leader.

A-92. Predetermined targets, including the FPL or PDF, are engaged on order or by SOP. The signal for calling for these fires is normally stated in the defense order. Control these predetermined targets by using arm-and-hand signals, voice commands, or pyrotechnic devices. Gunners fire the FPL or PDF at the sustained rate of fire unless the situation calls for a higher rate. When engaging other predetermined targets, the sustained rate of fire is also used unless a different rate is ordered.

METHODS OF FIRE CONTROL

A-93. The noise and confusion of battle may limit the use of some of these methods. Therefore, the leader must select a method or combination of methods that will accomplish the mission.

ORAL

A-94. The oral fire control method can be effective, but sometimes the leader may be too far away from the gunner, or the noise of the battle may make it impossible for him to hear. The primary means of the oral fire control method is the issuance of a fire command.

ARM-AND-HAND SIGNALS

A-95. Arm-and-hand signals are an effective fire control method when the gunner can see the leader. All gunners must know the standard arm-and-hand signals. The leader gets the gunner's attention and then points to the target. When the gunner returns the READY signal, the leader commands FIRE.

PREARRANGED SIGNALS

A-96. Prearranged signals are either visual or sound signals such as casualty-producing devices (rifle or claymore), pyrotechnics, whistle blasts, or tracers. These signals should be included in SOPs. If the leader wants to shift fire at a certain time, he gives a prearranged signal such as smoke or pyrotechnics. Upon seeing the signal, the gunner shifts his fire to a prearranged point.

PERSONAL CONTACT

A-97. In many situations, the leader must issue orders directly to individual Soldiers. Personal contact is used more than any other method by Infantry leaders. The leader must use maximum cover and concealment to keep from disclosing the position or himself.

RANGE CARDS

A-98. When using the range card method of fire control, the leader must ensure all range cards are current and accurate. Once this is accomplished, the leader may designate certain targets for certain weapons with the use of limiting stakes or with fire commands. He should also designate no-fire zones or restricted fire areas to others. The key factor in this method of fire control is that gunners must be well disciplined and pay attention to detail.

STANDING OPERATING PROCEDURES

A-99. SOPs are actions to be executed without command that are developed during the training of the squads. Their use eliminates many commands and simplifies the leader's fire control. SOPs for certain actions and commands can be developed to make gunners more effective. Some examples follow:

- **Observation.** The gunners continuously observe their sectors.
- **Fire.** Gunners open fire without command on appropriate targets that appear within their sectors.
- **Check.** While firing, the gunners periodically check with the leader for instructions.
- **Return Fire.** The gunners return enemy fire without order, concentrating on enemy automatic weapons.
- **Shift Fire.** Gunners shift their fires without command when more dangerous targets appear.
- **Rate of Fire.** When gunners engage a target, they initially fire at the rate necessary to gain and maintain fire superiority.
- **Mutual Support.** When two or more gunners are engaging the same target and one stops firing, the other increases the rate of fire and covers the entire target. When only one gunner is required to engage a target and the leader has alerted two or more, the gunner not firing aims on the target and follows the movements of the target. He does this to fire instantly in case the other machine gun malfunctions or ceases fire before the target has been eliminated.

FIRE COMMANDS

A-100. A fire command is given to deliver effective fire on a target quickly and without confusion. When the leader decides to engage a target that is not obvious to the squad, he must provide them with the information they need to effectively engage the target. He must alert the Soldiers; give a target direction, description, and range; name the method of fire; and give the command to fire. There are initial fire commands and subsequent fire commands.

A-101. It is essential that the commands delivered by the weapons squad leader are understood and echoed by the assistant gunner/gun team leader and the gunner. Table A-7 provides an example of the weapons squad fire commands and actions used by the weapons squad leader (WSL), assistant gunner (AG)/gun team leader (GTL), and gunner.

Table A-7. Example weapons squad fire commands and actions.

ACTION	WSL COMMANDS	AG/GTL COMMANDS AND ACTIONS	GUNNER ACTIONS	GUNNER RESPONSES
WSL or GTL identifies target within gun team's sector	"Light-skinned truck, 3 o'clock, 400 m, on my laser."	"Light-skinned truck, 3 o'clock, 400 m, on my laser." "Once on TGT engage."	Gunner looks for laser and identifies target. Gunner traverses and gets on target. Gunner engages target with correct rate of fire.	"TGT identified." "TGT acquired."
Gun team (or weapons SQD) go to bipod	"Gun 1-Bipod."	Repeats "Gun 1-Bipod" and identifies location for gun.	Gets down beside AG/GTL.	"Gun 1 up" once ready to fire.
Gun team go to tripod	"Gun 1-Tripod."	Repeats "Gun 1-Tripod" and lays down tripod (if not done) and prepares to lock gun on tripod.	Gunner picks up gun and places into tripod. He gets AG/GTL to lock it in. Once locked in, the AG/GTL collapses bipod legs.	"Gun 1 up" once ready to fire.
Barrel change	NA	"Gun 1 prepare for barrel change." "Gun 1 barrel change."	Fires one more burst. Waits for barrel change.	Repeats AG/GTL command. Once done, "Gun 1 up."
Displace gun	"Gun 1 out of action, prepare to move."	"Gun 1 out of action, prepare to move." Breaks down barrel bag, prepares to move.	Gunner takes gun off tripod, continues to orient towards target on bipod, and prepares to move.	"Gun 1, ready to move."
WSL identifies sector of fire for gun team(s) Day-marks w/tracer Night-marks with PEQ/tracer	"Gun 1, left, center, right sectors on my mark. Do you identify?" (Always marks left to right.)	Using binoculars identifies sectors and states, "Gun 1 identifies." Adjusts gunner onto target.	Gunner makes necessary adjustments, tells AG/GTL whether he identifies or not. Engages or makes further adjustments.	"Sector identified" to AG/GTL once he identifies.

Table A-7. Example weapons squad fire commands and actions (continued).

ACTION	WSL COMMANDS	AG/GTL COMMANDS AND ACTIONS	GUNNER ACTIONS	GUNNER RESPONSES
WSL or AG/GTL gives or adjusts rate of fire	"Gun 1, sustained __ seconds, engage."	Echoes command, starts count. Tells gunner to fire. Keeps count between bursts and ensures gun does not fire out of turn.	Gunner echoes command, also counts and fires when AG/GTL gives command to fire.	Echoes rate of fire "Sustained __ seconds."
WSL changes gun team(s) sector of fire or shift fire	"Gun 1, shift fire, target # (or) right/left sector." Marks sector same as above.	Echoes command to shift; identifies new target/sector. Adjusts gunner, alerts WSL once the gunner has shifted.	Gunner echoes command, makes necessary adjustment, acquires new target. Confirms with AG/GTL that all is OK. Engages new sector when told.	Echoes command with AG/GTL. "Shift fire to TGT #__." Once identified, "Sector/target identified."
Talking the gun teams (ensuring one gun fires during the other gun's interval and visa versa).	WSL gives gun teams the rate of fire. (As long as they are keeping correct interval, they should "talk" themselves.)	Repeats rate of fire and maintains proper count, telling gunner when to fire. Adjusts rate of fire off of lead gun.	Repeats rate of fire command, keeps own count. Fires when told to fire. Adjusts rate of fire off of lead gun.	"Sustained __ seconds."
Lift fire	"Lift fire, lift fire, lift fire." Or "Gun 1, lift fire."	Repeats command to gunner, ensures gunner lifts fire.	Repeats command. Ceases all fire onto the objective. Maintains overwatch and scans objective until told to reengage or go out of action.	Echoes "lift fire."
Round count	If need to know round count, prompt "Gun 1, round count."	AG/GTL continuously links rounds and gives WSL round count every 100. "Gun 1, 200 rounds."	Gunner echoes round count to ensure it is heard.	"Gun 1, 200 rounds."
"Watch and shoot" or "Traverse and search"	"Gun 1, watch and shoot." "Gun 1, traverse and search."	Repeats command, searches objective for targets of opportunity within sector.	Repeats command, searches objective for targets of opportunity in sector. Confirms target with AG/GTL before engaging.	"Gun 1, watch and shoot." "Gun 1, traverse and search."

INITIAL FIRE COMMANDS

A-102. Initial fire commands are given to adjust onto the target, change the rate of fire after a fire mission is in progress, interrupt fire, or terminate the alert.

ELEMENTS

A-103. Fire commands for all direct-fire weapons follow a pattern that includes similar elements. There are six elements in the fire command for the machine gun: alert; direction; description; range; method of fire; and command to open fire. The gunners repeat each element of fire command as it is given.

Alert

A-104. This element prepares the gunners for further instructions. The leader may alert both gunners in the squad and may have only one fire, depending upon the situation. To alert and have both gunners fire, the leader announces FIRE MISSION. If he desires to alert both gunners but have only one fire, he announces GUN NUMBER ONE, FIRE MISSION. In all cases, upon receiving the alert, the gunners load their machine guns and place them on FIRE.

Direction

A-105. This element indicates the general direction to the target and may be given in one or a combination of the following methods.

Oral

A-106. The leader orally gives the direction to the target in relation to the position of the gunner (for example, FRONT, LEFT FRONT, RIGHT FRONT).

Pointing

A-107. The leader designates a small or obscure target by pointing with his finger or aiming with a weapon. When he points with his finger, a Soldier standing behind him should be able to look over his shoulder and sight along his arm and index finger to locate the target. When aiming his weapon at a target, a Soldier looking through the sights should be able to see the target. Leaders may also use lasers in conjunction with night vision devices to designate a target to the gunner.

Tracer Ammunition

A-108. Tracer ammunition is a quick and sure method of designating a target that is not clearly visible. When using this method, the leader should first give the general direction to direct the gunner's attention to the target area. To prevent the loss of surprise when using tracer ammunition, the leader does not fire until he has given all elements of the fire command except the command to fire. The leader may fire his individual weapon. The firing of the tracer(s) then becomes the last element of the fire command, and it is the signal to open fire.

> **NOTE:** Soldiers must be aware that with the night vision device, temporary blindness ("white out") may occur when firing tracer ammunition at night or when exposed to other external light sources. Lens covers may reduce this effect.

Reference Points

A-109. Another way to designate obscure targets is to use easy-to-recognize reference points. All leaders and gunners must know terrain features and the terminology used to describe them (see FM 3-25.26, *Map Reading and Land Navigation*). When using a reference point, the word "reference" precedes its description. This is done to avoid confusion. The general direction to the reference point should be given.

Description

A-110. The target description creates a picture of the target in the minds of the gunners. To properly apply their fire, the Soldiers must know the type of target they are to engage. The leader should describe it briefly. If the target is obvious, no description is necessary.

Range

A-111. The leader always announces the estimated range to the target. The range is given, so the gunner knows how far to look for the target and what range setting to put on the rear sight. Range is announced in meters. However, since the meter is the standard unit of range measurement, the word "meters" is not used. With machine guns, the range is determined and announced to the nearest hundred or thousand (for example, THREE HUNDRED, or ONE THOUSAND).

Method of Fire

A-112. This element includes manipulation and rate of fire. Manipulation dictates the class of fire with respect to the weapon. It is announced as FIXED, TRAVERSE, SEARCH, or TRAVERSE AND SEARCH. Rate controls the volume of fire (sustained, rapid, and cyclic). Normally, the gunner uses the sustained rate of fire. The rate of fire is omitted from the fire command. The method of fire for the machine gun is usually 3- to 5-round bursts (M249) and 6- to 9-round bursts (M60/M240B).

Command to Open Fire

A-113. When fire is to be withheld so surprise fire can be delivered on a target or to ensure that both gunners open fire at the same time, the leader may preface the command to commence firing with AT MY COMMAND or AT MY SIGNAL. When the gunners are ready to engage the target, they report READY to the leader. The leader then gives the command FIRE at the specific time desired. If immediate fire is required, the command FIRE is given without pause and the gunners fire as soon as they are ready.

SUBSEQUENT FIRE COMMANDS

A-114. Subsequent fire commands are used to make adjustments in direction and elevation, to change rates of fire after a fire mission is in progress, to interrupt fires, or to terminate the alert. If the gunner fails to properly engage a target, the leader must promptly correct him by announcing or signaling the desired changes. When these changes are given, the gunner makes the corrections and resumes firing without further command.

A-115. Adjustments in direction and elevation with the machine gun are always given in meters; one finger is used to indicate 1 meter and so on. Adjustment for direction is given first. For example: RIGHT ONE ZERO METERS or LEFT FIVE METERS. Adjustment for elevation is given next. For example: ADD FIVE METERS or DROP ONE FIVE METERS. These changes may be given orally or with arm-and-hand signals.

- Changes in the rate of fire are given orally or by arm-and-hand signals.
- To interrupt firing, the leader announces CEASE FIRE, or he signals to cease fire. The gunners remain on the alert. They resume firing when given the command FIRE.
- To terminate the alert, the leader announces CEASE FIRE, END OF MISSION.

DOUBTFUL ELEMENTS AND CORRECTIONS

A-116. When the gunner is in doubt about any element of the fire command, he replies, SAY AGAIN RANGE, TARGET. The leader then announces THE COMMAND WAS, repeats the element in question, and continues with the fire command.

A-117. When the leader makes an error in the initial fire command, he corrects it by announcing CORRECTION, and then gives the corrected element. When the leader makes an error in the subsequent fire command, he may correct it by announcing CORRECTION. He then repeats the entire subsequent fire command.

ABBREVIATED FIRE COMMANDS

A-118. Fire commands do not need not be complete to be effective. In combat, the leader gives only the elements necessary to place fire on a target quickly and without confusion. During training, however, he should use all of the elements to get gunners in the habit of thinking and reacting properly when a target is to be engaged. After the gunner's initial training in fire commands, he should be taught to react to abbreviated fire commands, using one of the following methods.

Oral

A-119. The leader may want to place the fire of one machine gun on an enemy machine gun and quickly tells the gunner to fire on that gun.

Hand-and-Arm Signals

A-120. Battlefield noise and the distance between the gunner and the leader often make it necessary to use arm-and-hand signals to control fire (Figure A-19). When an action or movement is to be executed by only one of the gunners, a preliminary signal is given to that gunner only. The following are commonly used signals for fire control:

- **Ready.** The gunner indicates that he is ready to fire by yelling UP or having the assistant gunner raise his hand above his head toward the leader.
- **Commence Firing or Change Rate of Firing.** The leader brings his hand (palm down) to the front of his body about waist level, and moves it horizontally in front of his body. To signal an increase in the rate of fire, he increases the speed of the hand movement. To signal slower fire, he decreases the speed of the hand movement.
- **Change Direction or Elevation.** The leader extends his arm and hand in the new direction and indicates the amount of change necessary by the number of fingers extended. The fingers must be spread so the gunner can easily see the number of fingers extended. Each finger indicates 1 meter of change for the weapon. If the desired change is more than 5 meters, the leader extends his hand the number of times necessary to indicate the total amount of change. For example, *right nine* would be indicated by extending the hand once with five fingers showing and a second time with four fingers showing for a total of nine fingers.
- **Interrupt or Cease Firing.** The leader raises his arm and hand (palm outward) in front of his forehead and brings it downward sharply.
- **Other Signals.** The leader can devise other signals to control his weapons. A detailed description of arm-and-hand signals is given in FM 21-60.

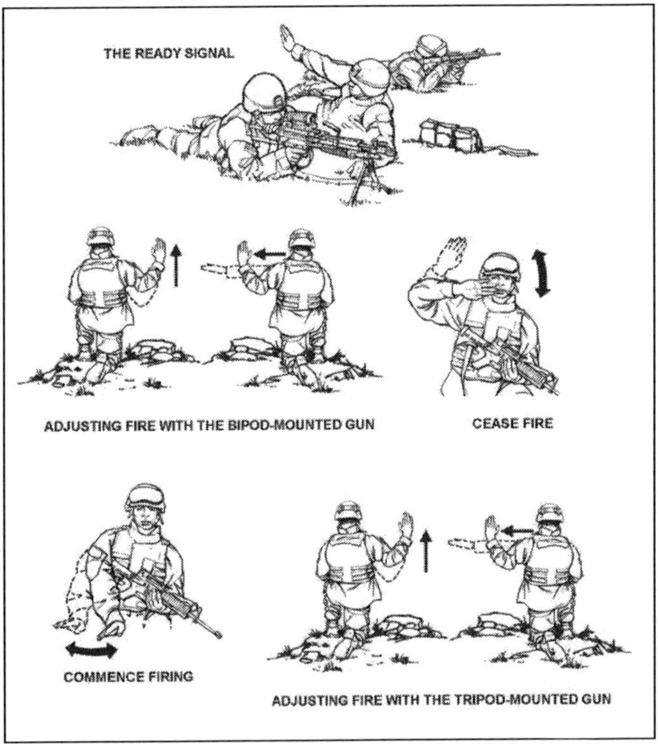

Figure A-19. Hand-and-arm signals.

SECTION V — MACHINE GUN USE

A-121. Despite their post-Civil War development, modern machine guns did not exhibit their full potential in battle until World War I. Although the machine gun has changed, the role of the machine gun and machine gunner has not. The mission of machine guns in battle is to deliver fires when and where the leader wants them in both the offense and defense. Machine guns rarely, if ever, have independent missions. Instead, they provide their unit with accurate, heavy fires to accomplish the mission.

TACTICAL ORGANIZATION OF THE MACHINE GUN

A-122. The accomplishment of the platoon's mission demands efficient and effective machine gun crews. Leaders consider the mission and organize machine guns to deliver firepower and fire support to any area or point needed to accomplish the assigned mission.

A-123. Infantry platoons will normally have an organic weapons squad that consists of a weapons squad leader and two gun teams. Depending on the unit's organization or the platoon's mission, there could be additional machine gun teams attached or organic to the platoon.

A-124. The weapons squad consists of a weapons squad leader and machine gun teams. Each machine gun team has a gunner, assistant gunner, and ammunition bearer. In some units the senior member of the gun team is the gunner. In other units the assistant gunner is the senior gun team member who also serves as the gun team leader. Table A-8 illustrates equipment carried by the weapons squad. Table A-9 illustrates the duty positions within the weapons squad and gives possible duty descriptions and responsibilities. The tables serve to show possible position and equipment use only. Individual unit SOPs and available equipment dictate the exact role each weapons squad member plays within his squad.

Table A-8. Example weapons squad equipment by position.

	Weapons Squad Leader	Assistant Gunner/ Gun Team Leader	Gunner	Ammunition Bearer
Weapon	M4 (w/ 7 mags*)	M4 (w/ 7 mags*)	M240B (50-100 rounds)	M4 (w/ 7 mags)
Day Optic	ACOG	ACOG	M145	M68/ACOG
Laser	PEQ-2	PEQ-2	PEQ-2	PAQ-4/PEQ-2
Additional Equipment	3x magnifier**	3x magnifier** Spare barrel(s)***	3x magnifier**	Tripod T&E
M240 Ammunition	100 rounds	300 rounds	100 rounds	300 rounds
Miscellaneous	Whistle Pen gun flare** Other shift signals** VS-17 panel Binoculars****	M9 pistol Cleaning kit Binoculars****	M9 pistol Cleaning kit CLP for 72 hours*****	NA

*WSL and AG/GTL load tracer rounds (4:1 mix) in magazines for marking targets.

**3x magnifier, flares, and shift signals are readily accessible at all times.

***Spare barrel(s) marked by relative age with ¼ pieces of green tape on carrying handle.
Oldest barrel=2 parallel strips Second newest barrel=1 strip Newest barrel=no tape

****Binoculars carried in the assault pack or in suitable pouch on vest (mission dependent).

*****Gunners always carry enough CLP for 72 hours of operations.

Table A-9. Example weapons squad duty positions and responsibilities.

Weapons Squad Leader	Senior squad leader within the platoon. Responsible for all training and employment of the machine guns. The WSL's knowledge, experience, and tactical proficiency influence the effectiveness of the squad.
Assistant Gunner/ Gun Team Leader	AG/GTL is a team leader with the responsibilities of a fire team leader.
	GTL is responsible for his team members and all the gun equipment.
	GTL and his team will be tactically proficient and knowledgeable on this FM and applicable FMs and TMs that apply to the machine gun.
	GTL assists the WSL on the best way to employ the M240B.
	GTL enforces field discipline while the gun team is employed.
	GTL leads by example in all areas. He sets the example in all things.
	GTL assists the WSL in all areas. He advises him of any problems either tactical or administrative.
	AG is responsible for all action concerning the gun.
	AG/GTL calls the ammunition bearer if ammunition is needed or actively seeks it out if the ammunition bearer is not available. Constantly updates the WSL on the round count and serviceability of the M240B.
	When the gun is firing, AG/GTL spots rounds and makes corrections to the gunner's fire. Also watches for friendly troops to the flanks of the target area or between the gun and the target.
	If the gunner is hit by fire, AG/GTL immediately assumes the roll of the gunner.
	AG/GTL is always prepared to change the gun's barrel (spare barrel is always out when the gun is firing). Ensures the hot barrel is not placed on live ammunition or directly on the ground when it comes out of the gun.
Machine Gunner	If second in the gun team's chain of command, he is always fully capable of taking the GTL position.
	Primary responsibility is to the gun. Focused on its cleanliness and proper function. Immediately reports any abnormalities to the GTL or WSL.
	If necessary for gunner to carry M240B ammunition, carries it in on his back so the AG/GTL can access it without stopping the fire of the gun.
	Always carries the necessary tools for the gun to be properly cleaned, along with a sufficient amount of oil for the gun's proper function.
Ammunition Bearer	The AB is the rifleman/equipment bearer for the gun team.
	Normally the newest member of the gun team. Must quickly learn everything he can, exert maximum effort at all times, and attempt to outdo his gun team members in every situation.
	Follows the gunner without hesitation. During movement moves to the right side of the gunner and no more than one 3-5 meters rush away from the gun.
	During firing, pulls rear security and if the gunner comes under enemy fire, provides immediate suppression while the gun moves into new position.
	Responsible for the tripod and T&E mechanism. They must always be clean and ready for combat. Responsible for replacing them, if necessary.

SECURITY

A-125. Security includes all command measures to protect against surprise, observation, and annoyance by the enemy. The principal security measures against ground forces include employment of security patrols and detachments covering the front flanks and rear of the unit's most vulnerable areas. The composition and strength of these detachments depends on the size of the main body, its mission, and the nature of the opposition expected. The presence of machine guns with security detachments augments their firepower to effectively delay, attack, and defend, by virtue of their inherent firepower.

A-126. The potential of air and any potential ground attacks on the unit demands every possible precaution for maximum security while on the move. Where this situation exists, the machine gun crew must be thoroughly trained in the hasty delivery of antiaircraft fire and of counterfires against enemy ground forces.

The distribution of the machine guns in the formation is critical. The machine gun crew is constantly on the alert, particularly at halts, ready to deliver fire as soon as possible. If the leader expects a halt to exceed a brief period, he carefully chooses machine gun positions to avoid unduly tiring the machine gun crew. If he expects the halt to extend for a long period, he can have the machine gun crew take up positions in support of the unit. The crew covers the direction from which he expects enemy activity as well as the direction from which the unit came. The leader selects positions that permit the delivery of fire in the most probable direction of enemy attack, such as valleys, draws, ridges, and spurs. He chooses positions that offer obstructed fire from any potential enemy locations.

MACHINE GUNS IN THE OFFENSE

A-127. Successful offensive operations result from the employment of fire and movement. Each is essential and greatly depends upon the other. Without the support of covering fires, maneuvering in the presence of enemy fire can result in disastrous losses. Covering fires, especially those that provide fire superiority, allow maneuvering in the offense. However, fire superiority alone rarely wins battles. The primary objective of the offense is to advance, occupy, and hold the enemy position.

MACHINE GUN AS A BASE OF FIRE

A-128. Machine gun fire from a support-by-fire (SBF) position must be the minimum possible to keep the enemy from returning effective fire. Ammunition must be conserved so the guns do not run out of ammunition.

A-129. The weapon squad leader positions and controls the fires of all machine guns in the element. Machine gun targets include key enemy weapons or groups of enemy targets either on the objective or attempting to reinforce or counterattack. In terms of engagement ranges, machine guns in the base-of-fire element may find themselves firing at targets within a range of 800 meters. The nature of the terrain, desire to achieve some standoff, and the other factors of METT-TC prompt the leader to the correct tactical positioning of the base-of-fire element.

A-130. The machine gun delivers an accurate, high-volume rate of lethal fire on fairly large areas in a brief time. When accurately placed on the enemy position, machine gun fires secure the essential element of fire superiority for the duration of the firing. Troops advancing in the attack should take full advantage of this period to maneuver to a favorable position from where they can facilitate the last push against the enemy. In addition to creating enemy casualties, machine gun fire destroys the enemy's confidence and neutralizes his ability to successfully engage the friendly maneuver element.

A-131. There are distinct phases of rates of fire employed by the base of fire element:

- Initial heavy volume (rapid rate) to gain fire superiority.
- Slower rate to conserve ammunition (sustained rate) while still preventing effective return fire as the assault moves forward.
- Increased rate as the assault nears the objective.
- Lift and shift to targets of opportunity.

A-132. All vocal commands from the leaders to change the rates of fire are accompanied simultaneously by hand-and-arm signals.

A-133. Machine guns in the SBF role should be set in and assigned a primary and alternate sector of fire as well as a primary and alternate position.

A-134. Machine guns are suppressive fire weapons used to suppress known and suspected enemy positions. Therefore, gunners cannot be allowed to empty all of their ammunition into one bunker simply because that is all they can identify at the time.

A-135. The SBF position, not the assault element, is responsible for ensuring there is no masking of fires. The assault element might have to mask the SBF line because they have no choice on how to move. It is the SBF gunner's job to continually shift fires, or move gun teams or the weapons squad to support the assault and prevent any masking.

A-136. Shift and shut down the weapon squad gun teams one at a time, not all at once. M203 and mortar or other indirect fire can be used to suppress while the machine guns are moved to where they can shoot.

A-137. Leaders must take into account the SDZ of the machine guns when planning and executing the lift and or shift of the SBF guns. The effectiveness of the enemy on the objective will play a large role in how much risk should be taken with respect to the lifting or shifting of fires.

A-138. Once the SBF line is masked by the assault element, fires are shifted and or lifted to prevent enemy withdrawal or reinforcement.

MACHINE GUN WITH THE MANEUVER ELEMENT

A-139. Under certain terrain conditions, and for proper control, machine guns may join the maneuver or assault unit. When this is the case, they are assigned a cover fire zone or sector.

A-140. The machine guns seldom accompany the maneuver element. The gun's primary mission is to provide covering fire. The machine guns are only employed with the maneuver element when the area or zone of action assigned to the assault or company is too narrow to permit proper control of the guns. The machine guns are then moved with the unit and readied to employ on order from the leader and in the direction needing the supporting fire.

A-141. When machine guns move with the element undertaking the assault, the maneuver element brings the machine guns to provide additional firepower. These weapons are fired from a bipod, in an assault mode, from the hip, or from the underarm position. They target enemy automatic weapons anywhere on the unit's objective. Once the enemy's automatic weapons have been destroyed (if there are any), the gunners distribute their fire over their assigned zone or sector. In terms of engagement ranges, the machine gun in the assault engages within 300 meters of its target and frequently at point-blank ranges.

A-142. Where the area or zone of action is too wide to allow proper coverage by the platoon's organic machine guns, the platoon can be assigned additional machine guns or personnel from within the company. This may permit the platoon to accomplish its assigned mission. The machine guns are assigned a zone or a sector to cover and they move with the maneuver element.

M 249 MACHINE GUN IN THE OFFENSE

A-143. In the offense, M249s target any enemy-supporting weapons being fired from fixed positions anywhere on the squad's objective. When the enemy's supporting weapons have been destroyed, or if there are none, the machine gunners distribute their fire over that portion of the objective that corresponds to their team's position.

MEDIUM MACHINE GUNS IN THE OFFENSE

A-144. In the offense the platoon leader has the option to establish his base of fire element with one or two machine guns, the M249 light machine gun, or a combination of the weapons. The platoon sergeant or weapons squad leader may position this element and control its fires when the platoon scheme of maneuver is to conduct the assault with the Infantry squads. The M240B machine gun, when placed on a tripod, provides stability and accuracy at greater ranges than the bipod, but it takes more time to maneuver the machine gun should the need arise. The machine gunners target key enemy weapons until the assault element masks their fires. They can also be used to suppress the enemy's ability to return accurate fire, or to hamper the maneuver of the enemy's assault element. They fix the enemy in position and isolate him by cutting off his avenues of reinforcement. They then shift their fires to the flank opposite the one being assaulted and continue to target any automatic weapons that provide enemy support, and engage any enemy counterattack. M240B fires also can be used to cover the gap created between the forward element of the friendly assaulting force and terrain covered by indirect fires when the indirect fires are lifted and shifted. On signal, the machine gunners and the base-of-fire element displace to join the assault element on the objective.

MK 19 AND M2 IN THE OFFENSE

A-145. The MK 19 and M2 can be used as part of the base-of-fire element to assist the friendly assault element by suppressing enemy bunkers and lightly-armored vehicles. Even if ammunition fired from the guns is not powerful enough to destroy enemy vehicles, well-aimed suppressive fire can keep the enemy buttoned up and unable to place effective fire on friendly assault elements. The MK 19 and M2 are particularly effective in preventing lightly-armored enemy vehicles from escaping or reinforcing. Both vehicle mounted weapons can fire from a long range stand-off position, or be moved forward with the assault element.

MACHINE GUNS IN THE DEFENSE

A-146. The platoon's defense centers on its machine guns. The platoon leader sites the rifle squad to protect the machine guns against the assault of a dismounted enemy formation. The machine gun provides the necessary range and volume of fire to cover the squad front in the defense.

A-147. The primary requirement of a suitable machine gun position in the defense is its effectiveness in accomplishing specific missions. The position should be accessible and afford cover and concealment. Machine guns are sited to protect the front, flanks, and rear of occupied portions of the defensive position, and to be mutually supporting. Attacking troops usually seek easily-traveled ground that provides cover from fire. Every machine gun should therefore have three positions: primary, alternate, and supplementary. All of these positions should be chosen by the leader to ensure his sector is covered and that the machine guns are protected on their flanks.

A-148. The leader sites the machine gun to cover the entire sector or to overlap sectors with the other machine guns. The engagement range may extend from over 1,000 meters where the enemy begins his assault to point-blank range. Machine gun targets include enemy automatic weapons and command and control elements.

A-149. Machine gun fire is distributed in width and depth in a defensive position. The leader can use machine guns to subject the enemy to increasingly devastating fire from the initial phases of his attack, and to neutralize any partial successes the enemy might attain by delivering intense fires in support of counterattacks. The machine gun's tremendous firepower enables the unit to hold ground. This is what makes them the backbone or framework of the defense.

M249 MACHINE GUN IN THE DEFENSE

A-150. In the defense, the M249 adds increased firepower without the addition of manpower. Characteristically, M249s are light, fire rapidly, and have more ammunition than the rifles in the squad they support. Under certain circumstances, the platoon leader may designate the M249 machine gun as a platoon weapon.

MEDIUM MACHINE GUNS IN THE DEFENSE

A-151. In the defense, the medium machine gun provides sustained direct fires that cover the most likely or most dangerous enemy dismounted avenues of approach. It also protects friendly units against the enemy's dismounted close assault. The platoon leader positions his machine guns to concentrate fires in locations where he wants to inflict the most damage to the enemy. He also places them where they can take advantage of grazing enfilade fires, stand-off or maximum engagement range, and best observation of the target area. Machine guns provide overlapping and interlocking fires with adjacent units and cover tactical and protective obstacles with traversing or searching fires. When final protective fires are called for, machine guns (aided by M249 fires) place an effective barrier of fixed, direct fire across the platoon front. Leaders position machine guns to—

- Concentrate fires where they want to kill the enemy.
- Fire across the platoon front
- Cover obstacles by direct fire.
- Tie in with adjacent units.

MK 19 AND M2 IN THE DEFENSE

A-152. In the defense, MK 19 and M2 machine guns may be fired from the vehicle mount or dismounted from the vehicle and mounted on a tripod at a defensive fighting position designed for the weapon system.

A-153. These guns provide sustained direct fires that cover the most likely enemy mounted avenue of approach. Their maximum effective range enables them to engage enemy vehicles and equipment at far greater ranges than the platoon's other direct fire weapons.

A-154. When mounted on the tripod, the M2 and MK 19 are highly accurate to their maximum effective range and predetermined fires can be planned for likely high pay off targets. The trade off is these weapon systems are relatively heavy, and take more time to move.

A-155. These guns are not as accurate when mounted on vehicles as they are when fired from the tripod-mounted system. They are, however, more easily maneuvered to alternate firing locations should the need arise.

AMMUNITION PLANNING

A-156. Leaders must carefully plan for the rates of fire to be employed by machine guns as they relate to the mission and the amount of ammunition available. The weapons squad leader must fully understand the mission the amount of available ammunition and the application of machine gun fire needed to fully support all key events of the mission. Planning will ensure the guns do not run out of ammunition.

A-157. A mounted platoon might have access to enough machine gun ammunition to support the guns throughout any operation. A dismounted platoon with limited resupply capabilities has to plan for only the basic load to be available. In either case, leaders must take into account key events the guns must support during the mission. They must plan for the rate of machine gun fire needed to support the key events, and the amount of ammunition needed for the scheduled rates of fire.

A-158. The leader must make an estimate of the total amount of ammunition needed to support all the machine guns. He must then adjust the amount of ammunition used for each event to ensure enough ammunition is available for all phases of the operation. Examples of planning rates of fire and ammunition requirements for a platoon's machine guns in the attack follow.

KNOW RATES OF FIRE

A-159. Leaders and gunners must know how much ammunition is required to support the different rates of fire each platoon machine gun and assault weapon will require. Coupling this knowledge with an accurate estimate of the length of time and rates of fire their guns are scheduled to fire will ensure enough ammunition resources to cover the entire mission. As part of an example of the planning needed to use M240Bs in support-by-fire roles, the rates of fire for the M240B are listed in Table A-10.

Table A-10. M240B rates of fire.

Sustained	• 100 rounds per minute • Fired in 6- to 9-round bursts • 4-5 seconds between bursts (barrel change every 10 minutes)
Rapid	• 200 rounds per minute • Fired in 10- to 12-round bursts • 2-3 seconds between bursts (barrel change every 2 minutes)
Cyclic	• 650-950 rounds per minute • Continuous burst (barrel change every minute)

Ammunition Requirement

A-160. Leaders must calculate the number of rounds needed to support every machine gun throughout all phases of the operation. Ammunition must be allocated for each key event and to support movement with suppressive fires. For example, in the following list, key events are given for a platoon using two M240Bs in a support-by-fire position. Figure A-20 illustrates steps the leader must take to accurately estimate the ammunition required.

Calculating Ammunition Requirement for Two M240Bs in a Support-by-Fire Position

Identify Key Events
Breach chain link fence.
Counterattack 3-4 light-skinned enemy vehicles with 30 enemy passengers.
Consolidte and reorganize.

Allocate Ammunition to Each Key Event
Support the breach with rapid rate of fire for 30 seconds prior to breach and 30 seconds after (1 minute total) = 200 rounds per gun.
Defeat counterattack = 100 rounds per gun.
Consolidate and reorganize = 200 rounds per gun.

Support Movement With Suppressive Fires
Use sustained rate of fire (100 rounds per minute) to support movement to breach (10 minutes) = 1,000 rounds per gun.
Use sustained rate of fire (100 rounds per minute) to support movement from first objective to the next (5 minutes) = 500 rounds per gun.

Add Everything Together
Breach (200) + counterattack (100) + consolidate and reorganize (200) + first movement (1,000) + second movement (500) = 2,000 rounds per gun.
2 machine guns = 4,000 total rounds needed.

Analyze and Adjust if Necessary
"Is this too much ammunition? What do I have right now?"
Look at key events: "Can we flex there? No."
"How much fire have I planned for during my movements?'

- Sustained rate of fire = 100 rounds per minute (6- to 9-round burst every 4-5 seconds) × 10 minutes = 1,000 rounds per gun planned.
- "Do I need that many rounds?"

Work Backwards

- "I now want one 9-round burst every 10 seconds to support movements."
- 60 seconds divided by 10 seconds = 6 bursts per minute × 9-round burst = 54 rounds per minute × 10 minutes = 540 rounds per gun.
- This a difference of 460 rounds per gun from the original plan.
- New round count = 1,540 rounds per gun or 3,080 total rounds.

Total savings of 920 rounds by adding 5 seconds between suppressive fire bursts during movements.

Figure A-20. Example of ammunition requirement calculation.

Appendix B

Shoulder-Launched Munitions and Close Combat Missile Systems

Shoulder-launched munitions (SLM) and Close Combat Missile Systems (CCMS) are employed by the Infantry platoon to destroy enemy field fortifications or disable enemy vehicles at ranges from 15 to 3,750 meters. They can engage targets in assault, support-by-fire, and defensive roles, and are the Infantry platoon's highest casualty-producing organic weapons when used against armored enemy vehicles. This appendix addresses SLM and CCMS use by the Infantry platoon and discusses their capabilities and limitations.

SECTION I — MUNITIONS

B-1. SLM and CCMS are used against field fortifications, enemy vehicles, or other similar enemy targets. SLM are issued to Infantry Soldiers as rounds of ammunition in addition to their assigned weapons. While Javelins are organic to the Infantry weapons squad, tube-launched, optically-tracked, wire-guided (TOW) missile weapon systems are found in the assault platoons in the Infantry battalion's weapon company. This section discusses the specific types of SLM and CCMS the Infantry platoon or squad will employ. Section II discusses their employment considerations. Section III discusses safety. For complete information read FM 3-23.25, *Shoulder Launched Munitions*; FM 3-22.37, *Javelin Medium Antiarmor Weapon System*; FM 3-22.34, *Tow Weapon System*; and FM 3-22.32, *Improved Target Acquisition System, M41.*

SHOULDER-LAUNCHED MUNITIONS

B-2. SLM include the M136 AT4; the M72A3 light antiarmor weapon (LAW) and improved M72A7 LAW; and the XM141 bunker defeat munition (BDM). The XM141 has also been referred to as the shoulder-launched multipurpose assault weapon-disposable (SMAW-D). Table B-1 lists select SLM specifications.

B-3. All SLM are lightweight, self-contained, single-shot, disposable weapons that consist of unguided free flight, fin-stabilized, rocket-type cartridges packed in expendable, telescoping launchers (except the AT4 which does not telescope) that also serve as storage containers. The only requirement for their care is a visual inspection. SLM can withstand extreme weather and environmental conditions, including arctic, tropical, and desert climates.

B-4. SLM increase the lethality and survivability of the Infantry Soldier and provide him a direct fire capability to defeat enemy personnel within armored platforms. BDM provides the Soldier a direct fire capability to defeat enemy personnel located within field fortifications, bunkers, caves, masonry structures, and lightly armed vehicles and to suppress enemy personnel in lightly armored vehicles.

B-5. The individual Soldier will use SLM to engage threat combatants at very close ranges—across the street or from one building to another. The Soldier may employ SLM as a member of a support- by-fire element to incapacitate threat forces that threaten the assault element. When the assault element clears a building, the leader may reposition the SLM gunner inside to engage a potential counterattack force.

Table B-1. Shoulder-launched munitions.

SHOULDER-LAUNCHED MUNITION	M136 AT4	M72A3 LAW	M72A7 IMPROVED LAW	XM141 BDM (SMAW-D)
FIELD MANUAL	FM 3-23.25	FM 3-23.25	FM 3-23.25	FM 3-23.25
CARRY WEIGHT	14.8 lbs 6.7 kg	5.5 lbs 2.5 kg	8.0 lbs 3.6 kg	15.7 lbs 7.2 kg
LENGTH: CARRY EXTENDED	 102.0 cm N/A	 67 cm 100 cm	 75.5 cm 98 cm	 79.2 cm 137.1 cm
CALIBER	84-mm	66-mm	60-mm	83-mm
MUZZLE VELOCITY	290 m/s 950 f/s	144.8 m/s 475 f/s	200 m/s 656 f/s	217 m/s 712 f/s
OPERATING TEMPERATURE	-40° to 60° C -40° to 140° F	-40° to 60° C -40° to 140° F	-40° to 60° C -40° to 140° F	-32° to 49° C -20° to 120° F
MAXIMUM EFFECTIVE RANGE	300 m	Stationary 200 m Moving 165 m	220 m	500 m
MAXIMUM RANGE	2,100 m	1,000 m	1,400 m	2,000 m
MINIMUM ARMING RANGE	10 m	10 m	25 m	15 m

M136 AT4

B-6. The M136 AT4 is a lightweight, self-contained, SLM designed for use against the improved armor of light armored vehicles. It provides lethal fire against light armored vehicles, and has some effect on most enemy field fortifications.

Ammunition

B-7. The AT4 is a round of ammunition with an integral, rocket-type cartridge. The cartridge consists of a fin assembly with tracer element; a point detonating fuze; and a high-explosive antitank (HEAT) warhead (Figure B-1).

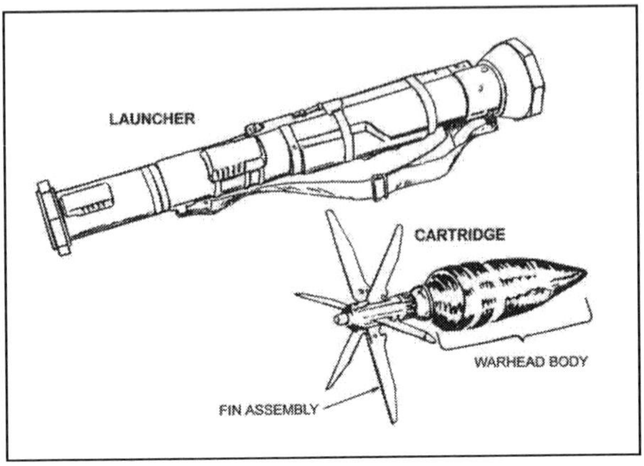

Figure B-1. M136 AT4 launcher and HEAT cartridge.

M72-SERIES LIGHT ANTITANK WEAPON (LAW)

B-8. The M72 LAWs used by Infantry platoons today are the M72A3 and M72A7. They are lightweight and self-contained SLM consisting of a rocket packed in a launcher (Figure B-2). They are man-portable, and may be fired from either shoulder. The launcher, which consists of two tubes, one inside the other, serves as a watertight packing container for the rocket and houses a percussion-type firing mechanism that activates the rocket.

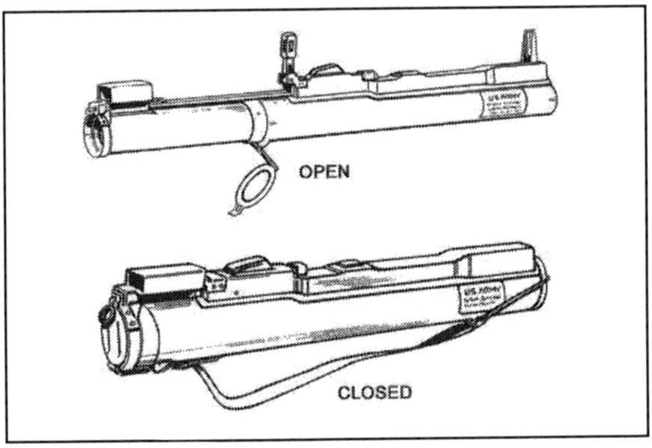

Figure B-2. M72A3 LAW.

M72A3

B-9. The M72A3 contains a nonadjustable propelling charge and a 66-mm rocket. Every M72A3 has an integral HEAT warhead in the rocket's head (or body) section (Figure B-3). Although the M72A3 is mainly employed as an antiarmor weapon, it may be used with limited success against secondary targets such as gun emplacements, pillboxes, buildings, or light vehicles.

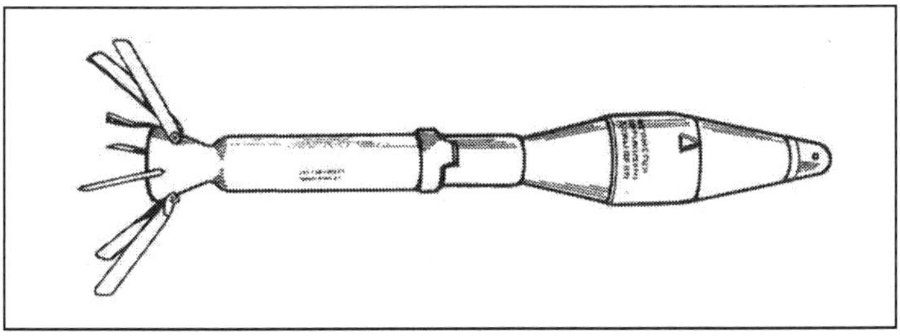

Figure B-3. M72A3 LAW 66-mm high-explosive antiarmor rocket.

Improved M72A7 LAW

B-10. The M72A7 is the Improved LAW currently employed by Infantry platoons. It is a compact, lightweight, single-shot, disposable weapon optimized to defeat lightly armored vehicles at close combat ranges (Figure B-4). The M72A7 offers enhanced capabilities beyond that of the original M72-series. The Improved M72 consists of a 60mm unguided rocket prepackaged at the factory in a telescoping, throw-away launcher. The system performance improvements include a higher velocity rocket motor that extends the weapon effective range, increased lethality warhead, lower and more consistent trigger release force, rifle-type sight system, and better overall system reliability and safety. The weapon contains a 60-mm rocket and an integral HEAT warhead. The warhead is designed to penetrate 150 millimeters of homogenous armor and is optimized for maximum fragmentation behind light armor, Infantry fighting vehicle(s) (IFV), and urban walls.

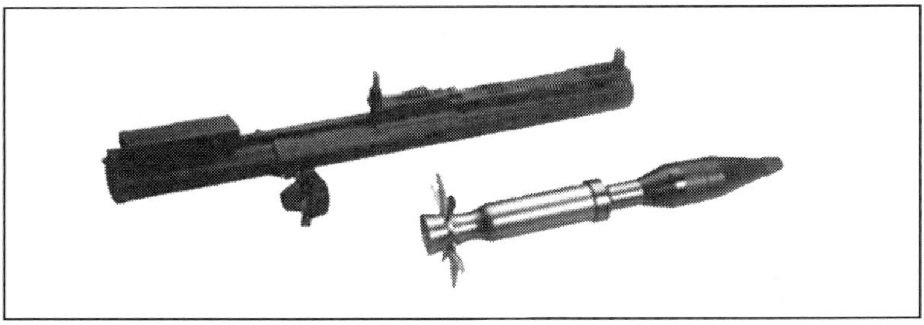

Figure B-4. Improved M72A7 LAW with rocket.

XM141 Bunker Defeat Munition (SMAW-D)

B-11. The XM141 BDM was developed to defeat enemy bunkers and field fortifications (Figure B-5). The XM141 is a disposable, lightweight, self-contained, man-portable, shoulder-fired, high explosive multipurpose munition.

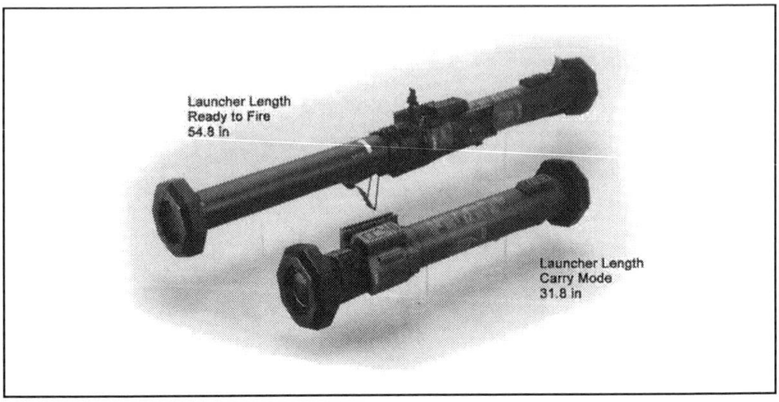

Figure B-5. XM141 bunker defeat munition.

Ammunition

B-12. The XM141 utilizes the 83-mm high explosive dual purpose (HEDP) assault rocket (Figure B-6). The 83-mm HEDP assault rocket warhead consists of a dual mode fuze, and 2.38 pounds of A-3 explosive.

B-13. Warhead function, in quick or delay mode, is automatically determined by the fuze when the rocket impacts a target. The XM141 is fired at hard or soft targets without any selection steps required by the gunner. This automatic feature assures that the most effective kill mechanism is employed. Warhead detonation is instantaneous when impacting a hard target, such as a brick or concrete wall or an armored vehicle. Impact with a softer target, such as a sandbagged bunker, results in a fuze time delay that permits the rocket to penetrate into the target before warhead detonation.

B-14. The XM141 BDM can destroy bunkers, but is not optimized to kill the enemy soldiers within masonry structures in urban terrain or armored vehicles. The XM141 BDM can penetrate masonry walls, but multiple rounds may be necessary to deliver sufficient lethality against enemy personnel behind the walls.

B-15. The XM141 has been used with great success in destroying personnel and equipment in enemy bunkers, field fortifications, and caves in recent operations.

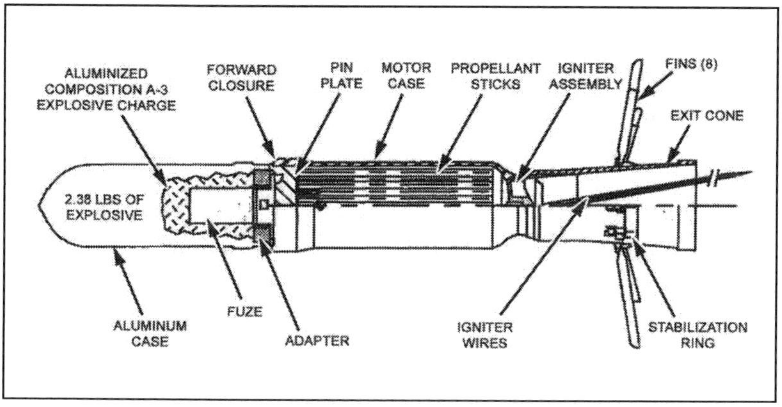

Figure B-6. XM141 high-explosive dual purpose assault rocket.

CLOSE COMBAT MISSILE SYSTEMS

B-16. CCMS are used primarily to defeat main battle tanks and other armored combat vehicles. In the current force, this category of weapons includes the TOW and the Javelin. The TOW and Javelin provide overmatch antitank fires during the assault and provide extended range capability for engaging armor during both offense and defense. These systems have a moderate capability against bunkers, buildings, and other fortified targets commonly found during combat in urban areas. The TOW's bunker buster round is capable of destroying the majority of urban targets.

JAVELIN

B-17. The Javelin is a fire-and-forget, shoulder-fired, man-portable CCMS that consists of a reusable M98A1 command launch unit (CLU) and a round (Figure B-7). The CLU houses the daysight, night vision sight (NVS), controls, and indicators. The round consists of the missile, the launch tube assembly (LTA), and the battery coolant unit (BCU). The LTA serves as the launch platform and carrying container for the missile. See FM 3-22.37 for complete information regarding the Javelin's technical specifications, care, maintenance, operation, gunnery skills training, training aids, and safety.

B-18. The Javelin CCMS' primary role is to destroy enemy armored vehicles out to 2,000 meters. The Javelin can be employed in a secondary role of providing fire support against point targets such as bunkers and crew-served weapons positions. In addition, the Javelin CLU can be used alone as an aided vision device for reconnaissance, security operations, and surveillance. When Bradley fighting vehicles are part of a combined-arms team, the Javelin becomes a secondary antiarmor weapons system. It supports the fires of tanks and TOWs, covers secondary armor avenues of approach, and provides observation posts with an antiarmor capability. The Javelin gunner should be able to engage up to three targets in two minutes, making him very effective against any armor threat.

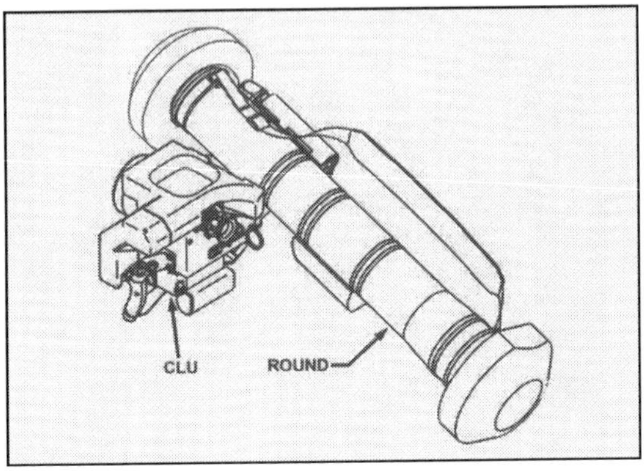

Figure B-7. Javelin close combat missile system.

Command Launch Unit

B-19. The M98A1 CLU is the reusable portion of the Javelin system. It contains the controls and indicators. The CLU provides further utility to the Infantry platoon by allowing accurate surveillance out to two kilometers in both day and night. CLUs have been used to spot and destroy enemy snipers in hidden positions over 1,000 meters away.

B-20. Tables B-2 through B-4 list the Javelin's capabilities and features, the physical characteristics of the CLU, and the physical characteristics of the round.

Table B-2. Javelin capabilities and features.

Javelin Missile System	Surface attack guided missile and M98A1 command launch unit
Type of System	Fire and forget
Crew	One- to three-Soldier teams based on TO&E
Missile modes	Top attack (default)
	Direct attack
Ranges	Top attack mode minimum effective engagement: 150 meters
	Direct attack mode minimum effective engagement range: 65 meters
	Maximum effective engagement range (direct attack and top attack modes): 2,000 meters
Flight Time	About 14 seconds at 2,000 meters
Backblast Area	Primary danger zone extends out 25 meters at a 60-degree (cone-shaped) angle
	Caution zone extends the cone-shaped area out to 100 meters
Firing From Inside Enclosures	Minimum room length: 15 feet
	Minimum room width: 12 feet
	Minimum room height: 7 feet

Table B-3. Physical characteristics of the command launch unit.

M98A1 Command Launch Unit	**With battery, carrying bag, and cleaning kit**	
	Weight: 14.16 lb. (6.42 kg)	
	Length: 13.71 in (34.82 cm)	
	Height: 13.34 in (33.88 cm)	
	Width: 19.65 in (49.91 cm)	
Sights	**Daysight**	
	Magnification: 4X	
	Field-of-view (FOV): 4.80° x 6.40°	
	Night Vision Sight	
	Wide field-of-view (WFOV) magnification: 4.2X	
	WFOV: 4.58° x 6.11°	
	Narrow field-of-view (NFOV) magnification: 9.2X	
	NFOV: 2.00° x 3.00° (approximately)	
Battery Type	**Lithium Sulfur Dioxide ($LiSO_2$) BA-5590/U (Nonrechargeable)**	
	Number required: 1	
	NSN: 6135-01-036-3495	
	Weight: 2.2 lbs. (1.00 kg)	
	Life:	4.0 hrs below 120°F (49°C)
		3.0 hrs between 50°F to 120°F (10°C to 49°C)
		1.0 hrs between -20°F to 50°F (-49°C to 10°C)
		0.5 hrs above 120°F (49°C)

Table B-4. Physical characteristics of the round.

Complete Round (Launch tube assembly with missile and BCU)	Weight: 35.14 lb. (15.97 kg)
	Length: 47.60 in (120.90 cm)
	Diameter with end caps: 11.75 in (29.85 cm)
	Inside diameter: 5.52 in (14.00 cm)
Battery Coolant Unit	Weight: 2.91 lb. (1.32 kg)
	Length: 8.16 in (20.73 cm)
	Width: 4.63 in (11.75 cm)
	Battery type: lithium, nonrechargeable
	Battery life: 4 min of BCU time
	Battery coolant gas: argon

Missile

B-21. The Javelin missile consists of the guidance section, the mid-body section, the warhead, the propulsion section, and the control actuator section. A discussion of the guidance section and warhead follows.

Guidance Section

B-22. The guidance section provides target tracking and flight control signals. It is the forward section of the missile and includes the seeker head section and the guidance electronics unit.

Warhead Section

B-23. The Javelin missile uses a dual charged warhead (Figure B-8) that contains a precursor charge and main charge.

- **Precursor Charge**. The precursor charge is an HE antitank shaped charge. Its purpose is to cause reactive armor on the target to detonate before the main charge reaches the armor. Once the reactive armor is penetrated, the target's main hull is exposed to the warhead's main charge. If the target is not equipped with reactive armor, the precursor provides additional explosives to penetrate the main armor.
- **Main Charge**. The main charge is the second charge of a dual-charge warhead and is also an HE antitank shaped charge. The primary warhead charge is designed to penetrate the target's main armor to achieve a target kill.

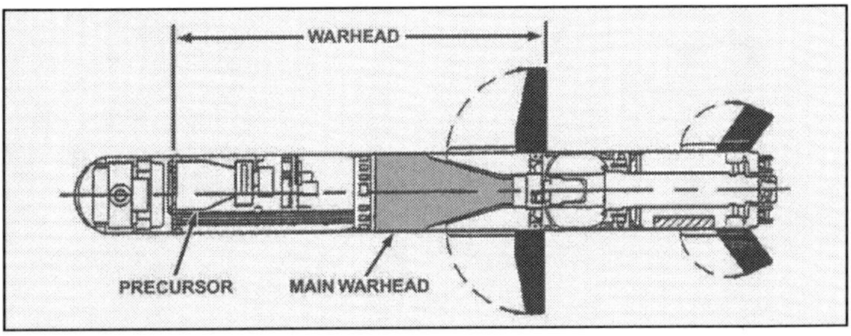

Figure B-8. Javelin missile warhead.

Capabilities and Limitations

B-24. The Javelin has some unique capabilities that provide the unit with an effective antiarmor weapon system. However, the Infantry leader should also understand the system's limitations in order to effectively employ this system (Table B-5).

Table B-5. Javelin capabilities and limitations.

	Capabilities	Limitations
Firepower	• Maximum effective range is 2,000 meters. • Fire-and-forget capability. Missile imaging infrared (I2R) system gives missile ability to guide itself to the target when launched by the gunner. • Two missile flight paths: ▪ Top attack – impacts on top of target. ▪ Direct attack – impacts on front, rear, or flank of target. • Gunner can fire up to three missiles within 2 minutes. • Dual-shaped charge warhead can defeat any known enemy armor. • NVS sees little degradation of target image. • Countermeasures used by enemy are countered by the NVS filter.	• CLU sight cannot discriminate targets past 2,000 meters. • NVS cool-down time is from 2.5 to 3.5 minutes. • Seeker's cool-down time is about 10 seconds. • BCU life, once activated, is only about 4 minutes. • FOV can be rendered useless during limited visibility conditions (rain, snow, sleet, fog, haze, smoke, dust, and night). Visibility is limited by the following: ▪ Day FOV relies on daylight to provide the gunner a suitable target image; limited visibility conditions may block sun. ▪ NVS uses the infrared naturally emitted from objects. *Infrared crossover* is the time at both dawn and dusk that terrain and targets are close enough in temperature to cause targets to blend in with their surroundings. ▪ *Natural clutter* occurs when the sun heats objects to a temperature close enough to surrounding terrain that it causes a target to blend in with terrain. ▪ *Artificial clutter* occurs when there are man-made objects that emit large amounts of infrared (for example, burning vehicles). ▪ Heavy fog reduces the capability of the gunner to detect and engage targets. • Flight path of missile is restricted in wooded, mountainous, and urban terrain. • Gunner must have line of sight for the seeker to lock onto a target.
Maneuver	• Man-portable. • Fire-and-forget capability allows gunner to shoot and move before missile impact. • Soft launch capability allows it to be fired from inside buildings and bunkers. • Maneuverable over short distances for the gunners.	• Weight of Javelin makes maneuvering slow over long distances. • The Javelin round is bulky and restricts movement in heavily-wooded or vegetative terrain.
Protection	• Passive infrared targeting system used to acquire lock-on cannot be detected. • Launch motor produces a small signature. • Fire-and-forget feature allows gunner to take cover immediately after missile is launched.	• Gunner must partially expose himself to engage the enemy. • CLU requires a line of sight to acquire targets.

TUBE-LAUNCHED, OPTICALLY-TRACKED, WIRE-GUIDED (TOW) MISSILE WEAPON SYSTEM

B-25. The Infantry TOW weapon system consists of the Improved Target Acquisition System (ITAS) launcher, which has tracking and control capabilities, and the missile, which is encased in a launch container. The launcher is equipped with self-contained, replaceable units.

B-26. The TOW is designed to destroy enemy tanks, fortifications, and other materiel targets. Its line-of-sight launcher initiates, tracks, and controls the missile's flight through command-link wire-transmitted guidance signals. It can be employed in all weather conditions as long as the gunner can see the target through the ITAS. The TOW also provides a long-range assault capability against heavily fortified bunkers, pillboxes, and gun emplacements.

B-27. The current versions of the TOW missile can destroy targets at a minimum range of 65 meters and a maximum range of 3,750 meters. The TOW 2B missile can destroy targets at a minimum range of 200 meters and a maximum range of 3,750 meters. TOW missiles in development are being produced to effectively engage enemy targets out to 4,500 meters.

Missile System Configurations and Types

B-28. The TOW CCMS consists of multiple configurations with numerous types of missiles. These configurations mainly consist of minor modified work orders that are transparent to the operator and are continually updated. All configurations use the same basic airframe, aerodynamic control system, command-link wire, and missile electronics designs. The current missile types are listed below.

- **Improved TOW**. The ITOW missile has an improved 5-inch warhead from the original TOW missile that includes extended probes for greater standoff and penetration. It can destroy targets at a minimum range of 65 meters and a maximum range of 3,750 meters.
- **TOW 2**. The TOW 2 missile has a full-caliber 6-inch warhead that includes an extended probe. In addition to the infrared radiator of the ITOW missile, TOW 2 has a second infrared radiator to provide hardened system performance against battlefield obscurants and countermeasures. The second radiator is called the thermal beacon and provides link compatibility with the electro-optical infrared nightsight, which is part of the TOW 2 launcher system.
- **TOW 2A**. The TOW 2A adds a small explosive charge in the tip of the extended probe that causes enemy reactive armor to detonate prematurely, thus allowing the TOW 2A's warhead to penetrate the main armor.
- **TOW 2B**. The TOW 2B has an entirely different warhead and kill mechanism than the previous TOW missiles. It is a top-attack missile (fly over/shoot down) that defeats enemy armor at its most vulnerable point—the top deck of the turret and hull. The TOW 2B has a tandem warhead that fires two explosively formed projectiles down through the thin upper deck armor of the enemy vehicle. The gunner tracks the target the same as any other TOW missile with the crosshairs on center mass, but the missile automatically flies 2.25 meters above the line of sight (LOS). When the missile senses that it is directly above the target (by means of the target's shape and magnetic field), it automatically fires its warhead. The TOW 2B missile can destroy targets at a minimum range of 288 meters when fired from the ground mount and 200 meters when fired from the HMMWV or BFV. The TOW 2B has a maximum range of 3,750 meters whether ground- or vehicle-mounted.
- **TOW 2B GEN 1**. The TOW 2B GEN 1 is similar to the TOW 2B but includes the addition of the GEN 1 Counter Active Protection System (CAPS), which is used to defeat enemy active protection systems.
- **TOW 2B Aero**. The TOW 2B Aero is an extended range version of the TOW 2B missile with an aerodynamic nose and has an effective range of 4,500 meters (Figure B-9). This longer range (compared to the 3,750 meter range of the previous TOW missiles) allows a TOW crew to fire well beyond the weapons range of its targeted vehicle.
- **TOW 2B Aero With GEN 1, 2, and 3A CAPS**. These versions of TOW 2B Aero have the addition of different generations of CAPS to defeat an enemy target's active protection system, allowing the TOW 2B missile to successfully engage any armored vehicle up to 4,500 meters (Figure B-9).
- **TOW Bunker Buster**. The TOW Bunker Buster (BB) replaces the TOW 2A warhead with a fragmenting bulk charge for non-armor targets (Figure B-10). The TOW BB has a range of 3,750 meters. Its missile is capable of defeating bunkers, breaching masonry walls, and engaging targets in support of urban operations.

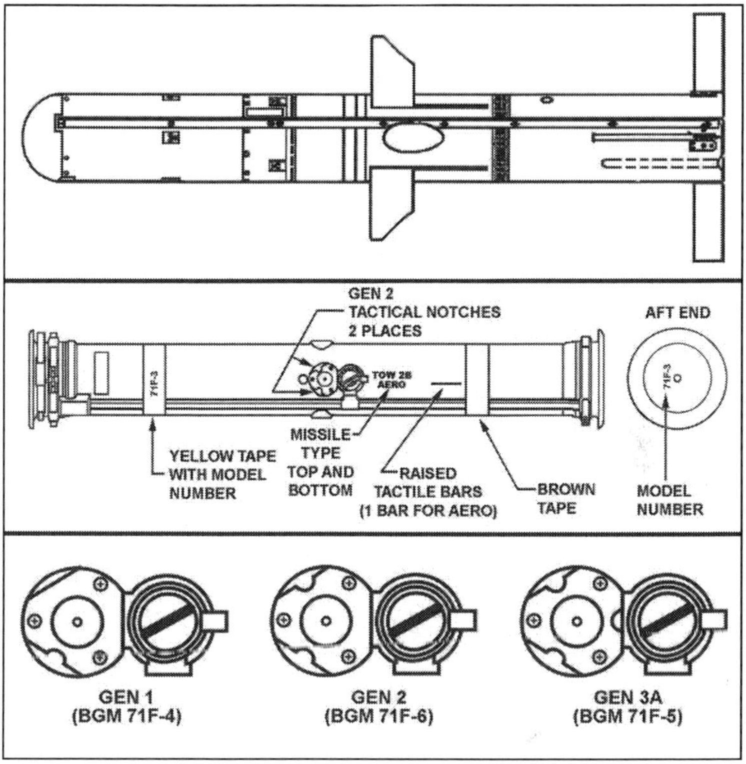

Figure B-9. TOW 2B Aero missile with identification.

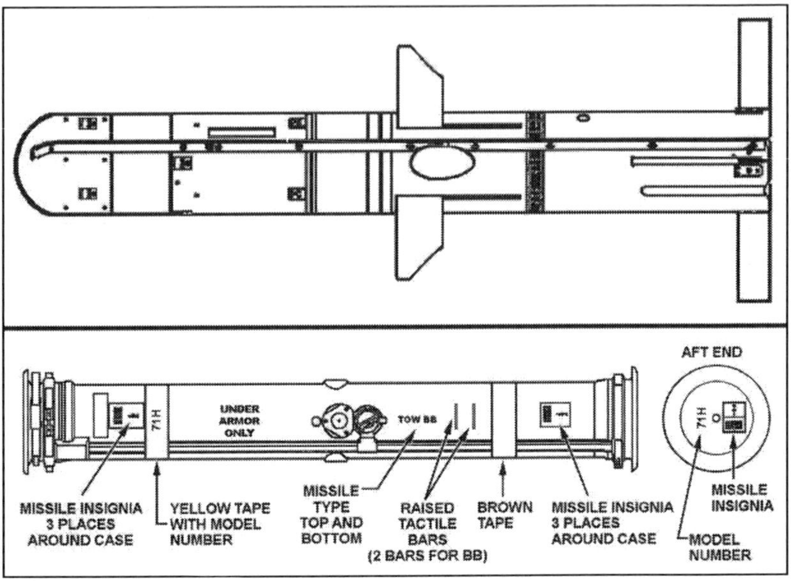

Figure B-10. TOW bunker buster (BB) missile and identification.

M41 Improved Target Acquisition System (ITAS)

B-29. The ITAS is primarily a mounted system that utilizes the M1121 HMMWV as the carrier vehicle. The M1121 HMMWV is a one-vehicle (1 1/4-ton truck) combat system that is air transportable, versatile, maintainable, and survivable. The vehicle carries one complete launcher system, seven encased missiles, and a three-man crew. The tactical or training situation may demand that the crew dismount the carrier and employ the ITAS in the dismounted or tripod configuration.

B-30. The M41 ITAS fires all existing and future versions of the TOW family of missiles. The ITAS provides for the integration of both the day sight and NVS into a single housing and for automatic boresighting. It has embedded training (for sustainment training) and advanced built-in test/built-in-test equipment (BIT/BITE), which provides fault detection and isolation.

B-31. The automatic missile tracking and control capabilities of the ITAS provide a high first-round-hit probability. To operate the system, the gunner places the track gates on the target, fires the missile, and centers the crosshairs on the target image until missile impact. The optical tracking and command functions within the system guide the missile to the target as long as the gunner keeps the crosshairs on target.

B-32. The ITAS provides the Infantry platoon with advanced optics during daylight and limited visibility to aid in surveillance and target acquisition in both defensive and offensive operations.

B-33. The ITAS can be vehicle-mounted or ground-emplaced (tripod-mounted) for operation. Missiles can be launched from either operational mode. The entire system can be carried by a single crew for short distances. Moving it over long distances without the vehicle will require two crews, which causes two systems to be out of operation at the same time. The vehicle-mounted launcher is more mobile and can be quickly prepared for use. The launcher can be assembled and disassembled without the use of tools.

SECTION II — EMPLOYMENT CONSIDERATIONS

B-34. The objective of the Army's warfighting doctrine is to concentrate decisive combat power at the right time and place, by massing fires rather than by massing forces, and by presenting the enemy with multiple threats. This section discusses SLM and CCSM employment considerations. A lethal mix of CCMs and SLM provide the Infantry unit with the flexibility to employ multiple systems designed to deliver maximum direct fire lethality and destroy enemy formations at both long range and in close combat. At close combat range (15-300 meters), SLM provide Soldiers with the ability to deliver direct fire lethality at very close proximity to the enemy. At extended range (300-4,500 meters), a mix of Javelin and TOW provides the Infantry leader with overwhelming combat overmatch. These weapons serve as key components by applying overlapping and interlocking fires to achieve synergy and mutual support for his maneuver force.

B-35. For a better understanding of how SLM and CCMS fit into the Infantry platoon's fire plan, see Chapter 3.

URBAN OPERATIONS AND FIELD FORTIFICATIONS

B-36. Operations in complex terrain and urban environments alter the basic nature of close combat. History tells us that engagements are more frequent and occur more rapidly when engagement ranges are close. Studies and historical analyses have shown that only 5 percent of all targets are more than 100 meters away. About 90 percent of all targets are located 50 meters or less from the identifying Soldier. Few personnel targets will be visible beyond 50 meters. Engagements usually occur at 35 meters or less.

B-37. Soldiers employ SLM in the short, direct fire, close-quarter engagement range of close combat. Their use is preferable in urban areas where other direct fire (M1 Abrams and M2/M3 BFV) and indirect fire systems (artillery and mortars) and CAS are incapable of operating due to risks of fratricide and collateral damage. In close combat, Soldiers employ SLM against a wide variety of targets. These include: personnel armed with individual and crew served weapons fighting from armored platforms (T-72s, BTRs, BRDMs); light armored personnel carriers and Infantry fighting vehicles (BMP1-3 and M113); modified personnel/Infantry vehicles; lightly armed vehicles; and enemy in fortified positions, behind walls, inside caves and masonry buildings, and within earthen bunkers.

B-38. CCMS teams provide overwatching antitank fires during the attack of a built-up area. They are best employed in these types of areas along major thoroughfares and in upper floors of buildings or roofs to attain long-range fields of fire. Because the minimum engagement distance limits firing opportunities in the confines of densely built-up areas, CCMS may not be the weapon of choice in the urban environment (FM 3-06.11). Urban area hazards include, fires caused by both friendly and enemy forces that may cause target acquisition and lock-on problems, clutter on the battlefield that may cause lock-on problems, and line-of-sight communications that may be limited by structures. CCMS unique flight path forces the gunner to think in three dimensions. Other urban environment hazards include overhead obstacles such as street signs, light poles, and wires, which could impede the missile's flight path.

SHOULDER-LAUNCHED MUNITIONS IN THE BUNKER DEFEAT ROLE

B-39. The current inventory of the M136 AT4 and the XM141 BDM in combination with the M72-series LAW provide the Infantry squad the capability to incapacitate personnel within earth and timber bunkers, masonry buildings, and light armored vehicles. However, neither system is fully capable of fire-from-enclosure.

B-40. SLM that can be safely fired from an enclosure to incapacitate personnel within earth and timber bunkers, masonry buildings, and light armored vehicles are currently being developed to increase the lethality, survivability, and mobility of the SLM gunner.

ENGAGEMENT OF FIELD FORTIFICATIONS AND BUILDINGS WITH SLM

B-41. The M72-series LAW and AT4 have proven to have only limited success inflicting casualties against enemy troops in field fortifications and buildings. The XM141 BDM was designed to enhance the destruction of these fortifications and enemy personnel inside them. The BDM's warhead contains a dual mode fuse that automatically adjusts for the type of target on impact. For soft targets, such as sandbagged bunkers, the XM141 warhead automatically adjusts to delayed mode and hits the target with very high kinetic energy. The warhead is propelled through the barrier and into the fortification or building where the fuze detonates the warhead and causes much greater damage. Soldiers should not expect to severely damage fortified targets with M72 LAWs or AT4s. However, if the recommendations shown in Table B-6 are used, Soldiers may be able to gain a temporary advantage.

Table B-6. Effects of the AT4 and M72A3 LAW on field fortifications or bunkers.

AIM POINT	EFFECT WHEN AT4 OR M72A3 IS FIRED AT AIM POINT	RECOMMENDED FIRING TECHNIQUE
Firing Port or Aperture	Rounds fired into firing ports or apertures may not have the desired effect on the enemy. The rounds may detonate against the rear wall of the position, causing little structural damage to the position or to the equipment or personnel within, unless they are hit directly. The AT4 produces less effect than the M72A3 LAW.	Coordinate fire: Fire CCM at a point 6 to 12 inches from the front edge of the firing ports in the berm. Fire small arms at the bunker or position to prevent personnel within from returning fire.
Berm	Firing at the berm causes the round to detonate outside the fighting position or inside the berm itself, creating only dust, a small hole in the berm, or minor structural damage to the position, but little damage to personnel or equipment unless they are hit directly. The AT4 produces less effect than the M72A3 LAW.	Firing the AT4 and LAW at berms should be avoided because of the negligible effects.
Window	The round may travel completely through the structure before detonating. If not, it creates dust and causes minor structural damage to the rear wall, but little damage to personnel or equipment, unless they are hit directly. The AT4 produces less effect than the M72A3 LAW.	Fire 6 to 12 inches from the sides or bottom of a window. CCMs explode on contact with brick or concrete, creating an opening whose size is determined by the type of round used.
Wall	The round detonates on contact, creating dust and causing a small hole and minor structural damage, but little damage to personnel or equipment, unless they are hit directly. Overpressure from the round entering the structure may temporarily incapacitate enemy personnel.	The M72-series LAW may be used to create a loophole, which is a hole large enough to throw hand grenades through. The AT4 produces less effect than the M72-series LAW.
Corner	Corners are reinforced and thus harder to penetrate than other parts of a wall. Any CCM round will detonate sooner on a corner than on a less dense surface. Detonation should occur in the targeted room, creating dust and causing overpressure, which can temporarily incapacitate personnel inside the structure near the point of detonation. The AT4 causes more overpressure than the M72-series LAW.	Avoid targeting corners because of the negligible effects.

CCMS Engagement Considerations

B-42. Urban engagement considerations for CCMS include engagement distance, thermal crossover, backblast, weapon penetration, and breaching structural walls. Details follow. TOW systems should always seek to engage at maximum range. If within 1,000 meters of an enemy, the flight time of the TOW missile would likely be greater than the flight time of a main gun tank round.

- **Engagement Distance.** The Javelin missile has a minimum engagement distance (150 meters in the attack mode and 65 meters in the direct attack mode), which limits its use in built-up areas. The TOW 2B has a minimum range of 200 meters and a maximum range of 3,750, which limits its use in built-up areas.
- **Crossover.** Sometimes the Javelin seeker or TOW round will not be able to distinguish between the background and the target because the two have the same temperature (crossover).
- **Time.** When a gunner comes across a target of opportunity, he may not be able to take advantage of it. The cool down time of the Javelin's NVS is 2.5 to 3.5 minutes. Javelin seeker cool down takes about 10 seconds. Once the BCU is activated, the gunner has a maximum of 4 minutes to engage the target before the battery coolant unit is depleted.
- **Backblast.** The soft launch capability of the Javelin enables the gunner to fire from inside buildings because there is little overpressure or flying debris.
- **Weapon Penetration.** The dual charge Javelin warhead penetrates typical urban targets. The direct attack mode is selected when engaging targets in a building. Enemy positions or bunkers in the open closer than 150 meters are engaged using the direct attack mode. Positions in the open farther than 150 meters are engaged using either the top or direct attack mode, depending on the situation.
- **Breaching Structural Walls.** The Javelin and TOW (except for the TOW BB) are not effective when breaching structural walls. Antitank guided missiles (ATGMs) are not designed to breach structural walls effectively. All CCMS are designed to produce a small hole, penetrate armor, and deliver the explosive charge. Breaching calls for the creation of a large hole. CCMS are better used against armored vehicles or for the destruction of enemy-fortified fighting positions.

ANTIARMOR ROLE

B-43. In the past decade, there has been a revolution in armor technology. Research and new developments have come from Europe, the United States, and Israel. These improvements are also becoming much more common in third world armies. In addition, many older tanks and other armored fighting vehicles are being retrofitted with improved armor protection. These advanced armor configurations improve the vehicles' survivability against all weapons. They are specifically designed to protect against HEAT warheads and essentially fall into four categories: reactive, laminated, composite, and appliqué. Improved armor types include:

- **Reactive Armor.** Reactive armor comes in several varieties, but the principle is essentially the same on all. The armor consists of blocks of explosives sandwiched between two metal plates and bolted on the outside of the vehicle. Small-arms and artillery shrapnel will not set off the blocks. However, when a HEAT round strikes the block, the explosive ignites and blows outwards. The blast and the moving steel plates disperse and deflect the jet of the HEAT warhead, dramatically reducing its ability to penetrate armor.
- **Laminated Armor.** Laminated armor consists of flat layers of steel armor plates with layers of ceramics, fiberglass, or other nonmetallic materials in between. This armor is highly effective against all types of weapons, but is difficult and expensive to manufacture. Vehicles with laminated armor are characterized by flat, slab sides, such as on the M1 Abrams and the German Leopard II.
- **Composite Armor.** Composite armor consists of a nonmetallic core (usually some kind of ceramic) around which the rest of the steel of the hull or the turret is molded. This is much more effective than conventional steel armor against all types of weapons, but less so than laminated armor.
- **Appliqué Armor.** Appliqué armor is essentially extra plates mounted or welded on top of the hull or turret of a vehicle. They can be made of any material, but are frequently made of ceramic or laminated materials. Like reactive armor, appliqué armor is an easy and cost-effective way of improving the protection of older vehicles.

Exploiting Armored Vehicle Weaknesses

B-44. Because they are designed mainly for offensive operations against other armored vehicles (Figure B-11), armored vehicles usually have their heaviest armor in front. All vehicles are vulnerable to repeated hits on their flanks and rear, though the flank offers the largest possible target. Firers should always aim center of mass to increase the probability of a hit. The older the vehicle model, the less protection it has against SLM and CCMS. Newer versions of older vehicle models may use bolt-on (appliqué) armor to improve their survivability. Reactive armor usually covers the forward-facing portions and sides of the vehicle and can defeat shaped-charge weapons such as the SLM. When reactive armor detonates, it disperses metal fragments to 200 meters. SLM cause only a small entry hole in an armored vehicle target, though some fragmentation or spall may occur.

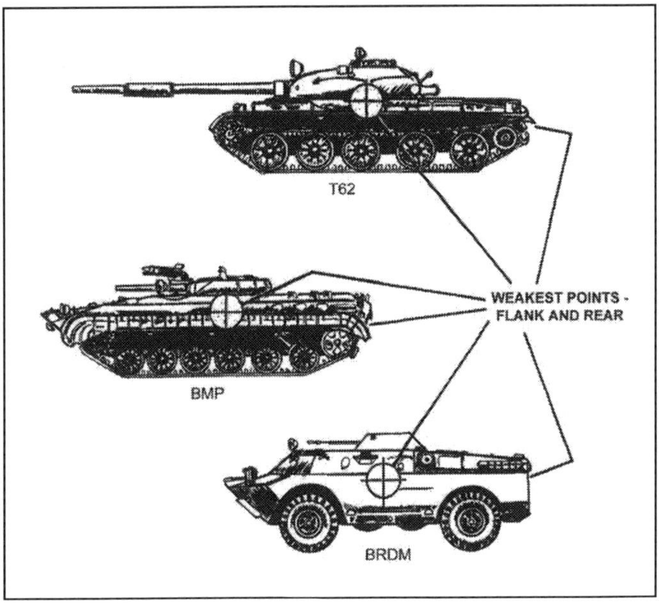

Figure B-11. Armored vehicle weak points.

B-45. Natural or man-made obstacles can be used to force the armored vehicle to slow, stop, or change direction. This pause enables the firer to achieve a first-round hit. If he does not achieve a catastrophic kill on the first round, he or another firer must be ready to engage the target vehicle immediately with another round.

B-46. The white area in Figure B-12 shows the most favorable direction of attack when the turret is facing to the front. The gray area shows the vehicle's principal direction of fire and observation when the turret is facing to the front). Volley fires can greatly degrade the additional protection that appliqué and reactive armors provide to the target vehicle.

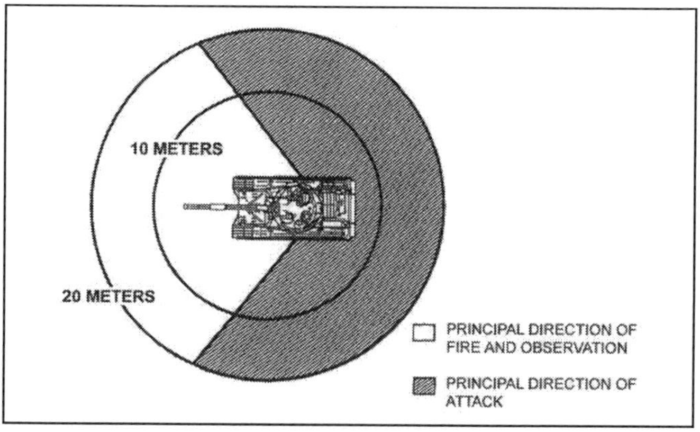

Figure B-12. Limited visibility of armored vehicles.

B-47. Armored vehicle kills are classified according to the level of damage achieved (Table B-7).

Table B-7. Armored vehicle kills.

Type Of Kill	Part of Vehicle Damaged or Destroyed	Capability After Kill
Mobility Kill	Suspension (track, wheels, or road wheels) or power train (engine or transmission) has been damaged.	Vehicle cannot move, but it can still return fire.
Firepower Kill	Main armament has been disabled.	Vehicle can still move, so it can get away.
Catastrophic Kill	Ammunition or fuel storage section has been hit by more than one round.	Vehicle completely destroyed.

SHOULDER-LAUNCHED MUNITIONS IN THE ANTIARMOR ROLE

B-48. When Soldiers employ the M136 AT4 and M72-series LAW to defeat threat armored vehicles, it requires Soldiers to engage threat vehicles using single or paired shots. Gunners require positions that allow engagement against the flank or rear of the target vehicles. They must seek covered and concealed positions from where targets can be engaged. However, the M136 AT4 cannot be fired safely from within an enclosure because it denies the protection offered by enclosed fighting positions and masonry buildings. FM 3-23.25 advises firing the M136 AT4 and XM141 BDM from an enclosure under combat conditions only when no other tactical option exists due to the risk of both auditory and non-auditory injury.

SLM Warhead Effects on Armor

B-49. SLM warheads have excellent armor penetration ability and lethal after-armor effects (especially the AT4 and M72A7). The extremely destructive shaped-charge explosives can penetrate more than 14 inches (35.6 centimeters) of rolled homogeneous armor (RHA). Types of warhead armor effects follow and are illustrated in Figure B-13.

- **Impact.** The nose cone crushes; the impact sensor activates the fuze.
- **Ignition.** The fuze element activates the electric detonator. The booster detonates, initiating the main charge.

- **Penetration.** The main charge fires and forces the warhead body liner into a directional gas jet that penetrates armor plate.
- **Spalling (After-Armor Effects).** The projectile fragments and incendiary effects produce blinding light and highly destructive results.

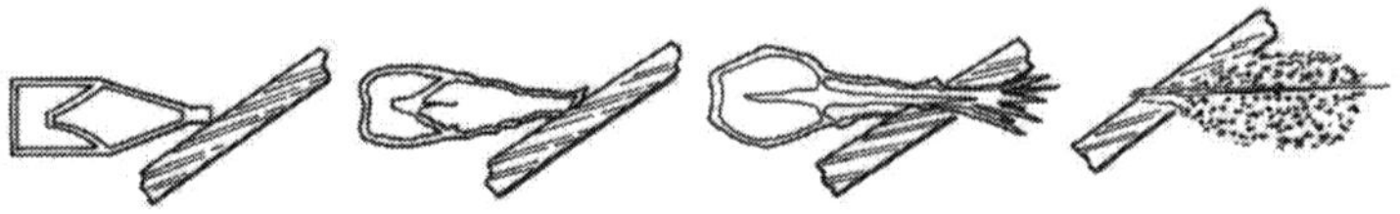

Figure B-13. Effects of SLM warheads on armor targets.

Engagement of Other Vehicles

B-50. The M72-series LAW proves more effective against light vehicles. The M136 AT4 proves more effective against armored vehicles. Non-armored vehicles such as trucks, cars, and boats are considered soft targets. Firing along their length offers the greatest chance of a kill, because this type of shot is most likely to hit their engine block or fuel tank.

Methods OF Engagement

B-51. The four engagement methods for SLM include single, sequence, pair, and volley firing. The leader evaluates the situation on the ground to determine which of these methods to use. Regardless of whether they are used singly or in combination, communications are needed as well. The methods of engagement are rehearsed IAW unit SOP.

Single Firing

B-52. A single Soldier with one SLM may engage an armored vehicle, but this is not the preferred method of engagement. Several SLM are normally required to effectively kill an armored vehicle. A single gunner firing one round must hit a vital part of the target in order to do damage (Figure B-14). A single firer can engage targets out to 225 meters with the LAW, or 300 meters with the AT4 (when he knows the actual range).

Figure B-14. Single firing.

Sequence Firing

B-53. A single firer, equipped with two or more SLM prepared for firing, engages the target. After engaging with the first round and observing the impact, the firer adjusts his point of aim. He then engages with another round until he destroys the target or runs out of rounds (Figure B-15).

Figure B-15. Sequence firing.

Pair Firing

B-54. Two or more firers, equipped with two or more SLM prepared for firing, engage a single target. Before firing, the first firer informs the others of the estimated speed and distance to the target. If the impact of his round proves his estimate to be correct, the other firers engage the target until it is destroyed. If the impact of the round proves his estimate to be incorrect, the second firer informs the others of his own estimate, and then he engages the target. This continues until the target is destroyed or all rounds are expended (Figure B-16).

Figure B-16. Pair firing.

Volley Firing

B-55. Two or more firers can engage a single target when the range is known. These firers engage the target at the same time on a prearranged signal such as a command, whistle, mine, or TRP. This can be the most effective means of engagement as it places the most possible rounds on one target at one time, increasing the possibility of a kill (Figure B-17).

Figure B-17. Volley firing.

TOW Countermeasures to Improved Armor

B-56. TOW crews can expect to be issued a mix of TOW missile types on the battlefield, with widely varying capabilities. Gunners and leaders must be familiar with the different missile types and their respective capabilities. The proper type of missile must be chosen for each type of target (Table B-8).

B-57. TOW crews must strive harder than ever to find positions where they can engage enemy vehicles from the flank. Modern tanks with reactive armor have become increasingly difficult to kill from the front.

Table B-8. Missile selection priority chart.

THREAT VEHICLE-TYPE TARGETS	SELECTION PRIORITY			
	First	Second	Third	Fourth
Tanks with appliqué armor	TOW 2B	TOW 2A	TOW 2	ITOW
Tanks with explosive reactive armor	TOW 2B	TOW 2A	TOW 2	ITOW
Tanks without appliqué/ reactive armor	TOW 2B	TOW 2A	TOW 2	ITOW
Light armored personnel carriers	TOW 2	TOW 2A	TOW 2B	ITOW
Light armored wheeled vehicles	TOW 2	TOW 2A	TOW 2B	ITOW
Antiaircraft vehicles	TOW 2	TOW 2A	TOW 2B	ITOW
Armored vehicles in hull defilade positions	TOW 2B	TOW 2A	TOW 2	ITOW
Bunkers/fortifications	TOW BB	TOW 2	TOW 2A	ITOW

Antiarmor Ambush Role

B-58. Antiarmor ambushes are usually conducted to destroy small groups of armored vehicles, force the enemy to move more slowly and cautiously, or force the enemy into a choke point. Units conducting an antiarmor ambush can use Javelins or TOWs for this purpose. The Javelin and TOW have a slow rate of fire, so other weapons systems must be prepared to engage the vehicles while the Javelin gunners attach the CLU to new rounds or the TOW gunners load new rounds. The Javelin's 2,000-meter range and the TOWs 3,750 meter range allow flexibility in choosing ambush positions. In addition to fires into the kill zones, the Javelin and TOW can be employed in a security role to guard high-speed avenues of approach, to slow or stop enemy reinforcements, or to destroy vehicles attempting to flee the kill zone (Figure B-18).

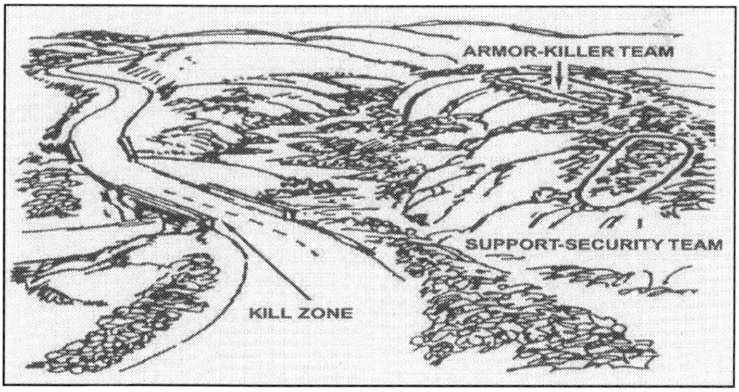

Figure B-18. Antiarmor ambush.

OFFENSIVE OPERATIONS

B-59. CCMS contribute to offensive operations by providing long-range fires that destroy enemy armor and protect the force from armored counterattacks. In the absence of armored targets, CCMS can engage enemy fortifications and hovering helicopters. CCMS are normally used in a support-by-fire role during offensive operations. The primary consideration for such employment is the availability of appropriate fields of fire and the armored threat. CCMS crews can effectively protect flanks against armored threats and can also provide overwatch for unit movement (Figure B-19).

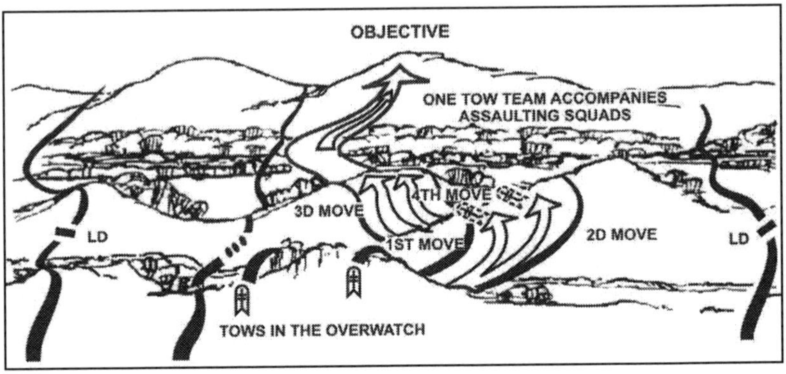

Figure B-19. TOW supporting offensive operations.

DEFENSIVE OPERATIONS

B-60. During planning, the leader considers the enemy armor threat, then positions antiarmor weapons accordingly to cover armor avenues of approach. He also considers the fields of fire, tracking time, and minimum engagement distance of each weapon. The section leader or squad leader selects a primary position and sector of fire for each antiarmor weapon. He also picks alternate and supplementary positions for them. Each position should allow flank fire and have cover and concealment. The leader should integrate the ITAS into his limited visibility security and observation plan. The squad leader selects the fighting position and assigns the sector of fire. Considering the fundamentals of antiarmor employment will greatly improve the crew's survivability. ITAS crews must coordinate with adjacent units to ensure security. The TOW's 3,750-meter maximum range makes it difficult for the enemy to engage the crew with direct fire, which forces the enemy to deploy earlier than intended. The gunner prepares a range card for his primary position. If time permits, he also prepares them for his alternate and supplementary positions (Table B-9).

B-61. Reserve forces armed with SLM may be employed to assist counterattacks to regain key positions. They are also used to block enemy penetrations, to meet unexpected enemy thrusts, and to provide support by fire to endangered friendly units during disengagements and withdrawals. In the event defensive positions are in danger of being overrun by enemy armored vehicles, SLM may be used against armored vehicles and lightly armored vehicles posing an immediate threat, including light tanks. The maximum range provides leaders with greater flexibility in positioning each round and provides a means of achieving overlapping sectors of fire for increased survivability.

Table B-9. Personnel duties.

Tasks to be Performed	Section Sergeant	Team Leader	Gunner/ Assistant Gunner
Integrate CCMS into the platoon tactical plan:			
• Select general weapons positions.	X		
• Assign sectors of fires.	X		
• Coordinate mutual support.	X		
• Coordinate with adjacent units.	X		
Reconnoiter for and select tentative CCMS firing positions (primary, alternate, and supplementary) and routes between positions.	X		
Supervise continual preparation and improvement of positions.	X	X	
Coordinate security for the CCMS teams.	X		
Inspect the selection of tentative firing positions, confirm or make adjustments.	X	X	
Supervise preparation of range card.	X	X	
Control movement of gunners between positions.	X	X	
Issue fire commands to gunners.	X	X	
Coordinate resupply and collection of extra rounds carried in platoon.	X		
Identify enemy avenues of approach.	X		
Prepare fighting position (primary, alternate, supplementary).		X	
Prepare range card.		X	X
Designate target reference points.	X		
Prestock rounds.		X	X
Prepare round for firing.			X
React to fire commands.			X
Engage targets.			X

SECTION III — SAFETY

B-62. Leaders must employ SLM/CCMS to effectively minimize danger to friendly Soldiers caused by the surface danger zone (SDZ) or backblast danger zones. They must weigh the risk of firing the missile in close proximity to friendly assault forces against the need to suppress or destroy enemy fortifications or vehicles from the support-by-fire or assault position. This section discusses SLM and CCMS safety.

SLM

B-63. Figures B-20 through B-27 and Table B-10 illustrate surface danger zone (SDZ) and backblast danger zone information for SLM. See DA PAM 385-63, *Range Safety*, and FM 3-23.25, *Shoulder-Launched Munitions*, for more specific information regarding this and other safety-specific information.

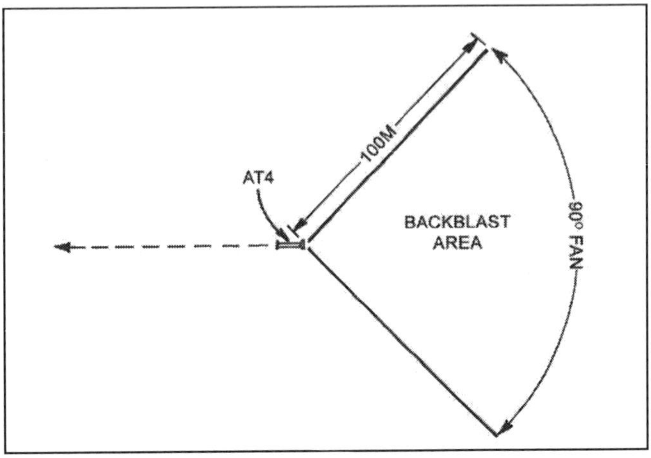

Figure B-20. M136 AT4 backblast danger area.

Table B-10. AT4 SDZ criteria in meters.

Type	Distance X	Minimum Range to Target	Area A	Area B	Area F[2]	
					Danger Zone Depth	Caution Area Depth
84-mm HEAT M136[1]	2,100	50	227	488	5[3]	95[4]
9-mm Trainer, M939	1,600	N/A	N/A	N/A	N/A	N/A

NOTES:
[1]Increased dud rates may occur when firing HE (M136) at impact angles of 10 degrees or less.
[2]Area F is 90-degree angle (45 degrees left and right) of rearward extension of launcher target line.
[3]Danger zone occupation could result in fatalities or serious casualties including, severe burns, eye damage, or permanent hearing loss. The hazards are baseplate fragments, debris, fireball, high noise levels, and overpressure.
[4]Caution area is an extension of the primary danger area. Occupation of this area could also result in severe casualties due to backblast, debris, high noise levels, and possible baseplate fragments. Primary danger area and caution area are conditions that may not be modified.

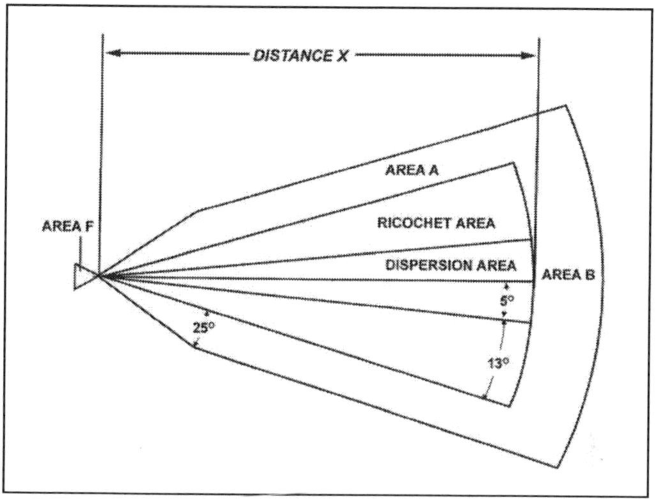

Figure B-21. SDZ for firing AT4.

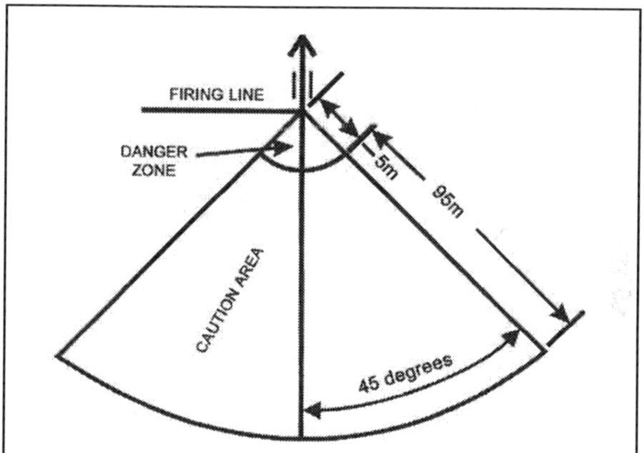

Figure B-22. SDZ area F for firing AT4.

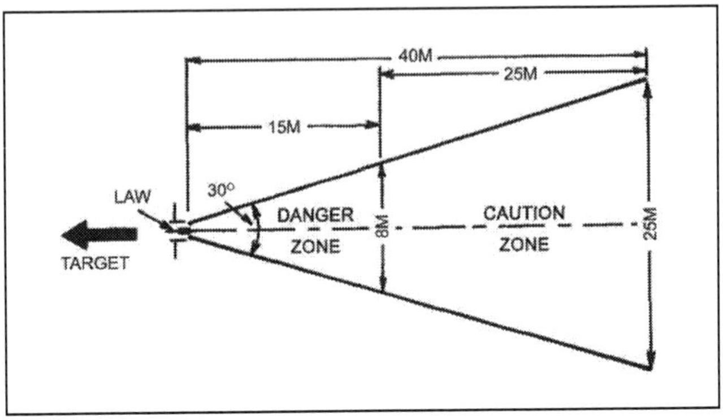

Figure B-23. M72A2/3 LAW backblast area.

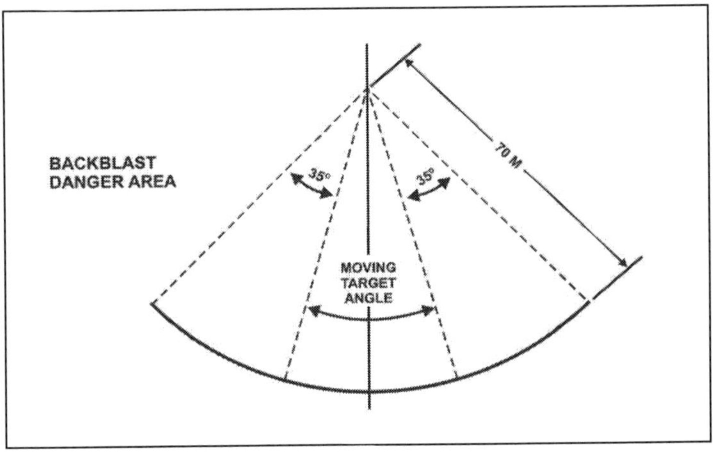

Figure B-24. Improved LAW backblast danger area.

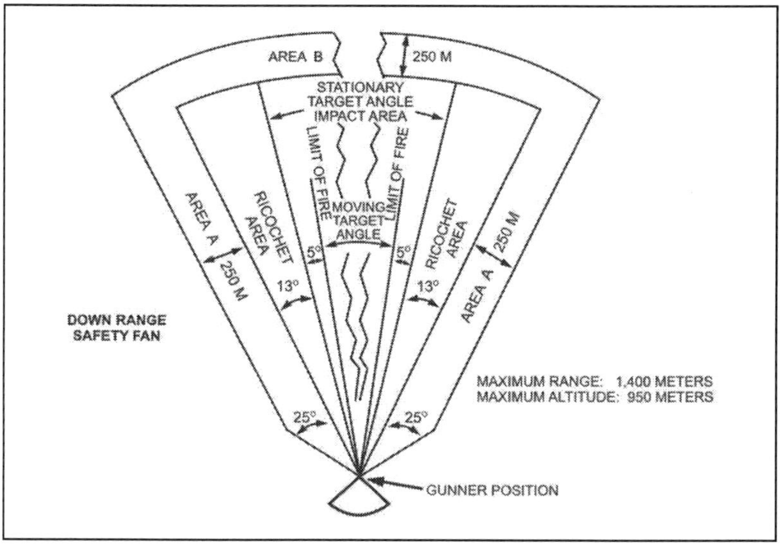

Figure B-25. SDZ for firing Improved LAW.

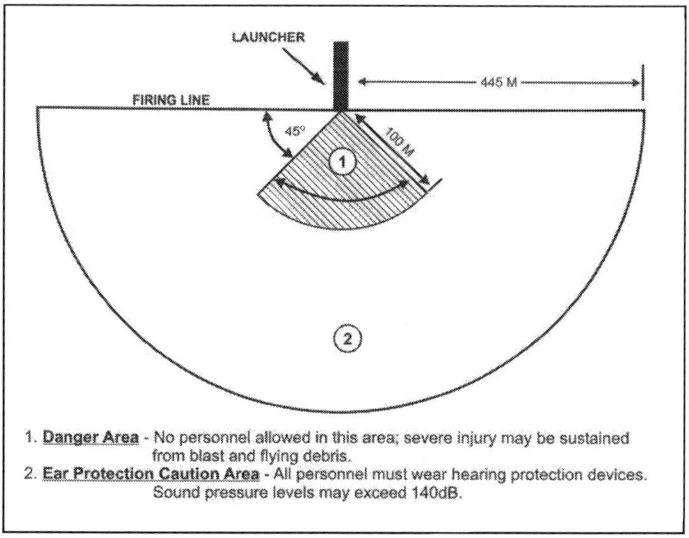

Figure B-26. XM141 BDM backblast danger area.

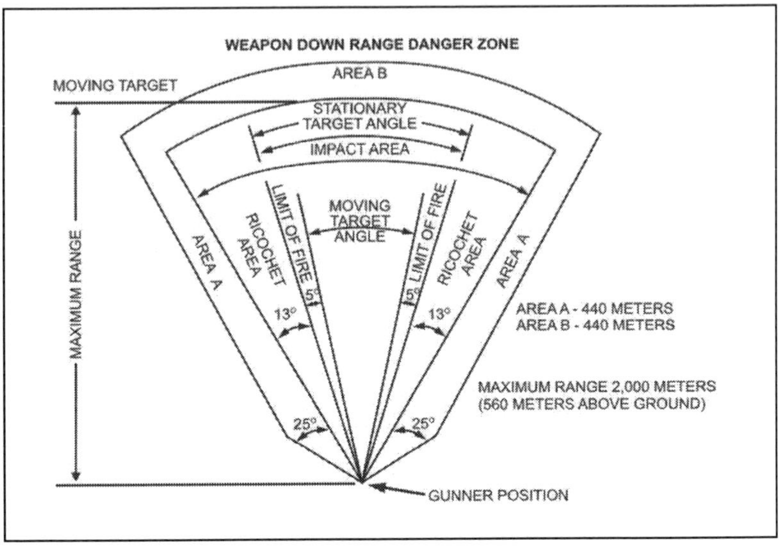

Figure B-27. SDZ for firing XM141 BDM.

COMBAT SAFETY FOR ALL SLM

B-64. Combat safety rules and procedures include all those that apply to training with the following modifications.

Engagement From an Enclosure

B-65. Firing from an enclosure creates unique hazards. Therefore, before positioning Soldiers in enclosures (combat only), leaders must consider several factors that affect safety. Only in combat, *when no other tactical option exists*, should the M136 AT4 and XM141 be fired from an enclosure. If it must be employed this way, the enclosure must meet the following minimum requirements. The M72-series LAW has been rated safe for use from an enclosure, but only when the enclosure meets the following minimum requirements.

- **Construction.** The building must be sturdily constructed to reduce structural damage that would occur in a weakly constructed enclosure such as one made of wood or stucco.
- **Size of Enclosure.** Minimum measurements for the enclosure are as follows:
  - AT4 and XM141 – minimum room size 17 by 24 feet; minimum ceiling height 8 feet (*combat only*).
  - LAW – minimum room size 12 by 15 feet.
- **Ventilation to the Rear and Sides.** To allow for backblast, at least 20 square feet of ventilation (for a standard 3-foot by 7-foot doorway) must be provided directly behind the firer. Doors and windows should be removed beside and behind the position to increase ventilation and reduce overpressure, noise, and blast effects. On the front wall, windows and doors should be reinforced, rather than removed, because removing would draw attention to the position. Reinforcing the windows also helps protect the firer from enemy direct-fire weapons.
- **Objects and Debris.** Any objects or debris to the rear of the weapon must be removed to prevent them from flying around the room and possibly injuring personnel as a result of the backblast.

- **Muzzle Clearance.** Muzzle clearance must be at least 6 inches.
- **Weapon Clearance.** Properly positioning the weapons within the enclosure is vital to the safety and survival of all personnel in the enclosure. The weapons should be positioned so no walls are within 5 meters to the rear or side of the weapon.
- **Non-Firing Personnel Positions.** If any other Soldiers are present, they must avoid standing in corners or near walls and must remain forward of the rear of the launcher.

Engagement from a Fighting Position

B-66. The M72-series LAW, M136 AT4, and SMAW-D can be fired from the standard Infantry fighting position. However, to increase accuracy and reduce danger to friendly Soldiers, the area to the rear of the firing position must have no walls, large trees, or other obstructions within 5 meters (5 1/2 yards). Ensuring the absence of such obstructions avoids deflection of weapon backblast onto the firer or into the position.

- **Individual Infantry Fighting Position.** The Soldier must lean against the rear wall and ensure that the venturi or the rear of the weapon protrudes past the rear of the position.
- **Two-Soldier Infantry Fighting Position.** Nonfiring personnel must remain clear of the backblast area. These positions should be constructed and sited so none are located in another position's backblast danger zone.
- **Modified Firing Position.** A modified firing position may be constructed to the side of the two-Soldier fighting position. Firing from a modified position reduces the possibility of injury to the firer or the other Soldier in the fighting position, while still offering the firer protection from enemy return fire.

OVERHEAD FIRE

B-67. SLM should not be fired over the heads of friendly Soldiers, unless the Soldiers have adequate protection against direct impact or other hazards.

JAVELIN

B-68. Figure B-28 shows the Javelin backblast danger area and SDZ. The primary danger area is a 60-degree sector, with the apex of the sector at the aft end of the missile launch motor.

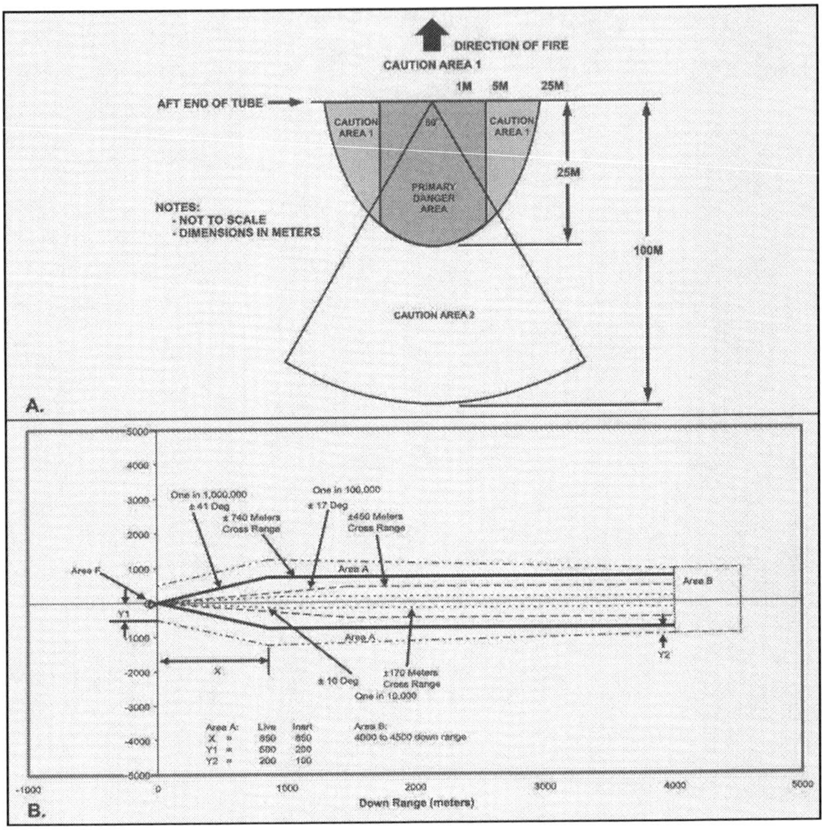

Figure B-28. Javelin backblast area and surface danger zone.

Firing From Enclosures

B-69. The Javelin can be fired from inside a building. However, the room from which it is fired must be at least 7 feet high, 12 feet wide, and 15 feet deep.

- **Debris.** Debris and loose objects are cleared from behind the launch site when firing within a confined area.
- **Venting.** When possible, doors and windows are opened to allow the backblast and overpressure to escape.
- **Structural Damage.** Escaping gases from the missile's first-stage motor are hot and flammable. The materials that can easily catch fire are removed before firing (for example, some types of curtains and throw rugs).
- **Hearing Protection.** All personnel within 25 meters of the Javelin must wear hearing protection.

- **Face Shield.** The face shield protects the gunner's face. It is possible to damage the face shield absorber between the indentation and the CLU main housing. If this part of the face shield is missing, the gunner must switch from firing the Javelin with the right eye to the left eye.

TOW

B-70. When firing from either a hasty or improved fighting position, the gunner must take into consideration obstructions directly to his front, to his rear, and to the sides of the fighting position.

FIRING LIMITATIONS

E-1. Some conditions may limit the firing and engagement capabilities of the TOW. The following information should be considered before engaging targets. (See TM 9-1425-450-12, *Operator and Organizational Maintenance Manual for TOW 2 Weapon System, Guided Missile System M220A2,* for updated firing limitations.)

- **Firing Over Bodies of Water.** Maximum and limited range firing over water varies according to missile type. If the range is less than 1,100 meters, the missile's range is not affected. However, if it is wider than 1,100 meters it can reduce the range of the TOW. A TOW position should be as high above and as far back from the water as the tactical situation allows. The squad or section leader should analyze his sector as soon as the position is occupied to determine if water will affect the employment of the TOW. Signals being sent through the command-link wires are shorted out when a large amount of wire is submerged in water.
- **Firing Over Electrical Lines.** If the command-link wires make contact with a live high-voltage power line, personnel can be injured or control of the missile could be lost. The launcher electronics may also be damaged. In addition to power lines, other high-voltage sources include street cars, electric train ways, and some moving target trolleys on training ranges.
- **Firing in Windy Conditions.** Gusty, flanking, or quartering winds can cause the launch tube to vibrate and spoil the tracking performance. The effect is similar to driving in a strong crosswind. Strong winds can move the missile around during flight, but as long as the crosshairs are kept on the center mass of the target, the weapon system itself can compensate for wind effects.
- **Firing Through Smoke and Area Fires.** Smoke can obscure the line of sight and hide the target when using the daysight tracker. A smooth tracking rate should be maintained as the target disappears into a smoke cloud so the missile will still be on target or very close as the vehicle goes out the other side of the smoke cloud. (This technique should be practiced during field tracking exercises.) A fire can burn through the command-link wire, causing loss of control of the missile.
- **Firing From Bunkers and Buildings.** In accordance with DA Pam 385-63, TOWs will not be fired from buildings, bunkers, or within 100 meters of a vertical or nearly vertical backstop without the approval of the commanding general.
- **Clearance Requirements.** The TOW muzzle must have at least nine inches of clearance at the end of the launch tube so the wings and control surfaces of the missile will not be damaged when they extend after clearing the launch tube. The muzzle of the launch tube must extend beyond any enclosure, window sill, or aperture. It must also have at least 30 inches of clearance between the line of sight and any obstruction from 500 to 900 meters downrange. A 30-inch line-of-sight clearance ensures a high probability the missile will not strike the ground on the way to the target (Figure B-29).
- **Firing TOW Bunker Buster Missile.** The missile warhead arms after launcher is between 35 and 65 meters. There is a very remote possibility of a TOW BB missile airburst 43 meters from launch platform. The probability of an inadvertent warhead detonation resulting in shrapnel injury to an exposed crewmember is also very remote. The crew is protected from shrapnel during firing from Stryker ATGM vehicles. The TOW BB is not currently fired from a HMMWV.

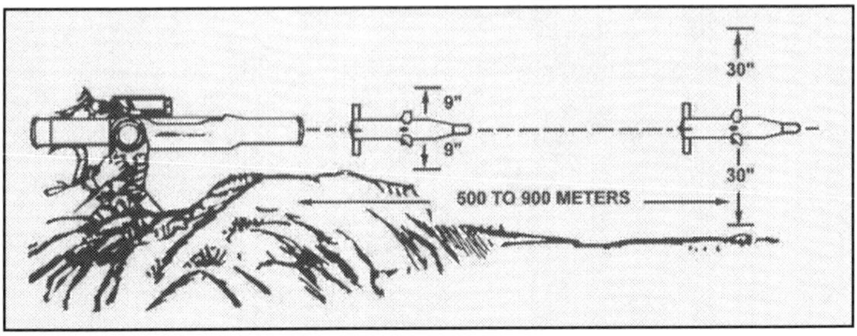

Figure B-29. Clearance requirements.

SURFACE DANGER ZONE

B-71. The surface danger zone for any firing range consists of a firing area, a target area, impact area, and danger areas surrounding these locations (Figure B-30). An additional area for occupation by personnel during firings may also be required. The shape and size of the surface danger zone varies with the type of missile or rocket being fired. (Refer to DA Pam 385-63 for dimensions.)

- **Primary Danger Area.** The primary danger area is a 90-degree cone with a 50-meter radius. The apex of the cone is centered at the rear of the missile launcher. Serious casualties or fatalities are likely to occur to anyone in the area during firing. Hazards include launch motor blast, high noise levels, overpressure, and debris.
- **Caution Area 1.** The caution area 1 extends in a radial pattern from each side of the primary danger area to the firing line with a radius of 50 meters. Permanent hearing damage could occur to personnel without adequate hearing protection in this area during firing. The hazards are high noise levels and overpressure.
- **Caution Area 2.** The caution area 2 is an extension of the primary danger area with the same associated hazards and personnel protection required. The radius of this area is 75 meters.
- **200-Meter Zone.** The 200-meter zone is the danger area for aerial firings 15.25 meters or more above ground level.

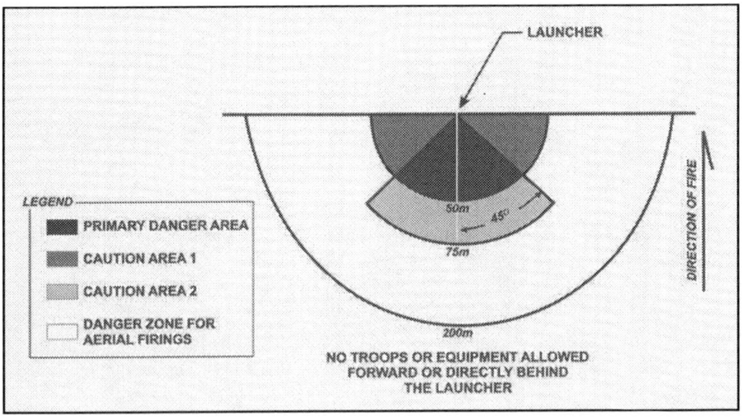

Figure B-30. Surface danger zone for firing basic TOW, TOW 2A, and TOW 2B missiles.

Firing Angle Limitations

B-72. Azimuth and elevation firing angles are limited by the traversing unit, the vehicle, and other external restrictions. All elevation angles are referenced to the horizontal plane of the traversing unit. Azimuth angles are referenced to the long axis of the vehicle and depend on whether the launch tube points over the front or rear of the vehicle. The other reference line is the LOS from the TOW to the target.

B-73. When the TOW is tripod-mounted, a 360-degree lateral track is possible, because the traversing unit is not restricted in azimuth. Mechanical stops limit the elevation angle coverage to 20 degrees below and 30 degrees above the horizontal plane. Before the missile is fired, the LOS angle should be estimated at the expected time of launch and throughout the expected missile flight time. The firing position should be changed or a different target selected if an expected line-of-sight angle exceeds the firing limitation angle.

B-74. Firing angle limitations of TOW carriers are as illustrated in Figure B-31.

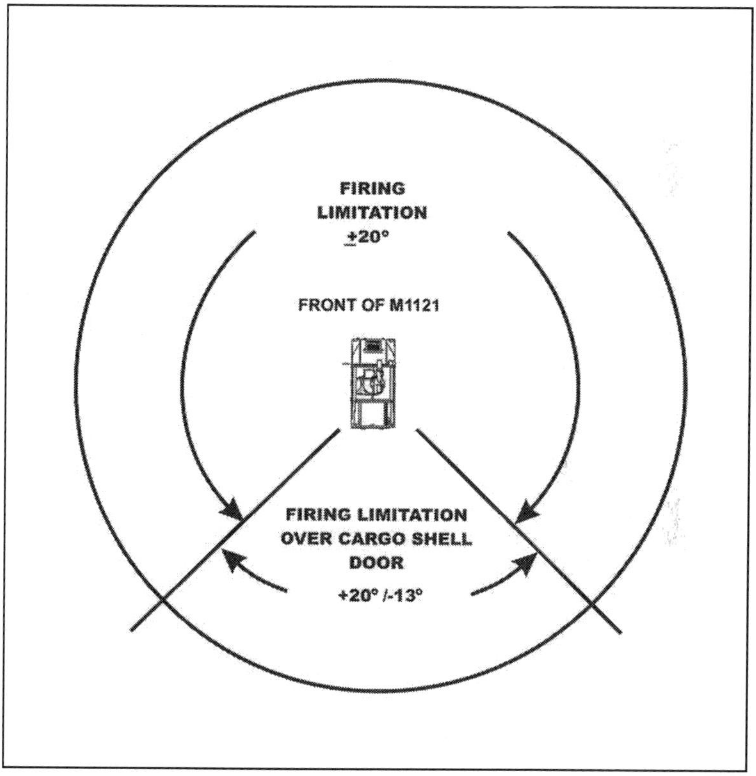

Figure B-31. M1121-mounted TOW firing angle limitations.

Appendix C
Fire Planning

Fire planning is the continual process of selecting targets on which fires are prearranged to support a phase of the concept of operation. Fire planning is accomplished concurrently with maneuver planning at all levels. Leaders conduct fire planning to suppress, isolate, obscure, neutralize, destroy, deceive, or disrupt known, likely, or suspected targets, and to support the actions of the maneuver element. Fires are planned for all phases of an operation.

SECTION I — FIRE PLANNING

C-1. Fire planning starts as soon as the leader gets a mission. Once begun, fire planning continues through the operation's completion. The primary aim of fire planning is to develop how fire is to be massed, distributed and controlled to best support the leader's concept of operation.

C-2. Fires are either *targets of opportunity, or planned targets.* Targets of opportunity are not planned in advance, but are engaged as they present themselves in accordance with established engagement criteria and rules of engagement. Planned targets are ones on which fires are prearranged, although the degree of this prearrangement may vary. The degree of prearrangement influences the time it takes to receive fires. The greater the prearrangement—the faster the reaction time. The subject of this section is planned fires.

C-3. Planned targets are categorized as *scheduled,* or *on-call.* Scheduled fires are fired in accordance with a pre-established time schedule and sequence. On-call targets are fired in response to a request for fires. Priority targets are a special type of on-call target. Priority targets have quick reaction times because the firing unit has guns set on a specific target when not engaged in other fire missions.

C-4. To be effective fires must be integrated and synchronized in time, space, and purpose over the entire concept of operation. Integration means all available assets are planned and used throughout an operation. Synchronization means that these assets are sequenced in time, space, and purpose in an optimal manner, producing complementary and reinforcing effects for the maneuver element.

On 14 May 1945 during the Ryukyus Campaign in Okinawa after three days of heavy fighting, the companies of 1st Battalion, 305th IN, 77th ID were reduced to the size of platoons, led by corporals and sergeants. Despite loses, the commander decided to continue its advance. In order to achieve surprise, the morning attack began without preparatory fires. The rifle companies moved over the LD at 0800 hours and advanced 200 yards with out a shot being fired by the enemy. Surprise had been achieved, but the enemy quickly recovered and achieved fire superiority by pouring machine gun and mortar fire on the attacking units, stopping their advance. Two of the enemy positions along ridge were destroyed by mortar fire but the troops were still unable to move with out being met by enemy fire. Determined not to loose ground already gained, the battalion commander ordered the 81-mm mortar platoon to place suppressive fires in front of the lead company. Placing fire only 50 yards in front of the troops, he kept moving the barrage ahead as troops advanced.

The battalion's mortar PL went forward to the lead elements, and after a hasty visual recon decided to use two mortars on the mission. He adjusted one mortar about 50 yards in front of the company and the second about 100 yards in front of the company. One fired at a range of 700; the other at a range of 750 yards.

At these ranges two turns of the elevating crank would move the impact of the round about 25 yards.

The lead company slowly resumed its advance, moving behind this curtain of mortar fire. The enemy moved back into their cave positions to get out of the fire, becoming easy prey for flame throwers and satchel charges. Seven caves were taken care of in this fashion as the advance moved slowly – but continuously forward. Each mortar fired at a rate of about 10 rounds per minute. Some rounds fell as close as 25 yards to the troops, wounding three riflemen with fragments. Within 45 minutes ridge 59 was secured.

—*Suppressive Fires*

FIRE PLANNING PROCESS

C-5. Fire planning begins with the concept of fires. This essential component of the concept of operation complements the leader's scheme of maneuver detailing the leader's plan for direct and indirect preparatory and supporting fires. Fire planning requires a detailed knowledge of weapon characteristics and logistical capabilities of those providing the support. Although leaders may be augmented with personnel to assist in planning and controlling attached or supporting assets, the responsibility for planning and execution of fires lies with the leader. The leaders do not wait to receive the higher headquarters' plan to begin their own fire planning. Rather, he begins as soon as possible to integrate fires into his own concept of operation and the concept of operation of the higher headquarters.

C-6. Additional assets are allocated in either a command or support relationship (see Chapter 1). An example of a command relationship would be an attachment of a section from the weapons company. The leader relies on the senior representative from the organization to provide expertise when planning. An example of a support relationship would be direct support from the artillery battalion or from an attack aviation company. When planning fires or CAS from a supporting unit, the leader normally receives someone from that organization to assist them. For example, if the unit were to get close air support (CAS), a Soldier trained to control the CAS would probably be attached to assist the leader in his planning and execution.

C-7. Developing the concept of fire should be fairly straight forward during deliberate operations because of the ability to conduct reconnaissance, planning, and preparation. However, during hasty operations the unit may have to rely on its internal SOPs and more hands on control by the leader.

PLOT-CR

C-8. Leaders refine, or establish if required, timings and control to ensure these targets are initiated, adjusted, and shifted properly. If possible, the observer should locate where he can see assigned target. Leaders refine, or develop a detailed execution matrix assigning responsibility for each target to the leader or observer who is in the best position to control them should be developed. These Soldiers must know when each target, series, or group is fired. They must also understand what effect is desired on which enemy positions, and when to lift or shift the fires. Leaders may consider the use of pyrotechnic or other signals to ensure communication. Units' assigned responsibilities for executing fires continually refine and rehearse their actions. Responsibilities are further refined with the information contained in the categories contained in the memory aid PLOT:

Purpose

C-9. The purpose outlines how the target assists the maneuver element or contributes to the higher headquarters' concept of operation.

Location

C-10. An identified target is the target's proposed location given as a grid preferably with a known point. The target location is not the location of the enemy – it is where the leader (or the higher headquarters) thinks the enemy will be.

Observer

C-11. The observation plan is how the leader plans to monitor the battlefield to execute the target. He assigns primary and alternate observers with proposed locations where they can observe the target and associated triggers. Positioning is perhaps the most important aspect of the plan. Observers' positions must allow them to see the trigger for initiating fires as well as the target area and the enemy forces on which the target is oriented. The leader also must consider other aspects of observer capabilities, including available equipment, communication, and their security. This information is critical to the leader. If an enemy asset is critical enough to be designated as a target, then it must be adequately resourced with execution assets.

Trigger

C-12. A trigger is event- or time-oriented criteria used to initiate planned actions directed toward achieving surprise and inflicting maximum destruction on the enemy or a designated point (FM 1-02). Triggers can be a physical point on the ground, a laser or lazed spot, or an action or event that causes and action among friendly forces. When using triggers to control fires, leaders ensure they have allocated them to start, shift, and cease fires. There are two types of triggers: tactical; and technical. Tactical triggers cue the observer/ executor of the target to communicate to the firing agency to prepare to fire. In the offense tactical triggers are tied to a friendly maneuver event. In the defense, tactical triggers are usually tied to enemy actions. Technical triggers involve the actual firing of the target, taking into account the enemy rate of march, and the friendly munition's time of flight.

C-13. When using triggers in the defense it is important for subordinates to have a method, usually addressed in the unit's SOP, for marking triggers. The marking method should work during day and limited visibility operations.

TACTICAL USES OF PLANNED FIRES

C-14. Fires are used for many different tactical reasons. They include:

- Fire delivered before an attack to weaken the enemy position (FM 1-02).
- Supporting fires (covering fires). Supporting fires enable the friendly maneuver element to move by destroying, neutralizing, or suppressing enemy fires, positions, and observers.
- Final protection fires (FPF) is an immediately available prearranged barrier of fire designed to impede enemy movement across defensive lines or areas.
- Suppression.
- Obscuration.
- Counterbattery (indirect fires only). Counterbattery is fire to destroy or neutralize enemy artillery / mortars. These missions are normally controlled at higher level headquarters. Direct support artillery moves with supported units and aviation is used to destroy enemy fire support means and key enemy units and facilities. Counter battery radars are positioned to maintain radar coverage to ensure continuous coverage during rapid movement forward.
- Harassing fire is observed or predicted (unobserved) fire intended to disrupt enemy troop and vehicle movement, disturb their rest, and lower their morale.
- Illumination.

ECHELONMENT OF FIRE – PLANNED FIRES TECHNIQUE

C-15. Echelonment of fires is the schedule of fire ranging from the highest caliber munitions to the lowest caliber munitions. The purpose of echeloning fires is to maintain constant fires on the enemy while using the optimum delivery system. Leaders use REDs, SDZs, and MSDs to manage associated risks. In the defense, triggers are tied to the progress of the enemy as it moves through the AO, enabling the leader to engage the enemy throughout the depth of the sector. In the offense triggers are tied to the progress of the maneuver element as it moves toward the objective protecting the force and facilitating momentum up to the objective.

Defensive Echelonment

C-16. In the defense, echeloning fires are scheduled based on their optimum ranges to maintain continuous fires on the enemy, disrupting his formation and maneuver. Echelonment of fires in the defense places the enemy under increasing volumes of fire as he approaches a defensive position. Aircraft and long-range indirect fire rockets and artillery deliver deep supporting fires. Close supporting fires such as final protective fires (FPF) are closely integrated with direct fire weapons such as Infantry weapons, tank support, and antiarmor weapons systems. Figure C-1 illustrates an example of defensive echelonment.

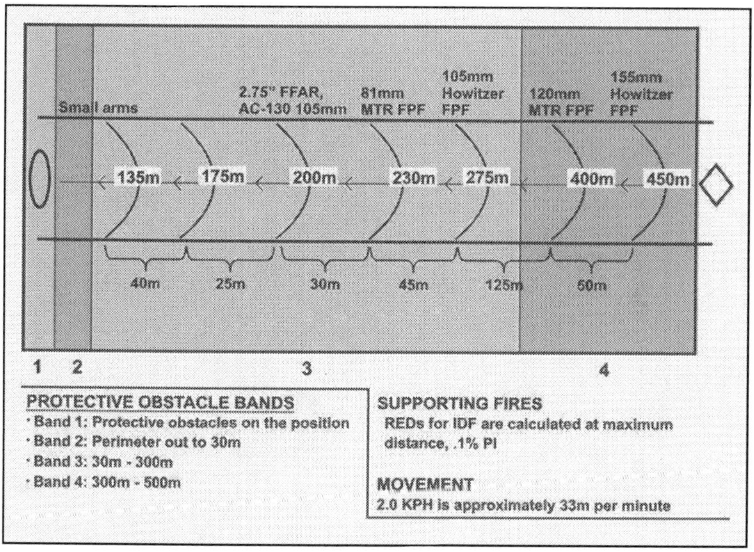

Figure C-1. Defensive echelonment of fires example.

Offensive Echelonment

C-17. In the offense, weapons are scheduled based on the point of a predetermined safe distance away from any maneuvering friendly troops. When scheduled effectively, fires provide protection for friendly forces as they move to and assault an objective. They also allow friendly forces to get in close with minimal casualties and prevent the defending enemy from observing and engaging the assault by forcing him to take cover. The overall objective of offensive scheduled fires is to allow the friendly force to continue the advance unimpeded (Figure C-2).

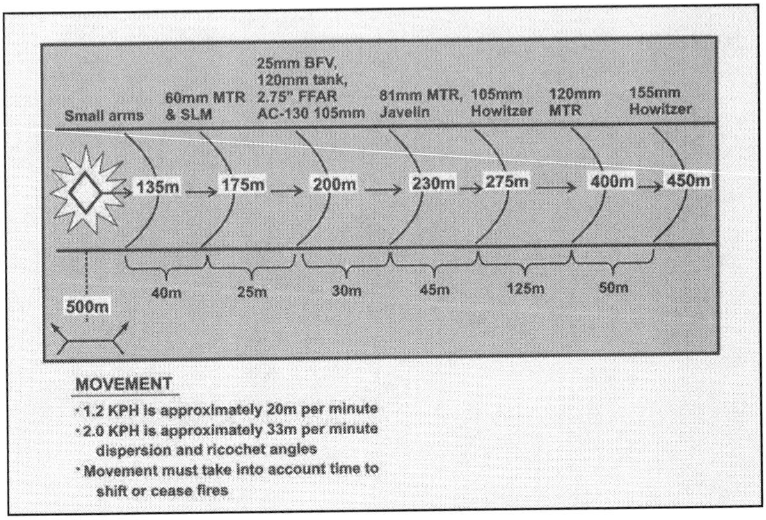

Figure C-2. Offensive echelonment of fires example.

C-18. As an example of echelonment of fires use during the conduct of a mission, consider an operation in which a platoon assaults an enemy position (Figures C-3 through C-6). As the lead elements of the unit approach the designated phase line en route to the objective, the leader orders the fire support officer (FSO) to begin the preparation. Observers track friendly movement rates and confirm them. Other fire support officers in the chain of command may need to adjust the plan during execution based on unforeseen changes to anticipated friendly movement rates.

C-19. As the unit continues its movement toward the objective, the first weapon system engages its targets. It maintains fires on the targets until the unit crosses the next phase line that corresponds to the RED of the weapon system being fired.

C-20. To maintain constant fires on the targets, the next weapon system begins firing before the previous weapon system ceases or shifts. This ensures no break in fires, enabling the friendly forces' approach to continue unimpeded. However, if the unit rate of march changes, the fire support system must remain flexible to the changes.

C-21. The FSO shifts and engages with each delivery system at the prescribed triggers, initiating the fires from the system with the largest RED to the smallest. Once the maneuver element reaches the final phase line, the FSO ceases the final indirect fire system or shifts to targets beyond the objective to cease all fires on the objective. Direct fire assets in the form of supporting fires are also maintained until the final assault, then ceased or shifted to targets beyond the objective.

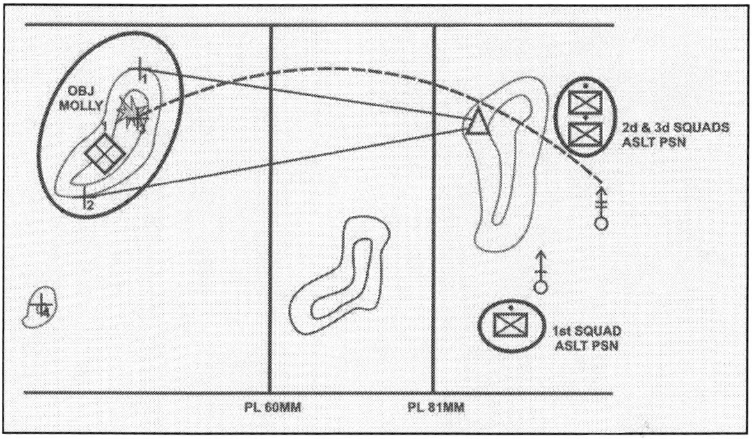

Figure C-3. 81-mm mortars begin firing.

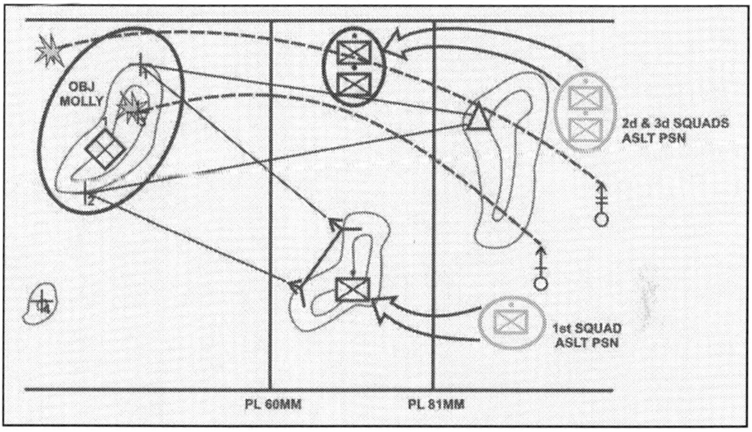

Figure C-4. 81-mm mortars shift, 60-mm mortars and supporting fires begin.

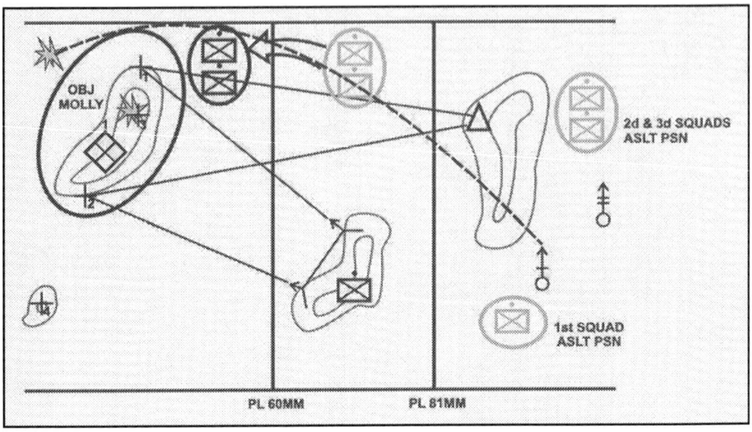

Figure C-5. 60-mm mortars shift.

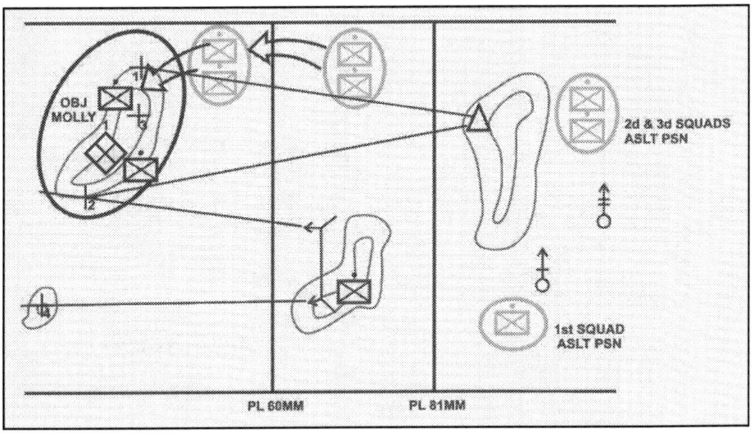

Figure C-6. Supporting fires shift for final assault.

FIRE PLANNING FOR THE DEFENSE

C-22. To develop a defensive fire plan, the leader—

- Assigns primary and secondary sectors from primary and alternate position to each subordinate.
- Designates unit point or area targets and other control measures, such as target reference points (TRPs), to coordinate the fire when more than one subordinate is firing into the same engagement area or sector.
- Receives target information from subordinates (normally provided on sector sketches and/or individual weapon range cards). The leader reviews this target information to insure that fire is equally distributed across the entire unit's sector and that sufficient control measures are established.
- Completes the unit's fire plan and gives a sketch to his higher headquarters.

C-23. In the defense, fires are planned in three locations – in front of the unit's position, on the position (FPF), and behind the position. Figure C-7 shows fires masses in front of a company-sized position. Fire plans are best developed using the seven steps of engagement area development technique:

(1) Identify likely enemy avenues of approach.

(2) Identify the enemy scheme of maneuver.

(3) Determine where to kill the enemy.

(4) Emplace weapon systems.

(5) Plan and integrate obstacles.

(6) Plan and integrate indirect fires.

(7) Conduct an engagement area rehearsal.

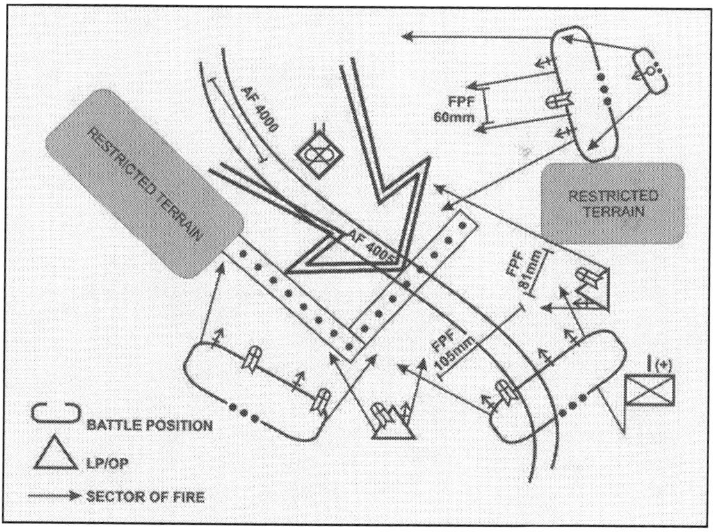

Figure C-7. Company defensive fire plan sketch.

C-24. The engagement area (EA) is the place where the leader intends to destroy an enemy force using the massed fires of all available weapons. The success of any engagement depends on how effectively the leader can integrate the obstacle and indirect fire plans with his direct fire plan in the EA to achieve the unit's purpose. Completing the steps of EA development is not a lengthy process. Particularly at the Infantry platoon level, EA development can occur rapidly without an elaborate decision making process.

SQUAD FIRE PLANNING

C-25. The squad leaders make two copies of their sector sketches. One copy goes to the platoon leader; the other remains at the position. The squad leaders draw sector sketches as close to scale as possible, showing the elements contained in Figure C-8.

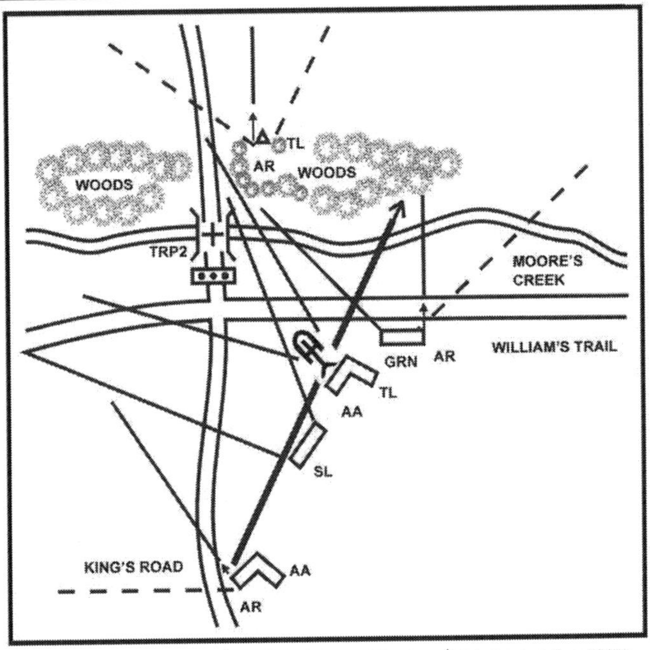

Main terrain features in the sector and the range to each.	M240B machine gun FPL or PDF (if applicable).	Reference points and TRPs in the sector.
Each primary position.	M249 SAW FPLs or PDFs.	OP locations.
EA or primary and secondary sectors of fire covering each position.	Type of weapon in each position.	Dead space.
MELs for all weapons systems.	MELs for Javelin (if applicable) and AT4s.	Obstacles.
		Indirect fire targets.

Figure C-8. Squad sector sketch.

PLATOON FIRE PLANNING

C-26. Squad leaders prepare their sketches and submit them to the platoon leader. The platoon leader combines all sector sketches (and possibly separate range cards) to prepare a platoon sector sketch. A platoon sector sketch is drawn as close to scale as possible that includes a target list for direct and indirect fires. One copy is submitted to the company commander, one copy is given to the PSG, and one copy is maintained by the platoon leader. As a minimum, the platoon sector sketch should show the elements contained in Figure C-9.

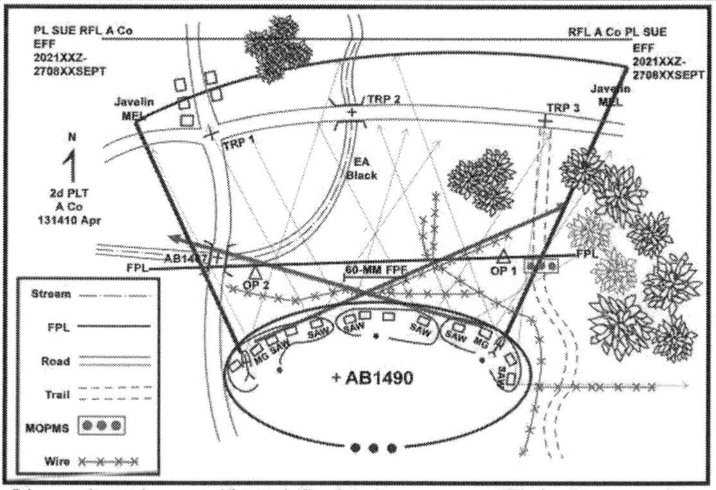

Primary and secondary sectors of fire or EAs.	Mines and other obstacles.	MELs for all weapons systems.
Primary, alternate, and supplementary squad positions.	M240B and M249 FPLs or PSFs.	OPs.
	Position and sector of flanking units.	TRPs.
Javelin, M240B, and M249 positions with primary directions of fire.	Indirect fire target locations and FPF location (if applicable).	Priority engagement by weapon system and crew.

Figure C-9. Platoon sector sketch.

FINAL PROTECTIVE LINE

C-27. The final protective line (FPL) is a line of fire selected where an enemy assault is to be checked by interlocking fire from all available weapons and obstacles (FM 1-02). The FPL consists of all available measures, to include protective obstacles, direct fires, and indirect fires. The FPF targets the highest type of priority targets and takes precedence over all other fire targets. The FPF differs from a standard priority target in that fire is conducted at the maximum rate until the mortars are ordered to stop, or until ammunition is depleted. If possible, the FPF should be registered.

C-28. If Soldiers are in well-prepared defensive positions with overhead cover, an FPF can be adjusted very close to the friendly positions, just beyond bursting range. If required, the leader can even call for artillery fires right on the unit's position using proximity or time fuzes for airbursts. Table C-1 shows indirect fire mortar weapon system characteristics that should be used when planning the FPF.

Table C-1. Normal FPF dimensions for each number of mortars.

Weapon	Number of Tubes	Width (meters)	Depth (meters)	Risk Estimated Distance, .1% PI	Risk Estimated Distance, 10% PI
MORTARS					
120 mm	4	300	75	400m	100m
120 mm	2	150	75		
81 mm	4	150	50	230m	80m
81 mm	2	75	50		
60 mm	2	60	30	175m	65m

FIRE PLANNING FOR THE OFFENSE

C-29. Offensive fire planning follows the same methodology as defensive fire planning within constraints of the situation. The main difference is that offensive fire planning always includes the synchronization between the base of fire and the maneuver element. Inevitably, the leader's plan will not be as detailed as the defensive plan, but the presence of a maneuver element requires a baseline of planning and control to ensure fire support is effective and efficient.

C-30. The leader must plan how he will engage known or suspected enemy targets, where friendly suppressive fire may be needed, and how he will control the unit's fires against both planned targets and targets of opportunity. Fire planning should include a thorough analysis of the type of threat expected. This will aid the supporting friendly element in tailoring the weapon and ammunition requirements to suit the situation.

C-31. Offensive fire planning supports four phases: planning and preparation, approach to the objective, actions on the objective, and follow-through. The degree of completeness and centralization of offensive fire planning depends on the time available to prepare for the offensive. Fires are planned in four locations on the battlefield – short of the LD / LC, LD / LC to the objective, on the objective, and behind the objective. Table C-2 lists planning considerations for each of the four locations.

Table C-2. Planning considerations.

Phase	Plan Fires to:
1) Planning and Preparation (Short of the LD / LC)	• Support unit in assembly areas. • Support unit's movement to the LD / LC. • Disrupt enemy reconnaissance forces. • Disrupt enemy defensive preparations. • Disrupt enemy spoiling attacks.
2) Approach to the Objective (LD / LC to the Objective)	• Begin echeloning fires for maneuver units. • Suppress and obscure for friendly breaching operations. • Suppress and obscure enemy security forces throughout movement. • Provide priority of fires to lead element. • Screen / guard exposed flanks.
3) Actions on the Objective (On the Objective)	• Fires to block enemy reinforcements. • Fires to suppress enemy direct fire weapons. • Suppress and obscure point of penetration. • Suppress and obscure enemy observation of friendly forces.
4) Follow Through (Beyond The Objective)	• Disrupt movement of enemy reinforcements during the assault. • Block avenues of enemy approach. • Disrupt enemy withdraw. • Screen friendly forces from enemy counterattacks during the assault. • Consolidate objective after the assault.

C-32. For simplicities, offensive fire planning is divided into two categories – preparatory and supporting fires. The concept of fires will have artillery and mortars in support of an attack to gain and maintain fire superiority on the objective until the last possible moment. When this indirect fire lifts, the enemy should be stunned and ineffective for a few moments. Take full advantage of this period by doing any or all of the following:

- Combat Vehicles. Vehicles used in the attack, or as fire support, continue to give close support.
- Maintaining Fire Superiority. Small-arms fire from local and internal SBF is continued as long as possible.
- Maneuver Elements. Assaulting troops must try to fire as they advance. Troops must observe fire discipline, as in many cases fire control orders will not be possible. They must not arrive at the objective without ammunition.
- Audacity. Where the ground and vegetation do not prohibit movement, leading sections should move very quickly over the last 30 or 40m to the enemy positions to minimize exposure.

C-33. When planning fires for the offense, leaders verify the fire element's task organization and ensure there exists plans and coordinating measures for the attack, exploitation, pursuit, and contingency plans. Leaders develop or confirm with the responsible level authority that supporting systems are positioned and repositioned to ensure continuous fires throughout the operation. Mutual support of fire systems promotes responsive support and provides the commanders of maneuver units freedom of action during each critical event of the engagement or battle

C-34. There exists a diverse variety of munitions and weapon systems, direct and indirect, to support close offensive operations. To effectively integrate fire support, the leader must understand the mission, the commander's intent, the concept of operations, and the critical tasks to be accomplished. The leader plans fires to focus on enemy capabilities and systems that must be neutralized. Critical tasks include:

- Continuous in-depth support (accomplished by proper positioning of systems).
- Isolating enemy forces.
- Softening enemy defenses by delivering effective preparatory fires.

- Suppressing and obscuring enemy weapon systems to reduce enemy standoff capabilities.
- Interdicting enemy counterattack forces, isolating the defending force, and preventing its reinforcement and resupply.

SECTION II — TARGET EFFECTS PLANNING

C-35. Not only must fire support planners determine what enemy targets to hit, and when, but must also decide how to attack each enemy target. Leaders should consider all the aspects of target effects when planning fires. Although this section is specific to mortars, the following concepts generally apply to most indirect fires.

HIGH-EXPLOSIVE AMMUNITION

C-36. When mortar rounds impact they throw fragments in a pattern that is never truly circular, and may even travel irregular, based on the round's angle of fall, the slope of the terrain, and the type soil. However, for planning purposes, each mortar high explosive (HE) round is considered to have a circular lethal bursting area. Figure C-10 shows a scale representation of the lethal bursting areas of mortar rounds.

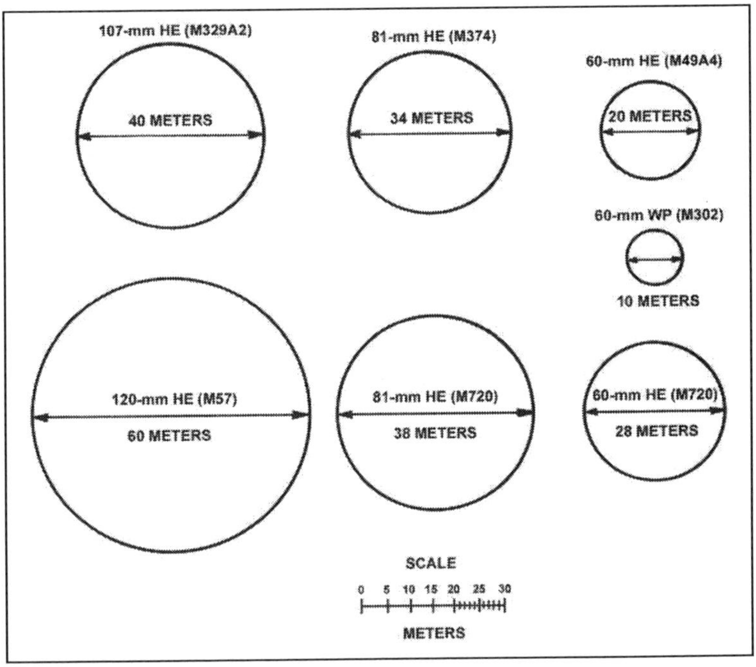

Figure C-10. Comparison of lethal bursting areas of U.S. mortar rounds.

FUZE SETTINGS

C-37. The decision concerning what fuze setting to use depends on the position of the enemy.

C-38. Exposed enemy troops that are standing up are best engaged with impact (IMP) or near surface burst (NSB) fuze settings. The round explodes on, or near, the ground. Shell fragments travel outward perpendicular to the long axis of the standing target (Figure C-11).

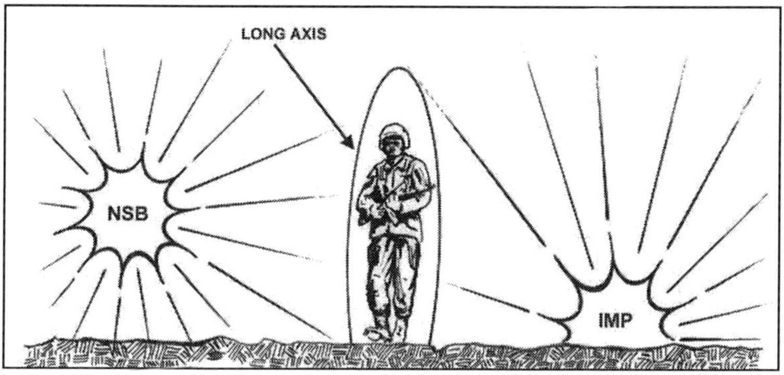

Figure C-11. Standing targets.

C-39. If exposed enemy troops are lying prone, the proximity (PRX) fuze setting is most effective. The rounds explode high above the ground, and the fragments coming downward are once again traveling perpendicular to the long axis of the targets (Figure C-12).

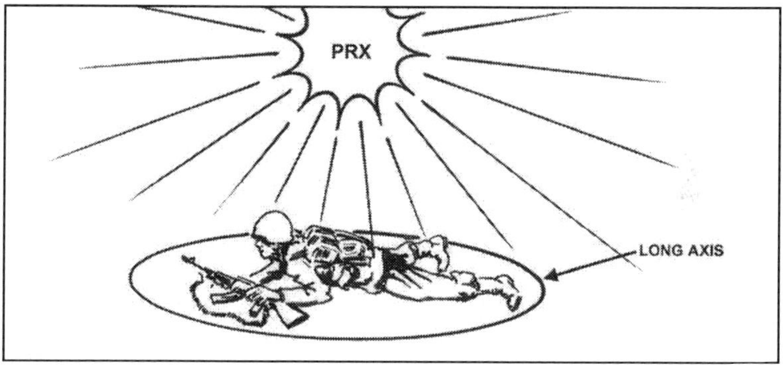

Figure C-12. Prone targets.

C-40. The PRX setting is also the most effective if the enemy is in open fighting positions, without overhead cover. Even PRX settings will not always produce effects if the positions are deep (Figure C-13).

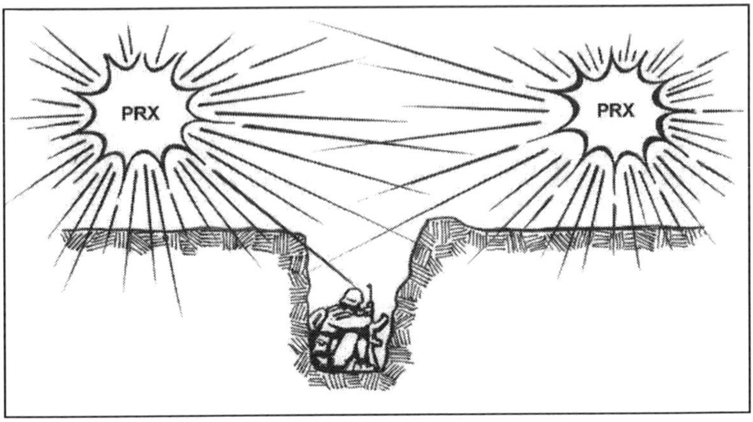

Figure C-13. Targets in open fighting positions.

C-41. The DLY fuze setting is most effective when the enemy is below triple canopy jungle or in fighting positions with overhead cover. Light mortars will have little effect against overhead cover. Even medium mortars have limited effect. Heavy mortars can destroy a bunker or enemy troops beneath jungle canopy with a hit or near-miss (Figure C-14).

Figure C-14. Targets beneath triple canopy jungle.

EFFECTS OF COVER ON HIGH-EXPLOSIVE ROUNDS

C-42. Enemy forces will normally be either standing or prone. They maybe in the open or protected by varying degrees of cover. Each of these changes the target effects of mortar fire.

C-43. Surprise mortar fire is always more effective than fire against an enemy that is warned and seeks cover. Recent studies have shown that a high casualty rate can be achieved with only two rounds against an enemy platoon standing in the open. The same studies required 10 to 15 rounds to duplicate the casualty rate when the platoon was warned by adjusting rounds and sought cover. If the enemy soldiers merely lay prone, they significantly reduce the effects of mortar fire. Mortar fire against standing enemy forces is almost twice as effective as fire against prone targets.

C-44. Proximity fire is usually more effective than surface-burst rounds against targets in the open. The effectiveness of mortar fire against a prone enemy is increased by about 40 percent by firing proximity-fuzed rounds rather than surface-burst rounds.

C-45. If the enemy is in open fighting positions without overhead cover, proximity-fuzed mortar rounds are about five times as effective as impact-fuzed rounds. When fired against troops in open fighting positions, proximity-fuzed rounds are only 10 percent as effective as they would be against an enemy in the open. For the greatest effectiveness against troops in open fighting positions, the charge with the lowest angle of fall should be chosen. It produces almost two times as much effect as the same round falling with the steepest angle.

C-46. If the enemy has prepared fighting positions with overhead cover, only impact-fuzed and delay-fuzed rounds will have much effect. Proximity-fuzed rounds can restrict the enemy's ability to move from position to position, but they will cause few, if any, casualties. Impact-fuzed rounds cause some blast and suppressive effect. Delay-fuzed rounds can penetrate and destroy a position but must achieve a direct hit. Only the 120-mm mortar with a delay-fuze setting can damage a Soviet-style strongpoint defense. Heavy bunkers cannot be destroyed by light or medium mortar rounds.

SUPPRESSIVE EFFECTS OF HE MORTAR ROUNDS

C-47. Suppression from mortar is not as easy to measure as the target effect. It is the psychological effect produced in the mind of the enemy that prevents him from returning fire or carrying on his duties. Inexperienced or surprised Soldiers are more easily suppressed than experienced, warned Soldiers. Soldiers in the open are much more easily suppressed than those with overhead cover. Suppression is most effective when mortar fires first fall; as they continue, their suppressive effects lessen. HE rounds are the most suppressive, but bursting WP mixed with HE has a great psychological effect on the enemy.

C-48. If a 60-mm mortar round lands within 20 meters of a target, the target will probably be suppressed, if not hit.

C-49. If a 60-mm mortar round lands within 35 meters of a target, there is a 50 percent chance it will be suppressed. Beyond 50 meters, little suppression takes place.

C-50. If an 81-mm mortar round lands within 30 meters of a target, the target will probably be suppressed, if not hit.

C-51. If an 81-mm mortar round lands within 75 meters of a target, there is a 50 percent chance that the target will be suppressed. Beyond 125 meters, little suppression takes place.

C-52. If a heavy mortar round (proximity-fuzed) lands within 65 meters of target, the target will probably be suppressed, if not hit.

C-53. If a heavy mortar round (proximity-fuzed) lands within 125 meters of a target, there is a 50 percent chance the target will be suppressed. Beyond 200 meters, little suppression takes place. The 120-mm mortar round is better for suppression than the 107-mm, but both are excellent suppressive rounds.

ILLUMINATION, SMOKE, AND WHITE PHOSPHORUS

C-54. Illumination and obscuration missions are important functions for mortar platoons or sections. Atmospheric stability, wind velocity, and wind direction are the most important factors when planning target effects for smoke and white phosphorus (WP) mortar rounds. The terrain in the target area also effects smoke and WP rounds.

C-55. The bursting WP round provides a screening, incendiary, marking, and casualty-producing effect. It produces a localized, instantaneous smoke cloud by scattering burning WP particles.

C-56. The WP round is used mainly to produce immediate, close point obscuration. It can be used to screen the enemy's field of fire for short periods, which allows troops to maneuver against him. The 60-mm WP round is not sufficient to produce a long-lasting, wide-area smoke screen, but the much larger WP round from the heavy mortar is.

C-57. The bursting WP round can be used to produce casualties among exposed enemy troops and to start fires. The casualty-producing radius of the WP round is much less than that of the HE round. Generally, more casualties can be produced by firing HE ammunition than by firing WP. However, the WP burst causes a significant psychological effect, especially when used against exposed troops. A few WP mixed into a fire mission of HE rounds may increase the suppressive effect of the fire.

C-58. The WP rounds can be used to mark targets, especially for attack by aircraft. Base-ejecting smoke rounds, such as the 81-mm M819 RP round, produce a dispersed smoke cloud, normally too indistinct for marking targets.

C-59. The effects of atmospheric stability can determine whether mortar smoke is effective at all or, if effective, how much ammunition will be needed.

- During unstable conditions, mortar smoke and WP rounds are almost ineffective--the smoke does not spread but often climbs straight up and quickly dissipates.
- Under moderately unstable atmospheric conditions, base-ejecting smoke rounds are more effective than bursting WP rounds. The M819 RP round for the M252 mortar screens for over 2½ minutes.
- Under stable conditions, both RP and WP rounds are effective.
- The higher the humidity, the better the screening effects of mortar rounds.

C-60. The M819 RP round loses up to 35 percent of its screening ability if the ground in the target area is covered with water or deep snow. During extremely cold and dry conditions over snow, up to four times the number of smoke rounds may be needed than expected to create an adequate screen. The higher the wind velocity, the more effective bursting WP rounds are, and the less effective burning smoke rounds become.

C-61. If the terrain in the target area is swampy, rain-soaked, or snow-covered, then burning smoke rounds may not be effective. These rounds produce smoke by ejecting felt wedges soaked in red phosphorus. These wedges then burn on the ground, producing a dense, long-lasting cloud. If the wedges fall into mud, water, or snow, they can be extinguished. Shallow water can reduce the smoke produced by these rounds by as much as 50 percent. Bursting WP rounds are affected little by the terrain in the target area, except that deep snow and cold temperatures can reduce the smoke cloud by about 25 percent.

C-62. Although bursting WP rounds are not designed to cause casualties, the fragments of the shell casing and bits of burning WP can cause injuries. Burning smoke rounds do not cause casualties and have little suppressive effect.

ILLUMINATION

C-63. Illumination rounds can be used to disclose enemy formations, to signal, or to mark targets. There are illumination rounds available for all mortars.

C-64. The 60-mm illumination round available now is the standard cartridge, illuminating, M83A3. This round has a fixed time of delay between firing and start of the illumination. The illumination lasts for about 25 seconds, providing moderate light over a square kilometer.

C-65. The 60-mm illumination round does not provide the same degree of illumination as do the rounds of he heavier mortars and field artillery. However, it is sufficient for local, point illumination. The small size of the round can be an advantage where illumination is desired in an area but adjacent friendly forces to not want to be seen. The 60-mm illumination round can be used without degrading the night vision devices of adjacent units.

C-66. The medium and heavy mortars can provide excellent illumination over wide areas. The 120-mm mortar illumination round provides one million candlepower for 60 seconds.

C-67. The M203 40-mm grenade, as well as all mortars have the capability to deliver IR illumination rounds in addition to the more common white light.

SPECIAL ILLUMINATION TECHNIQUES

C-68. Following are three special illumination techniques that mortars have effectively used.

C-69. An illumination round fired extremely high over a general area will not always alert an enemy force that it is being observed. However, it will provide enough illumination to optimize the use of image intensification (starlight) scopes such as the AN/TVS-5 and the AN/TVS-4.

C-70. An illumination round fired to burn on the ground will prevent observation beyond the flare into the shadow. This is one method of countering enemy use of image intensification devices. A friendly force could move behind the flare with greater security.

C-71. An illumination round fired to burn on the ground can be used to mark targets during day or night. Illumination rounds have an advantage over WP as target markers during high winds. The smoke cloud from a WP round will quickly be blown downwind. The smoke from the burning illumination round will continue to originate from the same point, regardless of the wind.

CONSIDERATIONS WHEN USING THERMAL SIGHTS

C-72. Although illumination rounds may aid target acquisition when friendly forces are using image intensification devices (such as night vision devices), this is not so when thermal sights are used. As the illumination flares burn out and land on the ground, they remain as a distinct hot spot seen through thermal sights for several minutes. This may cause confusion, especially if the flare canisters are between the enemy and the friendly forces. WP rounds can also cause these hot spots that can make target identification difficult for gunners using thermal sights (tanks, BFV, TOW, or Javelin).

Appendix D

Vehicle Employment Considerations

Employing combat vehicles with Infantry platoons and squads increases their combat power. Combining combat vehicles and Infantry to achieve complementary and reinforcing effects has proven to be a significant advantage. Operations that integrate combat vehicles and Infantry forces combine the advantages of the vehicle's mobility, protection, firepower, and ability to use their information platform. They also increase the Infantryman's ability to operate in restricted and severely restricted terrain.

Infantry units conduct operations with a variety of combat vehicles. The principles for integrating combat vehicles with Infantry are similar regardless of the specific vehicle type. Combat vehicles that most often work with Infantry forces include the M1 Abrams tank, the M2 Bradley fighting vehicle (BFV), the Stryker Infantry carrier vehicle (ICV), and multiple versions of the assault high-mobility multipurpose wheeled vehicle (HMMWV). This appendix is written from the perspective of an Infantry platoon leader controlling a combat vehicle section or platoon. However, the technical and tactical information addressed in the following pages is also generally valid for Infantry platoons attached to mechanized/heavy units.

SECTION I — CAPABILITIES

D-1. The primary roles of the combat vehicles discussed in this appendix are to provide Infantry platoons with mobility to allow them to maneuver. Combat vehicles also provide bases of fire; protection, breaching capabilities, enhanced communication platforms, and a variety of sustainment assets that include re-supply and MEDEVAC capabilities. Effective integration of these forces provides complementary and reinforcing effects to Infantry and mounted forces.

PRINCIPLES OF EMPLOYMENT

D-2. There are three general principles for employing combat vehicles with Infantrymen:

(1) So the combat power capabilities of the vehicle can support the maneuver of the Infantry.

(2) So the combat power of the Infantry platoon can support the maneuver of combat vehicle sections or platoons.

(3) The wingman concept. To achieve mutual support, combat vehicles almost always work in this concept. The wingman concept is similar to the buddy team concept Infantrymen employ (operating in two-vehicle sections). Just like Infantrymen never fight alone, combat vehicles never operate without the mutual support and evacuation capability the combat vehicle wingman provides.

GENERAL EMPLOYMENT CONSIDERATIONS

D-3. Employment of combat vehicles requires thorough understanding and integration of the vehicle and the Infantry unit. The following paragraphs focus on general employment considerations.

Combat Vehicles Supporting the Infantry

D-4. Combat vehicles support Infantry units by leading Infantrymen in open terrain and providing them a protected, fast-moving assault weapons system. They suppress and destroy enemy weapons, bunkers, and tanks by fire and movement. They may provide transport when the enemy situation permits.

Mobility

D-5. The following is a list of the primary mobility functions that combat vehicles provide an Infantry platoon during combat operations:

- Assist opposed entry of Infantry into buildings or bunkers.
- Breach or reduce obstacles by fire.
- Provide mobility to the dismounted force.
- Provide enhanced communication platforms and multiple communications systems.
- Sustainment (MEDEVAC and re-supply).

Firepower

D-6. The following is a list of the primary firepower functions that combat vehicles provide an Infantry platoon during combat operations:

- Speed and shock effect to assist the Infantry in rapidly executing an assault.
- Lethal and accurate direct fire support (support by fire).
- Suppression of identified sniper positions.
- Heavy volume of suppressive fires and a mobile base of fire for the Infantry.
- Employment of technical assets (thermal viewers and range finders) to assist in target acquisition and ranging.
- Neutralization or suppression of enemy positions with direct fire as Infantry closes with and destroys the enemy.
- Attack by fire any other targets designated by the Infantry.
- Accurate direct fires even while the vehicle is moving at high speeds with stabilized gun systems.
- Destruction of enemy tanks and armored personnel carriers (APCs).

Protection

D-7. The following are ways that combat vehicles protect an Infantry platoon during combat operations:

- Dominate the objective during consolidation and reorganization to defeat a counterattack and protect Infantry forces.
- Protect the movement of advancing Infantry through open terrain with limited cover and concealment.
- Secure cleared portions of the objective by covering avenues of approach.
- Establish roadblocks or checkpoints.
- Provide limited obscuration with smoke grenades and smoke generators.
- Isolate objectives with direct fire to prevent enemy withdrawal, reinforcement, or counterattack.

Infantrymen Supporting Combat Vehicles

D-8. Infantrymen support vehicular forces by finding and breaching or marking antitank obstacles. They detect and destroy or suppress enemy antitank weapons. Infantrymen may designate targets for armored vehicles and protect them in close terrain.

Mobility

D-9. Mobility functions that Infantry provide to units with vehicles during combat operations include:

- Seize and retain terrain.
- Clear defiles and restrictive urban terrain ahead of vehicular forces.

Firepower

D-10. Firepower functions that Infantry provide to units with vehicles during combat operations include:

- Actions on the objective (clear trenches, knock out bunkers, enter and clear buildings).
- Employ AT systems (Javelin) to destroy armored threats.

Protection

D-11. Ways Infantry protect units with vehicles during combat operations include:

- Provide local security over dead space / blind spots that weapon systems on combat vehicles cannot cover.
- Consolidate and reorganize (perform EPW procedures and direct MEDEVAC).

TECHNICAL CAPABILITIES

D-12. Infantry leaders must have a basic understanding of the technical capabilities of combat vehicles. These include vehicle characteristics, firepower and protection.

VEHICLE CHARACTERISTICS

D-13. To win in battle, leaders must have a clear understanding of the capabilities and limitations of their equipment. The tank, Bradley, Stryker ICV, and assault HMMWV each have their own capabilities, limitations, characteristics, and logistical requirements. Even though their role to the Infantry is virtually the same, these vehicles provide support in different ways. To effectively employ combat vehicles, leaders must understand the specific capabilities and limitations of vehicles that may be attached/OPCONed to their unit. The following information is a brief overview of the combat vehicles' characteristics as they apply to combat power. Table D-1 displays vehicle characteristics. (*Specifics vary by vehicle and modifications.)

Table D-1. Mobility characteristics of combat vehicles.

	ASSLT HMMWV*	ICV*	BFV*	Tank*
Tracks/Wheels	Wheels	Wheels	Tracks	Tracks
Length	196.5"	275"	254"	312"
Width	86"	107"	126"	144"
Height	74" (without wpn)	104"	117"	96"
Weight	5,600 lbs	38,000 lbs	50,000 lbs	68.7 tons
Speed	78 mph	60 mph	42 mph	42 mph

Firepower

D-14. The weapons and ammunition of vehicular units are designed to defeat specific enemy targets, though many are multi-purpose. An Infantry leader with a basic understanding of these weapons and ammunition types will be able to better employ vehicular units to defeat the enemy. Table D-2 lists the basic weapons and ammunition types offered by vehicular units that generally support Infantry platoons.

Table D-2. Weapons, ammunition, and targets.

		ASSLT HMMWV		ICV		BFV		Tank	
		Weapon Ammo	**Target**	**Weapon Ammo**	**Target**	**Weapon Ammo**	**Target**	**Weapon Ammo**	**Target**
Blast Munition		**40mm MK 19** Max area: 2,212m Max point: 1,500m	Trucks, troops, bunkers, buildings	**40mm MK 19** Max area: 2,212m Max point: 1,500m	Trucks, troops, bunkers, buildings	**25mm (HE)** Max effective: 3,000m	Trucks, troops, bunkers, buildings	**120mm (HEAT)** Max effective: 3,000m	Trucks, troops, bunkers, buildings, APCs
K I N E T I C (AT)	**Cannon**	None	None	None	None	**25mm (sabot)** Max effective: 2,500m	APCs	**120mm (sabot)** Max effective: 3,000m	APCs, tanks
	Machine Gun	**M249 5.56mm** Max area: 800m Max point: 600m **M240B 7.62mm** (mounted) Max area: 1,100m Max point: 800m **M2 .50 caliber** Max area: 1,830m Max point: 1,200m	Troops, trucks, eqpmnt	**M2 .50 caliber** Max area: 1,830m Max point: 1,200m	Troops, trucks, equipment	**M240C 7.62mm*** Max effective: 900m	Troops, trucks	**M240C 7.62mm** Max effective: 900m **M2 .50 caliber** Max area: 1,830m Max point: 1,200m	Troops, trucks, eqmnt
TOW Missile		Max effective: 3,750m	Tanks	Max effective: 3,750m	Tanks, helicopters, bunkers	Max effective: 3,750m	Tanks, helicopters, bunkers	None	None

*The BFV does not have a heavy machine gun.

Protection

D-15. All combat vehicles offer varying degrees of protection from direct and indirect fire. Figure D-1 illustrates the generally-progressive degrees of protection offered by combat vehicles.

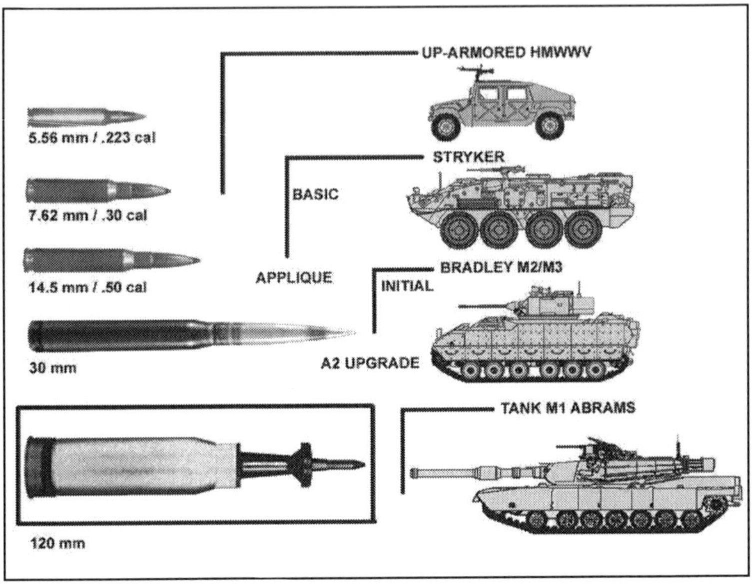

Figure D-1. Comparative levels of ballistic protection.

TANKS (M1)

D-16. The M1-series tank provides rapid mobility combined with excellent protection and highly lethal, accurate fires. They are most effective in generally open terrain with extended fields of fire.

Mobility Advantages

D-17. The tank's mobility comes from its capability to move at high speed both on and off road. The tank's ability to cross ditches; ford streams and shallow rivers; and push through small trees, vegetation, and limited obstructions allows effective movement in various types of terrain.

Mobility Disadvantages

D-18. Tanks consume large quantities of fuel. They are very noisy and must be started periodically in cold weather or when using thermal night sights and radios to ensure the batteries stay charged. The noise, smoke, and dust generated by tanks make it difficult for the Infantry in their vicinity to capitalize on stealth to achieve surprise. Tanks cannot cross bodies of water deeper than four feet without deep water fording kits or bridging equipment. Due to the length of the tank main gun, the turret will not rotate if a solid object such as a wall, post, or tree is blocking it. Tracked vehicles can also "throw track." This occurs when the track loses tension on the sprockets and/or support arms and the track becomes disconnected from the tank. Repairing the track can be a lengthy process.

Firepower Advantages

D-19. The tank's main gun is extremely accurate and lethal at ranges out to 4,000 meters. Tanks with stabilized main guns can fire effectively even when moving at high speeds cross-country. The tank remains the best antitank weapon on the battlefield. The various machine guns (M1 tank commander's caliber .50 and 7.62-mm coax and the loader's 7.62-mm MG) provide a high volume of supporting fires for the

Infantry. The target acquisition capabilities of the tank exceed the capability of all systems in the Infantry battalion. The thermal sight provides a significant capability for observation and reconnaissance. It can also be used during daylight hours to identify heat sources (personnel and vehicles), even through vegetation. The laser range finder provides an increased capability for the Infantry force to establish fire control measures (such as trigger lines and TRPs), and to determine exact locations.

Firepower Disadvantages

D-20. The normal, basic load for the tank's main gun is primarily armor piercing discarding sabots (APDS) antitank rounds. These rounds are not as effective against light armored or wheeled vehicles, bunkers, trench lines, buildings, or enemy personnel. They also present a safety problem when fired over the heads of exposed Infantrymen due to the discarded sabot pieces that fall to the ground. HE ammunition provides better destructive effects on the above-mentioned targets except enemy personnel, which the tank's machine guns are most effective against. The resupply of all tank ammunition is difficult and requires logistic support from the heavy battalion. The main gun of an M1A2 can only elevate +20 degrees and depress -9 degrees. Figure D-2 illustrates M1A2 fields of fire on the urban terrain.

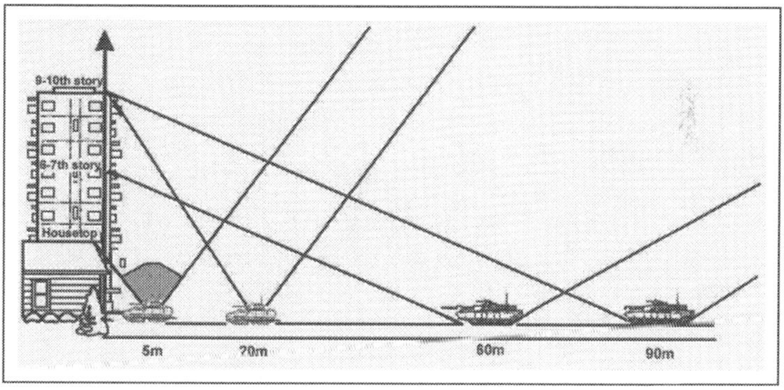

Figure D-2. M1A2 fields of fire on urban terrain.

Protection Advantages

D-21. Generally, tank armor provides excellent protection to the crew. Across the frontal 60-degree arc, the tank is impenetrable to all weapons except heavy AT missiles or guns and the main gun on enemy tanks. When fighting with the hatches closed, the crew is impenetrable to all small arms fire, artillery rounds (except a direct hit), and AP mines. The tank's smoke grenade launcher and on-board smoke generator provide rapid concealment from all but thermal observation.

Protection Disadvantages

D-22. The tank is most vulnerable to lighter AT weapons from the flanks, top, and especially the rear. The top is especially vulnerable to precision-guided munitions (artillery or air delivered). AT mines can also destroy or disable the vehicle. When fighting with hatches closed, the tank crew's ability to see, acquire, and engage targets (especially close-in Infantry) is greatly reduced.

Information Advantages

D-23. FBCB2, global positioning systems (GPSs), and inertial position navigation (POSNAV) systems allow today's tanks the mobility to virtually any designated location with greater speed and accuracy than ever before. Use of visual signals and the single channel ground/airborne radio system (SINCGARS) facilitates rapid and secure communication of orders and instructions. This capability allows tank crews to quickly mass the effects of their weapon systems while remaining dispersed to limit the effects of the enemy's weapons. On-board optics and sighting systems enable tank crews to acquire and destroy enemy tanks, armored vehicles, and fortifications using the main gun, and to suppress enemy positions, personnel, and lightly armored targets with the tank's machine guns

Information Disadvantages

D-24. Not all tanks are equipped with digitally enhanced systems (FBCB2). Additionally, at present, the situational awareness and enemy situation acquired by the FBCB2 cannot be easily shared with Infantry units on the ground.

M2 BRADLEY FIGHTING VEHICLE (BFV)

D-25. The M2 BFV provides good protection and mobility combined with excellent firepower to support Infantry units with direct fire.

Advantages

D-26. The mobility of the M2 is comparable to the tank. In addition to the three-man crew, the vehicle is designed to carry seven additional Infantrymen with a combat load.

Disadvantages

D-27. The M2 consumes significant quantities of fuel, but less than a M1. The BFV is louder than the M1, and like the M1, its engine must be started periodically in cold weather or when using the thermal night sight and radios to ensure the batteries stay charged. Like all heavy vehicles, the noise, smoke, and dust generated by the M2 makes it difficult for the Infantry to capitalize on its ability to move with stealth and to avoid detection when moving on the same approach. Improvised barricades, narrow streets and alleyways, or large amounts of rubble can block BFVs in an urban area. Heavy woods will restrict their movement in a rural area. The 25-mm cannon does not project out over the front of the Bradley like a tank, but it does protrude over the sides of the Bradley when the gunner is aiming at 3 o'clock or 9 o'clock. This will cause some problems for the Bradley when trying to negotiate narrow avenues of approach. Attaching and removing rucksacks to the exterior of the vehicle can be a lengthy process, and the rucksacks are exposed to enemy fire.

Firepower Advantages

D-28. The primary weapon on the M2 is the 25-mm chain gun that fires APDS, high explosive incendiary with tracer (HEI-T), and TPT. This weapon is extremely accurate and lethal against lightly armored vehicles, bunkers, trench-lines, and personnel at ranges out to 2,000 meters. The stabilized gun allows effective fires even when moving cross-country. The TOW provides an effective weapon for destroying enemy tanks or other point targets at extended ranges to 3,750 meters. The 7.62-mm coax provides a high volume of suppressive fires for self defense and supporting fires for the Infantry up to 800 meters. The combination of the stabilized turret, thermal sight, high volume of fire, and the reinforcing effects of weapons and ammunition makes the M2 an excellent suppression asset supporting Infantry assaults. The thermal sight provides a significant capability for observation and reconnaissance. It can also be used during the day to identify heat sources (personnel and vehicles) even through light vegetation. Figure D-3 shows the 25-mm supporting Infantry in an urban setting.

Figure D-3. BVF 25-mm infantry support.

Firepower Disadvantages

D-29. When operating the thermal sight with the M2engine off, a "clicking" sound can be heard at a considerable distance from the vehicle. The resupply of ammunition is difficult and requires external logistic support.

Protection Advantages

D-30. Overall, the M2 provides good protection. When fighting with the hatches closed, the crew is well protected from small-arms fire, fragmentation munitions, and AP mines. The M2 smoke-grenade launcher and on-board smoke generator provide rapid concealment from all but thermal observation.

Protection Disadvantages

D-31. The vehicle is vulnerable from all directions to any AT weapons and especially enemy tanks. AT mines can destroy or disable the vehicle. When the crew is operating the vehicle with the hatches open, they are vulnerable to small-arms fire.

Information Advantages

D-32. The target acquisition capabilities of the M2 exceed the capability of the other systems in the Infantry battalion. The thermal sight provides a significant capability for observation and reconnaissance. It can also be used during the day to identify heat sources (personnel and vehicles) even through light vegetation. Many models of the BFV are now equipped with the FBCB2.

Information Disadvantages

D-33. Bradley vehicle crewmen have poor all-round vision through their vision blocks. They are also easily blinded by smoke or dust. Therefore, the Bradley vehicle should not be approached while it is in contact because the crew may have difficulty seeing Infantryman outside of the vehicle. The Bradley commander (BC) must be informed where the dismounted Infantry are located to prevent any accidents on the battlefield.

STRYKER INFANTRY CARRIER VEHICLE (ICV)

D-34. There are two variants of the Stryker: the Infantry carrier vehicle (ICV); and the mobile gun system (MGS). The primary design of the Stryker is found in the basic ICV. This troop transport vehicle is quite capable of carrying nine Infantry Soldiers and their equipment, a crew of two, a driver, and a vehicle commander. There are eight configurations of the ICV that provide comprehensive sustainment. The eight ICV configurations include: command vehicle; reconnaissance vehicle; fire support vehicle; mortar carrier vehicle; antitank guided missile vehicle; engineer squad vehicle; medical evacuation vehicle; and nuclear, biological, and chemical reconnaissance vehicle. The MGS incorporates a 105-mm turreted gun and autoloader system. The Stryker can greatly reduce the amount of inventory and logistical support for combat brigades, while at the same time increasing the Infantry's ability to deploy.

Mobility

D-35. The Stryker vehicle enables the team to maneuver in close and urban terrain, provide protection in open terrain, and transport infantry quickly to critical battlefield positions.

Advantages

D-36. With 4x8- and 8x8-wheel drive, the Stryker is designed for all-weather use over all types of terrain and can ford hard-bottomed bodies of water to a depth of 67 inches. Stryker vehicles have a maximum speed of 60 miles per hour and a range of 300 miles on a tank of fuel. The vehicles are swift, easily maintainable, and include features designed for the safety of Soldiers. The Stryker's has run-flat tires that can be inflated or deflated from inside the vehicle to adapt to surfaces ranging from deep mud to hardtop. It also has a built-in fire suppression system, and a self-recovery winch. The vehicles run quieter than current armored personnel carriers, increasing their stealth. Steel-belted tires with run-flat liners enable vehicle mobility for 5 miles (8 km) with all tires flat.

Disadvantages

D-37. For vehicles weighing 10-20 tons, wheels are inferior to tracks in crossing sand, mud, and snow. Driving more than five miles on a flattened tire can cause a fire. Improvised barricades, narrow streets and alleyways, or large amounts of rubble can block Stryker vehicles in urban areas. Dense forests can block it in rural areas.

Firepower Advantages

D-38. The ICV has a remote weapon station with a universal soft mount cradle that can mount either a .50-caliber M2 machine gun, MK 19 40-mm grenade launcher, or M240B 7.62-mm machine gun. It is also armed with four M6 smoke grenade launchers. Stowed ammunition includes:

- 32 66-mm smoke grenades.
- 3,200 7.62-mm rounds.
- 2,000 .50 cal rounds *or* four hundred thirty MK 19 rounds.

D-39. Troops carry—

- 2,240 5.56-mm ball ammunition.
- 1,120 5.56-mm linked ammunition.

Firepower Disadvantages

D-40. The ICV loses some of the ammunition effects that tanks and Bradley fighting vehicles can provide to the Infantryman. For this reason the ICV can suppress light skinned vehicles, bunkers, buildings, and enemy Infantry, but is not as effective as a BFV or tank against enemy light-armored or armored vehicles.

Protection Advantages

D-41. The basic ICV provides armored protection for the two-man crew and a squad of nine Infantry Soldiers. The ICV's armor protection will stop .50-caliber bullets and protects against 152-mm airburst shells. The basic armor package on every Stryker vehicle is a steel hull that protects against 7.62-mm bullets; and a ceramic, added-on appliqué that gives protection against 14.5-mm machine guns. Hull floor plate and fuel tank armor protect from blast and fragment effects of antipersonnel mine detonations. Low silhouette and low noise output make the vehicle a difficult target to detect and engage.

Protection Disadvantages

D-42. The ICV is vulnerable to all AT fires and tanks. The effectiveness of RPG fire can be mitigated with a slat-armor application (cage) that causes a premature detonation of the RPG warhead away from the hull of the ICV.

Information

D-43. Just as with the tank and Bradley, the Stryker ICV vehicle crewmen have poor all-round vision through their vision blocks. They are also easily blinded by smoke or dust.

ASSAULT HIGH-MOBILITY MULTIPURPOSE WHEELED VEHICLE (HMMWV)

D-44. The HMMWV is a light, highly mobile, diesel-powered, four-wheel-drive vehicle equipped with an automatic transmission. Using components and kits common to the M998 chassis, the HMMWV can be configured as a troop carrier, armament carrier, TOW missile carrier, or a Scout vehicle.

MOBILITY ADVANTAGES

D-45. The HMMWV rests on a four-wheel chassis. Its four-wheel drive enables it to operate in a variety of terrain and climate conditions. It is capable of fording water up to 30 inches in depth, and can ford depths of up to 60 inches with the deep water fording kit. The HMMWV's size allows it to travel in the narrow streets of urban terrain with minimal damage to the infrastructure. Some models of the HMMWV (M1026, M1036, M1046, and M1114) employ a winch that aids in self recovery and recovery of similar vehicles.

Mobility Disadvantages

D-46. Although generally equipped with run-flat tires, the HMMWV's tires are very susceptible to enemy fire. HMMWVs have much less ability to breach obstacles than tracked vehicles. The HMMWV can be blocked by hasty and complex obstacles. I can also be easily rolled, especially with the armored M114.

Firepower Advantages

D-47. The HMMWV can employ a variety of weapon systems that offer excellent direct fire support to Infantry forces. The TOW, .M2, MK 19, M240B, and M249 can all be mounted in HMMWV models with turrets. The capabilities of these weapon systems are discussed in greater detail in Table D-2.

Firepower Disadvantages

D-48. In almost all instances, the HMMWV can only mount one weapon system. This makes it less effective than tanks or BFVs that can employ antitank and antipersonnel weapons simultaneously.

Protection Advantages

D-49. The M1114 is an up-armored HMMWV that provides ballistic, artillery, and mine blast protection to vehicle occupants. The M1114 can protect occupants from 7.62-mm assault rifle armor-piercing rounds and 155-mm artillery airburst, and provides 12 pounds front and 4 pounds rear antitank mine protection. Other protection features include complete perimeter ballistic protection, mine blast protection, and a turret shield for the gunner. Supplemental armor packages are now available for many models of the HMMWV. This armor has been shown to be effective against improvised explosive devices.

Protection Disadvantages

D-50. All models other than the M1114 offer extremely limited protection from direct or indirect fire. Leaders should not plan or direct the use of these vehicles for cover from enemy small arms, indirect fire, or rocket-propelled grenades. Gunners are exposed while manning their weapon system to direct and indirect fire. The lack of internal space causes difficulties if transporting a casualty.

Information Advantages

D-51. The HMMWV has a variety of features that make it excellent for gathering and managing information. The crew and passengers of the HMMWV generally have excellent situational awareness due to a large front windshield and large windows located on the door at each seat. HMMWVs can carry two SINCGARS-class FM radio systems. They can also employ a power amplifier to extend the communications range to 35 kilometers in open terrain. The HMMWV can be configured to carry many digital devices to include the FBCB2 and PLGRs. The weapon systems of the HMMWV can employ sites with night vision, thermal, and range-finding capabilities with high resolution and magnification in some systems.

Information Disadvantages

D-52. Many of the digital and electronic devices of the HMMWV require constant power sources. The need to start the HMMWV to keep the batteries charged can present a tactical problem if stealth is desired during an operation.

SIZE AND WEIGHT CONSIDERATIONS

D-53. Infantry leaders must consider the size and weight of combat vehicles operating in units before conducting an operation (Table D-3). Terrain that supports the movement of Infantrymen may or may not support the movement of combat vehicles. Structures of particular concern are bridges, overpasses, and culverts as structural failure could be deadly to the Soldiers in the vicinity. Many bridges in North America and Europe are marked with signs that state the load bearing capabilities of that structure. In other areas, Infantrymen should rely on route reconnaissance overlays that show the carrying capabilities of the routes being used. In the absence of such information, Infantry leaders should always use the cautious approach and avoid suspect infrastructure.

Table D-3. Vehicle size and weight classification.

Vehicle	Weight	Height (feet)	Width (inches)
M1 Tank	68.7 tons	10.14	143.75
BFV with reactive armor	33 tons	11.3	142.2
BFV without reactive armor	28 tons	11.3	130
Stryker ICV	38,000 lbs.	104	107
ASLT HMMWV	6,780 lbs.	74	85

SURFACE DANGER AREAS

D-54. Infantry leaders must consider the surface danger zones (SDZ) of combat vehicle weapon systems that are operating with their units. This information is crucial for the leaders to develop safe and effective direct fire control plans. Effective application of SDZs prevents fratricide and maximizes direct fire upon the enemy.

D-55. Each weapon system has a unique SDZ. SDZs are the minimum safe distances and angles that must be considered when operating in close proximity to weapon systems. SDZs take into consideration a round's maximum distance, lateral dispersion, and backblast (if applicable). This information allows leaders to plan for safe and effective maneuver of their forces. Reference Section III of this appendix for a detailed analysis of SDZs for weapon systems associated with combat vehicles in this appendix.

TACTICAL CAPABILITIES

D-56. Light Infantry units may have combat vehicle sections attached for combat operations. Table D-4 shows a list of tasks that these combat vehicle sections may perform while attached or under the operational control of Infantry units.

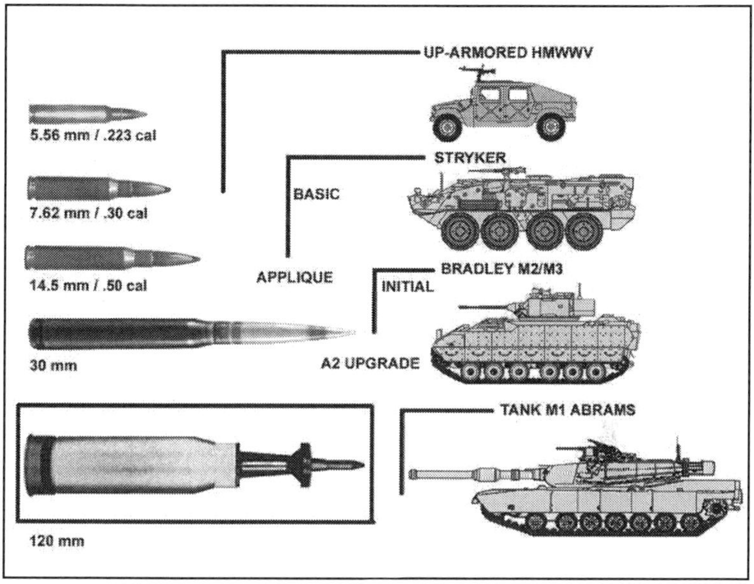

Figure D-1. Comparative levels of ballistic protection.

TANKS (M1)

D-16. The M1-series tank provides rapid mobility combined with excellent protection and highly lethal, accurate fires. They are most effective in generally open terrain with extended fields of fire.

Mobility Advantages

D-17. The tank's mobility comes from its capability to move at high speed both on and off road. The tank's ability to cross ditches; ford streams and shallow rivers; and push through small trees, vegetation, and limited obstructions allows effective movement in various types of terrain.

Mobility Disadvantages

D-18. Tanks consume large quantities of fuel. They are very noisy and must be started periodically in cold weather or when using thermal night sights and radios to ensure the batteries stay charged. The noise, smoke, and dust generated by tanks make it difficult for the Infantry in their vicinity to capitalize on stealth to achieve surprise. Tanks cannot cross bodies of water deeper than four feet without deep water fording kits or bridging equipment. Due to the length of the tank main gun, the turret will not rotate if a solid object such as a wall, post, or tree is blocking it. Tracked vehicles can also “throw track.” This occurs when the track loses tension on the sprockets and/or support arms and the track becomes disconnected from the tank. Repairing the track can be a lengthy process.

Firepower Advantages

D-19. The tank's main gun is extremely accurate and lethal at ranges out to 4,000 meters. Tanks with stabilized main guns can fire effectively even when moving at high speeds cross-country. The tank remains the best antitank weapon on the battlefield. The various machine guns (M1 tank commander's caliber .50 and 7.62-mm coax and the loader's 7.62-mm MG) provide a high volume of supporting fires for the

Figure D-3. BVF 25-mm infantry support.

Firepower Disadvantages

D-29. When operating the thermal sight with the M2engine off, a "clicking" sound can be heard at a considerable distance from the vehicle. The resupply of ammunition is difficult and requires external logistic support.

Protection Advantages

D-30. Overall, the M2 provides good protection. When fighting with the hatches closed, the crew is well protected from small-arms fire, fragmentation munitions, and AP mines. The M2 smoke-grenade launcher and on-board smoke generator provide rapid concealment from all but thermal observation.

Protection Disadvantages

D-31. The vehicle is vulnerable from all directions to any AT weapons and especially enemy tanks. AT mines can destroy or disable the vehicle. When the crew is operating the vehicle with the hatches open, they are vulnerable to small-arms fire.

Information Advantages

D-32. The target acquisition capabilities of the M2 exceed the capability of the other systems in the Infantry battalion. The thermal sight provides a significant capability for observation and reconnaissance. It can also be used during the day to identify heat sources (personnel and vehicles) even through light vegetation. Many models of the BFV are now equipped with the FBCB2.

Information Disadvantages

D-33. Bradley vehicle crewmen have poor all-round vision through their vision blocks. They are also easily blinded by smoke or dust. Therefore, the Bradley vehicle should not be approached while it is in contact because the crew may have difficulty seeing Infantryman outside of the vehicle. The Bradley commander (BC) must be informed where the dismounted Infantry are located to prevent any accidents on the battlefield.

STRYKER INFANTRY CARRIER VEHICLE (ICV)

D-34. There are two variants of the Stryker: the Infantry carrier vehicle (ICV); and the mobile gun system (MGS). The primary design of the Stryker is found in the basic ICV. This troop transport vehicle is quite capable of carrying nine Infantry Soldiers and their equipment, a crew of two, a driver, and a vehicle commander. There are eight configurations of the ICV that provide comprehensive sustainment. The eight ICV configurations include: command vehicle; reconnaissance vehicle; fire support vehicle; mortar carrier

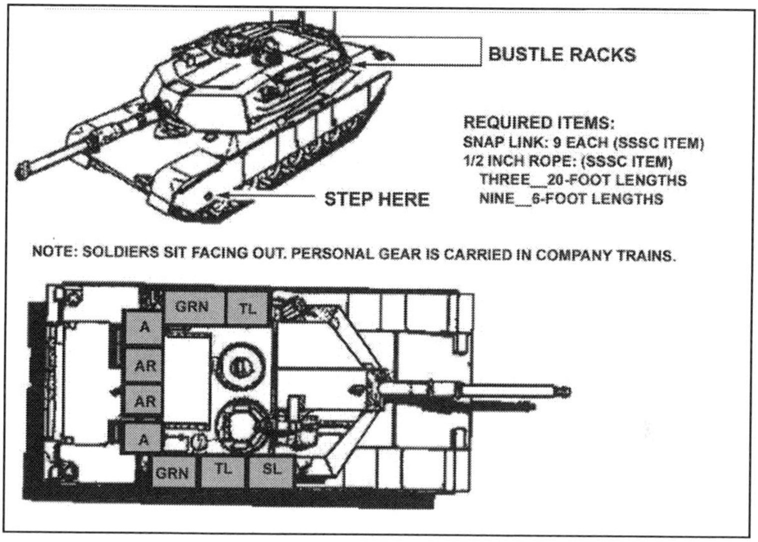

Figure D-4. Mounting and riding arrangements on an M1-series tank.

D-64. One Infantry squad can ride on the turret. Soldiers must mount in such a way that their legs cannot become entangled between the turret and the hull by an unexpected turret movement. Rope may be used as a field-expedient Infantry rail to provide secure handholds.

D-65. Everyone must ride to the rear of the smoke grenade launchers. This automatically keeps everyone clear of the coaxial machine gun and laser range finder.

D-66. The Infantry must always be prepared for sudden turret movement. Leaders should caution Soldiers about sitting on the turret blowout panels. This safety knowledge is critical because 250 pounds of pressure will prevent the panels from working properly. If there is an explosion in the ammunition rack, the panels blow outward to lessen the blast effect in the crew compartment.

D-67. If enemy contact is made, the tank should stop in a covered and concealed position and allow Infantry time to dismount and move away from the tank. This action needs to be practiced before movement.

D-68. The Infantry should not ride with anything more than their battle gear. Personal gear should be transported elsewhere.

Bradley Fighting Vehicle

D-69. The BFV is designed to carry six Infantrymen and a crew of three: a Bradley commander (BC), gunner (GNR), and driver (DVR). The troop compartment of the BFV carries six Infantrymen in combat gear. Rucksacks are generally carried on the outside of the vehicle. Prior to riding in the vehicle, Infantrymen who are not familiar with the BFV should be thoroughly trained on its exit points, fire drills, and rollover drills. The major difference in carrying capacity between the M2A1 and the M2A2/ODS/M2A3 is the seating configuration. The M2A1 has six individual seats, while the M2A2/ODS/M2A3 has two benches that are on the left and right sides of the troop compartment. Figures D-5 and D-6 illustrate the carrying capacity of the BFV-series combat vehicles.

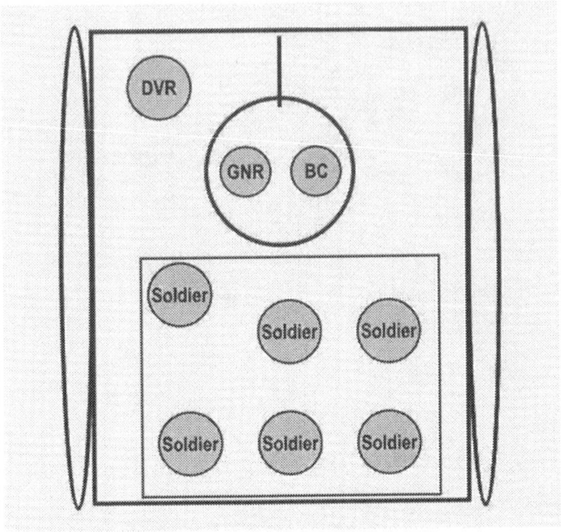

Figure D-5. M2A1 seating diagram.

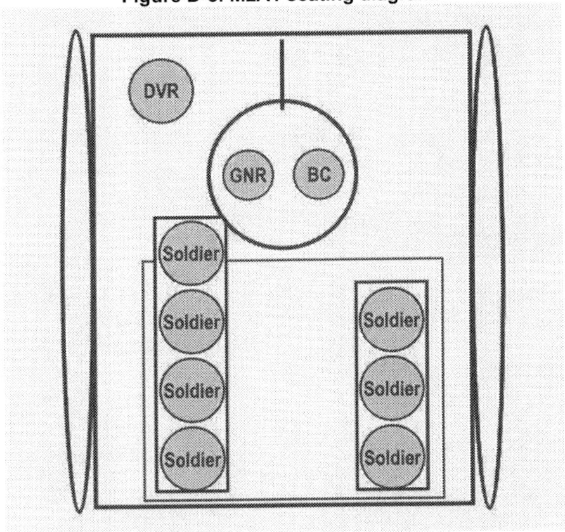

Figure D-6. M2A2, ODS, and M2A3 seating diagram.

Infantry Carrier Vehicle

D-70. The Stryker Infantry carrier vehicle is designed to carry a nine-man Infantry squad in combat gear, a driver, and a vehicle commander (VC). Rucksacks are generally carried on the outside of the ICV.

Infantrymen who are not familiar with the ICV should be thoroughly trained on its exit points, fire drills, and rollover drills prior to riding in the vehicle. Figure D-7 illustrates the carrying capacity of the ICV.

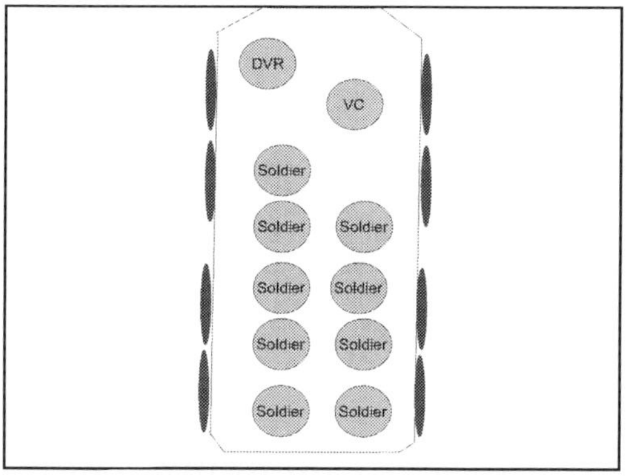

Figure D-7. Seating diagram for the ICV.

Assault HMMWV

D-71. The ASSLT HMMWV class of vehicles is designed to carry five Soldiers in combat gear, a truck commander (TC), a gunner, a driver, and two Soldiers in the rear passenger seats. Rucksacks are generally carried on the outside or in the rear cargo storage area of the ASSLT HMMWV. Infantrymen who are not familiar with the ASSLT HMMWV should be thoroughly trained on its exit points, fire drills, and rollover drills prior to riding in the vehicle. Figure D-8 illustrates the carrying capacity of the ASSLT HMMWV.

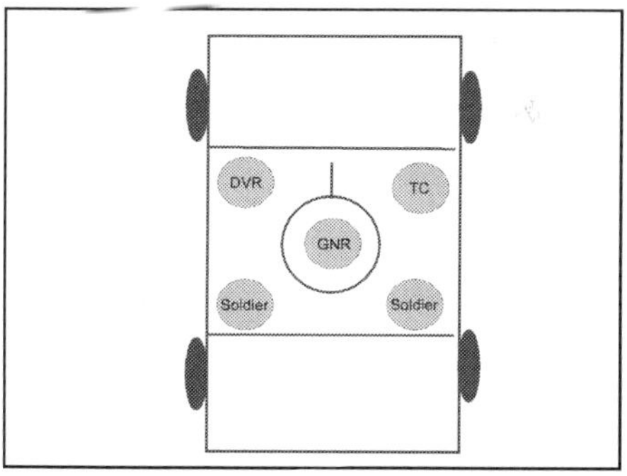

Figure D-8. Seating diagram for the ASSLT HMMWV.

SECTION II — OPERATIONS

D-72. The intent of this section is to familiarize leaders with conducting operations with combat vehicles. The section is divided under three subsections: plan, prepare, and execute.

PLAN

D-73. Employment of combat vehicles requires thorough understanding and integration of the vehicle with the Infantry unit. The following paragraphs focus on planning considerations for combat vehicles and dismounted Infantry integration.

TASK ORGANIZATION OPTIONS

D-74. A combat vehicle platoon or section would normally be OPCONed to an Infantry company during combined arms operations at the company team level. However, in the COE, Infantry platoons may receive combat vehicle platoons or sections to conduct operations. There are four basic techniques of task-organizing the combat vehicle section into the Infantry company for combat operations: combat vehicle platoon as a maneuver element; combat vehicle sections under Infantry control; combat vehicle sections under company and platoon control; and Infantry squads under combat vehicle control. This concept holds true for all combat vehicle units.

Combat Vehicle Platoon as a Maneuver Element

D-75. The combat vehicle platoon leader is responsible for maneuvering the vehicles IAW the company team commander's intent. Likely missions for the combat vehicles with this task organization are support by fire (SBF), or overwatch of the Infantry's movement. The combat vehicle platoon leader can choose to maneuver the platoon by sections to execute the mission. This maneuver provides greater flexibility in supporting the Infantry during the close fight.

Combat Vehicle Sections Under Infantry Platoon Control

D-76. Combat vehicles are broken down into two sections. Each section is placed under the OPCON of an Infantry platoon and maneuvered IAW the company team commander's intent. The commander relinquishes direct control of the combat vehicle maneuver to his subordinates. This technique is very effective in maintaining the same rate of progress between the combat vehicles and the Infantry. Leaders have the additional responsibility of maneuvering combat vehicles. The general lack of experience with combat vehicles and the overall battlefield focus of the leaders can affect this technique. This technique is best suited for when contact with the enemy is expected and close continuous support is required for movement or clearing buildings.

Combat Vehicle Sections Under Company and Platoon Control

D-77. Combat vehicle platoons can be broken down into two sections: one under company control; the other under platoon control. The selected maneuver Infantry platoon would have a combat vehicle section available to support the close fight. With this technique, the company team commander has a combat vehicle section to deploy. This task organization still allows support to the Infantry close fight while keeping additional support options in reserve for the commander to employ. The disadvantages to this technique are Infantry platoon leaders instead of the combat vehicle platoon leaders are maneuvering vehicles, and vehicles directly available to the company team commander are cut in half. This technique requires detailed planning, coordination, and rehearsals between the Infantry and combat vehicle sections.

Infantry Squads Under Combat Vehicle Platoon Control

D-78. The company team commander has the option of placing one or more Infantry squads under the OPCON of the combat vehicle platoon leader. He may also retain all combat vehicles under the control of the combat vehicle platoon leader, or place a combat vehicle section under the OPCON of an Infantry platoon leader. This provides the company team commander with a fourth maneuver platoon. It also involves the combat vehicle platoon leader in the fight. It can work well when a mobile reserve that needs Infantry protection is required.

Guidelines

D-79. None of the techniques described are inherently better than another one. The task organization must be tailored to accomplish the mission. Regardless of the technique selected, the following guidelines should be followed.

D-80. It is preferable for combat vehicles to operate as sections. This is an integral component of how combat vehicle units train and fight. If the company commander is controlling the combat vehicles, he needs to move forward to a position where he can effectively maneuver the combat vehicles in support of the Infantry.

D-81. Combat vehicles should be used to shield squads and teams (minus the unarmored versions of the ASSLT HMMWV) from building to building. As part of the maneuver plan, the leader of the forward element controls the combat vehicles.

D-82. The task organization should support the span of control. If the company commander is going to control the combat vehicles, there is no reason to task-organize the tanks by section under Infantry platoons.

D-83. Combat vehicles need Infantry support when the two elements are working together. Do not leave combat vehicles alone because they are not well suited to provide local security during the operation. Combat vehicles are extremely vulnerable to dismounted attack when operating in urban terrain. They are most vulnerable and need local security when Infantry are in the process of clearing buildings.

Risk Management

D-84. Infantry leaders must identify and implement controls to mitigate risks associated with conducting operations with combat vehicles. These risks are divided into two categories: tactical and accidental risk. Table D-7 contains a basic list of risks and control measures leaders should consider when conducting operations with combat vehicles. Table D-8 contains a list of possible accidental hazards and control measures.

Table D-7. Risk management matrix for tactical hazards.

Tactical Hazards	Control Measure
Enemy Direct Fire	Wear individual body armor (IBA), reinforce vehicle (sand bags), use proper scanning techniques, and engage in marksmanship training.
Enemy Indirect Fire	Practice mounted react to indirect fire drills, vary speed and distance to avoid a trigger from an enemy indirect fire system.
Mines	Maintain situational awareness (SA), maintain current obstacle overlay for AO, remain on cleared areas, be proficient in mine removal.
IEDs	Scan, use WARLOCK (anti-remote-detonation IED system), use up-armor, and avoid predictability.
Sniper Attacks	Scan, maintain SA, avoid predictability, use DVR techniques, engage in tactical movement (MVT) training.
Media Exploitation	Train leaders; refer to PAO; adhere to the ROE, Soldier's Creed, Law of War, and the Geneva Conventions.
VBIED	Gunner and Infantrymen riding on vehicles use proper scanning techniques, maintain SA, avoid predictability, use DVR techniques, and engage in tactical MVT training.
Ambush	Scan, maintain SA, avoid predictability, use DVR techniques, engage in tactical MVT training.

Table D-8. Risk management matrix for accidental hazards.

Accidental Hazards	Control Measure
Vehicle Collision	Ensure DVR is qualified and TC is alert.
Vehicle Fire	Conduct fire drills, keep fire extinguishers present and serviceable, perform proper PMCS.
Vehicle Rollover	Ensure DVR/TC/dismount situational awareness, train and rehearse with vehicles, know SOPs for communication between vehicle and dismounts. Secure loads.
Vehicle Striking Dismount	Train on high decibel danger zones and wear hearing protection.
Vehicle Malfunction	Perform proper PMCS, ensure BDR kit is available.
Hearing Damage	Train on high decibel danger zones and wear hearing protection.
Eye Damage	Verify eye protection during PCI, leaders enforce it during execution.
Burns	Be aware of TOW backblast and high heat exhaust zones, wear gloves when riding or operating equipment and weapons (changing barrels).
Falling From Moving Vehicle	Have proper load plan, use tie downs with snap links (M1 turret), wear seat belts (HMMWV, LMTV, 5-Ton), and ensure DVR is qualified.
Drowning After Water Entry	Train on vehicle exits and ensure Soldiers have passed the Combat Water Survival Test (CWST).
Fratricide by WPN System of Vehicle	Use day/night friendly recognition systems and proper fire control measures.
Disorientation	Ensure map is present, TC is briefed, and graphics are current.

D-85. Many Infantrymen are not familiar with the hazards that may arise during operations with combat vehicles. The most obvious of these include the dangers associated with main-gun fire, and the inability of combat vehicle crews to see people and objects near their vehicles. Leaders of heavy and Infantry units alike must ensure that their troops understand the following points of operational safety.

Discarding Sabot

D-86. Tank 120-mm sabot rounds and 25-mm BFV rounds discard stabilizing petals when fired, posing a downrange hazard for Infantry. The aluminum petals of the tank rounds are discarded in an area extending 70 meters to the left and right of the gun-target line out to a range of 1 kilometer (Figure D-9). The danger zone for plastic debris from BFV rounds extends 60 degrees to the left and right of the gun-target line, and out to 100 meters from the vehicle (Figure D-10). Infantrymen should not be in or near the direct line of fire for the tank main gun or BFV cannon unless they are under adequate overhead cover.

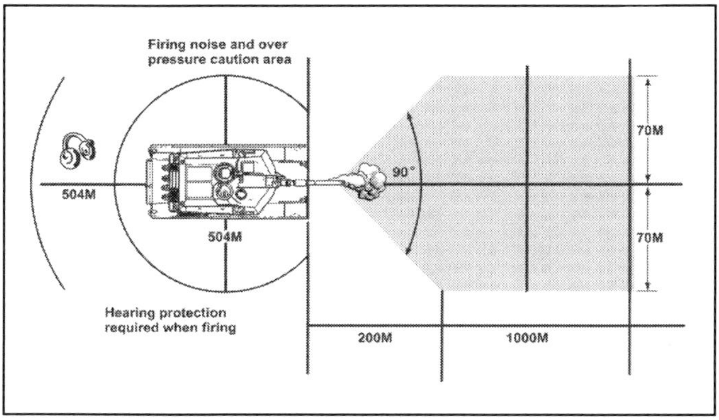

Figure D-9. M1 tank danger zone.

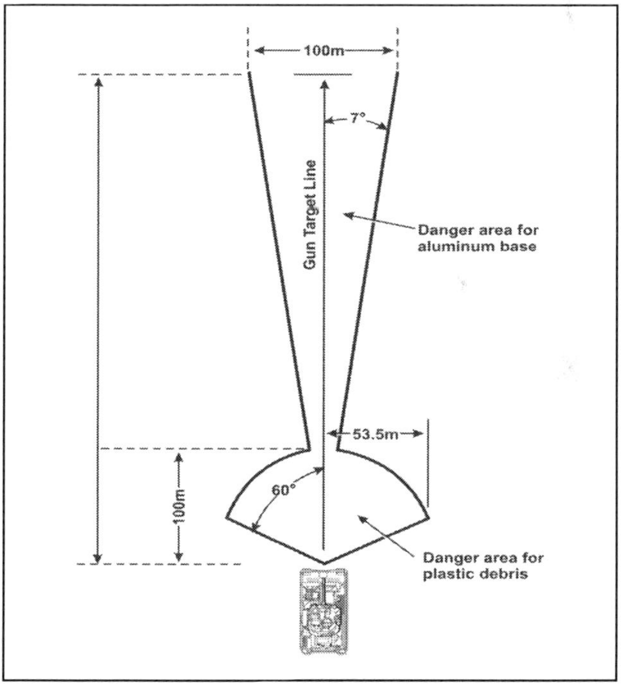

Figure D-10. BFV danger zone.

Ground Movement Hazards

D-87. Crewmen on combat vehicles have very limited abilities to see anyone on the ground to the side or rear of the vehicle. As a result, vehicle crews and dismounted Infantrymen share responsibility for avoiding the hazards this may create. Infantrymen must maintain a safe distance from heavy vehicles at all times. In addition, when they work close to heavy vehicles, Infantry Soldiers must ensure that the vehicle commander knows their location at all times, by establishing communication.

NOTE: Mounted and M1-series tanks are deceptively quiet and may be difficult for Infantrymen to hear as they approach. As noted, vehicle crews and Infantrymen share the responsibility of eliminating potential dangers in this situation.

M1 Exhaust Plume Hazard

D-88. M1-series tanks have an extremely hot exhaust plume that exits from the rear of the tank and angles downward. This exhaust is hot enough to burn skin and clothing. Infantrymen should therefore avoid the rear exhaust of the M1.

TOW Missile System

D-89. The TOW missile system can be employed on the BFV, the ASSLT HMMWV, and the ICV. The system has a dangerous area extending 75 meters to the rear of the vehicle in a 90-degree "cone." The area is divided into a 50-meter danger zone and a 25-meter caution zone (Figure D-11). In the 50-meter zone, serious casualties or fatalities are likely to occur from the blast and flying debris. Soldiers are safe in the 25-meter zone, provided they do not face the aft end of the launcher.

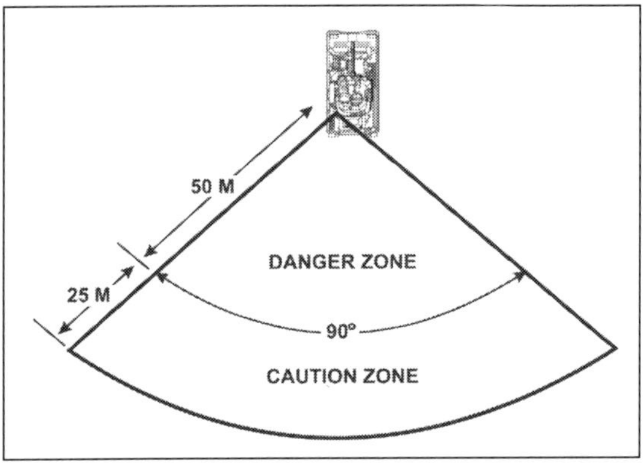

Figure D-11. BFV TOW backblast danger zone.

PREPARE

D-90. Key to planning operations with combat vehicles are rehearsals that gain the trust and confidence of vehicle crews and Infantryman.

REHEARSAL TECHNIQUES

D-91. A rehearsal is a session in which a staff or unit practices expected actions to improve performance during execution (FM 6-0). They are the cornerstone to any successful operation. Leaders are responsible to ensure that all combat vehicles attached to their units are incorporated into rehearsals. Rehearsals should include the tactical movement plan, and actions on the objective. Integration of combat vehicles is crucial because the relationship between vehicle crew men and Infantrymen may not be routine. Thorough rehearsals ensure that—

- Communications are established between the crewmen in the vehicles and Infantrymen prior to execution.
- Infantrymen are familiar with the technical capabilities and tactical movement of the vehicle.
- Vehicle crewmen understand the spatial relationship between the Infantrymen on the ground and their sectors of fire.
- Infantrymen understand the spatial relationship between the combat vehicles on the ground and their sectors of fire.

D-92. Following are five types of rehearsal techniques that can be used with combat vehicles: full-dress, reduced-force, terrain-model, sketch-map, and map.

Full-Dress Rehearsal

D-93. A full-dress rehearsal produces the most detailed understanding of the operation. It involves every participating Soldier, system, and combat vehicle. If possible, organizations execute full-dress rehearsals under the same conditions the force expects to encounter during an actual operation (weather, time of day, terrain—with use of live ammunition). The full-dress rehearsal is the most difficult to accomplish at higher echelons. At those levels, commanders develop a second rehearsal plan that mirrors the actual plan but fits the terrain available for the rehearsal. Mounted rehearsals involve actual movement of the combat vehicles along with the Infantrymen. Advantages of full-dress rehearsals include:

- Maintenance, communications, and weapon systems of the vehicles are checked during the rehearsal.
- Vehicle crewmen and Infantrymen gain a greater understanding of the battle space and spatial relationship of their operations.
- Leaders can ensure their graphic control measures are safe and effective.

D-94. The disadvantage of the full-dress rehearsal is it requires a larger area to conduct properly. Nevertheless, when METT-TC allows, leaders should always conduct a full-dress rehearsal.

Reduced-Force Rehearsal

D-95. A reduced-force rehearsal involves only key leaders of the organization and its subordinate units (squad leaders and vehicle commanders). It normally takes fewer resources than a full-dress rehearsal. Terrain requirements can be the same as for a full-dress rehearsal even though there are fewer participants. The platoon leader first decides the level of leader involvement. The selected leaders then rehearse the plan while traversing the actual or similar terrain. Leaders often use the reduced-force rehearsal technique to rehearse fire control measures for an engagement area during defensive operations. It may be used to prepare key leaders for a full-dress rehearsal, and may require developing a rehearsal plan that mirrors the actual plan, but fits the terrain of the rehearsal.

Terrain-Model Rehearsal

D-96. The terrain-model rehearsal takes less time and fewer resources than a full-dress or reduced-force rehearsal. (A terrain-model rehearsal takes proficient Soldiers to execute to standard.) It is the most popular rehearsal technique. An accurately-constructed terrain model helps subordinate leaders visualize the commander's intent and concept of operations. When possible, leaders place the terrain model where it overlooks the actual terrain of the area of operations (AO). However, if the situation requires more security, they place the terrain model on a reverse slope within walking distance of a point overlooking the AO. The model's orientation coincides with that of the terrain. The size of the terrain model can vary from small (using markers to represent units) to large (on which the participants can walk). A large model helps reinforce the participants' perception of unit positions on the terrain.

Sketch-Map Rehearsal

D-97. Leaders can use the sketch-map technique almost anywhere, day or night. Procedures are the same as for a terrain-model rehearsal, except the leader uses a sketch map in place of a terrain model. Effective sketches are large enough for all participants to see as each participant walks through execution of the operation. Participants move markers on the sketch to represent unit locations and maneuvers.

Map Rehearsal

D-98. A map rehearsal is similar to a sketch-map rehearsal, except the leader uses a map and operation overlay of the same scale used to plan the operation.

Exchange Information

D-99. Task organizations of units are likely to change during combat operations. When this occurs, some basic exchange information must occur to ensure success. First, an area must be chosen that provides security for the exchange to take place. The METT-TC may dictate the exchange must occur over FM or digital communications. However, when possible, leaders should meet and speak face to face. General exchange information includes:

- Number of personnel in the unit.
- Number of vehicles in the unit.
- Sensitive items list.
- Weapons capabilities.
- Logistical capability (particularly Class I, III, and V).
- Status/problems with logistics.
- Radio frequencies, call signs, and time hack.
- Graphics and overlays.
- Soldier uniform types.
- Day/night marking systems.
- Enemy situation updates.
- Terrain/route information.

Precombat Checks/Precombat Inspections

D-100. Infantry leaders may not always be proficient with the combat vehicles that are attached to their units for combat operations. Nevertheless, leaders are still responsible for ensuring that the combat vehicles and Soldiers in their unit are prepared to begin combat operations. Table D-9 contains a generic pre-execution checklist leaders can use to ensure that combat vehicles in their unit are prepared for combat operations.

Table D-9. Sample vehicle pre-execution checklist.

Vehicle Preparations	• Configured according to the secure load plan (personnel and equipment). • Vehicle refueled. • Water cans full, Class I stowed. • Equipment cleaned and stowed. • First-aid kit/combat-lifesaver bag complete and stowed. • Eye protection (sun, wind, dust goggles) stowed for exposed Soldiers. • Fire extinguisher secured and serviceable. • Slave cable secured and operational (at least one for each vehicle type). • One tow bar or recovery strap stowed for every two like-vehicle types. • Vehicle dispatched, technical manual (TM) present, vehicle tool kit stowed. • Basic load of ammunition stowed. • Rollover drill (water & land) complete. • CASEVAC drill complete. • Fire escape drill complete. • A basic Class IV load stowed (concertina wire, sandbags, pickets). • Battle damage repair kit (BDR) stowed. • Map of AO with current graphic control measures stowed.
Communications Equipment	• Radios operational, mounted, and secured; connections and receptacles cleaned and frequencies set. • Internal communication operational. • Extra hand microphones stowed. • Dismount kit for radios stowed. • Force XXI Battle Command, brigade and below (FBCB2); Blue Force Tracker (BFT); precision lightweight global positioning system receiver (PLGR); and inertial navigational system are operational, loaded with current graphics (if applicable), and communicating with other digital systems. • FM, integrated communications (ICOM), and communications checks are complete with higher, adjacent units, and subordinate units. • Vehicles' internal communication is operational. • Antennas present and operational, connections clean. • COMSEC (ANCD) equipment operational. • Telephones operational and stowed. • OE-254 complete, operational, and stowed. • All required nets entered and monitored.
CBRN	• M11 decontamination apparatus mounted and operational. • Hasty decontamination kit with DS-2 and nitrogen bottles stowed. • Automatic chemical alarm operational and mounted. • M256 kits stowed.
Optics	• Night-vision devices and binoculars cleaned, operational, and stowed for DVR/TC/GNR (night vision goggles [NVGs]) and driver's night vision block (VVS2 for BFV). • Weapons' optics operational, zeroed, clean, with extra batteries (if needed).
Maintenance	• Preventive maintenance checks (-10) and services conducted on all equipment. • DA Form 2404, *Equipment Inspection and Maintenance Worksheet*, completed on all equipment.
Firepower	• Weapons' mounts and turrets are operational and move freely. • Boresight complete (if needed). • All weapons cleaned and test-fired.

Security

D-101. Security must be maintained at all times during combat operations. Combat vehicles and Infantrymen provide complementary effects to one another with respect to security.

Combat Vehicles Securing Infantry

D-102. Combat vehicles can provide security to Infantrymen in many ways. In patrol bases and assembly areas, combat vehicles can use their weapon systems and night vision/thermal sights to provide early detection and a high volume of fire. During movement, combat vehicles can move to the front, rear, or flanks of the Infantry to provide protection from direct fire (tank, BFV, ICV, M1114 ASSLT HMMWV) and antipersonnel mines. They can also use their sights and weapon systems to detect and engage the enemy. On the objective, combat vehicles can dominate the terrain, provide security, and defeat a counterattack while the Infantrymen conduct actions on the objective.

Infantry Securing Combat Vehicles

D-103. Infantrymen can provide security to combat vehicles throughout an operation. In patrol bases and assembly areas, Infantrymen can secure the perimeter while combat vehicles conduct maintenance. During movement, Infantrymen can move to the front, rear, and flanks of combat vehicles to eliminate antiarmor threats and detect antitank mines. Infantrymen also clear defiles and other terrain that restrict the movement of combat vehicles. On the objective, Infantrymen can clear buildings, trenches, and bunkers while conducting EPW searches.

Sustainment

D-104. Infantry leaders should be aware of the robust logistical requirements of combat vehicles during combat operations. Normally, the leaders of attached vehicular units are responsible for bringing the majority of their logistical needs with them due to the austere and very different logistical support system of light Infantry units. Table D-10 provides leaders an overview of some logistical planning factors for combat operations.

Table D-10. Classes of supply considerations for combat vehicles.

Class I	Class I food requirements are determined based on the vehicular unit's personnel strength reports. This process may be complicated by unique mission requirements imposed on the team. This could include rapid changes in task organization or dispersion of subordinate team elements over a wide area.
Class II	Many Class II items required by tank and BFV crews such as specialized tools and flame retardant clothing may be difficult to obtain in a light organization. These items will usually come with the combat vehicles and should be checked by Infantry leaders.
Class III	The fuel and other POL products required by vehicular units are extremely bulky, so they present the greatest sustainment challenges in planning and preparing for light/heavy operations. Transportation support must be planned carefully. Planners must consider the placement of fuel heavy expanded mobility tactical trucks (HEMTTs) during all phases of the operation. Also, leaders must know their locations and the resupply plan. They must focus on general-use POL products such as lubricants that are not ordinarily used by light organizations. Vehicular units should stock their basic load of these items and make necessary resupply arrangements before attachment to the light Infantry unit.

Table D-10. Classes of supply considerations for combat vehicles (continued).

Class IV	Vehicular units do not have any unique requirements for barrier or fortification materials. The main consideration is any Class IV materials the vehicle commanders want may need loading and transport prior to attachment. Infantry leaders should be aware of the increased load capacity of combat vehicles and plan to utilize this asset to carry larger volumes of Class IV items such as sandbags, concertina wire, and pickets.
Class V	Along with POL products, ammunition for vehicular units presents the greatest transportation challenge in light/heavy operations. Class V requirements may include TOW missiles, 120-mm main gun rounds, 25-mm rounds, 40-mm MK19 rounds, .50 cal rounds, 7.62-mm link, 5.56-mm loose, and smoke grenades for smoke grenade launchers. Planning for Class V resupply should parallel that for Class III. Key considerations include anticipated mission requirements, and the availability of HEMTTs. Ammunition may be pre-stocked based on expected consumption rates.
Class VI	Vehicular unit operations create no unique requirements for personal demand items and sundries.
Class VII	Class VII consists of major end items. This includes entire vehicles such as a "float" tanks or BFVs units require as replacements for organic vehicles. The handling of these items requires thorough planning to determine transportation requirements and positioning in the scheme of the operation. Class VII items include smaller, but mission-essential items such as the boresight telescope for the BFV.
Class VIII	Vehicular units involved in light/heavy operations have no unique requirements for medical supplies. However, vehicular units may be capable of carrying more Class VIII supplies and provide standard/non-standard CASEVAC for combat operations.
Class IX	Class IX products (repair parts) are crucial to the sustainment of combat vehicles attached to Infantry units. Repair parts are essential during combat operations. Requirements for items on the team's parts load list (PLL) and ASL must be carefully considered before light/heavy operations begin. The vehicular unit may find it advantageous to prestock selected items in anticipation of its operational needs.

D-105. Combat vehicle sections attached to Infantry units may also receive resupply through a LOGPAC (logistical resupply) from their parent unit. These LOGPACs generally occur in the tailgate or service station method.

D-106. As directed by the commander or XO, the first sergeant establishes the company resupply point. He uses either the service station or tailgate method, and briefs each LOGPAC driver on which method to use. When he has the resupply point ready, the first sergeant informs the commander. The company commander then directs each unit or element to conduct resupply based on the tactical situation.

Service Station Method

D-107. The service station method allows vehicles with their squads to move individually or in small groups to a centrally-located resupply point (Figure D-12). Depending on the tactical situation, a vehicle, section, or platoon moves out of its position, conducts resupply operations, and then moves back into position. This process continues until the entire platoon has received its supplies. When using this method, vehicles enter the resupply point following a one-way traffic flow. Only vehicles that require immediate maintenance stop at the maintenance holding area. Vehicles move through each supply location. The crews rotate individually to eat, pick up mail and sundries, and refill or exchange water cans. When all platoon vehicles and crews have completed resupply, they move to a holding area. There, time permitting, leaders conduct a precombat inspection (PCI).

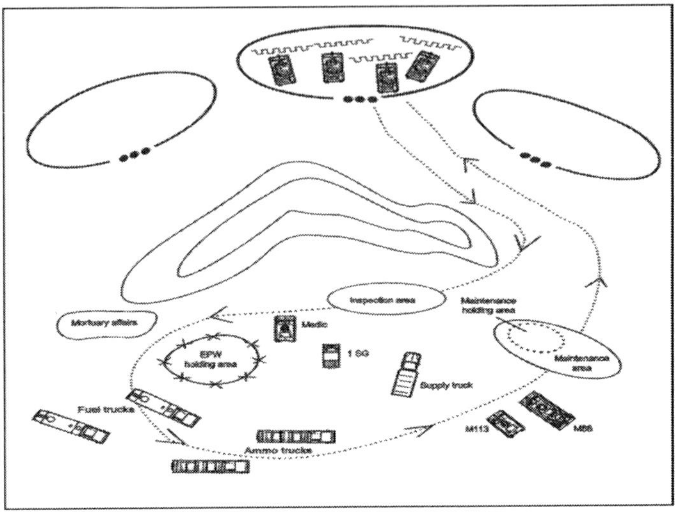

Figure D-12. Service station method.

Tailgate Method

D-108. In assembly areas, the first sergeant normally uses the tailgate method (Figure D-13). Combat vehicles remain in their vehicle positions or back out a short distance to allow trucks carrying Class III and V supplies to reach them. Individual Soldiers rotate through the feeding area. While there, they pick up mail and sundries and refill or exchange water cans. They also centralize and guard any EPW, and take Soldiers killed in action (KIA) and their personal effects to the holding area. Once there, the first sergeant assumes responsibility for them.

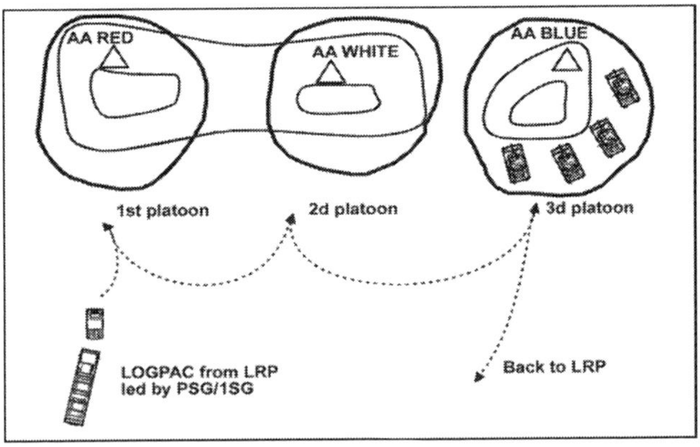

Figure D-13. Tailgate method.

Emergency Resupply

D-109. Occasionally (normally during combat operations), the unit might have such an urgent need for resupply that it cannot wait for a routine LOGPAC. Emergency resupply could involve CBRN equipment as well as Classes III, V, VIII, and water.

Prestock Resupply

D-110. In defensive operations, and at some other times, the unit will most likely need restocked supplies, also known as pre-positioned or "cached" resupply. Normally, the unit only pre-positions Class IV and V items, but they can also pre-position Class III supplies. However, they must refuel platoon vehicles before they move into fighting positions, while first occupying the battle position, or while moving out of their fighting position to refuel.

D-111. All levels must carefully plan and execute prestock operations. Every leader, down to vehicle commanders and squad leaders, must know the exact locations of prestock sites. During reconnaissance or rehearsals, they verify these locations. Leaders take steps to ensure the survivability of prestocked supplies. These measures include selecting covered and concealed positions and digging in the prestock positions. The leader must have a removal and destruction plan to prevent the enemy from capturing pre-positioned supplies.

D-112. During offensive operations, the unit can pre-position supplies on similar combat vehicles well forward on the battlefield. This works well if the unit expects to use a large volume of fire, with corresponding ammunition requirements, during a fast-moving operation.

MAINTENANCE AND RECOVERY

D-113. Recovery operations and maintenance are crucial components of the leader's plan when working with combat vehicles.

Maintenance

D-114. Leaders must plan for regular maintenance halts throughout extended operations. Combat vehicles require regular maintenance to perform consistently throughout combat operations. Combat vehicles can become non-mission capable (NMC) due to a number of variables including, direct and indirect enemy fire, mines and IEDs, vehicle accidents, and parts failure. Infantry leaders should enforce regular preventive maintenance checks and services (PMCS) of all combat vehicles attached to their unit. PMCS is operator-level maintenance conducted before, during, and after equipment operations. Comprehensive PMCS identifies actual and potential problems and ensures repairs are made in a timely manner to minimize vehicle downtime. Early detection and correction of these faults can decrease the possibility of the combat vehicle breaking down during combat operations and prevent minor faults from deteriorating into major faults. It is the vehicle crew's responsibility to conduct PMCS. It is the leader's job to ensure the PMCS is conducted regularly and to standard.

D-115. Leaders should plan vehicle security for the vehicle crews as they conduct PMCS, based on the enemy situation. Additionally, leaders should establish a maintenance rotation to ensure that all of their combat vehicles are not conducting maintenance at the same time. This will maximize the combat power of the unit. Leaders should also—

- Verify that all current and updated technical manuals and references are available or requisitioned for unit assigned equipment.
- Verify that all tools, POL, personnel, and other resources are available for PMCS.
- Observe operators performing PMCS at prescribed intervals.
- Review maintenance forms and reporting procedures for accuracy and completeness.
- Verify that the operator has correctly identified and corrected, or recorded, faults on DA Form 2404, *Equipment Inspection and Maintenance Worksheet*.
- Confirm that NMC faults are corrected before dispatch.

D-116. Leaders should also plan for the possibility of combat vehicles requiring maintenance at a level greater than the crew is equipped or trained to conduct. This often requires specially trained mechanics and equipment that is organic to the parent unit of the combat vehicle attachment. Leaders should plan for two possibilities. One, the maintenance team moves to the combat vehicles. This may require additional security and or escorts from the Infantry. Two, the combat vehicles must move to the maintenance team. Maintenance teams are often located at the parent unit's UMCP (unit maintenance collection point). Infantry leaders may have the responsibility of providing security or escort duties. Additionally, leaders should plan on the NMC vehicles to be absent from their task organization if a major maintenance fault is discovered.

Recovery Operations

D-117. Leaders are responsible for recovery operations that occur within their units. However, leaders should consult the senior officer or non-commissioned officer of the attached vehicular unit for the technical aspects of the recovery operation. Infantry leaders must have a thorough recovery plan that ensures their combat vehicles can be recovered throughout the operation. Recovery operations extricate damaged or disabled equipment and move it to locations where repairs can be made. Recovery is the primary responsibility of the using unit. The primary role of the Infantry during recovery operations is to provide security and assist with the recovery under supervision of the vehicle crew.

D-118. Recovery operations can be very dangerous. Recovery should be conducted under the supervision of the Infantry leader, using the experience and technical competence of the combat vehicle crew. The general rule in recovering a vehicle that is simply NMC in simple terrain is like vehicles can recover each other. For example, tanks recover tanks, and BFVs recover BFVs. However, there are vehicles specifically designed for recovery operations. These vehicles should be used if vehicles become stuck, flipped over, or severely damaged. The M-936 medium wrecker can be used to recover some wheeled vehicles, to include the assault HMMWV. The M88A1 medium recovery vehicle (MRV) is a full-tracked armored vehicle used to perform battlefield rescue and recovery missions. The M88A1 MRV performs hoisting, winching, and towing operations in support of recovery operations and evacuation of heavy tanks and other tracked combat vehicles. It has a fuel/defuel capability and is fully equipped to provide maintenance and recovery support for the main battle tank family and similar vehicles. These functions can be performed in all types of terrain during all weather conditions.

Appendix E

Helicopter Movement

Infantry platoons may conduct air movement operations to pick up patrols by helicopter, re-supply with helicopters, or evacuate casualties. This appendix discusses general helicopter information including, the five stages of an airmobile operation, how to organize the unit for a helicopter move, and how to select and secure a pickup zone.

SECTION I — CHARACTERISTICS OF HELICOPTERS

E-2. Helicopters most commonly used by Infantry platoons are the UH-60, Blackhawk and the CH-47, Chinook (Table E-1). See FM 90-4, *Air Assault Operations*, for information on air movement and air assault operations, and FM 3-21.38, *Pathfinder Operations*, for information on pathfinder operations.

Table E-1. Helicopter characteristics.

	UH-60A	UH-60L	CH-47D
Passenger capacity (seats in)	11	11	33
Passenger capacity (seats out)	18	18	60
Max cargo weight	8,500 lbs.	8,500 lbs.	26,000 lbs.
Cargo hook capacity	8,000 lbs.	9,000 lbs.	26,000 lbs. (center hook) 17,000 lbs (fore & aft hook) 25,000 lbs (fore & aft hook combined)
NOTE: Actual allowable cargo load (ACL) may be determined by ground and aviation unit commanders.			

CAPABILITIES

E-3. Under normal conditions, helicopters can climb and drop at steep angles. This allows them to fly from and into confines and unimproved areas. Other helicopter capabilities include—

- Transporting cargo as an internal load or external (sling) load and delivering to unit areas not supplied by any other means.
- Overflying or bypassing obstacles or enemy in order to reach objectives otherwise inaccessible.
- Flying at low altitudes to achieve surprise and deceive the enemy using hills and trees for cover and concealment.
- Operating under limited visibility conditions.

E-4. It is ALWAYS preferred to use a helicopter for loading or unloading of troops and equipment. If terrain prevents the helicopters from landing, troops and their combat equipment can be unloaded while hovering a short distance above the ground with troop ladders, rappelling ropes, or fast ropes. If the aircraft can hover low enough, Soldiers may jump out. The troop ladder (or in limited applications- a SPIES rope) can also be used to extract troops when the helicopter cannot land.

LIMITATIONS

E-5. The large amount of fuel used by helicopters may limit their range and allowable cargo load (ACL). Other helicopter limitations include:

- Extreme weather conditions such as fog, hail, sleet, ice, or winds (40 knots or more) and gusty winds (gusts up to 15 knots above a lull) will prevent the use of helicopters.
- Engine and rotor noise may compromise the secrecy of the mission.
- Limited size or number of suitable landing zones (LZs).
- The load-carrying capability of helicopters decreases with increases of pickup zone (PZ)/landing zone (LZ) altitude, humidity, and temperature.
- Vulnerability to enemy air defense systems and small arms fire.

SECTION II —AIRMOBILE OPERATIONS STAGES

E-6. There are five stages to an air movement operation (Figure E-1). The ground tactical plan is the key planning phase. All other planning is conducted in a backward manner from it. The five stages of this reverse planning sequence are—

(1) Ground tactical plan (GTP).
(2) Landing plan.
(3) Air movement plan.
(4) Loading plan.
(5) Staging plan.

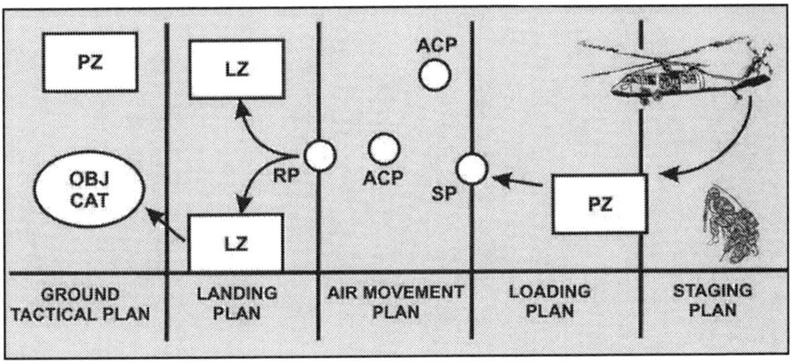

Figure E-1. Air movement through the five stages.

E-7. The ground tactical plan drives the entire mission. Convenience of landing considerations is subordinate to putting units on the ground where they can fight. The five plans tie together in this way:

- The ground tactical plan drives the sequence of arrival and amount of combat power onto the LZs.
- Combat power arriving at available LZs to accomplish the mission becomes the landing plan.
- Moving troops and equipment to LZs on the designated flight routes becomes the air movement plan.
- Getting troops and equipment from current friendly locations to the designated LZs dictate the loading plan and PZ locations.
- The PZ loading plan designates the requirements that become the staging plan to move friendly troops onto the PZ when and where needed.

GROUND TACTICAL PLAN

E-8. The ground tactical plan for an air movement operation contains the same essential elements as other Infantry missions, but differs in one area: it is prepared to capitalize on the speed and mobility of the aircraft to achieve surprise. Units are placed on or near the objective to immediately seize the objective.

The ground tactical commander, in accordance with doctrine and METT-TC, determines his ground tactical plan. The five stages of the reverse planning sequence cannot be developed independently. In addition to standard planning considerations for actions on the objective, the commander's plan should include—

- H-hour times.
- Primary and alternate LZ(s).
- Means of identifying LZ(s).
- Task organization.
- Chalk configurations.
- Special equipment required (such as kick-off bundles, ropes).
- Attack aviation assets available and missions.
- Suppression of enemy air defenses (SEAD).
- Landing formations.
- Offloading procedures.

LANDING PLAN

E-9. Unlike approaching an objective in armored vehicles, Soldiers in helicopters are most vulnerable when landing, and are potentially more vulnerable to enemy fire than if they were on the ground. Suppressive fires are employed to deny the enemy unhindered access to the landing forces, so the timing of fires is critical to the success of the landing.

E-10. The ground tactical commander's plan typically results in two types of landing plans: on the objective (within enemy small arms range), or away from the objective (outside of enemy small arms range). Landing away from the objective is the more common of the two landing plans. The mobility and speed of the helicopters further enables the unit to land to the rear of the objective and aid in the element of surprise and confusion during any subsequent assault. Table E-2 lists factors considered when constructing the landing plan. Regardless of the landing plan used, the Infantry platoon must land ready to fight.

Table E-2. Landing plan considerations.

Factors	Land away from the objective (outside of enemy small arms range) when...	Land on the objective (within enemy small arms range) when...
Mission	The mission is enemy force-oriented.	The mission is terrain-oriented.
Enemy	There is incomplete intelligence on enemy disposition.	There is precise intelligence on enemy dispositions.
Terrain	There is incomplete intelligence on terrain (especially LZs) and weather, or there are no suitable LZs on or near the objective.	There is precise intelligence on terrain (especially LZs) and weather, and there are suitable LZs on the objective.
Troops available	Conditions are not set.	Conditions are set and verified.
Time	There is time available to develop the situation.	Time is critical to secure the objective.
Intent	The unit plan is to arrive at the LZ prepared to move out quickly and ensure rapid advance on objective.	The unit has a plan to establish continuous suppression of any enemy fire immediately upon landing while aggressively assaulting to secure the objective.

E-11. Good PZs and LZs allow for helicopter insertion or extraction without exposing the unit or aircraft to unnecessary risks. Three-hundred-and-sixty-degree security must be maintained at all times. Preparatory and supporting fires are planned to suppress the enemy as the aircraft land on the LZ or the PZ. The control and distribution of all available means to suppress the enemy at a most vulnerable time is imperative. Fires should be focused along the base of the exit tree line (right door exit shoots at the right tree line). Regardless of threat data, suppressive fires are planned, although not necessarily executed, for every primary and alternate PZ or LZ. Whether a PZ or LZ, units establish a defensive posture and employ local security measures as required, shifting as necessary when chalks land or depart.

E-12. The ground tactical commander, in coordination with the supporting aviation unit, selects the location of helicopter PZs and LZs. There are many factors that leaders must consider when choosing appropriate LZs and PZs. These requirements are covered by aviation unit SOPs or are prearranged by the aviation unit commander in coordination with the pathfinder leader. The final decision concerning minimum landing zone requirements rests with the aviation unit commander. Among those factors considered is the number, type and landing formation of the helicopters, surface conditions, obstacles, ground slope, approach and departure route, atmospheric conditions, and type of loads.

Number, Type, and Landing Formation of the Helicopters

E-13. The number, type, and landing formation of helicopters determine the minimum landing space requirement and total size of the LZ and PZ. It may be necessary to have two PZs or LZs, or to land the necessary aircraft one at a time. Differing aircraft may have different landing point size requirements. A single UH-60 requires a touch down point (cleared area) of 50 meters in diameter without sling load, and 80 meters with sling load. A CH-47 requires a touchdown point of 80 meters in diameter without sling load, and 100 meters with sling load.

Surface Conditions

E-14. The surface at the landing point must be firm enough to keep helicopters from bogging down, raising too much dust, debris, or blowing snow. Troops remove loose debris that may damage the rotor blades or engines.

Obstacles

E-15. Helicopters should not land on a landing point that includes obstacles. An obstacle in this case is defined as any object or terrain feature (anything 18 inches high or deep) that could cause damage to the airframe or rotor system of the aircraft, or prevent safe landing. Objects or equipment placed on the PZ/LZ in conjunction with the operation (such as landing lights and slingloads) are not included. Obstructions (for example, rocks, stumps, and holes) that cannot be removed must be clearly marked. Methods of marking obstacles that cannot be cleared for both day and night must also be considered.

Ground Slope

E-16. When the slope is less than 7 percent (4 degrees), helicopters may land in any direction. Where ground slope is from 7 to 15 percent (4 to 8 degrees), aircraft must land and park sideslope or upslope. Helicopters with skids as landing gear may not land, but must terminate at a hover. If ground slope is greater than 15 percent (8 degrees), helicopters cannot land safely, and may sometimes hover to drop off Soldiers or supplies.

Approach and Departure Routes

E-17. The direction of departure and landing should be generally into the wind, over the lowest obstacle, and along the long axis of the LZ. If there is only one satisfactory approach direction because of obstacles or the tactical situation, most helicopters can land with a slight crosswind or tailwind. PZs or LZs should be free of tall trees, telephone and power lines, and similar obstructions on the approach and departure ends. Use an obstacle ratio of 10:1 when determining how much additional space is required for landing and take-off. A helicopter needs 100 meters of horizontal clearance from a 10-meter tree for takeoff or landing.

Atmospheric Conditions

E-18. As the humidity, altitude and temperature increase, the performance capability of aircraft decrease. This result in greater fuel consumption, lower ACLs, and larger LZ requirements. These limitations/considerations should be highlighted by aviation LNOs during planning.

TYPE OF LOAD

E-19. Most helicopters cannot take off or land vertically when fully loaded, so a larger LZ/PZ and better approach and departure routes may be required for fully loaded aircraft. LZs must be larger for aircraft delivering sling loads compared to aircraft delivering internal loads and Soldiers.

OTHER CONSIDERATIONS

E-20. Other considerations when selecting PZs and LZs include:

- Location in relation to objective.
- Ability of the unit to secure.
- Enemy location, capabilities, and strength.
- Cover and concealment.
- Identification from air.
- Weather and its effect.
- Visibility (darkness, fog, snow, dust, etc)

AIR MOVEMENT PLAN

E-21. Air movement involves flight operations from PZ, to LZ, and back. The Infantry leader and all chalk leaders should maintain the following items:

- A marked air route map.
- Compass/GPS.
- Watch synchronized with the flight crew and ground element.
- Air movement table, PZ sketch, and LZ sketch.
- Call signs and frequencies for all aviation and ground units involved in or around the operation.
- Backpack FM radio.

E-22. The air movement plan includes en-route security for the lift aircraft by attack aviation. It also includes, false insertions to deceive the enemy, suppression of enemy air defense positions along the flight route, and emergency procedures in the event an aircraft is lost en route due to maintenance or enemy fire.

E-23. To maximize operational control, aviation assets are designated as lifts, serials, and loads. A *lift* is all utility and cargo aircraft assigned to a mission. Each time all assigned aircraft pick up troops and/or equipment and set them down on the LZ, one lift is completed. The second lift is completed when all aircraft place their second load on the LZ. There may be times when a lift is too large to fly in one formation. In such cases, the lift is organized into a number of serials. A *serial* is a tactical group of two or more aircraft and separated from other tactical groupings within the lift by time or space. The use of serials may be necessary to maintain effective control of aviation assets when the capacity of available PZs or LZs is limited or to take advantage of available flight routes. The personnel and equipment designated to be moved a single aircraft is called a load or *chalk*. Each chalk must have a chalk leader who ensures that every man in his chalk gets on and off the helicopter, that everything is ready to load, and that everything gets loaded and unloaded correctly. The chalk leader should sit in the aircraft where he can best stay oriented during flight and where he can get off quickly at landing sites to control his men. Figure E-2 shows the relationship between a chalk, serial, and lift.

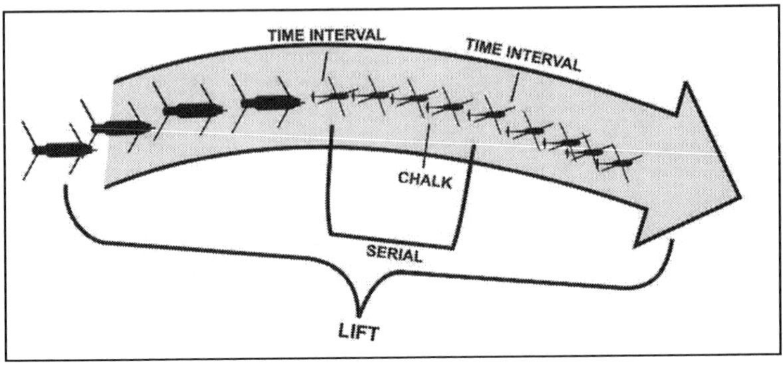

Figure E-2. Lifts, serials, and chalks.

LOADING PLAN

E-24. Air movement operations do not succeed on the PZ, but the failure of the mission can occur there. Therefore, PZs must be established to run efficiently. Assault forces are organized on the PZ, not the LZ. Every serial must be a self-contained force that understands what it must do on landing at either the primary or alternate LZ, and later in executing the ground tactical plan.

E-25. Before an Infantry platoon is lifted by helicopter, it must be organized for the move. The load (amount of men, weapons, equipment, and ammunition) that can be carried by a helicopter varies. It is based on the type of helicopter used, configuration of the helicopter, temperature, altitude of the PZ or LZ, humidity, and fuel load. What can be carried is the allowable cargo load (ACL). This is one of the main factors considered when planning aircraft loads. When the Infantry platoon is alerted for a movement by helicopter, the allowable cargo load will be given to the leader. The unit can then be organized into chalks/loads based on the given allowable cargo load of each type of aircraft. Page E-5 displays an example of a "Tadpole Diagram" (Figure E-3) that is used to plan and organize the chalks and loads.

LIFT	SER	CHLK	ARRIVAL TIME	PZ	T/0 TIME	INGRESS ROUTE	CHALK #1	CHALK #2	CHALK #3	CHALK #4	CHALK #5	TOTAL PAX	RMKS	LZ	LZ TIME	EGRESS ROUTE
1	1	1-5	H-3:00:00	ALEX	H-0:18:28	JUPITOR						90		AKBAR	H-HOUR	MARS
Unit							B	B	B	B	B					
#Pax							B	B	B	B	B					
Equipment																
1	2	6-10	H-3:00:00	ALEX	H-0:16:28	JUPITOR						72		AKBAR	H+00:02:00	MARS
Unit							B	B	D	TAC	TA					
#Pax							16	16	16	16	1XM338					
Equipment																
1	3	11-15	H-3:00:00	ALEX	H-0:14:28	JUPITOR						30		PHOENIX	h+00:04:00	MARS
Unit							B	C	C	C	C					
#Pax							18	18	18	18	18					
Equipment																
1	4	16-20	H-3:00:00	ALEX	H-0:12:28	JUPITOR						30		PHOENIX	h+00:06:00	MARS
Unit							C	C	C	C	C					
#Pax							18	18	18	18	18					
Equipment																
1	5	21-24	H-5:00:00	ALEX	H-0:10:28	JUPITOR						12		AKBAR	H+00:08:00	MARS
Unit							D	C	C	C						
#Pax							3	3	3	3						
Equipment							1XM1121	1XM1121	1XM1121	1XM1121						
										LIFT TOTAL PAX		354				
										LIFT TOTAL EQUIP	4XM1121 1XM338					

Figure E-3. Tadpole diagram.

E-26. The leader maintains the tactical integrity and self-sufficiency of each aircraft load as much as possible. He maintains tactical integrity by keeping squads and fire teams intact on chalks and the platoon intact within a serial. He maintains self-sufficiency by loading a machine gun and its ammunition and crew, or an entire antiarmor team on the same aircraft. Key men, weapons, and equipment should be cross-loaded among different aircraft. Platoon leaders and platoon sergeants should fly on separate helicopters. So should machine gun teams. This kind of cross-loading can prevent the loss of control or unit effectiveness in the event a helicopter is lost.

E-27. The leader prepares a load plan for the platoon that tells each man which aircraft he is to get in and who the chalk leader is.

E-28. The chalk leader tells each man in his chalk where to sit, what to do in case of emergency, and what to do when the aircraft lands.

STAGING PLAN

E-29. As part of the staging plan, Soldiers must mark obstacles on the PZ in both day and night operations. In daylight, troops use red panels or other easily seen objects and materials to mark obstacles. In night operations, units use signal lights to avoid security problems. Visible or infrared lights can be used, but the choice must be coordinated with the lift unit. In any case, pilots should be advised of obstacles whether marked or unmarked.

E-30. For a night operation, Soldiers can use flashlights, chemical lights, or expedient devices to show the direction of landing and to mark aircraft landing points. However, pilots cannot see blue or green chemical lights under aviator night vision goggles. Therefore, blue and green chemical lights should be used for Infantry staging purposes only. Always use red, orange, yellow, or infrared for aircraft positions.

E-31. There are many ways to mark a PZ or LZ at night. The inverted "Y" is one way. An inverted "Y" indicates the landing point of the lead aircraft and its direction of approach. The formation used by the aircraft will determine how to place the lights for other aircraft. Table E-3 lists examples of PZ markings during day and night operations.

E-32. Security on the PZ is of the utmost importance. It may be conducted by a separate unit that is not conducting the air movement. At a minimum, the Infantry platoon secures itself and maintains a high state of readiness while awaiting arrival of the aircraft.

E-33. Whenever possible, Infantry platoons should conduct "cold-load" rehearsals prior to conducting an air movement. This can be done on the actual aircraft (best method), or using field expedient methods. Chalk leaders arrange their chalk considering the last one to load the aircraft is the first one off. Soldiers are designated to open/close doors, secure and unload equipment, and understand the direction they will move or secure once getting off the aircraft. If the lift aircraft arrives at the LZ before execution of the mission, the chalk leader should conduct face-to-face coordination with the air crew. This is done to ensure everyone knows the PZ on-load and LZ off-load procedures. It also avoids confusion and speeds actions on the LZ, allowing the aircraft to spend minimal time on the ground. Information that should be coordinated include: which door(s) will be used to load and unload; actions if the aircraft takes enemy fire en route and on the ground; special safety considerations; crash procedures; location of the primary and alternate LZs; direction of landing; time warnings with hand and arm signals inside the aircraft; and any other special mission requirements.

Table E-3. Example PZ marking methods.

Position	Day	Night
PZ entry point for Infantry	NCOIC, signage	NCOIC, two blue chem lights
PZ control point	HMMWV and VS-17 panel	Green chem lights on antennae
Chalk stage points	Guides, signage	Guide, blue chem light
Lead touchdown points	VS-17 panel	Inverted "Y," infrared lights
Aircraft touchdown points	VS-17 panel	Red chem light per aircraft
Obstacles	FM communication	Red chem light ring around obstacle
Loads to be picked up	Hook-up teams stationed on loads	Swinging infrared chem light per load

SECTION III — SAFETY

E-34. Infantry leaders must enforce strict safety measures when working with helicopters. Measures include:

- Avoid the tail rotor. Never approach or depart to the rear of a helicopter except when entering or exiting a CH-47. Approach from 3 or 9 o'clock is preferred when using UH-60s.
- Keep a low body silhouette when approaching and departing a helicopter, especially on slopes.
- Keep safety belts fastened when helicopter is airborne.
- Keep muzzle pointing down and on safe.
- Keep all radio antennas down and secure.
- Keep hand grenades secured.
- Do not jump from a hovering helicopter until told to by an air crew member.

Appendix F

Obstacle Reduction and Employment

The Army defines mobility operations as "those activities that enable a force to move personnel and equipment on the battlefield without delays due to terrain or obstacles." Infantry units must be able to mass forces quickly at a chosen place and time to accomplish their assigned mission. Mobility is critical to achieving this situation. Mobility operations require the maintenance of force movement activities over great distances for extended periods of time. The Infantry platoon must be proficient in the reduction of obstacles to enable the movement of combat power through any obstacles while continuing to the objective.

Countermobility operations involve the augmentation of existing obstacles through the use of reinforcing obstacles that are integrated with direct- or indirect-fire systems. When employed effectively, this type of operation will disrupt, fix, turn, or block the enemy's ability to maneuver while giving the Infantry platoon opportunities to exploit enemy vulnerabilities. To be effective in countermobility operations, the Infantry platoon must be proficient in the employment of obstacles.

SECTION I — OBSTACLE TYPES AND CATEGORIES

F-1. An obstacle is any obstruction that is designed or employed by friendly or enemy forces to disrupt, fix, turn, or block the movement of the opposing force. Obstacles can impose additional losses in personnel, time, and equipment. It is therefore vital that Infantry leaders and Soldiers be knowledgeable in the various types of obstacles; not only to employ them effectively, but to reduce them when employed by enemy forces.

F-2. This appendix provides information on the types of obstacles (Section I), reduction of enemy obstacles (Section II), and employment of friendly obstacles (Section III). See FM 90-7, *Combined Arms Obstacle Integration*, for complete information on obstacles, and FM 20-32, *Mine/Countermine Operations*, for complete information on mine and countermine operations.

F-3. U.S. forces' employment of certain obstacles, booby traps, and antihandling devices are governed by the Law of Land Warfare and any applicable international laws. Rules governing their employment are also listed in the appropriate sections in this appendix.

F-4. There are four general types of obstacles. Each type is determined by its distinct battlefield purpose and the overall concept of the operation.

(1) *Protective obstacles* are employed to protect Soldiers, equipment, supplies, and facilities from enemy attacks or other threats.

(2) *Tactical obstacles* directly affect the opponent's maneuver in a way that gives the defending force a positional advantage.

(3) *Nuisance obstacles* impose caution on opposing forces. They disrupt, delay, and sometimes waken or destroy follow-on echelons.

(4) *Phony obstacles* deceive the attacking force concerning the exact location of real obstacles. They cause the attacker to question his decision to breach and may cause him to expend his reduction assets wastefully. Phony minefields are used to degrade enemy mobility and preservefriendly mobility. Intended to simulate live minefields and deceive the enemy, they are used when lack of time, personnel, or material prevents use of actual mines. They may also be used as gaps in live minefields. To be effective, a phony minefield must look like a live

minefield, so Soldiers must bury metallic objects or make the ground look as though objects are buried.

F-5. Obstacles are employed by both friendly and enemy forces. The two main categories of obstacles are:

(1) Existing obstacles.
(2) Reinforcing obstacles.

EXISTING OBSTACLES

F-6. Existing obstacles are those natural or cultural restrictions to movement that are part of the terrain. Existing obstacles can be reinforced into more effective obstacles. They are normally in defilade from enemy observation (located where observation and fires can prevent the opposing force from breaching them), and are difficult to bypass. Existing obstacles include steep slopes, escarpments, ravines, rivers, swamps, deep snow, trees, and built-up areas.

REINFORCING OBSTACLES

F-7. Reinforcing obstacles are used by both friendly and enemy forces to tie together, anchor, strengthen, and extend existing obstacles. Careful evaluation of the terrain to determine its existing obstructing or canalizing effect is required to achieve maximum use of reinforcing obstacles. Installation time and manpower are usually the two most important factors. The four types of reinforcing obstacles are:

(1) Land mines.
(2) Constructed obstacles.
(3) Demolition obstacles.
(4) Improvised obstacles.

LAND MINES

F-8. Land mines are explosive devices that are emplaced to kill, destroy, or incapacitate personnel/ equipment, and to demoralize an opposing force. A mine (or other explosive device) is detonated by the action of its target, the passage of time, or other controlled means (Figure F-1). There are two types of land-based mines: antitank (AT); and antipersonnel (AP). They can be employed in quantity to reinforce an existing obstacle within a specified area to form a minefield, or they can be used individually to reinforce nonexplosive obstacles such as wire. FM 20-32 is the primary reference for mine and countermine operations. See Section II for more information on reducing mine obstacles and Section III for more information on employing them.

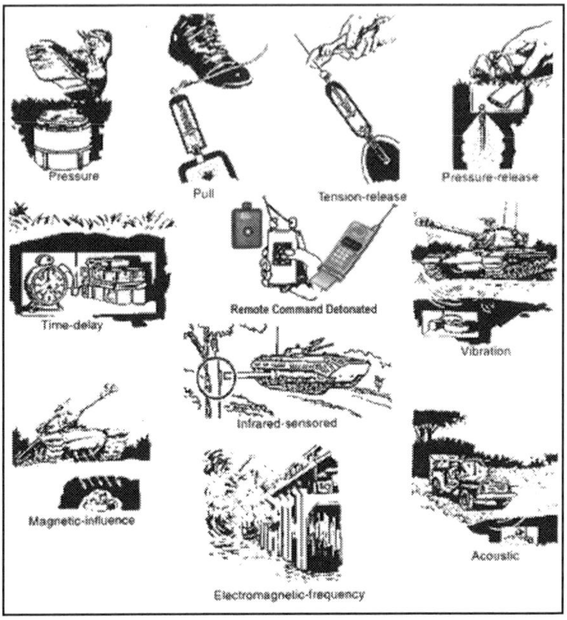

Figure F-1. Methods of actuating explosives.

CONSTRUCTED OBSTACLES

F-9. Units create constructed obstacles with manpower or equipment without the use of explosives. FM 5-34 covers constructed obstacles in detail. Examples of constructed obstacles include:

- **Ditches.** Ditches across roads and trails are effective obstacles. Large ditches in open areas require engineer equipment.
- **Log Hurdles.** Log hurdles act as "speed bumps" on roads.. They are easily installed and are most effective when used in conjunction with other obstacles.
- **Log Cribs.** A log crib is constructed of logs, dirt, and rocks. The logs are used to make rectangular or triangular cribs that are filled with dirt and rock. These are used to block narrow roads and defiles. Unless substantially built, log cribs will not stop tanks.
- **Log Posts.** Log posts embedded in the road and employed in depth can effectively stop tracked vehicles. If they are not high enough to be pushed out of the way, posts can cause a tracked vehicle to throw a track if it tries to climb over. If employed with wire and mines, they can also slow Infantry.
- **Wire Entanglements.** Wire entanglements impede the movement of dismounted Infantry, and in some cases, tracked and wheeled vehicles. Triple standard concertina is a common wire obstacle. However, there are other types, such as double apron, tanglefoot, and general-purpose barbed-tape obstacles. Figures F-2A and F-2B illustrate examples of wire and log obstacles. The materials used in constructing wire entanglements are relatively lightweight (compared to other obstacles) and inexpensive, considering the protection they afford.

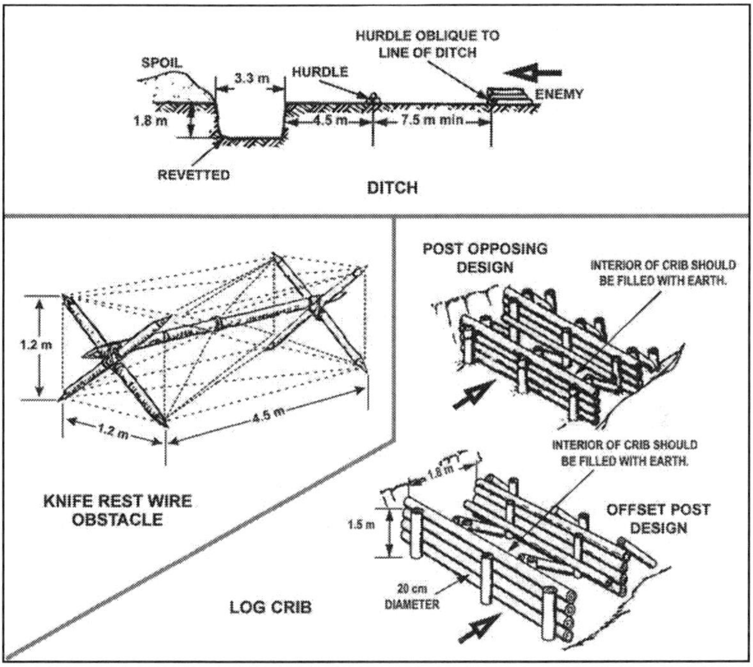

Figure F-2A. Constructed wire and log obstacles.

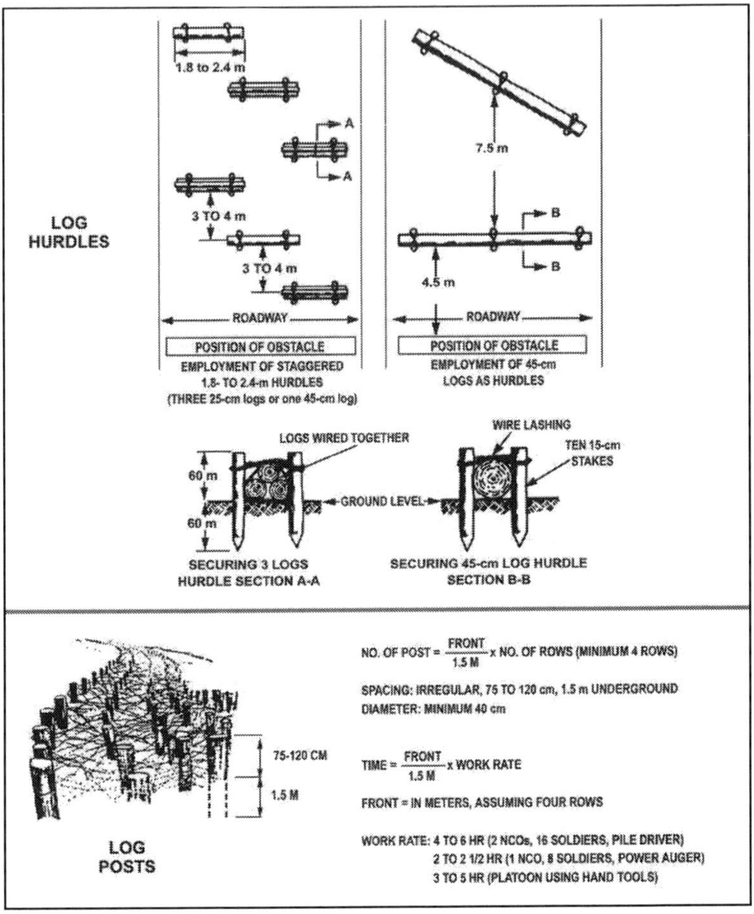

Figure F-2B. Constructed wire and log obstacles.

DEMOLITION OBSTACLES

F-10. Units create demolition obstacles by detonating explosives. FM 5-250, *Explosives and Demolitions*, covers demolitions in detail. There are many uses for demolitions, but some examples are road craters and abatis.

F-11. *Road craters* are effective obstacles on roads or trails if the areas on the flanks of the crater are tied into steep slopes or mined areas. Road craters can compel the opposing force to use earthmoving equipment, blade tanks, or mechanical bridging assets.

F-12. *Abatis* are only effective if large enough trees, telephone poles, or other similar objects are available to stop the opposing force. An abatis is an obstacle created by cutting down trees so their tops are crisscrossed and pointing toward the expected enemy direction. It is most effective for stopping vehicles in a forest or narrow movement routes. This obstacle may be reinforced with mines.

IMPROVISED OBSTACLES

F-13. Improvised obstacles are designed by Soldiers and leaders with imagination and ingenuity when using available material and other resources. An example of obstacles in urban terrain is shown in Figure F-3. Improvised obstacles include the following:

- **Rubble.** Rubble from selected masonry structures and buildings in a built-up area will limit movement through an area and provide fortified fighting positions.
- **Battle Damage.** Damaged vehicle hulks or other debris are used as roadblocks.
- **Flooding.** Flooded areas are created by opening floodgates or breaching levees.

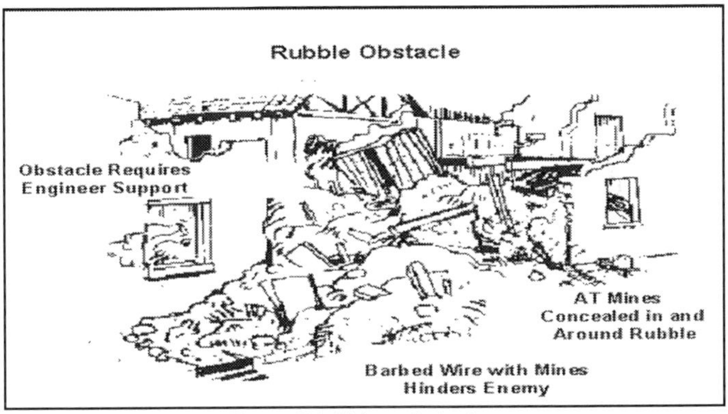

Figure F-3. Urban obstacles.

SECTION II — OBSTACLE REDUCTION

F-14. Suppress, obscure, secure, reduce, and assault (SOSRA) are the breaching fundamentals that must be applied to ensure success when breaching against a defending enemy. These obstacle reduction fundamentals will always apply, but they may vary based on the specific METT-TC situation.

BREACHING FUNDAMENTALS

SUPPRESS

F-15. Suppression is a tactical task used to employ direct or indirect fires or an electronic attack on enemy personnel, weapons, or equipment to prevent or degrade enemy fires and observation of friendly forces. The purpose of suppression during breaching operations is to protect forces reducing and maneuvering through an obstacle. Effective suppression is a mission-critical task performed during any breaching operation. Successful suppression generally triggers the rest of the actions at the obstacle. Fire control measures ensure that all fires are synchronized with other actions at the obstacle. Although suppressing the enemy overwatching the obstacle is the mission of the support force, the breach force should provide additional suppression against an enemy that the support force cannot effectively suppress.

Obscure

F-16. Obscuration must be employed to protect forces conducting obstacle reduction and the passage of assault forces. Obscuration hampers enemy observation and target acquisition by concealing friendly activities and movement. Obscuration smoke deployed on or near the enemy's position minimizes its vision. Screening smoke employed between the reduction area and the enemy conceals movement and reduction activities. It also degrades enemy ground and aerial observations. Obscuration must be carefully planned to provide maximum degradation of enemy observation and fires, but it must not significantly degrade friendly fires and control.

Secure

F-17. Friendly forces secure reduction areas to prevent the enemy from interfering with obstacle reduction and the passage of the assault force through lanes created during the reduction. Security must be effective against outposts and fighting positions near the obstacle and against overwatching units as necessary. The far side of the obstacle must be secured by fires or be occupied before attempting any effort to reduce the obstacle. The attacking unit's higher headquarters is responsible for isolating the breach area by fixing adjacent units, attacking enemy reserves in depth, and providing counterfire support.

F-18. Identifying the extent of the enemy's defenses is critical before selecting the appropriate technique to secure the point of breach. If the enemy controls the point of breach and cannot be adequately suppressed, the force must secure the point of breach before it can reduce the obstacle.

F-19. The breach force must be resourced with enough maneuver assets to provide local security against the forces that the support force cannot sufficiently engage. Elements within the breach force that secure the reduction area may also be used to suppress the enemy once reduction is complete. The breach force may also need to assault to the far side of the breach and provide local security so the assault element can seize its initial objective.

Reduce

F-20. Reduction is the creation of lanes through or over an obstacle to allow an attacking force to pass. The number and width of lanes created varies with the enemy situation, the assault force's size and composition, and the scheme of maneuver. The lanes must allow the assault force to rapidly pass through the obstacle. The breach force will reduce, proof (if required), mark, and report lane locations and the lane-marking method to higher command headquarters. Follow-on units will further reduce or clear the obstacle when required. Reduction cannot be accomplished until effective suppression and obscuration are in place, the obstacle has been identified, and the point of breach is secure.

Assault

F-21. A breaching operation is not complete until—

- Friendly forces have assaulted to destroy the enemy on the far side of the obstacle as the enemy is capable of placing or observing direct and indirect fires on the reduction area.
- Battle handover with follow-on forces has occurred, unless no battle handover is planned.

BREACHING ORGANIZATION

F-22. A commander or platoon leader organizes friendly forces to accomplish breaching fundamentals quickly and effectively. This requires him to organize support, breach, and assault forces with the necessary assets to accomplish their roles. For tactical obstacle breaches, platoons and squads are normally assigned as either one or part of the following forces (Table F-1).

SUPPORT FORCE

F-23. The support force's primary responsibility is to eliminate the enemy's ability to place direct or indirect fire on friendly force and interfere with a breaching operation. It must—

- Isolate the reduction area with fires and establish a support-by-fire position to destroy, fix, or suppress the enemy. Depending on METT-TC, this may be the weapons squad or the entire platoon.
- Mass and control direct and indirect fires to suppress the enemy and to neutralize any weapons that are able to bring fires on the breach force.
- Control obscuring smoke to prevent enemy-observed direct and indirect fires.

Breach Force

F-24. The breach force assists in the passage of the assault force by creating, proofing (if necessary), and marking lanes. The breach force may be a combined-arms force. It may include engineers, reduction assets, and enough maneuver forces to provide additional suppression and local security. The entire Infantry platoon may be part of the breach force. The breach force may apply portions of the following breaching fundamentals as it reduces an obstacle.

Suppress

F-25. The breach force must be allocated enough maneuver forces to provide additional suppression against various threats, including—

- Enemy direct-fire systems that cannot be effectively observed and suppressed by the support force due to the terrain or the masking of the support force's fires by the breach force as it moves forward to reduce the obstacle.
- Counterattacking and or repositioning forces that cannot be engaged by the support force.

Obscure

F-26. The breach force employs smoke pots, if necessary, for self-defense and to cover lanes while the assault force is passing.

Secure

F-27. The breach force secures itself from threat forces that are providing close-in protection of the obstacle. The breach force also secures the lanes through the tactical obstacles once they are created to allow safe passage of the assault force.

Reduce

F-28. The breach force performs its primary mission by reducing the obstacle. To support the development of a plan to reduce the obstacle, the composition of the obstacle system must be an information requirement. If the obstacles are formidable, the Infantry platoon will be augmented with engineers to conduct reduction. Without engineers and special equipment such as Bangalore torpedoes and line charges, mine fields must be probed.

Assault Force

F-29. The breach force assaults through the point of breach to the far side of an obstacle and seizes the foothold. The assault force's primary mission is to destroy the enemy and seize terrain on the far side of the obstacle to prevent the enemy from placing direct fires on the created lanes. The assault force may be tasked to assist the support force with suppression while the breach force reduces the obstacle.

F-30. The assault force must be sufficient in size to seize the point of penetration. Combat power is allocated to the assault force to achieve a minimum 3:1 ratio on the point of penetration. The breach and assault assets may maneuver as a single force when conducting breaching operations as an independent company team conducting an attack.

F-31. If the obstacle is defended by a small enemy force, assault and breach forces' missions may be combined. This simplifies C2 and provides more immediate combat power for security and suppression.

F-32. Fire control measures are essential because support and breach forces may be firing on the enemy when the assault force is committed. Suppression of overwatching enemy positions must continue and other enemy forces must remain fixed by fires until the enemy has been destroyed. The assault force must assume control for direct fires on the assault objective as support and breach force fires are ceased or shifted. Table F-1 illustrates the relationship between the breaching organization and breaching fundamentals.

Table F-1. Relationship between breaching organization and breaching fundamentals.

Breaching Organization	Breaching Fundamentals	Responsibilities
Support force	Suppress Obscure	Suppress enemy direct fire systems covering the reduction area. Control obscuring smoke. Prevent enemy forces from repositioning or counterattacking to place direct fires on the breach force.
Breach force	Suppress (provides additional suppression) Obscure (provides additional obscuration in the reduction area) Secure (provides local security) Reduce	Create and mark the necessary lanes in an obstacle. Secure the near side and far side of an obstacle. Defeat forces that can place immediate direct fires on the reduction area. Report the lane status/location.
Assault force	Assault Suppress (if necessary)	Destroy the enemy on the far side of an obstacle if the enemy is capable of placing direct fires on the reduction area. Assist the support force with suppression if the enemy is not effectively suppressed. Be prepared to breach follow-on and or protective obstacles after passing through the reduction area.

DETAILED REVERSE PLANNING

F-33. The platoon leader must develop the breaching plan using the following sequence when planning for a protective obstacle breach. The platoon leader can plan to breach wire, mine fields, trenches, and craters (Figure F-4).

- Reverse planning begins with actions on the objective.
- Actions on the objective drive the size and composition of the assault force.
- The size of the assault force determines the number and location of lanes to be created.
- The ability of the enemy to interfere with the reduction of the obstacle determines the size and composition of the security element in the breach force.
- The ability of the enemy to mass fires on the point of breach determines the amount of suppression and the size and composition of the support force.

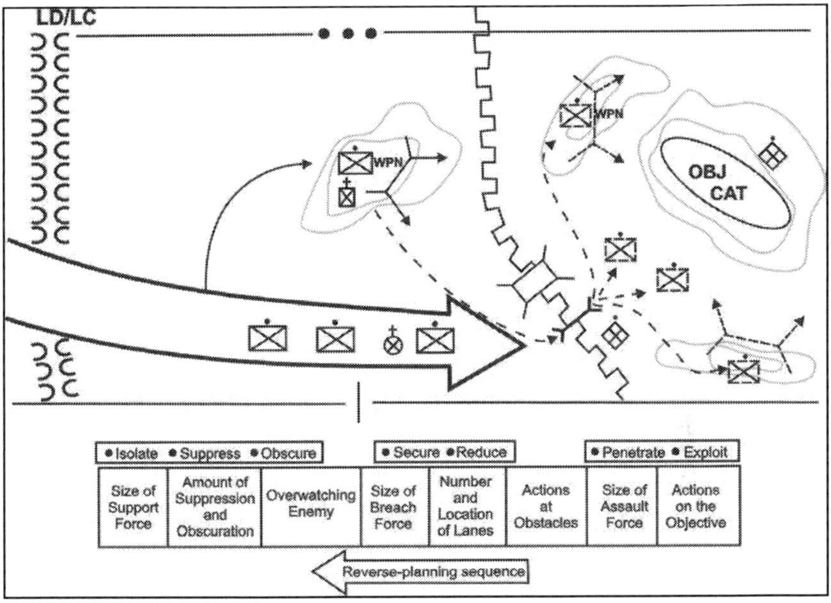

Figure F-4. Reverse planning.

F-34. The approved technique for conducting obstacle breaching operations is suppress, obscure, secure, reduce, assault (SOSRA). The section focuses specifically on platoon reduction techniques of land mines, construction obstacles, urban obstacles, and booby traps and expedient devices.

F-35. As part of reducing obstacles, units must also detect, report, proof, and mark.

F-36. *Detection* is the actual confirmation of the location of obstacles. It may be accomplished through reconnaissance. It can also be unintentional (such as a vehicle running into a mine or wire). Detection is used in conjunction with intelligence-gathering operations, bypass reconnaissance, and breaching/clearing operations. Specific detection methods for mines and booby traps are discussed further in this section.

F-37. Intelligence concerning enemy minefields is *reported* by the fastest means available. A SPOTREP should be sent to higher headquarters when the Infantry platoon or squad has detected a minefield or any other obstacle. This should be done whether they are sent on a specific minefield or obstacle reconnaissance mission, or if they encounter one in the course of normal operations. The SPOTREP should contain as much information as possible including the type, location, and size of the obstacle, and the results of any reduction efforts.

F-38. *Proofing* is normally done by engineers by passing a mine roller or another mine-resistant vehicle through the minefield to verify that a lane is free of mines. If the risk of live mines remaining in the lane does not exceed the risk of loss to enemy fires while waiting, proofing may not be practical. Some mines are resistant to specific breaching techniques. For example, magnetically fused mines may be resistant to some explosive blasts. So proofing should be done when the time available, the threat, and the mission allow. Proofing also involves verifying that other obstacles (such as wire) are free of explosive or injurious devices.

F-39. *Marking* breach lanes and bypasses is critical to obstacle reduction.

REDUCE A MINEFIELD

F-40. Most types of obstacles do not cause casualties directly. Minefields do have this potential, and will cause direct casualties if not reduced effectively. Buried mines are usually found in a highly prepared defense. When training for the reduction of surface-laid and buried minefields, always assume the presence of antihandling devices (AHDs) and trip wires until proven otherwise.

MINEFIELD DETECTION

F-41. The three types of minefield detection methods the platoon might employ are visual, physical (probing), and electronic.

Visual Detection

F-42. Visual detection is part of all combat operations. Soldiers should constantly be alert for minefields and all types of enemy obstacles. Soldiers visually inspect the terrain for the following obstacle indicators:

- Trip wires and wires leading away from the side of the road. They may be firing wires that are partially buried.
- Signs of road repair (such as new fill or paving, road patches, ditching, and culvert work).
- Signs placed on trees, posts, or stakes. Threat forces mark their minefields to protect their own forces.
- Dead animals or damaged vehicles.
- Disturbances in previous tire tracks or tracks that stop unexplainably.
- Odd features in the ground or patterns that are not present in nature. Plant growth may wilt or change color; rain may wash away some of the cover; the cover may sink or crack around the edges; or the material covering the mines may look like mounds of dirt.
- Civilians who may know where mines or booby traps are located in the residential area. Civilians staying away from certain places or out of certain buildings are good indications of the presence of mines or booby traps. Question civilians to determine the exact locations.
- Pieces of wood or other debris on a road. They may be indicative of pressure or pressure-release firing devices. These devices may be on the surface or partially buried.
- Patterns of objects that could be used as a sighting line. An enemy can use mines that are fired by command, so road shoulders and areas close to the objects should be searched.
- Berms may indicate the presence of an AT ditch.

Physical (Probing) Detection

F-43. Physical detection (probing) is very time-consuming and is used primarily for mine-clearing operations, self-extraction, and covert breaching operations. Detection of mines by visual or electronic methods should be confirmed by probing. Detailed probing instructions can be referenced in FM 21-75.

Electronic Detection

F-44. Electronic detection is effective for locating mines, but this method is time-consuming and exposes personnel to enemy fire. In addition, suspected mines must be confirmed by probing. As in probing, 20 to 30 minutes is the maximum amount of time an individual can use the detector effectively.

F-45. The AN/PSS-12 mine detector (Figure F-5) is very effective at finding metallic mines, but is less effective against low-metal mines. Employment and operation procedures for the AN/PSS-12 are discussed in FM 20-32. Technical data is available in TM 5-6665-298-10, *Operator's Manual for Mine Detecting Set AN/PSS-12*. The detector is handheld and identifies suspected mines by an audio signal in the headphones.

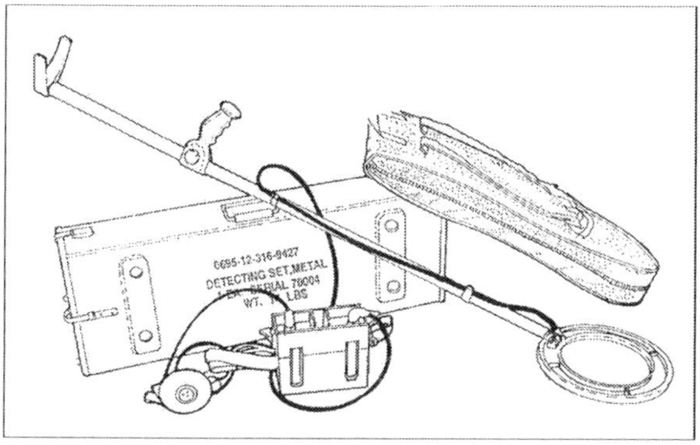

Figure F-5. AN/PSS-12 mine detector.

MINEFIELD REDUCTION AND CLEARING EQUIPMENT

F-46. Minefield reduction and clearing equipment is broken down into explosive, manual, mechanical, and electronic. While chiefly an engineer task, the platoon unit might need to reduce a minefield depending on the situation. The leader masses reduction assets to ensure it will successfully create as many lanes as necessary to ensure the rapid passage of the assault force through the obstacle system. If necessary, the leader must carefully plan and synchronize the creation of additional lanes to reduce the potential for fratricide with assaulting troops. The distance between lanes depends on the enemy, the terrain, the need to minimize the effects of enemy artillery, the direct-fire plan of the support force, C2, and the reduction-site congestion.

F-47. The breach force should be organized and equipped to use several different reduction techniques in case the primary technique fails. Additional reduction assets should be present to handle the unexpected. Normally, 50 percent more reduction assets than required for obstacle reduction are positioned with the breach force. Mechanical and electronic reduction techniques and equipment are employed by engineers and can be found in FM 20-32.

Explosive Minefield Reduction

F-48. FM 20-32 lists all explosive minefield reduction techniques and equipment. The different types of explosive minefield-reduction equipment that the platoon might use to breach obstacles are discussed below.

M1A1/M1A2 Bangalore Torpedo

F-49. The Bangalore torpedo (Figure F-6) is a manually emplaced, explosive-filled pipe designed as a wire breaching device that is also effective against simple pressure-activated AP mines. It is issued as a demolition kit and consists of 10 1.5-meter tubes, 10 connecting sleeves, and 1 nose sleeve. Each tube contains 4 kilograms of HE and weighs 6 kilograms. The kit clears a 1-by 15-meter lane.

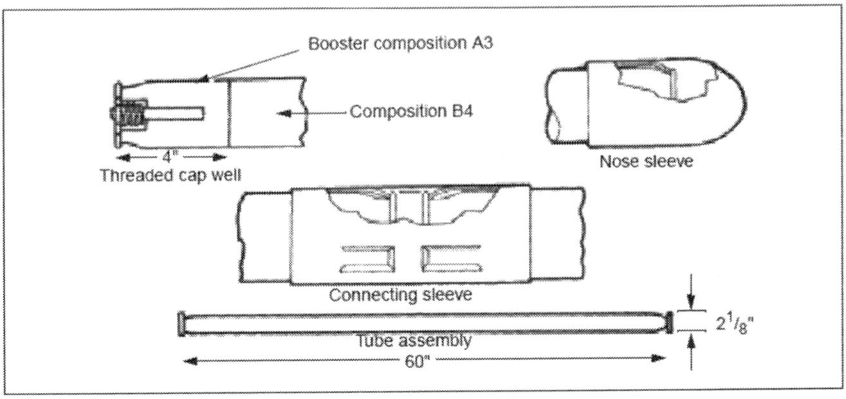

Figure F-6. Bangalore torpedo.

F-50. All torpedo sections have a threaded cap well at each end so they can be assembled in any order. The connecting sleeves are used to connect the torpedo sections together. An individual or pair of Soldiers connect the number of sections needed, and then push the torpedo through the AP minefield before priming the torpedo. A detailed reconnaissance is conducted before using the Bangalore torpedo to ensure trip wires have not been used. The Bangalore torpedo generates one short impulse. It is not effective against pronged, double-impulse, or pressure-resistant AP/AT mines.

WARNING

Do not modify the Bangalore torpedo. Cutting the Bangalore in half or performing any other modification could cause the device to explode.

Antipersonnel Obstacle Breaching System

F-51. The Antipersonnel Obstacle Breaching System (APOBS) (Figure F-7) is a man-portable device that is capable of quickly creating a footpath through AP mines and wire entanglements. It provides a lightweight, self-contained, two-man, portable line charge that is rocket-propelled over AP obstacles away from the obstacle's edge from a standoff position

F-52. For dismounted operations, the APOBS is carried in 25-kilogram backpacks by no more than two Soldiers for a maximum of 2 kilometers. One backpack assembly consists of a rocket-motor launch mechanism containing a 25-meter line-charge segment and 60 attached grenades. The other backpack assembly contains a 20-meter line-charge segment and 48 attached grenades.

F-53. The total weight of the APOBS is about 54 kilograms. It is capable of breaching a footpath about 0.6 by 45 meters and is fired from a 25-meter standoff.

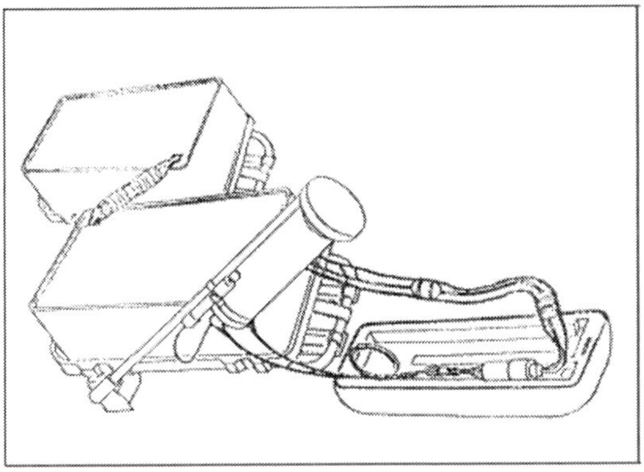

Figure F-7. Antipersonnel obstacle breaching system (APOBS).

Manual Minefield Reduction

F-54. Manual procedures are normally conducted by engineers (but can also be performed by Infantry units) and are effective against all obstacles under all conditions. Manual procedures involve dismounted Soldiers using simple explosives or other equipment to create a lane through an obstacle or to clear an obstacle. These procedures expose the Soldier and may be manpower and time-intensive. While mechanical and explosive reduction procedures are normally preferred, the Infantry platoon may have to use manual procedures for the following reasons:

- Explosive, mechanical, and electronic reduction assets are unavailable or ineffective against the type of obstacle.
- Terrain limitations.
- Stealth is required.

F-55. Different manual reduction techniques for surface-laid and buried minefields are discussed below.

Surface-Laid Minefield

F-56. First use grappling hooks from covered positions to check for trip wires in the lane. The limited range of the tossed hook requires the procedure to be repeated through the estimated depth of the obstacle. A demolition team then moves through the lane. The team places a line main down the center of the lane, ties the line from the explosive into the line main, and places blocks of explosive next to surface-laid mines. After the mines are detonated, the team makes a visual check to ensure that all mines were cleared before directing a proofing roller and other traffic through the lane. Members of the demolition team are assigned special tasks such as grappler, detonating-cord man, and demolitions man. All members should be cross-trained on all procedures. Demolitions are prepared for use before arriving at the point of breach (refer back to Table F-1). The platoon must rehearse reduction procedures until execution is flawless, quick, and technically safe. During reduction, the platoon will be exposed in the lane for five minutes or more depending on the mission, the minefield depth, and the Infantry platoon's level of training.

Buried Minefield

F-57. Manually reducing a buried minefield is extremely difficult to perform as part of a breaching operation. If mine burrows are not easily seen, mine detectors and probes must be used to locate mines. Mines are then destroyed by hand-emplaced charges. As an alternative, mines can be removed by using a grappling hook and, if necessary, a tripod (Figure F-8). Using a tripod provides vertical lift on a mine, making it easier to pull the mine out of the hole.

F-58. The leader organizes Soldiers into teams with distinct, rehearsed missions including grappling, detecting, marking, probing, and emplacing demolitions and detonating cord. The platoon is exposed in the obstacle for long periods of time.

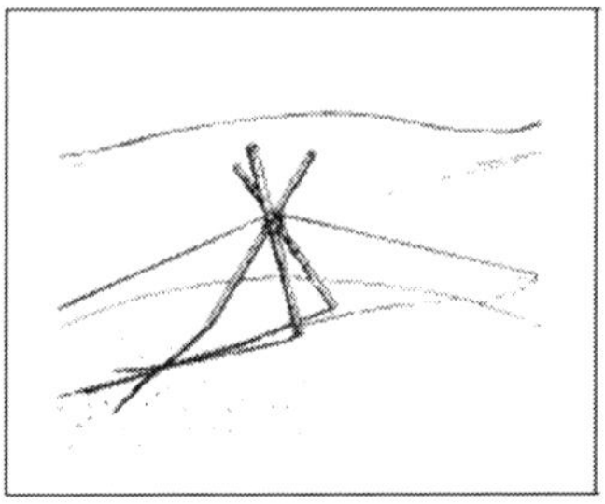

Figure F-8. Tripod.

Grappling Hook

F-59. The grappling hook is a multipurpose tool used for manual obstacle reduction. Soldiers use it to detonate mines from a standoff position by activating trip wires and AHDs. After the grapnel is used to clear trip wires in a lane, dismounted Soldiers can move through the minefield, visually locate surface-laid mines, and prepare mines for demolition. In buried minefields, Soldiers grapple and then enter the minefield with mine detectors and probes.

F-60. Multiple grapplers can clear a lane of trip wires quickly and thoroughly, but they must time their efforts and follow procedures simultaneously. A hit on a trip wire or a pressure fuse can destroy the grappling hook and the cord, so the platoon should carry extras.

F-61. There are two types of grappling hooks: hand-thrown; and weapon-launched.

F-62. **Hand-Thrown.** A 60+-meter light rope is attached to the grappling hook for hand throwing. The throwing range is usually no more than 25 meters. The excess rope is used for the standoff distance when the thrower begins grappling. The thrower tosses the grappling hook and seeks cover before the grappling hook and rope touch the ground in case their impact detonates a mine. He then moves backward, reaches the end of the excess rope, takes cover, and begins grappling. Once the grappling hook is recovered, the thrower moves forward to the original position, tosses the grapnel, and repeats the procedure at least twice. He then moves to the end of the grappled area and repeats this sequence through the depth of the minefield.

F-63. **Weapon-Launched.** A 150-meter lightweight rope is attached to a lightweight grappling hook that is designed to be fired from an M16-series rifle using an M855 cartridge. The grappling hook is pushed onto the rifle muzzle with the opening of the retrieval-rope bag oriented toward the minefield. The firer is located 25 meters from the minefield's leading edge and aims the rifle muzzle at a 30-to 40-degree angle for maximum range. Once fired, the grappling hook travels 75 to 100 meters from the firer's position. After the weapon-launched grappling hook (WLGH) has been fired, the firer secures the rope, moves 60 meters from the minefield, moves into a prone position, and begins to grapple. The WLGH can be used only once to clear a minefield, but it can be reused up to 20 times for training because blanks are used to fire it.

Demolitions

F-64. Different types of demolitions can be used for minefield obstacle reduction (Table F-2). FM 5-250 covers each different type of demolition available to support all Infantry missions. Demolitions are used differently against certain types of mines:

- **Pressure-Fused AP Mine.** Place at least a 1-pound charge within 15.2 centimeters of simple pressure-fused mines. Ensure that the charge is placed within 2.54 centimeters of blast-hardened mines.

- **Trip-Wire/Break-Wire-Fused AP Mine.** Place at least a 1-pound charge within 15.2 centimeters of the mine after the mine at the end of a trip wire has been located. Soldiers can use elevated charges if necessary against the Claymore and stake-type mines.
- **Influence-Fused AP Mine.** Do not use demolitions.
- **Command-Detonated Blast Mine.** Ensure that the observer is neutralized before approaching. Elevated charges can be used if necessary against Claymore mines.

Table F-2. Demolitions.

Item	Description
M183 Satchel Charge	Consists of 16 M112 (C4) charges and four priming assemblies. Total explosive weight of 20 pounds. Used primarily for breaching obstacles or demolishing structures when large charges are required. Also is effective on smaller obstacles such as small dragon's teeth.
M112 Charge	Consists of 1.25 pounds of C4 packed in an olive drab Mylar film container with a pressure-sensitive adhesive tape on one surface. Primarily used for cutting and breaching. Because of its ability to cut and be shaped, the M112 is ideally suited for cutting irregularly-shaped targets such as steel. The adhesive backing allows you to place the charge on any relatively flat surface.
Modernized Demolition Initiator (MDI)	MDI is a new family of nonelectric blasting caps and associated items. Components simplify initiation systems and improve reliability and safety. Components include the M11 high strength blasting cap, the M12 and M13 low strength blasting caps, and the M14 high strength time delay cap.
Detonating Cord	Consists of a core of HE (6.4 pounds of PETN per 1,000 feet) wrapped in a reinforced and waterproof olive drab coating. Can be used to prime and detonate single or multiple explosive charges simultaneously. Can be used in conjunction with the MDI components.

Marking and Crossing the Minefield

F-65. Effective lane marking allows the leader to project the platoon through the obstacle quickly with combat power and C2. It also gives the Infantry platoon or squad confidence in the safety of the lane and helps prevent unnecessary minefield casualties.

F-66. Once a footpath has been probed and the mines marked or reduced, a security team should cross the minefield to secure the far side. After the far side is secure, the rest of the unit should cross. If mines and any trip wires have been identified but not reduced, the mine and the line of the trip wire are marked along the ground surface, 12 inches before the trip wire (Figure F-9).

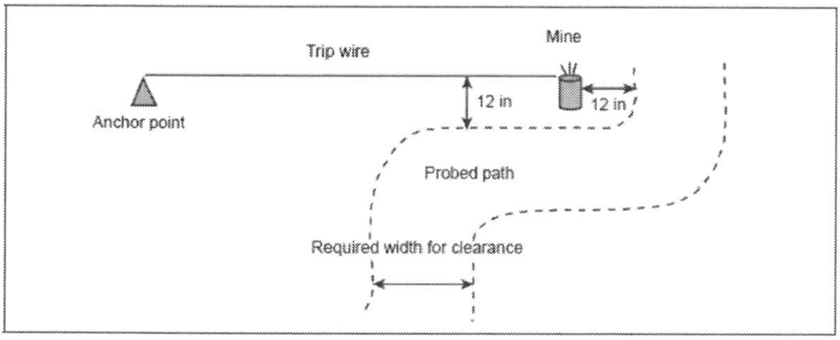

Figure F-9. Marking a footpath.

REDUCE A CONSTRUCTED OBSTACLE

F-67. Reduction methods for enemy wire and tank ditch obstacles are as follows.

REDUCE A WIRE OBSTACLE

F-68. The enemy uses wire and concertina obstacles to separate Infantry from tanks and to slow or stop the Infantry movement. His wire obstacles are similar to ours. On patrol, reducing a wire obstacle may require stealth and is conducted using wire cutters or by crawling under or crossing over the wire. It may not require stealth during an attack and can be accomplished with Bangalore torpedoes and wire cutters.

Cut the Wire

F-69. To cut through a wire obstacle with stealth—

- Cut only the lower strands and leave the top strand in place. That makes it less likely that the enemy will discover the gap.
- Cut the wire near a picket. To reduce the noise of a cut, have another Soldier wrap cloth around the wire and hold the wire with both hands. Cut part of the way through the wire between the other Soldier's hands and have him bend the wire back and forth until it breaks. If you are alone, wrap cloth around the wire near a picket, partially cut the wire, and then bend and break the wire.

F-70. To reduce an obstacle made of concertina—

- Cut the wire and stake it back to keep the breach open.
- Stake the wire back far enough to allow room to crawl through or under the obstacle.

Bangalore Torpedo

F-71. After the Bangalore torpedo has been assembled and pushed through the wire obstacle, prime it with either an electric or nonelectric firing system (Figure F-10). To prevent early detonation of the entire Bangalore torpedo if you hit a mine while pushing it through the obstacle, attach an improvised (wooden) torpedo section to its end. That section can be made out of any wooden pole or stick that is the size of a real torpedo section. Attach the nose sleeve to the end of the wooden section. Once the Bangalore torpedo has been fired, use wire cutters to cut away any wire not cut by the explosion.

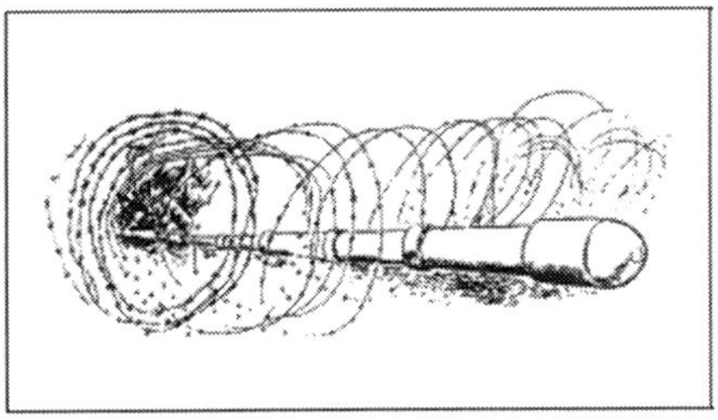

Figure F-10. Reducing wire obstacles with bangalore torpedos.

REDUCE AN URBAN OBSTACLE

F-72. Understanding how to employ and incorporate reduction techniques is an important part of urban operations. Gaining quick access to targeted rooms is integral to room clearing. Reduction teams need to be supported by fires or obscurants. Reduction operations should be performed during hours of limited visibility whenever possible. Reduction techniques vary based on construction encountered and munitions available.

F-73. The assault team's order of march to the breach point is determined by the method of reduction and its intended actions at the entry point. This preparation must be completed prior to or in the last covered and concealed location before reaching the entry point. Establishing an order of march aids the team leader with C2 and minimizes exposure time in open areas and at the entry point. One order of march technique is to number the assault team members one through four. The number-one man should always be responsible for frontal and door security. If the reduction has been conducted prior to its arrival, the assault team quickly moves through the entry point. If a reduction has not been made prior to its arrival at the entry point, depending on the type of breach to be made, the team leader conducts the reduction himself or signals forward the breach man or element. One option is to designate the squad leader as the breach man. If the breach man is part of the assault team, he will normally be the last of the four men to enter the building or room. This allows him to transition from his reduction task to his combat role. See FM 3-06.11, *Combined Arms Operations in Urban Terrain,* for more information on movement and breaching methods.

F-74. The three urban reduction methods discussed in this appendix are mechanical, ballistic, and explosive.

BREACH LOCATIONS

F-75. The success of the assault element often depends on the speed with which they gain access into the building. It is important that the breach location provide the assault element with covered or concealed access, fluid entry, and the ability to be overwatched by the support element.

Creating Mouseholes

F-76. Mouseholes provide a safe means of moving between rooms and floors. C4 plastic explosive can be used to create mouseholes when lesser means of mechanical reduction fail. Because C4 comes packaged with an adhesive backing or can be emplaced using pressure-sensitive tape, it is ideal for this purpose. When using C4 to blow a mousehole in a lath and plaster wall, one block or a strip of blocks should be placed on the wall from neck-to-knee height. Charges should be primed with detonating cord or MDI to obtain simultaneous detonation that will blow a hole large enough for a man to fit through.

Expedient Reduction Methods

F-77. Because the internal walls of most buildings function as partitions rather than load-bearing members, smaller explosive charges can be used to reduce them. When C4 or other military explosives are not available, one or more fragmentation grenades or a Claymore mine can be used to reduce some internal walls. These field-expedient reduction devices should be tamped to increase their effectiveness and to reduce the amount of explosive force directed to the rear. Take extreme care when attempting to perform this type of reduction because fragments may penetrate walls and cause friendly casualties. If walls are made of plaster or dry wall, mechanical reduction may be more effective.

Windows and Restrictive Entrances

F-78. Regardless of the technique used to gain entry, if the breach location restricts fundamental movement into the room or building, local or immediate support must be used until the assault team can support itself. For example, as a Soldier moves through a window and into the room, he may not be in a position to engage an enemy. Therefore, another window that has access to the same room may be used to overwatch the lead team's movement into the room. The overwatching element can come from the initial clearing team or from the team designated to enter the breach location second.

MECHANICAL REDUCTION

F-79. This method requires increased physical exertion by one or more Soldiers using hand tools such as axes, saws, crowbars, hooligan's tools, or sledgehammers to gain access. Although most Soldiers are familiar with these tools, practice on various techniques increases speed and effectiveness. The mechanical reduction is not the preferred primary breaching method because it may be time consuming and defeat the element of surprise. However, the ROE and situation may require the use of these tools, so Soldiers should be proficient in their use.

F-80. Typically, the order of movement for a mechanical breach is the initial assault team, followed by the breach man or element. At the breach point, the assault team leader brings the breach team forward while the assault team provides local security. After the reduction is conducted, the breach team moves aside and provides local security as the assault team enters the breach. See FM 3-06.11 for additional information concerning mechanical reduction and breaching.

F-81. When developing an urban operations mechanical breach kit SOP, Infantry units must consider their METL and the unit tactical SOP.

BALLISTIC REDUCTION

F-82. Ballistic reduction requires the use of a weapon firing a projectile at the breach point. Ballistic reduction is not a positive means of gaining entry and should not be considered the primary method for gaining initial entry into a structure. It may not supply the surprise, speed, and violence of action necessary to minimize friendly losses on initial entry. In certain situations, it may become necessary to use ballistic reduction as a back-up entry method. A misfire of an explosive charge or the compromise of the assault element during its approach to the target may necessitate the use of ballistic reduction as a means of initial entry into the structure. Ballistic reduction may have to be followed up with a fragmentation, concussion, or stun grenade before entry.

F-83. Once initial entry is gained, shotgun ballistic reduction may become the primary method for gaining access to subsequent rooms within the structure. Surprise is lost upon initial entry, and other reduction methods are often too slow, tending to slow the momentum of the assault team. If a door must be used for entry, several techniques can be used to open the door. Doors should be considered a fatal funnel because they are usually covered by fire, or may be booby-trapped. See FM 3-06.11 for more information concerning weapon employment and effects.

F-84. Unless a deliberate breach is planned, the platoon can employ a series of progressive reductions. An example is an attempt to open a door by using the doorknob first, then shotgun reduction, then explosive reduction as a final option. Mechanical reduction can be used to clean up a failed attempt of a shotgun or explosive reduction, but can also be used as the primary reduction technique. Based on the multiple situations that the complex urban environment presents, the leader needs latitude in his options.

Exterior Walls

F-85. For exterior walls, the use of a BFV or artillery piece in the direct fire role is ideal if the structure will support it and if the ROE will allow it. The BFV's 25-mm cannon is an effective reduction weapon when using HE rounds and firing a spiral firing pattern (Figure F-11). The main gun of an M1A1/A2 tank is very effective when using the high explosive antitank (HEAT) round. However, the armor-piercing discarding-sabot (APDS) round rarely produces the desired effect because of its penetrating power.

Figure F-11. Spiral firing pattern.

Doors, Windows, and Interior Walls

F-86. The 12-gauge shotgun breaching round is effective on doorknobs and hinges, while standard small arms (5.56 mm and 7.62 mm) have proven to be virtually ineffective for reducing obstacles. These should not be used except as a last resort because of their ricochet potential and shoot-through capability. Ballistic reduction of lightly-constructed interior walls by shotgun fire is normally an alternate means of gaining entry.

> **WARNING**
>
> **The fragmentation and ricochet effects of standard small arms (5.56 mm and 7.62 mm) as breaching rounds is unpredictable and considered extremely dangerous. Do not attempt in training.**

Rifle-Launched Entry Munitions

F-87. Rifle-launched entry munitions (RLEM) allow a remote ballistic reduction of an exterior door or window without having the assault or breaching element physically present at the entry point. This allows the assault element to assume a posture for entry in the last covered and concealed position before the breach. The RLEM firer is not normally part of the assault element but rather a part of the breaching or support element. This allows the RLEM to be fired from one position while the assault element waits in another position. In the event that the first round does not affect the reduction, the firer should prepare a second round for the reduction or a second firer should be prepared to engage the target.

> **WARNING**
>
> **The firer must be a minimum of 10 meters from the target to safely employ a 150-gram round.**

NOTE: Exact minimum safe distances for firers and assault elements have not been established for the 150-gram round.

Shotgun Reduction

F-88. Various shotgun rounds can be used for ballistic reduction. Breaching and clearing teams need to be familiar with the advantages as well as the disadvantages of each type of round. Leaders must consider the potential for over penetration on walls and floors in multi-story buildings to avoid potential fratricide incidents or killing of noncombatants.

- **Rifled Slugs.** Rifled slugs defeat most doors encountered, including some heavy steel doors. However, rifled slugs present a serious over penetration problem and could easily kill or injure anyone inside the room being attacked. Rifled slugs are excellent antipersonnel rounds and can be used accurately up to 100 meters.
- **Bird Shot.** Bird shot (number 6 through number 9 shot) is used in close-range work up to 15 meters. A 2 ¾-inch shell of number 9 shot typically contains an ounce of shot (though it can be loaded to 1 ½-ounce with an accompanied increase in recoil). The major advantage of bird shot is it does not over penetrate. Therefore, bird shot poses little hazard to fellow team members in adjoining rooms. When used at close range, bird shot offers the same killing potential as buckshot, especially in a full choke shotgun intended for dense shot patterns. Another advantage of bird shot is low recoil. This feature allows for faster recovery and quicker multi-target engagements. A disadvantage with bird shot is rapid-energy bleed-off that reduces penetration at medium and long ranges. Moreover, the small size of the individual pellets requires hits be made with a majority of the shot charge to be effective. A hit with one-third of the number 9 shot charge may not be fatal, unless the shot is at extremely close range. These disadvantages are negated when birdshot is fired from a full choke shotgun where it will produce a pattern that is quite small inside of 10 meters. Inside 5 meters, all of the shot will be clumped like a massive single projectile.
- **Buckshot.** Buckshot is used in close- to medium-range work, up to 30 meters. Because of its larger size, buckshot is more lethal than bird shot. A 2 ¾-inch shell of 00 buckshot contains nine .30-caliber balls. One .30-caliber ball of the 00 buckshot charge hit can prove fatal. Buckshot also retains its energy longer. Therefore, it is lethal at longer ranges than bird shot. A disadvantage of buckshot is over penetration. Because buckshot is typically loaded with heavier shot charges, it also has very heavy recoil. This problem becomes apparent when numerous shots have been taken and can result in fatigue.
- **Ferret Rounds.** Ferret rounds contain a plastic slug filled with liquid chemical irritant (CS). When shot through a door or wall (drywall or plywood), the plastic slug breaks up and a fine mist of CS is sprayed into the room. The effectiveness of one round is determined by the size of the room on the other side of the door or the wall and also the ventilation in that room.

F-89. When using the shotgun as an alternate reduction method to gain entry, shooters must consider the following target points on the door.

- **Doorknob.** Never target the doorknob itself because when the round impacts, the doorknob has a tendency to bend the locking mechanism into the doorframe. In most cases this causes the door to be bent in place and prevents entry into the room.
- **Locking Mechanism.** When attacking the locking mechanism, focus the attack on the area immediately between the doorknob and the doorframe. Place the muzzle of the shotgun no more than one inch away from the face of the door directly over the locking mechanism. The angle of attack should be 45 degrees downward and at a 45-degree angle into the doorframe. After breaching the door, kick it swiftly. This way, if the door is not completely open, a strong kick will usually open it. When kicking the door open, focus the force of the kick at the locking mechanism and close to the doorjamb. After the locking mechanism has been reduced, this area becomes the weakest part of the door.

- **Hinges.** The hinge breach technique is performed much the same as the doorknob reduction, except the gunner aims at the hinges. He fires three shots per hinge—the first at the middle, then at the top and bottom. He fires all shots from less than an inch away from the hinge. Because the hinges are often hidden from view, the hinge reduction is more difficult. Hinges are generally 8 to 10 inches from the top and bottom of the door. The center hinge is generally 36 inches from the top, centered on the door. Regardless of technique used, immediately after the gunner fires, he kicks the door in or pulls it out. He then pulls the shotgun barrel sharply upward and quickly turns away from the doorway to signal that the breach point has been reduced. This rapid clearing of the doorway allows the following man in the fire team a clear shot at any enemy who may be blocking the immediate breach site. See FM 3-06.11 for more information.

F-90. When the assault team members encounter a door to a "follow-on" room, they should line up on the side of the door that gives them a path of least resistance upon entering. When the door is encountered, the first Soldier to see it calls out the status of the door, OPENED, or CLOSED. If the door is open, Soldiers should never cross in front of it to give themselves a path of least resistance. If the door is closed, the number-one man maintains security on the door and waits for the number-two man to gain positive control of the number-one man. The number-one man begins the progressive breaching process by taking his nonfiring hand and checking the doorknob to see if it is locked. If the door is unlocked, the number-one man (with his hand still on the door) pushes the door open as he enters the room. If the door is locked, the number-one man releases the doorknob (while maintaining security on the door) and calls out for the breacher, BREACHER UP.

F-91. Once the breacher arrives at the door (with round chambered), he places the muzzle of the shotgun at the proper attack point, takes the weapon off safe, and signals the number-two man by nodding his head. At that time, the number-two man (with one hand maintaining positive control of the number-one man) takes his other hand (closest to the breacher) and forming a fist, places it within the periphery of the breacher and pumps his fist twice saying, READY BREACH. This action allows the breacher to see if a flashbang or grenade is to be used. Once the breacher defeats the door, he steps aside and allows the assault team to enter. He then either assumes the position of the number-four man if he is acting as a member of the assault team or remains on call as the breacher for any follow-on doors. He should keep the shotgun magazine full at all times. There may be numerous doors, and stopping to reload will slow the momentum of the assault.

NOTE: The shotgun should not be used as a primary assault weapon because of its limited magazine capacity and the difficulty of reloading the weapon.

Exterior Walls

F-92. One of the most difficult breaching operations for the assault team is reducing masonry and reinforced concrete walls. C4 is normally used for explosive reduction because it is safe, easy to use, and readily available. Engineers are usually attached to the platoon if explosive reduction operations are expected. The attached engineers will conduct the reduction themselves or provide technical assistance to the Infantrymen involved. The typical thickness of exterior walls is 15 inches or less, although some forms of wall construction are several feet thick. Assuming that all outer walls are constructed of reinforced concrete, a rule of thumb for reduction is to place 10 pounds of C4 against the target between waist and chest height. When detonated, this charge normally blows a hole large enough for a man to go through. On substandard buildings, however, a charge of this size could rubble the building. When explosives are used to reduce windows or doors, the blast should eliminate any booby traps in the vicinity of the window or doorframe. See FM 3-06.11 for information concerning demolitions.

Charge Placement

F-93. Place the charges (other than shape charges) directly against the surface that is to be reduced. When enemy fire prevents an approach to the wall, a potential technique is to attach the charge, untamped, to a pole and slide it into position for detonation at the base of the wall. Small-arms fire will not detonate C4 or TNT. Take cover before detonating the charge.

Tamping

F-94. Whenever possible, explosives should be tamped or surrounded with material to focus the blast to increase effectiveness. Tamping materials could be sandbags, rubble, desks, chairs, and even intravenous bags. For many exterior walls, tamping may be impossible due to enemy fire. An untamped charge requires approximately twice the explosive charge of a tamped charge to produce the same effect.

Second Charges

F-95. Charges will not cut metal reinforcing rods inside concrete targets. If the ROE permit, hand grenades should be thrown into the opening to clear the area of enemy. Once the area has been cleared of enemy, the reinforcing rods can be removed using special steel-cutting explosive charges or mechanical means.

Door Charges

F-96. Various charges can be utilized for explosive reduction of doors. Leaders must conduct extensive training on the use of the charges to get proper target feedback.

F-97. The general-purpose charge, rubber band charge, and the flexible linear charge are field-expedient charges that can be used to reduce interior and exterior doors. These charges give the breach element an advantage because they can be made ahead of time and are simple, compact, lightweight, and easy to emplace. See FM 3-06.11 for more information.

General-Purpose Charge

F-98. This charge is the most useful ready charge for reducing a door or other barrier. It can cut mild steel chain and destroy captured enemy equipment. To construct the general purpose charge—

- Take a length of detonation cord about 2 feet long. Using another length of detonation cord, tie two uli knots around the 2-foot long cord.
  - The uli knots need to have a minimum of six wraps and be loose enough for them to slide along the main line, referred to as an uli slider.
  - Trim the excess cord from the uli knots and secure them with tape.
- Cut a block of C4 explosive to a 2-inch square.
- Tape one slider knot to each side of the C4 block, leaving the length of detonation cord free to slide through the knots.

F-99. To place the charge, perform the following:

- To reduce a standard door, place the top loop of the charge over the doorknob. Slide the uli knots taped to the C4 so the charge is tight against the knob.
- Prime the loose ends of the detonation cord with an MDI firing system and detonate.

NOTE: To cut mild steel chain, place the loop completely around the chain link to form a girth hitch. Tighten the loop against the link by sliding the uli knots.

Rubber Band Charge

F-100. The rubber band charge is an easily fabricated lightweight device that can be used to remove the locking mechanism or doorknob from wooden/light metal doors, or to break a standard-size padlock at the shackle. To construct the rubber band charge—

- Cut a 10-inch piece of detonation cord and tie an overhand knot in one end.
- Using another piece of detonation cord, tie an uli knot with at least eight wraps around the first length of cord.
- Slide the uli knot tightly up against the overhand knot. Secure it in place with either tape or string.
- Loop a strong rubber band around the base of the uli knot tied around the detonation cord.
- Tie an overhand knot in the other end of the cord to form a pigtail for priming the charge.

F-101. To place the charge, attach the charge to the doorknob (or locking mechanism) by putting the loose end of the rubber band around the knob. The charge must be placed between the knob and the doorframe. This ensures the explosive is over the bolt that secures the door to the frame.

Flexible Linear Charge

F-102. The simplest field-expedient charge for reducing wooden doors is the flexible linear charge. See Tables F-3 and F-4 for charge use and system components. It can be made in almost any length and is easily carried until needed. It is effective against hollow-core, particle-filled, and solid wood doors. When detonated, the flexible linear charge cuts through the door near the hinges

F-103. To construct the flexible linear charge, lay out a length of double-sided contact tape with the topside adhesive exposed. Place the necessary number of strands of detonation cord down the center of the double-sided tape, pressing them firmly in place. Military detonation cord has 50 grains of explosives per foot and there are 7,000 grains in a pound. Most residential doors are 80 inches tall. Commercial doors are 84 inches tall. This must be considered when calculating the quantities of explosives, overpressure, and MSDs. For hollow-core doors, use a single strand; for particle-filled doors, use two strands; and for solid wood doors, use three strands. If the door type is unknown, use three strands. One of the strands must be cut about a foot longer than the others and should extend past the end of the double-sided tape. This forms a pigtail where the initiating system is attached once the charge is in place. Cover the strands of detonation cord and all the exposed portions of the double-sided tape with either sturdy single-sided tape or another length of double-sided tape. Roll the charge, starting at the pigtail, with the double-sided tape surface that is to be placed against the door on the inside.

F-104. At the breach site, place the charge straight up and down against the door tightly. If it is too short, place it so it covers at least half of the door's height. Prime and fire the charge from the bottom.

Table F-3. Charges.

Charge	Obstacle	Explosives Needed	Advantages	Disadvantages
Wall breach charge (satchel or U-shaped charge)	Wood, masonry, brick, and reinforced concrete walls	– Detonation cord – C4 or TNT	– Easy and quick to make – Quick to place on target	– Does not destroy rebar – High overpressure – Appropriate attachment methods needed – Fragmentation
Silhouette charge	Wooden doors (creates man-sized hole); selected walls (plywood, sheet-rock, CMU)	Detonation cord	– Minimal shrapnel – Easy to make – Makes entry hole to exact specifications	– Bulky; not easily carried
General purpose charge	Door knobs, mild steel chain, locks, and equipment	– C4 – Detonation cord	– Small, lightweight – Easy to make – Very versatile	Other locking mechanisms may make charge ineffective
Rubber strip charge	Wood or metal doors (dislodges doors from the frame); windows with a physical security system	– Sheet explosive – Detonation cord	– Small, lightweight – Quick to place on target – Uses small amounts of explosives	
Flexible linear charge	Wooden doors (widow cuts door along the length of the charge)	Detonation cord	– Small, lightweight – Quick to place on target – One man can carry several charges – Defeats most doors regardless of locking systems	Proper two-sided adhesive required
Doorknob charge	Doorknobs on wood or light metal doors	Detonation cord or flexible linear shaped charge	– Small, lightweight – Easily transported – Quick to place on door	Other locking mechanisms may make charge ineffective
Chain-link ladder charge	Chain link fence (rapidly creates a hole large enough to run through)	– C4 – Detonation cord	Cuts chain link quickly and effectively	Must stand to emplace it

Table F-4. Firing system components.

Firing System	Components
Time system	2 x M81 or M60; time fuze or M-14; 2 x M7 caps; detonation cord loop; red devil (detonation cord connector)
Command detonated	2 x M81; 2 x shock tube with caps (M11 or M12); detonation cord loop; red devil (desired length)
Delay system	1 x M81 or M60 (gutted); black adapter cap; direct shoot shock tube (NONEL); M11 MDI; detonation cord loop; red devil (STI may be used instead of a direct shoot with an M60)

Explosive Safety Factors

F-105. When employing explosives during breaching operations, leaders must consider three major safety factors: overpressure; missile hazard; and minimum safe distance requirements.

1 - Overpressure

F-106. Overpressure is the pressure per square inch (PSI) released from the concussion of the blast, both outside and into the interior of the building or room, that can injure, incapacitate, or kill.

2 - Missile Hazard

F-107. Missile hazards are fragmentation or projectiles sent at tremendous speed from the explosion area. This occurs from either the charge or target being breached.

3 - Minimum Safe Distance Requirements

F-108. When using explosives in the urban environment, Soldiers must consider the presence of noncombatants and friendly forces. Additionally, there are many hazardous materials located in the urban environment, including chemicals and construction materials. There is always a risk of secondary explosions and fires when employing explosive breaching techniques.

CAUTION

Always handle explosives carefully. Never divide responsibility for preparing, placing, priming, and firing charges. Always use proper eye and ear protection and cover exposed skin to prevent injuries. Explosives may produce hazardous fumes, flames, fragments, and overpressure. Use AR 385-63, FM 5-34, FM 5-250, and risk assessment to determine minimum safe distances (MSDs). Take into consideration whether the door is flush or receded when considering MSD.

REDUCE BOOBY TRAPS

F-109. Soldiers must be aware of the threat presented by booby traps that can be found in any operating environment in which the platoon might operate. The platoon must receive sufficient training to recognize locations and items that lend themselves to booby-trapping, striking a balance between what is possible and what is probable. See FM 20-32 for more information on booby traps and expedient devices.

F-110. When dealing with booby traps, the following rules and safety procedures can save lives:

- Suspect any object that appears to be out of place or artificial in its surroundings. Remember, what you see may well be what the enemy wants you to see. If you did not put it there, do not pick it up.
- Examine mines and booby traps from all angles, and check for alternative means of detonating before approaching them.
- Ensure that only one man works on a booby trap.
- Do not use force. Stop if force becomes necessary.
- Do not touch a trip wire until both ends have been investigated and all devices are disarmed and neutralized.
- Trace trip wires and check for additional traps along and beneath them.
- Treat all parts of a trap with suspicion, because each part may be set to actuate the trap.
- Wait at least 30 seconds after pulling a booby trap or a mine. There might be a delay fuse.
- Mark all traps until they are cleared.
- Expect constant change in enemy techniques.
- Never attempt to clear booby traps by hand if pulling them or destroying them in place is possible and acceptable.

F-111. Booby traps might be found in recently contested areas, so no items or areas that have not been cleared should be considered safe. By anticipating the presence of traps, it might be possible to isolate and bypass trapped areas. If this is not possible, employ countermeasures such as avoiding convenient and covered resting places along routes where mines or other explosive devices can be located. Collective training in booby-trap awareness and rapidly disseminating booby-trap incident reports to all levels is vital. This allows Soldiers to develop an understanding of the enemy's method of operation and a feel for what might or might not be targets.

INDICATIONS AND DETECTION

F-112. Successful detection depends on two things: being aware of what might be trapped and why, and being able to recognize the evidence of setting. The first requirement demands a well developed sense of intuition; the second, a keen eye. Intuition is gained through experience and an understanding of the enemy's techniques and habits. A keen eye is the result of training and practice in the recognition of things that might indicate the presence of a trap.

F-113. Detection methods depend on the nature of the environment. In open areas, methods used to detect mines can usually detect booby traps. Look for trip wires and other signs suggesting the presence of an actuating mechanism. In urban areas, mine detectors are probably of little use. The platoon will have to rely on manual search techniques and, if available, special equipment. The presence of booby traps or nuisance mines is indicated by—

- Disturbance of ground surface or scattered, loose soil.
- Wrappers, seals, loose shell caps, safety pins, nails, and pieces of wire or cord.
- Improvised methods of marking traps, such as piles of stones or marks on walls or trees.
- Evidence of camouflage, such as withered vegetation or signs of cutting.
- Breaks in the continuity of dust, paint, or vegetation.
- Trampled earth or vegetation; foot marks.
- Lumps or bulges under carpet or in furniture.

REDUCTION METHODS

F-114. Reducing booby traps and nuisance mines in AOs is done primarily by engineers, especially in secured areas. However, some booby traps may have to be cleared by Infantry Soldiers to accomplish a mission during combat. The method used to disarm a trap depends on many things including, time constraints, personnel assets, and the type of trap. A trap cannot be considered safe until the blasting cap or the detonation cord has been removed from the charge.

F-115. Use the safest method available to neutralize a trap. For example, if the firing device and the detonation cord are accessible, it is usually safer to cut the detonation cord. This method does not actuate the trap, but inserting pins in the firing device might. Unit resources or locally-manufactured or acquired aids are often used to clear traps. In areas with a high incidence of booby traps, assemble and reserve special clearing kits. Mark all booby traps found.

F-116. Nonexplosive traps are typically used in tropical or rain forest regions. Ideal construction materials abound and concealment in surrounding vegetation is relatively easy. No prescribed procedures exist for clearing nonexplosive traps. Each trap must be cleared according to its nature.

SECTION III — OBSTACLE EMPLOYMENT

F-117. Obstacles are used to reinforce the terrain. When combined with fires, they disrupt, fix, turn, or block an enemy force. Obstacles are used in all operations, but are most useful in the defense. Leaders must always consider what materials are needed and how long the obstacle will take to construct. See FM 5-34 for detailed instructions on specific types of obstacle construction methods.

F-118. A primary concern for the platoon in the defense is to supplement their fortified positions with extensive protective obstacles, both antipersonnel and antivehicle (particularly antipersonnel).

Antipersonnel obstacles, both explosive and nonexplosive, include all those mentioned in Section I (such as wire entanglements, antipersonnel mines, and field expedient devices), and are used to prevent enemy troops from entering a friendly position. Antipersonnel obstacles are usually integrated with fires and are close enough to the fortification for adequate surveillance by day or night, but beyond effective hand grenade range. Obstacles are also used within the position to compartmentalize the area in the event outer protective barriers are breached.

F-119. In the offense, the platoon/squad uses obstacles to—

- Aid in flank security.
- Limit enemy counterattack.
- Isolate objectives.
- Cut off enemy reinforcement or routes of withdrawal.

F-120. In the defense, the platoon uses obstacles to—

- Slow the enemy's advance to give the Infantry platoon more time to mass fires on him.
- Protect defending units.
- Canalize the enemy into places where he can more easily be engaged.
- Separate the enemy's tanks from his Infantry.
- Strengthen areas that are lightly defended.

MINES

U.S. NATIONAL POLICY ON ANTIPERSONNEL LAND MINES

On 16 May 1996, The President of the United States announced a national policy that eliminates or restricts the use of antipersonnel land mines, beginning with those that do not self-destruct, but eventually including all types. This policy is now in effect. It applies to all Infantry units either engaged in, or training for, operations worldwide.

Current U.S. policy allows the use of non-self-destructing antipersonnel land mines only along internationally recognized national borders or in established demilitarized zones, specifically for the defense of South Korea. Such mines must be within an area having a clearly marked perimeter. They must be monitored by military personnel and protected by adequate means to ensure the exclusion of civilians.

U.S. national policy also forbids U.S. forces from using standard or improvised explosive devices such as booby traps.

Except for South Korea based units and units deploying there for designated exercises, this policy specifically forbids all training on or actual employment of inert M14 and M16 antipersonnel land mines. Policy applies at the unit's home station and at Combat Training Centers, except in the context of countermine or de-mining training. No training with live M14 mines is authorized, and training with live M16 mines is authorized only for Soldiers actually on South Korean soil.

This policy does not affect the standard use of antivehicular mines. Nor does it affect training and use of the M18 Claymore mine in the command detonated mode.

When authorized by the appropriate commander, units may still use self-destructing antipersonnel mines such as the ADAM. Authorized units may also continue to emplace mixed minefields containing self-destructing antipersonnel land mines and antivehicular land mines such as MOPMS or Volcano.

The terms "mine", "antipersonnel obstacle", "protective minefield", or "minefield" contained in this FM should not be construed to mean an obstacle that contains non-self-destructing antipersonnel land mines or booby traps. Also, all references to antipersonnel mines and the employment of minefields should be considered in accordance with national policy that limits the use of non-self-destructing antipersonnel land mines.

F-121. Mines are one of the most effective tank and personnel killers on the battlefield. The type of minefield that a platoon or squad most commonly emplaces is the hasty protective.

F-122. It is important to distinguish the difference between the types of minefield and the means of emplacement. Volcano, Modular Pack Mine System (MOPMS), standard-pattern, and row mining are not types of minefields; they are just some of the means used to emplace tactical, nuisance, and protective minefields. They may also be the method of emplacement that is replicated by a phony minefield. Land-based mines and munitions are hand-emplaced, remote-delivered, air-delivered, or ground-delivered (Table F-5). FM 20-32 provides detailed instructions on the installation and removal of U.S. mines and firing devices.

Table F-5. Mine delivery methods.

Delivery Method	Characteristics
Hand-emplaced	Require manual arming and are labor-, resource-, and transport-intensive.
Remote- and Air-delivered	Require less time and labor; however, they are not as precisely placed as hand-emplaced mines and munitions.
Ground-delivered	Less resource-intensive than hand-emplaced mines. They are not precisely placed; however, the minefield boundaries are.

Scatterable Mines

F-123. SCATMINEs are laid without regard to a classical pattern. They are designed to be delivered remotely by aircraft, artillery, missile, or a ground dispenser. All U.S. SCATMINEs have a limited active life and self-destruct (SD) after that life has expired. The duration of the active life varies with the type of mine and the delivery system.

F-124. SCATMINEs enable minefield emplacement in enemy-held territories, contaminated territories, and in most other areas where it is impossible for engineers or the platoon to emplace conventional minefields. They may be used to support the platoon's mission by turning, fixing, disrupting, and blocking the enemy. However they are used, they must be planned and coordinated to fit into the overall obstacle plan. Characteristics of AP SCATMINE systems are listed in Table F-6. Table F-7 lists AT SCATMINE characteristics. SCATMINE placement authority is shown in Table F-8.

Table F-6. Characteristics of AP SCATMINE systems.

Mine	Delivery System	DODIC	Arming Time	Fuse	Warhead	AHD	SD Time	Explosive Weight	Mine Weight	Number of Mines
M67	155-mm artillery (ADAM)	D502	within 1 min after ground impact	trip wire	bounding frag	20%	4 hr	21 g Comp A5	540 g	36 per M731 projectile
M72	155-mm artillery (ADAM)	D501	within 1 min after ground impact	trip wire	bounding frag	20%	48 hr	21 g Comp A5	540 g	36 per M692 projectile
BLU 92/B	USAF (Gator)	K291 K292 K293	2 min	trip wire	blast frag	100%	4 hr 48 hr 15 days	540 g Comp B4	1.44 kg	22 per CBU 89/B dispenser
M77	MOPMS	K022	2 min	trip wire	blast frag	0%	4 hr (recycle up to 3 times)	540 g Comp B4	1.44 kg	4 per M131 dispenser
Volcano	Ground/ air	K045	2 min	trip wire	blast frag	0%	4hr 48 hr 15 days	540 g Comp B4	1.44 kg	1 per M87 canister

Table F-7. AT SCATMINE characteristics.

Mine	Delivery System	DODIC	Arming Time	Fuse	Warhead	AHD	SD Time	Explosive Weight	Mine Weight	Number of Mines
M73	155-mm artillery (RAAM)	D503	within 1 min after ground impact	magnetic	M-S plate	20%	48 hr	585 g RDX	1.7 kg	9 per M718 projectile
M70	155-mm artillery (RAAM)	D509	within 1 min after ground impact	magnetic	M-S plate	20%	4 hr	585 g RDX	1.7 kg	9 per M741 projectile
BLU 91/B	USAF (Gator)	K291 K292 K293	2 min	magnetic	M-S plate	NA	4 hr 48 hr 15 days	585 g RDX	1.7 kg	72 per CBU 89/B dispenser
M76	MOPMS	K022	2 min	magnetic	M-S plate	NA	4 hr (recycle up to 3 times)	585 g RDX	1.7 kg	17 per M131 dispenser
Volcano	Ground/ air	K045	2 min 30 sec	magnetic	M-S plate	NA	4 hr 48 hr 15 days	585 g RDX	1.7 kg	5 per M87 canister; 6 per M87A1 canister

Table F-8. SCATMINE emplacement authority.

Scatterable Mine System	Emplacement Authority
Ground- or artillery-delivered, with self-destruct time greater than 48 hours (long duration).	The corps commander may delegate emplacement authority to division level, which may further delegate to brigade level.
Ground- or artillery-delivered, with self-destruct time of 48 hours or less (short duration).	The corps commander may delegate emplacement authority to division level, which may further delegate to brigade level, which may further delegate to battalion level.
Aircraft-delivered (Gator), regardless of self-destruct time.	Emplacement authority is normally at corps, theater, or army command level, depending on who has air-tasking authority.
Helicopter-delivered (Volcano), regardless of self-destruct time.	Emplacement authority is normally delegated no lower than the commander who has command authority over the emplacing aircraft.
MOPMS when used strictly for a protective minefield.	Emplacement authority is usually granted to the company or base commander. Commanders at higher levels restrict MOPMS use only as necessary to support their operations.

Modular Pack Mine System (MOPMS), Man-Portable

F-125. The man-portable, 162-pound, suitcase-shaped MOPMS dispenses a total of 21 mines (17 antitank mines and 4 antipersonnel mines). It propels them in a 35-meter, 180-degree semicircle from the container. Mines are dispensed on command using the M71 remote control unit (RCU) or an electronic initiating device such as the M34 blasting machine. When dispensed, an explosive propelling charge at the bottom of each tube expels mines through the container roof (Figure F-12). The Infantry platoon can use MOPMS to create a protective minefield or to close lanes in tactical obstacles. The safety zone around one container is 55 meters to the front and sides, and 20 meters to the rear. MOPMS has a duration of 4 hours, which can be extended up to three times for a total of 16 hours. Once mines are dispensed, they cannot be recovered or reused. If mines are not dispensed, the container may be disarmed and recovered for later use. The RCU can also self-destruct mines on command, allowing a unit to counterattack or withdraw through the minefield. The RCU can control up to 15 MOPMS containers or groups of MOPMS containers from a distance of 300 to 1,000 meters.

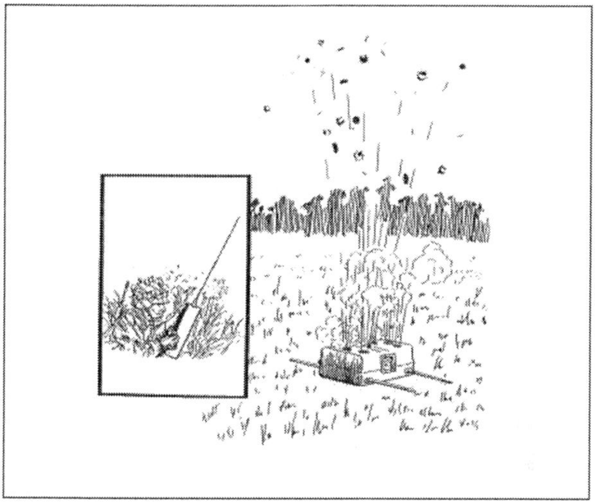

Figure F-12. MOPMS.

Hornet

F-126. The Hornet is a man-portable, nonrecoverable, AT/antivehicular, off-route munition made of lightweight material (35 pounds) that one person can carry and employ. It is capable of destroying vehicles by using sound and motion detection methods. It will automatically search, detect, recognize, and engage moving targets by using top attack at a standoff distance up to 100 meters. It can be a stand-alone tactical obstacle or can reinforce other conventional obstacles. It disrupts and delays the enemy, allowing long-range, precision weapons to engage more effectively. This feature is particularly effective in non-line-of-sight (LOS) engagements. It is normally employed by combat engineers, Rangers, and SOF. The remote control unit (RCU) is a handheld encoding unit that interfaces with the Hornet when the remote mode is selected at the time of employment. After encoding, the RCU can be used to arm the Hornet, reset its SD times, or destroy it. The maximum operating distance for the RCU is 2 kilometers.

CONVENTIONAL MINES

F-127. Conventional mines are hand-emplaced mines that require manual arming. This type of mine laying is labor-, resource-, and transport-intensive. Soldiers emplace conventional mines within a defined, marked boundary and lay them individually or in clusters. They record each mine location so the mines can be recovered. Soldiers can surface lay or bury conventional mines and may place AHDs on AT mines. FM 21-75 has complete information on emplacement of conventional AT mines.

NOTE: U.S. Soldiers can surface lay or bury AT mines and munitions and can place AHDs on hand-emplaced AT mines. Some countries employ conventional AP mines (with or without AHDs), but U.S. forces are not authorized to employ conventional AP mines (except on the Korean peninsula).

Antitank Mines

F-128. The M15 and M21 AT mines are used by U.S. forces. They are shown in Figure F-13. Their characteristics are listed in Table F-9.

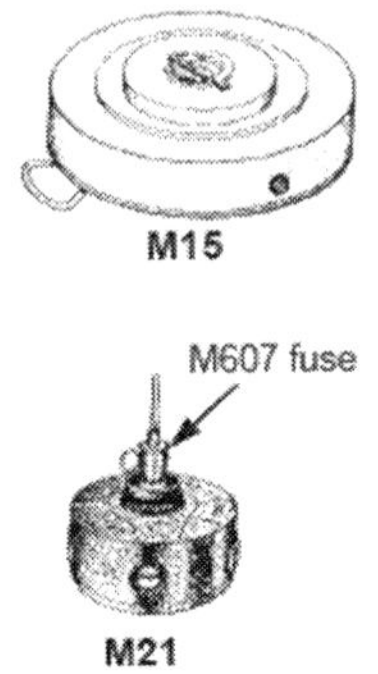

Figure F-13. Antitank (AT) mines.

Table F-9. Characteristics of AT mines.

Mine	DODIC	Fuse	Warhead	AHD	Explosive Weight	Mine Weight	Mines per Container
M15 with M603 fuse	K180	pressure	blast	yes	9.9 kg	13.5 kg	1
M15 with M624 fuse	K180 (mine) K068 (fuse)	tilt rod	blast	yes	9.9 kg	13.5 kg	1
M21	K181	tilt rod or pressure	SFF	yes*	4.95 kg	7.6 kg	4
*Conventional AHDs will not couple with this mine. However, the M142 multipurpose firing device can be emplaced under this mine.							

Antipersonnel Mines

F-129. The M14 and M16 AP mines are used by U.S. forces on the Korean peninsula. They are also used by many other countries. These mines are shown in Figure F-14. Their characteristics are listed in Table F-10.

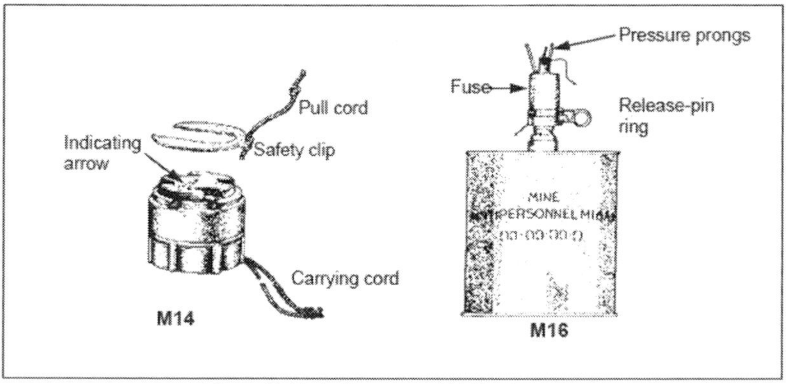

Figure F-14. Antipersonnel (AP) mines.

Table F-10. Characteristics of AP mines.

Mine	DODIC	Fuse	Warhead	AHD	Explosive Weight	Mine Weight	Mines per Container
M14	K121	pressure	blast	no	28.4 g	99.4 g	90
M16-series	K092	pressure or trip wire	bounding frag	no	450 g	3.5 kg	4

SPECIAL-PURPOSE MUNITIONS

F-130. Special-purpose munitions that the platoon might employ include the M18A1 Claymore and the selectable lightweight attack munition (SLAM).

M18A1 CLAYMORE

F-131. The M18A1 Claymore (Figure F-15) is a fragmentation munition that contains 700 steel balls and 682 grams of composition C4 explosive. It weighs 1.6 kilograms and is command detonated.

F-132. When employing the Claymore with other munitions or mines, separate the munitions by the following minimum distances:

- Fifty meters in front of or behind other Claymores.
- Three meters between Claymores that are placed side by side.
- Ten meters from AT or fragmentation AP munitions.
- Two meters from blast AP munitions.

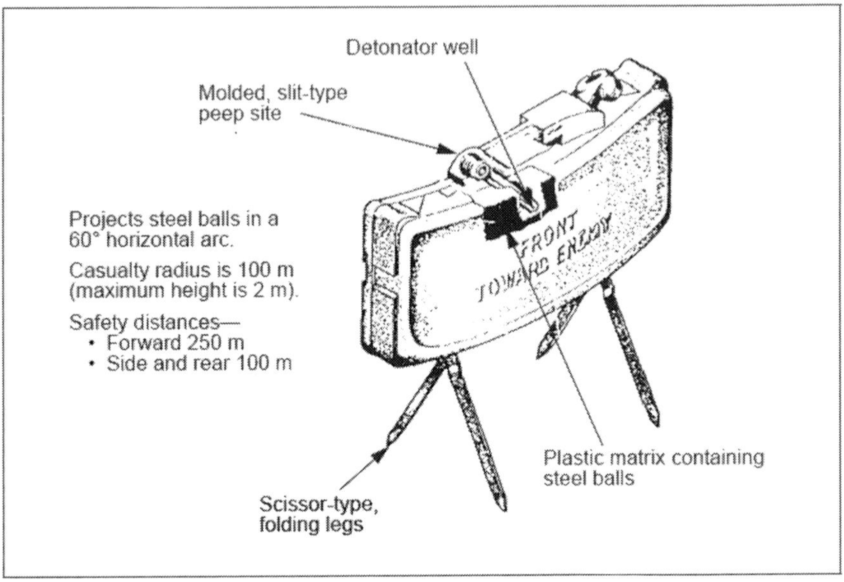

Figure F-15. M18A1 Claymore.

SLAM

F-133. The M4 SLAM is a multipurpose munition with an antitamper feature (Figure F-16). It is compact and weighs only 1 kilogram. It is easily portable and is intended for use against APCs, parked aircraft, wheeled or tracked vehicles, stationary targets (such as electrical transformers), small (less than 10,000-gallon) fuel-storage tanks, and ammunition storage facilities. The explosive formed penetrator (EFP) warhead can penetrate 40 millimeters of homogeneous steel. The SLAM has two models (the self-neutralizing [M2] and self-destructing [M4]). The SLAM's four possible employment methods include: bottom attack, side attack, timed demolition, and command detonation.

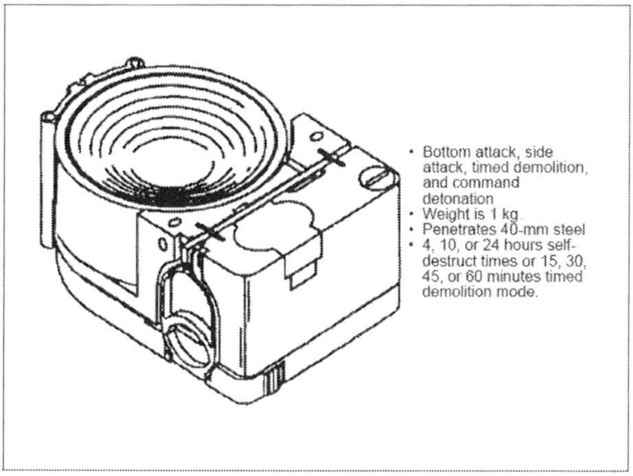

Figure F-16. SLAM.

M93 HORNET

F-134. The M93 Hornet is an AT and antivehicular off-route munition made of lightweight material (35 pounds) that one person can carry and employ (Figure F-17). It is a nonrecoverable munition capable of destroying vehicles through the use of sound and motion detection. It will automatically search, detect, recognize, and engage moving targets by using top attack mode at a standoff distance up to 100 meters from the munition.

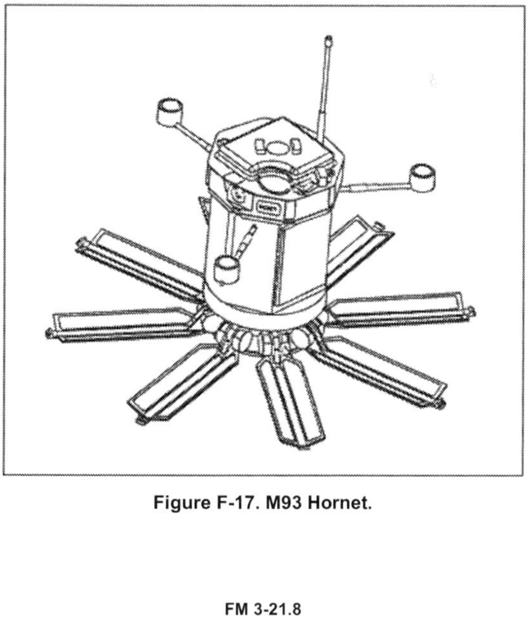

Figure F-17. M93 Hornet.

Hasty Protective Minefields

F-135. Neither AP nor AT mines are used in isolation. The majority of mine composition is designed against the most severe close-combat threat and the likelihood of that threat. The MOPMS automatically dispenses a mix of AT and AP mines.

F-136. In the defense, platoons lay hasty protective minefields to supplement weapons, prevent surprise, and give early warning of enemy advance. A platoon can install hasty protective minefields, but only with permission from the company commander. Conventional hasty protective minefields are reported to the company commander and recorded on DA Form 1355-1-R, *Hasty Protective Row Minefield Record.* The minefield should be recorded before the mines are armed. The leader puts the minefield across likely avenues of approach, within range of and covered by his organic weapons. If time permits, the mines should be buried to increase effectiveness, but they may be laid on top of the ground in a random pattern. The leader installing the minefield should warn adjacent platoons and tell the company commander of the minefield's location. When the platoon leaves the area (except when forced to withdraw by the enemy), it must remove the minefield (if it uses recoverable mines) or transfer the responsibility for the minefield to the relieving platoon leader. Only metallic mines are used in conventional hasty protective minefields. Booby traps are not used in hasty protective minefields because they delay removal of recoverable mines. The employing Infantry platoon must make sure that the minefield can be kept under observation and covered by fire at all times.

F-137. After requesting and receiving permission to lay the minefield, the Infantry platoon leader reconnoiters to determine exactly where to place the mines. While the Soldiers are placing the mines, the Infantry platoon leader finds an easily identifiable reference point in front of the platoon's position. A tree stump is used as the reference point in sample DA Form 1355-1-R shown in Figure F-18. The platoon leader records the minefield. The row of mines closest to the enemy is designated A, and the succeeding rows are B, C, and so on.

F-138. The ends of a row are shown by two markers. They are labeled with the letter of the row and number 1 for the right end of the row and number 2 for the left end of the row. The rows are numbered from right to left, facing the enemy. The marker can be a steel picket or wooden stake with a nail or a can attached so it can be found with a metallic mine detector.

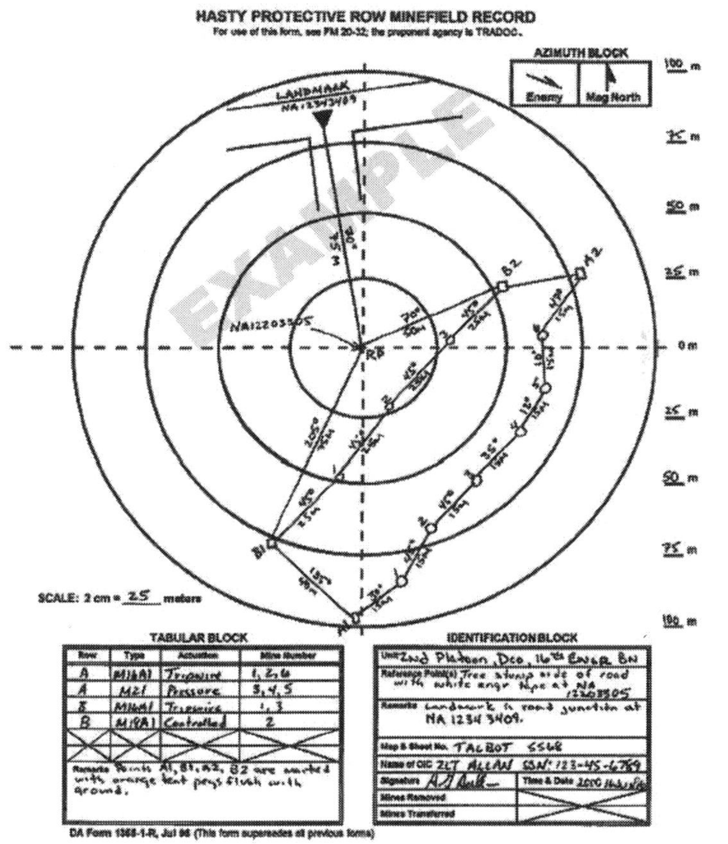

Figure F-18. Sample DA Form 1355-1-R (hasty protective row minefield record).

F-139. The platoon leader places a marker at B-1 and records the azimuth and distance from the reference point to B-1 on DA Form 1355-1-R.

F-140. Next, from B-1 the platoon leader measures the azimuth and distance to a point 15 to 25 paces from the first mine in row A. He places a marker at this point and records it as A-1. The platoon leader then measures the distance and azimuth from A-1 to the first mine in row A and records the location of the mine. He then measures the distance and azimuth from the first mine to the second, and so on until all mine locations have been recorded as shown. The platoon leader gives each mine a number to identify it in the tabular block of DA Form 1355-1-R. When the last mine location in row A is recorded, the platoon leader measures an azimuth and distance from the last mine to another arbitrary point between 15 and 25 paces beyond the last mine. He places a marker here and calls it A-2. The platoon leader follows the same procedure with row B.

F-141. When the platoon leader finishes recording and marking the rows, he measures and records the distance and azimuth from the reference point to B-2 to A-2. If antitank mines are being used, it is recommended that they be used at the A-2/B-2 markers, because their large size facilitates retrieval.

F-142. The platoon leader now ties in the reference point with a permanent landmark that he found on the map. He measures the distance and the azimuth from this landmark to the reference point. The landmark might be used to help others locate the minefield should it be abandoned. Finally, he completes the form by filling in the tabular and identification blocks.

F-143. While the platoon leader is tying in the landmark, the Soldiers arm the mines nearest the enemy first (Row A). The platoon leader reports that the minefield is completed and keeps DA Form 1355-1-R. If the minefield is transferred to another platoon, the gaining platoon leader signs and dates the mines transferred block and accepts the form from the previous leader. When the minefield is removed, the form is destroyed. If the minefield is left unattended or abandoned unexpectedly, the form must be forwarded to the company commander. The company commander forwards it to be transferred at battalion to more permanent records.

F-144. When retrieving the recoverable mines, the Soldiers start at the reference point and move to B-1, using the azimuth and distances as recorded. They then move from B-1 to the first mine in row B. However, if B-1 is destroyed, they move from the reference point to B-2 using that azimuth and distance. They will now have to shoot the back azimuth from B-2 to the last mine. The stakes at A-1, B-1, A-2, and B-2 are necessary because it is safer to find a stake when traversing long distances than to find a live mine.

WIRE OBSTACLES

F-145. The platoon normally employs wire obstacles as part of the protective obstacle plan in the defense. Wire obstacles include barbed-wire, triple-standard concertina, four-strand cattle fences, and tanglefoot. Construction methods for two of the more common wire obstacles that the platoon employs, triple standard concertina, and tanglefoot, are shown in Figures F-19 through F-23. See FM 5-34 for more information on these and other wire obstacles.

TRIPLE STANDARD CONCERTINA FENCE

F-146. The most common wire entanglement a platoon or squad may build is the triple standard concertina fence. It is built of either barbed wire concertina or barbed tape concertina. There is no difference in building methods. The material and labor requirements for a 300-meter triple standard concertina fence are—

- Long pickets – 160.
- Short pickets – 4.
- Barbed wire, 400-meter reels – 3.
- Rolls of concertina – 59.
- Staples – 317.
- Man-hours to erect – 30.

F-147. First, lay out and install pickets from left to right (facing the enemy). Put the long pickets five paces apart, and the short (anchor) pickets two paces from the end of the long pickets (Figure F-19). The enemy and friendly picket rows are offset and are placed 3 feet apart. Now lay out rolls of concertina. Place a roll in front of the third picket on the enemy side, and two rolls to the rear of the third picket on the friendly side. Repeat this step every fourth picket thereafter. Install the front row concertina and horizontal wire (Figure F-20). Place the concertina over the pickets. Install the rear row of concertina and horizontal wire. Install the top row of concertina and join the rear horizontal wire (Figure F-21).

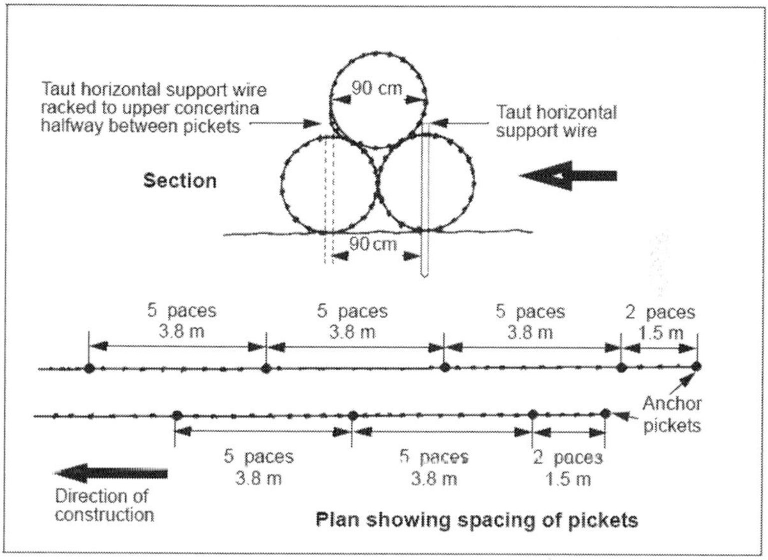

Figure F-19. Triple standard concertina fence.

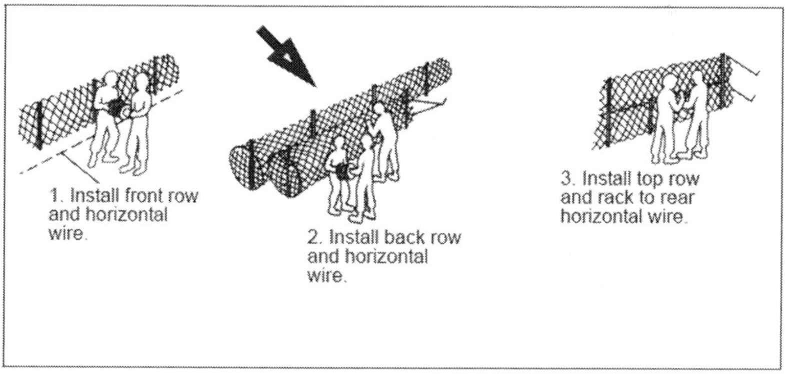

Figure F-20. Installing concertina.

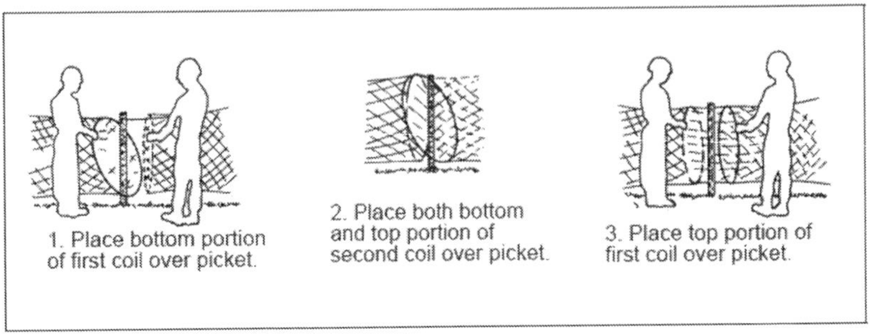

Figure F-21. Joining concertina.

CONCERTINA ROADBLOCK

F-148. The concertina roadblock is placed across roadways and designed to block wheeled or tracked vehicles. The roadblock is constructed of 11 concertina rolls or coils placed together, about 10 meters in depth, reinforced with long pickets five paces apart. The rolls or coils should not be tautly bound allowing them to be dragged and tangled around axles, tank road wheels, and sprockets. Additionally, wire is placed horizontally on top of the concertina rolls or coils (Figure F-22).

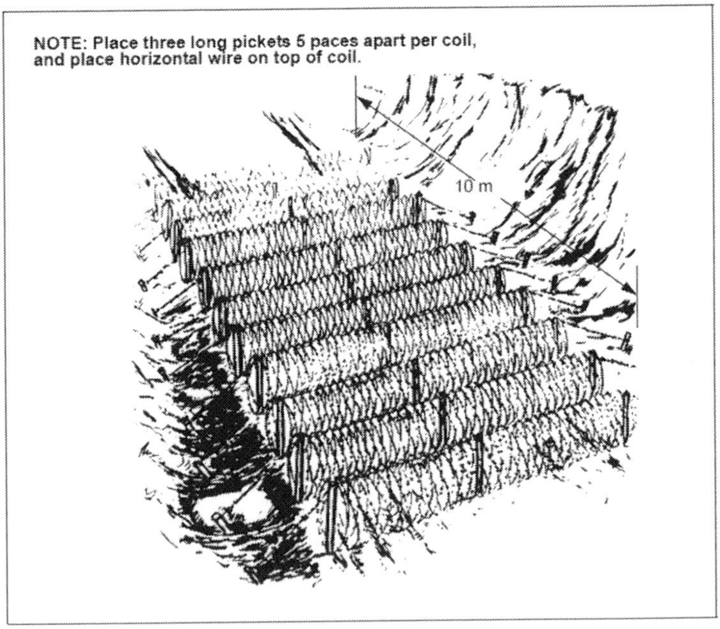

Figure F-22. Eleven-row antivehicular wire obstacle.

TANGLEFOOT

F-149. Tanglefoot is used where concealment is essential and to prevent the enemy from crawling between fences and in front of emplacements (Figure F-23). The obstacle should be employed in a minimum width of 32 feet. The pickets should be placed at irregular intervals of 2 ½ feet to 10 feet. The height of the barbed wire should vary between 9 to 30 inches. Tanglefoot should be sited in scrub, if possible, using bushes as supports for part of the wire. On open ground, short pickets should be used.

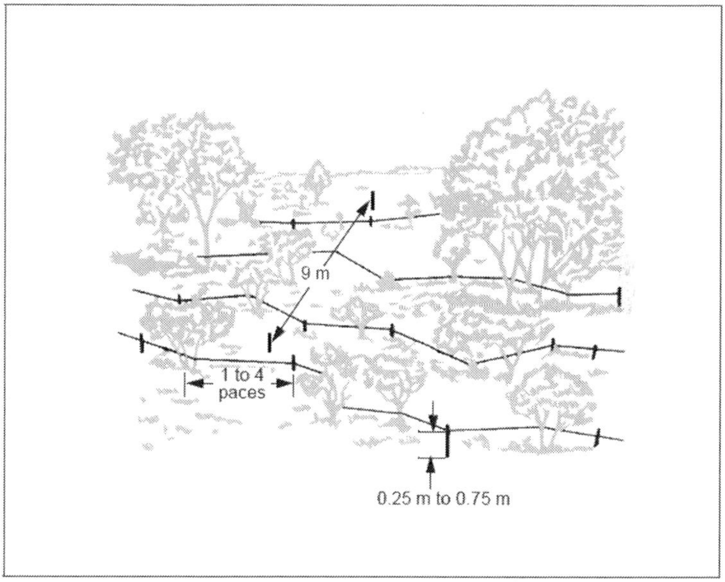

Figure F-23. Tanglefoot.

Appendix G

Other Small Unit Organizations

As part of full spectrum operations, the Infantry platoon can expect to conduct missions with other types of Infantry platoons (within and outside of their own Infantry battalion), combat arms units, and combat support units. To aid the Infantry platoon leader, this appendix briefly discusses the structure, capabilities, and limitations of—

- The Infantry battalion scout platoon
- Infantry battalion mortar platoon
- Infantry battalion sniper section
- Bradley platoon and squad
- Stryker platoon and squad
- Maneuver company fire support team (FIST)
- Combat engineer support
- Air defense assets
- Tank platoon

INFANTRY BATTALION SCOUT PLATOON

G-1. The Infantry battalion scout platoon serves as the forward "eyes and ears" for the battalion commander. The primary mission of the scout platoon is to conduct reconnaissance and security to answer CCIR, normally defined within the battalion's intelligence, surveillance, reconnaissance (ISR) plan. The scout platoon can conduct route, zone, and area reconnaissance missions. The platoon can also conduct limited screening operations and can participate as part of a larger force in guard missions.

G-2. The scout platoon is organized into a platoon headquarters and three squads of six men each. Each squad leader is responsible for controlling his squad's movement and intelligence collection requirements. He reports critical intelligence information obtained by his squad to the scout platoon leader or battalion TOC.

G-3. In either offensive or defensive operations, the commander may deploy his scout platoon to conduct screening operations of the battalion's front, flank, or rear. The scout platoon may also occupy outposts from which it can relay critical information to the TOC concerning enemy composition, disposition, and activities.

INFANTRY BATTALION MORTAR PLATOON

G-4. The primary role of the Infantry battalion mortar platoon is to provide immediate, responsive indirect fires in support of the maneuver companies or battalion. The battalion mortar platoon consists of a mortar platoon headquarters, a mortar section that contains the fire direction center (FDC), and four mortar squads. The platoon's FDC controls and directs the mortar platoon's fires. Infantry battalion mortar sections are equipped with 120-mm and 81-mm mortars, but only have the capability to man 50 percent of these mortars at any one time.

G-5. The mortar platoon provides the commander with the ability to shape the Infantry's close fight with indirect fires that—

- Provide close supporting fires for assaulting Infantry forces in any terrain.
- Destroy, neutralize, suppress, or disrupt enemy forces and force armored vehicles to button up.

- Fix enemy forces or reduce the enemy's mobility and canalize his assault forces into engagement areas.
- Deny the enemy the advantage of defile terrain and force him into areas covered by direct fire weapons.
- Optimize indirect fires in urban terrain.
- Significantly improve the Infantry's lethality and survivability against a close dismounted assault.
- Provide obscuration for friendly movement.

G-6. Each mortar system is capable of providing three primary types of mortar fires:

(1) High explosive (HE) rounds are used to suppress or destroy enemy Infantry, mortars, and other supporting weapons. HE is also used to interdict the movement of men, vehicles, and supplies in the enemy's forward area. Bursting white phosphorus (WP) rounds are often mixed with HE rounds to enhance their suppressive and destructive effects.

(2) Obscuration rounds are used to conceal friendly forces as they maneuver or assault and to blind enemy supporting weapons. Obscurants can also be used to isolate a portion of the enemy force while it is destroyed piecemeal. Some mortar rounds use bursting WP to achieve this obscuration. Bursting WP may be used to mark targets for engagement by other weapons, usually aircraft, and for signaling.

(3) Illumination rounds, to include infra-red illumination, are used to reveal the location of enemy forces hidden by darkness. They allow the commander to confirm or deny the presence of the enemy without revealing the location of friendly direct fire weapons. Illumination fires are often coordinated with HE fires to expose the enemy and to kill or suppress him.

INFANTRY BATTALION SNIPER SECTION

G-7. The primary mission of the sniper section in combat is to support combat operations by delivering precise long-range fire on selected targets. Snipers create casualties among enemy troops, slow enemy movement, lower enemy morale, and add confusion to their operations. They can engage and destroy high payoff targets. The secondary mission of the sniper section is collecting and reporting battlefield information. The sniper section is employed in all types of operations. This includes offensive, defensive, stability operations and civil support operations in which precision fire is delivered at long ranges. It also includes combat patrols, ambushes, countersniper operations, forward observation elements, military operations in urbanized terrain, and retrograde operations in which snipers are part of forces left in contact or as stay-behind forces.

COMPOSITION OF SNIPER SECTION

G-8. The Sniper section has 10 enlisted personnel: a section leader, 3 long range sniper rifle systems, and 3 standard sniper rifle systems. There are three sniper teams in the sniper section organized with a sniper, observer, and security. As a result, the sniper section can effectively employ three sniper teams at any one time. When necessary, the commander can employ up to five ad hoc sniper teams for limited duration missions by employing two man teams. Sniper teams can be task organized to any unit in the battalion or employed directly under battalion control. Snipers are most effective when leaders in the supported unit understand capabilities, limitations and tactical employment of sniper teams. See FM 3-21.10, *The Infantry Rifle Company,* and Appendix F for additional information on sniper team employment.

MECHANIZED INFANTRY RIFLE PLATOON AND SQUAD (BRADLEY)

G-9. BFV-equipped infantry rifle platoons and rifle squads normally operate as part of a larger force. They provide their own suppressive fires either to repel enemy assaults or to support their own maneuver. During close combat, platoon leaders consider the following to determine how to employ the BFVs.

- Support the rifle squads with direct fires.
- Provide mobile protection to transport rifle squads to the critical point on the battlefield.

- Suppress or destroy enemy infantry fighting vehicles and other lightly armored vehicles.
- Destroy enemy armor with TOW fires.

CAPABILITIES

G-10. The Bradley platoon's effectiveness is enhanced because of the lethality of its weapons systems and the rifle squad. To employ the platoon effectively, the platoon leader capitalizes on its strengths. The BFV-equipped mechanized infantry platoon can—

- Assault enemy positions.
- Assault with small arms and indirect fires to deliver rifle squads to tactical positions of advantage.
- Use 25-mm cannon and 7.62-mm machine gun fire to effectively suppress or destroy the enemy's infantry.
- Block dismounted avenues of approach.
- Seize and retain key and decisive terrain.
- Clear danger areas and prepare positions for mounted elements.
- Conduct mounted or dismounted patrols and operations in support of security operations.
- Develop the situation with Soldiers (three rifle squads) and equipment (25-mm cannon, TOW, and 7.62-mm coaxial machine gun).
- Establish strong points to deny the enemy important terrain or flank positions.
- Infiltrate enemy positions.
- Overwatch and secure tactical obstacles.
- Repel enemy attacks through close combat.
- Conduct assault breaches of obstacles.
- Participate in air assault operations.
- Destroy light armor vehicles using direct fire from the BFV.
- Employ 25-mm cannon fire to fix, suppress, or disrupt the movement of fighting vehicles and antiarmor systems up to 2,500 meters.
- Use TOW fires to destroy tanks and fighting vehicles out to 3,750 meters.
- Use Javelin fires to destroy tanks and fighting vehicles out to 2,000 meters.
- Operate in a chemical, biological, radiological, or nuclear (CBRN) environment.
- Participate in stability operations.

LIMITATIONS

G-11. BFV-equipped Infantry rifle platoons have the following limitations:

- Increased maintenance requirements.
- Increased fuel requirements.
- Size of vehicle limits maneuverability in restricted terrain.
- Load noise signature.
- Limited crew situational awareness.

ORGANIZATION

G-12. The mechanized infantry rifle platoon is equipped with four BFVs and can fight mounted or with rifle squads on the ground. Figure G-1 illustrates the BFV-equipped mechanized infantry rifle platoon organization. The platoon can fight as unified mutually supporting maneuver elements or as two distinct maneuver elements—one mounted and one dismounted. The platoon must prepare to fight in a variety of operational environments. Once the rifle squads have dismounted, the mounted element provides a base of fire for the rifle squads as they close with and destroy the enemy.

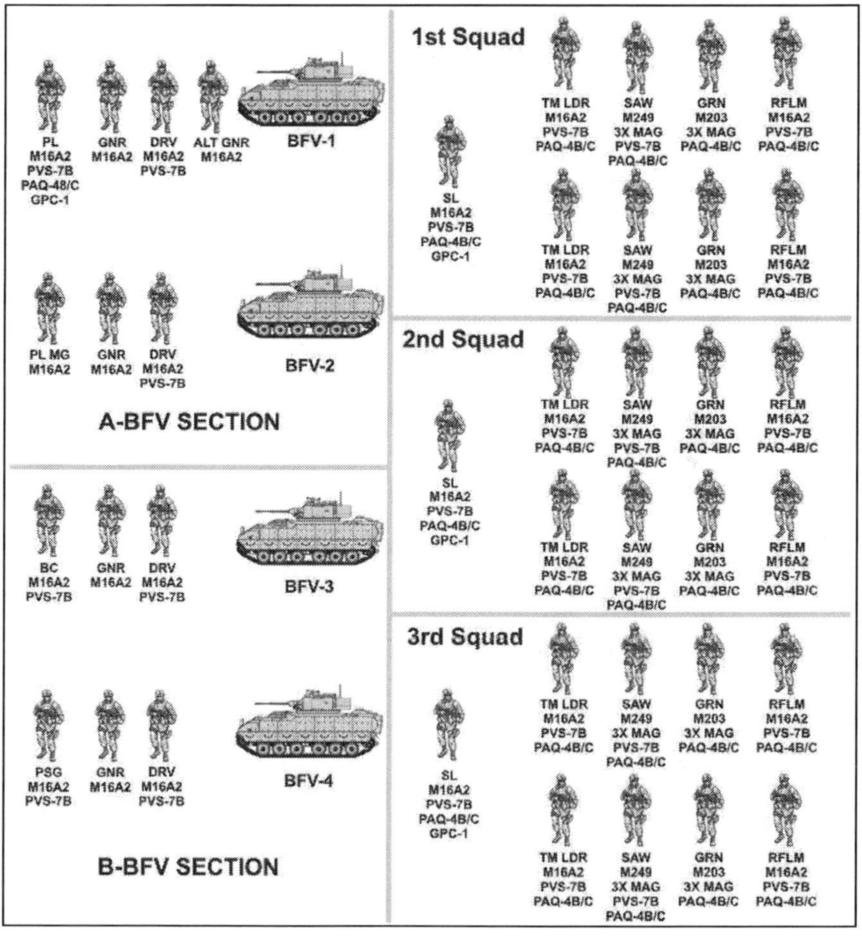

Figure G-1. Bradley fighting vehicle platoon organization.

STRYKER BRIGADE COMBAT TEAM INFANTRY RIFLE PLATOON AND SQUAD

G-13. The Army organized the Stryker brigade combat team (SBCT) in response to the need for a force that can deploy rapidly as an "early responder" to a crisis area anywhere in the world.

CAPABILITIES

G-14. The platoon combines the effects of the Infantry squads, the weapons squad, and the direct fires from the Infantry carrier vehicle (ICV). This includes Javelin fire-and-forget antitank missile fires. Protection is afforded by the vehicle and the ability of the vehicle to protect the infantrymen from small-arms fire and fragmentation before dismounting. The SBCT infantry platoon equipped with the ICV can—

- Use the mobility of the ICV to transport the infantry squads to a position of advantage under the protection of the vehicle.
- Operate in a mounted or dismounted role. .
- Destroy light armor vehicles and personnel using direct fire.
- Employ fires from the vehicle to destroy, suppress, or fix personnel and light infantry fighting vehicles.
- Destroy tanks and fighting vehicles with CCMS fires out to 2,000 meters (Javelin).
- Block dismounted avenues of approach.
- Protect obstacles and prevent enemy breaching operations.
- Establish strong points to deny the enemy key terrain or flank positions.
- Conduct assault breaches of obstacles.
- Clear danger areas and prepare positions for mounted elements.
- Assault enemy positions.
- Augment the ICV, mobile gun system (MGS), and tank antiarmor fires.
- Move over terrain not trafficable by other wheeled vehicles with the infantry squads.
- Infiltrate enemy positions.
- Conduct mounted or dismounted patrols and operations in support of security operations.
- Conduct air assault operations.

Limitations

G-15. The ICV-equipped infantry platoon has the following limitations:

- Platoon ICVs are vulnerable to enemy antiarmor fires.
- Platoon infantry squads are vulnerable to small arms and indirect fires.
- The pace of dismounted offensive operations is limited to the foot speed of the infantryman.
- The ICV poses a variety of difficulties in water-crossing operations, including the requirement for either adequate fording sites or a bridge with sufficient weight classification.
- A Soldier's load increases as a result of additional digital equipment and increased battery requirements.
- Inherent in a situation as an "early responder" is the difficulty in obtaining supplies for ongoing operations, especially with long lines of communication (LOC) and resupply in an underdeveloped area of operation. This situation is compounded because the unit may operate forward of the debarkation point and with threats to the LOCs, the routes may not be secure.

ORGANIZATION

G-16. The SBCT Infantry platoon has three elements: the platoon headquarters (Figure G-2), the mounted element, and the infantry squad element. The SBCT Infantry platoon is equipped with four ICVs. The ICV is a fully mobile system capable of operating in conjunction with infantry and other elements of the combined arms team. Each ICV has a vehicle commander (VC) and driver that operate the vehicle (Figure G-2). The PSG or a senior squad leader is included in the mounted section as the fourth VC and serves as one of the section leaders. The dismounted element (Figure G-3) consists of the platoon headquarters, three rifle squads, and a weapons squad.

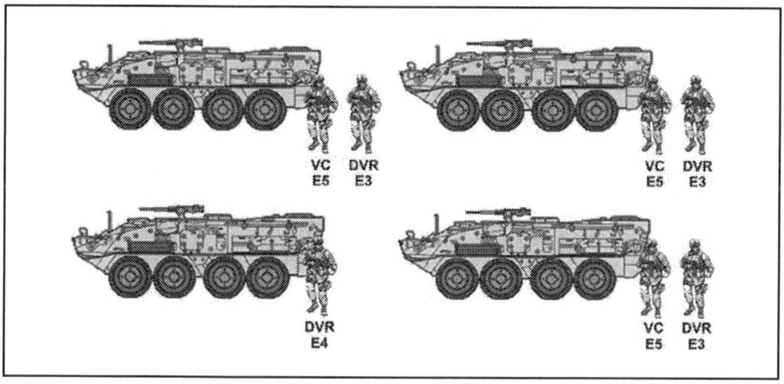

Figure G-2. Mounted element organization.

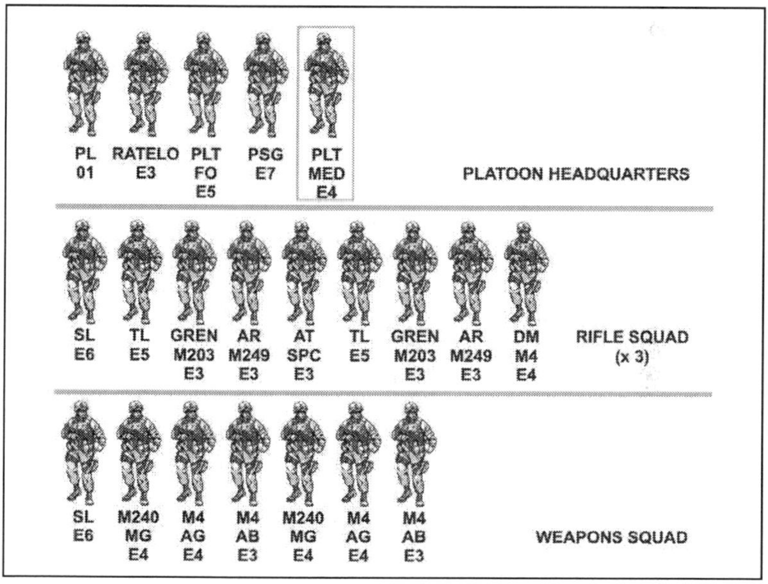

Figure G-3. Dismounted element organization.

Platoon Forward Observer Duties and Responsibilities

G-17. As the platoon's fire support representative, the primary duty of the FO is to locate targets and call for and adjust indirect fire support. Additional responsibilities include the following:

- Refine or submit key targets for inclusion in the company fire plan.
- Prepare, maintain, and use situation maps.

- Establish and maintain communications with company FIST.
- Advise the platoon leader as to the capabilities and limitations of available indirect fire support.
- Report battlefield intelligence.
- Laser designate targets when required.

MANEUVER COMPANY FIRE SUPPORT TEAM (FIST) FIRE REQUEST CHANNELS

G-18. The FIST serves as the net control system (NCS) on the company fire support net. The FIST relays the call for fire to supporting artillery on a digital net or sends the fire mission to the mortar platoon or section. The command net allows the FIST to monitor unit operations. It links the FIST to the commander and platoon leaders for planning and coordination. This net is also an alternate means the platoon leader can use to contact the company commander when primary means fail.

Quick Fire Channel

G-19. A quick fire channel is established to directly link an observer (or other target executor) with a weapon system (Figure G-4). Quick fire channels may be either voice or digital nets. Within a maneuver brigade, quick fire channels are normally established on FA or mortar nets. These channels are designed to expedite calls for fire against high profile targets (HPTs) or to trigger preplanned fires. Quick fire channels may also be used to execute fires for critical operations or phases of the battle. Examples include linking a combat observation and laser team (COLT) with a battery or platoon FDC for counter reconnaissance fires or an AN-TPQ-37 radar with the multiple launch rocket system (MLRS) battery FDC for counterfires. Copperhead missions can best be executed by using quick fire channels.

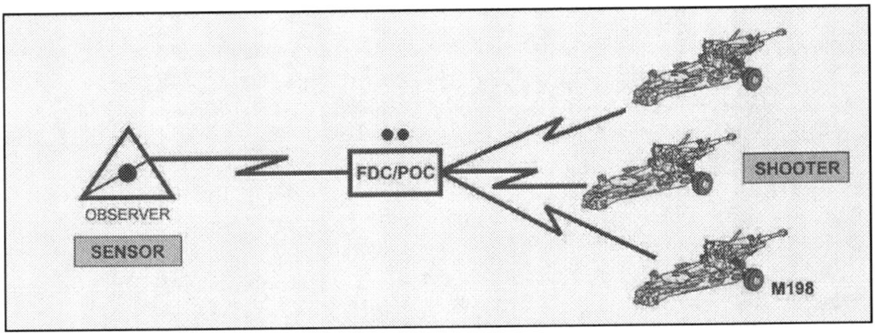

Figure G-4. Quick fire channel illustrating sensor-to-shooter link.

COMBAT ENGINEER SUPPORT

G-20. The light engineer organization is tailored to fight as part of the combined arms team in the Infantry. It focuses on mobility and provides limited countermobility and survivability engineer support. A light engineer unit can be task-organized to provide the necessary engineer functions to fight the battle.

ENGINEER PLATOON

G-21. An engineer platoon may be task-organized to a battalion or company based on the higher commander's analysis of METT-TC. The engineer platoon can be employed to accomplish most engineer missions. However they may require external support for them to conduct continuous operations for more than 48 hours. Figure G-5 shows an example of an engineer platoon.

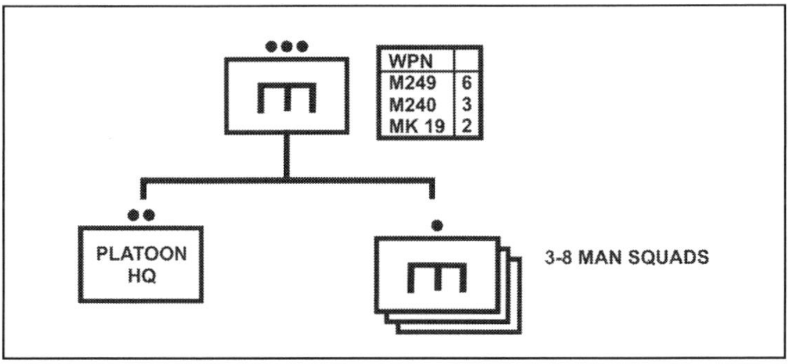

Figure G-5. Engineer platoon.

Sapper Squad

G-22. A sapper squad may be task-organized to a company and executes engineer tasks to support the company mission. Task organization is based on the battalion commander's analysis of METT-TC. The squad is the smallest engineer element that can be employed with its own organic C2 assets. Therefore, it can accomplish tasks such as reconnaissance, manual breaching, demolitions, or route clearance as part of a platoon or company mission. Depending on METT-TC, the engineer may receive augmentation of engineer equipment such as a small earth excavator (SEE) or other specialized engineer equipment. Figure G-6 shows an example of a sapper squad.

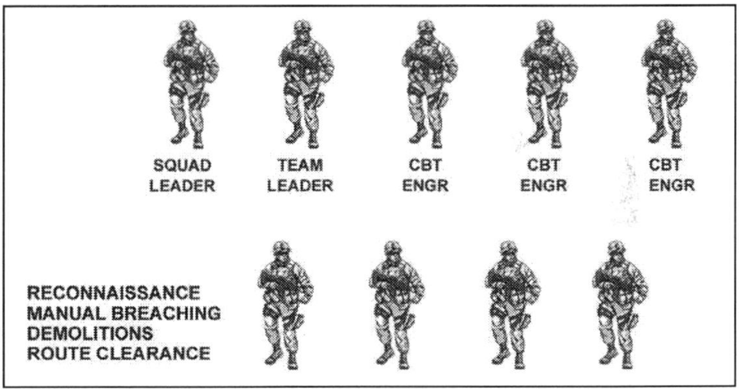

Figure G-6. Sapper squad.

Capabilities and Missions

G-23. The mission of engineers corresponds to those missions normally conducted by Infantry units. Engineer units can operate in restricted terrain such as forests, jungles, mountains, and urban areas. Because of their austere nature, once they are employed, light engineers have the same tactical mobility as the Infantry. To compensate for this, they train to operate in a decentralized manner. Like their supported maneuver force, they operate best under conditions of limited visibility.

Capabilities

G-24. The engineer's focus is mobility. They are experts in supporting infiltrations, air assaults, parachute assaults, ambushes, and raids. In this role, the engineer may conduct covert breaches, route reconnaissance, and obstacle reduction. He may also identify potential enemy counterattack routes and establish countermobility measures such as using scatterable mines (SCATMINEs) to protect the force. Engineers train in Infantry skills and are able to move undetected when close to the enemy.

Missions

G-25. Engineer missions fit into one of three categories: mobility, countermobility, and survivability. Table G-1 shows the tasks included in each of these categories. Depending on METT-TC, an engineer platoon or squad might be attached to a company. Engineers conduct reconnaissance, evaluate obstacles, and use demolitions and field expedients.

Table G-1. Engineer missions.

MOBILITY	COUNTERMOBILITY	SURVIVABILITY
Breach obstacles. Clear minefields. Clear routes. Cross gaps expediently. Construct combat roads or trails.	Construct obstacles to turn, fix, block, or disrupt enemy forces.	Construct crew-served weapons and vehicle fighting positions.

SURVIVABILITY

G-26. Engineer units may also be employed in survivability operations to assist in protecting friendly units by helping to prepare areas such as defensive positions. They may employ their blades to help prepare positions for systems such as mortars, C2, and key weapons. Units should prepare their areas for the arrival of the blades by marking the positions, identifying leaders to supervise position construction, and designating guides for the blade movement between positions.

G-27. Engineer units might employ a small earth excavator to aid in position construction. A SEE has a backhoe, bucket loader, handled hydraulic rock drill, chain saw, and pavement breaker. The SEE can dig positions for individual, crew-served, and AT weapons or for Stinger missile teams. It can also be employed to dig in ammunition pre-stock positions.

AIR AND MISSLE DEFENSE

G-28. Air defense systems that may operate in and adjacent to the Infantry platoon AO are the Avenger, man-portable air defense systems (MANPADS), and Linebacker (Table G-2). All systems can operate as MANPADS Stinger teams. Although other short-range air defense (SHORAD) systems support divisional units, the Infantry platoon is most likely to be supported by the Avenger or a MANPADS Stinger team. The Stinger is also fired from the Avenger and is designed to counter high-performance, low-level, ground attack aircraft; helicopters; and observation and transport aircraft.

Table G-2. Air defense systems.

Man-Portable System 	Personnel: 2-man crew Basic load: 6 missiles with M998 HMMWV Acquisition/range: Visual Engagement range: 5 km Engagement altitude: 3 km+ Mutual support: 2 km+
Bradley Linebacker 	Personnel: 4-man crew Basic load: 10 missiles (4 ready to fire, 6 stowed) Acquisition/range: Visual/thermal Engagement range: 5 km (Stinger); 2,500-m 25-mm; 900-m coax Engagement altitude: 3 km+ Mutual support: 3 km Emplacement time: Fire on the move Reload time: 4 minutes
Avenger 	Personnel: 2-man crew Basic load: 8 ready-to-fire missiles, 250 rounds .50 cal Acquisition/range: Visual/FLIR 9-10 km, laser range finder Engagement range: 5 km+, .50 cal range: 1,800 m Rate of fire: 1,025 rpm Engagement altitude: 3 km+ Mutual support: 3 km Emplacement time: 6 minutes, can remote operations out to 50 meters

AVENGER AND MANPADS STINGERS

G-29. The Avenger's combined arms mission is to provide protection to combat forces and other critical assets from attack. The Avenger is designed to counter hostile cruise missiles, unmanned aircraft systems, low-flying, high-speed, fixed-wing aircraft, and helicopters attacking or transiting friendly airspace. The Avenger provides the battalions with highly mobile dedicated air defense firepower. It is equipped with two standard vehicle-mounted launchers (SVMLs). Each carries four Stinger missiles. The Avenger has the following capabilities:

- A modified fire control subsystem and SVMLs that allow the Avenger to shoot on the move.
- An unobstructed, 360-degree field of fire that can engage at elevations between -10 and +70 degrees.
- A .50 cal machine gun that affords a measure of self-protection by providing additional coverage of the Stinger missile's inner launch boundary.
- A sensor package (forward-looking infrared radar [FLIR], carbon dioxide, eye-safe laser range finder, and a video autotracker) that provides target acquisition capability in battlefield obscuration at night and in adverse weather.
- Two-man crew can remain in the vehicle or remotely control the platform from a separate fighting position.
- Shoot-on-the-move and slew-to-cue capability.
- System maintains dismounted Stinger missile capability in event of launcher system damage, failure, or static mode.

G-30. The MANPADS Stinger Missile System employs a two-man crew that consists of a crew chief and a gunner. The MANPADS team normally has assigned transportation. Unit leaders must carefully consider the consequences before separating a Stinger team from its vehicle. Stinger teams operating away from their vehicles are limited in their ability to haul extra missiles to their firing point.

EARLY WARNING ALERTS

G-31. If SHORAD units are operating in the area, the platoon may receive early warning alerts from its elements. The SHORAD radar teams can broadcast an early warning of enemy air activity that will filter down to the platoon via the brigade, battalion, and company command nets. If METT-TC factors permit, the SHORAD platoon provides voice early warning directly to the battalions.

EMPLOYMENT OF AIR DEFENSE SYSTEMS

G-32. In offensive situations, air defense elements accompany the main attack. They may maneuver with the battalion's lead companies orienting on low-altitude air avenues of approach. When the unit is moving or in a situation that requires short halts, air defense elements should remain within the platoon's organic weapons systems maximum ranges to assure mutual support. The Stinger gunners (MANPADS) can dismount to provide air defense when the unit reaches the objective or pauses during the attack. In the defense, air defense elements may establish BPs based on available intelligence preparation of the battlefield (IPB) information and the company commander's scheme of maneuver.

Weapons Control Status

G-33. The weapons control status (WCS) describes the relative degree of control in effect for air defense fires. It applies to all weapons systems. The WCS is dictated in the battalion OPORD and may be updated based on the situation. The three levels of control are:

- **Weapons Free.** Crews can fire at any air target not positively identified as friendly. This is the least restrictive WCS level.
- **Weapons Tight.** Crews can fire only at air targets positively identified as hostile according to the prevailing hostile criteria.
- **Weapons Hold.** Crews are prohibited from firing except in self-defense or in response to a formal order. This is the most restrictive control status level.

TANK PLATOON

G-34. The tank platoon is the smallest maneuver element within a tank company. Organized to fight as a unified element, the platoon consists of four main battle tanks organized into two sections. The platoon leader (Tank 1) and platoon sergeant (Tank 4) are the section leaders. Tank 2 is the wingman in the platoon leader's section; Tank 3 is the wingman in the platoon sergeant's section (Figure G-7).

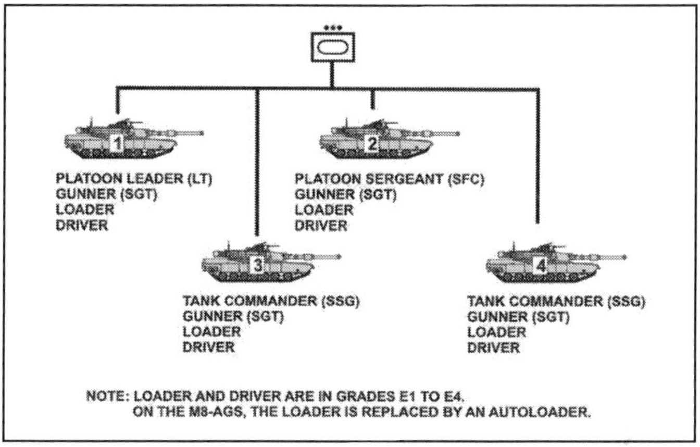

Figure G-7. Tank platoon organization.

G-35. The tank platoon is organic to tank companies and armored cavalry troops. The platoon may be cross-attached to a number of organizations, commonly a mechanized infantry company, to create company teams. It may also be placed under operational control (OPCON) of a light infantry battalion.

G-36. Under battlefield conditions, the wingman concept facilitates control of the platoon when it operates in sections. The concept requires that one tank orient on another tank on either its left or right side. In the absence of specific instructions, wingmen move, stop, and shoot when their leaders do. In the tank platoon, Tank 2 orients on the platoon leader's tank, while Tank 3 orients on the platoon sergeant's tank. The platoon sergeant (PSG) orients on the platoon leader's tank (Figure G-8).

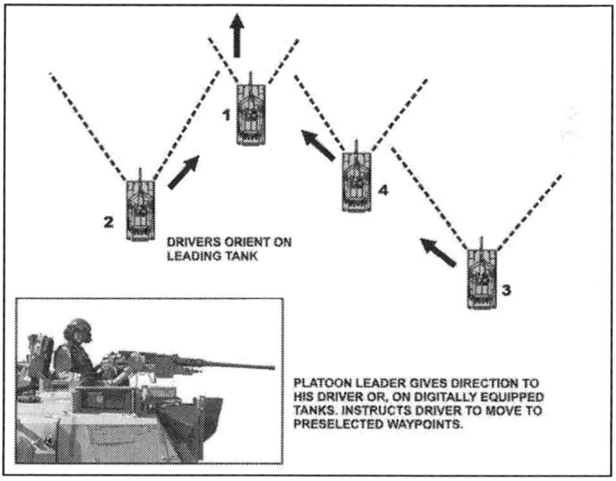

Figure G-8. The tank wingman concept.

Appendix H
Security

Security is the measures taken by the platoon to protect it against all acts designed to impair its effectiveness. Security measures are an inherent aspect of all military operations and can be moving or stationary.

SECTION I — SECURITY FUNDAMENTALS

H-1. Infantry platoons conduct local security measures. They may also be tasked to provide security measures for larger units (called the main body). Measures include screen, guard, cover, and area security. These tasks are executed in the larger unit's security zone (front, flank, or rear of the main body). The application of these security measures is founded on the enduring doctrine found in FM 22-6, *Guard Duty*. Leaders given these tasks or participating in the task of a larger unit must, at a minimum, understand their engagement criteria and whether or not to become decisively engaged.

- **Local security** consists of low-level security operations conducted near a unit to prevent surprise by the enemy (FM 1-02). Local security measures are the same as those outlined for *exterior guards* in FM 22-6.
- **Screen** is a form of security operations that primarily provides early warning to the protected force. (FM 1-02) A screen consists of a combination of observation posts and security patrols.
- **Guard** is a term with a dual meaning; the difference is the size element referred to. When used to refer to individuals, a *guard* is the individual responsible to keep watch over, protect, shield, defend, warn, or any duties prescribed by general orders and/or special orders. Guards are also referred to as a sentinels, sentries, or lookouts (FM 22-6). When used in reference to units, a *guard* is a tactical mission task where the guard force protects the main body by fighting to gain time while observing and preventing the enemy's observation and direct fire against the main body. (FM 1-02) Units conducting a guard mission cannot operate independently because they rely upon the fires and warfighting functions of the main body. Guards consist of a combination of OPs, battle positions, combat patrols, reconnaissance patrols, and movement to contact for force protection.
- **Cover** is a form of security operations with the primary task is to protect the main body. This is executed by fighting to gain time while also observing and preventing the enemy's ground observation and direct fire against the main body. (FM 1-02) Ordinarily only battalion -sized element and larger have the assets necessary to conduct this type of security operation.
- **Area security** is a form of security operations conducted to protect friendly forces, installations, routes, and actions within a specific area. (FM 1-02) During conventional operations (major theater of war scenarios) area security refers the security measures used in friendly controlled areas. Many of the tasks traditionally associated with stability operations and small scale contingencies fall within the scope of area security. These include road blocks, traffic control points, route security, convoy security, and searches.

H-2. The screen, guard, and cover are the security measures used primarily by battalion-sized units to secure themselves from conventional enemy units. These measures, respectively, contain increasing levels of combat power and provide increasing levels of security for the main body. Along with the increase of combat power, there is an increase in the unit's requirement to fight for time, space, and information on the enemy. Conceptually, the measures serve the same purpose as the local security measures by smaller units. For example, a battalion will employ a screen for early warning while a platoon will emplace an OP. The purpose is the same—early warning—only the degree and scale of the measures are different.

H-3. Local and area security are related in that they both focus on the enemy threat within a specified area. Again, the difference is one of degree and scale. Local security is concerned with protecting the unit from enemy in the immediate area, whereas area security is concerned with enemy anywhere in the leader's area of operation (AO).

SECURITY FUNDAMENTALS

H-4. The techniques employed to secure a larger unit are generally the same as those of traditional offensive and defensive operations. It is the application of those techniques that differ. Table H-1 lists the most common techniques used, information required to execute the operation, and the principles used to employ them.

Table H-1. Security fundamentals.

Principles of Security Operations	Techniques Used to Perform Security Operations	Information Required from Controlling Headquarters
• Three General Orders • Provide early and accurate warning • Provide reaction time and maneuver space • Orient on the force / facility being secured • Perform continuous reconnaissance • Maintain enemy contact	• Observation post • Combat outpost • Battle position • Patrols • Combat formations • Movement techniques • Infiltration • Movement to contact • Dismounted, mounted, and air insertion • Roadblocks • Checkpoints • Convoy and route security • Searches	• Trace of the security area (front, sides, and rear boundaries), and initial position within the area • Time security is to be established • Main body size and location • Mission, purpose and commander's intent of the controlling headquarters • Counterreconnaissance and engagement criteria • Method of movement to occupy the area (zone reconnaissance, infiltration, tactical road march, movement to contact; mounted, dismounted, or air insertion) • Trigger for displacement and method of control when displacing. • Possible follow-on missions

LOCAL SECURITY

A unit must be protected at all times from surprise. Exterior guards are utilized to protect a unit from surprise and to give the unit time to prepare to counter any threat. Guards must be alert for surprise by ground, airborne, and air attacks; to provide early warning of chemical, biological, radiological, and nuclear (CBRN) attack or contamination; and to protect supplies and supply installations. If the unit is moving, security may vary from observation to the use of security patrols. During short halts, guards, small security detachments, and forward patrols are used to provide all-round security. For stationary positions in combat or hostile areas, unit commanders use exterior guards to establish a surveillance system to operate day and night throughout the unit area. The commander may use guards, listening posts, observation posts, patrols, aerial observers, and any other available means. The guards may have any number of special devices to assist them in performing their duties. These may include CBRN detection devices, electronic detection devices, infrared or other night vision devices, trip flares and antipersonnel mines, noisemaking devices, or any other device to provide early warning to the guard and unit.

Local Security—FM 22-6, Guard Duty. 17 September 1971.

H-5. Local security prevents a unit from being surprised and is an important part of maintaining the initiative. Local security includes any local measure taken by units against enemy actions. It involves avoiding detection by the enemy or deceiving the enemy about friendly positions and intentions. It also includes finding any enemy forces in the immediate vicinity and knowing as much about their positions and intentions as possible. The requirement for maintaining local security is an inherent part of all operations. Table H-2 lists a sample of active and passive local security measures.

Table H-2. Active and passive security measures.

Active and Passive Security Measures	
Active Measures (moving)	-Combat formations, movement techniques, movement to contact, spoiling attacks - Moving as fast as conditions allow to prevent enemy detection and adaptation - Skillful use of terrain
Active Measures (stationary)	Outside the perimeter - Observation posts, security patrols - Battle positions, combat patrols, reconnaissance patrols - Employing early warning devices - Establishing roadblocks / checkpoints Inside the perimeter - Establishing access points (entrance and exits) - Establishing the number and types of positions to be manned - Establishing readiness control (REDCON) levels - Designating a reserve / response force - Establishing stand-to measures
Passive measures	-Camouflage, cover and concealment, and deception measures (see appendix X) - Signal security - Noise and light discipline

SCREEN

H-6. A screen primarily provides early warning to the main body. A unit performing a screen observes, identifies, and reports enemy actions. Screen is defensive in nature but not passive in execution. It is employed to cover gaps between forces, exposed flanks, or the rear of stationary or moving forces.

Generally, a screening force fights only in self-defense. However, it may engage enemy reconnaissance elements within its capability (counterreconnaissance). A screen provides the least amount of protection of any security mission. It does not have the combat power to develop the situation. It is used when the likelihood of enemy contact is remote, the expected enemy force is small, or the friendly main body needs only a minimum amount of time once it is warned to react effectively

H-7. Screen tasks are to—

- Provide early warning of threat approach.
- Provide real-time information, reaction time, and maneuver space to the protected force.
- Maintain contact with the main body and any security forces operating on its flanks.
- Maintain continuous surveillance of all avenues of approach larger than a designated size into the area under all visibility conditions.
- Allow no enemy ground element to pass through the screen undetected and unreported.
- Maintain contact with enemy forces and report any activity in the AO.
- Destroy or repel all enemy reconnaissance patrols within its capabilities.
- Impede and harass the enemy within its capabilities while displacing.
- Locate the lead elements of each enemy advance guard and determine its disposition, composition and strength, and capabilities.

Stationary Screen

H-8. When tasked to conduct a stationary screen (Figure H-1), the leader first determines likely avenues of approach into the main body's perimeter. The leader determines the location of potential OPs along these avenues of approach. Ideally, the leader assigns OPs in depth if he has the assets available. If necessary, he identifies additional control measures (such as threat named areas of interest [NAIs], phase lines, TRPs, or checkpoints) to assist in controlling observation, tracking of the enemy, and movement of his own forces. The unit conducts mounted and foot patrols to cover ground between OP that cannot be observed from OPs. Once the enemy is detected from an OP, the screening force may engage him with indirect fires. This prevents the enemy from penetrating the screen line and does not compromise the location of the OP. If enemy pressure threatens the security of the screening force, the unit reports the situation to the controlling headquarters and requests permission to displace to a subsequent screen line or follow-on mission.

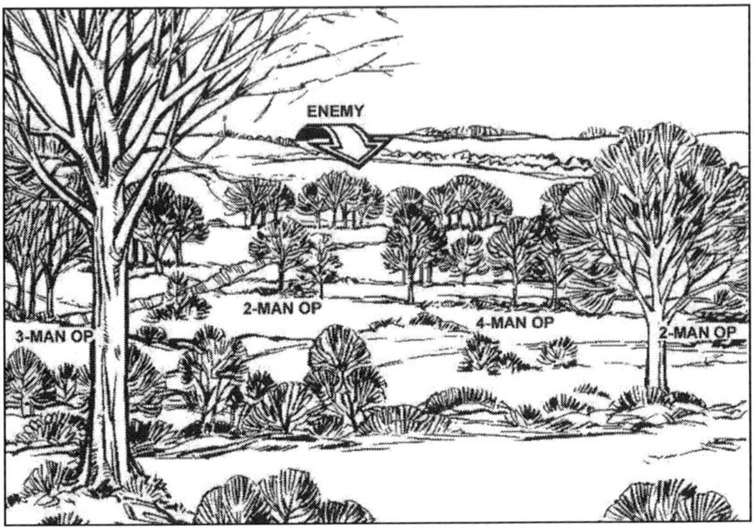

Figure H-1. Squad-sized stationary screen.

Moving Screen

H-9. Infantry platoons may conduct a moving screen to the flanks or rear of the main body force. The movement of the screen is tied to time and distance factors associated with the movement of the friendly main body.

H-10. Responsibilities for a moving flank screen begin at the front of the main body's lead combat element and end at the rear of the protected force. In conducting a moving flank screen, the unit either occupies a series of temporary OPs along a designated screen line to overwatch the main body, or if the main body is moving too fast, continues to move while maintaining surveillance. The screening force uses one or more of the three basic movement techniques to control movement along the screened flank (traveling, traveling overwatch, and bounding overwatch).

GUARD

H-11. A guard differs from a screen in that a guard force contains sufficient combat power to defeat, cause the withdrawal of, or fix the lead elements of an enemy ground force before it can engage the main body with direct fires. A guard force uses all means at its disposal, including decisive engagement, to prevent the enemy from penetrating the security zone. It operates within the range of the main body's indirect fire weapons, deploying over a narrower front than a comparable-size screening force to permit concentrating combat power. The three types of guard operations are: advance; flank; and rear guard.

H-12. Infantry platoons as part of a company can be assigned a guard mission conduct all of the measures associated with a screen. Additionally, they —

- Destroy the enemy advance guard.
- Cause the enemy main body to deploy, and then report its disposition, composition and strength, and capabilities.

AREA SECURITY

H-13. Area security is used by battalion-sized units and above to secure their area of operations (AO) from smaller enemy units (special purpose forces, guerrillas).

H-14. During area security operations civilians will be present. Therefore, commanders must ensure Soldiers understand the current ROE. However, leaders are always responsible for protecting their forces and consider this responsibility when applying the rules of engagement. Restrictions on conducting operations and using force must be clearly explained and understood by everyone. Soldiers must understand that their actions, no matter how minor, may have far-reaching positive or negative effects. They must realize that both friendly or hostile media and psychological operations organizations can quickly exploit their actions, especially the manner in which they treat the civilian population.

H-15. Leaders executing area security measures in a densely populated area must carefully assess the effect of imposing a degree of control on both traffic and pedestrians. For instance, during the rush hour period, however efficient the traffic control point (TCP), a crowd of impatient civilians or cars and trucks can quickly build-up and precipitate the very situation that the TCP leader is trying to avoid.

H-16. Population and resource control operations will cause inconvenience and disruption to all aspects of community life. Therefore, it is important that members of the civil community appreciate the purpose of such operations. In particular, they must understand that the control measures are protective and not punitive. All personnel involved in operations designed to ensure security must be thoroughly conversant with their duties and responsibilities. They must be able to work quickly and methodically to prevent delay and disruption to legitimate activities. They must also work to avoid unnecessary damage to personnel, vehicles, and property. To achieve their purpose they must be thorough. Leaders, at all levels, must ensure that adequate security is in place to counter all assessed risks.

SECTION III — OBSERVATION POSTS

H-17. The OP, the primary means of maintaining surveillance of an assigned avenue or NAI, is a position from where units observe the enemy and direct and adjust indirect fires against him. From the OP, Infantry platoons send SALUTE reports to their controlling headquarters when observing enemy activity.

TYPES OF OPS

H-18. OPs can be executed either mounted or dismounted. As they are complementary, if possible they should be used in combination.

H-19. The main advantage of a dismounted OP is that it provides maximum stealth hopefully preventing the enemy from detecting it. The two main disadvantages are that it has limited flexibility, taking time to displace and limited firepower to protect itself if detected.

H-20. The main advantages of a mounted OP are the flexibility that comes from vehicle mobility as well as the additional combat power resident in the vehicle's optics, communications, weapons, and protection. The main disadvantage is that vehicles are inherently easier to detect and can prevent the unit from accomplishing its mission.

POSITIONING OF OPS

H-21. Based on the specific METT-TC, leaders may array OPs linearly or in depth (Figures H-2 and H-3). Depth is the preferred technique for maintaining contact with a moving enemy along a particular avenue of approach. Linear placement is optimal when there is no clear avenue of approach or the enemy is not moving.

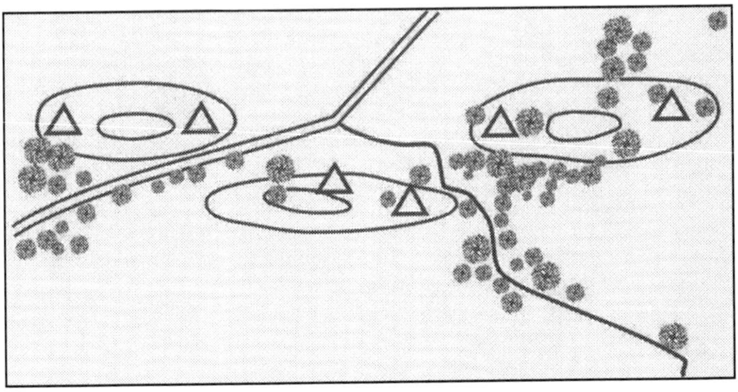

Figure H-2. Linear positioning of OPs.

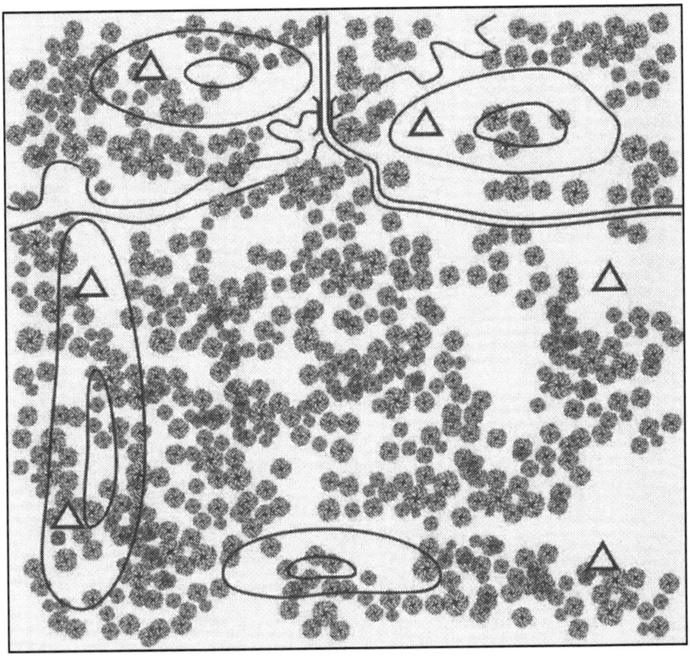

Figure H-3. In-depth positioning of OPs.

SELECTING AND SECURING THE OP

H-22. Based on guidance from the controlling headquarters, the leader selects the general location for the unit's OPs after conducting METT-TC analysis. From his analysis, he determines how many OPs he must establish. He also decides where they must be positioned to allow long-range observation along the avenues of approach assigned and to provide depth through the sector. Leaders assigned a specific OP select its exact position when they get on the actual ground. See Figure H-4 for example of OP selection in urban terrain. OPs should have the following characteristics:

- Covered and concealed routes to and from the OP. Soldiers must be able to enter and leave their OP without being seen by the enemy.
- Unobstructed observation of the assigned area or sector. Ideally, the fields of observation of adjacent OPs overlap to ensure full coverage of the sector.
- Effective cover and concealment. Leaders select positions with cover and concealment to reduce their vulnerability on the battlefield. Leaders may need to pass up a position with favorable observation capability but with no cover and concealment to select a position that affords better survivability. This position should not attract any attention or skyline the observer.

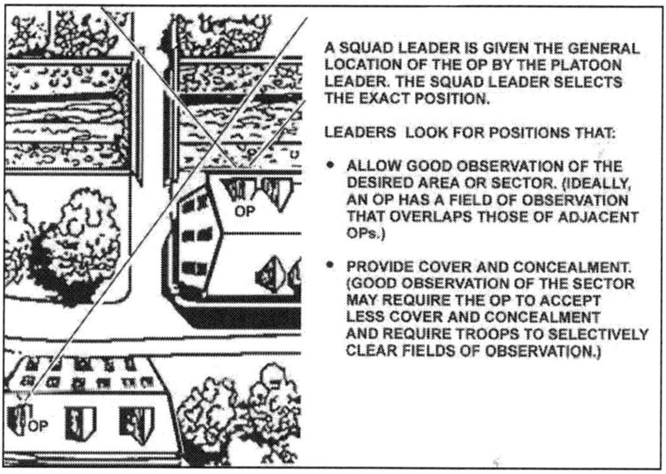

Figure H-4. Selection of OP location.

OP SECURITY

H-23. Small teams are extremely vulnerable in an OP. Their best self-defense is not to be seen, heard, or otherwise detected by the enemy. They employ active and passive local security measures.

OCCUPYING THE OP

H-24. The leader selects an appropriate technique to move to the observation post or screen line based on his analysis of METT-TC. (Infiltration, zone reconnaissance, movement to contact [mounted, dismounted, or air insertion], using traveling, traveling overwatch, or bounding overwatch.)

MANNING AND EQUIPMENT AT THE OP

H-25. At least two Soldiers are required to operate an OP. One man establishing security, recording information, and reporting to higher while the other observes. These men switch jobs every 20-30 minutes because the efficiency of the observer decreases with time. Three or more Soldiers are required to increase security. For extended periods of time (12 hours or more), the unit occupies long-duration OPs by squad-sized units. Essential equipment for the OP includes the following:

- Map of the area.
- Compass / GPS.
- Communications equipment.
- Observation devices (binoculars, observation telescope, thermal sights, and/or night vision devices).
- SOI extract.
- Report formats contained in the SOP.
- Weapons.
- Protective obstacles and early warning devices.
- Camouflage, cover and concealment, and deception equipment as required.

DRAWING A OP SECTOR SKETCH

H-26. Once the leader has established the OP he prepares a sector sketch. This sketch is similar to a fighting position sketch but with some important differences. Figure H-5 shows an example OP sector sketch. At a minimum, the sketch should include:

- A rough sketch of key and significant terrain.
- The location of the OP.
- The location of the hide position.
- The location of vehicle fighting and observation positions.
- Alternate positions (hide, fighting, observation).
- Routes to the OP and fighting positions.
- Sectors of observation.
- Direct and indirect fire control measures.

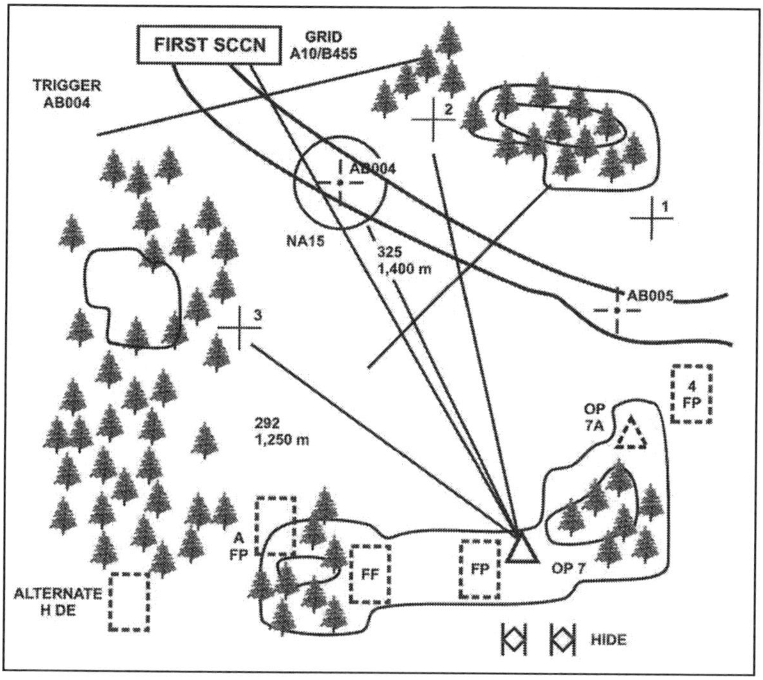

Figure H-5. Example OP sector sketch.

SECTION IV — TRAFFIC CONTROL POINTS (CHECKPOINTS)

H-27. Checkpoint (CP): As defined by FM 1-02 is a place where military police check vehicular or pedestrian traffic in order to enforce circulation control measures and other laws, orders, and regulations. The CP is primarily a military police task; however, while conducting area security, Infantry platoons are frequently employed to establish and operate CPs (Figure H-6).

H-28. Although similar, the CP should not to be confused with a roadblock or blocking position. Roadblocks are designed to prevent all access to a certain area by both wheeled and pedestrian traffic for a variety of purposes. The CP should also not be confused with an OP which is established to collect information.

H-29. When conducting checkpoint operations, Soldiers need the following support:

- Linguists that are familiar with the local language and understand English.
- HN police or a civil affairs officer.
- Wire / Sandbags.
- Signs to reduce misunderstandings and confusion on the part of the local populace
- Lighting.
- Communications equipment.
- Handheld translation devices.

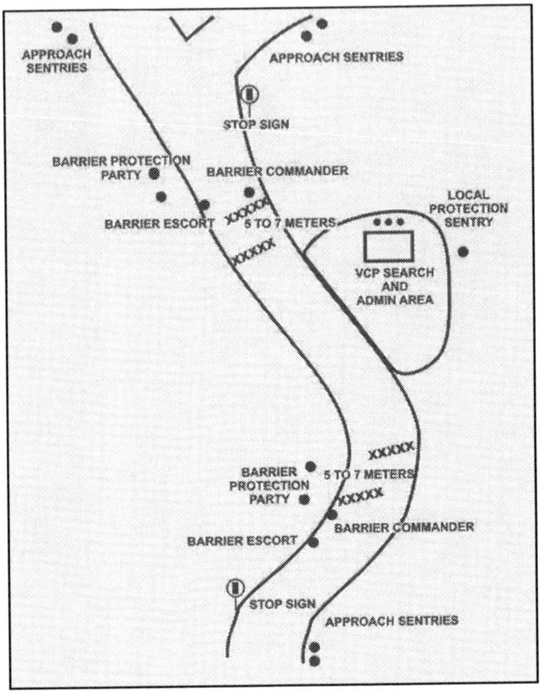

Figure H-6. Example check-point sketch.

TYPES OF CPS

H-30. There are two types of CPs: deliberate; and hasty.

DELIBERATE CP

H-31. A deliberate CP is permanent or semi-permanent. It is established to control the movement of vehicles and pedestrians, and to help maintain law and order. They are typically constructed and employed to protect an operating base or well-established roads. Like defensive positions, deliberate CPs should be continuously improved. Deliberate CPs—

- Control all vehicles and pedestrian traffic so crowds cannot assemble, known offenders or suspected enemy personnel can be arrested, curfews can be enforced, deter illegal movement, prevent the movement of supplies to the enemy, and deny the enemy contact with the local inhabitants.
- Dominate the area of responsibility around the CP. This includes maintaining law and order by local patrolling to prevent damage to property or injury to persons.
- Collect information.

HASTY CP

H-32. A hasty CP differs from a deliberate CP in that they are not, in most cases, pre-planned. A hasty CP will usually be activated as part of a larger tactical plan or in reaction to hostile activities (for example, bomb, mine incident, or sniper attack), and can be lifted on the command of the controlling headquarters. A hasty CP will always have a specific task and purpose. Most often used to avoid predictability and targeting by the enemy. It should be set up to last from five minutes to up to two hours using an ambush mentality. The short duration reduces the risk of the enemy organizing an attack against the checkpoint. The maximum time suggested for the CP to remain in place would be approximately eight hours, as this may be considered to be the limit of endurance of the units conducting the CP and may invite the CP to enemy attacks.

H-33. Characteristics of a hasty checkpoint (Figure H-7) include:

- Located along likely enemy avenues of approach.
- Achieve surprise.
- Temporary.
- Unit is able to carry and erect construction materials without additional assistance.
- Uses vehicles as an obstacle between the vehicles and personnel, and reinforces them with concertina wire.
- Soldiers are positioned at each end of the checkpoint.
- Soldiers are covered by mounted or dismounted automatic weapons.
- Assault force/response force is concealed nearby to attack or assault in case the site is attacked.

H-34. The hasty CPs success is brought about by swift and decisive actions. In many cases, there may be no clear orders before the CP is set up. Leaders must rely on common sense and instinct to determine which vehicles or pedestrians to stop for questioning or searching. They are moved quickly into position, thoroughly conducted, and just as swiftly withdrawn when lifted or once the threat has passed.

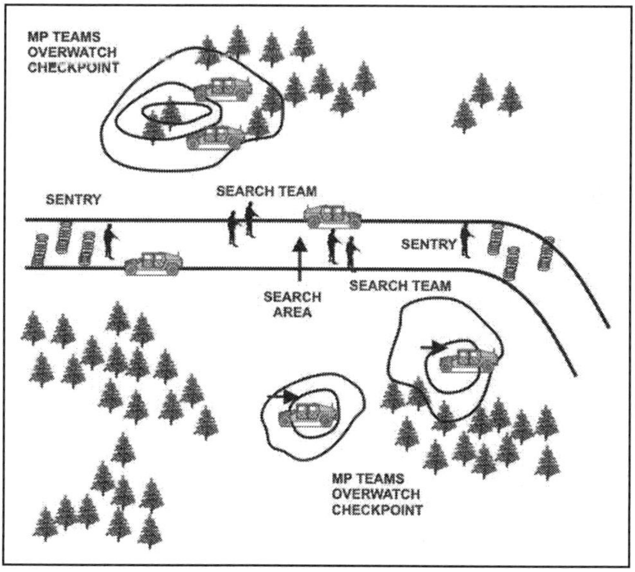

Figure H-7. Hasty check point example.

PHYSICAL LAYOUT

H-35. A checkpoint should consist of four areas: canalization zone, turning or deceleration zone, search zone, and safe zone (Figure H-8).

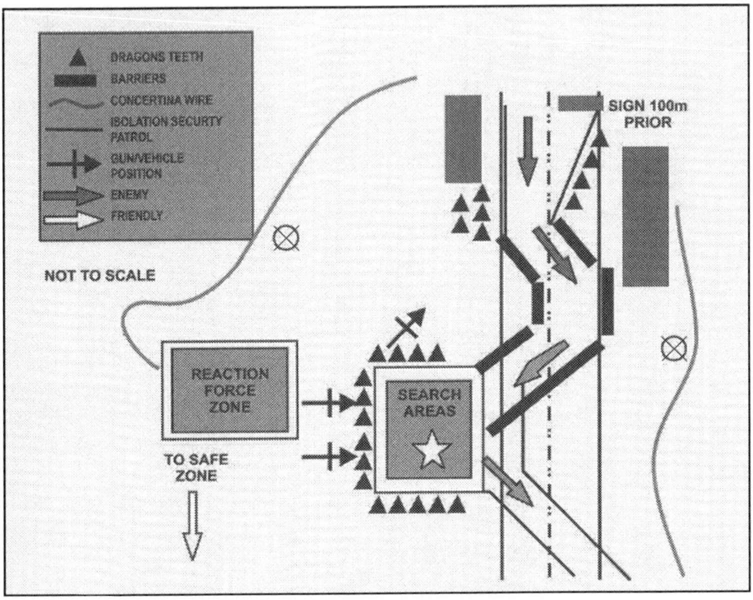

Figure H-8. Four zones of a CP.

H-36. The CP should be sited in such a position as to prevent persons approaching the site from bypassing it or turning away from the CP without arousing suspicion. Ideal sites are where vehicles have already had to slow down. It should be remembered that on country roads vehicles will need extra room to slow down and halt, (particularly large heavy vehicles). The sighting of the CP must take into consideration the type and number of vehicles expected to be using that part of the road where the CP will be sited. Areas where there are few road networks enhance the CP effectiveness.

H-37. The site should allow for a vehicle escape route and include plans to destroy a hostile element that uses such a route. If the checkpoint is completely sealed off, enemy forces may attempt to penetrate it by attempting to run over obstacles or personnel.

H-38. Location should make it difficult for a person to turn around or reverse without being detected. Soldiers establish hasty checkpoints where they cannot be seen by approaching traffic until it is too late for approaching traffic to unobtrusively withdraw. Effective locations on which to set up hasty checkpoints include—

- Bridges (near either or both ends, but not in the middle).
- Defiles, culvert, or deep cuts (either end is better than in the middle).
- Highway intersections (these must be well organized to reduce the inherent danger).
- The reverse slope of a hill (hidden from the direction of the main flow of traffic).
- Just beyond a sharp curve.

Canalization Zone

H-39. The canalization zone uses natural obstacles and/or artificial obstacles to canalize the vehicles into the checkpoint.

- Place warning signs out forward of the checkpoint to advise drivers of the checkpoint ahead (at least 100 meters).
- Canalize the vehicles so they have no way out until they have the consent of personnel controlling the checkpoint.
- This zone encompasses the area from maximum range to maximum effective range of your weapon systems. It usually consists of disrupting and/or turning obstacles.

Turning or Deceleration Zone

H-40. The search element establishes obstacles and an overwatch force to control each road or traffic lane being blocked. The turning or deceleration zone forces vehicles into making a rapid decision. The vehicle can decelerate, make slow hard turns, or maintain speed and crash into a series of obstacles. The road or traffic lanes should be blocked by means of obstacles positioned at either end of the CP. See Appendix F for a discussion of obstacles. These obstacles should be such as to be quickly and easily moved in case of emergencies. They should be sited so as to extend the full width of a traffic lane and staggered to force vehicles to slow to negotiate an 'S' turn (Figure H-9). Stop signs should also be erected ahead of the obstacles and at night illuminated by means of a light or lantern.

H-41. Ensure that vehicles are stopped facing an obstacle (berm, tank, or wall) that is capable of stopping a slow moving truck. Some obstacles will have to be improvised. Examples of these include:

- Downed trees.
- Beirut toothpick – nails driven through lumber.
- Caltrops placed across the road.
- Debris, rubble, large rocks.
- Abatis.
- Road cratering.
- Dragon's teeth, tetrahedrons, concrete blocks.
- Mines.
- Prepared demolitions.
- Concertina wire.

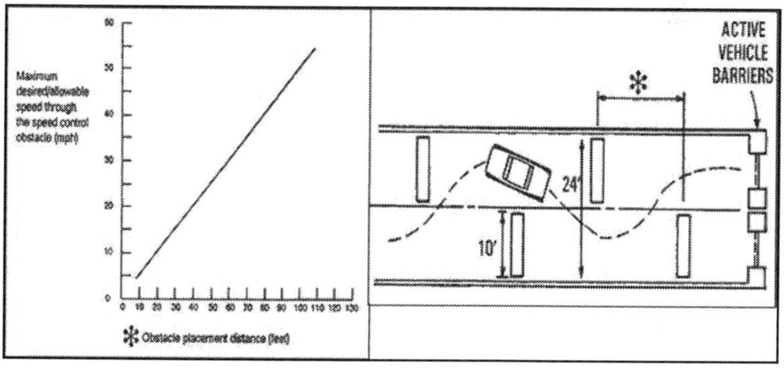

Figure H-9. Controlling vehicle speed through obstacle placement and serpentine placement.

SEARCH ZONE

H-42. The search zone is a relatively secure area where personnel and vehicles are positively identified and searched. A decision is made to confiscate weapons and contraband, detain a vehicle, or allow it to pass. The area is set up with a blocking obstacle that denies entry/exit without loss of life or equipment. When searching:

- Isolate the vehicle being checked from other cars by an obstacle of some type, which is controlled by a Soldier.
- Emplace an overwatch position with a crew-served weapon in an elevated position to cover the vehicle, particularly the driver. The crew-served weapon should be mounted on a T/E and tripod.

H-43. The search zone is further subdivided into three subordinate areas:

- **Personnel search zone** - where personnel are positively identified, searched, and/or detained. This may include partitioned or screened areas to provide privacy, especially when searching women and children. Use female Soldiers to search women, if available.
- **Vehicle search zone** - where vehicles are positively identified, and searched.
- **Reaction force zone** - where a reaction force is located to reinforce the checkpoint and immediately provide assistance using lethal and non-lethal force. Additionally, engineers, and EOD personnel may be co-located here to assist in analyzing and diffusing/destroying ammunition, demolitions, and/or booby traps. This element is organized and equipped to conduct close combat. This element engages in accordance with the established engagement criteria and ROE. This element has a position which allows it to overwatch the CP as well as block or detain vehicles that try to avoid the CP.

H-44. When establishing these zones, consider the following:

- Weapons' surface danger zones (SDZs), geometry.
- 360 degree security.
- Rapid removal of detainees and vehicles.
- Capabilities and skill level of all attachments.
- Potential suicide

H-45. Placing the search area to the side of the road permits two-way traffic. If a vehicle is rejected, it is turned back. If vehicle is accepted for transit, it is permitted to travel through the position. If the vehicle is a threat, the CP leader determines whether to attack or apprehend.

H-46. When confronted by a potentially threatening vehicle:

- The search element alerts the CP leader, moves to a safe/fortified position, and may engage or allow the vehicle to pass based on leader instructions and ROE.
- If the vehicle passes through the escape lane, the leader may direct the assault element to engage the vehicle based on ROE.

SAFE ZONE

H-47. The safe zone is the assembly area for the checkpoint that allows personnel to eat, sleep, and recover in relative security.

TASK ORGANIZATION

H-48. The basic organization of a CP includes a security element, a search element, an assault element, and a C2 element. The actual strength and composition of the force is determined by the nature of the threat, road layout, type of checkpoint required, and the anticipated number of vehicles to be processed. Table H-3 details typical duties of these elements as well as a general list of Do's and Don'ts.

Table H-3. Task organization.

C2
Overall Responsibility
- Exercises C2
- Maintains communications with controlling HQ
- Maintains a log of all activities
- Coordinates RIP as required
- Coordinates linkups as required
- Coordinates the role of civil authorities
- Coordinates local patrols.
- Integrates reserve / QRF
- If available, the C2 element should have a vehicle for patrolling, for moving elements, or administrative actions
Security Element
- Provides early warning to the CP through local security measures
- Prevent ambush
- Able to reinforce position is necessary
- Observes and reports suspicious activity
- Monitors traffic flow up to and through the checkpoint
Search Element
-Halts vehicles at the checkpoint.
- Guides vehicles to search area
- Conducts vehicle searches: passenger, cargo
- Conducts personnel searches: male, female
- Directs cleared vehicles out of the CP
- Detains personnel as directed
Assault Element
- Destroys escaping vehicles and personnel
- Able to reinforce position as necessary
(Soldiers occupy support by fire positions beyond the actual CP)
Do
- Speak to driver - driver speaks to occupants
- Have the driver open all doors and compartments before Soldier conducts search of vehicle
- Ask politely to follow your instructions
- Speak naturally and no louder than necessary
- Allow driver to observe the search
- All vehicle occupants are required to exit the vehicle
- Be courteous when searching
- Use scanners and metal detectors when possible
- Stay calm and make a special effort to be polite
- Maintain a high standard of dress, military bearing, and stay in uniform
Don't
- Be disrespectful or give any hint of dislike
- Put your head or arm in vehicle or open the door without permission
- Shout or show impatience
- Frisk women or tell them to put their hands up
- Become involved in a heated argument
- Use force as directed by unit ROE
- Become careless or sloppy in appearance

C2 Element

H-49. The C2 element controls the operation. The C2 element normally consists of a leader, his RTO and runner.

H-50. The leader normally establishes a headquarters / administrative area to synchronize the efforts of the subordinate activities. The headquarters and security element should be sited centrally and in a position which facilitates control of the obstacles. The headquarters area should be secure and sufficiently large to incorporate an administrative area and vehicle search area. Depending on the threat, this area should have sufficient cover or survivability positions should be built.

H-51. The CP should have communication to their controlling headquarters by radio. A spare radio and batteries should be supplied to the CP. Radio and telephone checks are carried out as per the unit's SOP using signal security measures. Communications within the site should be undertaken using whatever means are available.

Civil Authority Assistance

H-52. The closest liaison must be maintained between the CP leader and the senior policeman. Policemen at a CP are employed to assist in the checking and searching of vehicles and personnel, to make arrests when necessary. Police are ideally employed on the scale of one officer for each lane of traffic. These civil authorities should attend rehearsals. As the degree of threat increases, police officers should be on stand by to move with the patrol to the CP site. Wherever possible, it should be the responsibility of the military to command and control the CP while the police control the search aspects.

H-53. The leader must understand the guidance from his chain of command on contingencies that occur outside of the CP area that might require forces from the CP. The CP, unless otherwise ordered, is the primary task. If an incident occurs in the vicinity of the checkpoint that is likely to require manpower and affect the efficient operation of the CP, the leader should seek guidance from his higher headquarters.

H-54. Sequence of events for establishing the CP include:

- Leader's reconnaissance.
- Establish support by fire positions (and fighting positions as required).
- Establish blocking positions (entrance and exit).
- Establish search area for personnel and vehicles.
- Establish holding area (if required).
- Establish an area for C2 and admin.

Security Element

H-55. The nature of the CP makes it particularly vulnerable to enemy attack. Protection should therefore be provided for overall position as well as those of subordinate positions. Concealed sentries should also be positioned on the approaches to the CP to observe and report approaching traffic, and to prevent persons or vehicles from evading the CP. When available, early warning devices or radar may be used to aid guards on the approaches to the CP.

H-56. The security element stays alert for any change of scenery around the checkpoint. Crowds gathering for no apparent reason or media representatives waiting for an event are all indicators that something may happen.

Escalation Of Force

H-57. Escalation of Force (EOF) is a sequential action that begins with non-lethal force measures that could escalate to lethal force measures to protect the force. Infantrymen at the CP must ensure they follow ROE and EOF guidance when reacting to situations.

Search Element

Vehicle Searches

H-58. Two members of the search team position themselves at both rear flanks of the vehicle undergoing a search, putting the occupants at a disadvantage. These Soldiers maintain eye contact with the occupants once they exit the vehicle and react to any threat attempts by the occupants during the vehicle search.

H-59. The actual search is conducted by two Soldiers. One Soldier conducts interior searches; the other performs exterior searches. They instruct the occupants (with interpreters if available) to exit the vehicle during the interior search and instruct the driver to watch the vehicle search. Once the interior search is complete, they escort the driver to the hood of the vehicle and instruct him to open it. After the engine compartment has been examined, they instruct the driver to open the other outside compartments (tool boxes, gas caps, trunks). The driver removes any loose items that are not attached to the vehicle for inspection. Members of the search team rotate positions to allow for mental breaks.

H-60. Soldiers use mirrors and metal detectors to thoroughly search each vehicle for weapons, explosives, ammunition, and other contraband. Depending on the threat level, the vehicle search area provides blast protection for the surrounding area.

Personnel Searches

H-61. Soldiers may be required to conduct personnel searches at the checkpoints. Every attempt should be made for host nation authorities to conduct, or at least observe, searches of local nationals. Additionally, leaders must plan for same-gender searches. Personnel searches are conducted only when proper authorization has been obtained, usually from higher HQ, according to the ROE, Status of Forces Agreement (SOFA), or host nation agreements. This does not preclude units from searching individuals that pose a threat to U.S. or other friendly forces.

H-62. Units may have to detain local nationals who become belligerent or uncooperative at checkpoints. The OPORD and the ROE must address the handling of such personnel. In any case, self-protection measures should be planned and implemented according to the orders from higher HQ.

H-63. Searches of local nationals should be performed in a manner that preserves the respect and dignity of the individual. Special consideration must be given to local customs and national cultural differences. In many cultures it is offensive for men to touch or even talk to women in public. Searchers must be polite, considerate, patient, and tactful. Leaders must make every effort not to unnecessarily offend the local population. Such situations can have a very negative impact on peace operations and can quickly change popular opinion toward U.S. and other friendly forces.

H-64. Each captive is searched for weapons and ammunition, items of intelligence value, and other inappropriate items. Use of digital cameras will record any evidence of contraband.

H-65. When possible, conduct same gender searches. However, this may not always be possible due to speed and security considerations. If females are not available, use medics or NCOs with witnesses. Perform mixed gender searches in a respectful manner using all possible measures to prevent any action that could be interpreted as sexual molestation or assault. The on-site supervisor carefully controls Soldiers doing mixed-gender searches to prevent allegations of sexual misconduct.

H-66. Soldiers conduct individual searches in search teams that consist of the following:

- **Searcher:** A searcher is the Soldier that actually conducts the search. He is in the highest-risk position.
- **Security:** Security includes at least one Soldier to provide security. He maintains eye contact with the individual being searched.
- **Observer:** The observer is a leader that has supervisory control of the search operation. He also provides early warning for the other members of the team.

H-67. The two most common methods that are used to conduct individual searches are the frisk search, and the wall search.

- **Frisk search:** This method is quick and adequate to detect weapons, evidence, or contraband. However, it is more dangerous because the searcher has less control of the individual being searched.
- **Wall search:** This method affords more safety for the searcher because the individual is searched in a strained, awkward position. Any upright surface, such as a wall, vehicle, tree, or fence may be used.

H-68. If more control is needed to search an uncooperative individual, the search team places the subject in the kneeling or prone position.

SECTION V — CONVOY AND ROUTE SECURITY

H-69. Convoy security missions are conducted when insufficient friendly forces are available to continuously secure lines of communication in an AO. They may also be conducted in conjunction with route security missions. A convoy security force operates to the front, flanks, and rear of a convoy element moving along a designated route. Convoy security missions are offensive in nature and orient on the force being protected.

H-70. To protect a convoy, the security force must accomplish the following critical tasks:

- Reconnoiter and determine the trafficability of the route the convoy will travel.
- Clear the route of obstacles or positions from where the threat could influence movement along the route.
- Provide early warning and prevent the threat from impeding, harassing, containing, seizing, or destroying the convoy.
- Protect the escorted force from enemy contact
- React decisively to enemy contact

H-71. Company-sized units and larger organizations usually perform convoy or route security missions. Convoy security provides protection for a specific convoy. Route security aims at securing a specific route for a designated period of time, during which multiple convoys may use the route. These missions include numerous tasks such as reconnaissance, security, escorting, and establishing a combat reaction force. These tasks become missions for subordinate units. The size of the unit performing the convoy or route security operation depends on many factors, including the size of the convoy, the terrain, and the length of the route. For example, an Infantry platoon can escort convoys, perform route reconnaissance, and establish traffic control points along main supply routes.

ORGANIZATION OF FORCES

H-72. During convoy security operations, the convoy security commander and Infantry leader must establish and maintain security in all directions and throughout the platoon. As noted, several factors, including convoy size affect this disposition. The key consideration is whether the unit is operating as part of a larger escort force or is executing the escort mission independently. Additional METT-TC considerations include the employment of rifle squads during the mission (fire teams ride in escorted vehicles).

H-73. The unit should also be reinforced with engineers to reduce obstacles along the route. The higher headquarters should coordinate additional ISR assets to support the security mission. Unmanned aircraft systems (UASs) or aerial reconnaissance should reconnoiter the route in advance of the unit's lead elements.

H-74. When the platoon executes a convoy escort mission independently, the convoy commander and platoon leader disperse Infantry in vehicles throughout the convoy formation to provide forward, flank, and rear security. Engineer assets, if available, should be located near the front to respond to obstacles. At times, engineer assets may be required to move ahead of the convoy with scouts to proof the convoy route. In some independent escort missions, variations in terrain along the route may require the unit to operate using a modified traveling overwatch technique. In it, one section leads the convoy while the other trails the convoy. Dispersion between vehicles in each section is sufficient to provide flank security. The terrain may not allow the trail section to overwatch the movement of the lead section.

H-75. When sufficient forces are available, the convoy security should be organized into four elements: reconnaissance element; screen element; escort element; and a reaction element (Figure H-10). The Infantry platoon may be assigned any one of the four tasks, but as a general rule, probably cannot be assigned all four.

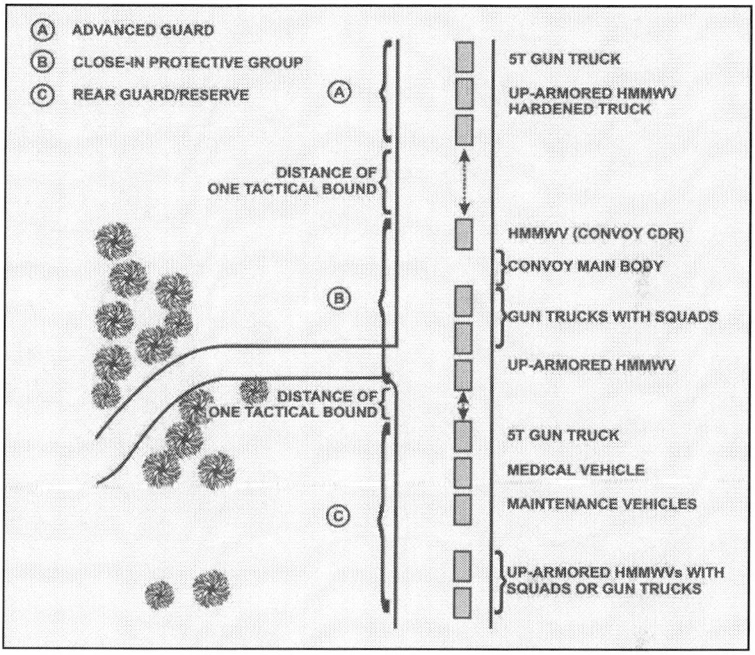

Figure H-10. Convoy escort organization.

ADVANCED GUARD

H-76. The advance guard reconnoiters and proofs the convoy route. The advanced guard element performs tasks associated with movement to contact and zone / route reconnaissance forward of the convoy. It searches for signs of enemy activity such as ambushes and obstacles. This element focuses on identifying enemy forces able to influence the route, route trafficability, or refugees or civilian traffic that may disrupt movement. Engineers are attached to the unit to assist reconnoitering and classifying bridges, fords, and obstacles along the route. The advanced guard normally operates from 3 to 4 kilometers ahead of the main body of the convoy. If available, UASs or aerial reconnaissance should precede the reconnaissance element by 5 to 8 kilometers dependent on the terrain and visibility conditions.

H-77. Within its capabilities, the advanced guard attempts to clear the route and provides the convoy commander with early warning before the arrival of the vehicle column. In some cases, an individual vehicle, a squad, or a platoon-sized element may be designated as part of the advanced guard and may receive additional combat vehicle support (tank with a mine plow, or mine roller). The leader plans for integrating engineer assets to aid in breaching point-type obstacles. Command-detonated devices and other improvised explosive devises (IEDs) pose a major threat during route reconnaissance.

Flank and Rear Guard/Screen

H-78. This element performs a guard or screen, depending on the amount of combat power allocated, providing early warning and security to the convoy's flanks and rear (unit may utilize outposts). The leader must develop graphic control measures to enable a moving flank screen centered on the convoy. The guard / screen's purpose is to prevent observation for employment of effective indirect fires and identify combat elements prior to a direct fire engagement against the convoy. These elements gain and maintain contact with threat reconnaissance and combat elements, employing fires (direct and indirect) to suppress and guiding reaction or escort elements to defeat or destroy the threat force. Units use a combination of OPs or battle positions on terrain along the route.

H-79. The rear guard follows the convoy (Figure H-11). It provides security in the area behind the main body of the vehicle column, often moving with medical and recovery assets. Again, an individual vehicle or the entire unit may make up this element.

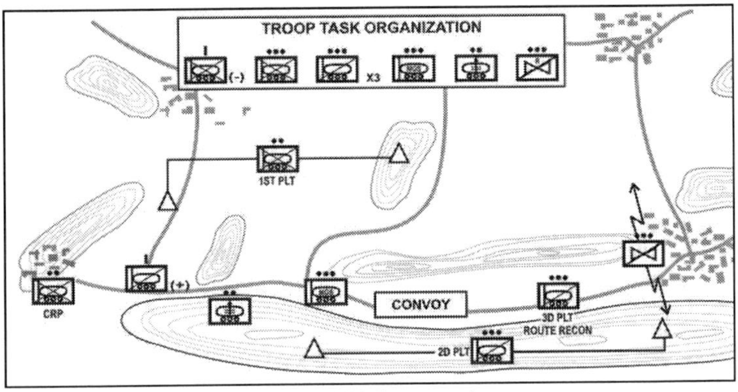

Figure H-11. Rear guard.

Escort Element

H-80. The escort element provides close-in protection to the convoy. The convoy may be made of many types of vehicles, including military sustainment and C2 as well as civilian trucks and buses. The escort element may also provide a reaction force to assist in repelling or destroying threat contact. The unit assigned the escort mission to provide local security throughout the length of the convoy. The escort element defeats close ambushes and marks bypasses or breaches obstacles identified by reconnaissance as necessary. If the reaction force is not available in sufficient time, the escort element may be required to provide a reaction force to defeat far ambushes or block attacking threat forces. The Infantry platoon or squad may perform a convoy escort mission either independently or as part of a larger unit's convoy security mission. Aviation units may also be a part of the escort force and the leaders of both ground and air must be able to quickly contact each other.

Reaction Force

H-81. The reaction force provides firepower and support to the elements above in order to assist in developing the situation or conducting a hasty attack. It may also perform duties of the escort element. The reserve will move with the convoy or be located at a staging area close enough to provide immediate interdiction against the enemy.

COMMAND AND CONTROL

H-82. Because of the task organization of the convoy escort mission, C2 is especially critical. The relationship between the Infantry platoon or squad and the convoy commander must provide unity of command and effort if combat operations are required during the course of the mission. In most cases, the unit will execute the escort mission under the control of the security force commander, who is usually under OPCON or attached to the convoy commander.

H-83. The leader should coordinate with the security force commander or the escorted unit to obtain or exchange the following information:

- Time and place of linkup and orders brief.
- Number and type of vehicles to be escorted.
- High value assets within the convoy.
- Available weapon systems, ammunition, and ordnance (crew served, squad, and individual).
- Vehicle maintenance status and operating speeds.
- Convoy personnel roster.
- Unit's or escorted unit SOP, as necessary.
- Rehearsal time / location.

H-84. It is vital that the convoy commander issues a complete OPORD to all convoy vehicle commanders before executing the mission. This is important because the convoy may itself be task-organized from a variety of units, and some vehicles may not have tactical radios. The order should follow the standard five-paragraph OPORD format (Table H-4), but special emphasis should be placed —

- Route of march (including a strip map for each vehicle commander).
- Order of march.
- Actions at halts.
- Actions in case of vehicle breakdown.
- Actions on contact.
- Chain of command.
- Communication and signal information.

Table H-4. Convoy OPORD example.

Task Organization	
SITUATION	**SERVICE AND SUPPORT**
Enemy: • Activity in the last 48 hours • Threats • Capabilities Friendly: • Units in the area or along the route • ROE Light and Weather Data: • Effects of light and weather on the enemy and on friendly forces • BMNT, sunrise, high temp, winds, sunset, EENT, moonrise, % illumination, low temp	MEDEVAC procedures: • 9-line MEDEVAC request • Location of medical support/combat lifesavers • Potential PZ/LZ locations Maintenance procedures: • Location of maintenance personnel • Location and number of tow bars • Recovery criteria • Stranded vehicle procedures
MISSION	**COMMAND AND SIGNAL**
Task and purpose of the movement mission statement	Convoy commander Sequence of command Location of convoy commander Call signs of every vehicle/unit in the convoy Convoy frequency MEDEVAC frequency Alternate frequencies
EXECUTION **Commander's intent** **End-state** **Concept of the operation (concept sketch or terrain model)** **Task to maneuver units** **Fires** **CAS**	

Coordinating instructions:
• Timeline
  o Marshal
  o Rehearsals
  o Convoy briefing
  o Inspections
  o Initiate movement
  o Rest halts
  o Arrival time
• Order of movement/bumper numbers and individual manifest
• Movement formation
• Speed/catch-up speed
• Interval (open areas and in built-up areas)
• Weapons orientation, location of key weapons systems
• Route
• Checkpoints
• Actions on contact
• Actions on breakdowns
• Actions at the halt (short halt and long halt)	

REACTING TO ENEMY CONTACT

H-85. As the convoy moves to its new location, the enemy may attempt to harass or destroy it. This contact will usually occur in the form of an ambush, often with the use of a hastily-prepared obstacle. The safety of the convoy rests on the speed and effectiveness with which escort elements can execute appropriate actions on contact. Based on the factors of METT-TC, portions of the convoy security force such as the unit may be designated as a reaction force. The reaction force performs its escort duties, conducts tactical movement, or occupies an AA (as required) until enemy contact occurs and the convoy commander gives it a reaction mission.

ACTIONS AT AN AMBUSH

H-86. An ambush is one of the more effective ways to interdict a convoy. Reaction to an ambush must be immediate, overwhelming, and decisive. Actions on contact must be planned for and rehearsed so they can be executed quickly.

H-87. In almost all situations, the unit will take several specific, instantaneous actions when it reacts to an ambush (Figures H-12 and H-13). However, if the convoy is moving fuel and other logistics, the best method might be to suppress the enemy, continue to move and report. These steps, illustrated in include the following:

- As soon as they encounter an enemy force, the escort vehicles take action toward the enemy. They seek covered positions between the convoy and the enemy and suppress the enemy with the highest volume of fire permitted by the ROE. Contact reports are submitted to higher headquarters as quickly as possible.
- The convoy commander retains control of the convoy vehicles and continues to move them along the route at the highest possible speed.
- Convoy vehicles, if armed, may return fire only if the escort has not positioned itself between the convoy and the enemy force.
- Leaders may request that any damaged or disabled vehicles be abandoned and pushed off the route.
- The escort leader uses SPOTREPs to keep the convoy security commander informed. If necessary, the escort leader or the convoy commander requests support from the reaction force and or calls for and adjusts indirect fires.

NOTE: Fire support for areas behind the forward line of troops is planned and coordinated on an area basis (such as a base operations center, base cluster operations center, or rear area operations center). This planning may provide fire support to main supply routes (MSRs) or other routes. Convoy commanders are responsible for the fire support plans for their convoy and for ensuring escort security leaders are familiar with the plan.

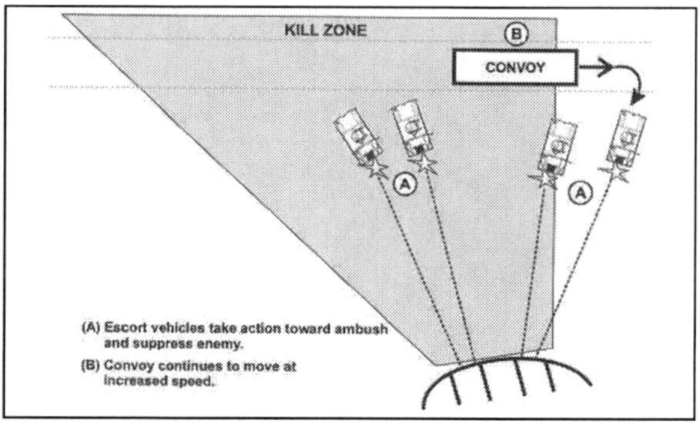

Figure H-12. Convoy escort actions toward ambush.

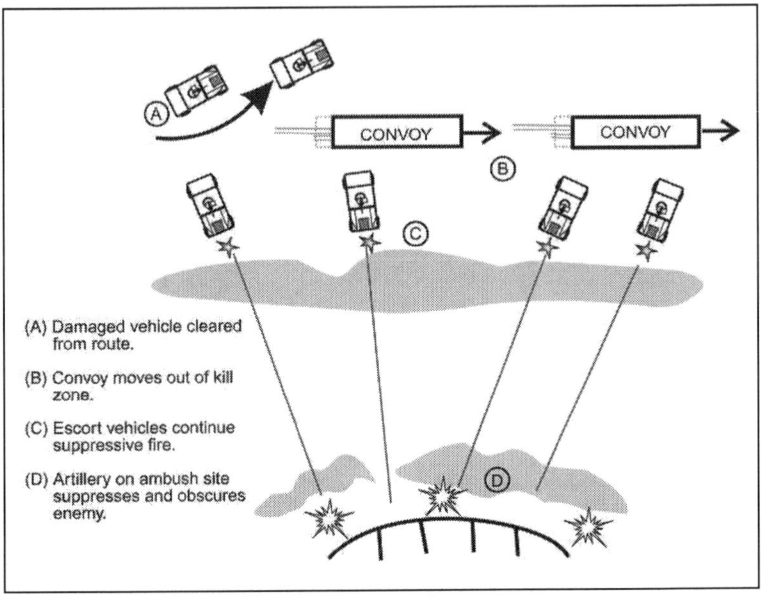

Figure H-13. Convoy continues to move.

H-88. Once the convoy is clear of the kill zone, the escort element executes one of the following COAs:

- Continues to suppress the enemy as combat reaction forces move to support (Figure H-14).
- Uses the Infantry to assault the enemy (Figure H-15).
- Breaks contact and moves out of the kill zone.
- Request immediate air support to cut-off escape routes.

H-89. In most situations, Infantry platoons or squads will continue to suppress the enemy or execute an assault. Contact should be broken only with the approval of the controlling commander.

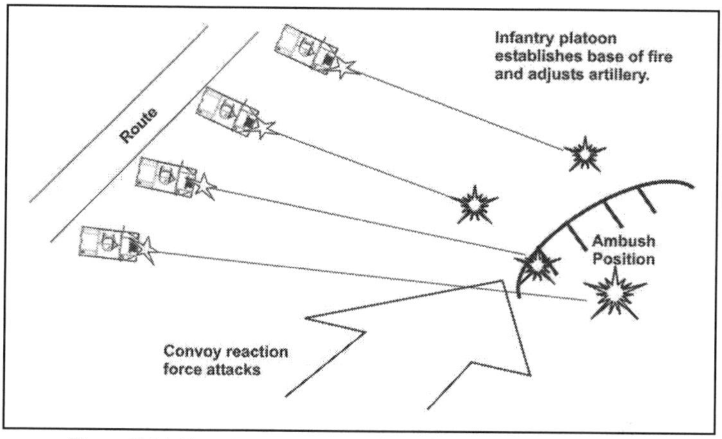

Figure H-14. Escort suppresses ambush for reaction force attack.

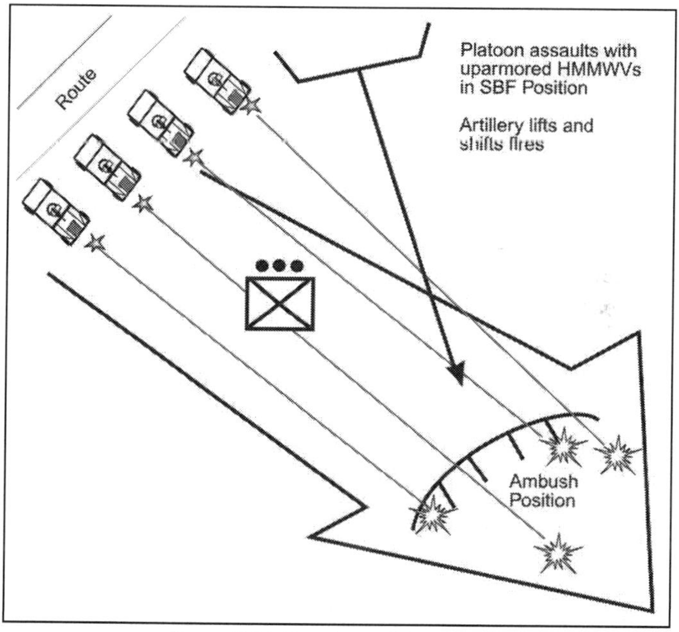

Figure H-15. Escort assaults ambush.

Actions at an Obstacle

H-90. Obstacles are a major impediment to convoys. The purpose of reconnaissance ahead of a convoy is to identify obstacles and either breach them or find bypasses. In some cases the enemy or its obstacles may avoid detection by the reconnaissance element.

H-91. Obstacles can be used to harass the convoy by delaying it. If the terrain is favorable, the obstacle may stop the convoy altogether. Obstacles may also be used to canalize the convoy to set up an enemy ambush. When an obstacle is identified, the convoy escort faces two problems: reducing or bypassing the obstacle, and maintaining protection for the convoy. Security becomes critical, and actions at the obstacle must be accomplished very quickly. The convoy commander must assume that the enemy is covering the obstacle with direct- and indirect-fire weapons systems.

H-92. To reduce any time the convoy is halted and to reduce its vulnerability, the following actions should occur when the convoy escort encounters a point-type obstacle:

- The lead element identifies the obstacle and directs the convoy to make a short halt to establish security. The convoy escort overwatches the obstacle and requests the breach element force to move forward (Figure H-16).
- The convoy escort maintains 360-degree security of the convoy and provides overwatch as the breach force reconnoiters the obstacle in search of a bypass.

H-93. Once all reconnaissance is complete, the convoy commander determines which of the following COAs he will take:

- Bypass the obstacle.
- Breach the obstacle with assets on hand.
- Breach the obstacle with reinforcing assets.

H-94. The convoy security commander relays a SPOTREP and requests support by combat reaction forces, engineer assets (if they are not part of the convoy), and aerial reconnaissance elements. Artillery units are alerted to prepare to provide fire support.

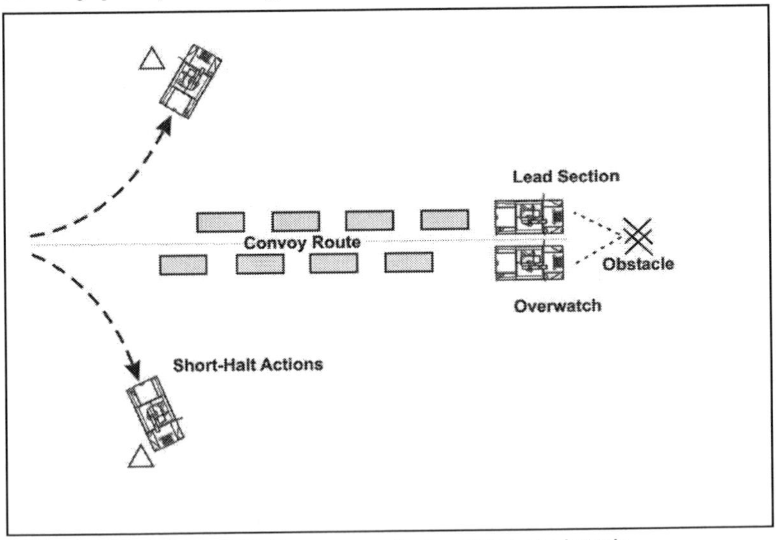

Figure H-16. Convoy escort overwatches an obstacle.

H-95. Obstacles may be in the form of unexploded ordnance (UXO), or uncharted minefields. If the convoy encounters UXO or mines, the convoy security commander should identify, mark, report, and bypass.

Actions During Halts

H-96. During a short halt, the convoy escort remains alert for possible enemy activity. If the halt is for any reason other than an obstacle, the following actions should be taken.

H-97. The convoy commander signals the short halt and transmits the order via tactical radio. All vehicles in the convoy initially assume a herringbone formation.

H-98. If possible, escort vehicles are positioned up to 100 meters beyond the convoy vehicles that are just clear of the route. Escort vehicles remain at the ready, dismount the rifles teams or squads as required, and establish local security. Infantry security elements or escort vehicles must occupy terrain within small arms range that dominates the convoy route during halts.

H-99. When the order is given to move out, convoy vehicles reestablish movement formation, leaving space for escort vehicles. Once the convoy is in column, local security elements (if used) return to their vehicles, and the escort vehicles rejoin the column.

H-100. The convoy resumes movement.

Appendix I

Improvised Explosive Devices, Suicide Bombers, and Unexploded Ordnance

Improvised explosive devices (IEDs), car bombs, unexploded ordnance (UXO), and suicide bombers pose deadly and pervasive threats to Soldiers and civilians in operational areas all over the world. Infantrymen at all levels must know how to identify, avoid, and react to these hazards properly. Newly assigned leaders and Soldiers should read everything they can find on current local threats. They should also become familiar with unit SOP policies and other relevant information contained in locally produced Soldier handbooks and leader guidebooks.

This appendix introduces discussions of improvised explosive devices (IEDs), homicide bombers, and unexploded ordnance (UXO). It incorporates tactical-level countermeasures learned from recent combat operations.

SECTION I — IMPROVISED EXPLOSIVE DEVICES

I-1. IEDs are nonstandard explosive devices used to target U.S. Soldiers, civilians, NGOs, and government agencies. IEDs range from crude homemade explosives to extremely intricate remote-controlled devices. The devices are used to instill fear in U.S. Soldiers, coalition forces, and the local civilian population. Their employment is intended to diminish U.S. national resolve with mounting casualties. The sophistication and range of IEDs continue to increase as technology continues to improve and as terrorists gain experience.

TYPES

I-2. Some of the many types of IEDs follow.

Timed Explosive Devices

I-3. These can be detonated by remote control such as by the ring of a cell phone, by other electronic means, or by the combination of wire and either a power source or timed fuze (Figure I-1).

Impact Detonated Devices

I-4. These detonate after being dropped, thrown, or impacted in some manner.

Vehicle Bombs

I-5. These may include explosive-laden vehicles detonated with electronic command wire or wireless remote control, or with timed devices. They might be employed with or without drivers.

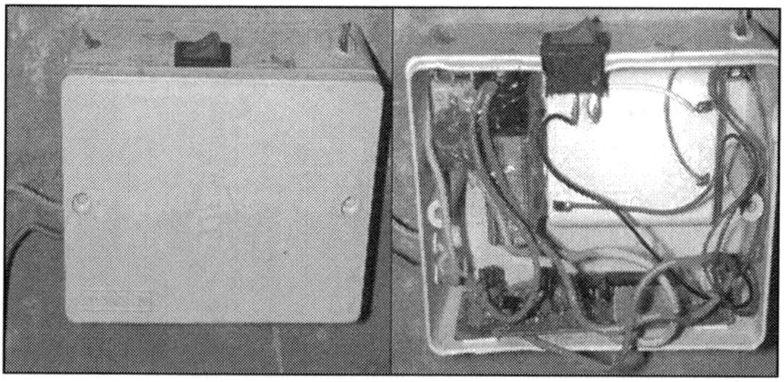

Figure I-1. Example of IED detonation device with explosive.

CHARACTERISTICS

I-6. Key identification features and indicators of suspected IEDs include—

- Exposed wire, cord, or fuze protruding from an object that usually has no such attachment.
- An unusual smell, sound, or substance emanating from an object.
- An item that is oddly light or heavy for its size.
- An object that seems out of place in its surrounding.
- An object or area locals are obviously avoiding.
- An threatening looking object covered with written threats or whose possessor uses verbal threats.
- An object that is thrown at personnel, facilities, or both.

INGREDIENTS

I-7. Anything that can explode will be used to make IEDs. Examples include:

- Artillery rounds containing high explosives or white phosphorous.
- Any type of mine (antitank or antipersonnel).
- Plastic explosives such as C4 or newer.
- A powerful powdered explosive.
- Ammonium nitrate (fertilizer) combined with diesel fuel in a container. (The truck bomb that destroyed the Oklahoma City Federal Building used ammonium nitrate and diesel fuel.)

CAMOUFLAGE

I-8. An IED can vary from the size of a ballpoint pen to the size of a water heater. They are often contained in innocent-looking objects to camouflage their true purpose. The type of container used is limited only by the imagination of the terrorist. However, containers usually have a heavy metal casing to increase fragmentation. Figure I-2 shows some of the types of camouflage that have been used to hide IEDs in Iraq. Some of the more commonly used containers include:

- Lead, metal, and PVC pipes with end caps (most common type).
- Fire extinguishers.
- Propane tanks.
- Mail packaging.
- Wood and metal boxes.

- Papier-mâché or molded foam or plastic "*rocks*," (containers that look like rocks, usually employed along desert roads and trails).
- Military ordnance, or rather modified military ordnance, which uses an improvised fuzing and firing system.

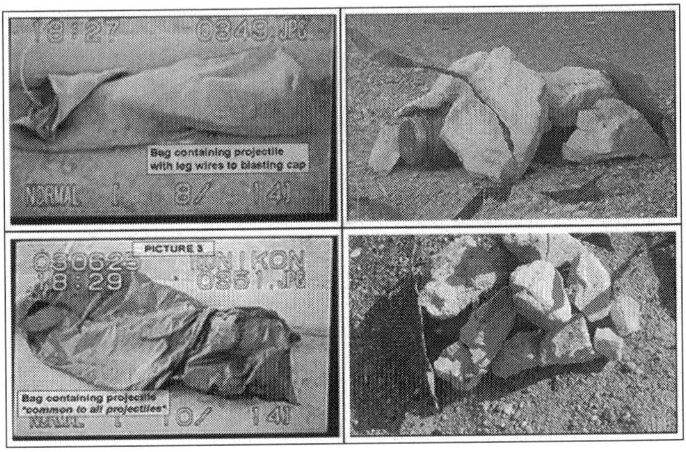

Figure I-2. Camouflaged UXO.

VEHICLE-BORNE DEVICES (CAR BOMBS)

I-9. Car bombs obviously use a vehicle to contain the device. The size of the device varies by the type of vehicle used. They can be packed into varying sizes of sedans, vans, or a large cargo trucks (Figure I-3). Larger vehicles can carry more explosives, so they cause more damage than smaller vehicles. Device functions, like package types, vary.

I-10. Signs of a possible car bomb include:

- A vehicle riding low, especially in the rear, especially if the vehicle seems empty. Explosive charges can also be concealed in the panels of the vehicle to distribute the weight of the explosives better.
- Suspiciously large boxes, satchels, bags, or any other type of container in plain view on, under, or near the front seat in the driver's area of the vehicle.
- Wire or rope-like material coming from the front of the vehicle that leads to the rear passenger or trunk area.
- A timer or switch in the front of a vehicle. The main charge is usually out of sight, and often in the rear of the vehicle.
- Unusual or very strong fuel-like odors.
- An absent or suspicious-behaving driver.

ATF	Vehicle Description	Maximum Explosives Capacity	Lethal Air Blast Range	Minimum Evacuation Distance	Falling Glass Hazard
	Compact Sedan	500 pounds 227 Kilos (In Trunk)	100 Feet 30 Meters	1,500 Feet 457 Meters	1,250 Feet 381 Meters
	Full Size Sedan	1,000 Pounds 455 Kilos (In Trunk)	125 Feet 38 Meters	1,750 Feet 534 Meters	1,750 Feet 534 Meters
	Passenger Van or Cargo Van	4,000 Pounds 1,818 Kilos	200 Feet 61 Meters	2,750 Feet 838 Meters	2,750 Feet 838 Meters
	Small Box Van (14 Ft. box)	10,000 Pounds 4,545 Kilos	300 Feet 91 Meters	3,750 Feet 1,143 Meters	3,750 Feet 1,143 Meters
	Box Van or Water/Fuel Truck	30,000 Pounds 13,636	450 Feet 137 Meters	6,500 feet 1,982 Meters	6,500 Feet 1,982 Meters
	Semi-Trailer	60,000 Pounds 27,273 Kilos	600 feet 183 Meters	7,000 Feet 2,134 Meters	7,000 Feet 2,134 Meters

Figure I-3. Vehicle IED capacities and danger zones.

EMPLOYMENT

I-11. IEDs have been used against the U.S. military throughout its history. Operation Enduring Freedom (Afghanistan) and Iraqi Freedom (OIF) have seen the use of IED attacks on a significant scale targeting U.S., coalition, and Iraqi Security forces, and civilian concentrations. Some threat TTPs might include:

- An IED dropped into a vehicle from a bridge overpass. An enemy observer spots a vehicle and signals a partner on the overpass when to drop the IED. Uncovered soft-top vehicles are the main targets. These IEDs are triggered either by timers or by impact (Figure I-4).
- An IED used in the top-attack mode and attached to the bottom of a bridge or overpass. This IED is command-detonated as a vehicle passes under it. This method gets around the side and undercarriage armor used on U.S. vehicles.
- An IED used with an ambush. Small arms, RPGs, and other direct-fire weapons supplement the IED, which initiates the ambush (Figures I-5, I-6, and I-7). Terrorists sometimes use deception measures such as dummy IEDs to stop or slow vehicles in the real kill zone.
- The driver of a suicide or homicide vehicle such as a taxicab feigns a breakdown and detonates the vehicle when Soldiers approach to help. The vehicle with IEDs might also run a checkpoint and blow up next to it.
- Suicide bombers sometimes approach U.S. forces or other targets and then self-detonate. Children might approach coalition forces wearing explosive vests.

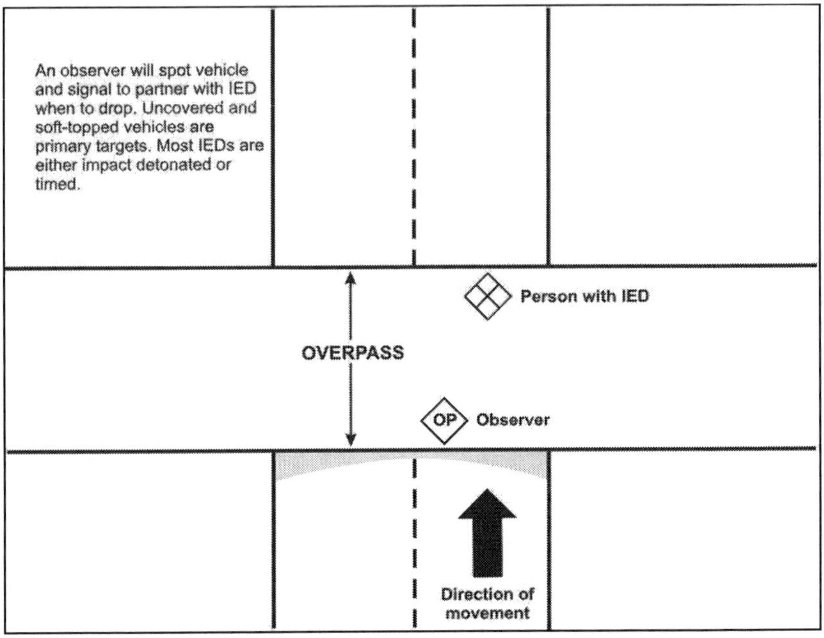

Figure I-4. Example of IED dropped into vehicles.

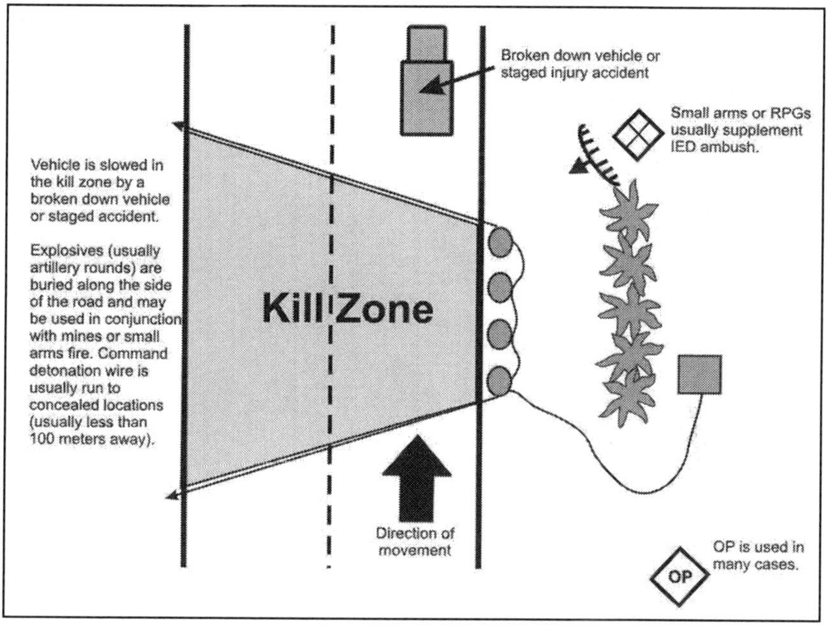

Figure I-5. Typical IED combination ambush.

Figure I-6. IED combination ambush in Iraq.

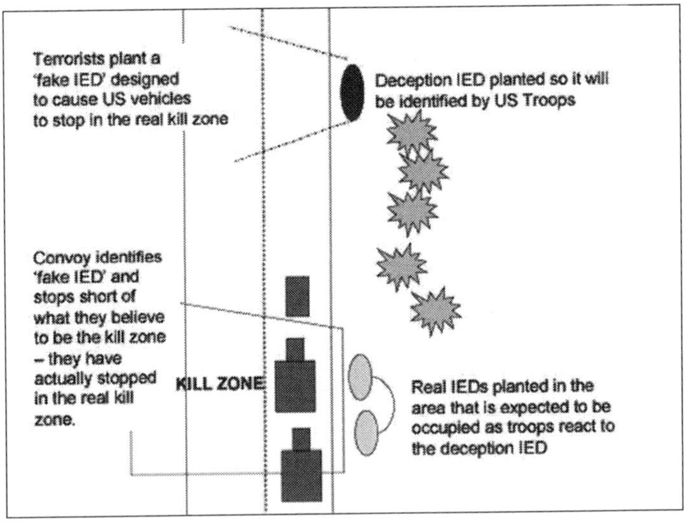

Figure I-7. Deception or fake IED used to stop convoy in kill zone.

COUNTERMEASURES

I-12. The enemy continues to adapt as friendly countermeasures evolve. Following are some measures used to counter an IED threat.

AVIATION SUPPORT

I-13. Operate with army aviation support when possible. Terrorists employing command-detonated IEDs generally rely on a quick escape after detonating an IED or executing an ambush. Recent trends have shown that OH-58D support deters attacks because terrorists are unable to break contact easily.

ALL-ROUND SECURITY

I-14. Remain alert. Maintain all-round security at all times. Scan rooftops and bridge overpasses for enemy activity.

CONVOY SECURITY

I-15. When possible, travel in large convoys. Vary road speed to disrupt the timing of command-detonated devices. However, terrorists often target convoys (or specific vehicles within convoys) with poor security postures. All occupants of convoy vehicles should have and keep their weapons pointed in an alert and defensive posture. Maintain a strong rear security element or a follow-on "shadow" trail security element. This force can more quickly be brought to bear on an enemy attacking the rear of a convoy. Use armed vehicles to speed ahead of a convoy to overwatch overpasses as the convoy passes. The lead vehicle in a convoy should have binoculars to scan the route ahead. All convoys should have extra tow bars or towing straps to recover broken-down vehicles quickly.

ADAPTATION

I-16. Be aware of evolving enemy tactics/procedures and be prepared to design countermeasures (Figure I-8). To the maximum possible extent, avoid becoming predictable. Vary routes, formations, speeds, and techniques.

TURNS

I-17. Avoid moving toward or stopping for an item in the roadway. Give wide clearance to items in the road. Turn to the outside of corners because terrorists often plant IEDs on the inside of turns to close the distance to the target. Turning to the outside also allows a longer field of view past the turn.

AUDIBLE SIGNALS

I-18. At night be aware of audible signals that can be used to communicate the approach of a convoy such as flares, gunfire, lights going off, or horns honking.

ENEMY OBSERVERS

I-19. Be alert for people who seem overly interested in your convoy, especially those using cell phones while watching your convoy.

UNUSUAL SILENCE

I-20. Be aware of unusually quiet areas. Often, local civilians have been warned of an enemy attack on coalition forces.

USE OF HEADLIGHTS

I-21. Do not use service drive headlights during the day. Having lights on during daylight makes the military vehicles stand out and easier to identify at a greater distance.

VEHICLE PROTECTION

I-22. Harden all vehicles.

OTHER TRAVELING PRECAUTIONS

I-23. Do not stop for broken down civilian vehicles, vehicle accidents, or wounded civilians along a convoy route.

CIVILIAN VEHICLE THREATS

I-24. Be alert to civilian vehicles cutting in and out or ramming vehicles in a convoy as if attempting to disrupt, impede, or isolate the convoy. Current ROE might permit you to fire warning shots or to engage threatening vehicles.

FIVES C's TECHNIQUE

I-25. Using the five C's (confirm, clear, call, cordon, control) technique helps to simplify both awareness and reaction to a suspected IED.

CONFIRM

I-26. The first step when encountering a suspected IED is to confirm that it is an IED. If Soldiers suspect an IED while performing 5- and 25-meter searches of their positions, they should act as if it could detonate at any moment, even if it turns out to be a false alarm. Using as few people as possible, troops should begin looking for telltale signs such as wires, protruding ordnance, or fleeing personnel.

CLEAR

I-27. If an IED is confirmed, the next step is to clear the area. The safe distance is determined by several factors: the tactical situation, avoidance of predictability, and movement several hundred meters away. Everyone within the danger zone should be evacuated. If more room is needed such as when the IED is vehicle-born, Soldiers should clear a wider area and continuously direct people away. Only explosive ordnance disposal (EOD) personnel or their counterparts may approach the IED. While clearing, avoid following a pattern and look out for other IEDs. If you find any more, reposition to safety and notify a ranking member on the scene.

CALL

I-28. While the area around the IED is being cleared, a nine-line IED/UXO report should be called in. The report is much like the nine-line MEDEVAC report. It includes the necessary information for the unit's TOC to assess the situation and prepare an appropriate response.

CORDON

I-29. After the area has been cleared and the IED has been called in, Soldiers should establish fighting positions around the area to prevent vehicle and foot traffic from approaching the IED. They assure the area is safe by checking for secondary IEDs. They use all available cover. The entire perimeter of the effected area should be secured and dominated by all available personnel. Available obstacles should be used to block vehicle approach routes. Scan near and far for enemy observers who might try to detonate the IED. Insurgents often try to hide where they can watch their target area and detonate at the best moment. To deter attacks, randomly check the people leaving the area.

CONTROL

I-30. Since the distance of all personnel from the IED directly affects their safety, Soldiers should control the site to prevent people from straying too close until the IED is cleared. No one may leave the area until the EOD gives the "all clear." While controlling the site, assure all Soldiers know the contingency plans in case they come under attack by any means, including direct-fire small arms or RPGs, or indirect fires.

SECTION II — SUICIDE BOMBERS

I-31. These are different from all other terrorist threats, and require specific guidance on actions, particularly the interpretation of the ROE.

DEFINITION

I-32. A suicide attack is so called because it is an attack that means certain death for the attacker. The terrorist knows that success depends on his willingness to die. He conducts this kind of attack by detonating a worn, carried, or driven portable explosive charge. In essence, the attacker is himself a precision weapon. Suicide bombers aim to cause the maximum number of casualties, or to assassinate a particular target. Stopping an ongoing suicide attack is difficult. Even if security forces stop him before he reaches his intended target, he can still activate the charge and kill or injure those around him at the time. An additional benefit is the simplicity of such an attack. Neither escape nor extraction is an issue. Nor is intelligence, for no one will be left to interrogate. The only way to prepare for a suicide attack is to train Soldiers to react immediately with competence and confidence. Soldiers should also train to avoid overreacting with unnecessary or inappropriate lethal force. The following are potential high-value targets for suicide bombers.

- High-signature forces such as uniformed military and security elements; military vehicles; civilian vehicles used for military purposes; military bases; checkpoints; patrols; liaison personnel; or supportive host nation personnel.
- Members and facilities of the international community such as ambassadors and other diplomats; embassy, U.N., and NGO buildings; and diplomatic vehicles and staffs.
- National and provincial leaders and government officials.
- Civilians in public places such as markets, shops, and cafes. Although civilians in these locations are seldom primary targets, some groups do attack them.

DELIVERY METHODS

I-33. The two main methods of employing devices are by person or by vehicle.

- A person-borne suicide bomb usually has a high-explosive and fragmentary effect and uses a command-detonated firing system such as a switch or button the wearer activates by hand. A vest, belt, or other specially modified clothing can conceal explosives with fragmentation (Figure I-8).
- A vehicle-borne suicide bomb uses the same methods and characteristics of other package or vehicle bombs, and is usually command detonated.

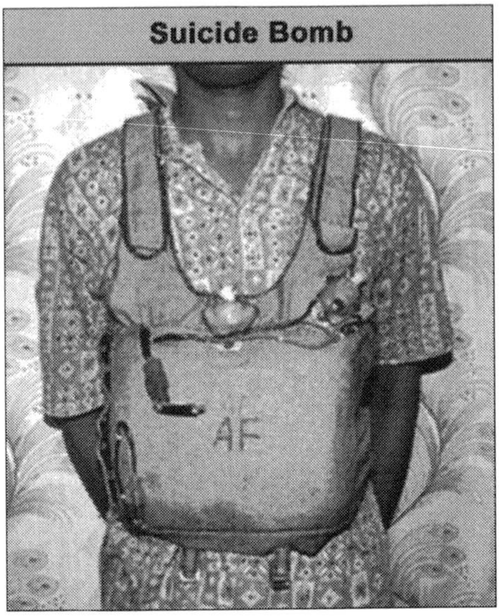

Figure I-8. Suicide bomber vest.

INDICATORS

I-34. Suicide bombers can be either gender and any age. For example, recent Palestinian bombers were female teenagers. You might be looking at a suicide bomber if you see someone who—

- Tries to blend in with the (target) environment.
- Wears ordinary, nondistinctive clothing, military or religious garb, or an oversized, bulky, or unseasonably heavy coat or jacket.
- Demonstrates fanatical religious beliefs by behaviors such as praying fervently, possibly loudly, in public.
- Has a shaved head (Muslim males); wears their hair short and their face clean shaven; or wears fragrance, which is unusual for an Arab man.
- Behaves nervously, that is, sweats, or glances about anxiously.
- Has religious verses from the Quran written or drawn onto their body, hands, or arms.
- (Islamic males) dresses as and pretends to be a woman.
- Carries a bag tightly, clutched close to the body, and in some cases squeezes or strokes it.

SPECIAL CONSIDERATIONS

I-35. Consider the following when dealing with potential suicide bombers:

- Most will try to detonate the device if they believe they have been discovered.
- Suicide bombers are of any nationality, not necessarily of direct Middle Eastern descent. They may simply sympathize with the terrorist group's cause(s).
- If you determine that a suspect is a suicide bomber, then you will probably have to use deadly force. Prepare for and expect a detonation. Shoot from a protected position from as far away as possible.

- Many suicide bombers use pressure-release-type detonation devices that they hold in their hands. They apply the pressure before they begin their final approach to the target. The explosive payload will detonate as soon as the bomber relaxes his grip, so it will go off even if you kill him.
- Some bombers also have a command-detonated system attached to their bomb, and a second person observes and tracks him to the target. This also allows the terrorists to control and detonate the bomb, even if the bomber dies or his trigger is destroyed or disabled.
- The suicide bomber may also use a timed detonation system, and again this works whether or not you kill him before he reaches his target.

COMPLICATIONS

I-36. Dealing with a suicide bomber is one of the toughest situations a Soldier can face. In just a few seconds he must identify the bomber, assess the situation, consider how to comply with the ROE, and act decisively. There is seldom time to think beyond that or to wait for orders. The only possible way to stop the bomber short of his target is to immediately incapacitate him with lethal force. Challenging him would probably cause him to trigger his device at once. The suicide bomber is trained and prepared to carry out his mission. Some experts believe that a suicide bomber considers himself already dead when setting out on an attack. The Soldier and leader must continually be aware that—

- A pressure release switch can detonate the device as soon as the bomber is shot.
- A device could be operated by remote control or timer even after the bomber is incapacitated.
- Another person observe and command-detonate the bomb.
- A second suicide bomber might be operating as a backup or to attack the crowd and assistance forces that normally gather after a detonation.

SECTION III — UNEXPLODED ORDNANCE

I-37. Unexploded ordinance (UXO) are made up of both enemy and friendly force ordnance that have failed to detonate. UXO sometimes pose no immediate threat, but they can cause injuries, loss of life, and damage to equipment if appropriate actions are not taken. UXO can be found on the battlefield, in urban areas, caves, and almost anywhere in an AO. UXO can be a result of a recent battle or war, or left over from past conflicts. During Operation Enduring Freedom (OEF), U.S. Soldiers, coalition forces, and the local population were in danger of encountering an estimated 10,000,000 pieces of UXO and mines left over from 23 years of war in Afghanistan. Soldiers in Bosnia and Soldiers fighting in Operation Iraqi Freedom have been exposed to an estimated 8 million antipersonnel mines and 2 million antitank mines, as well as UXO. Soldiers can expect to encounter UXO in any future conflict.

RECOGNITION

I-38. Soldiers' knowledge of UXO is essential to help prevent the risk of injury. Soldiers are generally familiar with the appearance of ammunition and munitions used in their own weapons. They seldom recognize what the actual projectile looks like once it has been fired, especially if it is discolored or deformed by impact. Also, Soldiers might not be able to easily recognize UXO from USAF-delivered weapons or from non-U.S. weapons. In general, leaders should caution their Soldiers against disturbing any unknown object on the battlefield.
I-1.

I-39. FM 3-100.38 provides detailed illustrations and identifying characteristics of the four categories of UXO, including projected, thrown, placed, and dropped.

PROJECTED ORDNANCE

I-40. Projected ordnance includes:

- Projectiles such as HE, chemical, illumination, and submunitions.
- Mortar rounds such as HE, chemical, WP, and illumination.

- Rockets such as self-propelled projectiles, no standard shape.
- Guided missiles such as missiles with guidance systems.
- Rifle grenades similar to mortars but fired from rifles.

THROWN ORDNANCE

I-41. Thrown ordnance including fragmentation, smoke, illumination, chemical, and incendiary hand grenades.

PLACED ORDNANCE

I-42. Placed ordnance includes:

- AP mines, generally small, of various shapes and sizes, and made of plastic, metal, or wood. They might have trip wires attached.
- AT mines, large, of various shapes and sizes, and made of plastic, metal, or wood. They might have antihandling devices.

DROPPED ORDNANCE

I-43. Dropped ordnance includes:

- Bombs, small to very large, with metal casings, tail fins, lugs, and fuzes. They may contain HE, chemicals, or other hazardous materials.
- Dispensers that look similar to bombs but may have holes or ports in them. Do not approach as sub-munitions might be scattered around.
- Very sensitive submunitions such as small bombs, grenades, or mines.

DANGER

DO NOT TRY TO TOUCH OR MOVE UXO. ORDNANCE FAILS FOR MANY REASONS, BUT ONCE FIRED OR THROWN, THE FUZING SYSTEM WILL LIKELY ACTIVATE. THIS MAKES THE ORDNANCE TOO UNSTABLE TO HANDLE. IF A ROUND FAILS TO FUNCTION INITIALLY, ANY SUBSEQUENT STIMULUS OR MOVEMENT MIGHT SET IT OFF.

IMMEDIATE ACTION

I-44. Many areas, especially previous battlefields, might be littered with a wide variety of sensitive and deadly UXO. Soldiers need to follow these precautions on discovering a suspected UXO:

- Do not move toward the UXO. Some types of ordnance have magnetic or motion-sensitive fuzing.
- Never approach or pick up UXO even if identification is impossible from a distance. Observe the UXO with binoculars if available.
- Send a UXO report to higher HQ (Figure I-9). Use radios at least 100 meters away from the ordnance. Some UXO fuzes might be set off by radio transmissions.
- Mark the area with mine tape or other obvious material at a distance from the UXO to warn others of the danger. Proper markings will also help EOD personnel find the hazard in response to the UXO report.
- Evacuate the area while carefully scanning for other hazards.
- Take protective measures to reduce the hazard to personnel and equipment. Notify local people in the area.

BOOBY TRAPS

I-45. Booby traps are typically hidden or disguised explosive devices rigged on common items to go off unexpectedly (Figure I-9). They may also be employed as antihandling devices on UXO, emplaced mines, or as improvised explosive devices (IED). Identify, mark, and report using the nine-line UXO incident report (Figure I-10). Field-expedient booby traps have also been employed with some success during most conflicts.

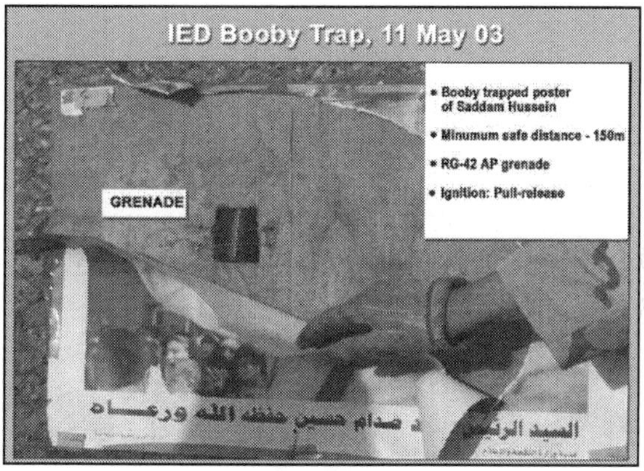

Figure I-10. Example booby trap.

1. *DTG:* Date and time UXO was discovered.
2. *Reporting Unit or Activity, and UXO Location:* Grid coordinates.
3. *Contact Method:* How EOD team can contact the reporting unit.
4. *Discovering Unit POC:* MSE, DSN phone number and unit frequency, or call sign.
5. *Type of UXO:* Dropped, projected, thrown, or placed, and number of items discovered.
6. *Hazards Caused by UXO:* Report the nature of perceived threats such as a possible chemical threat or a limitation of travel over key routes.
7. *Resources Threatened:* Report any equipment, facilities, or other assets threatened by the UXO.
8. *Impact on Mission:* Your current situation and how the UXO affects your status.
9. *Protective Measures:* Describe what you have done to protect personnel and equipment such as marking the area and informing local civilians.

Figure I-10. Nine-line UXO incident report.

Glossary

ACRONYMS AND ABBREVATIONS

Acronym/Term	*Definition*
AA	assembly area
AAR	after-action review
ACL	allowable cargo load
AD	air defense
ADA	air defense artillery
ADAM	air defense and missile
AG	assistant gunner
AHD	antihandling device
ALO	air liaison officer
ANCD	automated network control device
AO	area of operations
AP	Antipersonnel
APC	armored personnel carrier
APDS	armor piercing discarding sabot
APOBS	Antipersonnel Obstacle Breaching System
AR	automatic rifleman
AT	antitank
ATGM	antitank guided missile
BAS	battalion aid station
BB	bunker buster
BC	Bradley commander
BCU	battery coolant unit
BDM	bunker defeat munition
BFT	blue force tracker
BIT	built-in text
BITE	built-in test equipment
BMNT	beginning morning nautical twilight
BVF	M2 Bradley Fighting Vehicle
C2	command and control
CAPS	counter active protection system
CAS	close air support
CASEVAC	casualty evacuation
CBRN	chemical, biological, radiological, and nuclear
CCIR	commander's critical information requirements
CCM	close combat missile
CCMS	Close Combat Missile System

CCP	casualty collection point
CFL	coordinated fire line
CLU	command launch unit
COA	course of action
COE	contemporary operational environment
COP	common operating picture
CP	check point
CWST	combat survival water test
DA	Department of the Army
DD	Deparment of Defense form
DED	detailed equipment decontamination
DLIC	detachment left in contact
DoD	Department of Defense
DTD	detailed troop decontamination
DTG	date-time group
DVR	driver
EA	engagement area
ECM	electronic countermeasures
EENT	end of evening nautical twilight
EFP	explosive formed penetrator
EPW	enemy prisoner of war
ES2	enemy soldier and sensor
ETAC	enlisted tactical air controller
FBCB2	Force XXI Battle Command Brigade and Below System
FDC	fire direction center
FEBA	forward edge of the battle area
FFE	fire for effect
FIST	fire support team
FLIR	forward looking infared radar
FLOT	forward line of troops
FM	field manual
FO	forward observer
FOV	field of vision
FPF	final protective fire
FPL	final protective line
FRAGO	fragmentary order
FS	fire support
FSCL	fire support coordination line
FSCM	fire support coordination measures
FSE	fire support element
FSO	fire support officer

GNR	gunner
GPS	Global Positioning System
GSR	ground surveillance radar
GTA	graphic training aid
GTL	gun team leader
GTP	ground tactical plan
HE	high explosive
HEAT	high explosive antitank
HEDP	high explosive dual purpose
HEI-T	high explosive incendiary w/ tracer
HEMTT	heavy expanded mobility tactical truck
HMMWV	high-mobility multipurpose wheeled vehicle
HPT	high profile target
HQ	Headquarters
HUMINT	human intelligence
IBA	individual body armor
ICV	Infantry carrier vehicle
ICOM	ingetrated communication
IED	improvised explosive device
IMP	Impact
IMT	individual movement techniques
IPB	intelligence preparation of the battlefield
IR	information requirement
IRP	initial rally point
ISR	intelligence, surveillance, reconnaissance
ITAS	Improved Target Acquistion System
JSLIST	joint service lightweight integrated suit technology
JTAC	joint terminal air controller
KIA	killed in action
KPH	kilometers per hour
LACE	liquid, ammunition, casualty, and equipment
LAW	light antiarmor weapon
LC	line of contact
LCMR	lightweight counter-mortar radar
LD	line of departure
LOA	limit of advance
LOC	line of communication
LOGPAC	logistics package
LOS	line of sight
LOW	law of war
LTA	launch tube assembly

LZ	landing zone
MANPADS	Man-Portable Air Defense System
MBA	main battle area
MDI	modernized demolition initiator
MEDEVAC	medical evacuation
MEL	maximum engagement line
METTC-TC	mission, enemy, terrain, troops-time, civil
MGS	Mobile Gun System
MLO	multipurpose rain/snow/cb overboot
MOPMS	modular pack mine system
MOPP	mission-oriented protective posture
MR	Moonrise
MS	Moonset
MSD	minimum safe distance
MSR	main supply route
MVT	Movement
NAI	named area of interest
NCO	non-commissioned officer
NCS	net control system
NFA	no-fire area
NFV	narrow field of view
NLT	not later than
NMC	nonmission capable
NSB	near surface burst
NVD	night vision device
NVG	night vision goggles
NVS	night vision sight
OAKOC	observation and fields of fire, avenues of approach, key and decisive terrain, obstacles, cover and concealment
OP	observation post
OPCON	operational control
OPORD	operation order
OPTEMPO	operational tempo
ORP	objective rally point
OT	observer target
PA	physician's assistant
PCI	precombat inspection
PDF	principle direction of fire
PI	percent of incapacitation
PIR	priority intelligence requirement
PLD	probable line of deployment

PLGR	precision lightweight GPS receiver
PLL	parts load list
PMCS	preventive maintenance checks and services
PMM	preventive medicine measures
POL	petroleum, oil, and lubricants
POSNAV	position navigation
PRX	Proximity
PSG	platoon seargent
PSI	pressure per square inch
PZ	pickup zone
RCU	remote control unit
RED	risk estimate distance
RFA	restrictive fire area
RIP	relief in place
RLEM	rifle-launched entry munitions
ROE	rules of engagement
ROI	rules of interaction
RPG	rocket-propelled grenade
RRP	reentry rally point
RTO	radiotelephone operator
SA	situational awareness
SALUTE	size, activity, location, uniform, time, equipment
SBCT	Stryker brigade combat team
SBF	support by fire
SCATMINEs	scatterable mines
SD	self-destruct
SDM	squad designated marksman
SDZ	surface danger zone
SEAD	suppression of enemy air defenses
SEE	small earth excavator
SHORAD	short-range air defense
SINCGARS	Single Channel Ground/Airborne Radio System
SITEMP	situation template
SL	squad leader
SLAM	selectable lightweight attack munition
SLM	shoulder-launched munitions
SMAW-D	shoulder-launched multipurpose assault weapon (disposable)
SME	small earth excavator
SOFA	status of forces agreement
SOP	standing operating procedure
SOSRA	suppress, obscure, secure, reduce, and assault

SR	Sunrise
SSC	small-scale contingency
SS	Sunset
SU	situational understanding
SVML	standard vehicle-mounted launcher
TACP	tactical air control party
TC	truck commander
TCP	traffic control point
TL	team leader
TLP	troop-leading procedures
TM	technical manual
TOW	tube-launched, optically-tracked, wire-guided
TPT	trainer practice tracer
TRP	target reference point
TSOP	tactical standing operating procedure
TTP	tactics, techniques, and procedures
TWS	thermal weapons site
UAS	unmanned aircraft system
UCMJ	Uniform Code of Military Justice
UHF	ultra high frequency
UMCP	unit maintenance collection point
UXO	unexploded ordnance
VC	vehicle commander
WARNO	warning order
WCS	weapons control status
WFF	warfighting function
WFOV	wide field of view
WIA	wounded in action
WLGH	weapon-launched grappling hook
WP	white phosphorus
WSL	weapons squad leader
XO	executive officer

References

Army Regulations

AR 190-8, *Enemy Prisoners Of War, Retained Personnel, Civilian Internees and Other Detainees*. 01 October 1997.
AR 385-63, *Range Safety {MCO 3570.1B}*. 19 May 2003.

Army Training Evaluation Program Mission Training Plans

ARTEP 7-8-DRILL, *Battle Drills for the Infantry Rifle Platoon and Squad.* 25 June 2002.
ARTEP 7-8-MTP, *Mission Training Plan for the Infantry Rifle Platoon and Squad.* 29 September 2004.

Field Manuals

FM 3-21.10, *The Infantry Rifle Company*. 27 July 2006.
FM 1-02, *Operational Terms and Graphics*. 21 September 2004.
FM 6-22, *Army Leadership*. 12 October 2006.
FM 6-0, *Mission Command: Command and Control of Army Forces*. 11 August 2003.
FM 3-22.68, *Crew-Served Machine Guns, 5.56-mm and 7.62-mm*. 31 January 2003.
FM 3-0, *Operations*. 14 June 2001.
FM 3-90, *Tactics*. 04 July 2001.
FM 21-60, *Visual Signals*. 30 September 1987.
FM 6-30, *Tactics, Techniques, and Procedures for Observed Fire*. 16 July 1991.
FM 21-75, *Combat Skills of the Soldier*. 03 August 1984.
FM 21-18, *Foot Marches*. 01 June 1990 (with C1, 08 August 2005).
FM 5-19, *Composite Risk Management*. 21 August 2006.
FM 3-34.2, *Combined-Arms Breaching Operations*. 31 August 2000 (with C3, 11 October 2002).
FM 3-06, *Urban Operations*. 26 October 2006.
FM 3-06.11, *Combined Arms Operations in Urban Terrain*. 28 February 2002.
FM 90-7, *Combined Arms Obstacle Integration*. 29 September 1994 (with C1, 10 April 2003).
FM 5-103, *Survivability*. 10 June 1985.
FM 5-34, *Engineer Field Data*. 19 July 2005.
FM 3-23.25, *Shoulder-Launched Munitions*. 31 January 2006.
FM 3-22.37, *Javelin Medium Antiarmor Weapon System*. 23 January 2003.
FM 3-22.34, *Tow Weapon System*. 28 November 2003.
FM 3-22.32, *Improved Target Acquisition System, M41*. 08 July 2005.
FM 90-4, *Air Assault Operations*. 16 March 1987.
FM 20-32, *Mine/Countermine Operations*. 29 May 1998 (with C5, 01 April 2005).
FM 5-250, *Explosives and Demolitions*. 30 July 1998 (with C1, 30 June 1999).
FM 90-7, *Combined Arms Obstacle Integration*. 29 September 1994 (with C1, 10 April 2003).
FM 3-21.38, *Pathfinder Operations*. 25 April 2006.
FMI 2-91.4, *Intelligence Support to Operations in the Urban Environment*. 30 June 2005.
FM 7-8, *Infantry Rifle Platoon and Squad*. 22 April 1992 (with C1, 01 March 2001).
FM 3-21.10, *The Infantry Rifle Company*. 27 July 2006.
FM 2-21.20 (FM 7-20), *The Infantry Battalion*. 13 December 2006.
FM 3-22.9, *Rifle Marksmanship M16A1, M16A2/3, M16A4, and M4 Carbine*. 24 April 2003 (with C4, 13 September 2006).
FM 3-07, *Stability Operations and Support Operations*. 20 February 2003 (with C1, 30 April 2003).
FM 8-10-6, *Medical Evacuation in A Theater of Operations, Tactics, Techniques, and Procedures*. 14 April 2000.
FM 8-10.26, *Employment of the Medical Company (Air Ambulance)*. 16 February 1999 (with C1, 30 May 2002).
FM 3-22.65, *Browning Machine Gun, Caliber .50 Hb, M2*. 03 March 2005.
FM 3-22.27, *MK 19, 40-mm Grenade Machine Gun, MOD 3*. 28 November 2003 (with C1, 14 September 2006).

FM 3-25.26, *Map Reading and Land Navigation.* 18 January 2005 (with C1, 30 August 2006).
FM 22-6, *Guard Duty.* 17 September 1971 (with C1, 15 January 1975).
FM 7-8, *Infantry Rifle Platoon and Squad.* 22 April 1992 (with C1, 01 March 2001).
FM 3-100.38, *(Ux0) Multi-Service Tactics, Techniques And Procedures For Unexploded Operations {MCRP 3-17.2B; NTTP 3-02.4.1;AFTTP(I) 3-2.12}.* 16 August 2005.

Department of the Army Forms

DA Form 1355-1-R, *Hasty Protective Row Minefield Record.* September 2001.
DA Form 5517-R, *Standard Range Card.* February 1986.
DA Form 4137, *Evidence/Property Custody Document.* July 1976.
DA Form 2404, *Equipment Inspection and Maintenance Worksheet.* April 1979.
DA Form 2028, *Recommended Changes to Publications and Blank Forms.* February 1974.
DA Form 1156, *Casualty Feeder Card.* March 2006.

Department of Defense Forms

DD Form 2745, *Enemy Prisoner of War (EPW) Capture Tag.* May 1996.

Department of Defense Directives

Department of Defense Directive 2311.01E, *DoD Law of War Program.* 09 May 2006.

Department of the Army Pamphlets

DA PAM 385-63, *Range Safety.* 10 April 2003.

DA PAM 350-38, *Standards in Weapons Training.* 16 August 2004.

TRADOC Regulations

TRADOC Reg. 350-70, *Systems Approach to Training Management, Processes, and Procedures.* 09 March 1999.

TRADOC Reg. 385-2, *TRADOC Safety Program.* 27 January 2000 (with C1, 10 October 2000).

Technical Manuals

TM 9-1005-319-10, *Operator's Manual for Rifle, 5.56 mm, M16A2W/E; Rifle 5.56-mm, M16A3; Rifle, 5.56 mm, M16A4; Carbine, 5.56-mm, M4 W/E; Carbine, 5.56-mm, M4A1.* 01 October 1998.

TM 9-1010-221-10, *Operator's Manual for Grenade Launcher, 40-mm: M203.* 01 August 2001.

TM 9-1005-201-10, *Operator's Manual Machine Gun, 5.56-mm, M249 w/Equip.* 26 July 1991.

TM 9-1005-313-10, *Operator's Manual for Machine Gun, 7.62mm, M240B.* 11 November 2002 (with C2, 20 May 2005).

TM 9-1315-886-12, *Operator's and Unit Maintenance Manual for Launcher and Cartridge, 84 Millimeter: M136 (AT4).* 15 May 1990.

TM 9-1425-687-12, *Operator and Organizational Maintenance Manual for Javelin.* 24 May 2005.

TM 9-1425-688-12, *Operator and Maintenance Manual for Javelin Weapon System, M98A1.* 24 May 2005.

TM 9-1005-213-10, *Operator's Manual for Machine Guns, Caliber .50; M2, Heavy Barrel Flexible, W/E.* 01 June 2001.

TM 9-1010-230-10, *Operator's Manual for Machine Gun, 40-mm, MK 19, MOD 3.* 30 May 2001.

TM 9-1425-450-12, *Operator and Organizational Maintenance Manual for TOW 2 Weapon System, Guided Missile System M220A2.* 25 May 1983.

TM 5-6665-298-10, *Operator's Manual for Mine Detecting Set AN/PSS-12.* 01 April 2002.

GRAPHIC TRAINING AIDS

GTA 05-08-001, *Survivability Positions.* 01 August 1993.
GTA 07-06-001, *Fighting Position Construction Infantry Leader's Reference Card.* 01 June 1992.

JOINT PUBLICATIONS

JP 3-09, *Joint Fire Support.* 13 November 2006.

JP 3-09.3, *Joint Tactics, Techniques, and Procedures for Close Air Support (CAS).* 03 September 2003.

RECOMMENDED READING

These sources contain relevant supplemental information.

AR 190-11, *Physical Security of Arms, Ammunition, and Explosives.* 15 November 2006.

FM 3-11.5, *Multiservice Tactics, Techniques, and Procedures for Chemical, Biological, Radiological, and Nuclear Decontamination.* 04 April 2006.

FM 21-11, *First Aid for Soldiers.* 27 October 1988 (with C2, 4 December 1991.)

U.S. LAW

The War Crimes Act of 1996, U.S. Code 18 (1996), § 2441.

INTERNET WEB SITES

Doctrinal and Training Literature (ARTEPs, FMs, GTAs, STPs, TCs)

www.adtdl.army.mil or

https://akocomm.us.army.mil/usapa/doctrine/index.html

U.S. Army Publishing Agency (DA forms and pamphlets)

www.usapa.army.mil

Index

A

actions at danger areas (mounted), 3-42
actions on contact, 1-32 (illus)
adjust fire, 2-34
advance to objective, 7-15
aerial sustainment, 6-9 (illus)
after-action reviews, 1-34
air and missile defense, 4-8, G-9
  active air defense, 4-10
  early warning procedures, 4-9
alternating fire, 2-20
ambush, 7-26, 9-12
area fire, 2-20
area reconnaissance, 9-29
  actions from the ORP, 9-29
    multiple teams, 9-30
    observation plan, 9-29
    security element, 9-30
    short range, 9-30
    surveillance teams, 9-30
area of operation (AO), 7-12, 7-13 (illus)
assault a building, 7-38
  building entry, 7-39
  clear a room, 7-40, 7-42 (illus)
  clearing a "T" intersection, 7-45, 7-46 (illus)
  clearing hallway junctions, 7-43, 7-44 (illus)
  critical tasks, 7-36
  follow-through, 7-47
  isolate building, 7-37, 7-38 (illus)
  moving in building , 7-42, 7-43 (illus)
assaulting trench systems, 7-51
  clearing, 7-52
  entering, 7-51
  moving, 7-53
attack by fire, 7-36
attacking fortified positions, 7-48 through 7-53
  find, 7-48
  finish, 7-49
  fix, 7-49
  follow-through 7-53
attacks during limited visibility, 7-25

B

booby traps, I-14
briefings and orders (patrols), 9-37

C

call for fire, 2-27
camouflage (IED), I-2
camouflage, noise, and light discipline, 3-45
car bombs, I-3
casualty procedures, 6-10
  casualty evacuation (CASEVAC), 6-11, 6-12 (table)
  casualty report, 6-14 (illus), 6-15 (illus)
  initial care, 6-10
  killed in action (KIA), 6-15
  movement, 6-10, 6-12 (table)
  unit SOPs, 6-14
chemical, biological, radiological, and nuclear (CBRN) defense, 4-10
  CBRN defense tenets, 4-11
  CBRN reconnaissance, 4-11
  contamination avoidance, 3-11
  decontamination, 4-12, 4-13 (table)
  protection, 4-11
    MOPP levels, 4-11 (table)
civil support operations, 1-42
clearing a trench, 7-52
clearing hallway junctions, 7-43, 7-44 (illus)
clearing a "T" intersection, 7-45, 7-46 (illus)
close air support, 2-43
close combat, characteristics of 1-6
  courage, 1-6
  flexibility, 1-6
  mental stamina, 1-6
  physical and mental toughness, 1-6
close combat missile systems, B-6
  javelin, B-6
  TOW, B-7
combat engineer support, G-7
combat orders, 5-4
  fragmentary order (FRAGO), 5-5
  operations order (OPORD), 5-5
  warning order (WARNO), 5-4, 5-5 (table)
combat patrols, 9-7
  ambush, 9-8
  assault element, 9-8
  final preparations, 9-14, 9-15 (illus)
  L-shaped ambush, 9-13, 9-14 (illus)
  leader locations, 9-9
  linear ambush, 9-13 (illus)
  planning, 9-8 (illus)
  raid, 9-10 (illus)
  security, 9-8
  security element, 9-9
  support element, 9-9
  v-shaped ambush , 9-14 (illus)
combat power, 1-5, 1-35
  doctrinal hierarchy of operations, 1-37, 1-38 (illus)
  four critical functions, 1-35
combat vehicle and infantry platoon formations, 3-40
combat vehicle and infantry squad formations, 3-37
combined arms warfare, 1-21
  complimentary effects, 1-21, 1-22 (illus)
command and control, 5-1, 7-12
  fire support, 7-11
  intelligence, 7-11
  leadership, 5-1 (table)
  mission command, 5-2
  mission-oriented, 5-1
  movement and maneuver, 7-11
    protection, 7-11
    sustainment, 7-12

movment to contact, 7-18, 7-19 (illus)
plans and orders, 5-3
combat orders, 5-4
FRAGO, 5-5
OPORD, 5-5
WARNO, 5-4, 5-5 (table)
mission statement, 5-3

convoy and route security, H-19
command and control, H-22
escort element, H-21
flank and rear guard screen, H-21
organization of forces, H-19
reacting to enemy contact, H-23 (illus)
actions at ambush, H-24 (illus)
actions at obstacle, H-27 (illus)
actions during halts, H-28
reaction force, H-21

convoys, 3-42

counterattack, 7-30

creeping method, 2-40, 2-41 (illus)

crossing danger areas, 3-34

crossing large open areas, 3-35, 3-36 (illus)

crossing linear danger areas (platoon), 3-35 (illus)

crossing small open areas, 3-36, 3-37 (illus)

D

debriefing and reporting, 1-11

defensive operations, 1-39, 8-1
characteristics, 8-1
conduct engagement area rehearsal, 8-18 (illus)

defend a battle position, 8-25, 8-26 (illus)

defend a perimeter, 8-30, 8-31 (illus)

defend a reverse slope, 8-31, 8-32 (illus)

defend an area, 8-24 (illus)

defend a strong point, 8-28, 8-29 (illus)

defensive purposes, 1-40

defensive techniques, 8-23
delay, 8-48
determine where to kill the enemy, 8-16
disruption, 8-2
emplace weapons systems, 8-17
engagement area development, 8-16
fighting and survivability positions, 8-33 (illus)
fighting position materials, 8-37
fighting position principles, 8-34
fighting positions, types of 8-38
javelin, 8-42, 8-43 (illus)
machine gun, 8-41, 8-42 (illus)
one-soldier, 8-39 (illus)
three-soldier, 8-40, 8-41 (illus)
trenches, 8-43, 8-44 (illus)
two-soldier, 8-40 (illus)
weapons placement (defense), 8-22
flexibility, 8-2
massing effects, 8-2
occupation of and preparation of defensive positions, 8-19
coordination, 8-23
establishment of, 8-22
planning considerations, 8-6
command and control, 8-15
fire support, 8-13
intelligence, 8-15
movement and maneuver, 8-6
protection, 8-13
priority of work, 8-20
reconnaissance, security operations, and enemy preparatory fires, 8-2
retrograde defensive operation, 8-44
security, 8-2
security in the defense, 8-21
sequence of the defense, 8-2
approach of enemy attack, 8-4
consolidation and reorganization, 8-5
counterattack, 8-5
enemy assault, 8-4
occupation, 8-3
stay-behind operations, 8-49
retirement, 8-50
deliberate attack, 7-14, 7-15 (illus)

deployment, 7-9

direct fires, 3-4

direct fire and boundaries, 7-12

doctrinal hierarchy of operations, 1-37, 1-38 (illus)

E

echelonment of fire, C-3

employing fires, 2-1
avoid target overkill, 2-3
command and control, 2-1
destroy greatest threat, 2-2
develop contingencies, 2-4
employ best weapon, 2-3
mass the effects, 2-2
minimize friendly exposure, 2-4
plan for limited visibility, 2-4
prevent fratricide, 2-4

employing direct fire, 2-11
control measures, 2-11
enemy target types, 2-7
engagement area (EA), 2-15
engagement priority, 2-18
final protective fire, 2-17
final protective line (FPL), 2-16
fire control process, 2-12
fire patterns, 2-18
leader responsibilities, 2-12
maximum engagement line (MEL), 2-16
principle direction of fire, 2-17
restrictive fire line, 2-17
rules of engagement (ROE), 2-11
sector of fire, 2-15
target reference point (TRP), 2-13, 2-14 (illus)

terrain and threat-based fire control measures, 2-12, 2-13 (table)
threat-based fire control measures, 2-18
trigger, 2-19
weapons control status, 2-19
weapons-ready posture, 2-19
weapons safety posture, 2-19

employing indirect fires, 2-27
adjust fire, 2-34
call for fire, 2-27
creeping method, 2-40, 2-41 (illus)
deviation correction, 2-36
angle T, 2-37, 2-38 (illus)
range correction, 2-38
range spotting, 2-38
successive bracketing, 2-39 (illus)
deviation spotting, 2-34 (illus), 2-35, 2-36 (illus)
end of mission, 2-41
fire support coordination measures, 2-41
free-fire area, 2-42
no-fire area, 2-42
refinement and end of mission, 2-34
restrictive fire area, 2-42
hasty bracketing, 2-39 2-40 (illus)
message to observer, 2-32
repeat, 2-33
request splash, 2-33
required elements, 2-27
time on target, 2-33
time to target , 2-33
method of fire and control, 2-33
observer I.D. and WARNO, 2-27
adjust fire, 2-27
fire for effect, 2-27
fire when ready, 2-33
grid mission, 2-28 (table)
optional elements, 2-32
ammunition, 2-32
danger close, 2-32
method of engagement, 2-32
suppress, 2-27
target description, 2-31
target location methods, 2-28
grid mission, 2-28 (illus)
polar mission, 2-28, 2-29 (table) (illus)
shift from known point, 2-29, 2-30, 2-31 (table) (illus)
trajectory, 2-32

employment, machine gun, A-1

employment, vehicles, D-1

enemy contact at danger areas, 3-37

enemy prisoners of war (EPW) and retained/detained persons, 1-46, 6-15, 6-16 (illus), 6-17, 6-18 (illus)

engagement techniques, 2-20

engaging the enemy without endangering friendly troops, 2-9

entering a trench, 7-51

envelopment movement, 7-3, 7-4 (illus)

every soldier is a sensor (ES2), 1-11

exploitation, 7-24

F

feint, 7-31

fighting and survivability positions, 8-33 (illus)

fighting position, 8-37
materials, 8-37
principles, 8-34

fire commands, 2-21
fire planning, C-1
defense, C-8
offence, C-11
process, C-2
tactical uses of planned fires, C-3
echelonment, C-3
targets effects planning, C-13 (illus)
effects of HE rounds, C-16
HE ammunition, C-13 (illus)
illumination, smoke, white phosphorous, C-17
special illumination techniques, C-18
suppressive effects of HE mortar rounds, C-16
use of thermal sites, C-18

fire support, 7-11

fire support warfighting function, 1-4

fire team formations, 3-8 (table)
file, 3-9 (illus)
wedge, 3-8 (illus)

fire team weapons, 2-7

flexibility, 8-2

follow-through, 7-53

forms of maneuver, 7-3
envelopment, 7-3, 7-4 (illus)
frontal attack, 7-7, 7-8 (illus)
infiltration, 7-5, 7-6 (illus)
penetration, 7-6, 7-7 (illus)
turning, 7-4, 7-5 (illus)

forward observer (FO), 1-14

fragmentary order (FRAGO), 5-5

G

grenadier, 1-15

grid mission, 2-28 (table)

ground tactical plan, E-2

H

hand and arm signals, 2-24, 2-25 (illus)

hasty attack, 7-17
conduct, 7-17
task organization, 7-18

helicopter movement, E-1
airmobile operation stages, E-2
air movement plan, E-5
ground tactical plan, E-2
landing plan, E-3 (table)
loading plan, E-6, E-7 (illus)
capabilities, E-1
helicopter characteristics, E-1 (table)
limitations, E-1
safety, E-9
staging plan, E-8, E-9 (table)

high explosive (HE)
  ammunition, C-13
  effects of HE cover, C-16

human dimension, 1-43
  friction, 1-44
  skill, 1-44
  will, 1-44

I

identification of enemy approach, 8-16

illumination, smoke, white phosphorous, C-17
  special illumination techniques, C-18

improvised explosive devices (IEDs), I-1 (illus)
  booby traps, I-14
  camouflage, I-2
  car bombs, I-3
  characteristics, I-2
  countermeasures, I-8
  employment, I-4
  five c's technique, I-9
  ingredients, I-2
  suicide bombers, I-10
    complications, I-12
    definition, I-10
    delivery methods, I-10
    indicators, I-11
    special considerations, I-11
    types, I-1

individual leadership, 1-24
  accountability, 1-26
  authority, 1-25
  by example, 1-25 (illus)
  delegation, 1-26
  responsibility, 1-26

infantry, 1-1
  doctrine, 1-7
  forward observer, 1-14
  individual skills, 1-8
    communicate, 1-9
    move, 1-8
    shoot, 1-8
    survive, 1-9
    sustain, 1-9
  principles, 1-7
  tactics, 1-7

infantry battalion mortar platoon, G-1

infantry battalion scout platoon, G-1

infantry battalion sniper section, G-2

infantry fire team, 1-14
  automatic rifleman, 1-16
  grenadier, 1-15
  rifleman, 1-15
  team leader, 1-16

infantry platoon, 1-11
  platoon headquarters, 1-12
  platoon leader, 1-12
  platoon medic, 1-14
  platoon radiotelephone operator, 1-13
  platoon sergeant, 1-13

infantry squad, 1-17
  squad designated marksman, 1-18
  squad leader, 1-17

infantry weapons squad, 1-20
  close combat missile team, 1-21
  medium machine gun team, 1-20
  weapons squad leader, 1-21

infiltration maneuver, 7-5, 7-6 (illus)

information requirements (IR), 9-27, 9-28 (table)

intelligence, 7-11

intelligence warfighting function, 1-4

K

kill zone, 9-12

knocking out bunkers, 7-50

L

L-shaped ambush, 9-13, 9-14 (illus)

law of war, 1-44
  enemy prisoners of war and other detainees, 1-45
  principles, 1-45
  ten soldier rules, 1-46

leadership by example, 1-25 (illus)

linear ambush, 9-13 (illus)

M

M1-series tank , D-5

M2 Bradley fighting vehicle, D-7

machine gun employment, A-1
  applications of fire, A-13
  characteristics, A-5
  combat techniques of fire, A-5
  distribution, concentration, and rate of fire, A-16
  fire commands, A-23 (table)
  fire control, A-22

machine gun tech data, A-1 (table)
  M2.50 Cal, A-5 (illus)
  M240B, A-3, A-4 (illus)
  M249, A-2, A-3 (illus)
  MK 19, A-4 (illus)

machine gun use, A-28
  ammunition planning, A-33
  machine guns in the defense, A-32
  machine guns in the offense, A-30
  methods of fire control, A-22
  predetermined fires, A-19
    field expedients, A-21
    range card, A-20
    terminology, A-19
  security, A-29
  tactical organization, A-28 (table)
  target engagements by types of targets, A-14
  target engagement during limited visibility, A-16
  techniques, A-17

maintenance, 6-6

maneuver company fire support team (FIST) fire request channels, G-7

maneuver forms, 7-3 through 7-8 (illus)
  envelopment, 7-3, 7-4 (illus)
  frontal attack, 7-7, 7-8 (illus)
  infiltration, 7-5, 7-6 (illus)
  penetration, 7-6, 7-7 (illus)
  turning, 7-4, 7-5 (illus)

maneuver movement, 7-11

mechanized infantry rifle platoon and squad, M2, G-2

message to observer, 2-32

method of engagement, 2-32

method of fire and control, 2-33
mission command, 1-27
  disciplined initiative, 1-27
  nested purpose, 1-27
mounted tactical movement, 3-41
movement and maneuver, 7-11
  warfighting function, 1-4
movement formations, 3-2
  fire team, 3-8 (table)
  primary formations, 3-6, 3-7 (table)
    squad, 3-9
    wedge, 3-8
movement techniques, 3-19
  platoon, 3-20 (table)
    bounding overwatch, 3-26 (illus)
    traveling overwatch, 3-25 (illus)
  squad, 3-21 (illus)
    bounding overwatch, 3-22, 3-23 (illus)
    command and control, 3-27
    successive and alternative bounds, 3-20 (illus),
    traveling overwatch, 3-21 (illus)
movement to contact, 7-18, 7-19 (illus)
movement to line of departure, 7-9
movement with combat vehicles, 3-37
  actions at danger areas (mounted), 3-42
  combat vehicle and infantry platoon formations, 3-40
  combat vehicle and infantry squad formations, 3-37
  convoys, 3-42
  mounted tactical movement, 3-41
  other movement situations, 3-47
    by water, 3-48
    during limited visibility, 3-48
moving in a trench, 7-53

N

naval gunfire support, 2-46
navigation aids, 3-28
navigation, types of, 3-29
nested purpose, 1-27

O

observation posts, H-6
  occupying, H-8
  types and positioning, H-6
observed fire, 2-21
obstacles, 5-10, 5-11 (illus)
  avenues of approach, 5-11 (illus)
  cover and concealment, 5-13 (illus)
  key terrain, 5-12 (illus)
  observation and fields of fire, 5-12, 5-13
obstacle reduction and employment, F-1, F-27 (illus)
  mines, F-28
  special-purpose munitions, F-34
  wire, F-39
  reduction, F-6 (illus)
    booby traps, F-26
    breaching, F-6
    constructed obstacles, F-11
    minefields, F-11
    urban obstacles, F-17
  types and catagories, F-1
    exiting, F-2
    reinforcing, F-2
offensive and defensive combat, 1-1
  basic actions, 1-2
  individual leadership, 1-3
  tactical decision making, 1-3
  tactical maneuver, 1-2
  tactical principles, 1-2 (table)
  warfighting functions, 1-3
offensive operations, 1-38, 7-1
  ambush, 7-26
  attacking fortified positions, 7-48
    find, 7-48
    finish, 7-49
    fix, 7-49
  attacks during limited visibility, 7-25
  characteristics, 7-1
  command and control, 7-12
  counterattack, 7-30
  demonstration, 7-31
  enemy-oriented, 1-39
  exploitation, 7-24
  feint, 7-31
  fire support, 7-11
  follow-through, 7-53
  forms of maneuver, 7-3, 7-4 (illus)
    envelopment, 7-3, 7-4 (illus)
    frontal attack, 7-7, 7-8 (illus)
    infiltration, 7-5, 7-6 (illus)
    penetration, 7-6, 7-7 (illus)
    turning,7-4, 7-5 (illus)
  intelligence, 7-11
  movement and maneuver, 7-11
  movment to contact, 7-18, 7-19 (illus)
  movement to line of departure, 7-9
  offensive operation types, 1-39, 7-2
  planning considerations, 7-11
    protection, 7-11
    sustainment, 7-12
  platoon attacks, 7-14
    deliberate, 7-14, 7-15 (illus)
    hasty, 7-17
  purposes, 1-39
  pursuit, 7-24
  raid, 7-29, 7-30 (illus)
  sequence of offensive operations, 7-8
    assault, 7-10
    assembly area, 7-9
    consolidation and reorganization, 7-10 (table)
    deployment, 7-9
    maneuver, 7-9
    preparation, 7-8
    reconnaissance, 7-9
  spoiling attack, 7-30
  tactical enabling and infantry platoon actions, 1-39
  tactical tasks, 7-31
    attack by fire, 7-36
    clear, 7-32
    seize, 7-31
    support by fire, 7-35
    suppress, 7-34

terrain-oriented, 1-39
urban areas, 7-36
assault a building, 7-38, 7-42, 7-43, 7-46, (illus)
critical tasks, 7-36
find, 7-37
follow-through, 7-47
isolate building, 7-37, 7-38 (illus)
operational environment, 1-46
operations order (OPORD), 5-5
organizing reconnaissance patrols, 9-26 (illus)

P

patrols, 9-1
ambush, 9-10
actions from the ORP, 9-29
multiple teams, 9-30
observation plan, 9-29
security element, 9-30
short range, 9-30
surveillance teams, 9-30
area reconnaissance, 9-29
combat patrols, 9-7
planning, 9-8 (illus)
raid, 9-10 (illus)
departure from base, 9-5
initial planning and coordination, 9-3
organization, 9-2
password, 9-5
plan, 9-4 (illus)
preparations, 9-37
5 and 20 meter checks, 9-41
briefings and orders, 9-37
equipment, 9-38
exiting and entering a fixed base, 9-40
security checks, 9-40
purpose, 9-1
rally points, 9-6, 9-7 (illus)
signals, 9-5
types of patrols, 9-2
penetration, 7-8
plan and integrate indirect fires, 8-17
plan and integrate obstacles, 8-17
planning considerations, 8-1
defensive ops, 8-1
offensive ops, 7-1
sustainment ops, 8-1
platoon attacks, 7-14
deliberate attack, 7-14, 7-15 (illus)
advance to objective, 7-15
consolidate and reorganize, 7-17
isolate objective, 7-16
reconnaissance, 7-15
seize foothold and exploit objective, 7-17
site exploitation, 7-17
hasty attack, 7-17
conduct, 7-18
task organization, 7-18
platoon formations, 3-12, 3-14 (table)
column, 3-14, 3-15 (illus)
file, 3-18, 3-19 (illus)
line, 3-16 (illus)
vee, 3-17 (illus)
wedge, 3-17, 3-18 (illus)
platoon, infantry, 1-11
headquarters, 1-12
leader, 1-12
medic, 1-14, 6-2
radiotelephone operator, 1-13
sergeant, 1-13, 6-1
platoon movement techniques, 3-24
point fire, 2-20
post-patrol activities, 9-41
accounting for members, 9-41, 9-42 (illus)
patrol reports, 9-42 (illus)
reconnaissance patrols, 9-22
actions, 9-27 (illus)
control measures, 9-24
IR, 9-27, 9-28 (table)
observation plan, 9-28
organizing, 9-26 (illus)
security patrols, 9-22
route reconnaissance, 9-31
conduct, 9-31 (illus), 9-33 (illus)
moving element techniques, 9-33
stationary element techniques, 9-36
zone reconnaissance, 9-33
box method, 9-34, 9-35 (illus)
converging routes method, 9-35 (illus)
fan method, 9-34 (illus)
stationary element techniques, 9-36 (illus)
successive sector method, 9-36 (illus)
protection, 4-1, 7-11
air missile defense, 4-8
CBRN defense, 4-10
risk management and fratricide, 4-1
avoidance, 4-5
procedures, 4-1
responsibilities, 4-5
protection warfighting function, 1-5, 7-11
pursuit, 7-24

R

raid, 7-29, 7-30 (illus)
rally points, 9-6
range cards, 2-25, 2-26 (illus)
range correction, 2-38
range spotting, 2-38
reacting to enemy contact, H-23
actions at ambush, H-24 (illus)
actions at obstacle, H-27 (illus)
actions during halts, H-28
reaction force, H-21
reconnaissance, 7-9, 7-15
reconnaissance by fire, 2-21
reconnaissance patrols, 9-22
actions on objective, 9-27 (illus)
control measures, 9-24
security patrols, 9-22
reinforcing fire effects, 1-22, 1-23 (illus)
responsibility, 1-26
resupply operations, 6-6
aerial sustainment, 6-9
prestock resupply, 6-9
retirement, 8-50
risk management and fratricide, 4-1
avoidance, 4-5
procedures, 4-1
responsibilities, 4-5

route planning, 3-29
route reconnaissance, 9-31
conduct, 9-31 (illus), 9-33 (illus)
route selection and navigation, 3-28
execute route, 3-32
leg development, 3-31 (illus)
navigation aids, 3-28
planning, 3-29
types of navigation, 3-29
types of routes, 3-30
zone reconnaissance, 9-33
box method, 9-34, 9-35 (illus)
converging routes method, 9-35 (illus)
fan method, 9-34 (illus)
stationary element techniques, 9-36 (illus)
successive sector method, 9-36 (illus)
route, types of, 3-30

S

sector of fire, 2-15
security, 3-43, H-1
at halts, 3-44, 3-46 (table)
camouflage, noise, and light discipline, 3-44
convoy and route security, H-19
command and control, H-22
escort element, H-21
flank and rear guard screen, H-21
organization of forces, H-19
reacting to enemy contact, H-23 (illus)
reaction force, H-21
observation posts, H-6
occupying, H-8
types and positioning, H-6
traffic control checkpoints, H-10
physical layout, H-13
selecting and securing, H-7
task organization, H-15
types, H-11
security at halts, 3-44
security fundamentals, H-1
security in the defense, 8-21
seize foothold and exploit objective, 7-17
sequence of the defense, 8-2
approach of enemy attack, 8-4
consolidation and reorganization, 8-5
counterattack, 8-5
enemy assault, 8-4
occupation, 8-3
sequential fire, 2-21
shoulder-launched munitions (SLM) and close combat missile systems, B-1
close combat missile systems, B-6
antiarmor, B-16
javelin, B-6, B-7 (illus) (table)
urban field fortifications, B-14, B-15 (table)
defensive operations, B-23
employment considerations, B-14
offensive operations, B-23
safety, B-24 (illus)
javelin, B-30 (illus)
SLM B-24 (illus)
TOW, B-32 (illus)
urban operations and field fortifications, B-14, B-15 (illus)
site exploitation, 1-11, 7-17
situation, 1-44
human dimension, 1-44
soldier's load, 6-6
special illumination techniques, C-18
special purpose attacks, 7-26
ambush, 7-26
counterattack, 7-30
demonstration, 7-31
feint, 7-31
raid, 7-29, 7-30 (illus)
spoiling attack, 7-30
squad, 1-17
squad designated marksman, 1-18
squad formations, 3-9 (table)
column, 3-10 (illus)
file, 3-11 (illus)
line, 3-11 (illus)
movement techniques, 3-19, 3-20 (table)
weapons squad, 3-11
squad leader, 1-17, 6-1
stability operations, 1-41
stationary element techniques, 9-36 (illus)
stay-behind operations, 8-49
Stryker brigade combat infantry rifle platoon and squad, G-4
Stryker infantry carrier vehicle, D-8
successive bracketing, 2-39 (illus)
suicide bombers, I-10
complications, I-12
definition, I-10
delivery methods, I-10
indicators, I-11
special considerations, I-12
support by fire, 7-35
suppressive effects of HE mortar rounds, C-16
sustainment, 6-1 7-12
aerial sustainment, 6-9 (illus)
combat lifesaver, 6-2
combat load and basic load, 6-6
maintenance, 6-6
planning considerations, 6-2
emergency resupply, 6-8
platoon medic, 6-2
platoon sergeant, 6-1
resupply operations, 6-6
soldier's load, 6-6
squad leader, 6-1
supply class considerations, 6-4
support, types of, 6-3
sustainment planning, 6-3
transportation, 6-6
sustainment warfighting function, 1-5

T

tactical decision making, 1-3, 1-27, 1-28 (illus)
tactical enabling operations, 1-42
combined arms breach, 1-43

passage of lines, 1-43
reconnaissaince, 1-42
relief in place, 1-43
security, 1-42

tactical maneuver, 1-2

tactical movement, 3-1
actions at danger areas, 3-33
crossing danger areas, 3-34
crossing large open areas, 3-35, 3-36 (illus)
crossing linear danger areas (platoon), 3-35 (illus)
crossing small open areas, 3-36, 3-37 (illus)
enemy contact at danger areas, 3-37
control, 3-4
convoys, 3-42
fire team formations, 3-8 (table)
file, 3-9 (illus)
wedge, 3-8 (illus)
movement formations, 3-2, 3-8 (table)
movement techniques, 3-20 (table)
movement with combat vehicles, 3-37
actions at danger areas (mounted), 3-42
combat vehicle and infantry platoon formations, 3-40
combat vehicle and infantry squad formations, 3-37
convoys 3-42
mounted tactical movement. 3-41
other movement situations, 3-47
administrative, 3-47
by water, 3-48
limited visibility, 3-48
tactical road marches, 3-48
platoon formations, 3-12, 3-14 (table)
column, 3-14, 3-15 (illus)
file, 3-18, 3-19 (illus)
line, 3-16 (illus)
vee, 3-17 (illus)
wedge, 3-16, 3-17 (illus)
route selection and navigation, 3-28
security, 3-44
camouflage, noise, and light discipline, 3-45
enemy, 3-44
terrain, 3-45
squad formations, 3-9 (table)
column, 3-10 (illus)
file, 3-11 (illus)
line, 3-10, 3-11 (illus)
movement techniques, 3-19
platoon, 3-24
squad, 3-21
weapons squad, 3-11

tactical principles, 1-2 (table)

tactical questioning, 1-11

tactical tasks (offensive), 7-31
attack by fire, 7-36
clear, 7-32
seize, 7-31
support by fire, 7-35
suppress, 7-34

tactical uses of planned fires, C-3
echelonment, C-3

tank platoon, G-11 (illus)

target description, 2-23 (table)

targets effects planning, C-13
effects of cover on HE rounds, C-16
HE ammunition, C-13
illumination, smoke, and white phosphorous C-17 (illus)
special illumination techniques, C-18
suppressive effects of HE mortar rounds, C-16
time of suppression, 2-21

terrain, 3-3, 3-45, 7-13

terrain association, 3-29

thermal sites, use of, C-18

traffic control checkpoints, H-10
checkpoint types, H-11
deliberate, H-11
hasty, H-12
physical layout, H-13 (illus)
canalization zone, H-14
safe zone, H-15
search zone, H-15
turning or deceleration zone, H-14
task organization, H-15, H-16 (table)
C2 element, H-17
civil authority assistance, H-17
escalation of force, H-17
security element, H-17

trajectory, 2-7, 2-32

trenches, 8-43, 8-44 (illus)

troop-leading procedures, 1-29, 5-6
avenues of approach, 5-11 (illus)
complete plan, 5-18
conduct reconnaissance 5-18
course of action (COA) development, 5-15, 5-16 (table)
cover and concealment, 5-13 (illus)
initiate movement, 5-17
issue OPORD, 5-18, 5-21 (illus)
issue WARNO, 5-7
key terrain, 5-12 (illus)
make tentative plan, 5-8
observation and fields of fire, 5-12, 5-13 (illus)
obstacles, 5-10, 5-11 (illus)
receive mission, 5-6
supervise and refine, 5-19 (table)
squad orders, 5-20
terrain and weather analysis, 5-9, 1-10 (OCKA table)

turning movement ,7-4, 7-5 (illus)

U

unexploded ordnance (UXO), I-12
immediate action, I-13
incident report, I-14, I-15 (illus)
recognition, I-12

urban areas, 7-36
assault a building, 7-38
building entry, 7-39
clear a room, 7-40, 7-42 (illus)
clearing a "T" intersection, 7-45, 7-46 (illus)

clearing hallway junctions, 7-43, 7-44 (illus)
critical tasks, 7-36
moving in a building , 7-42, 7-43 (illus)
follow-through, 7-47
isolate building, 7-37, 7-38 (illus)

V

v-shaped ambush, 9-14 (illus)

vehicle employment, D-1
assault high mobility multi-purpose wheeled vehicle, D-10
Bradley fighting vehicle, D-7, D-8 (illus)
combat vehicles supporting infantry, D-2
infantrymen supporting combat vehicles, D-2
maintenance and recovery, D-28
operations, D-16
plan, D-17
prepare, D-21
risk management, D-18
protection, D-4
size and weight considerations, D-11 (table)
Stryker infantry carrier vehicle, D-8
tactical capabilities, D-11, D-12 (table)
tactical movement rates, D-12, D-13 (table)
tanks (M1), D-5
technical capabilities, D-3
fire power, D-3, D-4 (table)
vehicle characteristics, D-3 (table)

volley fire, 2-20

W

warfighting functions, 1-3
command and control, 1-5
fire support, 1-4
intelligence, 1-4
movement and maneuver, 1-4
protection, 1-5
sustainment, 1-5

warrior ethos and Army values, 1-9

weapons and munitions
characteristics, 2-5 (illus)
ammunition, 2-6
complementary and reinforcing effects, 2-8
enemy target types, 2-7
fire team weapons, 2-7
high explosive (HE) munitions, 2-6
lay, 2-5, 2-6 (illus)
trajectory, 2-7

weapons placement, (defense), 8-22

weapons squad, 1-20
close combat missile team, 1-21
medium machine gun team, 1-20

weapons squad leader, 1-21

weapons squad movement formations, 3-11

Z

zone reconnaissance, 9-33
box method, 9-34, 9-35 (illus)
converging routes method, 9-35 (illus)
fan method, 9-34 (illus)
stationary element techniques, 9-36 (illus)
successive sector method, 9-36 (illus)

QUICK REFERENCE CHECKLIST

Decide to Survive!

S - **S**ize up the situation, surroundings, physical condition, equipment.
U - **U**se all your senses
R - **R**emember where you are.
V - **V**anquish fear and panic.
I - **I**mprovise and improve.
V - **V**alue living.
A - **A**ct like the natives.
L - **L**ive by your wits.

1. Immediate Actions

a. Assess immediate situation. ***THINK BEFORE YOU ACT!***

b. Take action to protect yourself from nuclear, biological, or chemical hazards (Chapter IX).

c. Seek a concealed site.

d. Assess medical condition; treat as necessary (Chapter V).

e. Sanitize uniform of potentially compromising information.

f. Sanitize area; hide equipment you are leaving.

g. Apply personal camouflage.

h. Move away from concealed site, zigzag pattern recommended.

i. Use terrain to advantage, communication, and concealment.

j. Find a hole-up site.

2. Hole-Up-Site (Chapter I)

a. Reassess situation; treat injuries, then inventory equipment.

b. Review plan of action; establish priorities (Chapter VI).

c. Determine current location.

d. Improve camouflage.

e. Focus thoughts on task(s) at hand.

f. Execute plan of action. Stay flexible!

Recommend inclusion of this manual in the aviator's survival vest.

3. Concealment (Chapter I)

a. Select a place of concealment providing—

(1) Adequate concealment, ground and air.

(2) Safe distance from enemy positions and lines of communications (LOC).

(3) Listening and observation points.

(4) Multiple avenues of escape.

(5) Protection from the environment.

(6) Possible communications/signaling opportunities.

b. Stay alert, maintain security.

c. Drink water.

4. Movement (Chapters I and II)

a. Travel slowly and deliberately.

b. ***DO NOT*** leave evidence of travel; use noise and light discipline.

c. Stay away from LOC.

d. Stop, look, listen, and smell; take appropriate action(s).

e. Move from one concealed area to another.

f. Use evasion movement techniques (Chapter I).

5. Communications and Signaling (Chapter III)

a. Communicate as directed in applicable plans/orders, particularly when considering transmitting ***in the blind***.

b. Be prepared to use communications and signaling devices on short notice.

c. Use of communications and signaling devices may compromise position.

6. Recovery (Chapter IV)

a. Select site(s) IAW criteria in theater recovery plans.

b. Ensure site is free of hazards; secure personal gear.

c. Select best area for communications and signaling devices.

d. Observe site for proximity to enemy activity and LOC.

e. Follow recovery force instructions.

FM 21-76-1
MCRP 3-02H
NWP 3-50.3
AFTTP(I) 3-2.26

FM 21-76-1	U.S. Army Training and Doctrine Command Fort Monroe, Virginia
MCRP 3-02H	Marine Corps Combat Development Command Quantico, Virginia
NWP 3-50.3	Navy Warfare Development Command Newport, Rhode Island
AFTTP(I) 3-2.26	Headquarters Air Force Doctrine Center Maxwell Air Force Base, Alabama

29 JUNE 1999

Survival, Evasion, and Recovery

Multiservice Procedures for Survival, Evasion, and Recovery

Note: This *UNCLASSIFIED* publication is designed to provide Service members quick-reference survival, evasion, and recovery information. See Appendix B for the scope, purpose, application, implementation plan, and user information.

TABLE OF CONTENTS

Page

CHAPTER I EVASION

1. Planning ... I-1
2. Camouflage ... I-1
3. Shelters ... I-3
4. Movement ... I-3

CHAPTER II NAVIGATION

1. Stay or Move Considerations ... II-1
2. Navigation and Position Determination ... II-1
3. Travel Considerations ... II-10
4. River Travel ... II-10
5. Ice and Snow Travel ... II-11
6. Mountain Hazards ... II-12
7. Summer Hazards ... II-12
8. Dry Climates ... II-12

9. Tropical Climates II-13
10. Open Seas II-13
CHAPTER III RADIO COMMUNICATIONS AND SIGNALING
1. Radio Communications (Voice and Data) III-1
2. Signaling III-2
CHAPTER IV RECOVERY
1. Responsibilities IV-1
2. Site Selection IV-1
3. Site Preparation IV-1
4. Recovery Procedures IV-1
CHAPTER V MEDICAL
1. Immediate First Aid Actions V-1
2. Common Injuries and Illnesses V-9
3. Plant Medicine V-15
4. Health and Hygiene V-18
6. Rules for Avoiding Illness V-18
CHAPTER VI PERSONAL PROTECTION
1. Priorities VI-1
2. Care and Use of Clothing VI-1
3. Other Protective Equipment VI-2
4. Shelters VI-3
5. Fires VI-8
CHAPTER VII WATER
1. Water Requirements VII-1
2. Water Procurement VII-1
3. Water Preparation and Storage VII-7
CHAPTER VIII FOOD
1. Food Procurement VIII-1
2. Food Preparation VIII-9
3. Food Preservation VIII-11
CHAPTER IX INDUCED CONDITIONS
1. Nuclear Conditions IX-1
2. Biological Conditions IX-6
3. Chemical Conditions IX-7
APPENDIX A THE WILL TO SURVIVE A-1
APPENDIX B PUBLICATION INFORMATION B-2

THE CODE OF CONDUCT

ARTICLE I

I am an American, fighting in the forces which guard my country and our way of life. I am prepared to give my life in their defense.

ARTICLE II

I will never surrender of my own free will. If in command, I will never surrender the members of my command while they still have the means to resist.

ARTICLE III

If I am captured, I will continue to resist by all means available. I will make every effort to escape and aid others to escape. I will accept neither parole nor special favors from the enemy.

ARTICLE IV

If I become a prisoner of war, I will keep faith with my fellow prisoners. I will give no information or take part in any action which might be harmful to my comrades. If I am senior, I will take command. If not, I will obey the lawful orders of those appointed over me and will back them up in every way.

ARTICLE V

When questioned, should I become a prisoner of war, I am required to give name, rank, service number and date of birth. I will evade answering further questions to the utmost of my ability. I will make no oral or written statements disloyal to my country and its allies or harmful to their cause.

ARTICLE VI

I will never forget that I am an American, fighting for freedom, responsible for my actions, and dedicated to the principles which made my country free. I will trust in my God and in the United States of America.

Chapter I
EVASION

1. Planning

a. Review the quick reference checklist on the inside cover.

b. Guidelines for successful evasion include—

(1) Keeping a positive attitude.
(2) Using established procedures.
(3) Following your evasion plan of action.
(4) Being patient.
(5) Drinking water (***DO NOT*** eat food without water).
(6) Conserving strength for critical periods.
(7) Resting and sleeping as much as possible.
(8) Staying out of sight.

c. The following odors stand out and may give an evader away:

(1) Scented soaps and shampoos.
(2) Shaving cream, after-shave lotion, or other cosmetics.
(3) Insect repellent (camouflage stick is least scented).
(4) Gum and candy (smell is strong or sweet).
(5) Tobacco (odor is unmistakable).

d. Where to go (initiate evasion plan of action):

(1) Near a suitable area for recovery.
(2) Selected area for evasion.
(3) Neutral or friendly country or area.
(4) Designated area for recovery.

2. Camouflage

a. Basic principles:

(1) Disturb the area as little as possible.
(2) Avoid activity that reveals movement to the enemy.
(3) Apply personal camouflage.

b. Camouflage patterns **(Figure I-1)**:

(1) Blotch pattern.
(a) Temperate deciduous (leaf shedding) areas.
(b) Desert areas (barren).
(c) Snow (barren).

(2) Slash pattern.
(a) Coniferous areas (broad slashes).

(b) Jungle areas (broad slashes).
(c) Grass (narrow slashes).
(3) Combination. May use blotched and slash together.

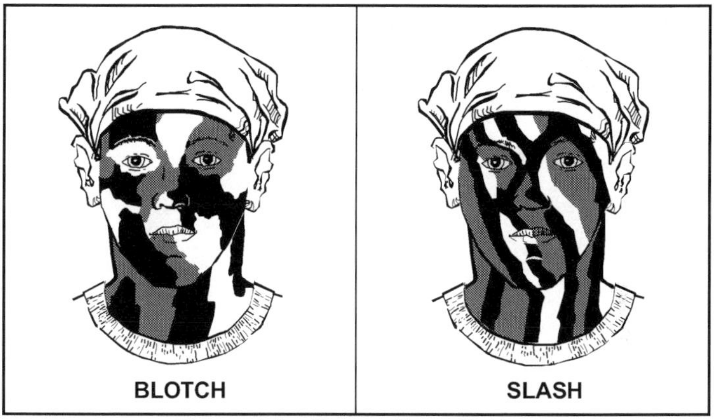

Figure I-1. Camouflage Patterns

c. Personal camouflage application follows:

(1) Face. Use dark colors on high spots and light colors on any remaining exposed areas. Use a hat, netting, or mask if available.

(2) Ears. The insides and the backs should have **2** colors to break up outlines.

(3) Head, neck, hands, and the under chin. Use scarf, collar, vegetation, netting, or coloration methods.

(4) Light colored hair. Give special attention to conceal with a scarf or mosquito head net.

d. Position and movement camouflage follows:
(1) Avoid unnecessary movement.
(2) Take advantage of natural concealment:
(a) Cut foliage fades and wilts, change regularly.
(b) Change camouflage depending on the surroundings.
(c) ***DO NOT*** select vegetation from same source.
(d) Use stains from grasses, berries, dirt, and charcoal.
(3) ***DO NOT*** over camouflage.
(4) Remember when using shadows, they shift with the sun.

(5) Never expose shiny objects (like a watch, glasses, or pens).

(6) Ensure watch alarms and hourly chimes are turned off.

(7) Remove unit patches, name tags, rank insignia, etc.

(8) Break up the outline of the body, ***"V"*** of crotch/armpits.

(9) Conduct observation from a prone and concealed position.

3. Shelters

a. Use camouflage and concealment.

b. Locate carefully—easy to remember acronym: ***BLISS***.

B - **B**lend **L** - **L**ow silhouette **I** - **I**rregular shape **S** - **S**mall **S** - **S**ecluded location

(1) Choose an area—

(a) Least likely to be searched (drainages, rough terrain, etc.) and blends with the environment.

(b) With escape routes (***DO NOT*** corner yourself).

(c) With observable approaches.

(2) Locate entrances and exits in brush and along ridges, ditches, and rocks to keep from forming paths to site.

(3) Be wary of flash floods in ravines and canyons.

(4) Conceal with minimal to no preparation.

(5) Take the direction finding threat into account before transmitting from shelter.

(6) Ensure overhead concealment.

4. Movement

a. A moving object is easy to spot. If travel is necessary—

(1) Mask with natural cover **(Figure I-2)**.

(2) Use the military crest.

(3) Restrict to periods of low light, bad weather, wind, or reduced enemy activity.

Figure I-2. Ground Movement

(4) Avoid silhouetting **(Figure I-3)**.

(5) At irregular intervals—

(a) ***STOP*** at a point of concealment.

(b) ***LOOK*** for signs of human or animal activity (smoke, tracks, roads, troops, vehicles, aircraft, wire, buildings, etc.). Watch for trip wires or booby traps and avoid leaving evidence of travel. Peripheral vision is more effective for recognizing movement at night and twilight.

(c) ***LISTEN*** for vehicles, troops, aircraft, weapons, animals, etc.

(d) ***SMELL*** for vehicles, troops, animals, fires, etc.

Figure I-3. Avoid Silhouetting

(6) Employ noise discipline; check clothing and equipment for items that could make noise during movement and secure them.

b. Break up the human shape or recognizable lines.

c. Route selection requires detailed planning and special techniques (irregular route/zigzag) to camouflage evidence of travel.

d. Some techniques for concealing evidence of travel follows:

(1) Avoid disturbing the vegetation above knee level.

(2) ***DO NOT*** break branches, leaves, or grass.

(3) Use a walking stick to part vegetation and push it back to its original position.

(4) ***DO NOT*** grab small trees or brush. (This may scuff the bark or create movement that is easily spotted. In snow country, this creates a path of snowless vegetation revealing your route.)

(5) Pick firm footing (carefully place the foot lightly but squarely on the surface to avoid slipping). ***TRY NOT TO***—

(a) Overturn ground cover, rocks, and sticks.

(b) Scuff bark on logs and sticks.

(c) Make noise by breaking sticks. (Cloth wrapped around feet helps muffle this.)

(d) Mangle grass and bushes that normally spring back.

(6) Mask unavoidable tracks in soft footing by—

(a) Placing tracks in the shadows of vegetation, downed logs, and snowdrifts.

(b) Moving before and during precipitation allows tracks to fill in.

(c) Traveling during windy periods.

(d) Taking advantage of solid surfaces (logs, rocks, etc.) leaving less evidence of travel.

(e) Patting out tracks lightly to speed their breakdown or make them look old.

(7) Secure trash or loose equipment—hide or bury discarded items. (Trash or lost equipment identifies who lost it.)

(8) Concentrate on defeating the handler if pursued by dogs.

e. Penetrate obstacles as follows:

(1) Enter deep ditches feet first to avoid injury.

(2) Go around chain-link and wire fences. Go under fence if unavoidable, crossing at damaged areas. ***DO NOT*** touch fence; look for electrical insulators or security devices.

(3) Penetrate rail fences, passing under or between lower rails. If impractical, go over the top, presenting as low a silhouette as possible **(Figure I-4)**.

(4) Cross roads after observation from concealment to determine enemy activity. Cross at points offering concealment such as bushes, shadows, bend in road, etc. Cross in a manner leaving your footprints parallel (cross step sideways) to the road. **(Figure I-5)**

(5) Use same method of observation for railroad tracks that was used for roads. Next, align body parallel to tracks with face down, cross tracks using a semi-pushup motion. Repeat for the second track. **(Figure I-6)**.

Figure I-4. Rail Fences

Figure I-5. Road Crossing

Figure I-6. Railroad Tracks

WARNING: If 3 rails exist, 1 may be electrified.

Chapter II
NAVIGATION

Assess the threat and apply appropriate evasion principles.

1. Stay or Move Considerations

a. Stay with the vehicle/aircraft in a non-combat environment.

b. Leave only when—

(1) Dictated by the threat.

(2) Are certain of your location, have a known destination, and have the ability to get there.

(3) Can reach water, food, shelter, and/or help.

(4) Convinced rescue is not coming.

c. Consider the following if you decide to travel:

(1) Follow the briefed evasion plan.

(2) Determine which direction to travel and why.

(3) Decide what equipment to take, cache, or destroy.

d. Leave information at your starting point (in a non-combat environment) that includes—

(1) Destination.

(2) Route of travel.

(3) Personal condition.

(4) Supplies available.

e. Consider the following for maps (in a combat environment):

(1) ***DO NOT*** write on the map.

(2) ***DO NOT*** soil the map by touching the destination.

(3) ***DO NOT*** fold in a manner providing travel information.

Note: **These actions may compromise information if captured.**

2. Navigation and Position Determination

a. Determine your general location by—

(1) Developing a working knowledge of the operational area.

(a) Geographic checkpoints.

(b) Man-made checkpoints.

(c) Previous knowledge of operational area.

(2) Using the ***Rate x Time = Distance*** formula.

(3) Using information provided in the map legend.

(4) Using prominent landmarks.

(5) Visualizing map to determine position.

b. Determine cardinal directions (north, south, east, and west) by—

(1) Using compass.

CAUTION: The following methods are ***NOT*** highly accurate and give only general cardinal direction.

(2) Using stick and shadow method to determine a true north-south line **(Figure II-1)**.

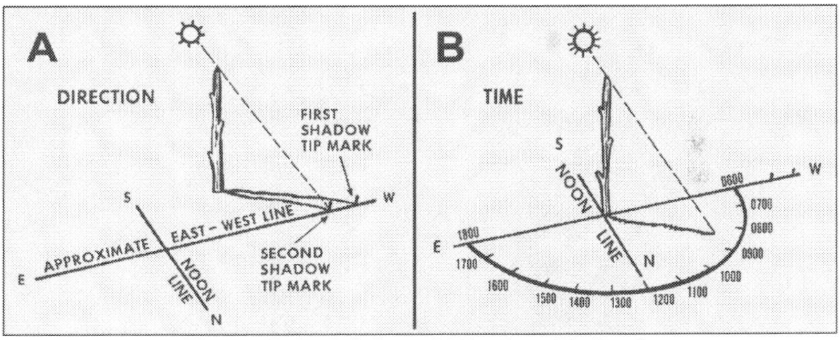

Figure II-1. Stick and Shadow Method

(3) Remembering the sunrise/moonrise is in the east and sunset/moonset is in the west.

(4) Using a wristwatch to determine general cardinal direction **(Figure II-2)**.

(a) Digital watches. Visualize a clock face on the watch.

(b) Northern Hemisphere. Point hour hand at the sun. South is halfway between the hour hand and 12 o'clock position.

(c) Southern Hemisphere. Point the 12 o'clock position on your watch at the sun. North is halfway between the 12 o'clock position and the hour hand.

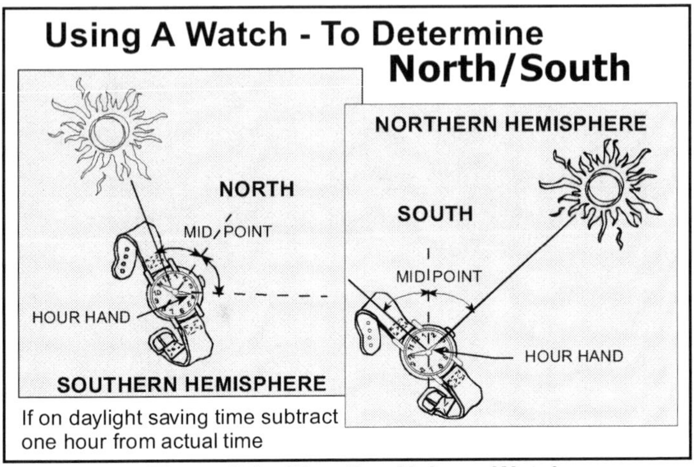

Figure II-2. Direction Using a Watch

(5) Using a pocket navigator **(Figure II-3)**—

(a) Gather the following necessary materials:

- Flat writing material (such as an MRE box).
- 1-2 inch shadow tip device (a twig, nail, or match).
- Pen or pencil.

(b) Start construction at sunup; end construction at sundown. Do the following:

- Attach shadow tip device in center of paper.
- Secure navigator on flat surface (***DO NOT*** move during set up period).
- Mark tip of shadow every 30 minutes annotating the time.
- Connect marks to form an arc.
- Indicate north with a drawn arrow.

Note: The shortest line between base of shadow tip device and curved line is a north-south line.

(c) Do the following during travel:

- Hold navigator so the shadow aligns with mark of present time (drawn arrow now points to true north).

(d) Remember the navigator is current for approximately 1 week.

CAUTION: The Pocket Navigator is ***NOT*** recommended if evading.

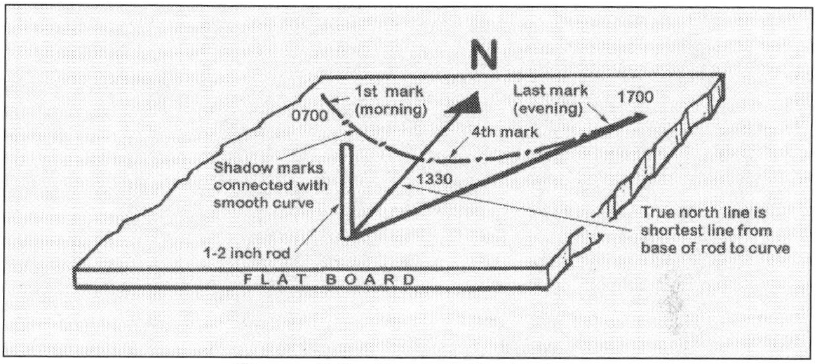

Figure II-3. Pocket Navigator

(6) Using the stars **(Figure II-4)** the—

(a) North Star is used to locate true north-south line.

(b) Southern Cross is used to locate true south-north line.

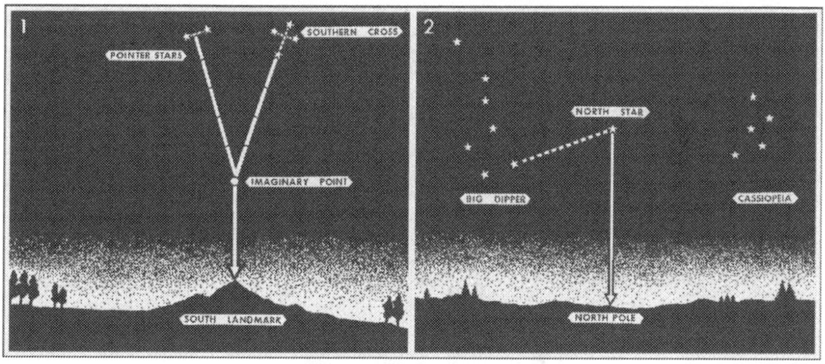

Figure II-4. Stars

c. Orient the map by—

(1) Using a true north-south line **(Figure II-5)**—

(a) Unfold map and place on a firm, flat, level nonmetallic surface.

(b) Align the compass on a true north-south line.

(c) Rotate map and compass until stationary index line aligns with the magnetic variation indicated in marginal information.

- Easterly (subtract variation from 360 degrees).
- Westerly (add variation to 360 degrees).

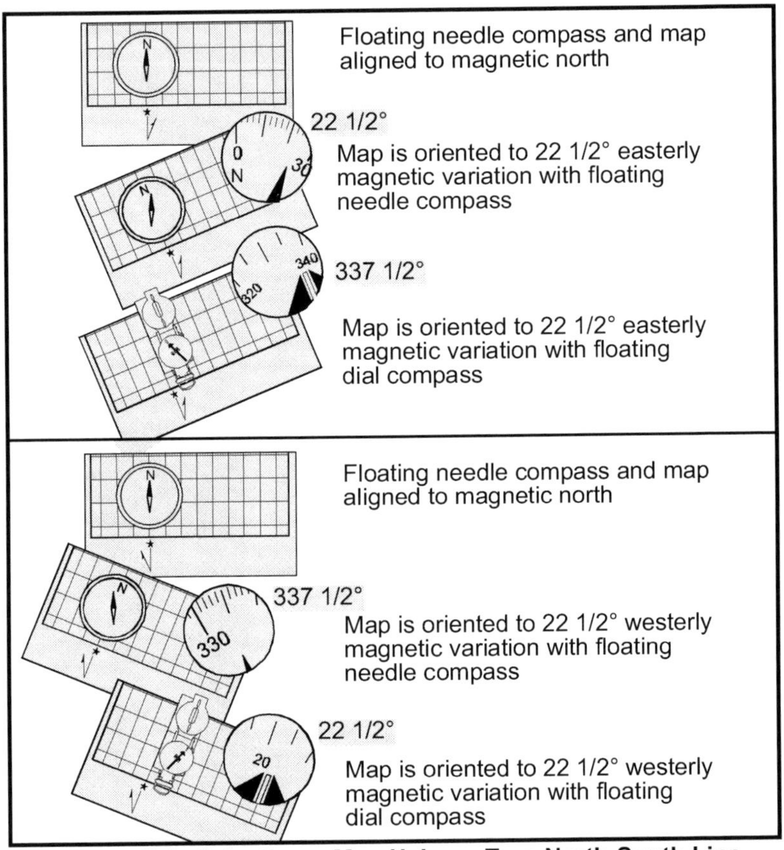

Figure II-5. Orienting a Map Using a True North-South Line

(2) Using a compass rose **(Figure II-6)**—

(a) Place edge of the lensatic compass on magnetic north line of the compass rose closest to your location.

(b) Rotate map and compass until compass reads 360 degrees.

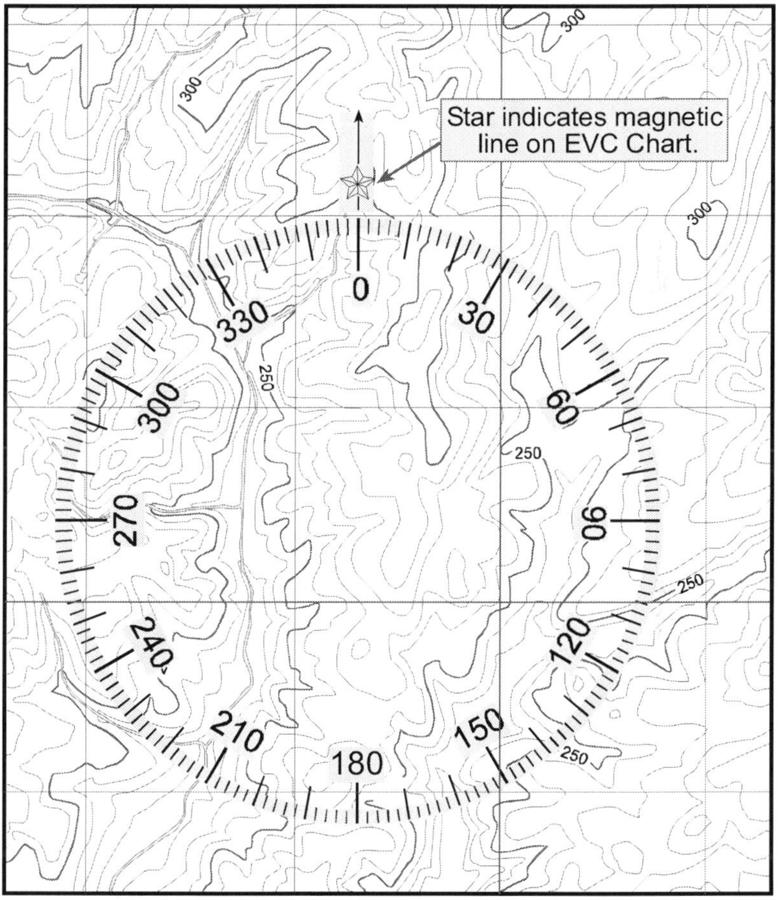

Figure II-6. Map Orientation with Compass Rose

(3) If there is **NO** compass, orient map using cardinal direction obtained by the stick and shadow method or the celestial aids (stars) method.

d. Determine specific location.

(1) Global Positioning System (GPS).

(a) ***DO NOT*** use GPS for primary navigation.

(b) Use GPS to confirm your position ***ONLY***.

(c) Select area providing maximum satellite reception.

(d) Conserve GPS battery life.

(2) Triangulation (resection) with a compass **(Figure II-7)**.

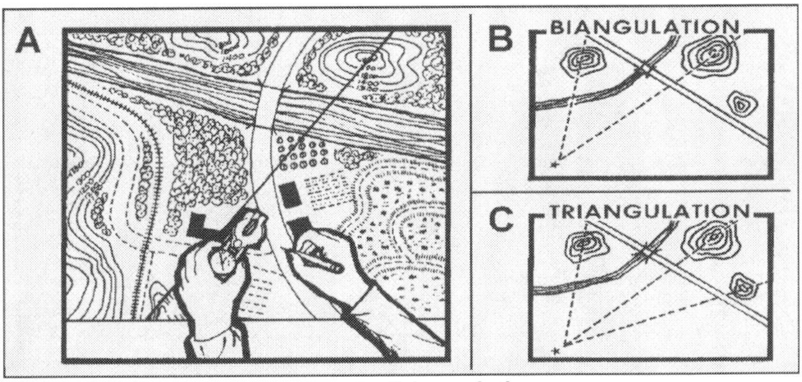

Figure II-7. Triangulation

(a) Try to use **3** or more azimuths.

(b) Positively identify a major land feature and determine a line of position (LOP).

(c) Check map orientation each time compass is used.

(d) Plot the LOP using a thin stick or blade of grass (combat) or pencil line (non-combat).

(e) Repeat steps **(b)** through **(d)** for other LOPs.

e. Use the compass for night navigation by—

(1) Setting up compass for night navigation **(Figure II-8)**.

(2) Aligning north-seeking arrow with luminous line and follow front of compass.

(3) Using point-to-point navigation.

f. Route selection techniques follow:

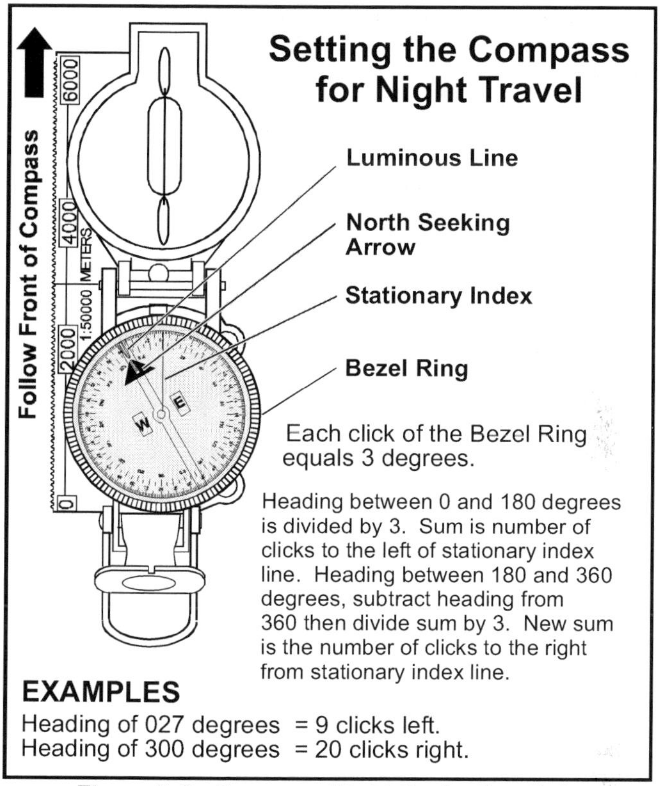

Figure II-8. Compass Night Navigation Setup

(1) Circumnavigation.

(a) Find a prominent landmark on the opposite side of the obstacle.

(b) Contour around obstacle to landmark.

(c) Resume your route of travel.

(2) Dogleg and 90 degree offset **(Figure II-9)**.

(3) Straight-line heading as follows:

(a) Maintain heading until reaching destination.

(b) Measure distance by counting the number of paces in a given course and convert to map units.

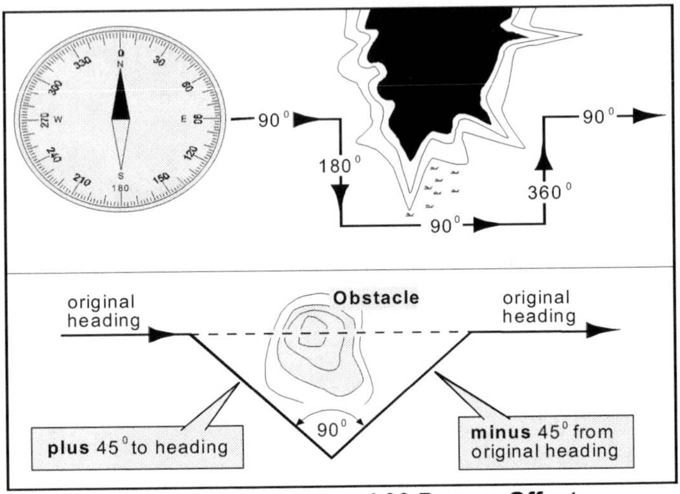

Figure II-9. Dogleg and 90 Degree Offset

•One pace is the distance covered each time the same foot touches the ground.

•Distances measured by paces are approximate (example in open terrain, 900 paces per kilometer [average], or example in rough terrain, 1200 paces per kilometer [average]).

(c) Use pace count in conjunction with terrain evaluation and heading to determine location. An individual's pace varies because of factors such as steep terrain, day/night travel, or injured/uninjured condition. Adjust estimation of distance traveled against these factors to get relative accuracy when using a pace count.

(4) Deliberate offset is—

(a) Used when finding a point on a linear feature (that is, road or river).

(b) Intentionally navigated to left or right of target so you know which way to turn at the linear feature.

(5) Point-to-point is same as straight line.

(a) Pick out landmarks on the heading and walk the trail of least resistance to a point.

(b) On reaching a point, establish another landmark and continue.

3. Travel Considerations

a. Pick the easiest and safest route (non-combat).

b. Maintain a realistic pace; take rest stops when needed.

c. Avoid overdressing and overheating.

d. Consider food and water requirements.

e. Take special care of feet (change socks regularly).

f. Pack equipment to prevent loss, damage, pack imbalance, and personal safety.

g. Go **around** obstacles, not over or through them.

h. Travel on trails whenever possible (non-combat).

i. Travel in forested areas if possible.

j. Avoid creek bottoms and ravines with ***NO*** escape in the event of heavy rains.

k. Consider the following for swamps, lakes, and unfordable rivers:

(1) Circumnavigate swamps, lakes, and bogs if needed.

(2) Travel downstream to find people and slower water.

(3) Travel upstream to find narrower and shallow water.

4. River Travel

River travel may be faster and save energy when hypothermia is not a factor. It may be a primary mode of travel and LOC in a tropical environment **(use with caution if evading)**.

a. Use flotation device (raft, log, bamboo, etc.).

b. Use a pole to move the raft in shallow water.

c. Use an oar in deep water.

d. Stay near inside edge of river bends (current speed is less).

e. Keep near shore.

f. Watch for the following ***DANGERS:***

(1) Snags.

(2) Sweepers (overhanging limbs and trees).

(3) Rapids (***DO NOT*** attempt to shoot the rapids).

(4) Waterfalls.

(5) Hazardous animals.

g. Consider using a flotation device when crossing rivers or large/deep streams.

5. Ice and Snow Travel

Travel should be limited to areas free of hazards.

a. ***DO NOT*** travel in—

(1) Blizzards.

(2) Bitterly cold winds.

(3) Poor visibility.

b. Obstacles to winter travel follow:

(1) Reduced daylight hours (***BE AWARE***).

(2) Deep soft snow (if movement is necessary, make snowshoes **[Figure II-10])**. Travel is easier in early morning or late afternoon near dusk when snow is frozen or crusted.

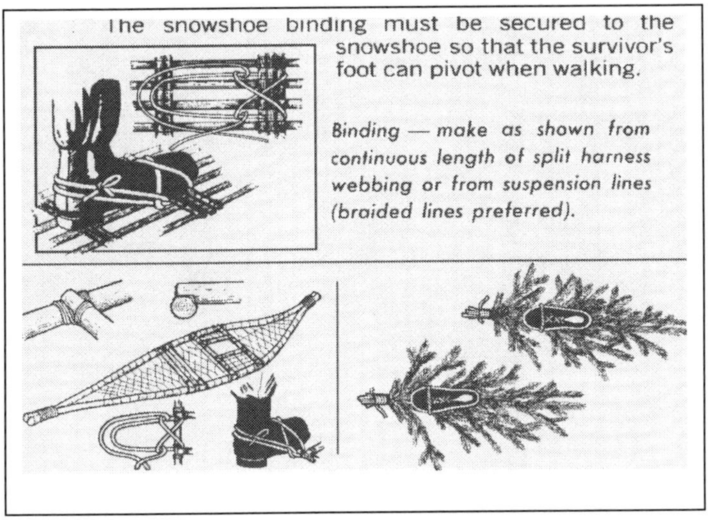

Figure II-10. Improvised Snowshoes

(3) Avalanche prone areas to avoid:

(a) Slopes 30-45 degrees or greater.

(b) Trees without uphill branches (identifies prior avalanches).

(c) Heavy snow loading on ridge tops.

(4) If caught in an avalanche, do the following:

(a) Backstroke to decrease burial depth.

(b) Move hand around face to create air pocket as moving snow slows.

(5) Frozen water crossings.

(a) Weak ice should be expected where—

- Rivers are straight.
- Objects protrude through ice.
- Snow banks extend over the ice.
- Rivers or streams come together.
- Water vapor rising indicates open or warm areas.

(b) Air pockets form when a frozen river loses volume.

(c) When crossing frozen water, distribute your weight by laying flat, belly crawling, or using snowshoes.

c. Glacier travel is hazardous and should be avoided.

6. Mountain Hazards

a. Lightning. Avoid ridge tops during thunderstorms.

b. Avalanche. Avoid areas prone to avalanches.

c. Flash floods. Avoid low areas.

7. Summer Hazards (see page II-10; paragraph 3, *Travel Considerations*, items h through k.)

(1) Dense brush.

(a) Travel on trails when possible (non-combat).

(b) Travel in forested areas if possible.

(c) Avoid creek bottoms and ravines with no escape in the event of heavy rains.

(2) Swamps, lakes, and unfordable rivers.

(a) Circumnavigate swamps, lakes, and bogs if needed.

(b) Travel downstream to find people and slower water.

(c) Travel upstream to find narrower and shallow water.

8. Dry Climates

a. ***DO NOT*** travel unless certain of reaching the destination using the water supply available.

b. Travel at dawn or dusk on hot days.

c. Follow the easiest trail possible (non-combat), avoiding—

(1) Deep sandy dune areas.

(2) Rough terrain.

d. In sand dune areas—

(1) Follow hard valley floor between dunes.

(2) Travel on the windward side of dune ridges.

e. If a sandstorm occurs—

(1) Mark your direction of travel.

(2) Sit or lie down in direction of travel.

(3) Try to get to the downwind side of natural shelter.

(4) Cover the mouth and nose with a piece of cloth.

(5) Protect the eyes.

(6) Remain stationary until the storm is over.

9. Tropical Climates

a. Travel only when it is light.

b. Avoid obstacles like thickets and swamps.

c. Part the vegetation to pass through. Avoid grabbing vegetation; it may have spines or thorns (**use gloves** if possible).

d. ***DO NOT*** climb over logs if you can go around them.

e. Find trails—

(1) Where **2** streams meet.

(2) Where a low pass goes over a range of hills.

f. While traveling trails—

(1) Watch for disturbed areas on game trails; they may indicate a pitfall or trap.

(2) Use a walking stick to probe for pitfalls or traps.

(3) ***DO NOT*** sleep on the trail.

(4) Exercise caution, the enemy uses the trails also.

10. Open Seas

a. Using currents—

(1) Deploy sea anchor **(Figure II-11)**. Sea anchor may be adjusted to make use of existing currents.

(2) Sit low in the raft.

(3) Deflate the raft slightly so it rides lower in the water.

b. Using winds—

(1) Pull in sea anchor.

(2) Inflate raft so it rides higher.

(3) Sit up in raft so body catches the wind.

(4) Construct a shade cover/sail **(Figure II-12)**. (Sail aids in making landfall.)

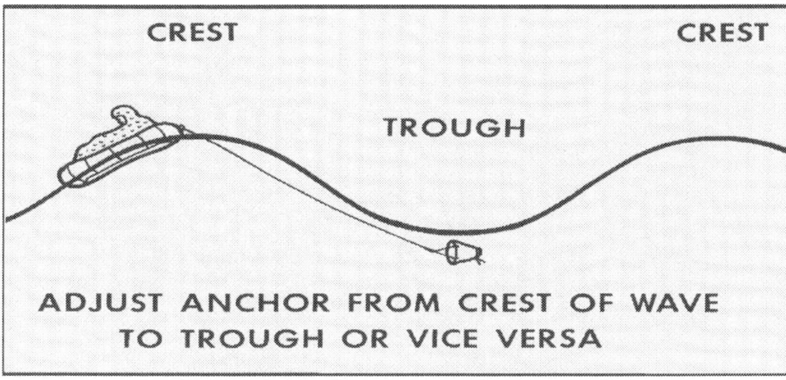

Figure II-11. Sea Anchor Deployment

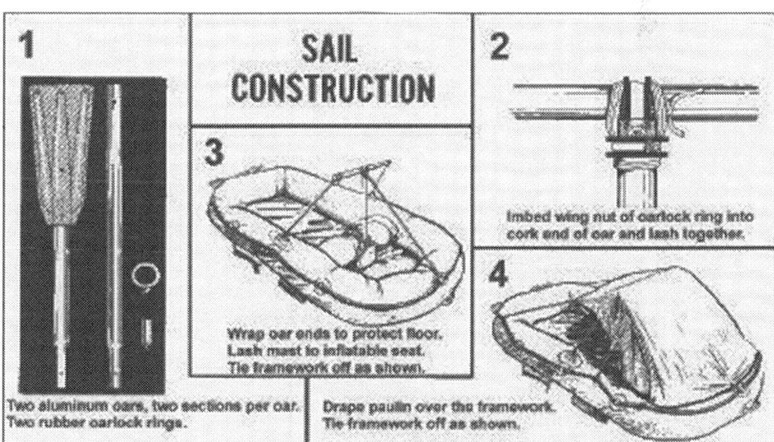

Figure II-12. Shade/Sail Construction

c. Making landfall. Indications of land are—

(1) Fixed cumulus clouds in a clear sky or in a cloudy sky where all other clouds are moving.

(2) Greenish tint in the sky **(in the tropics)**.

(3) Lighter colored reflection on clouds (open water causes dark gray reflections) (**in the arctic**).

(4) Lighter colored water (indicates shallow water).

(5) The odors and sounds.

(a) Odors from swamps and smoke.

(b) Roar of surf/bird cries coming from one direction.

(6) Directional flights of birds at dawn and at dusk.

d. Swimming ashore—

(1) Consider physical condition.

(2) Use a flotation aid.

(3) Secure all gear to body before reaching landfall.

(4) Remain in raft as long as possible.

(5) Use the sidestroke or breaststroke to conserve strength if thrown from raft.

(6) Wear footgear and at least **1** layer of clothing.

(7) Try to make landfall during the lull between the sets of waves (waves are generally in **sets** of **7**, from **smallest** to **largest**).

(8) In moderate surf.

(a) Swim forward on the back of a wave.

(b) Make a shallow dive just before the wave breaks to end the ride.

(9) In high surf.

(a) Swim shoreward in the trough between waves.

(b) When the seaward wave approaches, face it and submerge.

(c) After it passes, work shoreward in the next trough.

(10) If caught in the undertow of a large wave—

(a) Remain calm and swim to the surface.

(b) Lie as close to the surface as possible.

(c) Parallel shoreline and attempt landfall at a point further down shore.

(11) Select a landing point.

(a) Avoid places where waves explode upon rocks.

(b) Find a place where waves smoothly rush onto the rocks.

(12) After selecting a landing site—

(a) Face shoreward.

(b) Assume a sitting position with feet 2 or 3 feet lower than head to absorb the shock of hitting submerged objects.

e. Rafting ashore—

(1) Select landing point carefully.

(2) Use caution landing when the sun is low and straight in front of you causing poor visibility.

(3) Land on the lee (downwind) side of islands or point of land if possible.

(4) Head for gaps in the surf line.

(5) Penetrate surf by—

(a) Taking down most shade/sails.

(b) Using paddles to maintain control.

(c) Deploying a sea anchor for stability.

CAUTION: ***DO NOT*** deploy a sea anchor if traveling through coral.

f. Making sea ice landings on large stable ice flows. Icebergs, small flows, and disintegrating flows are dangerous **(ice can cut a raft)**.

(1) Use paddles to avoid sharp edges.

(2) Store raft away from the ice edge.

(3) Keep raft inflated and ready for use.

(4) Weight down/secure raft so it does not blow away

Chapter III

RADIO COMMUNICATIONS AND SIGNALING

Inventory and review the operating instructions of all communications and signaling equipment.

1. Radio Communications (Voice and Data)

a. Non-combat.

(1) Ensure locator beacon is operational.

(2) Follow standing plans for on/off operations to conserve battery use.

b. Combat.

(1) Turn off locator beacon.

(2) Keep it with you to supplement radio communications.

(3) Follow plans/orders for on/off operations.

c. Make initial contact as soon as possible or as directed in applicable plans/orders.

d. If no immediate contact, then as directed in applicable plans/orders.

e. Locate spare radio and batteries (keep warm and dry).

f. Transmissions.

(1) Use concealment sites (combat) that optimize line of site (LOS).

(2) Face recovery asset.

(3) Keep antenna perpendicular to intended receiver **(Figure III-1)**.

(4) ***DO NOT*** ground antenna (that is finger on antenna or attaching bolt, space blanket, vegetation, etc.).

(5) Keep transmissions short (3-5 seconds maximum). Use data burst if available.

(6) Move after each transmission (***ONLY*** in combat, if possible).

(7) If transmitting in the blind, ensure a clear LOS towards the equator.

(8) Use terrain masking to hinder enemy direction finding.

g. Listening (use reception times in applicable plans/orders or as directed by recovery forces).

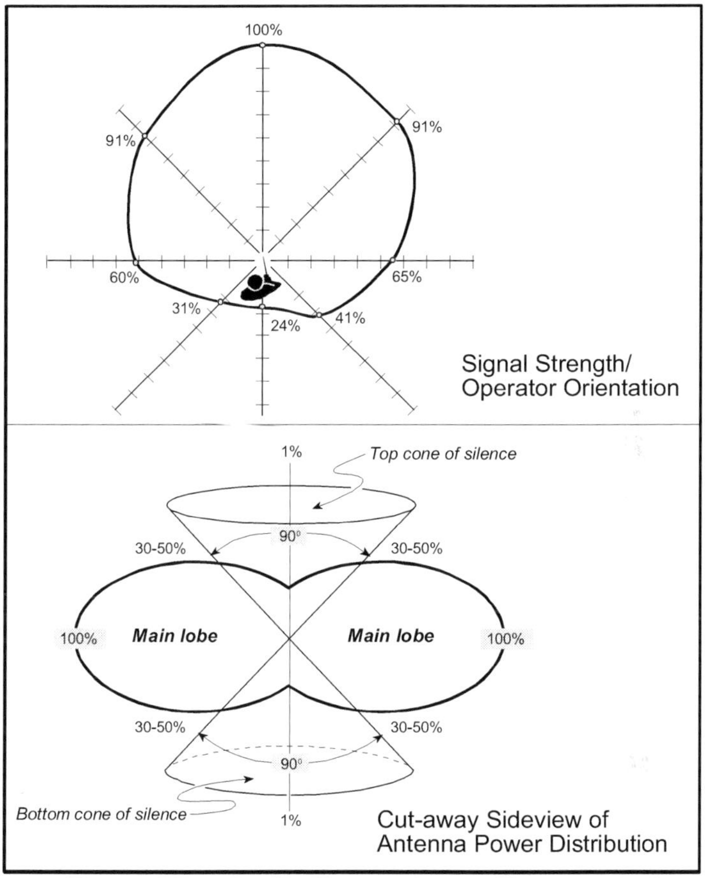

Figure III-1. Radio Transmission Characteristics

2. Signaling

a. Pyrotechnic signals.

(1) Prepare early (weather permitting).

(2) Use as directed in applicable plans/orders or as directed by recovery forces.

(3) Extend over raft's edge before activating.

b. Signal mirror **(Figure III-2)**.
   (1) Use as directed by recovery forces.
   (2) If no radio, use only with confirmed friendly forces.
   (3) Cover when not in use.

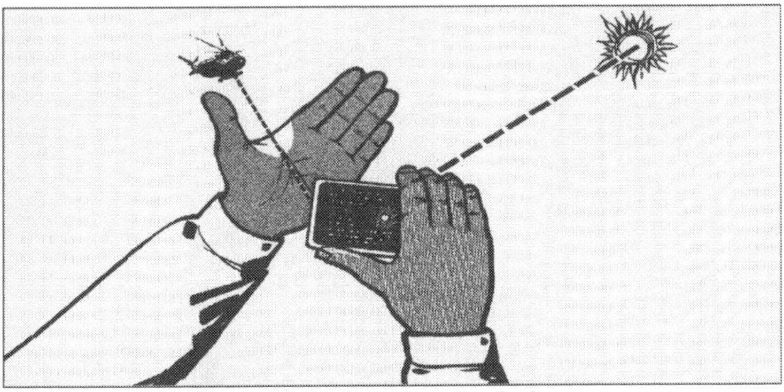

Figure III-2. Sighting Techniques

> **Note:** Make a mirror from any shiny metal or glass.

c. Strobe/IR lights.
   (1) Prepare early, consider filters and shields.
   (2) Use as directed by recovery forces.
   (3) Conserve battery life.

> **Note:** Produces one residual flash when turned off.

d. Pattern signals (use as directed in applicable plans/orders).
   (1) Materials:
      (a) Manmade (space blanket, signal paulin, parachute).
      (b) Natural use materials that contrast the color and/or texture of the signaling area (rocks, brush, branches, stomped grass).
   (2) Location.
      (a) Maximize visibility from above.
      (b) Provide concealment from ground observation.
   (3) Size (large as possible) and ratio **(Figure III-3)**.

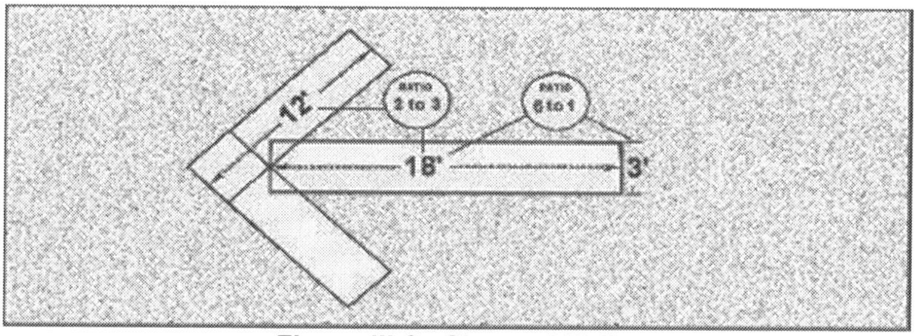

Figure III-3. Size and Ratio

(4) Shape (maintain straight lines and sharp corners).
(5) Contrast (use color and shadows).
(6) Pattern signals **(Figure III-4)**.

NO.	MESSAGE	CODE SYMBOL
1	REQUIRE ASSISTANCE	V
2	REQUIRE MEDICAL ASSISTANCE	X
3	NO or NEGATIVE	N
4	YES or AFFIRMATIVE	Y
5	PROCEEDING IN THIS DIRECTION	↑

Figure III-4. Signal Key

e. Sea dye marker.
   (1) ***DO NOT*** waste in rough seas or fast moving water.
   (2) Conserve unused dye by rewrapping.
   (3) May be used to color snow.

f. Non-combat considerations:

(1) Use a fire at night.

(2) Use smoke for day (tires or petroleum products for dark smoke and green vegetation for light smoke). **(Figure III-5)**

(3) Use signal mirror to sweep horizon.

(4) Use audio signals (that is, voice, whistle, and weapons fire).

Figure III-5. Smoke Generator

Chapter IV
RECOVERY

1. Responsibilities

a. Establish radio contact with recovery forces (if possible).

b. Maintain communication with recovery forces until recovered.

c. Be prepared to authenticate as directed in applicable plans/orders.

d. Follow recovery force instructions, be prepared to report—

(1) Enemy activity in the recovery area.

(2) Recovery site characteristics (slope, obstacles, size, etc.).

(3) Number in party/medical situation.

(4) Signal devices available.

e. If no radio, a ground-to-air signal may be your only means to effect recovery.

2. Site Selection

a. Locate area for landing pick-up, if practical (approximately 150 feet diameter, free of obstructions, flat and level).

b. Assess evidence of human activity at/near the site (in combat).

c. Locate several concealment sites around area (in combat).

d. Plan several tactical entry and exit routes (in combat).

3. Site Preparation

a. Pack and secure all equipment.

b. Prepare signaling devices (use as directed or as briefed).

c. Mentally review recovery methods (aircraft, ground, boat, etc.).

4. Recovery Procedures

a. Assist recovery force in identifying your position.

b. Stay concealed until recovery is imminent (in combat).

c. For a landing/ground recovery—

(1) Assume a non-threatening posture.

(2) Secure weapons and avoid quick movement.

(3) ***DO NOT*** approach recovery vehicle until instructed.

(4) Beware of rotors/propellers when approaching recovery vehicle, especially on sloping or uneven terrain. Secure loose equipment that could be caught in rotors/propellers.

d. For **hoist** recovery devices **(Figures IV-1 and IV-2)**—

(1) Use eye protection, if available (glasses or helmet visor).

(2) Allow metal on device to contact the surface before touching to avoid injury from static discharge.

(3) Sit or kneel for stability while donning device.

(4) Put safety strap under armpits.

(5) Ensure cable is in front of you.

(6) Keep hands clear of all hardware and connectors.

(7) ***DO NOT*** become entangled in cable.

(8) Use a thumbs up, vigorous cable shake, or radio call to signal you are ready.

(9) Drag feet on the ground to decrease oscillation.

(10) ***DO NOT*** assist during hoist or when pulled into the rescue vehicle. Follow crewmember instructions.

e. For **nonhoist** recovery (rope or unfamiliar equipment)—

(1) Create a ***"fixed loop"*** big enough to place under armpits **(Figure IV-3)**.

(2) Follow the procedures in "**d**" above.

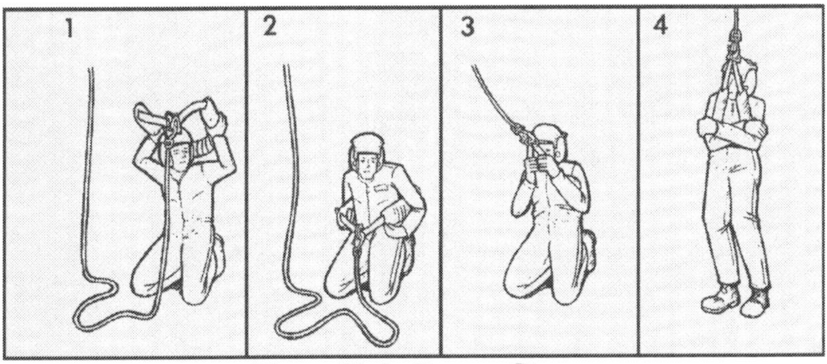

Figure IV–1. Rescue Strap

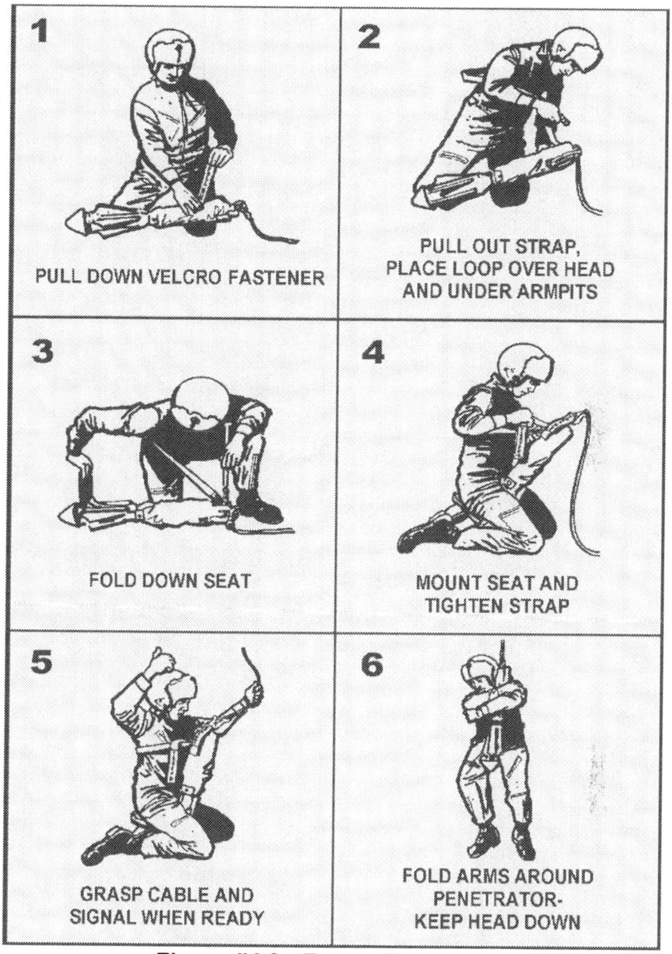

Figure IV-2. Forest Penetrator

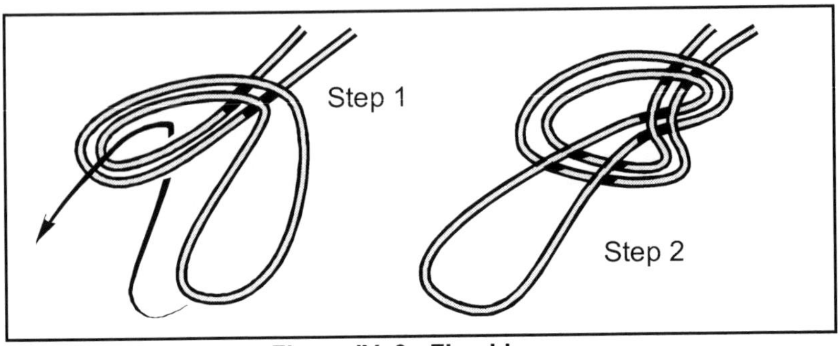

Figure IV–3. Fixed Loop

Chapter V
MEDICAL

WARNING: These emergency medical procedures are for survival situations. Obtain professional medical treatment as soon as possible.

1. Immediate First Aid Actions

Remember the *ABCs* of Emergency Care:
Airway Breathing Circulation

a. Determine responsiveness as follows:
   (1) If unconscious, arouse by shaking gently and shouting.
   (2) If no response—
      (a) Keep head and neck aligned with body.
      (b) Roll victims onto their backs.
      (c) Open the airway by lifting the chin **(Figure V-1)**.
      (d) Look, listen, and feel for air exchange.

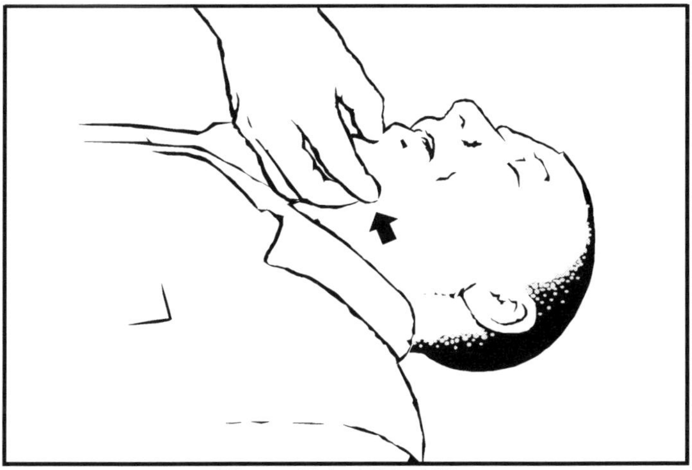

Figure V-1. Chin Lift

(3) If victim is not breathing—
(a) Check for a clear airway; remove any blockage.
(b) Cover victim's mouth with your own.
(c) Pinch victim's nostrils closed.
(d) Fill victim's lungs with **2** slow breaths.
(e) If breaths are blocked, reposition airway; try again.
(f) If breaths still blocked, give **5** abdominal thrusts:
•Straddle the victim.
•Place a fist between breastbone and belly button.
•Thrust upward to expel air from stomach.
(g) Sweep with finger to clear mouth.
(h) Try **2** slow breaths again.
(i) If the airway is still blocked, continue **(c)** through **(f)** until successful or exhausted.
(j) With open airway, start mouth to mouth breathing:
•Give **1** breath every 5 seconds.
•Check for chest rise each time.
(4) If victim is unconscious, but breathing—
(a) Keep head and neck aligned with body.
(b) Roll victim on side (drains the mouth and prevents the tongue from blocking airway).
(5) If breathing difficulty is caused by chest trauma, refer to **page V-7, paragraph 1d, *Treat Chest Injuries***.

CAUTION: ***DO NOT*** remove an impaled object unless it interferes with the airway. You may cause more tissue damage and increase bleeding. For travel, you may shorten and secure the object.

b. Control bleeding as follows:
(1) Apply a pressure dressing **(Figure V-2)**.
(2) If ***STILL*** bleeding—
(a) Use direct pressure over the wound.
(b) Elevate the wounded area above the heart.

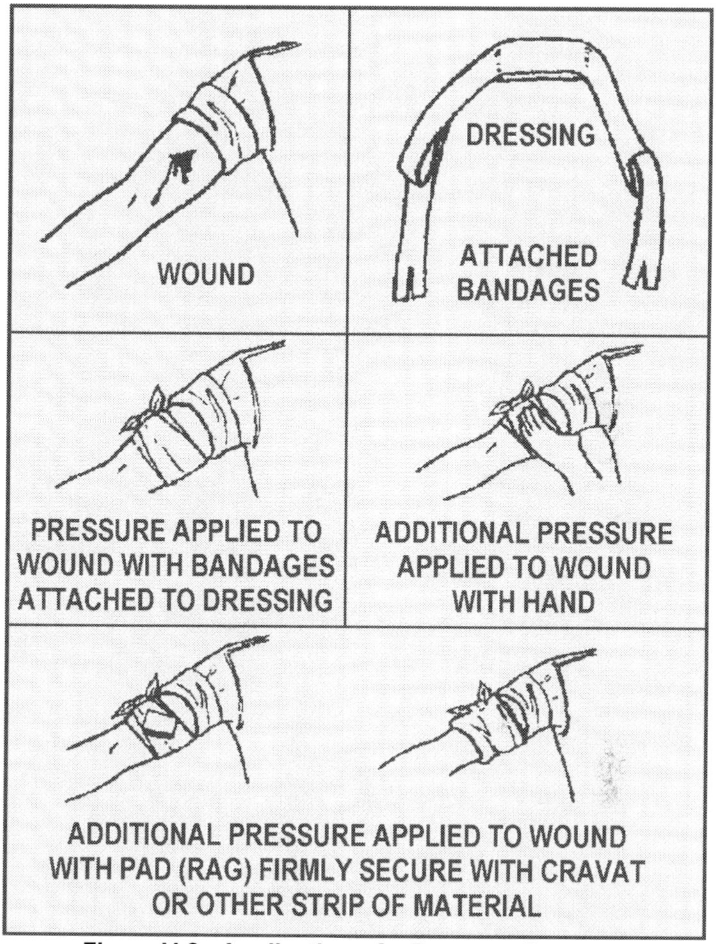

Figure V-2. Application of a Pressure Dressing

(3) If ***STILL*** bleeding—

(a) Use a pressure point between the injury and the heart **(Figure V-3)**.

(b) Maintain pressure for 6 to 10 minutes before checking to see if bleeding has stopped.

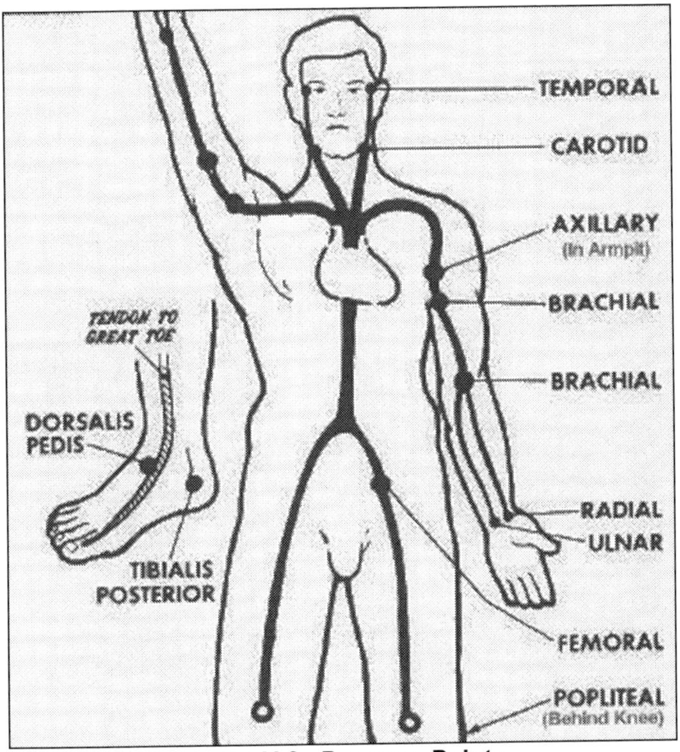

Figure V-3. Pressure Points

(4) If a limb wound is ***STILL*** bleeding—

CAUTION: Use of a tourniquet is a ***LAST RESORT*** measure. Use ***ONLY*** when severe, uncontrolled bleeding will cause loss of life. Recognize that long-term use of a tourniquet may cause loss of limb.

(a) Apply tourniquet (TK) band just above bleeding site on limb. A band at least 3 inches (7.5 cm) or wider is best.

(b) Follow steps illustrated in **Figure V-4**.

(c) Use a stick at least 6 inches (15 cm) long.

(d) Tighten only enough to stop arterial bleeding.

(e) Mark a ***TK*** on the forehead with the time applied.

(f) ***DO NOT*** cover the tourniquet.

CAUTION: The following directions apply ***ONLY*** in survival situations where rescue is ***UNLIKELY*** and ***NO*** medical aid is available.

(g) If rescue or medical aid is not available for over 2 hours, an attempt to ***SLOWLY*** loosen the tourniquet may be made 20 minutes after application. Before loosening—

- Ensure pressure dressing is in place.
- Ensure bleeding has stopped
- Loosen tourniquet ***SLOWLY*** to restore circulation.
- Leave loosened tourniquet in position in case bleeding resumes.

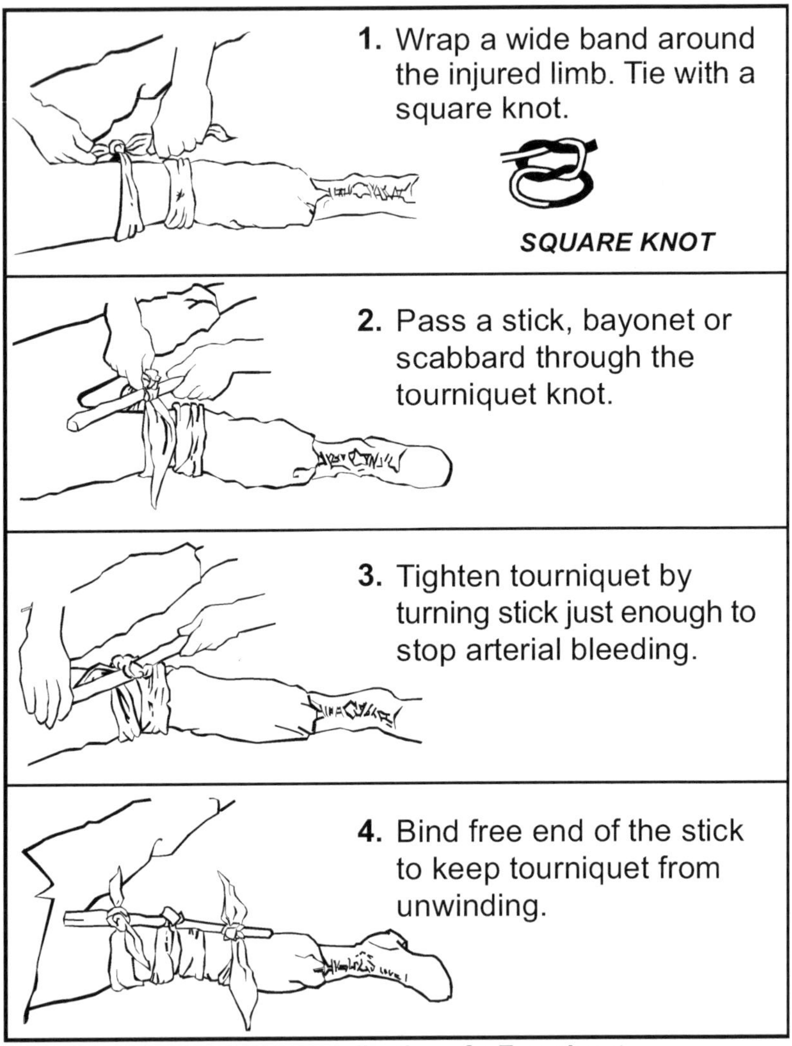

Figure V-4. Application of a Tourniquet

c. Treat shock. (Shock is difficult to identify or treat under field conditions. It may be present with or without visible injury.)

(1) Identify by one or more of the following:

(a) Pale, cool, and sweaty skin.

(b) Fast breathing and a weak, fast pulse.

(c) Anxiety or mental confusion.

(d) Decreased urine output.

(2) Maintain circulation.

(3) Treat underlying injury.

(4) Maintain normal body temperature.

(a) Remove wet clothing.

(b) Give warm fluids.

•***DO NOT*** give fluids to an unconscious victim.

•***DO NOT*** give fluids if they cause victim to gag.

(c) Insulate from ground.

(d) Shelter from the elements.

(5) Place conscious victim on back.

(6) Place very weak or unconscious victim on side, this will—

(a) Allow mouth to drain.

(b) Prevent tongue from blocking airway.

d. Treat chest injuries.

(1) Sucking chest wound. This occurs when chest wall is penetrated; may cause victim to gasp for breath; may cause sucking sound; may create bloody froth as air escapes the chest.

(a) ***Immediately*** seal wound with hand or airtight material.

(b) Tape airtight material over wound on ***3 sides only*** **(Figure V-5)** to allow air to escape from the wound but not to enter.

(c) Monitor breathing and check dressing.

(d) Lift untapped side of dressing as victim ***exhales*** to allow trapped air to escape, as necessary.

(2) Flail chest. Results from blunt trauma when ***3*** or ***more*** ribs are broken in ***2*** or more places. The flail segment is the broken area that moves in a direction opposite to the rest of chest during breathing.

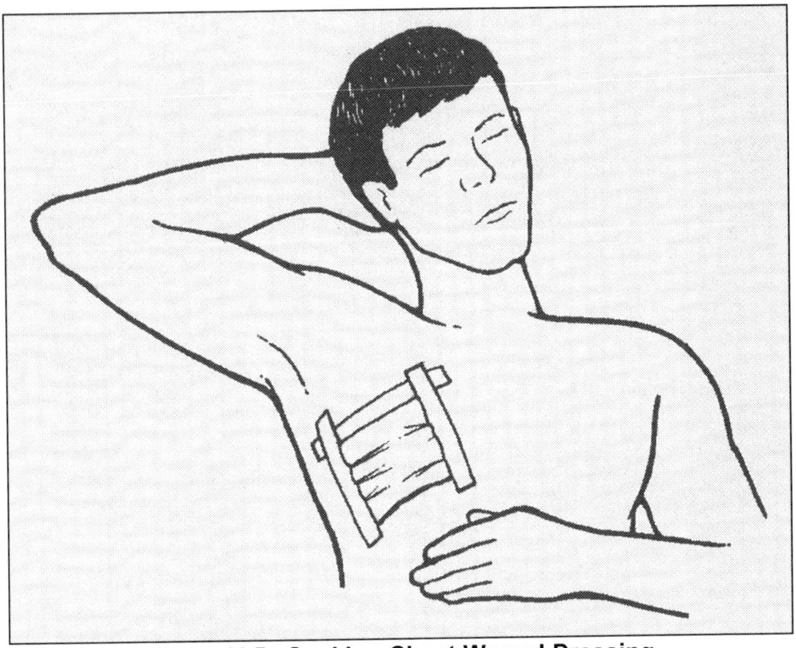

Figure V-5. Sucking Chest Wound Dressing

(a) Stabilize the flail segment as follows:

•Place rolled-up clothing or bulky pad over site.

•Tape pad to site

•***DO NOT*** wrap tape around chest.

(b) Have victim keep segment still with hand pressure.

(c) Roll victim onto side of flail segment injury (as other injuries allow).

(3) Fractured ribs.

(a) Encourage deep breathing (painful, but necessary to prevent the possible development of pneumonia).

(b) ***DO NOT*** constrict breathing by taping ribs.

e. Treat fractures, sprains, and dislocations.

(1) Control bleeding.

(2) Remove watches, jewelry, and constrictive clothing.

(3) If fracture penetrates the skin—

(a) Clean wound by gentle irrigation with water.
(b) Apply dressing over wound.
(4) Position limb as normally as possible.
(5) Splint in position found (if ***unable*** to straighten limb).
(6) Improvise a splint with available materials:
(a) Sticks or straight, stiff materials from equipment.
(b) Body parts (for example, opposite leg, arm-to-chest).
(7) Attach with strips of cloth, parachute cord, etc.
(8) Keep the fractured bones from moving by immobilizing the joints on both sides of the fracture. If fracture is in a joint, immobilize the bones on both sides of the joint.

CAUTION: Splint fingers in a slightly flexed position, ***NOT*** in straight position. Hand should look like it is grasping an apple.

(9) Use ***RICES*** treatment for 72 hours.
(a) **R**est.
(b) **I**ce.
(c) **C**ompression.
(d) **E**levation.
(e) **S**tabilization.
(10) Apply cold to acute injuries.
(11) Use 15 to 20 minute periods of cold application.
(a) ***DO NOT*** use continuous cold therapy.
(b) Repeat 3 to 4 times per day.
(c) Avoid cooling that can cause frostbite or hypothermia.
(12) Wrap with a compression bandage after cold therapy.
(13) Elevate injured area above heart level to reduce swelling.
(14) Check periodically for a pulse beyond the injury site.
(15) Loosen bandage or reapply splint if no pulse is felt or if swelling occurs because bandage is too tight.

2. Common Injuries and Illnesses

a. Burns.
(1) Cool the burned area with water.
(a) Use immersion or cool compresses.
(b) Avoid aggressive cooling with ice or frigid water.
(2) Remove watches, jewelry, constrictive clothing.

(3) ***DO NOT*** remove embedded, charred material that will cause burned areas to bleed.

(4) Cover with sterile dressings.

(5) ***DO NOT*** use lotion or grease.

(6) Avoid moving or rubbing the burned part.

(7) Drink ***extra*** water to compensate for increased fluid loss from burns. (Add ***1/4 teaspoon*** of ***salt*** [if available] to ***each quart*** of ***water***.)

(8) Change dressings when soaked or dirty.

b. Eye injuries.

(1) Sun/snow blindness (gritty, burning sensation, and possible reduction in vision caused by sun exposure).

(a) Prevent with improvised goggles. **(See Chapter VI, page VI-3, Figure VI-2.)**

(b) Treat by patching affected eye(s).

- Check after 12 hours.
- Replace patch for another 12 hours if not healed.

(c) Use cool compresses to reduce pain.

(2) Foreign body in eye.

(a) Irrigate with clean water from the ***inside*** to the ***outside*** corner of the eye.

(b) If foreign body is not removed by irrigation, improvise a small swab. Moisten and wipe gently over the affected area.

(c) If foreign body is ***STILL*** not removed, patch eye for 24 hours and then reattempt removal using steps **(a)** and **(b)**.

c. Heat injury.

(1) Heat cramps (cramps in legs or abdomen).

(a) Rest.

(b) Drink water. Add ***1/4 teaspoon*** of salt ***per quart***.

(2) Heat exhaustion (pale, sweating, moist, cool skin).

(a) Rest in shade.

(b) Drink water.

(c) Protect from further heat exposure.

(3) Heat stroke (victim disoriented or unconscious, skin is hot and flushed [sweating **may** or **may not** occur], fast pulse).

> **CAUTION:** Handle heat stroke victim gently. Shock, seizures, and cardiac arrest can occur.

(a) Cool as rapidly as possible (saturate clothing with water and fan the victim). Remember to cool the groin and armpit areas. (Avoid overcooling.)

(b) Maintain airway, breathing, and circulation.

d. Cold injuries:

(1) Frostnip and frostbite—

(a) Are progressive injuries.

•Ears, nose, fingers, and toes are affected first.

•Areas will feel cold and may tingle leading to—

••Numbness that progresses to—

•••Waxy appearance with stiff skin that cannot glide freely over a joint.

(b) Frostnipped areas rewarm with body heat. If body heat ***WILL NOT*** rewarm area in 15 to 20 minutes, then frostbite is present.

(c) Frostbitten areas are deeply frozen and require medical treatment.

CAUTION: In frostbite, repeated freezing and thawing causes severe pain and increases damage to the tissue. ***DO NOT*** rub frozen tissue. ***DO NOT*** thaw frozen tissue.

(2) Hypothermia—

(a) Is a progressive injury.

•Intense shivering with impaired ability to perform complex tasks leads to—

••Violent shivering, difficulty speaking, sluggish thinking go to—

•••Muscular rigidity with blue, puffy skin; jerky movements go to—

••••Coma, respiratory and cardiac failure.

(b) Protect victim from the environment as follows:

•Remove wet clothing.

•Put on dry clothing (if available).

•Prevent further heat loss.

••Cover top of head.

••Insulate from above and below.

•Warm with blankets, sleeping bags, or shelter.

•Warm central areas before extremities.

••Place heat packs in groin, armpits, and around neck.

••Avoid causing burns to skin.

CAUTION: Handle hypothermia victim gently. Avoid overly rapid rewarming which may cause cardiac arrest. Rewarming of victim with skin-to-skin contact by volunteer(s) inside of a sleeping bag is a survival technique but can cause internal temperatures of all to drop.

e. Skin tissue damage.

(1) Immersion injuries. Skin becomes wrinkled as in ***dishpan hands***.

(a) Avoid walking on affected feet.

(b) Pat dry; ***DO NOT*** rub. Skin tissue will be sensitive.

(c) Dry socks and shoes. Keep feet protected.

(d) Loosen boots, cuffs, etc., to improve circulation.

(e) Keep area dry, warm, and open to air.

(f) ***DO NOT*** apply creams or ointments.

(2) Saltwater sores.

(a) Change body positions frequently.

(b) Keep sores dry.

(c) Use antiseptic (if available).

(d) ***DO NOT*** open or squeeze sores.

f. Snakebite.

CAUTION: This snakebite treatment recommendation is for situations where medical aid and specialized equipment are not available.

(1) Nonpoisonous. Clean and bandage wound.

(2) Poisonous.

(a) Remove constricting items.

(b) Minimize activity.

(c) ***DO NOT*** cut the bite site; ***DO NOT*** use your mouth to create suction.

(d) Clean bite with soap and water; cover with a dressing.

(e) Overwrap the bite site with a tight (elastic) bandage **(Figure V-6)**. The intent is to slow capillary and venous blood flow but not arterial flow. Check for pulse below the overwrap.

(f) Splint bitten extremity to prevent motion.

(g) Treat for shock **(page V-7, paragraph 1c)**.

(h) Position extremity below level of heart.

(i) Construct shelter if necessary (let the victim rest).

(j) For conscious victims, force fluids.

g. Marine life.

(1) Stings.

(a) Flush wound with salt water (fresh water stimulates toxin release).

(b) Remove jewelry and watches.

(c) Remove tentacles and gently scrape or shave skin.

(d) Apply a steroid cream (if available).

(e) ***DO NOT*** rub area with sand.

(f) Treat for shock; artificial respiration may be required **(page V-1, paragraph 1a)**.

(g) ***DO NOT*** use urine to flush or treat wounds.

(2) Punctures.

(a) Immerse affected part in hot water or apply hot compresses for 30-60 minutes (as hot as victim can tolerate).

(b) Cover with clean dressing.

(c) Treat for shock as needed.

h. Skin irritants (includes poison oak and poison ivy).

(1) Wash with large amounts of water. Use soap (if available).

(2) Keep covered to prevent scratching.

i. Infection.

(1) Keep wound clean.

(2) Use iodine tablet solution or diluted betadine to prevent or treat infection.

(3) Change bandages as needed.

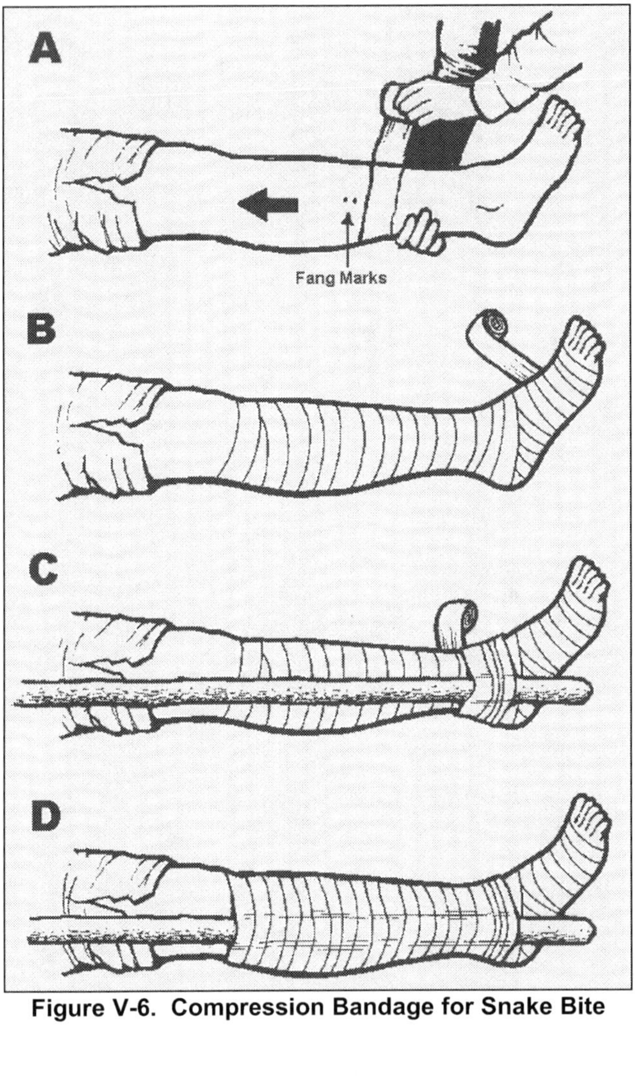

Figure V-6. Compression Bandage for Snake Bite

j. Dysentery and diarrhea.

(1) Drink ***extra*** water.

(2) Use a liquid diet.

(3) Eat charcoal. Make a paste by mixing fine charcoal particles with water. (It may relieve symptoms by absorbing toxins.)

k. Constipation (can be expected in survival situations).

(1) ***DO NOT*** take laxatives.

(2) Exercise.

(3) Drink **extra** water.

3. Plant Medicine

a. Tannin.

(1) Medical uses. Burns, diarrhea, dysentery, skin problems, and parasites. Tannin solution prevents infection and aids healing.

(2) Sources. Found in the outer bark of all trees, acorns, banana plants, common plantain, strawberry leaves, and blackberry stems.

(3) Preparation.

(a) Place crushed outer bark, acorns, or leaves in water.

(b) Leach out the tannin by soaking or boiling.

•Increase tannin content by longer soaking time.

•Replace depleted material with fresh bark/plants.

(4) Treatments.

(a) Burns.

•Moisten bandage with cooled tannin tea.

•Apply compress to burned area.

•Pour cooled tea on burned areas to ease pain.

(b) Diarrhea, dysentery, and worms. Drink strong tea solution (may promote voiding of worms).

(c) Skin problems (dry rashes and fungal infections). Apply cool compresses or soak affected part to relieve itching and promote healing.

(d) Lice and insect bites. Wash affected areas with tea to ease itching.

b. Salicin/salicylic acid.

(1) Medical uses. Aches, colds, fever, inflammation, pain, sprains, and sore throat (aspirin-like qualities).

(2) Sources. Willow and aspen trees **(Figure V-7)**.

(3) Preparation.

(a) Gather twigs, buds, or cambium layer (soft, moist layer between the outer bark and the wood) of willow or aspen.

(b) Prepare tea as described in paragraph **3a(3)**.

(c) Make poultice.

•Crush the plant or stems.

•Make a pulpy mass.

(4) Treatments.

(a) Chew on twigs, buds, or cambium for symptom relief.

(b) Drink tea for colds and sore throat.

(c) Use warm, moist poultice for aches and sprains.

•Apply pulpy mass over injury.

•Hold in place with a dressing.

c. Common plantain.

(1) Medical uses. Itching, wounds, abrasions, stings, diarrhea, and dysentery.

(2) Source. There are over 200 plantain species with similar medicinal properties. The common plantain is shown in **Figure V-7**.

(3) Preparation.

(a) Brew tea from seeds.

(b) Brew tea from leaves.

(c) Make poultice of leaves.

(4) Treatments.

(a) Drink tea made from seeds for diarrhea or dysentery.

(b) Drink tea made from leaves for vitamin and minerals.

(c) Use poultice to treat cuts, sores, burns, and stings.

d. Papain.

(1) Medical uses. Digestive aid, meat tenderizer, and a food source.

(2) Source. Fruit of the papaya tree **(Figure V-7)**.

(3) Preparation.

(a) Make cuts in **unripe** fruit.

(b) Gather milky white sap for its papain content.

(c) Avoid getting sap in eyes or wounds.

(4) Treatments.

(a) Use sap to tenderize tough meat.

(b) Eat **ripe** fruit for food, vitamins, and minerals.

e. Common Cattail.

(1) Medical uses. Wounds, sores, boils, inflammations, burns, and an excellent food source.

(2) Source. Cattail plant found in marshes **(Figure V-7)**.

(3) Preparation.

(a) Pound roots into a pulpy mass for a poultice.

(b) Cook and eat green bloom spikes.

(c) Collect yellow pollen for flour substitute.

(d) Peel and eat tender shoots (raw or cooked).

(4) Treatments.

(a) Apply poultice to affected area.

(b) Use plant for food, vitamins, and minerals.

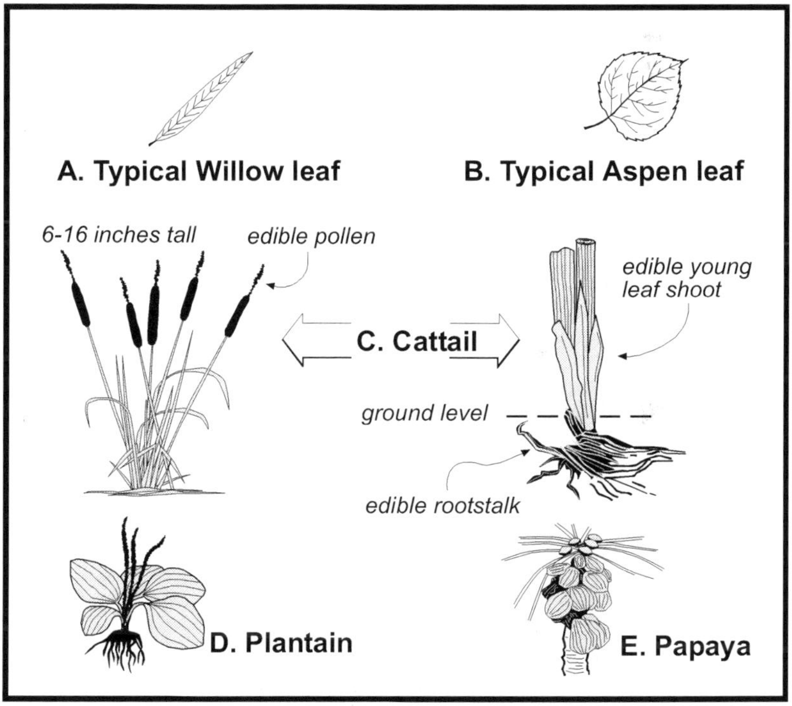

Figure V-7. Useful Plants

4. Health and Hygiene

a. Stay clean (daily regimen).

(1) Minimize infection by washing. (Use white ashes, sand, or loamy soil as soap substitutes.)

(2) Comb and clean debris from hair.

(3) Cleanse mouth and brush teeth.

(a) Use hardwood twig as toothbrush (fray it by chewing on one end then use as brush).

(b) Use single strand of an inner core string from parachute cord for dental floss.

(c) Use clean finger to stimulate gum tissues by rubbing.

(d) Gargle with salt water to help prevent sore throat and aid in cleaning teeth and gums.

(4) Clean and protect feet.

(a) Change and wash socks

(b) Wash, dry, and massage.

(c) Check frequently for blisters and red areas.

(d) Use adhesive tape/mole skin to prevent damage.

b. Exercise daily.

c. Prevent and control parasites.

(1) Check body for lice, fleas, ticks, etc.

(a) Check body regularly.

(b) Pick off insects and eggs (***DO NOT*** crush).

(2) Wash clothing and use repellents.

(3) Use smoke to fumigate clothing and equipment.

5. Rules for Avoiding Illness

a. Purify all water obtained from natural sources by using iodine tablets, bleach, or boiling for 5 minutes.

b. Locate latrines 200 feet from water and away from shelter.

c. Wash hands before preparing food or water.

d. Clean all eating utensils after each meal.

e. Prevent insect bites by using repellent, netting, and clothing.

f. Dry wet clothing as soon as possible.

g. Eat varied diet.

h. Try to get 7-8 hours sleep per day.

Chapter VI
PERSONAL PROTECTION

1. Priorities

a. Evaluate available resources and situation, then accomplish individual tasks accordingly.

b. First 24 hours in order of situational needs—

(1) Construct survival shelter according to selection criteria.

(2) Procure water.

(3) Establish multiple survival signals.

(4) Build Fire.

c. Second 24 hours in order of situational needs—

(1) Construct necessary tools and weapons.

(2) Procure food.

2. Care and Use of Clothing

a. Never discard clothing.

b. Wear loose and layered clothing.

(1) Tight clothing restricts blood flow regulating body temperature.

(2) Layers create more dead air space.

c. Keep entire body covered to prevent sunburn and dehydration in hot climates. When fully clothed, the majority of body heat escapes through the head and neck areas.

d. Avoid overheating.

(1) Remove layers of clothing before strenuous activities.

(2) Use a hat to regulate body heat.

(3) Wear a hat when in direct sunlight (in hot environment).

e. Dampen clothing when on the ocean in hot weather.

(1) Use salt water, ***NOT*** drinking water.

(2) Dry clothing before dark to prevent hypothermia.

f. Keep clothing dry to maintain its insulation qualities (dry damp clothing in the sun or by a fire).

g. If you fall into the water in the winter—

(1) Build fire.

(2) Remove wet clothing and rewarm by fire.

(3) Finish drying clothing by fire.

h. If no fire is available—
   (1) Remove clothing and get into sleeping bag (if available).
   (2) Allow wet clothes to freeze.
   (3) Break ice out of clothing.

i. Keep clothing clean (dirt reduces its insulation qualities). Examine clothing frequently for damage.
   (1) ***DO NOT*** sit or lie directly on the ground.
   (2) Wash clothing whenever possible.
   (3) Repair when necessary by using—
      (a) Needle and thread.
      (b) Safety pins.
      (c) Tape.

j. Improvised foot protection **(Figure VI-1)**.
   (1) Cut **2** to **4** layers of cloth into a 30-inch square.
   (2) Fold into a triangle.
   (3) Center foot on triangle with toes toward corner.

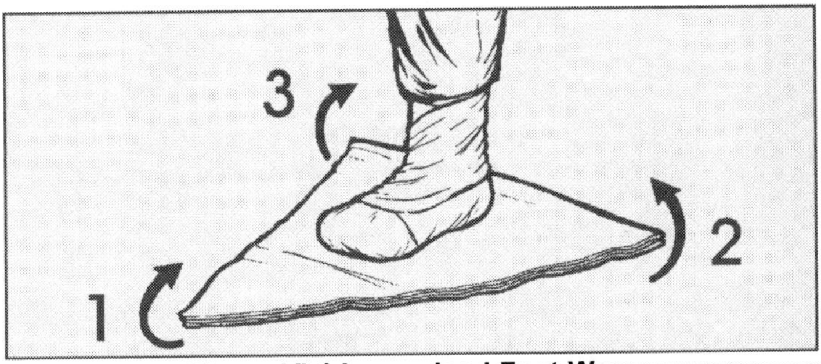

Figure VI-1 Improvised Foot Wear

   (4) Fold front over the toes.
   (5) Fold side corners, one at a time, over the instep.
   (6) Secure by rope, vines, tape, etc., or tuck into other layers of material.

3. Other Protective Equipment

a. Sleeping bag.
   (1) Fluff before use, ***especially*** at foot of bag.
   (2) Air and dry daily to remove body moisture.

(3) Improvise with available material, dry grass, leaves, dry moss, etc.

b. Sun and snow goggles **(Figure VI-2)**.

(1) Wear in bright sun or snow conditions.

(2) Improvise by cutting small horizontal slits in webbing, bark, or similar materials.

Figure VI-2. Sun and Snow Goggles

c. Gaiters **(Figure VI-3)**. Used to protect from sand, snow, insects, and scratches (wrap material around lower leg and top of boots).

Figure VI-3. Gaiters

4. Shelters

Evasion considerations apply.

a. Site selection.

(1) Near signal and recovery site.

(2) Available food and water.

(3) Avoid natural hazards:

(a) Dead standing trees.

(b) Drainage and dry river beds except in combat areas.

(c) Avalanche areas.

(4) Location large and level enough to lie down in.

b. Types.

(1) Immediate shelters. Find shelter needing minimal improvements **(Figure VI-4)**.

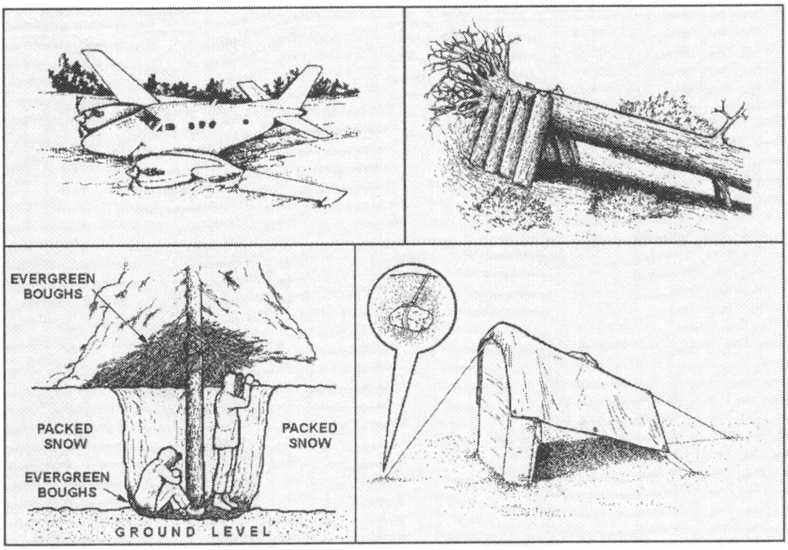

Figure VI-4. Immediate Shelters

(2) General shelter. Temperate climates require any shelter that gives protection from wind and rain.

(3) Thermal A Frame, Snow Trench, Snow Cave. **(Figures VI-5 through VI-7)**. Cold climates require an enclosed, insulated shelter.

(a) Snow is the most abundant insulating material.

(b) Air vent is required to prevent carbon monoxide poisoning when using an open flame inside enclosed shelters.

Note: As a general rule, unless you can see your breath, your snow shelter is too warm and should be cooled down to preclude melting and dripping.

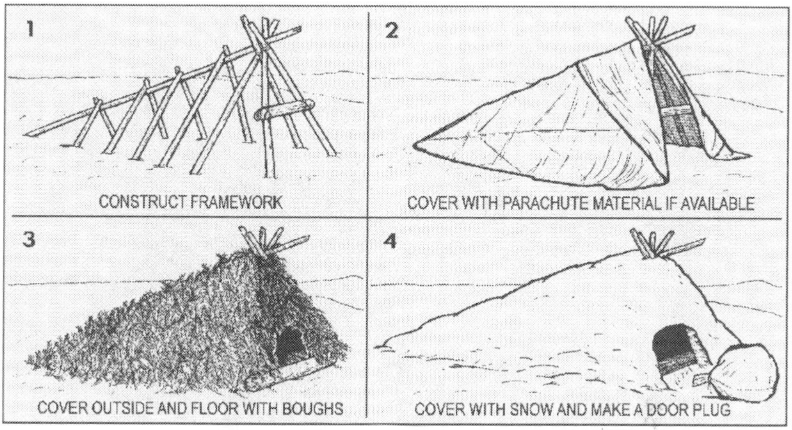

Figure VI-5. Thermal A Frame

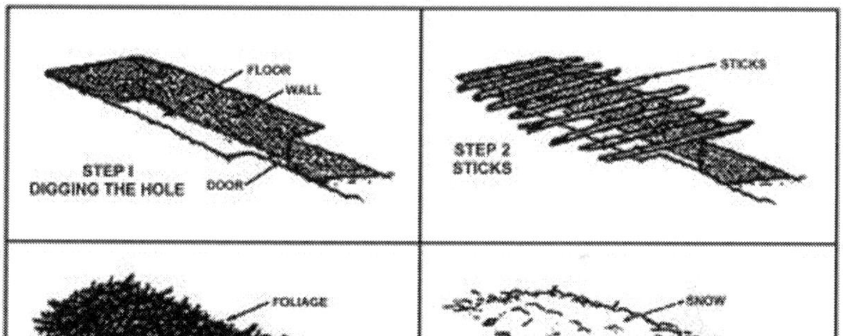

Figure VI-6. Snow Trench

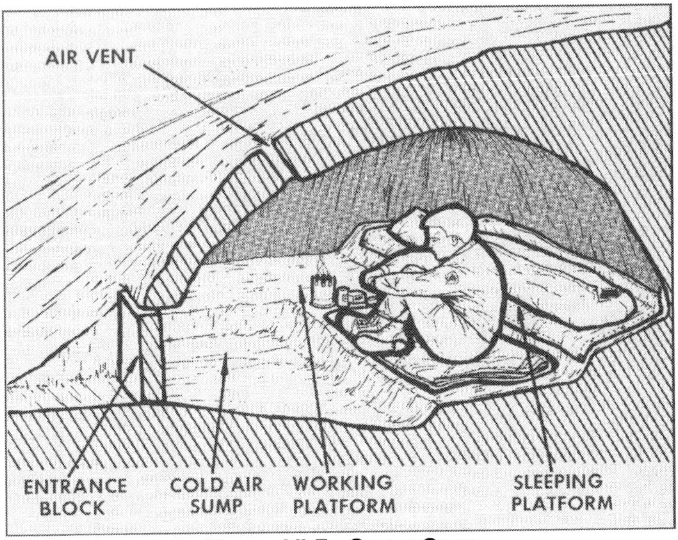

Figure VI-7. Snow Cave

(4) Shade shelter. Hot climates require a shade shelter to protect from ultraviolet rays **(Figure VI-8)**.

(a) To reduce the surface temperature, the shelter floor should be elevated or dug down (approximately 18 inches).

(b) For thermal protection, a minimum of **2** layers of material suspended 12-18 inches above the head is required. White is the best color to reflect heat (inner most layer should be of darker material).

(5) Elevated platform shelter **(Figure VI-9)**. Tropical/wet climates require enclosed, elevated shelter for protection from dampness and insects.

c. Shelter construction.

(1) Have entrance 45-90 degrees from prevailing wind.

(2) Cover with available material.

(a) If natural materials are used, arrange them in layers starting at the bottom with each layer overlapping the previous one. See **Figure VI-10** for an example.

Figure VI-8. Poncho/Parachute Shade Shelter

Figure VI-9. Elevated Platform Shelter

Figure VI-10. Shingle Method

(b) If using porous material like parachute, blankets, etc.—

- •Stretch as tight as possible
- •Use a 40–60 degree slope.
- •Use additional layers in heavy rain.

d. Shelter construction materials:

(1) Raft and raft parts.
(2) Vehicle or aircraft parts.
(3) Blankets, poncho, or parachute material.
(4) Sheet of plastic or plastic bag.
(5) Bark peeled off dead trees.
(6) Boughs, broad leaves, dry moss.
(7) Grass and sod.
(8) Snow.
(9) Sand and rocks.

e. Bed construction. Construct a bed to protect from cold, damp, ground using—

(1) Raft or foam rubber from vehicle seats.
(2) Boughs, leaves, or dry moss.

5. Fires

> **CAUTION:** Weigh hazards and risks of detection against the need for a fire.

a. Evasion considerations:

(1) Use trees or other sources to dissipate smoke.
(2) Use fires at dusk, dawn, or during inclement weather.
(3) Use fires at times when the local populace is cooking.

b. Fire building. The **3** essential elements for starting a fire are heat, fuel, and oxygen.

(1) Heat sources:

(a) Matches or lighter.
(b) Flint and steel (experiment with various rocks and metals until a good spark is produced).
(c) Sparks from batteries.
(d) Concentrated sunlight (use magnifying glass or flashlight reflectors).
(e) Pyrotechnics, such as flares **(last resort)**, etc.

(f) Friction method **(Figure VI-11)**. Without prior training, this method is difficult to master and requires a lot of time to build the device.

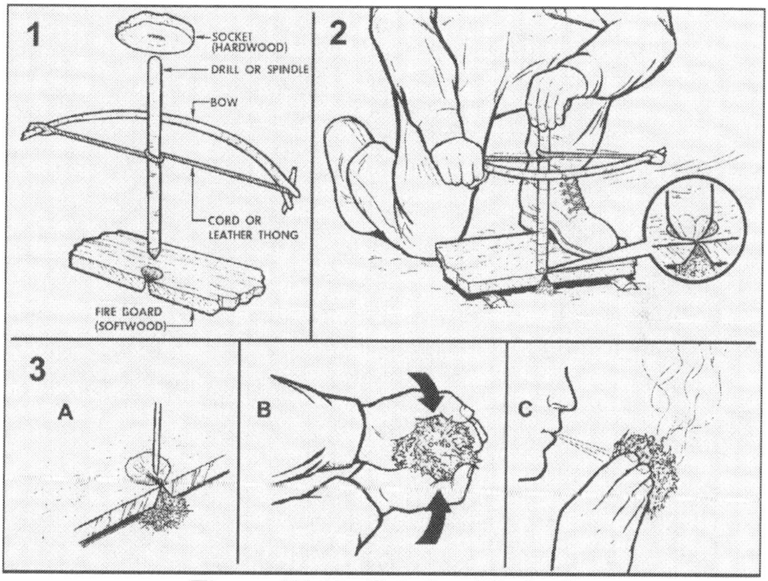

Figure VI-11. Friction Method

Note: If possible, carry a fire-starting device with you.

(2) Fuel is divided into **3** categories: tinder, kindling, and fuel. (Gather large amounts of each category before igniting the fire.)

(a) **Tinder**. Tinder must be very finely shaved or shredded to provide a low combustion point and fluffed to allow oxygen to flow through. (To get tinder to burn hotter and longer, saturate with Vaseline, Chapstick, insect repellant, aircraft fuel, etc.) Examples of tinder include—

- Cotton.
- Candle (shred the wick, not the wax).
- Plastic spoon, fork, or knife.
- Foam rubber.

•Dry bark.
•Dry grasses.
•Gun powder.
•Pitch.
•Petroleum products.

(b) **Kindling**. Kindling must be small enough to ignite from the small flame of the tinder. Gradually add larger kindling until arriving at the size of fuel to burn.

(c) **Fuel**. Examples of fuel include—

•Dry hardwood (removing bark reduces smoke).
•Bamboo (open chambers to prevent explosion).
•Dry dung.

c. Types. Fires are built to meet specific needs or uses.

(1) Tepee fire **(Figure VI-12)**. Use the tepee fire to produce a concentrated heat source for cooking, lighting, or signaling.

Figure VI-12. Tepee Fire

(2) Log cabin fire **(Figure VI-13)**. Use the log cabin fire to produce large amounts of light and heat, to dry out wet wood, and provide coals for cooking, etc.

Figure VI-13. Log Cabin or Pyramid Fires

(3) Sod fire and reflector (**Figure VI-14**). Use fire reflectors to get the most warmth from a fire. Build fires against rocks or logs.

CAUTION: ***DO NOT*** use porous rocks or riverbed rock—they may explode when heated.

Figure VI-14. Sod Fire and Reflector

(4) Dakota fire hole (Figure VI-15). Use the Dakota fire hole for high winds or evasion situations.

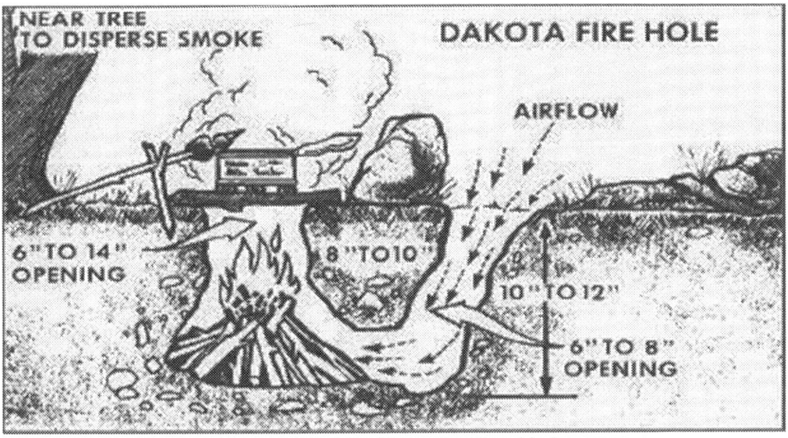

Figure VI-15. Dakota Fire Hole

(5) Improvised stoves **(Figure VI-16)**. These are very efficient.

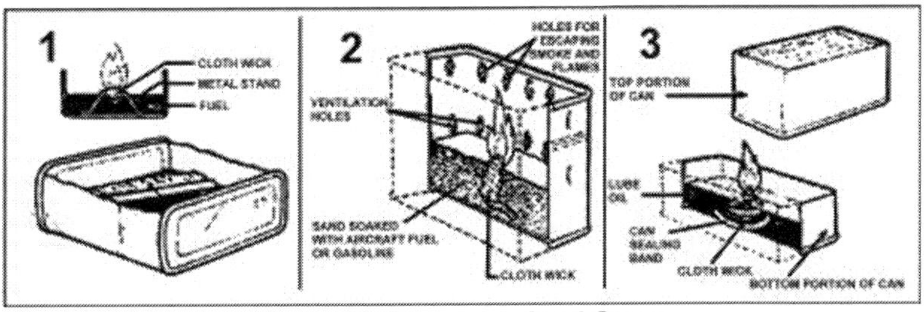

Figure VI-16. Improvised Stove

Chapter VII
Water

1. Water Requirements

Drink **extra** water. Minimum 2 quarts per day to maintain fluid level. Exertion, heat, injury, or an illness increases water loss.

Note: Pale yellow urine indicates adequate hydration.

2. Water Procurement

a. ***DO NOT*** drink—
   (1) Urine.
   (2) Fish juices.
   (3) Blood.
   (4) Sea water.
   (5) Alcohol.
   (6) Melted water from new sea ice.

b. Water sources:
   (1) Surface water (streams, lakes, and springs).
   (2) Precipitation (rain, snow, dew, sleet) **(FigureVII-1)**.
   (3) Subsurface (wells and cisterns).
   (4) Ground water (when no surface water is available) **(Figure VII-2)**.
      (a) Abundance of lush green vegetation.
      (b) Drainages and low-lying areas.
      (c) *"V"* intersecting game trails often point to water.
      (d) Presence of swarming insects indicates water is near.
      (e) Bird flight in the early morning or late afternoon might indicate the direction to water.
   (5) Snow or ice.
      (a) ***DO NOT*** eat ice or snow.
         - Lowers body temperature.
         - Induces dehydration.
         - Causes minor cold injury to lips and mouth.

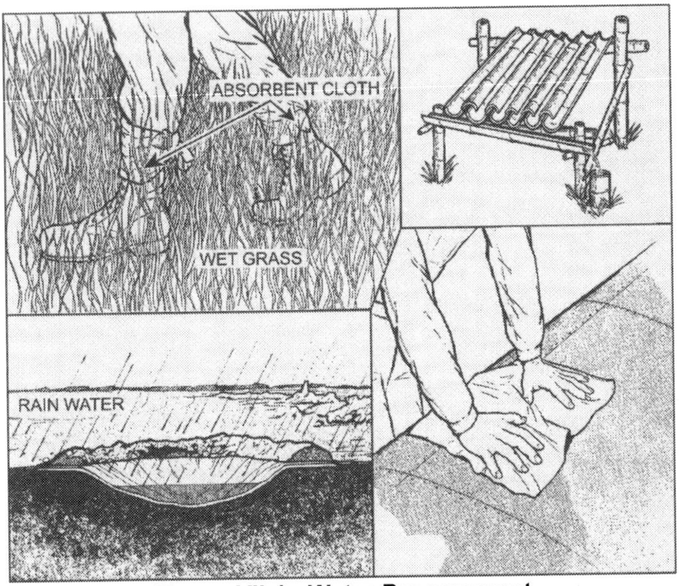

Figure VII-1. Water Procurement

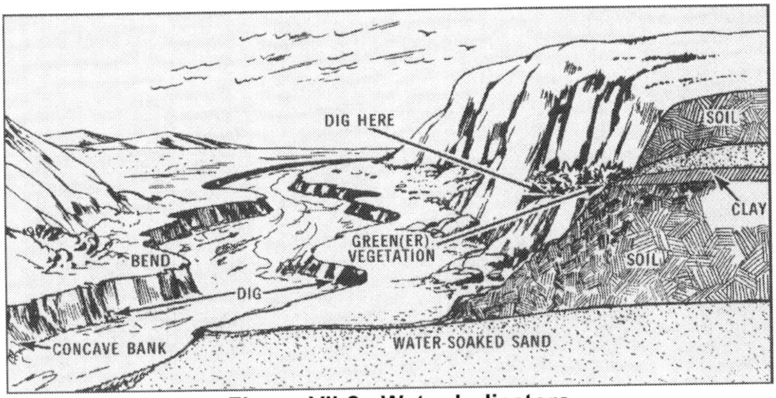

Figure VII-2. Water Indicators

(b) Melt with fire.
- •Stir frequently to prevent damaging container.
- •Speed the process by adding hot rocks or water.

(c) Melt with body heat.
- •Use waterproof container.
- •Place between layers of clothing.
- •***DO NOT*** place next to the skin.

(d) Use a water generator **(Figure VII-3)**.

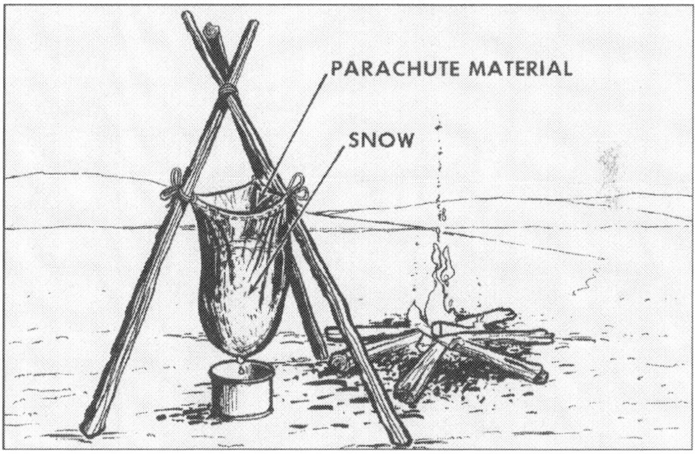

Figure VII-3. Water Generator

(6) Open seas.

(a) Water available in survival kits.

(b) Precipitation.
- •Drink as much as possible.
- •Catch rain in spray shields and life raft covers.
- •Collect dew off raft.

(c) Old sea ice or icebergs **(Table VII-1)**.

Table VII-1. Old Sea Ice or Icebergs

OLD SEA ICE	NEW SEA ICE
Bluish or blackish	Milky or grey
Shatters easily	Does not break easily
Rounded corners	Sharp edges
Tastes relatively salt-free	Tastes extremely salty

(7) Tropical areas.

(a) All open sources previously mentioned.

(b) Vegetation.

•Plants with hollow sections can collect moisture.

•Leaning Tree. Cloth absorbs rain running down tree and drips into container **(Figure VII-4)**.

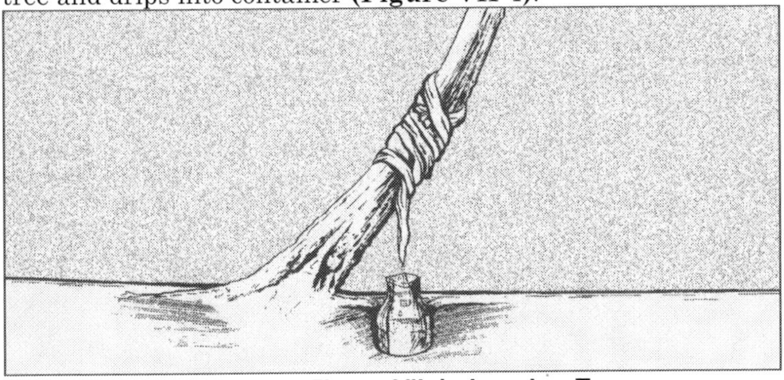

Figure VII-4. Leaning Tree

•Banana plants.

•Water trees (avoid milky sap).

••Tap before dark. Let sap stop running and harden during the daytime.

••Produce most water at night.

••For evasion situations, bore into the roots and collect water.

•Vines **(Figure VII-5A)**.

••Cut bark (***DO NOT*** use milky sap).

••If juice is clear and water like, cut as large a piece of vine as possible (cut the top first).

••Pour into hand to check smell, color, and taste to determine if drinkable.

••***DO NOT*** touch vine to lips.

••When water flow stops, cut off 6 inches of opposite end, water will flow again.

•Old bamboo.

••Shake and listen for water.

••Bore hole at bottom of section to obtain water.

••Cut out entire section to carry with you.
••Filter and purify.
•Green bamboo **(Figure VII-5B)**.

Figure VII-5 A and B. Water Vines and Green Bamboo

CAUTION: Liquid contained in green coconuts (**ripe** coconuts may cause diarrhea).

•Beach well. Along the coast, obtain water by digging a beach well **(Figure VII-6)**.

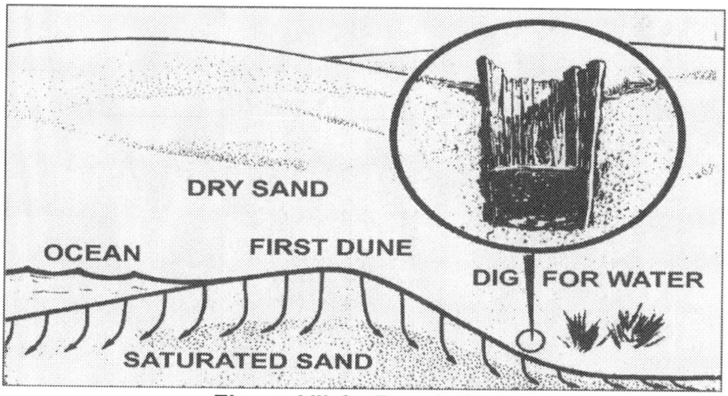

Figure VII-6. Beach Well

(8) Dry areas.

(a) Solar still **(Figure VII-7)**.

(b) Vegetation bag **(Figure VII-8)**.

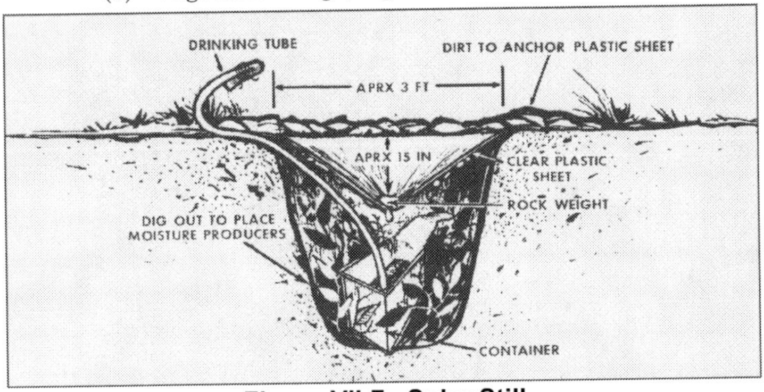

Figure VII-7. Solar Still

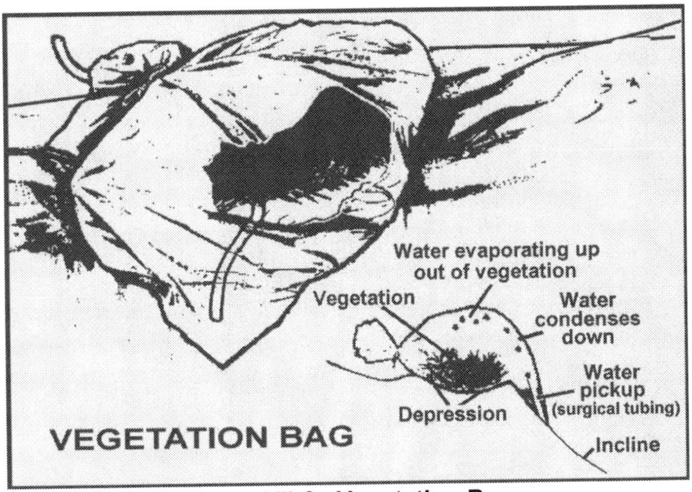

Figure VII-8. Vegetation Bag

(c) Transpiration bag **(Figure VII-9)**.

•Water bag must be clear.

•Water will taste like the plant smells.

(d) Seepage basin **(Figure VII-10)**.

> **CAUTION:** ***DO NOT*** use poisonous/toxic plants in vegetation/transpiration bags.

Figure VII-9. Transpiration Bag

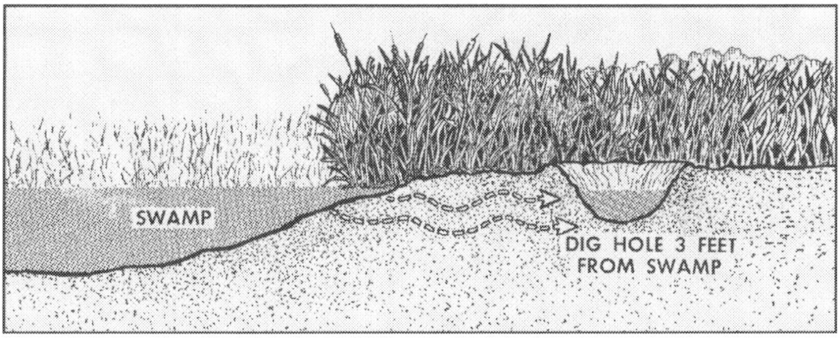

Figure VII-10. Seepage Basin

3. Water Preparation and Storage

a. Filtration. Filter through porous material (sand/charcoal).

b. Purification.

(1) Water from live plants requires no further treatment.

(2) Purify all other water.

(a) Boil at least 1 minute.

(b) Pour from one container to another to improve taste to aerate.

(c) Water purification tablets. Follow instructions on package.

c. Potable Water.

(1) If water cannot be purified, obtain water from a clear, cold, clean, and fast running source (if possible).

(2) Put in clear container and expose to the sun's ultraviolet rays to kill bacteria.

d. Storage. To prevent contamination, use a clean, covered or sealed container.

(1) Trash bag.

(2) Prophylactic.

(3) Section of bamboo.

(4) Flotation gear.

Chapter VIII
FOOD

1. Food Procurement

a. Sources and location.

(1) Mammals can be found where—

(a) Trails lead to watering, feeding, and bedding areas.

(b) Droppings or tracks look fresh.

(2) Birds can be found by—

(a) Observing the direction of flight in the early morning and late afternoon (leads to feeding, watering, and roosting areas).

(b) Listening for bird noises (indication of nesting areas

(3) Fish and other marine life locations **(Figure VIII-1)**.

Figure VIII-1. Fishing Locations

(4) Reptiles and amphibians are found almost worldwide.

(5) Insects are found—

(a) In dead logs and stumps.

(b) At ant and termite mounds.

(c) On ponds, lakes, and slow moving streams.

b. Procurement techniques.

(1) Snares—

(a) Work while unattended.

(b) Location:
•Trails leading to water, feeding, and bedding areas.
•Mouth of dens **(Figure VIII-2)**.

Figure VIII-2. Snare Placement

(c) Construction of simple loop snare.
•Use materials that will not break under the strain of holding an animal.
•Use a figure 8 (locking loop) if wire is used **(Figure VIII-3)**.
••Once tightened, the wire locks in place, preventing reopening, and the animal's escape.
•To construct a squirrel pole **(Figure VIII-4)** use simple loop snares.
•Make noose opening slightly larger than the animal's head **(3-finger** width for squirrels, **fist-sized** for rabbits**)**.
(d) Placement of snares (set as many as possible).
•Avoid disturbing the area.
•Use funneling (natural or improvised) **(Figure VIII-5)**.

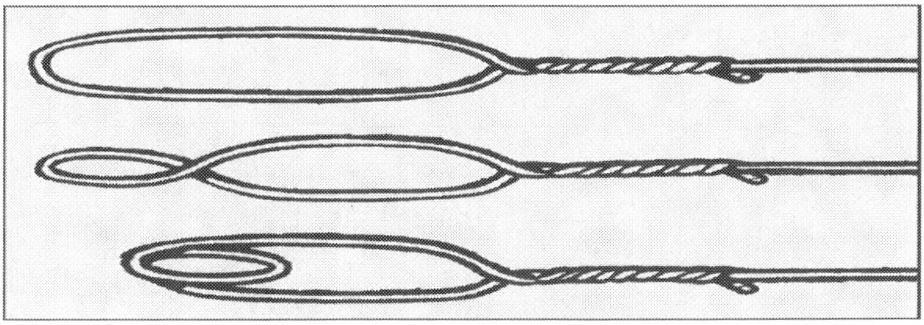

Figure VIII-3. Locking Loop

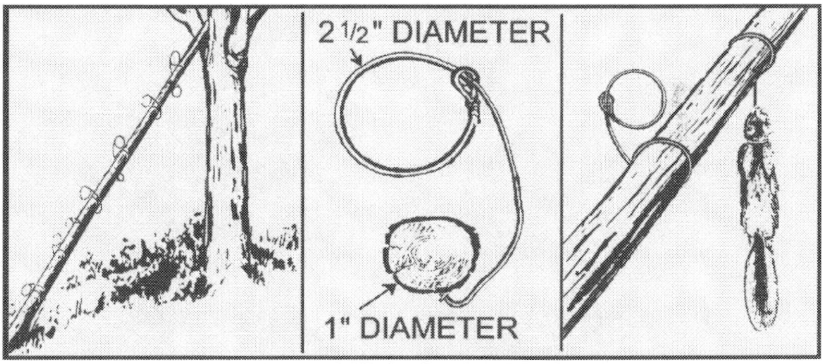

Figure VIII-4. Squirrel Pole

Figure VIII-5. Funneling

(2) Noose stick (easier and safer to use than the hands).

(3) Twist stick **(Figure VIII-6)**.

(a) Insert forked stick into a den until something soft is met.

(b) Twist the stick, binding the animal's hide in the fork.

(c) Remove the animal from the den.

(d) Be ready to **kill** the animal; **it may be dangerous**.

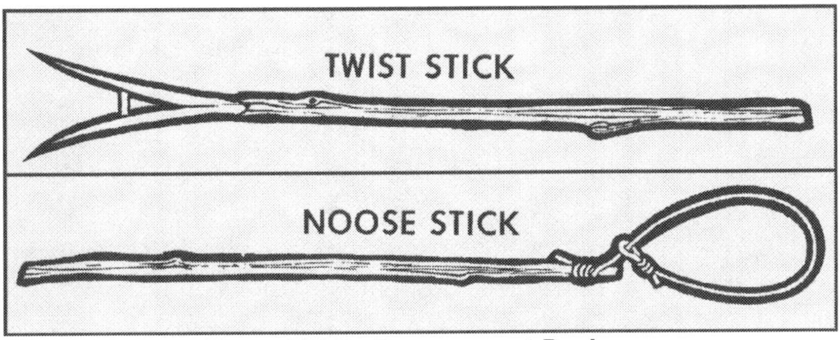

Figure VIII-6. Procurement Devices

(4) Hunting and fishing devices. (See **Figure VIII-7** for fishing procurement methods.)

(a) Club or rock.

(b) Spear.

(c) Slingshot.

(d) Pole, line, and hook.

(e) Net.

(f) Trap.

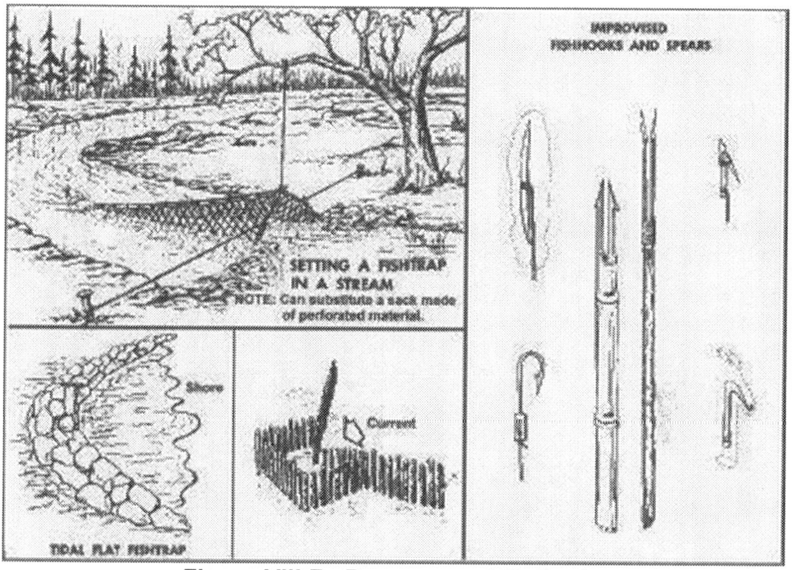

Figure VIII-7. Procurement Methods

(5) Precautions:

(a) Wear shoes to protect the feet when wading in water.

(b) Avoid reaching into dark holes.

(c) **Kill** animals before handling. Animals in distress may attract the enemy.

(d) ***DO NOT*** secure fishing lines to yourself or the raft.

(e) **Kill** fish before bringing them into the raft.

(f) ***DO NOT*** eat fish with—

- Spines.
- Unpleasant odor.
- Pale, slimy gills.
- Sunken eyes.
- Flabby skin.
- Flesh that remains dented when pressed.

(g) ***DO NOT*** eat fish eggs or liver (entrails).

(h) Avoid all crustaceans above the high tide mark.

(i) Avoid cone-shaped shells **(Figure VIII-8)**.

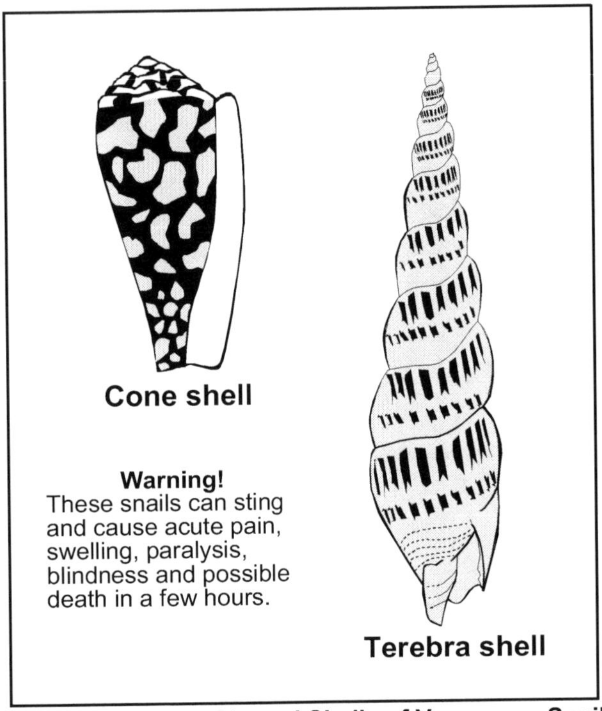

Figure VIII-8. Cone-Shaped Shells of Venomous Snails

(j) Avoid hairy insects; the hairs could cause irritation or infection.

(k) Avoid poisonous insects, for example:
- Centipedes.
- Scorpions.
- Poisonous spiders.

(l) Avoid disease carrying insects, such as—
- Flies.
- Mosquitoes.
- Ticks.

c. Plant Foods. ***Before using the following guide use your evasion chart to identify edible plants:***

> **Note:** If you cannot positively identify an edible plant and choose to try an unknown plant, these guidelines may help determine edibility.

(1) Selection criteria.

(a) Before testing for edibility, ensure there are enough plants to make testing worth your time and effort. Each part of a plant (roots, leaves, stems, bark, etc.) requires more than 24 hours to test. ***DO NOT*** waste time testing a plant that is not abundant.

(b) Test only **1** part of **1** plant at a time.

(c) Remember that eating large portions of plant food on an empty stomach may cause diarrhea, nausea, or cramps. ***Two*** good examples are ***green apples*** and ***wild onions***. Even after testing food and finding it safe, eat in moderation.

(2) Avoid plants with the following characteristics:

> **Note:** Using these guidelines in selecting plants for food may eliminate some edible plants; however, these guidelines will help prevent choosing potentially toxic plants.

(a) Milky sap (dandelion has milky sap but is safe to eat and easily recognizable).

(b) Spines, fine hairs, and thorns (skin irritants/contact dermatitis). ***Prickly pear*** and ***thistles*** are exceptions. ***Bracken fern fiddleheads*** also violate this guideline.

(c) Mushrooms and fungus.

(d) Umbrella shaped flowers (hemlock is eliminated).

(e) Bulbs (***only*** onions smell like onions).

(f) Grain heads with pink, purplish, or black spurs.

(g) Beans, bulbs, or seeds inside pods.

(h) Old or wilted leaves.

(i) Plants with shiny leaves.

(j) White and yellow berries. (Aggregate berries such as black and dewberries are always edible, test all others before eating.)

(k) Almond scent in woody parts and leaves.

d. Test procedures.

> **CAUTION:** Test all parts of the plant for edibility. Some plants have both edible and inedible parts. **NEVER ASSUME** a part that proved edible when cooked is edible raw, test the part raw before eating. The same part or plant may produce varying reactions in different individuals.

(1) Test only **1** part of a plant at a time.

(2) Separate the plant into its basic components (stems, roots, buds, and flowers).

(3) Smell the food for strong acid odors. Remember, smell alone does not indicate a plant is edible or inedible.

(4) ***DO NOT*** eat 8 hours before the test and drink only purified water.

(5) During the 8 hours you abstain from eating, test for contact poisoning by placing a piece of the plant on the inside of your elbow or wrist. The sap or juice should contact the skin. Usually 15 minutes is enough time to allow for a reaction.

(6) During testing, take ***NOTHING*** by mouth **EXCEPT** purified water and the plant you are testing.

(7) Select a small portion of a single part and prepare it the way you plan to eat it.

(8) Before placing the prepared plant in your mouth, touch a small portion (a pinch) to the outer surface of your lip to test for burning or itching.

(9) If after 3 minutes there is no reaction on your lip, place the plant on your tongue and hold it for 15 minutes.

(10) If there is no reaction, thoroughly chew a pinch and hold it in your mouth for 15 minutes **(*DO NOT* SWALLOW)**. If any ill effects occur, rinse out your mouth with water.

(11) If nothing abnormal occurs, swallow the food and wait 8 hours. If **any ill effects** occur during this period, **induce** vomiting and drink a water and charcoal mixture.

(12) If no ill effects occur, eat ¼ **cup** of the same plant prepared the same way. Wait another 8 hours. If no ill effects occur, the plant part as prepared is safe for eating.

CAUTION:
1. Ripe tropical fruits should be peeled and eaten raw. Softness, rather than color, is the best indicator of ripeness. Cook unripe fruits and discard seeds and skin.
2. Cook underground portions when possible to reduce bacterial contamination and ease digestion of their generally high starch content.
3. During evasion, you may not be able to cook. Concentrate your efforts on leafy green plants, ripe fruits, and above ground ripe vegetables not requiring significant preparation.

2. Food Preparation

Animal food gives the greatest food value per pound.

a. Butchering and skinning.

(1) Mammals.

(a) Remove the skin and save for other uses.

(a) One cut skinning of small game **(Figure VIII-9)**.

•Open the abdominal cavity.

•Avoid rupturing the intestines.

•Remove the intestines.

•Save inner organs (heart, liver, and kidneys) and all meaty parts of the skull, brain, tongue, and eyes.

(b) Wash when ready to use.

(c) If preserving the meat, remove it from the bones.

(d) Unused or inedible organs and entrails may be used as bait for other game.

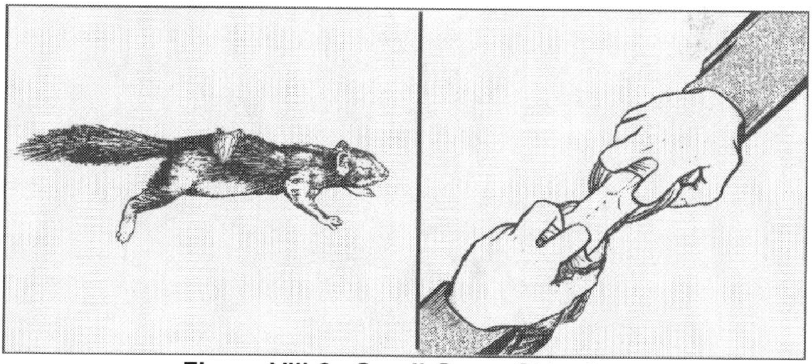

Figure VIII-9. Small Game Skinning

(2) Frogs and snakes.

(a) Skin.

(b) Discard skin, head with 2 inches of body, and internal organs.

(3) Fish.

(a) Scale (if necessary) and gut fish soon after it is caught.

(b) Insert knifepoint into anus of fish and cut open the belly.

(c) Remove entrails.

(d) Remove gills to prevent spoilage.

(4) Birds.

(a) Gut soon after killing.

(b) Protect from flies.

(c) Skin or pluck them.

(d) Skin scavengers and sea birds.

(5) Insects.

(a) Remove all hard portions such as the legs of grasshoppers or crickets. (The rest is edible.)

(b) Recommend cooking grasshopper-size insects.

CAUTION: Dead insects spoil rapidly, ***DO NOT*** save.

(6) Fruits, berries, and most nuts can be eaten raw.

b. Cooking.

CAUTION: To kill parasites, thoroughly cook all wild game, freshwater fish, clams, mussels, snails, crawfish, and scavenger birds. Saltwater fish may be eaten raw.

(1) Boiling (most nutritious method of cooking—drink the broth).

(a) Make metal cooking containers from ration cans.

(b) Drop heated rocks into containers to boil water or cook food.

(2) Baking.

(a) Wrap in leaves or pack in mud.

(b) Bury food in dirt under coals of fire.

(3) Leaching. Some nuts (acorns) must be leached to remove the bitter taste of tannin. Use one of the following leaching methods:

(a) First method:

•Soaking and pouring the water off.

•Crushing and pouring water through. Cold water should be tried first; however, boiling water is sometimes best.

•Discarding water.

(b) Second method:

•Boil, pour off water, and taste the plant.

•If bitter, repeat process until palatable.

(4) Roasting.

(a) Shake shelled nuts in a container with hot coals.

(b) Roast thinly sliced meat and insects over a candle.

3. Food Preservation

b. Keeping an animal alive.

c. Refrigerating.

(1) Long term.

(a) Food buried in snow maintains a temperature of approximately 32 degrees F.

(b) Frozen food will not decompose (freeze in meal-size portions).

(2) Short term.

(a) Food wrapped in waterproof material and placed in a stream remains cool in summer months.

(b) Earth below the surface, particularly in shady areas or along streams, is cooler than the surface.

(c) Wrap food in absorbent material such as cotton and re-wet as the water evaporates.

c. Drying and smoking removes moisture and preserves food.

(1) Use salt to improve flavor and promote drying.

(2) Cut or pound meat into thin strips.

(3) Remove fat.

(4) ***DO NOT*** use pitch woods such as fir or pine; they produce soot giving the meat an undesirable taste.

d. Protecting meat from animals and insects.
(1) Wrapping food.
(a) Use clean material.
(b) Wrap pieces individually.
(c) Ensure all corners of the wrapping are insect proof.
(d) Wrap soft fruits and berries in leaves or moss.
(2) Hanging meat.
(a) Hang meat in the shade.
(b) Cover during daylight hours to protect from insects.
(3) Packing meat on the trail.
(a) Wrap before flies appear in the morning.
(b) Place meat in fabric or clothing for insulation.
(c) Place meat inside the pack for carrying. Soft material acts as insulation helping keep the meat cool.
(d) Carry shellfish, crabs, and shrimp in wet seaweed.

e. ***DO NOT*** store food in the shelter; it attracts unwanted animals.

Chapter IX

INDUCED CONDITIONS

(NUCLEAR, BIOLOGICAL, AND CHEMICAL CONSIDERATIONS)

1. Nuclear Conditions

CAUTION: Radiation protection depends on time of exposure, distance from the source, and shielding.

a. Protection.

(1) **FIND PROTECTIVE SHELTER IMMEDIATELY!**

(2) Gather all equipment for survival (time permitting).

(3) Avoid detection and capture.

(a) Seek existing shelter that may be improved **(Figure IX-1)**.

Figure IX-1. Immediate Action Shelter

(b) If no shelter is available, dig a trench or foxhole as follows:

•Dig trench deep enough for protection, then enlarge for comfort **(Figure IX-2)**.

•Cover with available material.

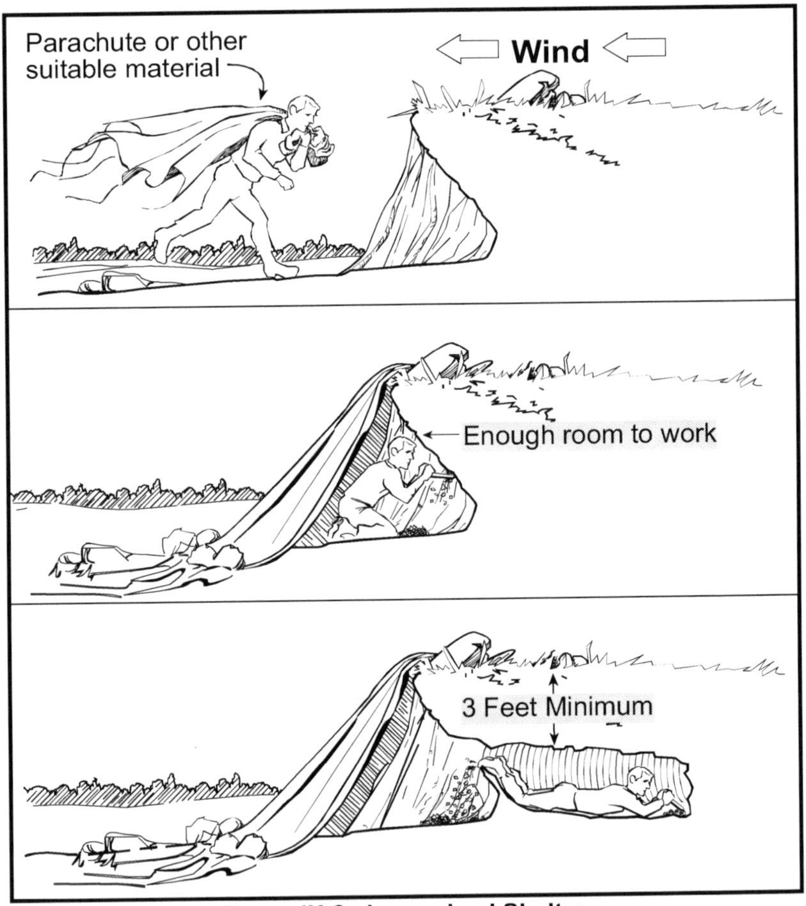

Figure IX-2. Improvised Shelter

(4) Radiation shielding efficiencies **(Figure IX-3)**.

NUCLEAR EXPLOSIONS: Fall flat. Cover exposed body parts. Present minimal profile to direction of blast. ***DO NOT*** *look at fireball!* Remain prone until blast effects are over.

SHELTER: Pick, as soon as possible, 5 minutes unsheltered is maximum!

Priority:
(1) Cave or tunnel covered with 3 or more feet of earth.
(2) Storm/storage cellars
(3) Culverts.
(4) Basements.
(5) Abandoned stone/mud buildings.
(6) Foxhole 4 feet deep (remove topsoil within 2 feet radius of foxhole lip).

RADIATION SHIELDING EFFICIENCIES

Iron/Steel	.7 inches	**Cinder Block**	5.3 inches
Brick	2.0 inches	**Ice**	6.8 inches
Concrete	2.2 inches	**Wood (Soft)**	8.8 inches
Earth	3.3 inches	**Snow**	20.3 inches

One thickness reduces received radiation dose by 1/2. Additional thickness added to any amount of thickness reduces received radiation dose by 1/2.

SHELTER SURVIVAL: Keep contaminated materials out of shelter.
Good Weather: Bury contaminated clothing outside of shelter (recover later).
Bad Weather: Shake strongly or beat with branches. Rinse and /or shake wet clothing. ***DO NOT*** *wring out!*

PERSONAL HYGIENE: Wash entire body with soap and *any* water; give close attention to fingernails and hairy parts.
No Water: Wipe all exposed skin surfaces with clean cloth or uncontaminated soil. Fallout/dusty conditions--keep entire body covered. Keep handkerchief/cloth over mouth and nose. Improvise goggles. ***DO NOT*** *smoke!*

DAILY RADIATION TIME TABLE for NO RATE METER

4-6	Complete isolation	**9-12**	2-4 hours exposure per day
3-7	Brief exposure (30 minutes maximum)	**13**	Normal movement.
8	Brief exposure (1 hour maximum)		

Figure IX-3. Radiation Shielding Efficiencies

(5) Leave contaminated equipment and clothing near shelter for retrieval after radioactive decay.

(6) Lie down, keep warm, sleep, and rest.

b. Substance:

(1) Water. Allow no more than 30 minutes exposure on **3d** day for water procurement.

(a) Water sources (in order of preference):

- Springs, wells, or underground sources are **safest**.
- Water in pipes/containers in abandoned buildings.
- Snow (**6** or **more inches below** the surface during the fallout).

•Streams and rivers (filtered before drinking).

•Lakes, ponds, pools, etc.

•Water from below the surface (***DO NOT*** stir up the water).

•Use a seep well.

(b) Water preparation **(Figures IX-4 and IX-5)**.

•Filtering through earth removes 99 percent of radioactivity.

•Purify all water sources.

(2) Food.

(a) Processed foods (canned or packaged) are preferred; wash and wipe containers before use.

(b) Animal foods.

•Avoid animals that appear to be sick or dying.

•Skin carefully to avoid contaminating the meat.

•Before cooking, cut meat away from the bone, leaving at least 1/8 inch of meat on the bone.

•Discard all internal organs.

•Cook all meat until **very well** done.

(c) Avoid.

•Aquatic food sources (use only in extreme emergencies because of high concentration of radiation).

•Shells of all eggs (contents will be safe to eat).

•Milk from animals.

(d) Plant foods (in order of preference).

•Plants whose edible portions grow underground (for example, potatoes, turnips, carrots, etc.). Wash and remove skin.

•Edible portions growing above ground that can be washed and peeled or skinned (bananas, apples, etc.).

•Smooth skinned vegetables, fruits, or above ground plants that are not easily peeled or washed.

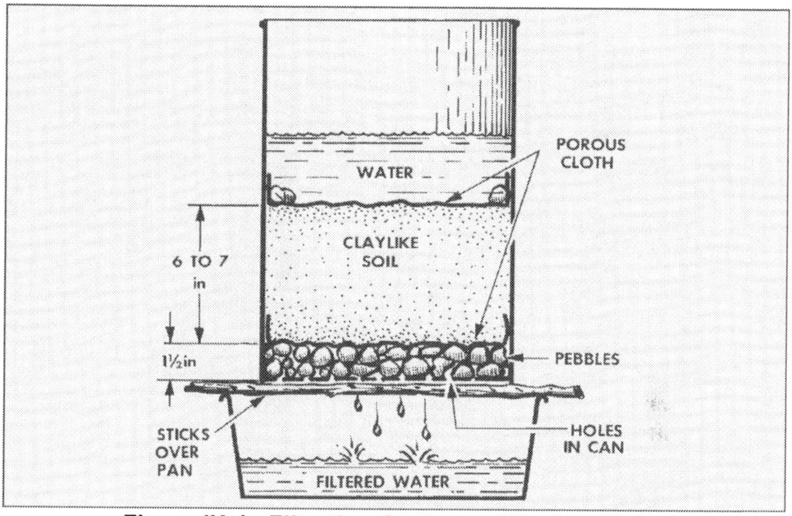

Figure IX-4. Filtration Systems, Filtering Water

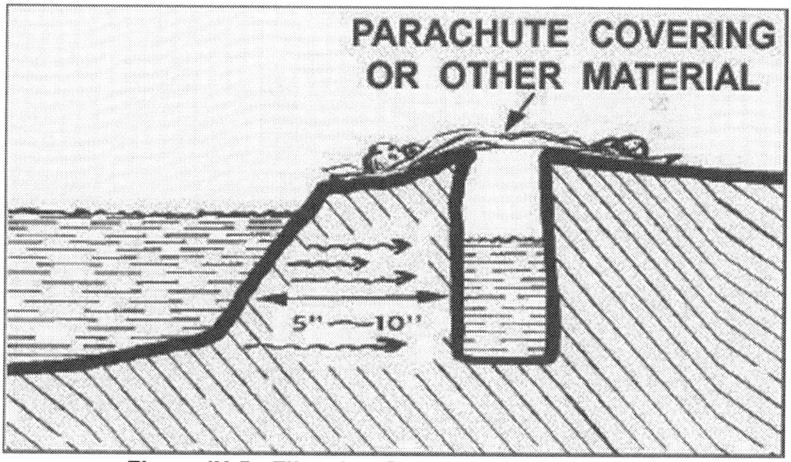

Figure IX-5. Filtration Systems, Settling Water

c. Self-aid:

(1) General rules:

(a) Prevent exposure to contaminants.

(b) Use personal hygiene practices and remove body waste from shelter.

(c) Rest, avoid fatigue.

(d) Drink liquids.

(2) Wounds.

(a) Clean affected area.

(b) Use antibacterial ointment or cleaning solution.

(c) Cover with clean dressing.

(d) Watch for signs of infection.

(3) Burns.

(a) Clean affected area.

(b) Cover with clean dressing.

(4) Radiation sickness (nausea, weakness, fatigue, vomiting, diarrhea, loss of hair, radiation burns).

(a) Time is required to overcome.

(b) Rest.

(c) Drink fluids.

(d) Maintain food intake.

(e) Prevent additional exposure.

2. Biological Conditions

a. Clues which may alert you to a biological attack follow:

(1) Enemy aircraft dropping objects or spraying.

(2) Breakable containers or unusual bombs, particularly those bursting with little or no blast, and muffled explosions.

(3) Smoke or mist of unknown origin.

(4) Unusual substances on the ground or vegetation; sick looking plants or crops.

b. Protection from biological agents follow:

(1) Use protective equipment.

(2) Bathe as soon as the situation permits.

(3) Wash hair and body thoroughly with soap and water.

(4) Clean thoroughly under fingernails.

(5) Clean teeth, gums, tongue, and roof of mouth frequently.

c. Survival tips for biological conditions follow:

(1) Keep your body and living area clean.

(2) Stay alert for clues of biological attack.

(3) Keep nose, mouth, and skin covered.

(4) Keep food and water protected. Bottled or canned foods are safe if sealed. If in doubt, boil food and water for 10 minutes.

(5) Construct shelter in a clear area, away from vegetation, with entrance 90 degrees to the prevailing wind.

(6) If traveling, travel crosswind or upwind (taking advantage of terrain to stay away from depressions).

3. Chemical Conditions

a. Detecting.

(1) Smell. Many agents have little or no odor.

(2) Sight. Many agents are colorless:

(a) Color. Yellow, orange, or red smoke or mist.

(b) Liquid. Oily, dark patches on leaves, ground, etc.

(c) Gas. Some agents appear as a mist immediately after shell burst.

(d) Solid. Most solid state agents have some color.

(3) Sound. Muffled explosions are possible indications of chemical agent bombs.

(4) Feel. Irritation to the nose, eyes, or skin and/or moisture on the skin are danger signs.

(5) Taste. Strange taste in food or water indicates contamination.

(6) General indications. Tears, difficult breathing, choking, itching, coughing, dizziness.

(7) Wildlife. Presence of sick or dying animals.

b. Protection against chemical agents follows:

(1) Use protective equipment.

(2) Avoid contaminated areas.

(a) Exit contaminated area by moving crosswind.

(b) Select routes on high ground.

(c) Avoid cellars, ditches, trenches, gullies, valleys, etc.

(d) Avoid woods, tall grasses, and bushes as they tend to hold chemical agent vapors.

(e) Decontaminate body and equipment as soon as possible by—

- Removing. Pinch-blotting.
- Neutralizing. Warm water.
- Destroying. Burying.

c. Self-aid in chemically contaminated areas.

(1) If a chemical defense ensemble is available—

(a) Use all protective equipment.

(b) Follow antidote directions when needed.

(2) If a chemical defense ensemble is not available—

(a) Remove or tear away contaminated clothing.

(b) Rinse contaminated areas with water.

(c) Improvise a breathing filter using materials available (T-shirt, handkerchief, fabric, etc.).

d. Tips for the survivor:

(1) ***DO NOT*** use wood from a contaminated area for fire.

(2) Look for signs of chemical agents around water sources before procurement (oil spots, foreign odors, dead fish, or animals).

(3) Keep food and water protected.

(4) ***DO NOT*** use plants for food or water in contaminated areas.

Appendix A
THE WILL TO SURVIVE

ARTICLE VI CODE OF CONDUCT

I will never forget that I am an American fighting for freedom, responsible for my actions, and dedicated to the principles which made my country free. I will trust in my God and in the United States of America.

1. Psychology of Survival

a. Preparation—

(1) Know your capabilities and limitations.

(2) Keep a positive attitude.

(3) Develop a realistic plan.

(4) Anticipate fears.

(5) Combat psychological stress by—

(a) Recognizing and anticipating existing *stressors* (injury, death, fatigue, illness, environment, hunger, isolation).

(b) Attributing normal reactions to existing *stressors* (fear, anxiety, guilt, boredom, depression, anger).

(c) Identifying signals of distress created by *stressors* (indecision, withdrawal, forgetfulness, carelessness, and propensity to make mistakes).

b. Strengthen your will to survive with—

(1) The Code of Conduct.

(2) Pledge of Allegiance.

(3) Faith in America.

(4) Patriotic songs.

(5) Thoughts of return to family and friends.

c. Group dynamics of survival include—

(1) Leadership, good organization, and cohesiveness promote high morale:

(a) Preventing panic.

(b) Creating strength and trust in one another.

(c) Favoring persistency in overcoming failure.

(d) Facilitating formulation of group goals.

(2) Taking care of your buddy.

(3) Working as a team.

(4) Reassuring and encouraging each other.

(5) Influencing factors are—
   (a) Enforcing the chain of command.
   (b) Organizing according to individual capabilities.
   (c) Accepting suggestions and criticism.

2. Spiritual Considerations

a. Collect your thoughts and emotions.
b. Identify your personal beliefs.
c. Use self-control.
d. Meditate.
e. Remember past inner sources to help you overcome adversity.
f. Pray for your God's help, strength, wisdom, and rescue.
   (1) Talk to your God.
   (2) Give thanks that God is with you.
   (3) Ask for God's help.
   (4) Pray for protection and a positive outcome.
g. Remember scripture, verses, or hymns; repeat them to yourself and to your God.
h. Worship without aid of written scripture, clergy, or others.
i. Forgive—
   (1) Yourself for what you have done or said that was wrong.
   (2) Those who have failed you.
j. Praise God and give thanks because—
   (1) God is bigger than your circumstances.
   (2) God will see you through (no matter what happens).
   (3) Hope comes from a belief in heaven and/or an after-life.
k. Trust.
   (1) Faith and trust in your God.
   (2) Love for family and self.
   (3) Never lose hope.
   (4) Never give up.
l. With other survivors—
   (1) Identify or appoint a religious lay leader.
   (2) Discuss what is important to you.
   (3) Share scriptures and songs.
   (4) Pray for each other.
   (5) Try to have worship services.

(6) Write down scriptures and songs that you remember.

(7) Encourage each other while waiting for rescue, remember—

(a) Your God loves you.

(b) Praise your God.

Appendix B
Publication Information

1. Scope

This UNCLASSIFIED multiservice tactics, techniques, and procedures publication is designed to assist Service members in a survival situation regardless of geographic location.

2. Purpose

This publication provides Service members a quick reference, weatherproof, pocket-sized guide on basic survival, evasion, and recovery information.

3. Application

The target audience for this publication is any Service member requiring basic survival, evasion, and recovery information.

4. Implementation Plan

Participating Service command offices of primary responsibility (OPRs) will review this publication, validate the information, and reference and incorporate it in Service and command manuals, regulations, and curricula as follows:

Army. The Army will incorporate the procedures in this publication in US Army training and doctrinal publications as directed by the commander, US Army Training and Doctrine Command (TRADOC). Distribution is in accordance with the Department of Army Form 12-99-R.

Marine Corps. The Marine Corps will incorporate the procedures in this publication in US Marine Corps training and doctrinal publications as directed by the commanding general, US Marine Corps Combat Development Command (MCCDC). Distribution is in accordance with Marine Corps Publication Distribution System.

Navy. The Navy will incorporate these procedures in US Navy doctrinal and training publications as directed by the commander, Navy Warfare Development Command (NWDC). Distribution is in accordance with Military Standard Requisitioning and Issue Procedures Desk Guide and Navy Standing Operating Procedures Publication 409.

Air Force. Air Force units will validate and incorporate appropriate procedures in accordance with applicable governing directives. Distribution is in accordance with Air Force Instructions 33-360.

5. User Information

a. The TRADOC-MCCDC-NWDC-AFDC Air Land Sea Application (ALSA) Center developed this publication with the joint participation of the approving Service commands. ALSA will review and update this publication as necessary.

b. This publication reflects current joint and Service doctrine, command and control (C2) organizations, facilities, personnel, responsibilities, and procedures. Changes in Service protocol, appropriately reflected in joint and Service publications, will likewise be incorporated in revisions to this document.

c. We encourage recommended changes for improving this publication. Key your comments to the specific page and paragraph and provide a rationale for each recommendation. Send comments and recommendation directly to—

Army

Commander
US Army Training and Doctrine Command
ATTN: ATDO-A
Fort Monroe VA 23651-5000
DSN 680-3153 COMM (757) 727-3153

Marine Corps

Commanding General
US Marine Corps Combat Development Command
ATTN: C42
3300 Russell Road
Quantico VA 22134-5001
DSN 278-6234 COMM (703) 784-6234

Navy

Commander
Navy Warfare Development Command (Det Norfolk)
ATTN: ALSA Liaison Officer
1530 Gilbert Street
Norfolk VA 23511-2785
DSN 262-2782 COMM (757) 322-2782
E-mail: ndcjoint@nctamslant.navy.mil

Air Force

HQ Air Force Doctrine Center
ATTN: DJ
216 Sweeney Boulevard Suite 109
Langley AFB VA 23665-2722
DSN 754-8091 COMM (757) 764-8091
E-mail Address: afdc.dj@langley.af.mil

ALSA

ALSA Center
ATTN: Director
114 Andrews Street
Langley AFB, VA 23665-2785
DSN 575-0902 COMM (757) 225-0902
E-mail: alsadirector@langley.af.mil

Made in United States
Orlando, FL
04 November 2026

90204533R00378